NONLINEAR FUNCTIONAL ANALYSIS
AND ITS APPLICATIONS

PROCEEDINGS OF SYMPOSIA
IN PURE MATHEMATICS
Volume 45, Part 1

Nonlinear Functional Analysis and Its Applications

Felix E. Browder, Editor

AMERICAN MATHEMATICAL SOCIETY
PROVIDENCE, RHODE ISLAND

PROCEEDINGS OF THE SUMMER RESEARCH INSTITUTE
ON NONLINEAR FUNCTIONAL ANALYSIS AND ITS APPLICATIONS
HELD AT THE UNIVERSITY OF CALIFORNIA
BERKELEY, CALIFORNIA
JULY 11–29, 1983

with support from the National Science Foundation, Grant MCS 82-19023

1980 *Mathematics Subject Classification*
Primary 34-XX, 35-XX, 46-XX, 47HXX, 49AXX, 58-XX, 76D05; 30-XX, 31-XX,
32H15, 40AXX, 42B30, 45-XX, 47D05, 49G05, 49H05, 51M10, 52A07,
54-XX, 65-XX, 73CXX, 76-XX, 81-XX, 92AXX.
Secondary 06F20, 26D10, 53-XX, 55M25, 57H10, 70F15, 70H15, 90A15, 90C25.

Library of Congress Cataloging-in-Publication Data
Main entry under title:

Nonlinear functional analysis and its applications.
　　(Proceedings of symposia in pure mathematics; v. 45)
　　1. Nonlinear functional analysis—Congresses. I. Browder, Felix E.
III. Series.
QA321.5.N66　　1986　　　　　　　　515.7　　　　　　　　85-28725
ISBN 0-8218-1467-2 (set)
ISBN 0-8218-1471-0 (part 1)
ISBN 0-8218-1472-9 (part 2)

　　COPYING AND REPRINTING. Individual readers of this publication, and nonprofit libraries acting for them, are permitted to make fair use of the material, such as to copy an article for use in teaching or research. Permission is granted to quote brief passages from this publication in reviews, provided the customary acknowledgment of the source is given.
　　Republication, systematic copying, or multiple reproduction of any material in this publication (including abstracts) is permitted only under license from the American Mathematical Society. Requests for such permission should be addressed to the Executive Director, American Mathematical Society, P.O. Box 6248, Providence, Rhode Island 02940.
　　The appearance of the code on the first page of an article in this book indicates the copyright owner's consent for copying beyond that permitted by Sections 107 or 108 of the U.S. Copyright Law, provided that the fee of $1.00 plus $.25 per page for each copy be paid directly to the Copyright Clearance Center, Inc., 21 Congress Street, Salem, Massachusetts 01970. This consent does not extend to other kinds of copying, such as copying for general distribution, for advertising or promotional purposes, for creating new collective works, or for resale.

Copyright ©1986 by the American Mathematical Society. All rights reserved.
Printed in the United States of America.
The American Mathematical Society retains all rights
except those granted to the United States Government.
The paper used in this book is acid-free and falls within the guidelines
established to ensure permanence and durability.

Contents

PART 1

Foreword	xi
A note on the regularity of solutions obtained by global topological methods	
J. C. ALEXANDER	1
Asymmetric rotating wave solutions of reaction-diffusion equations	
J. C. ALEXANDER	7
Parabolic evolution equations with nonlinear boundary conditions	
HERBERT AMANN	17
Nonlinear oscillations with minimal period	
ANTONIO AMBROSETTI	29
On steady Navier–Stokes flow past a body in the plane	
CHARLES J. AMICK	37
On a contraction theorem and applications	
IOANNIS K. ARGYROS	51
Dual variational principles for eigenvalue problems	
GILES AUCHMUTY	55
Generalized elliptic solutions of the Dirichlet problem for n-dimensional Monge–Ampère equations	
ILYA J. BAKELMAN	73
Existence and containment of solutions to parabolic systems	
PETER W. BATES	103
Periodic Hamiltonian trajectories on starshaped manifolds	
HENRI BERESTYCKI, JEAN-MICHEL LASRY, GIANNI MANCINI and BERNHARD RUF	109
Integrability of nonlinear differential equations via functional analysis	
M. S. BERGER, P. T. CHURCH and J. G. TIMOURIAN	117
Extinction of the solutions of some quasilinear elliptic problems of arbitrary order	
F. BERNIS	125
On a system of degenerate diffusion equations	
M. BERTSCH, M. E. GURTIN, D. HILHORST and L. A. PELETIER	133
Pointwise continuity for a weak solution of a parabolic obstacle problem	
MARCO BIROLI	141
A note on iteration for pseudoparabolic equations of fissured media type	
MICHAEL BÖHM	147

Homogenization of two-phase flow equations
  ALAIN BOURGEAT ... 157

Some variational problems with lack of compactness
  HAIM BREZIS ... 165

Degree theory for nonlinear mappings
  FELIX E. BROWDER ... 203

Construction of periodic solutions of periodic contractive evolution systems from bounded solutions
  RONALD E. BRUCK ... 227

An abstract critical point theorem for strongly indefinite functionals
  A. CAPOZZI and D. FORTUNATO ... 237

Uniqueness of positive solutions for a sublinear Dirichlet problem
  ALFONSO CASTRO ... 243

Applications of homology theory to some problems in differential equations
  KUNG-CHING CHANG ... 253

On the continuation method and the method of monotone iterations
  PHILIPPE CLÉMENT ... 263

Existence theorems for superlinear elliptic Dirichlet problems in exterior domains
  CHARLES V. COFFMAN and MOSHE M. MARCUS ... 271

A global fixed point theorem for symplectic maps and subharmonic solutions of Hamiltonian equations on tori
  C. CONLEY and E. ZEHNDER ... 283

Harmonic maps from the disk into the Euclidean N-sphere
  JEAN-MICHEL CORON ... 301

Nonlinear semigroups and evolution governed by accretive operators
  MICHAEL G. CRANDALL ... 305

A theorem of Mather and the local structure of nonlinear Fredholm maps
  JAMES DAMON ... 339

Remarks on S^1 symmetries and a special degree for S^1-invariant gradient mappings
  E. N. DANCER ... 353

Abstract differential equations, maximal regularity, and linearization
  G. DA PRATO ... 359

Positive solutions for some classes of semilinear elliptic problems
  DJAIRO G. DE FIGUEIREDO ... 371

Elliptic and parabolic quasilinear equations giving rise to a free boundary: the boundary of the support of the solutions
  JESUS ILDEFONSO DIAZ ... 381

An index theory for periodic solutions of convex Hamiltonian systems
  I. EKELAND ... 395

CONTENTS

An extension of the Leray–Schauder degree for fully nonlinear elliptic problems
    P. M. FITZPATRICK and JACOBO PEJSACHOWICZ   425

Nonlinear special manifolds for the Navier–Stokes equations
    C. FOIAŞ and J. C. SAUT   439

Regularity criteria for weak solutions of the Navier–Stokes system
    YOSHIKAZU GIGA   449

A strongly nonlinear elliptic problem in Orlicz–Sobolev spaces
    JEAN-PIERRE GOSSEZ   455

Nontrivial solutions of semilinear elliptic equations of fourth order
    YONG-GENG GU   463

Approximation in Sobolev spaces and nonlinear potential theory
    LARS INGE HEDBERG   473

Some free boundary problems for predator-prey systems with nonlinear diffusion
    JESÚS HERNÁNDEZ   481

On positive solutions of semilinear periodic-parabolic problems
    PETER HESS   489

The topological degree at a critical point of mountain-pass type
    HELMUT HOFER   501

On a conjecture of Lohwater about asymptotic values of meromorphic functions
    J. S. HWANG   511

Parametrix of $\Box_b$
    CHISATO IWASAKI   521

Applications of Nash–Moser theory to nonlinear Cauchy problems
    NOBUHISA IWASAKI   525

Nonlinear multiparametric equations: structure and topological dimension of global branches of solutions
    J. IZE, I. MASSABÓ, J. PEJSACHOWICZ and A. VIGNOLI   529

PART 2

Remarks on the Euler and Navier–Stokes equations in $\mathbf{R}^2$
    TOSIO KATO   1

Nonlinear equations of evolution in Banach spaces
    TOSIO KATO   9

Geometrical properties of level sets of solutions to elliptic problems
    BERNHARD KAWOHL   25

Remarks about St. Venant solutions in finite elasticity
    DAVID KINDERLEHRER   37

Nonexpansive mappings in product spaces, set-valued mappings and k-uniform rotundity
    W. A. KIRK   51

A new operator theoretic algorithm for solving first order scalar quasilinear equations
    YOSHIKAZU KOBAYASHI   65

A general interpolation theorem of Marcinkiewicz type
    HIKOSABURO KOMATSU   77

An easy proof of the interior gradient bound for solutions to the prescribed mean cuvature equation
    N. KOREVAAR   81

On application of the monotone iteration scheme to wave and biharmonic equations
    PHILIP KORMAN   91

Unilateral obstacle problem for strongly nonlinear second order elliptic operators
    RÜDIGER LANDES and VESA MUSTONEN   95

Product formula, imaginary resolvents and modified Feynman integral
    MICHEL L. LAPIDUS   109

Quasilinear elliptic equations with nonlinear boundary conditions
    GARY M. LIEBERMAN   113

Some L^p inequalities and their applications to fixed point theorems of uniformly Lipschitzian mappings
    TECK-CHEONG LIM   119

Some remarks on the optimal control of singular distributed systems
    J. L. LIONS   127

Global continuation and complicated trajectories for periodic solutions of a differential-delay equation
    JOHN MALLET-PARET and ROGER D. NUSSBAUM   155

On quasiconvexity in the calculus of variations
    PAOLO MARCELLINI   169

Asymptotic growth for evolutionary surfaces of prescribed mean curvature
    PAOLO MARCELLINI and KEITH MILLER   175

L^2-decay of solutions of the Navier–Stokes equations in the exterior domain
    KYÛYA MASUDA   179

On the index and the covering dimension of the solution set of semilinear equations
    P. S. MILOJEVIĆ   183

Pointwise potential estimates for elliptic obstacle problems
    UMBERTO MOSCO   207

On singularities of solutions to nonlinear elliptic systems of partial differential equations
    JINDŘICH NEČAS   219

Uniqueness, nonuniqueness and related questions of nonlinear elliptic and parabolic equations
    WEI-MING NI   229

Global solutions of the hyperbolic Yang–Mills equations and their sharp asymptotics
    STEPHEN M. PANEITZ   243

Surjectivity of generalized ϕ-accretive operators
    SEHIE PARK and JONG AN PARK   255

Approximation-solvability of periodic boundary value problems via the A-proper mapping theory
    W. V. PETRYSHYN   261

On a generalization of the Fuller index
    A. J. B. POTTER   283

Minimax methods for indefinite functionals
    PAUL H. RABINOWITZ   287

Nonlinear semigroups, holomorphic mappings and integral equations
    SIMEON REICH   307

Rearrangement of functions and reverse Jensen inequalities
    CARLO SBORDONE   325

Bifurcation and even-like vector fields
    STEVE J. SCHIFFMAN and JAY H. WOLKOWISKY   331

Classical and quantized invariant wave equations—progress and problems
    IRVING SEGAL   341

Resonance and the stationary Navier–Stokes equations
    VICTOR L. SHAPIRO   359

On removable point singularities of coupled Yang–Mills fields
    L. M. SIBNER   371

Two-function minimax theorems and variational inequalities for functions on compact and noncompact sets, with some comments on fixed-point theorems
    S. SIMONS   377

On approximating fixed points
    S. P. SINGH and B. WATSON   393

Symmetry breaking and nondegenerate solutions of semilinear elliptic equations
    JOEL A. SMOLLER and ARTHUR G. WASSERMAN   397

A generalized Palais-Smale condition and applications
    MICHAEL STRUWE   401

Existence and nonuniqueness of solutions of a noncoercive elliptic variational inequality
    ANDRZEJ SZULKIN   413

Fixed point, minimax, and Hahn-Banach theorems
  Wataru Takahashi 419

Remarks on the Euler equations
  R. Temam 429

Infinite-dimensional dynamical systems in fluid mechanics
  R. Temam 431

Existence of symmetric homoclinic orbits for systems of Euler–Lagrange equations
  J. F. Toland 447

Graphs with prescribed curvature
  Neil S. Trudinger 461

Navier–Stokes equations for compressible fluids: global estimates and periodic solutions
  Alberto Valli 467

Weak and strong singularities of nonlinear elliptic equations
  Laurent Véron 477

Regularity of weak solutions of the Navier–Stokes equations
  Wolf von Wahl 497

A monotone convergence theorem in abstract Banach spaces
  Ioan I. Vrabie 505

On a class of strongly nonlinear Dirichlet boundary-value problems: beyond Pohožaev's results
  Pierre A. Vuillermot 521

Topological degree and global bifurcation
  Jeffrey R. L. Webb and Stewart C. Welsh 527

Generalized idea of Synge and its applications to topology and calculus of variations in positively curved manifolds
  S. Walter Wei 533

On extending the Conley–Zehnder fixed point theorem to other manifolds
  Alan Weinstein 541

L^p-energy and blow-up for a semilinear heat equation
  Fred B. Weissler 545

A geometric theory of bifurcation
  Jay H. Wolkowisky 553

Some problems on degenerate quasilinear parabolic equations
  Zhuoqun Wu 565

Quasi-homogeneous microlocal analysis for nonlinear partial differential equations
  Masao Yamazaki 573

Foreword

The two volumes *Nonlinear Functional Analysis and Its Applications*, published in the series Proceedings of Symposia in Pure Mathematics (vol. 45, parts 1 and 2), are the result of the thirty-first Summer Research Institute of the American Mathematical Society held at the University of California at Berkeley from July 11 to July 29, 1983. This institute was partially supported by a grant from the National Science Foundation, and organized by an Organizing Committee consisting of Haim Brezis, Felix Browder (Chairman), Tosio Kato, J.-L. Lions, Louis Nirenberg, and Paul Rabinowitz.

The purpose of the institute was to present and develop research on an international basis in nonlinear functional analysis and its applications, especially in the study of boundary value problems for nonlinear partial differential equations and corresponding problems in geometry and mathematical physics. Major topics which were covered in a series of expository lectures as well as research talks included: Minimax methods in the calculus of variations, existence theory for variational problems without compactness, theories of degree of mapping, inverse function theorems of Nash–Moser type, nonlinear semigroup theory, nonlinear equations of evolution, nonlinear problems of control theory, periodic solutions of Hamiltonian systems, generalizations of the Morse theory, nonlinear partial differential equations in gauge field theory, the theory of Feigenbaum cascades, the study of the Navier–Stokes equations, nonlinear elliptic equations in differential geometry, and a variety of topics concerning nonlinear elliptic boundary value and eigenvalue problems, bifurcation theory, nonlinear hyperbolic equations, nonlinear conservation laws, nonlinear Hamilton–Jacobi equations, and an even wider variety of physical applications.

There were 13 series of expository lectures totaling 39 hours of lectures which summarized main directions and methods in current research. In addition, there were 115 one-hour research lectures.

A total of 203 mathematicians registered for the Institute, twenty of whom were students. The international character of the Institute is reflected in the national origins of the participants. Twenty-two countries not in North America were represented by the following numbers of participants: Africa (1), Australia (3), Belgium (1), Brazil (1), China (4), Czechoslovakia (1), England (4), France (22),

Israel (2), Italy (14), Japan (8), Korea (1), Netherlands (2), New Zealand (1), Poland (1), Rumania (1), Scotland (1), Spain (4), Sweden (2), Switzerland (3), and West Germany (12).

One final comment may be in order. As compared to an earlier volume, *Nonlinear Functional Analysis*, which appeared in the Proc. Sympos. Pure Math. series 15 years ago, the present volume as well as the Institute itself represent a much more forceful emphasis upon applications as opposed to general theory. This reflects, in my view, a major shifting of focal emphasis in the field as well as in the tastes of different organizing committees. Though new conceptual advances are being made and new general methods are being developed, they tend on the whole to be much more closely linked with particular domains of application. In part, this represents the process of assimilation of the general theories developed in previous decades, as well as a mood of distrust of general theories which move too far from the context of applications. This probably reflects a more general process going on in the mathematical world at large, but it can be seen in a transparent way in the present context. No one can predict with greater security than their own self-confidence whether this tendency is irreversible or part of a broad pendular swing, and no one can determine to anyone else's satisfaction whether it is due to the internal processes of mathematical development or to the pressures arising from the external institutional context in which mathematics is being done today. Suffice it to record the facts in their broad outline so that we can look at them in the classical spirit of non-attachment.

FELIX E. BROWDER

NONLINEAR FUNCTIONAL ANALYSIS
AND ITS APPLICATIONS

A Note on the Regularity of Solutions Obtained by Global Topological Methods

J. C. ALEXANDER[1]

Over the past decade or so, there have been a number of results establishing "global" families of solutions of functional-analytic operators. The paper that set the ideas and terminology is Rabinowitz [6], although the classic work of Leray–Schauder [5] also fits in the framework. More recently, this author and other workers have found it useful to consider global families of dimension greater than one. One way of expressing and proving the results is via cohomology theory. The solution set of the operator, considered as a topological space, is shown to carry some kind of nonzero Čech cohomology class. The dimension of the class is a lower bound on the dimension of the solution set. If the solution set is to be connected, as in Rabinowitz's paper, the class is one dimensional. In the classic existence results, the class is zero dimensional; i.e., the solution set is not empty. In virtually all cases the underlying space on which the operator acts is a Banach space (possibly finite dimensional). This is more than a technical point; in general the results are false in Fréchet spaces.

The purpose of this note is to consider what happens when "one" operator acts on an ordered family of Banach spaces. The result is motivated by some work on differential operators; however, it was felt it would be worthwhile to isolate a general functional analytic result.

Consider the following motivating and prototypical example. Let $\mathscr{C}^{r+\alpha}$ be the Hölder space of $\mathscr{C}^{r+\alpha}$ functions on some domain with the $\mathscr{C}^{r+\alpha}$ topology. Suppose P is a parameter space. Suppose $\mathscr{F}$ is an integral operator which inverts a differential operator and which defines a compact operator $\mathscr{F}_r\colon P \times \mathscr{C}^{r+\alpha} \to \mathscr{C}^{r+\alpha}$ for each r. Under suitable standard assumptions, the solution set $\mathscr{S}_r$ of $\mathscr{F}_r$ in $P \times \mathscr{C}^{r+\alpha}$ carries a nonzero cohomology class, and of course $\mathscr{S}_r \subset \mathscr{S}_{r-1}$. Thus

---

1980 *Mathematics Subject Classification.* Primary 47H99, 35B65, 35B32; Secondary 58C30, 34A34.
[1] Partially supported by NSF.

© 1986 American Mathematical Society
0082-0717/86 $1.00 + $.25 per page

there are global families with any arbitrary degree of differentiability. However the techniques do not directly yield any information about $\mathscr{S}_\infty = \cap_r \mathscr{S}_r$. Even with regularity theorems which guarantee that any solution in $\mathscr{S}_r$ is automatically in $\mathscr{S}_\infty$, it is not clear that the set $\mathscr{S}_\infty$, with the limit topology (in this example, the C^∞ topology), has desired topological properties such as connectivity. For connectivity, one can use point-set topological arguments (Alexander [1]); however, for higher-dimensional results these are not available.

In this note we establish that global results carry over in considerable generality to $\mathscr{S}_\infty$. Stated abstractly, and with categorical and cohomological machinery at hand, the result admits a proof of a very few lines (a phenomenon which occurs elsewhere with the cohomological formulation—it speaks to the correctness of the formulation). However the abstract functional analytic formulation loses some immediacy, and we assume the reader is familiar with the general cohomological formulation of such global results. In particular, we begin with the conclusion of most topological arguments.

To set the result, consider a sequence of Banach spaces $\mathscr{X}_r$ with continuous linear inclusions $i_r : \mathscr{X}_r \to \mathscr{X}_{r-1}$. The obvious examples are Hölder spaces and Sobolev spaces. Let $\mathscr{P}$ be a locally compact parameter space (perhaps a single point) with a metric so that bounded sets have compact closure. Let $j_r : \mathscr{P} \times \mathscr{X}_r \to \mathscr{P} \times \mathscr{X}_{r-1}$ be the product of i_r and the identity of $\mathscr{P}$. We suppose there is a family of nonlinear operators $\mathscr{F}_r : \mathscr{P} \times \mathscr{X}_r \to \mathscr{X}_r$ such that $i_r \mathscr{F}_r = \mathscr{F}_{r-1} j_r$. The usual situation is that the $\mathscr{F}_r$ all come from a single analytic operator $\mathscr{F}$.

For an explicit example, consider the following (Auchmuty [3], Alexander–Auchmuty [2], Auchmuty [4]). Let (r, θ) be polar coordinates on the disk or annulus. Consider the equation

$$D\nabla^2 u = c(\partial u/\partial \theta) + f_\beta(u), \quad u = 0 \text{ on boundary},$$

where u is an n-vector-valued function, D is a nonsingular $(n \times n)$ diagonal matrix, $f_\beta(u)$ is a nonlinear function with $f_\beta(0) = 0$, and c, β are parameters. Thus $\mathscr{P} = R^2$. The equation comes from chemical kinetics. The term $f_\beta(u)$ expresses the kinetics. Solutions u correspond to rotating wave solutions with wave speed c.

Let

$$\mathscr{G}_r u = \text{Green}\big(D^{-1}(c(\partial u/\partial \theta) + f_\beta(u))\big)$$

where Green is convolution with the appropriate Green's function. If $u \in \mathscr{C}^{r+\alpha}$ then $\partial u/\partial \theta \in \mathscr{C}^{r-1+\alpha}$ and $\mathscr{G}_r u \in \mathscr{C}^{r+1-\alpha}$. Let $\mathscr{F}_r = i_r \mathscr{G}_r : \mathscr{C}^{r+\alpha} \to \mathscr{C}^{r+\alpha}$. It is shown under certain transversality conditions on $\partial f_\beta(u)/\partial u$, that there are global Hopf bifurcation branches, i.e., global connected branches of fixed points of $\mathscr{F}_r$ for any r. The solutions are C^∞, but the bifurcation results do not guarantee topological properties of the set of solution in $\mathscr{C}^\infty$—in particular that they form connected branches. That they do form connected branches in the $\mathscr{C}^\infty$ topology follows from the result of this note.

The global results for a single $\mathscr{F}_r$ can all be stated in the following general form. Let

$$\mathscr{S}_r = \{(p, x) \in \mathscr{P} \times \mathscr{X}_r : \mathscr{F}_r(p, x) = x\}$$

denote the fixed point set, or solution set, of $\mathscr{F}_r$. There is also a closed subset $\mathscr{S}_0 \subset \mathscr{S}$ (possibly empty) of trivial solutions. It is technically convenient to adjoin a point at infinity (with neighborhood basis the complements of bounded sets); we let $\hat{\mathscr{X}}_r, \hat{\mathscr{S}}_r, \hat{\mathscr{S}}_r^0$ etc. denote sets with the point at infinity adjoined. The sets have the properties:

(i) every bounded set in $\mathscr{S}_r$ has compact closure (thus $\hat{\mathscr{S}}_r/\hat{\mathscr{S}}_r^0$ is compact),

(ii) there is a space $\mathscr{Y}$ and closed subspace $\mathscr{Y}^0$ and a continuous map $e_r: \hat{\mathscr{S}} \to \mathscr{Y}$ with $e_r(\hat{\mathscr{S}}^0) \subset \mathscr{Y}^0$ (in applications to date, $\mathscr{Y}$ is a sphere and $\mathscr{Y}^0$ a point, or in some cases $\mathscr{Y} \subset \mathscr{P}$),

(iii) there is a reduced Čech cohomology class $\gamma \in \check{H}^*(\mathscr{Y}/\mathscr{Y}^0)$ such that $e_r^*\gamma$ is not zero in $\check{H}^*(\hat{\mathscr{S}}/\hat{\mathscr{S}}^0)$.

This last property is the consequence of some analytic or degree-theoretic assumptions on $\mathscr{F}_r$. All other global topological conclusions about $\mathscr{S}_r$, such as connectivity or dimension, follow from (iii).

As for the system of $\mathscr{F}_r$, we make the following assumptions:

(iv) $j_r(\mathscr{S}_r^0) \subset \mathscr{S}_{r-1}^0$ (i.e. $\mathscr{S}_r^0 \subset \mathscr{S}_{r-1}^0 \cap \mathscr{X}_r$),

(v) $j_r(\mathscr{S}_r) \subset \mathscr{S}_{r-1}$ (i.e. $\mathscr{S}_r \subset \mathscr{S}_{r-1} \cap \mathscr{X}_r$),

and moreover, if $\mathscr{T} \subset \mathscr{S}_r$ is unbounded, so is $j_r\mathscr{T} \subset \mathscr{S}_{r-1}$. Under this condition $j_r: \mathscr{S}_r \to \mathscr{S}_{r-1}$ extends to a continuous $\hat{j}_r: \hat{\mathscr{S}}_r \to \hat{\mathscr{S}}_{r-1}$. We further assume:

(vi) the spaces $\mathscr{Y}, \mathscr{Y}^0$ are independent of r and $e_r = e_{r-1}j_r: \mathscr{S}_r \to \mathscr{Y}$ (this last condition can be weakened to a homotopy condition, but this is unnecessary for the applications).

Let $\mathscr{S}_\infty = \cap \mathscr{S}_r$, $\mathscr{S}_\infty^0 = \cap \mathscr{S}_r^0$ with the limit topologies. A sequence in $\mathscr{S}_\infty$ converges if and only if it converges in each $\mathscr{S}_r$. A set in $\mathscr{S}_\infty$ is bounded if and only if it is bounded in each $\mathscr{S}_r$. We adjoin a point at infinity as above.

THEOREM. *The natural map* $e_\infty: \mathscr{S}_\infty \to \mathscr{Y}$ *extends to a map* $\hat{e}_\infty: \hat{\mathscr{S}}_\infty \to \mathscr{Y}$. *The space* $\hat{\mathscr{S}}_\infty/\hat{\mathscr{S}}_\infty^0$ *is compact and the class* $\hat{e}_\infty^*\gamma \in \check{H}^*(\hat{\mathscr{S}}_\infty/\hat{\mathscr{S}}_\infty^0)$ *is not zero.*

Thus the same global topological consequences can be drawn concerning $\mathscr{S}_\infty$ as for the $\mathscr{S}_r$.

PROOF. For any sequence of spaces $\mathscr{X}_r$ with continuous maps $j_r: \mathscr{X}_r \to \mathscr{X}_{r-1}$, the inverse limit $\lim \mathscr{X}_r \subset \prod \mathscr{X}_r$ and it equals $\{z = (z_r): j_r z_r = z_{r-1}\}$. For such a sequence denote the point in $\lim \mathscr{X}_r$ by $\lim z_r$. Let $k_r: \lim \mathscr{X}_r \to \mathscr{X}_r$ be the continuous map $\lim z_r \mapsto z_r$. The topology on $\lim \mathscr{X}_r$ is the topology of pointwise convergence; i.e. a sequence $z^i = \lim z_r^i$ converges to $z^0 = \lim z_r^0$ if and only if each $z_r^i \to z_r^0$. This is precisely the topology on $\mathscr{S}_\infty$.

Given any sequence $\lim z_r^i \in \mathscr{S}_\infty$, by (v) there is a subsequence $\lim z_r^i$ such that either $|z_r^i| \to \infty$ for each r or $|z_r^i|$ is bounded for each r. In the first case

$\lim z_r^i \to \infty$ and in the second case, by a diagonal argument, we can find a further subsequence $\lim z_r^k$ which converges. This establishes that $k_r \colon \mathscr{S}_\infty \to \mathscr{S}_r$ extends to $k_r \colon \hat{\mathscr{S}}_\infty \to \hat{\mathscr{S}}_r$ and also that $\hat{\mathscr{S}}_\infty$ is compact.

Let $\mathscr{E}$ be the space $\lim(\hat{\mathscr{S}}_r/\hat{\mathscr{S}}_r^0)$ and $e \colon \mathscr{E} \to \mathscr{Y}/\mathscr{Y}^0$ the natural map. For general categorical reasons (or directly by construction), there is a continuous map $f \colon \hat{\mathscr{S}}_\infty \to \mathscr{E}$ such that $e_\infty = ef$. By the limit property of Čech cohomology, $e^*\gamma \in \check{H}^*(\mathscr{E})$ is not zero. The theorem is proved if we show that f is a homeomorphism. Let $s_r \in \hat{\mathscr{S}}_r/\hat{\mathscr{S}}_r^0$ be the point $\{\hat{\mathscr{S}}_r^0\}$. By (iv) $f^{-1} \lim s_r = \hat{\mathscr{S}}_\infty^0$. Thus f induces an inclusion $f \colon \hat{\mathscr{S}}_\infty/\hat{\mathscr{S}}_\infty^0 \to \mathscr{E}$. To show surjectivity, let $z = \lim z_r \in \mathscr{E}$. If all the $z_r = s_r = \{\hat{\mathscr{S}}_r^0\}$, then z is in the image of f. If only a finite number of the $z_r = s_r$, then $z' = \lim z_r$ exists in $\lim \hat{\mathscr{S}}_r$ and $fz' = z$. The theorem is proved.

For the usual constructions, conditions (iv)–(vi) are automatically satisfied, except for the condition on bounded sets in (v). However, for operators $\mathscr{F}_r$ which make functions more regular, such as integral operators, and in particular for operators which invert parabolic and elliptic differential operators, for which Schauder type estimates are available, (v) is satisfied.

PROPOSITION. *Suppose $\mathscr{F}_r = i_{r+1}\mathscr{G}_r$ where $\mathscr{G}_r \colon \mathscr{P} \times \mathscr{X}_r \to \mathscr{X}_{r+1}$ takes bounded sets to bounded sets. Then condition* (v) *holds.*

PROOF. Suppose $\mathscr{T} \subset \mathscr{P} \times \mathscr{X}_r$ is unbounded, but $j_r\mathscr{T} \subset \mathscr{P} \times \mathscr{X}_{r-1}$ is bounded. There exists a sequence $(p_i, x_i) \in \mathscr{T}$ such that either $|p_i| \to \infty$ or $|x_i| \to \infty$. If $|p_i| \to \infty$ then the sequence $j_r(p_i, x_i) = (p_i, i_r x_i)$ is unbounded, so it must be that $|x_i| \to \infty$. However

$$i_r x_i = i_r \mathscr{F}_r(p_i, x_i) = \mathscr{F}_{r-1} j_r(p_i, x_i) = i_r \mathscr{G}_{r-1} j_r(p_i, x_i)$$

so that $x_i = \mathscr{G}_{r-1} j_r(p_i, x_i) \in \mathscr{G}_{r-1}\mathscr{T}$, and thus, the $|x_i|$ are bounded. The proposition is proved.

Note that $\mathscr{G}_r$ need not take all bounded sets to bounded sets, but only those in $\mathscr{S}_r$.

Two final comments: First, the set $\mathscr{S}_\infty$ is a subset of the Fréchet space $\mathscr{X}_\infty$, and it is possible to phrase the result totally within the context of a single Fréchet space. The hypotheses of such a phrasing would involve the relation between an operator $\mathscr{F}_\infty$ and the individual seminorms of the Fréchet space. In this there would be a resemblance to results of the Nash–Moser type. However, the present conditions are much stronger than those of Nash–Moser results, and the result is definitely not a global Nash–Moser result. For the present result, however it is phrased, there really are operators operating on some Banach spaces, and it seemed best to phrase the result in that context.

Second, the result extends to operators defined on open subsets of Banach spaces, if one technical point is valid. To wit, the assumption concerning unbounded $\mathscr{T}$ in (v) must hold for unbounded $\mathscr{T}$ in the closure of $\mathscr{S}_r$, not just $\mathscr{S}_r$ itself.

References

1. J. C. Alexander, *A primer on connectivity* (Proc. Conf. on Fixed Point Theory, 1980, E. Fadell and G. Fournier, eds.), Lecture Notes in Math., vol. 886, Springer-Verlag, 1981, pp. 455–483.

2. J. C. Alexander and J. F. G. Auchmuty, *Global bifurcation of waves*, Manuscripta Math. **27** (1979), 159–166.

3. J. F. G. Auchmuty, *Bifurcating waves*, Bifurcation Theory and Applications in Scientific Disciplines (O. Gurel and O. E. Rössler, eds.), Ann. New York Acad. Sci., vol. 316, 1979, pp. 263–278.

4. _____, *Bifurcation analysis of reaction-diffusion equations*. V: *Rotating waves on a disk*, Partial Differential Equations and Dynamical Systems (W. E. Fitzgibbon, ed.), Res. Notes in Math., vol. 101, Pitman, Boston, 1984, pp. 173–181.

5. J. Leray and J. Schauder, *Topologie et équations fonctionelles*, Ann. Sci. École Norm. Sup. **51** (1934), 45–78.

6. P. H. Rabinowitz, *Some global results for nonlinear eigenvalue problems*, J. Funct. Anal. **7** (1971), 183–200.

UNIVERSITY OF MARYLAND

Asymmetric Rotating Wave Solutions of Reaction-Diffusion Equations

J. C. ALEXANDER[1]

Introduction. J. F. G. Auchmuty [5, 6] has shown there exist travelling wave solutions of reaction-diffusion equations with rotational spatial symmetry. The motivating examples are from chemical kinetics with the Brusselator (Lefever–Prigogine [12]) as the prototypical example. The machinery is bifurcation theory. Branches of wave solutions bifurcate, via Hopf bifurcation, from a stationary solution. The Hopf bifurcations are global in the sense of Alexander–Yorke [3]; this is discussed in Alexander–Auchmuty [1] and used in Auchmuty [6].

The waves investigated by Auchmuty respect the spatial symmetry of the system. That is, they rotate uniformly around the center. The purpose here is to show that asymmetric waves exist. More precisely we observe that the global bifurcation branches persist under perturbations of the system. Since the topological machinery is so straightforward, proving existence of global branches is perhaps easier than proving a local result, say with an implicit function theorem.

The reason for establishing such perturbed global branches is twofold. Obviously bifurcation should obtain for any physically or numerically meaningful example. Auchmuty mentions that his work was motivated in part by numerical studies of Erneux and Herschkowitz-Kaufman [8]. Secondly, we establish travelling wave solutions for systems without rotational symmetry, and asymmetric solutions (uncountably many) for symmetric systems. We illustrate with examples.

After setting up the problem we sketch that part of Auchmuty's machinery that we need for our results. Then we establish a general result, which in the last section we apply to several examples.

1980 *Mathematics Subject Classification*. Primary 35B32, 35B20, 92A20; Secondary 47H10, 35K22.
[1] Partially supported by the NSF.

The formulation. In this section we review the functional analytic formulation of the problem, and put it in a form suitable for the machinery of bifurcation theory. The material in the next two sections is basically from Auchmuty [6] to which the reader is referred for more details.

Consider a smooth bounded domain Ω with axial symmetry in n-dimensional Euclidean space. Let θ be the coordinate which denotes the rotational angle with respect to some reference direction, and let r denote an $(n-1)$-dimensional complementary "radial coordinate." Thus $\Omega = \{(r,\theta): \theta \in S^1, r \in \Omega^\perp \subset R^{n-1}\}$. Auchmuty has explicitly considered the cases of Ω an annular ring ($n = 2$, $0 < r_0 \leq r \leq r_1$) (Auchmuty [5]) and a disk ($n = 2$, $0 \leq r \leq r_1$) (Auchmuty [6]) but it is clear the analysis applies also, for example, to rotationally symmetric 3-dimensional bodies such as cylinders (which Auchmuty has also considered), balls, ellipsoids, toroids,... (in which case r denotes a 2-dimensional "coordinate").

Consider a system of m (chemical) reactants; let $u_i(r, \theta, t)$ denote the concentration of the ith reactant at time t at the point $(r, \theta) \in \Omega$. Let $u = u(r, \theta, t)$ denote the column vector of the u_i. The reaction-diffusion equation satisfied by u is the vector equation

$$(1) \qquad \partial u/\partial t = f_\beta(u) + D \cdot \nabla^2 u.$$

Here $\partial/\partial t$ and the Laplacian operate componentwise, D is a diagonal matrix with diagonal entries diffusivities $D_i > 0$, and f_β expresses the point (or well-mixed) chemical kinetics of the system. Thus $f_\beta(u)$ is a smooth m-vector function of $(u_1, \ldots, u_m)$. The parameter β is a parameter of the kinetics which we isolate explicitly as a bifurcation parameter. We assume $f_\beta(0) = 0$, so $u = 0$ is a stationary point of the point system.

For example, the Brusselator has two chemicals X, Y ($m = 2$) which satisfy the point kinetics

$$\dot{X} = A - (B+1)X + X^2Y; \qquad \dot{Y} = BX - X^2Y.$$

These equations have the unique stationary point $X = A$, $Y = B/A$ and we let $u_1 = X - A$, $u_2 = Y - B/A$. The parameters A, B are auxiliary parameters; usually A is fixed, and B plays the role of a bifurcation parameter. Thus for the Brusselator, (1), written out in components, reads

$$\frac{\partial u_1}{\partial t} = A - (B+1)(u_1 + A) + (u_1 + A)^2(u_2 + B/A) + D_1 \nabla^2 u_1,$$

$$\frac{\partial u_2}{\partial t} = B - (u_1 + A)^2(u_2 + B/A) + D_2 \nabla^2 u_2.$$

For a broad expository treatment of such types of chemical reactions, see Nicolis-Prigogine [14].

A solution u of (1) is a rotating wave if it has the form $u(r, \theta, t) = v(r, \theta - ct)$, where v is a function of the space variables alone. Here c is a nonzero constant wave speed. Let $\zeta = \theta - ct$; then v satisfies

(2) $$D\nabla^2 v + c\frac{\partial v}{\partial \zeta} + f_\beta(v) = 0.$$

Thus c has become a parameter.

Let $J_\beta = (\partial f_{\beta,i}/\partial u_j)_u = 0$ denote the Jacobian of f at $u = 0$. The formal linearization of (2) at $u = 0$ is thus

(3) $$D\nabla^2 v + c\frac{\partial v}{\partial \zeta} + J_\beta \cdot v = 0.$$

We consider (2) (and thus (1)) with Neumann or Dirichlet boundary conditions:

(4a) $$\partial v/\partial \nu = 0 \quad \text{on } \partial\Omega,$$

(4b) $$v = 0 \quad \text{on } \partial\Omega$$

respectively. We mostly discuss (4a) here, the case (4b) is similar but slightly easier.

Let $\mathscr{X}_1, \mathscr{X}_0$ be two function spaces over Ω such that the Laplacian operator $\mathscr{F}_0$: $\mathscr{X}_1 \to \mathscr{X}_0$ is Fredholm of index 0; e.g. $\mathscr{X}_1$ is the Sobolev space $\mathscr{H}^{2,2}$ and $\mathscr{X}_0$ is $\mathscr{L}^2$. Alternatively, $\mathscr{X}_1$ and $\mathscr{X}_0$ could be appropriate Hölder spaces. Let $\mathscr{X}$ be such that $\mathscr{X}_1 \subset \mathscr{X} \subset \mathscr{X}_0$ and (i) for $v \in \mathscr{X}$, $\partial v/\partial \zeta \in \mathscr{X}_0$ and depends continuously on v, and (ii) the inclusion $\mathscr{X}_1 \subset \mathscr{X}$ is compact. In particular, we can let $\mathscr{X} = \mathscr{C}^1(\Omega)$.

For Neumann conditions (4a), the kernel and cokernel of $\mathscr{F}_0$ are both the linear subspace $\mathscr{K}$ of constant functions. For $w \in \mathscr{X}_i$ ($i = 1, 2$), let $\langle w \rangle = |\Omega|^{-1}\int_\Omega w$ denote the average of w. The operator $\mathscr{F}: \mathscr{X}_1 \to \mathscr{X}_0$ defined by $\mathscr{F}w = \mathscr{F}_0 w + \langle w \rangle$ is an isomorphism; its inverse is convolution with a Green's kernel. We form the operator $\mathscr{G} = \mathscr{G}_{c,\beta}: \mathscr{X} \to \mathscr{X}$ defined by (suppressing notation for inclusions)

(4) $$\mathscr{G}(v) = \mathscr{F}^{-1}\left(D^{-1}\left(c\frac{\partial v}{\partial \zeta} + f_\beta(v)\right) + \langle v \rangle\right).$$

The following are then verified: (i) v is a fixed point of $\mathscr{G}$ if and only if $v \in \mathscr{X}$ satisfies (2); (ii) $\mathscr{G}$ is a compact continuous operator in c, β, v; (iii) $\mathscr{G}$ has a Fréchet derivative given at 0 by

(5) $$\mathscr{L}v = \mathscr{L}_{c,\beta}v = \mathscr{F}^{-1}\left(D^{-1}\left(c\frac{\partial v}{\partial \zeta} + J_\beta \cdot v\right) + \langle v \rangle\right)$$

which depends continuously on c, β, where J_β is the Jacobian defined above. Thus in particular v satisfies (5) if and only if v satisfies the linear equation (3).

The case (4b) is slightly easier, since there is no kernel or cokernel, and the average $\langle w \rangle$ need not be used. Auchmuty actually uses a slightly different operator to eliminate the kernel; the reduction to (3) is unchanged.

The linear bifurcation analysis. The first step in the bifurcation analysis is to determine when (3) has nontrivial solutions. To this end, Auchmuty considers the scalar eigenvalue problem on Ω,

(6) $$-\nabla^2 G = \lambda G.$$

with respective boundary conditions

(7a) $$\partial G/\partial \nu = 0 \quad \text{on } \partial\Omega,$$
(7b) $$G = 0 \quad \text{on } \partial\Omega.$$

The important fact about (6) is that the Laplacian separates so we can write $G(r,\theta) = R(r)\Theta(\theta)$ and (6) has solutions $R_{(l)}(r)e^{\pm i|l|\theta}$. Here (l) is meant symbolically as an n-index for the eigenfunctions; it consists of an $(n-1)$-index $(l^\perp)$ and a nonnegative integer $|l|$. The eigenvalue $\lambda = \lambda(l)$. For example, for Ω a disk or annulus, $l = (p,q)$ with $p = |l|$, $\lambda(l) = q^2$. The radial function $R_{(l)}(r)$ is a sum of Bessel functions $\mathcal{J}_p(qr)$ and $\mathcal{Y}_p(qr)$. For a ball in 3-dimensional space, r is the pair (ρ, ψ) of radial and colatitudinal spherical coordinates, $l = (|l|, p, q)$ and $R_{(l)}(\rho, \psi)$ is the product of a spherical Bessel function $\rho^{-1/2}\mathcal{J}_{p+1/2}(qr)$ and an associated Legendre function $p_q^{|l|}(\psi)$. Again $\lambda(l) = q^2$.

The following result completely determines the nontrivial periodic solutions of the linearized operator.

THEOREM (AUCHMUTY). *(3) and (4) have nontrivial periodic solutions if and only if there is an eigenvalue $\lambda(l)$ of (6), (7) with $|l| > 0$ such that the $(m \times m)$-matrix*

(8) $$\lambda(l)D - J_{\beta_0}$$

has a pair of conjugate purely imaginary eigenvalues $\pm i|l|c_0$. Let $\gamma = (\gamma_1, \ldots, \gamma_m)^T$ and $\bar{\gamma}$ be the corresponding eigenvectors of (8). Then the associated solutions of (3), (4) are

(9) $$R_{(l)}(r)e^{\pm i|l|\zeta}\gamma.$$

Thus the real and imaginary parts of (9) are two linearly independent real solutions of (3), (4); these have components

$$|\gamma_i|R_{(l)}(r)\cos(l\zeta - \arg \bar{\gamma}_i), \quad |\gamma_i|R_{(l)}(r)\sin(l\zeta - \arg \bar{\gamma}_i),$$

respectively. The dimension of the space of solutions equals the geometric multiplicity of null space of the operator $\mathcal{L}_{c,\beta} - I$.

Rotational symmetry is imperative for this theorem. The reason is that the eigenfunctions for the Laplacian form a smooth basis for $\mathcal{L}^2(\Omega)$ (with appropriate boundary conditions) that have a simple dependence on ζ. That is, there is complete control over the derivative $\partial/\partial\zeta$ in the operator $\mathcal{L}$. We illustrate later with the ellipse the necessity of rotational symmetry.

Note that the wave speed c is not arbitrary, but is determined by an eigenvalue condition. Recall that $\zeta = \theta - ct$. Thus the integer $|l|$ is the wave number of the solution. Note that the geometric multiplicity of solutions is always finite (since there are only a finite number of (l) with $\lambda(l) \leq L$ for any L) and is even (as one expects for bifurcations of periodic solutions), and for any β_0, c_0 it equals twice the sum of the geometric multiplicities of ilc_0 as eigenvalues of (8), as (l) ranges over indices (counted with multiplicity) such that $\lambda(l)$ is an eigenvalue of (6), (7). Generically the multiplicity is two.

Global bifurcation. There is next the transition from the linear to the nonlinear problem. With the machinery of bifurcation theory in place, this is a short step. We need a transversality condition. Assume

(T): there are $\varepsilon > 0$, $\delta > 0$ such that when $\max(|c - c_0|, |\beta - \beta_0|) \leq \delta$ then

$$|\mathscr{L}_{c,\beta} v - v| \geq \varepsilon \cdot \max(|c - c_0|, |\beta - \beta_0|) \cdot |v|,$$

where $|v|$ is the norm in $\mathscr{X}$.

This is implied for example if, as β varies, the continuation of all the purely imaginary eigenvalues $\pm i|l|c_0$ of (8) cross the imaginary axis transversally (see Alexander–Yorke [3], Fitzpatrick [9]). If quantitative information about the dependence on β of the real part of the relevant eigenvalues is available, the parameters δ, ε of (T) can be estimated.

THEOREM (ALEXANDER–AUCHMUTY). *If the transversality condition* (T) *holds, and the dimension of the null space of $\mathscr{G}_{c_0,\beta_0} - I$ equals $2r$ with r odd, then a global branch of nontrivial solutions of* (3), (4) *bifurcates from the trivial solution at c_0, β_0.*

We remark that it is also meaningful to consider one or more of the D_i as the bifurcation parameter rather than β. In particular, multiplying the matrix D by k^2 corresponds to shrinking the size of Ω by the factor k. For the Brusselator, or any 2-chemical system, the corresponding transversality condition always holds. This is because if the eigenvalues of $\lambda k^2 D - J$ are $\alpha(k) \pm i\gamma(k)$, with $\alpha(1) = 0$, then

$$\alpha(k) = \tfrac{1}{2} \operatorname{Tr}(\lambda k^2 D - J) = \tfrac{1}{2}\lambda(k^2 - 1) \operatorname{Tr} D,$$

which has derivative $\lambda \operatorname{Tr} D$ at $k = 1$.

Perturbed systems. Suppose $\mathscr{G}_{c,\beta}^{(\kappa)}: \mathscr{X} \to \mathscr{X}$ is a perturbation of (2) such that (i) $\mathscr{G}_{c,\beta}^{(\kappa)}(v)$ is continuous and compact in c, β, v, (ii) $\mathscr{G}_{c,\beta}^{(\kappa)}(0) = 0$, (iii) $\mathscr{G}_{c,\beta}^{(\kappa)}$ has a Fréchet derivative at 0, denoted $\mathscr{L}_{c,\beta}^{(\kappa)}$, which depends continuously on c, β. Let c_0, β_0 be as in the previous theorem, and let δ, ε be as in condition (T). Let $\Sigma = \{(c, \beta): \max(|c - c_0|, |\beta - \beta_0|) = \delta\}$.

THEOREM *Suppose for $(c, \beta) \in \Sigma$ that $|\mathscr{L}_{c,\beta}^{(\kappa)} - \mathscr{L}_{c,\beta}| < \varepsilon$. Then there exists a global branch of fixed points of $\mathscr{G}_{c,\beta}^{(\kappa)}$ which bifurcates from the trivial solution at some point $\bar{c}_0$, $\bar{\beta}_0$ with $|c_0 - \bar{c}_0| < \delta$, $|\beta_0 - \bar{\beta}_0| < \delta$.*

PROOF. Let $\mathscr{S}_d$ be the space of nonlinear compact continuous operators $\mathscr{X} \to \mathscr{X}$ with no nontrivial fixed points on the disk $\{v \in X: |v| \leq d\}$ for some $d > 0$. The map $\sigma_0: (c, \beta) \mapsto \mathscr{G}_{c,\beta}$ is a map $\Sigma \to \mathscr{S}_d$ for some $d > 0$. The assumptions of the theorem of the previous section guarantee that σ_0 represents the nontrivial element in the fundamental group $\pi_\perp(\mathscr{S}_d) \approx \mathbf{Z}/2\mathbf{Z}$ and this in turn implies that global bifurcation occurs (this is precisely the way the machinery is set up in Alexander–Fitzpatrick [2]). The assumptions of the present theorem guarantee that the map $\sigma_\kappa: (c, \beta) \mapsto \mathscr{G}_{c,\beta}^{(\kappa)}$ is homotopic to σ_0 and thus also represents the nontrivial homotopy element. Thus bifurcation has to occur as claimed. The proof is complete.

Note that the perturbation estimates are not on $\mathscr{G}^{(\kappa)}$ itself, but only on its Fréchet derivative at $v = 0$. Away from $v = 0$, $\mathscr{G}^{(\kappa)}$ can differ from $\mathscr{G}$ by an arbitrarily large amount.

We remark that although the theorem is stated locally in $\mathscr{L}^{(\kappa)}$ there is a continuation property for $\mathscr{L}^{(\kappa)}$. That is, if $\mathscr{L}^{(\kappa)}$ is continuously changed, the bifurcation persists beyond the bounds guaranteed by the theorem. There are only a few ways the bifurcation point can disappear. It can amalgamate with, and cancel, another bifurcation point which is connected to the bifurcation point in question by a global branch, it can wander off to infinity or the boundary of the domain of definition, or the period of the solutions of the linearized problem at the bifurcation point can grow to infinity.

Applications. The type of perturbation allowed in the previous theorem is quite general. Any small time-independent change is permitted, and global branches of rotating wave solutions exist. In this section we consider the possibilities. Any combination of the possibilities discussed below is permitted. We do not attempt detailed estimates of the size of perturbations permitted. Such estimates depend on the values of δ, ε of condition (T) and can be worked out in detail in any particular case. All of the informal statements below can be formalized in the form: There exists δ_1, $\varepsilon_1 > 0$ such that if for $|u| < \varepsilon_1, \ldots$ is perturbed by no more than δ_1 in the ... topology from its form in (1) or (2), and c_0, β_0 are as above, then bifurcation occurs from a point $\bar{c}_0$, $\bar{\beta}_0$ within distance δ of c_0, β_0. Moreover, as noted in the previous section, bifurcation will generally occur for much larger perturbations.

(I) The most obvious change is an alteration of the kinetics f. A small $\mathscr{C}^1$ change in f effects a small change in the operator.

(II) A second obvious change is an alteration of the diffusion operator. For example, the diffusion could be nonlinear or could depend on position. The requirements are that the corresponding operator $\mathscr{G}^{(\kappa)}$ of equation (4) can be formed satisfying properties (i)–(iii) (i.e. there exist an invertible $\mathscr{F}^{(\kappa)}$ of some sort), and that the Fréchet derivative $\mathscr{L}^{(\kappa)}$ at $v = 0$, be close to $\mathscr{L}$.

(III) As a special case, the diffusivity matrix D can be C^0 perturbed. In particular, it can be spatially dependent.

(IV) A less obvious change is a perturbation in the coordinate system (r, ζ). Let $h\colon \Omega \to \Omega$ be a C^1 diffeomorphism which is C^1 close to the identity. Then $(r', \zeta') = (rh, \zeta h)$ is another coordinate system for Ω which is close to the original one. Since

$$\frac{\partial}{\partial \zeta'} = \frac{\partial \zeta}{\partial \zeta'} \frac{\partial}{\partial \zeta} + \frac{\partial r}{\partial \zeta'} \frac{\partial}{\partial r}$$

(where the partials denote the appropriate block matrices), the operators (2) and (3) with $\partial/\partial \zeta'$ substituted for $\partial/\partial \zeta$ are small perturbations of (2) and (3) and thus exhibit bifurcation. In the case of boundary conditions (4a), it is necessary that the diffeomorphism h satisfy a first-order condition at the boundary in order to preserve the boundary condition. The requirement is that h preserve the normal direction at the boundary. In particular, the condition is satisfied if h is conformal at the boundary.

This type of perturbation is interesting because the coordinate system is not involved in (1). It is introduced in (2) in order to define a rotating wave. The hypersurfaces ζ = const. define the "wave front" and the spacing between the hypersurfaces governs the speed of the wave, since the wave propagates at constant ζ speed. Also the axis $r = 0$ is the center of rotation of the wave. Thus near any bifurcation branch of symmetric rotating wave solutions of (1) we obtain a continuum (one for each diffeomorphism close to the identity) of rotating wave solutions which have the geometry of the associated coordinate system.

As a particular example, let Ω be the disk $\{z \in C, |z| \leq 1\}$. The conformal diffeomorphisms of the disk are given by the complex analytic maps

$$h(z) = e^{i\delta} \frac{z - \eta}{1 - \eta z}$$

for $|\eta| < 1$ and δ real (Schwarz' theorem). The $e^{i\delta}$ is an uninteresting rotation. However, for each η near 0, we obtain a rotating wave solution of (1) which has η as its center and circular arcs as its wavefronts.

(V) Finally Ω itself can be perturbed. Such a change effects a change in Green's function which inverts $\mathscr{F}$. There are variational formulas (Bergmann–Schiffer [7], Hadamard [11], see also Garabedian [10]) which show that a C^1 perturbation in the boundary of Ω effects a perturbation in $\mathscr{G}$ and $\mathscr{L}$ and thus we obtain wave solutions for all regions C^1 close to Ω.

Consider, as an explicit example, the case of an annular ring bounded by two confocal ellipses. The case that the ellipses are circles is the subject of Auchmuty [5], Alexander–Auchmuty [1]. Let the foci be at the points $(\pm h, 0)$ in the (x, y) plane. The coordinates of choice are elliptic coordinates (ξ, η) with

$$x = h \cosh \xi \cos \eta, \qquad y = h \sinh \xi \sin \eta.$$

The coordinate curves are confocal ellipses (ξ = const.) and hyperbolae (η = const.). The domain $\Omega^\perp = \Omega^\perp(\xi_0, \xi_1) = \{\xi : \xi_0 \leq \xi \leq \xi_1\}$. Equation (6) becomes

$$(10) \qquad \frac{2}{h^2(\cosh 2\xi - \cos 2\eta)} \left(\frac{\partial^2 G}{\partial \xi^2} + \frac{\partial^2 G}{\partial \eta^2} \right) = -\lambda G,$$

which separates, if $G = \psi(\xi)\phi(\eta)$, into two Mathieu equations

$$(11) \qquad \frac{d^2\phi}{d\eta^2} + \left(a - \frac{\lambda h^2}{2} \cos 2\eta \right) \phi = 0,$$

$$(12) \qquad \frac{d^2\psi}{d\xi^2} - \left(a - \frac{\lambda h^2}{2} \cosh 2\xi \right) \psi = 0,$$

where a is a separation constant. Equation (11) imposes a relation between a and λh^2, and given λh^2, the separation constant a can assume one of a countable number of nonnegative values a_n, b_n with respective eigenvalues $ce_n(\eta, \lambda^2 h/4)$,

$se_n(\eta, \lambda^2 h/4)$ (Mathieu functions). As $h \to 0$, $a_n \to n^2$, $b_n \to n^2$,

$$ce_n\left(\eta, \frac{\lambda^2 h}{4}\right) \to \cos n\eta, \qquad se_n\left(\eta, \frac{\lambda^2 h}{4}\right) \to \sin n\eta,$$

but for $h > 0$, $a_n \neq b_n$.

A complete set of eigenfunctions of (10) are products of Mathieu functions

$$Ge_n\left(\xi, \frac{\lambda h^2}{4}\right) ce_n\left(\eta, \frac{\lambda h^2}{4}\right), \quad Fey_n\left(\xi, \frac{\lambda h^2}{4}\right) ce_n\left(\eta, \frac{\lambda h^2}{4}\right) \qquad (n \geq 0),$$

$$Se_n\left(\xi, \frac{\lambda h^2}{4}\right) se_n\left(\eta, \frac{\lambda h^2}{4}\right), \quad Gey_n\left(\xi, \frac{\lambda h^2}{4}\right) se_n\left(\eta, \frac{\lambda h^2}{4}\right) \qquad (n \geq 1),$$

(for notation and a discussion, see e.g. McLachlan [13]). As $h \to 0$, the functions Ge_n, Se_n, Fey_n, Gey_n, suitably rescaled, converge to Bessel functions. As usual, the boundary conditions determine conditions on λ.

The point is that there is a complete and explicit theory of the eigenvalue problem (6), (7). However, the linear bifurcation theorem does not apply. The rotational symmetry necessary for that theorem is broken, the eigenvalues of (6), (7) are not doubly degenerate, and the ζ dependence is not of the form $e^{\pm i|l|\zeta}$. There is no convenient expression for the derivatives ce'_n, se'_n and the derivative $\partial/\partial \eta$ cannot be controlled. However by the perturbation theorem, we know branches exist for small $h > 0$.

It is also possible to formulate this example as a perturbation of the elliptic operator. If we use the semi-major axis $\rho = h \cosh \xi$ as a variable instead of ξ, the Laplacian $\nabla^2 v$ becomes

$$(13) \qquad \frac{2}{2\rho^2 - h^2 - h^2 \cos 2\zeta}\left[(\rho^2 - h^2)\frac{\partial^2 v}{\partial \rho^2} + \rho\frac{\partial v}{\eta \rho} + \frac{\partial^2 v}{\partial \eta^2}\right].$$

As $h \to 0$, (13) becomes the expression for the Laplacian in polar coordinates. Thus for small enough $h > 0$, bifurcation occurs and there are rotating wave solutions on the elliptic annular region; the wavefronts are hyperbolic arcs.

Finally there is the general question, unaddressed in this paper, of how the bifurcation branches behave as the perturbation is continuously varied. Do the global brances themselves, as well as the bifurcation points, vary in some reasonable way? This can be answered in terms of higher-dimensional cohomology, and will be discussed in general in another paper.

References

1. J. C. Alexander and J. F. G. Auchmuty, *Global bifurcation of waves*, Manuscripta Math. **27** (1979), 159–166.

2. J. C. Alexander and P. M. Fitzpatrick, *The homotopy of certain spaces of nonlinear operators, and its relation to global bifurcation of the fixed points of parametrized condensing operators*, J. Functional Analysis **34** (1979), 87–106.

3. J. C. Alexander and J. A. Yorke, *Global bifurcation of periodic orbits*, Amer. J. Math. **100** (1978), 37–53.

4. _____, *Calculating bifurcation invariants as elements in the homotopy of the general linear group*, J. Pure Appl. Algebra **13** (1978), 1–8.

5. J. F. G. Auchmuty, *Bifurcating waves*, Bifurcation theory and applications in scientific disciplines (O. Gurel and O. E. Rössler, eds.), Ann. N. Y. Acad. Sci., Vol. 316, 1979, 263–278.

6. _____, *Bifurcation analysis of reaction-diffusion equations. V: Rotating waves on a disc*, Partial Differential Equations and Dynamical Systems (W. E. Fitzgibbon, ed.), Res. Notes in Math. no. 101, Pitman 1984, pp. 173–181.

7. S. Bergman and M. Schiffer, *Kernel functions and elliptic differential equations in mathematical physics*, Academic Press, 1953.

8. T. Erneux and M. Herschkowitz-Kaufman, *Rotating waves as asymptotic solutions of a model chemical reaction*, J. Chem. Phys. **66** (1977), 248–253.

9. P. M. Fitzpatrick, *Bifurcation, linearization and homotopy*, J. Nonlinear Anal. (to be published)

10. P. R. Garabedian, *Partial differential equations*, Wiley, 1964.

11. J. Hadamard, *Mémoire sur le problème d'analyse relatif à l'équilibre des plaques elastiques encastrées*, Mem. Acad. Sci. **33** (1908), 1–128.

12. R. Lefever and I. Prigogine, *Symmetry-breaking instabilities in dissipative systems*. II, J. Chem. Phys. **48** (1968), 1695–1700.

13. N. W. McLachlan, *Theory and applications of Mathieu functions*, Oxford Univ. Press, 1947.

14. G. Nicolis and I. Prigogine, *Self-organization in non-equilibrium systems*, Wiley, 1977.

DEPARTMENT OF MATHEMATICS, AND INSTITUTE FOR PHYSICAL SCIENCE AND TECHNOLOGY, UNIVERSITY OF MARYLAND

Parabolic Evolution Equations with Nonlinear Boundary Conditions

HERBERT AMANN

1. Introduction. In this note we consider semilinear parabolic systems of the form

(1.1)
$$\frac{\partial u}{\partial t} + \mathscr{A}u = f(x,t,u) \quad \text{in } \Omega \times (0,\infty),$$
$$\mathscr{B}u = g(x,t,u) \quad \text{on } \partial\Omega \times (0,\infty),$$
$$u(x,0) = u_0(x) \quad \text{on } \Omega.$$

Here Ω denotes a bounded domain in $\mathbf{R}^n$ with a smooth boundary and $(\mathscr{A}, \mathscr{B})$ is a regular elliptic system operating on N-vector valued functions $u: \Omega \to \mathbf{R}^N$. Instead of writing down the precise hypotheses upon $(\mathscr{A}, \mathscr{B})$ we consider the following simple example which indicates what we have in mind:

(1.2)
$$\frac{\partial u}{\partial t} - M\Delta u = f(x,t,u) \quad \text{in } \Omega \times (0,\infty),$$
$$M\frac{\partial u}{\partial \nu} = g(x,t,u) \quad \text{on } \partial\Omega \times (0,\infty),$$
$$u(\cdot,0) = u_0 \quad \text{on } \Omega.$$

Here $N = 2$ and Δ denotes the diagonal Laplacian (that is, $\Delta u = (\Delta u^1, \Delta u^2)$ for $u = (u^1, u^2)$), and $M := \begin{bmatrix} \alpha & \gamma \\ 0 & \beta \end{bmatrix}$, where the "diffusion" coefficients α and β are positive and the "drift" coefficient—describing the drift of the quantity u^1 in the direction of ∇u^2—is arbitrary. Thus, if $\gamma \neq 0$, (1.2) is a strongly coupled nonlinear parabolic system with nonlinear boundary conditions (if g is a nonlinear function of u, of course). Here ν denotes the outer normal on $\Gamma := \partial\Omega$.

Problems of the above type—in particular so-called chemotaxis problems of the type of example (1.2)—occur in many mathematical models of phenomena studied in physics, chemistry, biology, population dynamics, etc. Despite the great

---

1980 *Mathematics Subject Classification.* Primary 35K55; Secondary 47H20.

amount of work which in recent years has been done in the mathematical study of so-called "reaction-diffusion equations" not much seems to be known for reaction-diffusion equations with nonlinear boundary conditions.

Of course, there is the classical work summarized in the books by Ladyženskaya–Solonnikov–Ural'ceva [12] and by Friedman [8]. These results are very precise. But they are essentially restricted to one equation and require somewhat restrictive structure conditions. More recently problems with nonlinear boundary conditions have been studied in the framework of monotone operator theory (e.g. [4, 6, 10]). However these methods are not applicable to problem (1.2) if $|\gamma| > 2\sqrt{\alpha\beta}$. Finally there is some recent work by Alikakos [2], Pao [16], and Leung [13] on problems with nonlinear boundary conditions. However all of these results apply only to particular situations and depend heavily on the classical results given in [8, 12].

It is the purpose of this note to indicate a quite different approach in the spirit of abstract evolution equations. Thus it applies to very general systems of arbitrary order and does not impose any structure conditions upon the nonlinearities. In order to motivate this approach we recall briefly how one can handle problem (1.1) in the case of homogeneous boundary conditions $g = 0$. In this case one interprets (1.1) as an abstract evolution equation in a Banach space X (usually an L_p-space) of the form

$$\dot{u} + Au = f(t, u), \quad t > 0, \quad u(0) = u_0,$$

where A is an unbounded linear operator whose domain consists of those functions (in an appropriate Sobolev space) satisfying the boundary conditions $\mathcal{B}u = 0$. Then $-A$ is the infinitesimal generator of an analytic semigroup $U(t) := e^{-tA}$, $t \geq 0$, and problem (1.1) is essentially equivalent to the integral equation

$$(1.3) \qquad u(t) = U(t)u_0 + \int_0^t U(t - \tau)f(\tau, u(\tau))\, d\tau, \qquad t \geq 0.$$

This integral equation, which is a consequence of the "variations-of-constants formula", is the basis for the proof of local and global existence theorems and for the study of the asymptotic behavior of the solutions, that is, for the study of stability questions, etc.

Now in the case of nonlinear boundary conditions there is no obvious generalization of the variations-of-constants formula (1.3). This seems to be the main difficulty in dealing with problems of the type (1.1). It is the main purpose of this note to show that we can obtain a kind of variations-of-constants formula also in the presence of nonlinear boundary conditions, which can be used in much the same way as the integral equation (1.3).

2. Heuristic derivation of a variations-of-constants formula. Let W, X, and Y be Banach spaces such that $W \hookrightarrow X$, that is, W is continuously imbedded in X. Moreover we suppose that

$$(2.1) \qquad \mathcal{A} \in \mathcal{L}(W, X), \qquad \mathcal{B} \in \mathcal{L}(W, Y),$$

where $\mathscr{L}(X, Y)$ denotes the Banach space of all continuous linear operators mapping X into Y. Furthermore, we suppose that

(2.2) $\quad\quad\quad\quad\quad\quad\quad\quad \mathscr{B}$ is surjective

and

(2.3) $\quad\quad A := \mathscr{A}|W_{\mathscr{B}}\colon W_{\mathscr{B}} := \ker(\mathscr{B}) \to X \quad$ is an isomorphism.

Observe that $P := A^{-1}\mathscr{A} \in \mathscr{L}(W, W_{\mathscr{B}})$ satisfies $P^2 = P$ and $W_{\mathscr{A}} := \ker(\mathscr{A}) = \ker(P)$. Hence it follows that W is the topological direct sum of $W_{\mathscr{A}}$ and $W_{\mathscr{B}}$,

(2.4) $\quad\quad\quad\quad\quad\quad\quad\quad W = W_{\mathscr{A}} \oplus W_{\mathscr{B}},$

and that

(2.5) $\quad\quad\quad\quad \mathscr{R} := (\mathscr{B}|W_{\mathscr{A}})^{-1}\colon Y \to W_{\mathscr{A}} \quad$ is an isomorphism.

Now we consider the abstract linear evolution problem

(2.6) $$\begin{aligned} \dot{u} + \mathscr{A}u &= f(t), \\ \mathscr{B}u &= g(t), \\ u(0) &= u_0, \end{aligned} \quad 0 < t < \infty,$$

where $f \in C(\mathbf{R}^+, X)$ and $g \in C(\mathbf{R}^+, Y)$. According to (2.4) we can decompose a solution u of (2.6) (whose existence we presuppose) as

$$u(t) = v(t) + w(t) \in W_{\mathscr{A}} \oplus W_{\mathscr{B}}.$$

Then, by (2.6),

(2.7) $$\begin{aligned} \dot{w} + \mathscr{A}w &= f(t) - \dot{v}(t), \\ \mathscr{B}v &= g(t), \\ w(0) &= u_0 - v(0). \end{aligned}$$

Since, by (2.5), $v(t) = \mathscr{R}g(t)$, we formally obtain from (2.7) the initial value problem

(2.8) $$\begin{aligned} \dot{w} + Aw &= f(t) - (\mathscr{R}g)^{\cdot}(t), \quad t > 0, \\ w(0) &= u_0 - v(0). \end{aligned}$$

Now we suppose that

(2.9) $\quad\quad$ $-A$ is the infinitesimal generator of a strongly continuous analytic semigroup $U(t) := e^{-tA}$, $t \geq 0$, on X.

Then we can represent the solution w of (2.8) by means of the variations-of-constants formula

$$\begin{aligned} w(t) &= U(t)[u_0 - v(0)] + \int_0^t U(t - \tau)[f(\tau) - (\mathscr{R}g)^{\cdot}(\tau)]\, d\tau \\ &= U(t)[u_0 - v(0)] - \mathscr{R}g(t) + U(t)\mathscr{R}g(0) \\ &\quad + \int_0^t U(t - \tau)f(\tau)\, d\tau - \int_0^t DU(t - \tau)\mathscr{R}g(\tau)\, d\tau, \end{aligned}$$

where we integrated formally by parts. Since $\mathscr{R}g(t) = v(t)$, we obtain *by formal manipulations* that every solution of (2.6) is given by

$$(2.10) \quad u(t) = U(t)u_0 + \int_0^t U(t-\tau)f(\tau)\,d\tau - \int_0^t DU(t-\tau)\mathscr{R}g(\tau)\,d\tau,$$

where $DU(t) = -Ae^{-tA} = -AU(t)$ denotes the derivative of the semigroup for $t > 0$. This relation is our *generalized variations-of-constants formula* we were looking for.

In the case of the semilinear evolution problem

$$(2.11) \quad \begin{aligned} \dot{u} + \mathscr{A}u &= f(t,u), \quad 0 < t < \infty, \\ \mathscr{B}u &= g(t,u), \\ u(0) &= u_0 \end{aligned}$$

we obtain from (2.10) formally the integral equation

$$(2.12) \quad u(t) = U(t)u_0 + \int_0^t U(t,\tau)f(\tau, u(\tau))\,d\tau - \int_0^t DU(t-\tau)\mathscr{R}g(\tau, u(\tau))\,d\tau$$

for $0 \le t < \infty$. Letting

$$(2.13) \quad E \hookrightarrow X \quad \text{and} \quad E_0 := X \times Y,$$

we define

$$(2.14) \quad V(t) \in \mathscr{L}(E_0, E)$$

by $V(t)(x, y) := U(t)x - DU(t)\mathscr{R}y$ for $t > 0$ and

$$(2.15) \quad F: \mathbf{R}^+ \times E \to E_0$$

by $F(t, u) := (f(t, u), g(t, u))$. Here the space E has to be chosen appropriately so that (2.13) and (2.14) are satisfied and such that the nonlinear map (2.15) has "nice continuity properties". Then we can rewrite the integral equation (2.12) in the form

$$(2.16) \quad u(t) = U(t)u_0 + \int_0^t V(t-\tau)F(\tau, u(\tau))\,d\tau, \quad 0 \le t < \infty.$$

Observe that the family $\{V(t) | t > 0\}$ possesses the "evolution property"

$$(2.17) \quad U(t)V(s) = V(t+s) \quad \text{for } s, t > 0.$$

For this reason (2.16) is said to be an *integral-evolution equation* in the pair of Banach spaces (E, E_0).

It is well known that there exists a constant c such that

$$\|DU(t)\|_{\mathscr{L}(X)} = \|AU(t)\|_{\mathscr{L}(X)} \le ct^{-1}$$

for $0 < t < 1$ and that, in general, this estimate cannot be improved. Hence DU has a nonintegrable singularity at $t = 0$ if it is considered as a map from $(0, \infty)$ into $\mathscr{L}(X)$. However the operator $\mathscr{R}$ maps Y into $W \hookrightarrow X$, that is, $\mathscr{R}$ will have regularizing properties in general. Thus it may well be the case that, in more concrete situations, it is possible to balance the smoothing properties of $\mathscr{R}$ against the nonintegrable singularity of DU at $t = 0$ by a judicious choice of the spaces E (and perhaps also E_0) so that V satisfies an estimate of the form

$$\|V(t)\|_{\mathscr{L}(E_0, E)} \leq ct^{-\alpha}, \quad 0 < t < 1,$$

for some $\alpha \in [0, 1)$. In this case the integral-evolution equation (2.16) can be given a meaning, independently of the above formal derivation, which would be very hard to justify in general.

In other words, having found—by formal manipulations—a possible candidate for a nonlinear variations-of-constants formula for problem (2.11), namely formula (2.12), we can now study this equation directly.

We can try to find an appropriate set-up such that (2.16) has a unique solution. It remains then, of course, the problem to relate this solution to the original differential equation (2.11).

It is clear that this approach is, in principle, very general. Namely in the formal derivation of the fundamental variations-of-constants formula (2.10) we have only used the assumptions (2.1)–(2.3) and (2.9). These hypotheses are satisfied practically for every regular elliptic boundary value problem $(\mathscr{A}, \mathscr{B})$ of arbitrary order.

3. A chemotaxis problem. In order to show that the general approach indicated above works we restrict our consideration in this note to problem (1.2). The general case will be dealt with in another paper.

In the following we let $n < p < \infty$ and $W_p^s := W_p^s(\Omega, \mathbf{R}^2)$. Then we let

$$W := W_p^2, \quad X := L_p, \quad Y := W_p^{1-1/p}(\Gamma, \mathbf{R}^2)$$

and we define $\mathscr{A}$ and $\mathscr{B}$ by

$$\mathscr{A}: W \to X, \quad u \mapsto -M\Delta u + u$$

and

$$\mathscr{B}: W \to Y, \quad u \mapsto M(\partial u/\partial \nu),$$

respectively. Then, by means of the well-known L_p-theory for nonhomogeneous elliptic boundary value problems [14] and the triangular structure of M, it is not difficult to see that the conditions (2.1)–(2.3) and (2.9) are satisfied. Moreover, it follows from [14, Theorem 4.2] that

(3.1) $$\mathscr{R} \in \mathscr{L}\left(W_p^{s-1-1/p}(\Gamma, \mathbf{R}^2), W_p^s\right)$$

for every $s \in [0, 2]$ with $s - 1/p \notin \mathbf{Z}$.

Now let $s, \sigma \in \mathbf{R}$ with

(3.2) $$n/p < \sigma < s < 1 + 1/p$$

be fixed and let $E := W_p^\sigma$. Extending slightly some results due to Grisvard [9, Théorème 7.5], it can be shown that

$$W_p^t = \left(L_p, W_{p,\mathscr{B}}^2\right)_{t/2, p}, \quad 0 < t < 1 + 1/p, t \neq 1,$$

where $W_{p,\mathscr{B}}^2 := \{u \in W_p^2 | \mathscr{B}u = 0\}$ and $(\cdot, \cdot)_{\theta, p}$ denotes the "real interpolation functor" with parameters $\theta \in (0, 1)$ and $p \in (1, \infty)$ (e.g. [5, 18]). Since, up to equivalent norms, $W^2_{p,\mathscr{B}} = D(A)$, where $D(A)$ denotes the domain of A endowed with the graph norm, it is not difficult to see that the semigroup e^{-tA} restricts to a strongly continuous analytic semigroup on W^t_p, $0 < t < 1 + 1/p$, $t \neq 1$, whose infinitesimal generator is the W^t_p-realization of $-A$, that is, the restriction of $-A$ to $\{u \in W^2_{p,\mathscr{B}} | Au \in W^t_p\}$. Using these facts and interpolation techniques it can be shown that there exist positive constants c and ω_0 such that

(3.3) $$\|AU(t)\|_{\mathscr{L}(W^s_p, W^\sigma_p)} \leq ct^{-\alpha}e^{-\omega_0 t}, \qquad t > 0,$$

where $\alpha := 1 - (s - \sigma)/2$. Thus, since

$$L_p(\Gamma, \mathbf{R}^2) \hookrightarrow W^{s-1-1/p}_p(\Gamma, \mathbf{R}^2),$$

we deduce from (3.2) and (3.3) that there is a constant c such that

(3.4) $$\|DU(t)\mathscr{R}\|_{\mathscr{L}(L_p(\Gamma, \mathbf{R}^2), W^\sigma_p)} \leq ct^{-\alpha}e^{-\omega_0 t}$$

for $t > 0$, where $\alpha := 1 - (s - \sigma)/2$. Moreover, by interpolation, it follows that there is a constant c such that

(3.5) $$\|U(t)\|_{\mathscr{L}(L_p, W^\sigma_p)} \leq ct^{-\sigma/2}e^{-\omega_0 t}, \qquad t > 0.$$

Hence, letting $E_0 := L_p \times L_p(\Gamma, \mathbf{R}^2)$ we obtain from (3.4) and (3.5) the existence of constants c and $\omega \in (0, \omega_0)$ such that

(3.6) $$\|V(t)\|_{\mathscr{L}(E_0, E)} \leq ct^{-\alpha}e^{-\omega t}, \qquad t > 0.$$

Suppose now that

$$f \in C^{0,1-}(\bar{\Omega} \times \mathbf{R}^2, \mathbf{R}^2) \quad \text{and} \quad g \in C^{0,1-}(\Gamma \times \mathbf{R}^2, \mathbf{R}^2),$$

where $u \in C^{0,1-}$ means that u is continuous in the first variable and locally Lipschitz continuous with respect to the second variable. Then it is easily seen that

(3.7) $$\hat{f} \in C^{1-}(C(\bar{\Omega}, \mathbf{R}^2), C(\bar{\Omega}, \mathbf{R}^2))$$

and

(3.8) $$\hat{g} \in C^{1-}(C(\Gamma, \mathbf{R}^2), C(\Gamma, \mathbf{R}^2)),$$

respectively, where $\hat{f}$ denotes the substitution operator induced by f, that is, $\hat{f}(u)(x) := f(x, u(x))$ for $x \in \bar{\Omega}$ and $u: \bar{\Omega} \to \mathbf{R}^2$, and $\hat{g}$ is the substitution operator induced by g. Since, by (3.2) and $p > n$,

(3.9) $$E \hookrightarrow C(\bar{\Omega}, \mathbf{R}^2) \quad \text{and} \quad W^{\sigma-1/p}_p(\Gamma, \mathbf{R}^2) \hookrightarrow C(\Gamma, \mathbf{R}^2),$$

and since

(3.10) $$\gamma \in \mathscr{L}(E, W^{\sigma-1/p}_p(\Gamma, \mathbf{R}^2)),$$

where γ denotes the trace operator, it follows from (3.7)–(3.10) and from

$$C(\bar{\Omega}, \mathbf{R}^2) \hookrightarrow L_p(\Omega, \mathbf{R}^2) \quad \text{and} \quad C(\Gamma, \mathbf{R}^2) \hookrightarrow L_p(\Gamma, \mathbf{R}^2)$$

that

(3.11) $$F := (\hat{f} + \text{id}, \hat{g} \circ \gamma) \in C^{1-}(E, E_0).$$

Now (3.6) and (3.11) and the fact that $\{U(t)|t \geq 0\}$ restricts to a strongly continuous semigroup on E imply that the integral evolution equation (2.16) is well defined in the pair of Banach spaces (E, E_0). Hence this equation can be solved—locally—by means of standard Banach fixed point arguments. Moreover, by means of standard continuation arguments it can be shown that (2.16) has for each $u_0 \in E$ a unique solution $u(\cdot, u_0)$ defined on a maximal interval $[0, t^+(u_0))$ and that

$$\lim_{t \to t^+(u_0)} \|u(t, u_0)\|_E = \infty \quad \text{if } t^+(u_0) < \infty.$$

Finally it can be shown that $u(t, u_0)$ depends continuously on $u_0 \in E$ and that the map $(t, u_0) \mapsto u(t, u_0)$ defines a local semiflow on E, due to the fact that f and g do not depend on $t \in \mathbf{R}^+$.

It remains to be shown how the solution $u(\cdot, u_0)$ of the integral-evolution equation (2.16) is related to problem (1.2). To see this we introduce the formally adjoint elliptic system $(\mathscr{A}^\#, \mathscr{B}^\#)$ of $(\mathscr{A}, \mathscr{B})$ by

$$\mathscr{A}^\# := -M^T \Delta, \qquad \mathscr{B}^\# := M^T \partial/\partial \nu,$$

where M^T denotes the transpose of the matrix M. Then it is not too difficult to show that $u(\cdot, u_0)$ satisfies the integral relation

$$\int_0^T \langle -\dot{\varphi}(t) + \mathscr{A}^\# \varphi(t), u(t, u_0) \rangle dt$$
$$= \int_0^T \langle \varphi(t), \hat{g}(u(t, u_0)) \rangle_\Gamma dt + \int_0^T \langle \varphi(t), \hat{f}(u(t, u_0)) \rangle dt + \langle \varphi(0), u_0 \rangle$$

for every $T \in (0, t^+(u_0))$ and every $\varphi \in C^1([0, T], W_{p'}^2)$ satisfying $\varphi(T) = 0$ and $\mathscr{B}^\# \varphi(t) = 0$ for $t \in [0, T]$. Here $\langle \, , \, \rangle$ denotes the duality pairing between $L_{p'}$ and L_p, $p' := p/(p-1)$, and $\langle \cdot, \cdot \rangle_\Gamma$ denotes the duality pairing between $L_{p'}(\Gamma, \mathbf{R}^2)$ and $L_p(\Gamma, \mathbf{R}^2)$. Thus $u(\cdot, u_0)$ is a weak solution of the parabolic initial value problem (1.2). Using this fact and the fact that we know from the integral-evolution equation that

$$[t \mapsto \hat{f}(u(t, u_0))] \in C([0, t^+(u_0)), C(\overline{\Omega}))$$

and

$$[t \mapsto \hat{g}(u(t, u_0))] \in C([0, t^+(u_0)), C(\Gamma)),$$

we can apply classical arguments (e.g. [12]) to show that $u(\cdot, u_0)$ is in fact a regular solution of (1.2) (in a sense we do not specify here).

In summary we obtain the following theorem, where $\|\cdot\|_{\sigma, p}$ denotes the norm in W_p^σ.

THEOREM 1. *Suppose that $\alpha, \beta > 0$ and that*

$$f \in C^{0,1-}(\overline{\Omega} \times \mathbf{R}^2, \mathbf{R}^2) \quad \text{and} \quad g \in C^{0,1-}(\Gamma \times \mathbf{R}^2, \mathbf{R}^2),$$

and let $n < p < \infty$ and $1 < \sigma < 1 + 1/p$. Then the semilinear parabolic system
$$\frac{\partial u}{\partial t} - \begin{bmatrix} \alpha & \gamma \\ 0 & \beta \end{bmatrix} \Delta u = f(x, u) \quad \text{in } \Omega \times (0, \infty),$$
$$\begin{bmatrix} \alpha & \gamma \\ 0 & \beta \end{bmatrix} \frac{\partial u}{\partial \nu} = g(x, u) \quad \text{on } \Gamma \times (0, \infty)$$
has for each $u_0 \in W_p^\sigma(\Omega, \mathbf{R}^2)$ a unique solution $u(\cdot, u_0)$ on a maximal interval of existence $[0, t^+(u_0))$ such that $u(0, u_0) = u_0$ and

(3.12) $\qquad\qquad \|u(t, u_0)\|_{\sigma, p} \to \infty \quad \text{as } t \to t^+(u_0)$

if $t^+(u_0) < \infty$. Moreover the map $(t, u_0) \mapsto u(t, u_0)$ defines a local semiflow on W_p^σ such that bounded orbits are relatively compact.

The last assertion follows from the fact that W_p^s is compactly imbedded in W_p^σ, if $s > \sigma$, and that it is not difficult to show that bounded orbits are in fact bounded in W_p^s for $\sigma < s < 1 + 1/p$.

REMARKS. (a) It should be observed that there are no growth restrictions whatsoever for the nonlinearities f and g.

(b) Theorem 1 can be extended to parabolic systems of arbitrary even order provided the corresponding elliptic system $(\mathscr{A}, \mathscr{B})$ satisfies L_p-estimates in the sense of Agmon-Douglis-Nirenberg [1] and there is a "Green's formula" involving an appropriate formally adjoint system $(\mathscr{A}^\#, \mathscr{B}^\#)$. These conditions are always satisfied if $N = 1$ and also for large classes of strongly coupled systems if $N > 1$.

(c) The above results can also be extended to time-dependent problems of the form

(3.13)
$$\frac{\partial u}{\partial t} + \mathscr{A}(t)u = f(t, u),$$
$$\mathscr{B}(t)u = g(t, u),$$
$$u(\cdot, 0) = u_0,$$

provided there exists a "parabolic evolution operator" in the sense of Kato-Tanabe [11]. Of course, in the nonautonomous case the solutions do not generate a semiflow.

(d) The above results can be extended to systems of the form (3.13), where the nonlinearities f and g depend also on derivatives of u (in relation to the orders of $\mathscr{A}(t)$ and $\mathscr{B}(t)$).

(e) Except for the relative compactness of bounded orbits the above results can also be extended to unbounded domains which are uniformly regular in the sense of Browder [7].

The general situations indicated above will be considered in forthcoming publications.

4. Global existence. In this section we give a further indication for the usefulness of the variations-of-constants formula (2.10) or (2.16), respectively. Namely, we give a simple proof for the global existence of $u(\cdot, u_0)$ if f and g

satisfy appropriate growth restrictions and *if it is a priori known that $u(\cdot, u_0)$ is bounded in some weak norm*.

THEOREM 2. *Let the hypotheses of Theorem 1 be satisfied and suppose, in addition, that there exist constants c and δ such that*

$$|f(x, \xi)| \leq c\left(1 + |\xi|^\delta\right) \quad \forall (x, \xi) \in \overline{\Omega} \times \mathbf{R}^2$$

and

$$|g(x, \xi)| \leq c(1 + |\xi|) \quad \forall (x, \xi) \in \Gamma \times \mathbf{R}^2.$$

Moreover, suppose that there is a $p_0 \in (1, p]$ such that $1 < \delta < 1 + \sigma p_0/n$ and

(4.1) $$\sup\left\{\|u(t, u_0)\|_{0, p_0} \mid 0 \leq t < t^+(u_0)\right\} < \infty.$$

Then $t^+(u_0) = \infty$ and the orbit through u_0 is bounded in W_p^σ.

PROOF. Let $u(t) := u(t, u_0)$ for $0 \leq t < t^+(u_0)$. Since $u \in C([0, t^+(u_0)), W_p^\sigma)$ is a solution of the integral equation (2.16) it follows that there is a constant c such that

(4.2) $$\|u(t)\|_{\sigma, p} \leq \|U(t)u_0\|_{\sigma, p} + \int_0^t \|U(t - \tau)\|_{\mathscr{L}(L_p, W_p^\sigma)} c\left(1 + \||u(\tau)|^\delta\|_{0, p}\right) d\tau$$
$$+ \int_0^t \|DU(t - \tau)\mathscr{R}\|_{\mathscr{L}(L_p(\Gamma, \mathbf{R}^2), W_p^\sigma)} c\left(1 + \|u(\tau)\|_{\rho, p}\right) d\tau,$$

where $1/p < \rho < \sigma$. In the last integral we used the fact that the trace operator maps W_p^ρ continuously into $W_p^{\rho - 1/p}(\Gamma, \mathbf{R}^2) \hookrightarrow L_p(\Gamma, \mathbf{R}^2)$.

Observe that $\||u(\tau)|^\delta\|_{0, p} = \|u(\tau)\|_{0, \delta p}^\delta$. Moreover, since W_p^ρ can be characterized as a (real or complex) interpolation space between L_p and W_p^2 (if $\rho \neq 1$ or $\rho = 1$, respectively), and by using standard imbedding theorems (e.g. [5, 18]), it is not difficult to show that there exists a constant $c := c(\theta, p_0, p_1, q, s_0, s_1, t)$ such that

(4.3) $$\|u\|_{t, q} \leq c\|u\|_{s_1, p_1}^\theta \|u\|_{s_0, p_0}^{1-\theta}$$

provided $0 < \theta < 1$, $1 < p_0, p_1, q < \infty$, $0 \leq t, s_0, s_1 < \infty$ satisfy

$$s_0 \neq s_1, \quad 1/q \leq (1 - \theta)/p_0 + \theta/p_1$$

and

$$t - n/q < (1 - \theta)s_0 + \theta s_1 - n((1 - \theta)/p_0 + \theta/p_1).$$

From this we deduce that

$$\|u\|_{0, \delta p}^\delta \leq c\|u\|_{\sigma, p}^{\delta\theta} \|u\|_{0, p_0}^{\delta(1-\theta)},$$

where $\theta \in (0, 1)$ can be chosen such that $\delta\theta < 1$. Furthermore, (4.3) implies the estimate

$$\|u\|_{\rho, p} \leq c\|u\|_{s, p}^\eta \|u\|_{0, p_0}^{1-\eta}$$

for an appropriate $\eta \in (0, 1)$. Thus, letting $\lambda := \max\{\delta\theta, \eta\}$ and using (3.4) and (3.5), we deduce from (4.1) and (4.2) the existence of a constant c such that

$$\|u(t)\|_{\sigma,p} \leq c\left\{e^{-\omega_0 t}\|u_0\|_{\sigma,p} + 1 + \int_0^t (t-\tau)^{-\alpha} e^{-\omega(t-\tau)}\|u(\tau)\|_{\sigma,p}^\lambda \, d\tau\right\}.$$

Hence there exists a constant c such that

$$\sup_{0 \leq t < t^+(u_0)} \|u(t)\|_{\sigma,p} \leq c\left(1 + \sup_{0 \leq t < t^+(u_0)} \|u(t)\|_{\sigma,p}^\lambda\right),$$

which implies

$$\sup_{0 \leq t < t^+(u_0)} \|u(t)\|_{\sigma,p} < \infty.$$

Now the assertion follows from (3.12). □

REMARKS. (a) It is clear that the above proof works also in the general situations mentioned in the remarks to Theorem 1.

(b) If $g = 0$, we can let $0 < \sigma < 2$. Then, by choosing σ close to 2, it follows that $t^+(u_0) = \infty$, provided $1 < \delta < 2p_0/n$ and the L_{p_0}-norm of $u(\cdot, u_0)$ is uniformly bounded. Thus we obtain a very simple and very flexible proof for some related results due to Alikakos [3], Massat [15], and Rothe [17].

(c) Theorem 2 is not optimal. Indeed, suppose that (4.1) is true and that there are constants c, λ, and κ such that

$$|f(x, \xi)| \leq c\left(1 + |\xi|^\lambda\right) \quad \forall (x, \xi) \in \overline{\Omega} \times \mathbf{R}^2$$

and

$$|g(x, \xi)| \leq c\left(1 + |\xi|^\kappa\right) \quad \forall (x, \xi) \in \Gamma \times \mathbf{R}^2,$$

and such that $1 < \lambda < 1 + 2p_0/n$ and $1 < \kappa < 1 + p_0/n$. Then it can be shown that $t^+(u_0) < \infty$ and that $u(\cdot, u_0)$ is uniformly bounded in W_p^σ. For this one has to replace the space $E_0 := L_p \times L_p(\Gamma, \mathbf{R}^2)$ by $W_p^{-\alpha}(\Omega, \mathbf{R}^2) \times W_p^{-\beta}(\Gamma, \mathbf{R}^2)$ for appropriate choices of α and β. Details will be given elsewhere.

REFERENCES

1. S. Agmon, A. Douglis and L. Nirenberg, *Estimates near the boundary for solutions of elliptic partial differential equations satisfying general boundary conditions.* I, II, Comm. Pure Appl. Math. **12** (1959), 623–727; **17** (1964), 35–92.

2. N. D. Alikakos, *Regularity and asymptotic behaviour for the second order parabolic equation with nonlinear boundary conditions*, J. Differential Equations **39** (1981), 311–344.

3. _____, L^p *bounds of solutions of reaction-diffusion equations*, Comm. Partial Differential Equations **4** (1979), 827–868.

4. V. Barbu, *Nonlinear semigroups and differential equations in Banach spaces*, Nordhoof, Leyden, 1976.

5. J. Bergh and J. Loefstroem, *Interpolation spaces. An introduction*, Springer-Verlag, 1976.

6. H. Brézis, *Monotonicity methods in Hilbert spaces and some applications to nonlinear partial differential equations*, Contributions to Nonlinear Functional Analysis (E. H. Zarantonello, ed.), Academic Press, New York, 1971, pp. 101–156.

7. F. E. Browder, *On the spectral theory of elliptic operators.* I, Math. Ann. **142** (1961), 22–130.

8. A. Friedman, *Partial differential equations of parabolic type*, Prentice-Hall, Englewood Cliffs, N. J., 1964.

9. P. Grisvard, *Équations différentielles abstraites*, Ann. Sci. École Norm. Sup. Sér. 4 **2** (1969), 311–395.

10. J. Hernandez, *Some existence and stability results for solutions of reaction-diffusion systems with nonlinear boundary conditions*, Nonlinear Differential Equations: Invariance, Stability, and Bifurcation (P. de Mottoni, ed.), Academic Press, New York, 1981, pp. 161–173.

11. T. Kato and H. Tanabe, *On the abstract evolution equation*, Osaka Math. J. **14** (1962), 107–133.

12. O. A. Ladyženskaja, V. A. Solonnikov and N. N. Ural'ceva, *Linear and quasilinear equations of parabolic type*, Amer. Math. Soc., Providence, R. I., 1968.

13. A. Leung, *A semilinear reaction-diffusion prey-predator system with nonlinear coupled boundary conditions: equilibrium and stability*, Indiana Univ. Math. J. **31** (1982), 223–241.

14. J.-L. Lions and E. Magenes, *Problemi ai limiti non omogenei* (V), Ann. Scuola Norm. Sup. Pisa **16** (1962), 1–44.

15. P. Massatt, *Obtaining L^∞ bounds from L^p bounds for solutions of various parabolic partial differential equations*, Proc. Internat. Sympos. on Dynamical Systems, 1981 (to appear).

16. C. V. Pao, *Reaction diffusion equations with nonlinear boundary conditions*, Nonlinear Anal. **5** (1981), 1077–1094.

17. F. Rothe, *Uniform bounds from bounded L_p-functionals in reaction-diffusion equations*, J. Differential Equations **45** (1982), 207–233.

18. H. Triebel, *Interpolation theory, function spaces, differential operators*, North-Holland, Amsterdam, 1978.

UNIVERSITÄT ZÜRICH, SWITZERLAND

Nonlinear Oscillations with Minimal Period

ANTONIO AMBROSETTI[1]

1. Introduction. Let $H \in C^1(\mathbf{R}^{2n}; \mathbf{R})$ (this is assumed throughout) and let J be the simplectic matrix $\begin{bmatrix} 0 & -\mathrm{Id} \\ \mathrm{Id} & 0 \end{bmatrix}$, Id = identity in $\mathbf{R}^n$; set $H' = \mathrm{grad}\, H$, $\dot z = dz/dt$, and consider the Hamiltonian system

$$-J\dot z = H'(z). \tag{1}$$

In this paper we deal with the following problem: *Find periodic solutions of* (1) *having a prescribed number* $T > 0$ *as **minimal** period*.

Among others, the above problem is connected with the existence of multiple solutions of (1) on a Hamiltonian surface $H = \mathrm{const}$ (cf. [9, 2, 4]).

From the abstract point of view the solutions of (1) are critical points of functionals which are indefinite (cf., for example, [11]), and the difficulty lies in finding critical points with suitable *minimal* properties.

Many of the results discussed here are related to those of [1].

Other references are [6, 10].

2. Nonexistence. In this section we start with some nonexistence results. Denote by $\langle\, \cdot\, ,\, \cdot\, \rangle$ and $|\cdot|$ the Euclidean scalar product and the Euclidean distance in $\mathbf{R}^{2n}$, respectively. Recall that, for every periodic solution $z(t)$ of (1), $H(z)$ is constant and denoted by H_z. Let

$$S_z = \{ x \in \mathbf{R}^{2n} : H(x) = H_z \}, \qquad \mu_z = \max\{ |H'(x)| : x \in S_z \}.$$

LEMMA 1. *Suppose there exists $\beta \neq 0$ such that*

$$\langle H'(x), x \rangle \geq \beta H(x) \qquad \forall x \in \mathbf{R}^{2n}, \tag{2}$$

and let z be a T-periodic solution of (1). *Then*

$$\mu_z^2 T \geq 2\pi\beta H_z.$$

1980 *Mathematics Subject Classification.* Primary 35C15, 34C25.
[1] Supported in part by Ministero P.I. (Italy), Research on "Calcolo delle Variazioni".

PROOF. Let $T = 2\pi\lambda$ and $u(t) := z(\lambda t)$. Then u is a 2π-periodic solution of

(3) $$-J\dot{u} = \lambda H'(u)$$

and $H_u = H_z$. From (3) and (2) one has

(4) $$\int_0^{2\pi} \langle \dot{u}, Ju \rangle = \lambda \int_0^{2\pi} \langle H'(u), u \rangle \geq \lambda\beta \int_0^{2\pi} H(u) = 2\pi\lambda\beta H_z.$$

Using the Wirtinger inequality and (3) we get

(4') $$\int_0^{2\pi} \langle \dot{u}, Ju \rangle \leq \int_0^{2\pi} |\dot{u}|^2 = \lambda^2 \int_0^{2\pi} |H'(u)|^2 \leq 2\pi\lambda^2\mu_z^2.$$

Combining (4) and (4'), we obtain the lemma.

Consequently, we can deduce lower bounds for the periods of possible solutions of (1).

THEOREM 2. *Let H be convex, β-homogeneous ($\beta \neq 0$), and $0 < H(x) \leq (1/\beta)|x|^\beta \, \forall x \neq 0$. Let $z \not\equiv 0$ be a solution of (1). Then*

$$T \geq 2\pi\beta^\gamma H_z^\gamma, \quad \gamma = (2-\beta)/\beta.$$

PROOF. Since $\langle H'(x), x \rangle = \beta H(x)$ and $z \not\equiv 0$, it follows that $\forall x \in S_z$ $\langle H'(x), x \rangle = \beta H_z > 0$; in particular, $H'(x) \neq 0$ for such x. Moreover, from the convexity of H we deduce

(5) $$\min_{S_z} |x| = \min_{S_z} \langle x, H'(x)/|H'(x)| \rangle$$

because the right side is the minimal distance from 0 to the tangent hyperplane of S_z at x. Since H is β-homogeneous, (5) gives

$$\min_{S_z} |x| = \beta(H_z/\mu_z).$$

Further, $H(x) \leq |x|^\beta/\beta$ implies

$$\min_{S_z} |x| \geq (\beta H_z)^{1/\beta}$$

and, hence,

(5') $$1/\mu_z \geq (\beta H_z)^{(1-\beta)/\beta}.$$

Using Lemma 1 and (5'), one has $T \geq 2\pi\beta H_z \mu_z^{-2} \geq 2\pi\beta^\gamma H_z^\gamma$.

COROLLARY [7]. *Suppose H is convex, 2-homogeneous, and $0 < H(x) \leq \frac{1}{2}|x|^2$ $\forall x \neq 0$. Then for the periods of any possible solution of (1), $T \geq 2\pi$.*

3. Existence. In this section we give some positive answers to our problem, under the main assumption that H is convex. More precisely we assume:

(A1) $H \in C^1(\mathbf{R}^{2n}; \mathbf{R})$ is convex;

(A2) there are $\beta > 1$, $a_1, a_2 > 0$ such that $\forall x \in \mathbf{R}^{2n}$ one has $a_1|x|^\beta \leq H(x) \leq a_2|x|^\beta$.

Let
$$G(y) := \sup\{\langle x, y \rangle - H(x) : x \in \mathbf{R}^{2n}\}.$$

By (A1), (A2), $G \in C^1(\mathbf{R}^{2n}; \mathbf{R})$ is convex and

(6) $\qquad b_1|y|^\alpha \leq G(y) \leq b_2|y|^\alpha, \qquad 1/\alpha + 1/\beta = 1.$

Moreover, using the duality relation $G(y) + H(x) = \langle x, y \rangle$, $H'(x) = y$ (i.e., $G'(y) = x$), we can deduce the following properties from (A2) and (2):

(6') $\qquad G(y) \leq \langle G'(y), y \rangle \leq \alpha G(y) \qquad \forall y \in \mathbf{R}^{2n}.$

Before we specify the abstract setting, we rescale, for convenience, the time t, looking for 2π-periodic solutions of (3). If z is such a solution, $z(t/\lambda)$ is a T-periodic solution of (1) for $T = 2\pi\lambda$. Of course, T is minimal whenever 2π is the minimal period of z.

Let

(7) $\qquad E = \left\{ u \in L^\alpha(0, 2\pi; \mathbf{R}^{2n}) : \int u = 0 \right\},$ [2]

and let $L: E \to E^*$ be the inverse of the densely defined operator $u \to -J\dot{u}$. L is continuous, selfadjoint and compact. For u in E we set

(7') $\qquad f(u) = \lambda \int G(u) - \frac{1}{2} \int (u, L(u)).$

If u is any critical point of f in E, then $z = G'(u)$ defines a 2π-periodic solution of (3) ("dual action principle" [5]).

The advantage of such a device is that the quadratic part in f is a compact perturbation of the term $\int G(u)$, which behaves like $\|u\|_E^\alpha$. In fact, it is easy to find critical points of f (and hence solutions of (3)) in this framework. More precisely, we distinguish:

Case $\beta < 2$. f is bounded from below, and there is a critical point $\bar{u} \in E$ such that $f(\bar{u}) = \min_E f < 0$. In particular, $\bar{u} \neq 0$; since $\int \bar{u} = 0$, it is nonconstant and gives rise to a nonconstant solution z of (3).

Case $\beta > 2$. Now $\inf_E f = -\infty$ and $\sup_E f = +\infty$. However, it is possible to see [8] that if H satisfies (A1), (A2) and (2), then the following "mountain-pass" theorem [3] can be applied:

M-P THEOREM. *Let $f \in C^1(E; \mathbf{R})$ satisfy*

(PS) *every u_j such that $f(u_j)$ is bounded and $f'(u_j) \to 0$ has a converging subsequence;*

(f1) *there are $\alpha, \rho > 0$ such that $f(u) \geq 0 \ \forall \|u\| \leq \rho$ and $f(u) \geq \alpha \ \forall \|u\| = \rho$;*

(f2) *there is $e \neq 0$ such that $f(e) \leq 0$.*

---

[2] From now on all the integrals are evaluated from 0 to 2π. Moreover, $(\cdot, \cdot)$ denotes the pairing in E.

Then f has a critical point $u^* \neq 0$ such that

(8) $$c^* = f(u^*) = \inf_{p \in \Gamma} \{\max\{f(p(s)): 0 \leq s \leq 1\}\},$$

where $\Gamma = \{p \in C([0,1]; E): p(0) = 0 \text{ and } f(p(1)) < 0\}$.

REMARK. The M-P Theorem as stated is slightly different with respect to [3]. But the proof can be easily modified to cover the present form. In fact, (f2) implies $\Gamma \neq \varnothing$, and (f1) implies that every $p \in \Gamma$ intersects the sphere $\|u\| = \rho$. The remainder is the same.

Concerning the minimal period, in the subquadratic case (i.e., $\beta < 2$) we easily see that $f(\bar{u}) = \min f$
implies 2π is the minimal period of $\bar{u}$. Precisely, one has

THEOREM 3. [6]. *If (A1), (A2) hold with $\beta < 2$, then, for all $T > 0$, (1) has a periodic solution with T as minimal period.*

The case $\beta > 2$ requires an additional argument, which we report in a general setting.

Let $f \in C^1(E; \mathbf{R})$, let E be a Banach space, and suppose there is a subset $M \subset E - \{0\}$ such that:

(f3) $u^* \in M$;
(f4) $f(su) \to -\infty$ as $s \to +\infty \ \forall u \in M$;
(f5) $(f'(su), u) > 0 \ (< 0) \forall s < 1 \ (> 1)$.

Clearly, (f5) implies $f(u) = \max\{f(su): s \geq 0\}$.

LEMMA 4. *Suppose f satisfies (f1), (f2) and (PS). Moreover assume there exists $M \subset E - \{0\}$ such that (f3)–(f5) hold. Then for c^* given by (8), $c^* = \inf_M f := m$.*

PROOF. Since $u^* \in M$, then $c^* = f(u^*) \geq m$. Suppose $m < c^*$ and let $\hat{u} \in M$ be such that $f(\hat{u}) < c^*$. By (f4) we take $\sigma > 0$ such that $f(\sigma \hat{u}) < 0$. Setting $\hat{p}(s) := s\sigma\hat{u}$, we have $\hat{p} \in \Gamma$ and, hence,

$$c^* \leq \max\{f(s\sigma\hat{u}): 0 \leq s \leq 1\}.$$

But from (f5) we infer $\max\{f(s\sigma\hat{u}): 0 \leq s \leq 1\} = f(\hat{u})$; thus $c^* \leq f(\hat{u}) = m$.

Coming back to our specific problem, we will show how Lemma 4 allows us to prove the minimality of the periods for the corresponding u^*. Let E and f be given by (7) and (7'), respectively, and let

$$M = \{u \in E - \{0\}: (f'(u), u) = 0\}$$

Every $u \in E - \{0\}$ with $f'(u) = 0$ belongs to M. In particular (f3) holds. For all $u \in M$ it results that $\int(u, Lu) = \lambda \int (G'(u), u)$; hence $\int (u, Lu) > 0$ and (f4) follows from $f(su) \geq b_1 \lambda s^\alpha \int |u|^\alpha - \tfrac{1}{2}s^2 \int(u, Lu)$. As for (f5), we shall impose

(A3) $\quad \langle G'(sy), y \rangle - s\langle G'(y), y \rangle > 0 \ (< 0) \qquad \forall 0 < s < 1 \ (s > 1).$

THEOREM 5. *If* (A1)–(A3) *and* (2) *hold with* $\beta > 2$, *then, for all* $T > 0$, (1) *has a solution with* T *as minimal period*.

PROOF. As remarked before, f in (9) satisfies (PS) and (f1)–(f5). Let u^* be the critical point such that (8) holds. If u^* has period $T \leq \pi$, then we let $v^*(t) := u^*(t/2)$. Note that $v^* \in E$ and

$$\text{(10)} \quad \int G(v^*) = \int G(u^*), \quad \int (G'(sv^*), v^*) = \int (G'(su^*), u^*),$$

$$\int (v^*, Lv^*) = 2 \int (u^*, Lu^*).$$

We use (10) to get

$$(f'(sv^*), v^*) = \lambda \int (G'(su^*), u^*) - 2s \int (u^*, Lu^*)$$

$$= \lambda \left[\int (G'(su^*), u^*) - 2s \int (G'(u^*), u^*) \right] := \lambda h(s).$$

Note that

$$h(s) = \lambda (f'(su^*), u^*) - s \int (G'(u^*), u^*) < \lambda (f'(su^*), u^*).$$

In particular, $h(s) < 0 \ \forall s \geq 1$. Moreover, from (6)–(6') one has

$$\text{(11)} \quad h(s) \geq b_1 s^{\alpha-1} \int |u^*|^\alpha - 2s \int (G'(u^*), u^*).$$

Since $1 < \alpha < 2$, $h(s) > 0$ for $s > 0$ small enough. Then (A3) implies $\exists s^*$, $0 < s^* < 1$, such that $h(s^*) = 0$, namely, $s^*v^* \in M$. Now $s^* < 1$ and (f5) imply $f(s^*v^*) < f(v^*)$, and (10) gives

$$f(v^*) = f(u^*) - \frac{1}{2} \int (u^*, Lu^*) < f(u^*).$$

Then $f(s^*v^*) < f(u^*) = c^*$, a contradiction, because $s^*v^* \in M$ and $c^* = \inf_M f$ by Lemma 4.

Theorem 5 slightly improves the result of [1]. A similar result has been obtained in [12] by different arguments.

4. Remarks. (i) Examples of Hamiltonians for which (A3) holds are given in [1]. Here we take the second-order system

$$\text{(12)} \quad \ddot{x} + Qx + U'(x) = 0, \quad x \in \mathbf{R}^n,$$

where Q is a symmetric matrix with eigenvalues $\omega_1 \leq \cdots \leq \omega_n$, $\omega_1 > 0$, and U is convex, β-homogeneous, $\beta > 2$. We will show that, *for all* $T < 2\pi/\omega_n$, (12) *has a solution with minimal period* T.

Let V be the convex conjugate of U; V is α-homogeneous. Set $E = L^\alpha(0, 2\pi; \mathbf{R}^n)$, and, for $\lambda < \omega_n^{-1}$, let A be the inverse of $x \to -\ddot{x} - Qx$. As in §3 the 2π-solutions of (12) correspond to the critical points $u \in E$ of

$$\Phi(u) = \lambda \int V(u) - \frac{1}{2} \int (u, Au).$$

It is easy to check that Φ satisfies the assumptions of the M-P Theorem. Let $c^* = \Phi(u^*)$ be as in (8). We will show directly that 2π is the minimal period of u^*. In fact, if u^* has period π, its Fourier expansion will be $u^* = \sum_Z u_{2k} e^{i2kt}$ ($u_{-j} = \bar{u}_j$), and we set $v^* = \sum u_{2k} e^{ikt}$. By direct computations one finds (cf. Lemma 4.2 in [1])

(13) $$K' := \int (v^*, Av^*) > \int (u^*, Au^*) := K.$$

Let

$$C := \lambda \int V(v^*) = \lambda \int V(u^*).$$

Since V is α-homogeneous and $\Phi'(u^*) = 0$, we get

(14) $$K = \int (u^*, Au^*) = \lambda \int (V'(u^*), u^*) = \alpha\lambda \int V(u^*) = \alpha C.$$

In particular, we remark that $K' > K > 0$. Then from

$$\Phi(sv^*) = s^\alpha C - \tfrac{1}{2} s^2 K',$$

we obtain that $p(s) := sv^*$ (after reparametrization) belongs to Γ. By a direct computation one gets $(d/ds)\Phi(sv^*) = 0$ for $s^{\alpha-2} = K'/K$, and

$$\max_p \Phi = \left(\frac{1}{\alpha} - \frac{1}{2}\right) K^{2\theta} K'^{1-\alpha\theta} \quad \left(\theta = \frac{1}{2-\alpha}\right).$$

By (13) we get

$$\max_p \Phi < \left(\frac{1}{\alpha} - \frac{1}{2}\right) K^{(2-\alpha)\theta} = \left(\frac{1}{\alpha} - \frac{1}{2}\right) K = \Phi(u^*) = c^*,$$

a contradiction with respect to the variational characterization of c^* in (8). This proves the claim.

(ii) The same direct argument can be used to obtain existence of solutions with any minimal period for (1) provided (A1), (A2), (2) hold, and a_2/a_1 is small enough.

(iii) Very little is known for nonconvex Hamiltonians. See [10].

References

1. A. Ambrosetti and G. Mancini, *Solutions of minimal period for a class of convex Hamiltonian systems*, Math. Ann. **255** (1981), 405–421.
2. _____, *On a theorem by Ekeland and Lasry concerning the number of periodic Hamiltonian trajectories*, J. Differential Equations **43** (1981), 1–6.
3. A. Ambrosetti and P. H. Rabinowitz, *Dual variational methods in critical point theory and applications*, J. Funct. Anal. **14** (1973), 349–387.
4. H. Berestycki, J. M. Lasry, G. Mancini and B. Ruf, *Existence of multiple periodic orbits on star-shaped Hamiltonian surfaces* (to appear) 1983.
5. F. Clarke, *Periodic solutions to Hamiltonian inclusions*, J. Differential Equations **40** (1981), 1–7.
6. F. Clarke and I. Ekeland, *Hamiltonian trajectories having prescribed minimal period*, Comm. Pure Appl. Math. **33** (1980), 103–116.

7. C. B. Croke and A. Weinstein, *Closed curves on convex hypersurfaces and periods of nonlinear oscillations*, Invent. Math. **64** (1981), 199–202.

8. I. Ekeland, *Periodic solutions of Hamiltonian equations and a theorem of P. Rabinowitz*, J. Differential Equations **34** (1979), 523–534.

9. I. Ekeland and J. M. Lasry, *On the number of Hamiltonian trajectories for a Hamiltonian flow on a convex energy surface*, Ann. of Math. (2) **112** (1980), 283–319.

10. M. Girardi and M. Matzeu, *Some results on solutions of minimal periods to superquadratic Hamiltonian systems*, Nonlinear Anal. **7–5** (1983), 475–482.

11. P. H. Rabinowitz, *Periodic solutions of Hamiltonian systems*, Comm. Pure Appl. Math. **11** (1978), 157–184.

12. D. Shitao, *On periodic solutions of Hamiltonian systems of second order*, preprint, 1983.

UNIVERSITY OF VENICE, ITALY

On Steady Navier–Stokes Flow Past a Body in the Plane

CHARLES J. AMICK

1. Introduction. In this note, we consider the problem of finding a solution (p, w) of the steady Navier–Stokes equations in an exterior plane domain Ω:

(1) $$-\nu\Delta w + (w \cdot \nabla)w = -\nabla p,$$
(2) $$\nabla \cdot w = 0 \quad \text{in } \Omega,$$
(3) $$w = 0 \quad \text{on } \Gamma,$$
(4) $$w(x) \to w_\infty \quad \text{as } |x| \to \infty.$$

Here $\nu > 0$ is the given kinematic viscosity, Γ is the boundary of a smooth compact set $K \subset \mathbf{R}^2$, $\Omega = \mathbf{R}^2 \setminus K$ is the exterior of K, and $w_\infty = (u_\infty, v_\infty) \in \mathbf{R}^2 \setminus \{0\}$ is a given constant vector. We assume that K is simply-connected, so that Γ has only one component. The scalar-valued function p denotes the pressure, while $w = (u, v)$ is the velocity vector.

In 1933, Leray [1] approached the problem (1)–(4) by solving a sequence of approximate problems:

(5) $$-\nu\Delta w + (w \cdot \nabla)w = -\nabla p,$$
(6) $$\nabla \cdot w = 0 \quad \text{in } \Omega_R,$$
(7) $$w = 0 \quad \text{on } \Gamma,$$
(8) $$w = w_\infty \quad \text{on } \{|x| = R\},$$

where Ω_R denotes the annular region $\{(x_1, x_2) \in \Omega : |x|^2 \equiv x_1^2 + x_2^2 < R^2\}$. We assume that the body K lies within the unit circle, that the origin is contained in K, and that $R \geq 2$ above. Leray proved the existence of a solution (p^R, w^R) to (5)–(8) with the additional property that

(9) $$\int_{\Omega_R} |\nabla w^R|^2 < \text{const},$$

1980 *Mathematics Subject Classification.* Primary 76D05; Secondary 30Q10, 35B40, 35B50.

© 1986 American Mathematical Society
0082-0717/86 $1.00 + $.25 per page

where the constant is independent of $R \geq 2$. Standard theory gives the existence of a subsequence $\{(p^R, w^R)\}$ which converges on compact subsets of $\bar{\Omega}$ to a solution (p, w) of (1)–(3), and

(10) $$\int_\Omega |\nabla w|^2 \leq \text{const}.$$

Without loss of generality, we may assume that each w^R has been extended to equal w_∞ outside $\{|x| = R\}$. In order to show that the limiting velocity w satisfies (4), one must prove the uniform convergence of the w^R on the *unbounded* set $\bar{\Omega}$; unfortunately, this has still not been done for any ν or w_∞. A method completely different from (5)–(8) has been used [2] to prove the existence of solutions to (1)–(4) when the Reynolds number is sufficiently small; that is, when ν is sufficiently large or w_∞ is sufficiently small. This method does not appear to extend to large Reynolds number, and so the best hope of solving (1)–(4) for general Reynolds number is still that of Leray.

Another reason for the interest in the approximate solution w^R of Leray is that their three-dimensional counterparts work. More precisely, let Ω denote the exterior of a compact set $K \subset \mathbf{R}^3$, and let (p^R, w^R) denote solutions of (5)–(8) where Ω_R and w_∞ have their obvious meaning. The solutions w^R satisfy (9) again, and a subsequence converges uniformly on compact subsets of $\bar{\Omega}$ to a solution (p, w) of (1)–(3). If we extend the w^R to equal w_∞ outside $\{|x| = R\}$, then $w^R - w_\infty$ has compact support in $\mathbf{R}^3$. The standard estimate

(11) $$\int_{\mathbf{R}^3} |\varphi|^6 \leq \text{const} \left(\int_{\mathbf{R}^3} |\nabla \varphi|^2 \right)^3$$

for functions of compact support, gives $|w^R - w_\infty|_{L^6(\Omega)} \leq \text{const}$, and the same estimate is inherited by the limiting velocity field w:

(12) $$\int_\Omega |w - w_\infty|^6 \leq \text{const} \left(\int_\Omega |\nabla w|^2 \right)^3 \leq \text{const}.$$

The use of (12) with the Navier–Stokes equations yields $w(x) \to w_\infty$ as $|x| \to \infty$ [3, 4]. It is also known that if (p, w) is any solution (that is, not necessarily the limit of Leray approximants) of (1)–(3) in $\mathbf{R}^3$ with finite Dirichlet norm (10), then there exists a constant vector w_∞ such that (12) holds, $w(x) \to w_\infty$ as $|x| \to \infty$, $|w(x) - w_\infty| = O(|x|^{-1})$, and the flow has the expected wake properties [5].

In the case of two dimensions, estimates relating compactly supported functions φ to their derivative $|\nabla \varphi|$ do exist, but they are not as useful as (11):

$$\int_\Omega \frac{|\varphi|^2}{|x|^2 (\log |x|)^2} \leq \text{const} \int_\Omega |\nabla \varphi|^2$$

whence

(13) $$\int_\Omega \frac{|w^R - w_\infty|^2}{|x|^2 (\log |x|)^2} \leq \text{const} \int_\Omega |\nabla w^R|^2 \leq \text{const}.$$

It follows that the limiting velocity w satisfies (1)–(3),

$$\int_\Omega |\nabla w|^2 < \infty,$$

and

(14) $$\int_\Omega \frac{|w - w_\infty|^2}{|x|^2 (\log|x|)^2} \leq \text{const.}$$

Unfortunately, (14) does not reveal anything about the behavior of w at infinity since if it holds for one w_∞, then it holds *for all* vectors $w_\infty \in \mathbf{R}^2$. This prevents one from immediately applying the methods for the three-dimensional exterior problem to the two-dimensional one.

A considerable step forward towards solving (1)–(4) came in two papers of Gilbarg and Weinberger [6, 7], who exploited maximum principles for certain physical quantities. In this note, we shall pursue certain of the approaches in these two papers, and will introduce a new quantity which satisfies a useful maximum principle. In [6], it was shown that p^R and w^R are bounded in Ω_R, independently of R. (Actually, the authors only showed this on $\Omega_{R/2}$, but similar arguments carry over to the whole annulus.) Since the w^R converge uniformly on compact sets to w, it follows that $w \in L_\infty(\Omega)$. It was then proved (Theorem 3 in §4) that if (p, w) satisfies (1)–(4), has finite Dirichlet norm (10), and has $w \in L_\infty(\Omega)$, then there exists a constant vector $\tilde{w}_\infty \in \mathbf{R}^2$ such that

(15) $$\lim_{r \to \infty} \int_0^{2\pi} |w(r, \theta) - \tilde{w}_\infty|^2 \, d\theta = 0.$$

However, it was not shown that $w_\infty = \tilde{w}_\infty$ or that $w(x) \to \tilde{w}_\infty$ pointwise as $|x| \to \infty$. The authors also showed that the pressure p had a limit at infinity, and the vorticity $\omega \equiv u_{x_2} - v_{x_1}$ was $o(|x|^{-3/4})$ as $|x| \to \infty$. In the second paper [7], Gilbarg and Weinberger considered a solution (not necessarily that constructed by Leray) of (1)–(3) with finite Dirichlet norm (10), and showed that either the velocity field w is bounded in Ω or becomes unbounded (but is $o(\log|x|)$) as $|x| \to \infty$. We shall show in §4 that only the former can occur, whence Theorem 3 is applicable and (15) always holds.

In §2, we introduce certain quantities and maximum principles which will be needed later. In §3, we consider the approximate solution (p^R, w^R) of Leray, and show that their Dirichlet norm is uniformly bounded away from zero (Theorem 2); this is to be contrasted with Stokes flow (Theorem 1) for which the Dirichlet norm goes to zero as $R \to \infty$. In §4, we consider solutions of (1)–(3) with finite Dirichlet norm, and prove in Theorem 4 that $w \in L_\infty(\Omega)$. As noted earlier, this gives us the result (15) above. In Theorem 5, we show that a certain quantity $\gamma(x) \equiv p(x) + \frac{1}{2}|w(x)|^2 - \psi(x)\omega(x)$ satisfies $\gamma(x) \to \frac{1}{2}|\tilde{w}_\infty|^2$ as $|x| \to \infty$. Here ψ is the Stokes stream function, $\omega = u_{x_2} - v_{x_1}$ is the vorticity, and $\tilde{w}_\infty$ is as in (15). We use this result for γ to show that the vorticity $\omega(x)$ is $o(|x|^{-1})$ in sectors away from the x_1-axis. We also prove the pointwise convergence of $|w(x)|$ to $|\tilde{w}_\infty|$ in such sectors.

In order to simplify certain proofs, we shall assume in §3 that the body K is symmetric about the x_1-axis. More detailed proofs of the theorems in §§3 and 4, and additional results such as $w(x) \to \tilde{w}_\infty$ as $|x| \to \infty$ and the existence of a wake, may be found in [8].

2. Maximum principles. Standard theory ensures that a solution (p, w) of (1)–(2) is real-analytic in Ω, and similarly for (p^R, w^R) in $\Omega_R \cup \{|x| = R\}$. Define the *vorticity* ω of a velocity field $w = (u, v)$ by $\omega = u_{x_2} - v_{x_1}$. If (p, w) satisfies the steady Navier–Stokes equations (5)–(6) in Ω_R, then a calculation yields

(16) $$-\nu \Delta \omega + w \cdot \nabla \omega = 0 \quad \text{in } \Omega_R,$$

and so ω satisfies a two-sided maximum principle: The maximum and minimum values of ω on a compact subset $\overline{U}$ of $\overline{\Omega}_R$ are taken on ∂U. We have omitted an R superscript on the terms w and ω in (16), and continue to do so in the important displayed equations (17)–(20).

If we have a point $\tilde{x} \in \Omega_R$ where $\nabla \omega(\tilde{x}) \neq 0$, then the level curve of ω through $\tilde{x}$ is (locally) a real-analytic curve. If $\nabla \omega(\tilde{x}) = 0$, there may not be a single level curve through $\tilde{x}$. More precisely, there exists a $\delta > 0$ such that the set

$$\{x \in \Omega_R : |x - \tilde{x}| < \delta, \omega(x) = \omega(\tilde{x})\} \setminus \{x = \tilde{x}\}$$

consists of an even number of real-analytic arcs with $\tilde{x}$ in each of their closure. If $\nabla \omega(\tilde{x}) \neq 0$, then this even number is two. The use of the maximum principle with (16) ensures that either ω is identically a constant in Ω_R or $\nabla \omega$ can only vanish at isolated points in Ω_R. We claim that ω cannot be identically a constant C, say, in Ω_R. Indeed,

$$\int_{\Omega_R} \omega \, dx_1 \, dx_2 = \int_{\Omega_R} (u_{x_2} - v_{x_1}) \, dx_1 \, dx_2 = -\oint_{\{|x| = R\}} u \, dx_1 + v \, dx_2 = 0$$

since $w = (u, v) = (u_\infty, v_\infty)$ on $\{|x| = R\}$. Hence, $C = 0$, and so the flow is irrotational in Ω_R. But then each component of w is harmonic in Ω_R. Now u takes its maximum or minimum on $\{|x| = R\}$, and so the maximum principle ensures that $|\partial u/\partial r| \neq 0$ on $\{|x| = R\}$. However, the conjugate harmonic function v is constant there and so $0 = |\partial v/\partial \theta| = |\partial u/\partial r| \neq 0$. This contradiction ensures that ω is not constant in Ω_R.

Define the *total-head pressure* Φ of a pressure and velocity pair (p, w) by $\Phi = p + \frac{1}{2}|w|^2$. A calculation from (5)–(6) yields

(17) $$-\nu \Delta \Phi + w \cdot \nabla \Phi = -\nu \omega^2 \quad \text{in } \Omega_R,$$

whence Φ satisfies a one-sided maximum principle: The maximum value of Φ is taken on ∂U. Equations (16) and (17) were critical to the analysis in [6] and [7]. We now introduce a new quantity by $\gamma(x) = \Phi(x) - \psi(x)\omega(x)$, where ψ denotes a stream function:

$$\psi(x) = \int_{y_0}^{x} u \, dx_2 - v \, dx_1.$$

We are unaware of any physical significance of γ or of its use in other fluid-mechanics problems. Here y_0 is a fixed point on Γ and the line integral is taken over a smooth curve connecting y_0 to the point $x = (x_1, x_2)$. Note that $w = (u, v) = (\psi_y, -\psi_x)$. A calculation using (5)–(6) yields the following relationships between γ and the vorticity ω:

(18a) $$\gamma_{x_1} = \nu\omega_{x_2} - \psi\omega_{x_1},$$
(18b) $$\gamma_{x_2} = -\nu\omega_{x_1} - \psi\omega_{x_2}$$ in Ω_R,

or, equivalently,

(19a) $$\omega_{x_1} = -(\nu\gamma_{x_2} + \psi\gamma_{x_1})/(\nu^2 + \psi^2),$$
(19b) $$\omega_{x_2} = (\nu\gamma_{x_1} - \psi\gamma_{x_2})/(\nu^2 + \psi^2)$$ in Ω_R.

Another calculation from (16) and (18) gives

(20) $$-\nu\Delta\gamma + (w + q) \cdot \nabla\gamma = 0 \quad \text{in } \Omega_R,$$

where $q = \{-2\nu^2 w + 2\nu\psi(-v, u)\}/(\nu^2 + \psi^2)$. It follows from (20) that γ satisfies a two-sided maximum principle and has the same type of level curve structure described earlier for the vorticity. Equations (18) and (19) show that $\nabla\omega$ and $\nabla\gamma$ vanish at the same points in Ω_R which, if there are any, we know to be isolated. If $\nabla\omega(\tilde{x}) \neq 0$, then the level curve through $\tilde{x}$ is locally a real-analytic curve, and the maximum principle applied to (16) ensures that the normal derivative of ω to this curve is one-signed. It follows from (18) that the tangential derivative to this curve of γ is one-signed, whence γ is monotone (increasing or decreasing) as one moves along level curves of ω. In this respect, ω and γ behave like conjugate harmonic functions. We shall combine these properties of the functions ω and γ with certain average values of integrals arising from (9) or (10) to derive the main results in the following two sections.

3. The approximate solutions of Leray. In this section, we consider the solutions (p^R, w^R) of (5)–(8). We remind the reader of the bound (9) on the Dirichlet norm. In addition, Ω_R, $R \geq 2$, *is symmetric about the x_1-axis* and Γ is contained in the unit disc, while u^R, p^R, Φ^R, and γ^R are even functions of x_2 and v^R and ω^R are odd functions of x_2. In §1, we noted that the three-dimensional version of (1)–(4) has been solved by the approximation scheme of Leray while the two-dimensional problem remains unresolved. Another difference between the two- and three-dimensional problems may be seen in the Stokes equations:

(21) $$-\nu\Delta w = -\nabla p,$$
(22) $$\nabla \cdot w = 0$$ in Ω,
(23) $$w = 0 \quad \text{on } \Gamma,$$
(24) $$w \to w_\infty \quad \text{as } |x| \to \infty.$$

The first equation arises from (1) if one assumes the viscous term is dominant. A natural method for solving (21)–(24) would be to solve (21)–(23) in the annular domains Ω_R with the boundary condition (24) imposed on $\{|x| = R\}$. This problem has a unique solution (p_s^R, w_s^R) in either two or three dimensions, and the estimate

$$\int_{\Omega_R} |\nabla w_s^R|^2 \leq \text{const} \tag{25}$$

holds. If we use estimates of the form (12) for the *three-dimensional* problem, then one can easily show that the (p_s^R, w_s^R) converge to the unique solution of (21)–(24). The problem in two dimensions is made more curious by a result of Finn and Noll [9]: If (p, w) satisfies (21)–(23) in an exterior domain $\Omega \subset \mathbf{R}^2$ and $|w(x)| = o(\log |x|)$ as $|x| \to \infty$, then $w \equiv 0$ in Ω. Hence, (21)–(24) has no solution in two dimensions if $w_\infty \neq 0$. Since the Leray approximants w_s^R satisfy (25), a subsequence converges on compact sets to a solution (p_s, w_s) of (21)–(23) with finite Dirichlet norm.

One can easily show that any solution of (21)–(23) with finite Dirichlet norm has $w(x) = o(\log |x|)$ at infinity, whence $w_s \equiv 0$ by [9]. Let us extend each w_s^R to equal w_∞ outside $\{|x| = R\}$. Since $w_s^R = 0$ on $\Gamma = \partial \Omega$, one can show that each w_s^R belongs to the Hilbert space $H(\Omega)$, the completion of $\{\varphi \in C_0^\infty(\Omega \to \mathbf{R}^2): \text{div } \varphi = 0 \text{ in } \Omega\}$ in the Dirichlet norm. (This might seem surprising since each w_s^R is nonzero at infinity but it may be proved easily.) Equation (25) ensures that $w_s^R \to w_s = 0$ in $H(\Omega)$ as $R \to \infty$. There are two possibilities: either $w_s^R \to 0$ in $H(\Omega)$ or

$$\limsup_{R \to \infty} \int_{\Omega_R} |\nabla w_s^R|^2 > 0.$$

Since w_s^R converges pointwise to zero on compact subsets of $\overline{\Omega}$, the latter possibility can occur by the Dirichlet norm being concentrated near the outer boundary $\{|x| = R\}$ and becoming small on compact sets. The following theorem shows that $w_s^R \to 0$ in $H(\Omega)$. Theorem 2 shows that this cannot occur for the Navier–Stokes equations.

THEOREM 1. *Let $\nu > 0$ and $w_\infty \in \mathbf{R}^2$ be given. Let (p_s^R, w_s^R) denote the unique solution of the Stokes equation in the exterior two-dimensional domain Ω:*

$$-\nu \Delta w_s^R = -\nabla p_s^R, \tag{26}$$
$$\nabla \cdot w_s^R = 0 \tag{27}$$
$$\text{in } \Omega_R,$$
$$w_s^R = 0 \quad \text{on } \Gamma, \tag{28}$$
$$w_s^R = w_\infty \quad \text{on } \{|x| = R\}. \tag{29}$$

Then

$$\int_{\Omega_R} |\nabla w_s^R|^2 \leq \frac{\text{const}|w_\infty|^2}{\log R}, \quad R \geq 2,$$

and the constant is independent of R, w_∞, and ν.

PROOF. Let $\tau \in C^\infty(\mathbf{R})$ with $\tau' \in C_0^\infty(\tfrac{1}{2}, 1)$, and $\tau(r) \equiv 0$ for $r \leq \tfrac{1}{2}$ and $\tau(r) \equiv 1$ for $r \geq 1$. Define $\mu(r) = \tau(\log r / \log R)$ so that $\mu(r) = 0$ when $r \leq \sqrt{R}$. We shall drop the R and s super- and subscripts in our calculations, and we may rotate the axes so that $w_\infty = (u_\infty, 0)$ without loss of generality. Define the solenoidal velocity field $A = (A_1, A_2)$ by

$$A_1 = \frac{\partial}{\partial x_2}\{u_\infty x_2 \mu(|x|)\}, \quad A_2 = -\frac{\partial}{\partial x_1}\{u_\infty x_2 \mu(|x|)\},$$

set $\tilde{w} = w - A$, and note that $\tilde{w} = 0$ on $\partial \Omega_R$. If we combine this with the equation $-\nu \Delta \tilde{w} - \nu \Delta A = -\nabla p$, then

$$\int_{\Omega_R} |\nabla \tilde{w}|^2 = -\int_{\Omega_R} \nabla \tilde{w} : \nabla A,$$

and so

(30) $$\int_{\Omega_R} |\nabla \tilde{w}|^2 \leq \int_{\Omega_R} |\nabla A|^2.$$

An explicit calculation yields

$$|\nabla A(x)|^2 \leq \frac{\text{const}|w_\infty|^2}{|x|^2 (\log R)^2}, \quad x \in \Omega_R,$$

where $|x|^2 = x_1^2 + x_2^2$, and the constant is independent of R and w_∞. The use of this gives

$$\int_{\Omega_R} |\nabla \tilde{w}|^2 \leq \int_{\Omega_R} |\nabla A|^2 \leq \frac{\text{const}|w_\infty|^2}{\log R},$$

and the representation $w = \tilde{w} + A$ completes the proof. Q.E.D.

REMARK. We did not need to assume any symmetry of Ω in Theorem 1, and so the result holds for arbitrary Ω.

The following theorem shows that the Dirichlet norms of Leray's approximate solutions are bounded away from zero as $R \to 0$, and this is clearly due to the presence of the nonlinear term $(w \cdot \nabla)w$ in (5).

The proof depends crucially on using (18) and certain level curves where $\omega^R = 0$ to show that the total-head pressure Φ^R is monotone on certain arcs connecting Γ (where $w^R = 0$) to the exterior boundary $\{|x| = R\}$ where $w^R = w_\infty$. In remarks following the theorem, we show why such a method fails for the Stokes equations (26)–(29).

THEOREM 2. *Let $\nu > 0$ and $w_\infty \in \mathbf{R}^2 \setminus \{0\}$ be given, and let (p^R, w^R) denote any solution of (5)–(8).*

(a) *There exist Jordan arcs $J_i(\cdot\,; R)$, $i = 1, 2$, with $J_i(0; R) \in \Gamma$ and $J_i(1; R) \in \{|x| = R\}$ such that $\omega^R(J_i(t; R)) = 0$, $t \in [0, 1]$, $i = 1, 2$, and*

$$\Phi^R(J_1(t; R)) > \Phi^R(J_1(s; R)), \quad \Phi^R(J_2(t; R)) < \Phi^R(J_2(s; R))$$

if $s, t \in [0, 1]$ and $t > s$.

(b)
$$0 < \liminf_{R \to \infty} \int_{\Omega_R} |\nabla w^R|^2 \leq \limsup_{R \to \infty} \int_{\Omega_R} |\nabla w^R|^2 \leq \text{const.}$$

PROOF. (a) The proof of (a) is technical and depends on the relations (18)–(19) and the maximum principle for the vorticity. The details may be found in [**8**].

(b) We shall drop the R superscript on functions during the proof. The upper bound for the Dirichlet norm comes from (9), and so we need only prove the lower bound. Assume the contrary, so that there is a subsequence, which we relabel as $\{w^R\}$, such that

$$\int_{\Omega_R} |\nabla w^R|^2 \to 0 \quad \text{as } R \to \infty. \tag{31}$$

We shall divide the proof into convenient steps.

(i) Equation (31) allows us to assume that (p^R, w^R) and as many derivatives as we wish converge to zero uniformly on compact subsets of $\bar{\Omega}$; this follows from (31) and the usual regularity theory for (5)–(8). For each $r \in (1, R)$, let a bar over a function denote the mean average over circles, e.g.,

$$\bar{p}^R(r) = \frac{1}{2\pi} \int_0^{2\pi} p^R(r, \theta) \, d\theta, \quad r \in [1, R].$$

Note that (5)–(6) gives

$$\frac{\partial}{\partial r} p^R = \frac{\nu}{r} \frac{\partial}{\partial \theta} \omega^R + \frac{1}{r} \left\{ u^R \frac{\partial}{\partial \theta} v^R - v^R \frac{\partial}{\partial \theta} u^R \right\} \tag{32}$$

for all $(r, \theta) \in (1, R) \times [0, 2\pi]$. Hence,

$$\frac{d}{dr} \bar{p}^R(r) = \frac{1}{2\pi r} \int_0^{2\pi} \left[u^R(r, \theta) \frac{\partial}{\partial \theta} v^R(r, \theta) - v^R(r, \theta) \frac{\partial}{\partial \theta} u^R(r, \theta) \right] d\theta$$

$$= \frac{1}{2\pi r} \int_0^{2\pi} \left[\{u^R(r, \theta) - \bar{u}^R(r)\} \frac{\partial}{\partial \theta} v^R(r, \theta) \right. \tag{33}$$

$$\left. - \{v^R(r, \theta) - \bar{v}^R(r)\} \frac{\partial}{\partial \theta} u^R(r, \theta) \right] d\theta.$$

The use of Wirtinger's inequality gives

$$\left| \frac{d}{dr} \bar{p}^R(r) \right| \leq \frac{1}{2\pi r} \int_0^{2\pi} \left\{ \left(\frac{\partial}{\partial \theta} u^R \right)^2 + \left(\frac{\partial}{\partial \theta} v^R \right)^2 \right\} d\theta \tag{34}$$

$$\leq \frac{1}{2\pi} \int_0^{2\pi} r |\nabla w^R|^2 \, d\theta, \quad r \in (1, R).$$

Now

$$|\bar{p}^R(r) - \bar{p}^R(1)| \leq \int_1^R \left| \frac{d}{dr} \bar{p}^R(r) \right| dr \leq \frac{1}{2\pi} \int_{\Omega_R} |\nabla w^R|^2 \to 0 \quad \text{by (31)},$$

while $p^R \to 0$ on compact sets as noted before. Hence,

$$\limsup_{1 \leq r \leq R} |\bar{p}^R(r)| \to 0 \quad \text{as } R \to \infty. \tag{35}$$

(ii) We now show that

$$\int_{\Omega_{R/2}} |\nabla \omega^R|^2 \to 0 \quad \text{as } R \to \infty. \tag{36}$$

Let $\tau \in C^\infty(\mathbf{R})$ with $\tau' \in C_0^\infty(0,1)$, and $\tau(r) = 1$ for $r \leq 0$ and $\tau(r) = 0$ for $r \geq 1$. Define $\mu(r; R) = \tau(r - R/2)$. If we multiply (16) by $\mu(\cdot\,; R)\omega^R$ and integrate over Ω_R, then

$$-\nu \int_{\Omega_R} \mu \omega^R \Delta \omega^R = \nu \int_{\Omega_R} \mu |\nabla \omega^R|^2 - \frac{\nu}{2} \int_{\Omega_R} (\omega^R)^2 \Delta \mu + \{\text{B.T.}\}$$

$$= -\int_{\Omega_R} \mu \omega^R w^R \cdot \nabla \omega^R = -\frac{1}{2} \int_{\Omega_R} \mu \, \text{div}\big((\omega^R)^2 w^R\big)$$

$$= \frac{1}{2} \int_{\Omega_R} (\omega^R)^2 w^R \cdot \nabla \mu,$$

where {B.T.} denotes a boundary integral around Γ involving ω and its normal derivative. A result of Gilbarg and Weinberger ensures that $|w^R|$ is bounded on $\Omega_{3R/4}$, independently of R, while (31) ensures that

$$\int_{\Omega_R} (\omega^R)^2 \leq 2 \int_{\Omega_R} |\nabla w^R|^2 \to 0 \quad \text{as } R \to \infty$$

and also that the boundary terms vanish as $R \to \infty$. This proves (36).

Equations (5)–(6) yield

$$\frac{\partial}{\partial x_1} p^R = \nu \frac{\partial}{\partial x_2} \omega^R - u^R \frac{\partial}{\partial x_1} u^R - v^R \frac{\partial}{\partial x_2} u^R, \tag{37a}$$

$$\frac{\partial}{\partial x_2} p^R = -\nu \frac{\partial}{\partial x_1} \omega^R - u^R \frac{\partial}{\partial x_1} v^R - v^R \frac{\partial}{\partial x_2} v^R, \tag{37b}$$

and the use of (31) and (36) gives

$$\int_{\Omega_{R/2}} |\nabla p^R|^2 \to 0 \quad \text{as } R \to \infty. \tag{38}$$

(iii) Equations (31) and (38) give

$$\int_{R/4}^{R/2} \frac{dr}{r} \int_0^{2\pi} \left\{ \left|\frac{\partial}{\partial \theta} p^R(r, \theta)\right|^2 + \left|\frac{\partial}{\partial \theta} w^R(r, \theta)\right|^2 \right\} d\theta \to 0 \quad \text{as } R \to \infty,$$

and so there exists $\tilde{R} = \tilde{R}(R) \in (R/4, R/2)$ such that

(39) $$\int_0^{2\pi} \left\{ \left| \frac{\partial}{\partial \theta} p^R(\tilde{R}, \theta) \right|^2 + \left| \frac{\partial}{\partial \theta} w^R(\tilde{R}, \theta) \right|^2 \right\} d\theta \to 0 \quad \text{as } R \to \infty.$$

Say that the curves J_i intersect the circle $\{|x| = \tilde{R}\}$ at z_i, $i = 1, 2$, where $|z_i| = \tilde{R}$. Let $\tilde{z}_i = J_i(0)$ denote the points where these curves intersect the body Γ. Part (a) gives

(40) $$\Phi^R(z_1) = p^R(z_1) + \tfrac{1}{2}|w^R(z_1)|^2 \geq \Phi^R(\tilde{z}_1) = p^R(\tilde{z}_1),$$
$$\Phi^R(z_2) = p^R(z_2) + \tfrac{1}{2}|w^R(z_2)|^2 \leq \Phi^R(\tilde{z}_2) = p^R(\tilde{z}_2).$$

It follows from (35) and (39) that

$$\max_{|x|=\tilde{R}} |p^R(x)| \to 0 \quad \text{as } R \to \infty.$$

Since $p^R \to 0$ on Γ as $R \to \infty$, we have $p^R(\tilde{z}_2) \to 0$, whence $|w^R(z_2)| \to 0$ as $R \to \infty$. If we combine this with (39), then

(41) $$\max_{|x|=\tilde{R}} |w^R(x)| \to 0 \quad \text{as } R \to \infty.$$

(iv) Now

(42) $$\int_0^{2\pi} |w^R(R, \theta)| \, d\theta - \int_0^{2\pi} |w^R(\tilde{R}, \theta)| \, d\theta = \int_{\tilde{R}}^R dr \int_0^{2\pi} \frac{\partial}{\partial r} |w^R| \, d\theta$$
$$\leq \left(\int_{\tilde{R}}^R \frac{2\pi \, dr}{r} \right)^{1/2} \left(\int_{\Omega_R} |\nabla w^R|^2 \right)^{1/2} \to 0 \quad \text{as } R \to \infty.$$

It follows from (41) and (42) that

$$\int_0^{2\pi} |w^R(R, \theta)| d\theta \to 0 \quad \text{as } R \to \infty.$$

However, this is impossible since (8) gives $|w^R(R, \theta)| = |w_\infty|$ for all R, and so (31) is invalid. Q.E.D.

If we compare Theorems 1 and 2, we see that solutions to the Stokes equations behave quite differently from the Navier–Stokes equations. If (p_s, w_s) satisfies the Stokes equations $-\nu \Delta w_s = -\nabla p_s$ and $\nabla \cdot w_s = 0$, then p_s is harmonic as is the vorticity of ω_s. Indeed, the Stokes equations say that $p_s + i\nu\omega_s$ is an analytic function. Hence, along level curves of ω_s, the pressure p_s is monotone, and conversely. In the proof of Theorem 2, we needed a quantity γ, involving the pressure *and the velocity field*, which was monotone along level cures of ω.

We remind the reader that we have only shown a positive lower bound on the Dirichlet norm for the w^R, and not that the limiting velocity field is nontrivial. However, Theorem 2 is suggestive of that, and such a result is proved in [8].

4. Solutions of (1)–(3) with finite Dirichlet norm.

Throughout the section, we shall restrict attention to a pair (p, w) satisfying

$$
\begin{aligned}
-\nu\Delta w + (w \cdot \nabla)w &= -\nabla p, \\
\nabla \cdot w &= 0
\end{aligned}
\quad \text{in } \Omega, \tag{43, 44}
$$

$$w = 0 \quad \text{on } \Gamma, \tag{45}$$

$$\int_\Omega |\nabla w|^2 < \infty. \tag{46}$$

Here ν is a given positive constant, $\Omega = \mathbf{R}^2 \setminus K$, where K is a smooth, compact, simply connected set (not necessarily symmetrical), and we assume w is nontrivial. Standard theory ensures that (p, w) is real-analytic in Ω, and all the equations and the maximum principle in §2 hold with Ω_R replaced by Ω. Since w is nontrivial, the same is true of the vorticity ω, and so the points where $\nabla \omega = 0$ are isolated. We know from §1 that a solution of (43)–(46) can be found from the Leray approximants, but we shall not assume that our (p, w) has been so constructed.

The following theorem is due to Gilbarg and Weinberger [6].

THEOREM 3. *Let (p, w) satisfy (43)–(46), and assume that $w \in L_\infty(\Omega)$. Then there exists a constant vector $\tilde{w}_\infty \in \mathbf{R}^2$ such that*

$$\lim_{r \to \infty} \int_0^{2\pi} |w(r, \theta) - \tilde{w}_\infty|^2 \, d\theta = 0. \tag{47}$$

Furthermore, the pressure p has a limit at infinity and the vorticity ω is $o(|x|^{-3/4})$ as $|x| \to \infty$.

If (p, w) is only known to satisfy (43)–(46) then [7] shows that $w = o(\log|x|)$, $\omega = o(|x|^{-3/4}(\log|x|)^{1/8})$, and that p has a limit at infinity. This particular growth estimate for w is not surprising since (13) gives

$$\int_\Omega \frac{|w|^2}{|x|^2 (\log|x|)^2} \leq \text{const.}$$

We also remark that the best estimate one can expect for the vorticity is $O(|x|^{-1})$ inside the wake and $o(|x|^{-1})$ outside the wake.

The following theorem shows that solutions with finite Dirichlet norm have $w \in L_\infty(\Omega)$; recall that there is no assumption that Ω be symmetric.

THEOREM 4. *Assume that (p, w) satisfies (21)–(24).*

(a) *There exist Jordan arcs $M_i(t)$, $t \in [0, 1)$, $i = 1, 2$, with $M_i(0) \in \Gamma$ and $M_i(t) \to \infty$ as $t \to 1$ such that $\omega(M_i(t)) = 0$, $t \in [0, 1)$, and*

$$\Phi(M_1(t)) > \Phi(M_1(s)), \qquad \Phi(M_2(t)) < \Phi(M_2(s)),$$

where $s, t \in [0, 1)$ and $s < t$.

(b) $w \in L_\infty(\Omega)$.

PROOF. (a) The remark after Theorem 3 allows us to assume that $\omega(x) \to 0$ as $|x| \to \infty$. We begin by noting that ω is not one-signed at infinity; that is, there exist positive x and y with $|x|$ and $|y|$ arbitrarily large, such that $\omega(x) > 0$ and $\omega(y) < 0$. To see this, let us assume to the contrary that $\omega(x) \geq 0$ for all $|x| \geq N$, say. Since $\omega \geq 0$ on $\{|x| = N\}$, $\omega(x) \to 0$ as $|x| \to \infty$, and ω satisfies (16), the maximum principle yields $\omega > 0$ in the set $\{|x| > N\}$. Let $\varepsilon > 0$ denote the minimum value of ω on the circle $\{|x| = N + 1\}$. Now the set $\{x: |x| \geq N + 1$ and $\nabla\omega(x) = 0\}$ is at most countable, and so there exists an $\tilde{\varepsilon} \in (0, \varepsilon)$ such that $\tilde{\varepsilon}$ is never a critical value: If $|x| \geq N + 1$ and $\omega(x) = \tilde{\varepsilon}$, then $\nabla\omega(x) \neq 0$. Let $\tilde{x}$ lie in $\{|x| \geq N + 1\}$ and have $\omega(\tilde{x}) = \tilde{\varepsilon}$. Since $\omega(x) \to 0$ as $|x| \to \infty$ and $\omega(x) > \tilde{\varepsilon}$ on $\{|x| = N + 1\}$, the level curve through $\tilde{x}$ is a real-analytic closed Jordan curve S contained in $\{|x| \geq N + 1\}$. Equation (16) and the maximum principle ensure that the normal derivative of ω to S is one-signed, whence the tangential derivative of γ is one-signed by (18). This is a contradiction, and so it is impossible for $\omega(x) \geq 0$ for all $|x|$ sufficiently large. A similar argument holds for $\omega \leq 0$.

For each $x \in \Omega$ with $\omega(x) > 0$, let $U^+(x)$ denote the maximal connected open set containing x in Ω_R on which $\omega > 0$. Each U^+ is simply connected since any closed Jordan curve T in U^+ has int $T \subset U^+$. If not, then the interior of T would contain the body, and the exterior of T would be contained in Ω. But $\omega > 0$ on $\partial(\text{ext } T) = T$ and $\omega \to 0$ at infinity, whence $\omega(x) > 0$ for all large x. This is a contradiction, and so each U^+ is simply connected. One can show that $\partial U^+ \cup \{\infty\}$ is a closed Jordan curve. We note that $\partial U^+ \cap \Gamma \neq \emptyset$ else $\omega = 0$ on $\partial U^+ \subset \Omega$ whence $\omega \equiv 0$ in Ω by the maximum principle.

We claim there is a $U^+(\tilde{x})$ which is unbounded. To see this, we assume the contrary and derive a contradiction. Now there exists a sequence $\{x_n\}$ with $|x_n| \to \infty$ and $\omega(x_n) > 0$ as $n \to \infty$, and our hypothesis allows us to assume that $U_i \cap U_j = \emptyset$ if $i \neq j$, where $U_i = U^+(x_i)$. Since $\partial U_i \cap \Gamma \neq \emptyset$ for each i and since Γ is contained in the unit disc (by hypothesis in §1), we know that the sets $\partial U^i \cap \{|x| = 2\}$ are nonempty for all large i, and that the union over i is an infinite set. But this means that the real-analytic function $\theta \mapsto \omega(|x| = 2, \theta)$ has an infinite number of discrete zeros on $[0, 2\pi]$. This implies that $\omega(x) = 0$ if $|x| = 2$ whence $\omega \equiv 0$ in the region $\{|x| \geq 2\}$ by the maximum principle.

Since $U^+(\tilde{x})$ is unbounded, $\partial U^+ \cap \Gamma \neq \emptyset$, and $\omega > 0$ in U^+, we may use the arguments as for Theorem 2(a) to construct the M_i.

(b) The remark after Theorem 3 allows us to assume that $p(x)$ has a limit as $|x| \to \infty$. Since $\Phi(x) = p(x) + \frac{1}{2}|w(x)|^2$ and $\Phi(M_2(t))$ is nonincreasing, it follows that $|w(x)| \leq \text{const}$ if x lies on the Jordan arc M_2. Equation (46) gives

$$(48) \qquad \int_{2^n}^{2^{n+1}} \frac{dr}{r} \int_0^{2\pi} \left|\frac{\partial}{\partial \theta} w(r, \theta)\right|^2 d\theta \to 0 \quad \text{as } n \to \infty$$

and so there exists $r_n \in (2^n, 2^{n+1})$ such that

$$(49) \qquad \int_0^{2\pi} \left|\frac{\partial}{\partial \theta} w(r_n, \theta)\right|^2 d\theta \to 0 \quad \text{as } n \to \infty.$$

By connectedness, $M_2 \cap \{|x| = r_n\} \neq \varnothing$ for all n, and so there exists $\theta_n \in [0, 2\pi]$ such that $|w(r_n, \theta_n)| \leq \text{const}$, independently of n. If we combine this with (49), then
$$\max_{\theta \in [0, 2\pi]} |w(r_n, \theta)| \leq \text{const},$$
independently of n. Since the pressure has a limit at infinity, we have $|\Phi(x)| \leq \text{const}$ when $|x| = r_n$. If we apply (17) to the annular regions $\{r_n < |x| < r_n + 1\}$, then
$$\max_{r_n \leq |x| \leq r_{n+1}} \left\{ p(x) + \tfrac{1}{2}|w(x)|^2 \right\} = \max_{|x| = r_n, r_{n+1}} \Phi(x) \leq \text{const}.$$
Since p is bounded at infinity and the constant is independent of n, we have $w \in L_\infty(\Omega)$. Q.E.D.

The level curves and maximum principles used so far have exploited the equality of γ and Φ on curves where $\omega = 0$. The following theorem uses the function γ in its own right.

THEOREM 5. *Let (p, w) satisfy (43)–(46), and assume that the pressure has been normalized by $p(x) \to 0$ as $|x| \to \infty$. Then*

(a) $\quad \gamma(x) \equiv \Phi(x) - \psi(x)\omega(x)$
$$= p(x) + \tfrac{1}{2}|w(x)|^2 - \psi(x)\omega(x) \to \tfrac{1}{2}|\tilde{w}_\infty|^2 \quad \text{as } |x| \to \infty,$$
where $\tilde{w}_\infty$ is as in Theorem 3.

(b) *Assume that $|\tilde{w}_\infty| \neq 0$ and that the axes have been rotated so that $w_\infty = (u_\infty, 0)$. For every $\varepsilon \in (0, \pi/2)$, let*
$$S_\varepsilon = \{(r, \theta) \in (0, \infty) \times [-\pi, \pi] : r \geq 2, |\theta| \in (\varepsilon, \pi - \varepsilon)\}.$$
Then $\omega(x) = o(|x|^{-1})$ and $|w(x)| \to |\tilde{w}_\infty|$ as $x \to \infty$ in S_ε.

PROOF. (a) We shall only mention the main ideas and reserve the proof to [8]. Since w satisfies (47) and (49), one has
$$(50) \qquad \max_{\theta \in [0, 2\pi]} |w(r_n, \theta) - \tilde{w}_\infty| \to 0 \quad \text{as } n \to \infty,$$
and the use of this with the monotonicity of Φ along the M_i gives
$$|w(M_i(t)) - \tilde{w}_\infty| \to 0 \quad \text{as } M_i(t) \to \infty.$$
Since $\omega = 0$ on M_i, it follows that $\gamma(M_i(t)) \to \tfrac{1}{2}|\tilde{w}_\infty|^2$ as $t \to \infty$. The rest of the proof uses the monotonicity of γ along level curves of ω.

(b) Let $\tilde{\varepsilon} \in (0, \pi/2)$ be fixed. Since
$$\int_0^{2\pi} d\theta \int_{r_n}^{r_{n+1}} r \left| \frac{\partial}{\partial r} w(r, \theta) \right|^2 dr \to 0 \quad \text{as } n \to \infty$$
there exists $\varepsilon_n \in (\tilde{\varepsilon}/2, \tilde{\varepsilon})$ such that
$$\int_{r_n}^{r_{n+1}} r \left| \frac{\partial}{\partial r} w(r, \pm \varepsilon_n) \right|^2 dr, \quad \int_{r_n}^{r_{n+1}} r \left| \frac{\partial}{\partial r} w(r, \pm(\pi - \varepsilon_n)) \right|^2 dr \to 0 \quad \text{as } n \to \infty.$$

Combining this with (50) and the fact that $r_n \in (2^n, 2^{n+1})$ yields

$$\max_{x \in \partial S_n} |w(x) - \tilde{w}_\infty| \to 0 \quad \text{as } n \to \infty,$$

where $S_n = \{(r, \theta): r \in (r_n, r_{n+1}), |\theta| \in (\varepsilon_n, \pi - \varepsilon_n)\}$. The use of this with (a) gives

(51) $$\max_{x \in \partial S_n} |\psi(x)\omega(x)| \to 0 \quad \text{as } n \to \infty.$$

Since w satisfies (47) and $w_\infty = (u_\infty, 0) \neq 0$, one can show that the stream function $\psi(x_1, x_2)$ behaves like $u_\infty x_2$ at infinity; more precisely,

$$\lim_{x \in S_\varepsilon \to \infty} \frac{\psi(x_1, x_2)}{x_2} = u_\infty$$

for any $\varepsilon \in (0, \pi/2)$. In particular, $|\psi(x)| \geq \text{const}|x|$ if $x \in S_n$, where the constant is independent of n, and depends only on our choice of $\tilde{\varepsilon}$. Equation (51) gives

$$\max_{x \in \partial S_n} |x|\omega(x) \to 0 \quad \text{as } n \to \infty,$$

and so the maximum principle gives

$$\max_{x \in \overline{S}_n} |x| |\omega(x)| \to 0 \quad \text{as } n \to \infty.$$

Each $\overline{S}_n$ contains the set $\{(r, \theta): r \in [r_n, r_{n+1}], |\theta| \in [\tilde{\varepsilon}, \pi - \tilde{\varepsilon}]\}$, whence $\omega(x) = o(|x|^{-1})$ in $S_{\tilde{\varepsilon}}$ as $|x| \to \infty$.

Since $\psi(x)\omega(x) \to 0$ in $S_{\tilde{\varepsilon}}$ as $|x| \to \infty$, it follows from part (a) that $|w(x)| \to |\tilde{w}_\infty|$. Q.E.D.

If one uses the estimate $\omega(x) = o(|x|^{-1})$ in S_ε with the methods of Theorem 1 in [7], then $w(x) \to \tilde{w}_\infty$ as $|x| \to \infty$ in every S_ε. This will be proved in [8], where we shall prove the stronger result that $w(x) \to \tilde{w}_\infty$ as $|x| \to \infty$.

References

1. J. Leray, *Études de diverse équations intégrales non linéaires et de quelques problèmes que pose l'hydrodynamique*, J. Math. Pures Appl. **12** (1933), 1–82.
2. R. Finn and D. R. Smith, *On the stationary solution of the Navier–Stokes equations in two dimensions*, Arch. Rational Mech. Anal. **25** (1967), 26–39.
3. R. Finn, *On the steady-state solutions of the Navier–Stokes equations. III*, Acta Math. **105** (1961), 197–244.
4. O. A. Ladyzhenskaya, *The mathematical theory of viscous incompressible flow*, Gordon & Breach, New York, 1969.
5. K. I. Babenko, *On stationary solutions of the problem of flow past a body by a viscous incompressible fluid*, Math. Sb. **91** (1973), 3–26.
6. D. Gilbarg and H. F. Weinberger, *Asymptotic properties of Leray's solution of the stationary two-dimensional Navier–Stokes equations*, Russian Math. Surveys **29** (1974), 109–123.
7. _____, *Asymptotic properties of steady plane solutions of the Navier–Stokes equations with bounded Dirichlet integral*, Ann. Scuola Norm. Sup. Pisa (4) **5** (1978), 381–404.
8. C. Amick, *Stationary solutions of the Navier–Stokes equations in two dimensions* (to appear).
9. R. Finn and W. Noll, *On the uniqueness and non-existence of Stokes flow*, Arch. Rational Mech. Anal. **1** (1957), 97–106.

University of Chicago

On a Contraction Theorem and Applications

IOANNIS K. ARGYROS

Consider the equation

(1) $$x = y + \lambda B(x, x)$$

in a Banach space X, where $B: X \times X \to X$ is a bounded, symmetric bilinear operator, λ is a positive parameter and $y \in X$ is fixed. We announce a consequence of the contraction mapping principle which can be used to prove existence and uniqueness for (1). For the special cases of Chandrasekhar's equation [3]

(C) $$x(s) = 1 + \lambda x(s) \int_0^1 \frac{s}{s+t} x(t)\, dt$$

and the Anselone–Moore system [1]

(H) $$x_j(s) = y_j(s) + \lambda \int_0^1 L_{j1}(s,t) x_1(t) x_2(t)\, dt + \lambda \int_0^1 L_{j2}(s,t) \frac{1}{2} x_1^2(t)\, dt,$$

$$j = 1, 2,$$

our theorem yields existence and uniqueness for larger values of λ than previously known, as well as providing more accurate information on the location of solutions.

The principal new idea in our general theorem is the introduction of a second quadratic equation

(2) $$z = y + F(z, z)$$

for comparison with (1). The estimates on (C) and (H) are then obtained under suitable choices of F.

THEOREM. *Suppose $F: X \times X \to X$ is a bounded symmetric bilinear operator and that (2) has a solution z^* such that*

(3) $$\|z^*\| < \left[2\|B\|^{1/2}\left(\|B - F\|^{1/2} + \|B\|^{1/2}\right)\right]^{-1}.$$

1980 *Mathematics Subject Classification.* Primary 46B15.

© 1986 American Mathematical Society
0082-0717/86 $1.00 + $.25 per page

Then

(I) *Equation* (1) (*with* $\lambda = 1$) *has a unique solution* x^* *in the ball* $U(z^*, b) = \{|z|\, \|z - z^*\| < b\}$, *where*

$$b = [1 - 2\|B\|\, \|z^*\|]/2\|B\|;$$

(II) *Moreover*, $x^* \in \overline{U}(z^*, a)$, *where*

$$a = \left\{1 - 2\|B\|\, \|z^*\| - \left[(1 - 2\|B\|\, \|z^*\|)^2 - 4\|B\|\, \|B - F\|\, \|z^*\|^2\right]^{1/2}\right\}/2\|B\|.$$

For the choice $F = 0$ in (2), we have $z^* = y$ and (3) becomes

(4) $$\|y\| < (4\|B\|)^{-1}.$$

This choice of F yields Applications 1 and 2.

Application 1 (Chandrasekhar's Equation (C)). Here $X = C[0, 1]$, $y = 1$ and

$$\|B\| = \max_{0 \leq s \leq 1} \left|\lambda x(s) \int_0^1 \frac{s}{s + t}\, dt\right| = \lambda \log 2, \qquad \|y\| = 1,$$

so (4) becomes

(5) $$\lambda < [4 \log 2]^{-1} = .36067376\ldots$$

Condition (5) is the same as the restriction on λ in [3], but we prove the solution of (C) lies in a smaller ball (centered at y instead of 0).

Application 2 (Anselone-Moore system (4)). Let

$$X = C[0, 1] \times C[0, 1], \quad \text{with } \|x\| = \max(\|x_1\|, \|x_2\|),$$

where

$$x = \begin{bmatrix} x_1 \\ x_2 \end{bmatrix}, \quad y = \begin{bmatrix} y_1 \\ y_2 \end{bmatrix}, \quad B(x, x) = \begin{bmatrix} B_1(x, x) \\ B_2(x, x) \end{bmatrix}$$

and

$$B_j(x, x) = \lambda \int_0^1 L_{j1}(s, t) x_1(t) x_2(t)\, dt + \lambda \int_0^1 L_{j2}(s, t) \frac{1}{2} x_1^2(t)\, dt,$$

$$j = 1, 2.$$

Then

$$\|B\| \leq \lambda \max_{j=1,2} \left(\|L_{j1}\| + \frac{\|L_{j2}\|}{2}\right)$$

and (4) becomes

(6) $$\lambda < [4\|B\|\, \|y\|]^{-1}.$$

Condition (6) improves the restrictions on λ in [1] and with our theorem we obtain better bounds on the solution of (H), as in Application 1.

Another choice of F is obtained by assuming a solution x of (1) is known for a certain value of λ (satisfying (5) or (6), for example), and then choosing $F = \lambda B$ and $z^* = x^*$, in (2). We have

COROLLARY. *If* $2\lambda \|B\| \|z^*\| < 1$ *then there exists* $\lambda_1 > 0$ *such that*

$$\lambda \leq \lambda_1 < \left(4\|B\| \|z^*\|(1 - \lambda\|B\| \|z^*\|)\right)^{-1} = d$$

and (1) *(with* $\lambda = \lambda_1$*) has a unique solution* x^* *in* $U(z^*, b)$. *Moreover*, $x^* \in \overline{U}(z^*, a)$.

For the second choice of F, the Corollary yields

Application 3. The range of λ in (C) can be extended to $\lambda \approx .424059379$. Some characteristic values of λ, the norm of the solution corresponding to λ and the bound on λ_1 are given in Table 1.

λ	$\|z^*\|$	d
.35	1.44474532	.384363732
.4	1.59821923	.405244551
⋮	⋮	⋮
.424059378	1.700973716	.424059379
.424059379	1.700973721	.424059379

TABLE 1

The solution x^* is always such that

(7) $$\|x^*\| < 1/2\|B\|.$$

Numerical iteration [4] suggests that if $\lambda \geq .42406$ then (7) is violated, which implies that if the estimate on $\|x^*\|$ given in (7) is "best" possible then the Corollary provides the widest possible range for λ in this case.

ACKNOWLEDGEMENTS. The author wishes to express his gratitude to his advisors Douglas Clark and Edward Azoff for their help in the preparation of this paper.

REFERENCES

1. P. M. Anselone and R. H. Moore, *An extension of the Newton–Kantorovich method for solving nonlinear equations*, Tech. Rep. No. 520, U. S. Army Math. Research Center, Univ. of Wisconsin, Madison, 1965.

2. S. Chandrasekhar, *Radiative transfer*, Dover, New York, 1960.

3. L. B. Rall, *Quadratic equations in Banach space*, Rend. Circ. Math. Palermo **10** (1961), 314–332.

4. D. W. N. Stibbs and R. E. Weir, *On the H-functions for isotropic scattering*, Monthly Not. Roy. Astron. Soc. **119** (1959), 512–525.

UNIVERSITY OF GEORGIA

Dual Variational Principles for Eigenvalue Problems

GILES AUCHMUTY[1]

Introduction. In this paper we shall use the duality theory for nonconvex variational principles developed in [1] to describe and analyze some new variational principles for eigenvalues of linear and nonlinear operators. Particular attention will be paid to certain variational principles that are dual to Rayleigh's principle for finding the eigenvalues of selfadjoint compact linear operators or selfadjoint linear elliptic problems.

For these problems the duality theory leads to unconstrained variational problems. The functionals have critical values corresponding to each eigenvalue of the original operator and the minimizers are the corresponding eigenfunctions of norm depending on the eigenvalue.

The methods also provide existence theorems for various nonlinear eigenvalue problems.

In the next section, a brief description of the problems are given and in §3 the standard form for eigenvalue problems for operators of potential type is given. In §4 we describe a number of dual problems to Rayleigh's principle for the eigenvalues of a selfadjoint compact linear operator. These are analyzed in §5 where we show that a number of different unconstrained variational principles all provide similar information. In §§6 and 7, a similar analysis is done for selfadjoint linear elliptic eigenvalue problems. The last section indicates how these methods may be used to prove the existence of nontrivial solutions of nonlinear elliptic eigenvalue problems.

2. Duality theory for nonconvex variational principles. Let X, Y be real, locally convex Hausdorff topological vector spaces and $\overline{\mathbb{R}} = [-\infty, \infty]$. The primal problem ($\mathfrak{P}$) will be to find

(2.1) $$\alpha = \inf_{x \in X} f(x)$$

1980 *Mathematics Subject Classification.* Primary 49G05; Secondary 35P30.
[1] This research was partially supported by NSF grant MCS 8201889.

where $f: X \to \overline{\mathbb{R}}$ is a lower semicontinuous (l.s.c.) function. α is called the value of $(\mathfrak{P})$.

Let X^* (Y^*) be the dual spaces of X (Y), respectively, and let $\langle\,,\,\rangle$ denote the pairing between a space and its dual.

A function $L: X \times Y^* \to \overline{\mathbb{R}}$ is a Lagrangian of type I associated with $(\mathfrak{P})$ provided, for every x in X,

$$(2.2) \qquad f(x) = \sup_{y^* \in Y^*} L(x, y^*).$$

It is a Lagrangian of type II associated with $(\mathfrak{P})$ provided, for every x in X,

$$(2.3) \qquad f(x) = \inf_{y^* \in Y^*} L(x, y^*).$$

When L is a Lagrangian of type I, the standard dual variational principle $(\mathfrak{P}^*)$ is to find

$$(2.4) \qquad \alpha^* = \sup_{y^* \in Y^*} g(y^*)$$

where

$$(2.5) \qquad g(y^*) = \inf_{x \in X} L(x, y^*).$$

There is also an anomalous dual variational principle $(\mathfrak{P}^\otimes)$ of finding

$$(2.6) \qquad \alpha^\otimes = \inf_{y^* \in Y^*} h(y^*)$$

where

$$(2.7) \qquad h(y^*) = \sup_{x \in X} L(x, y^*).$$

When L is a Lagrangian of type II, the dual variational principle $(\mathfrak{P}^*)$ is to find

$$(2.8) \qquad \alpha^* = \inf_{y^* \in Y^*} g(y^*)$$

where g is defined by (2.5).

The theory of these dual principles is described in Auchmuty [1] where many examples of dual problems are developed. In this paper we shall study the application of these ideas to elliptic eigenvalue problems and eigenvalue problems for nonlinear compact operators of potential type.

In this paper, the functionals of interest have the form

$$(2.9) \qquad f(x) = f_1(\Lambda x) - f_2(x)$$

or

$$(2.10) \qquad f(x) = k_2(x) - k_1(\Lambda x).$$

Here

($\mathfrak{A}$1) $f_1: Y \to \overline{\mathbb{R}}$, $f_2: X \to \overline{\mathbb{R}}$, $k_1: Y \to \overline{\mathbb{R}}$ and $k_2: X \to \overline{\mathbb{R}}$ are proper, lower semicontinuous, convex functions, and

($\mathfrak{A}$2) $\Lambda: X \to Y$ is a continuous linear map.

When f has either of the forms (2.9) or (2.10) there are certain natural Lagrangians associated with f.

When f is given by (2.9), the Lagrangians of type I and II, respectively, are

$$(2.11) \qquad L(x, y^*) = \langle \Lambda x, y^* \rangle - f_1^*(y^*) - f_2(x)$$

and

$$(2.12) \qquad \mathfrak{L}(x, x^*) = -\langle x, x^* \rangle + f_1(\Lambda x) + f_2^*(x^*).$$

The anomalous dual variational principle ($\mathfrak{P}^\circledast$) is to minimize h on Y^* where, from (2.7),

$$(2.13) \qquad h(y^*) = f_2^*(\Lambda^* y^*) - f_1^*(y^*).$$

The dual principle ($\mathfrak{P}^*$) associated with (2.12) is to minimize g on Y^* where, from (2.5),

$$(2.14) \qquad g(x^*) = f_2^*(x^*) - (f_1 \circ \Lambda)^*(x^*).$$

When Λ and Λ^* are homeomorphisms, these two dual principles are equivalent.

Here, and throughout this paper, f_i^* will denote the polar functional of f_i defined as in Ekeland and Temam [2, Chapter 1, Definition 4.1]. Similarly, any other undefined terms will have the definitions given in [2].

When f has the form (2.10), the Lagrangians of type I and II are, respectively,

$$(2.15) \qquad L(x, x^*) = \langle x, x^* \rangle - k_2^*(x^*) - k_1(\Lambda x)$$

and

$$(2.16) \qquad \mathfrak{L}(x, y^*) = \langle \Lambda x, y^* \rangle + k_2(x) + k_1(-y^*).$$

The dual principles ($\mathfrak{P}^\circledast$) and ($\mathfrak{P}^*$), respectively, associated with these Lagrangians, are to minimize

$$(2.17) \qquad h(x^*) = (k_1 \circ \Lambda)^*(x^*) - k_2^*(x^*)$$

and

$$(2.18) \qquad g(y^*) = k_1^*(-y^*) - k_2^*(-\Lambda^* y^*)$$

on X^* and Y^* respectively.

One notes that these dual principles, in each case, are again problems of minimizing the difference of two convex functionals. For a number of basic results on the relationships between these primal and dual variational principles see [1, especially §7].

LEMMA 2.1. *Suppose f has the form (2.9) [or (2.10)] and ($\mathfrak{A}1$)–($\mathfrak{A}2$) hold. If X is a reflexive, locally convex, Hausdorff, topological vector space, then ($\mathfrak{P}$) and ($\mathfrak{P}^{\circledast\circledast}$) [or ($\mathfrak{P}^{**}$)] are the same.*

PROOF. From (2.13) one observes that ($\mathfrak{P}^\circledast$) is the problem of minimizing h on Y^* and h has the form (2.9) with X replaced by Y^*, Λ by Λ^*.

Using this construction again, one has that ($\mathfrak{P}^{\otimes\otimes}$) is the problem of minimizing $H: X^{**} \to \overline{\mathbb{R}}$ where

$$H(z) = f_1^{**}(\Lambda^{**}z) - f_2^{**}(z).$$

Since ($\mathfrak{A}1$) holds, one has $f_1^{**}(y) = f(y)$ for all y in Y and similarly for f_2. Since X is reflexive, X^{**} is isomorphic to X and if x is in X, then $\Lambda^{**}x = \Lambda x$ for all x in X.

Thus ($\mathfrak{P}^{\otimes\otimes}$) is equivalent to minimizing H on X where $H(x) = f_1(\Lambda x) - f_2(x)$ and this is ($\mathfrak{P}$).

The proof in case (2.10) is similar.

3. Eigenvalue problems for operators of potential type. Assume here that $X = H$ is a Hilbert space and $\mathfrak{f}: H \to \mathbb{R}$ is a convex, weakly continuous functional which is bounded on bounded subsets of H. Let $\langle \, , \, \rangle$ denote the inner product on H.

Define $f: H \to \overline{\mathbb{R}}$ by

(3.1) $$f(x) = \mathfrak{X}_r(x) - \mathfrak{f}(x)$$

where

$$\mathfrak{X}_r(x) = \begin{cases} 0 & \text{if } \|x\| \leq r, \\ \infty & \text{if } \|x\| > r \end{cases}$$

is the indicator functional of the closed unit ball B_r in H.

THEOREM 3.1. *Under the above assumptions, f attains a finite minimum value α on H. If f is minimized at $\hat{x}$, then there is a constant $\lambda \geq 0$ such that*

(3.2) $$\lambda \hat{x} \in \partial \mathfrak{f}(\hat{x}).$$

PROOF. By assumption, $\mathfrak{f}$ is bounded on B_r so f is bounded below on B_r and hence on H. Let $\alpha = \inf_{x \in H} f(x)$; then α is finite and the minimum is attained since $\mathfrak{f}$ is weakly continuous and B_r is weakly compact.

If $\hat{x}$ minimizes f on H, then for any $x \in B_r$ one has $\mathfrak{f}(\hat{x}) \geq \mathfrak{f}(x)$.

If $\|\hat{x}\| < r$, this implies $\{0\} = \partial \mathfrak{f}(\hat{x})$ and thus (3.2) holds with $\lambda = 0$.

If $\|\hat{x}\| = r$ and $y \in \partial \mathfrak{f}(\hat{x})$, then $\langle x - \hat{x}, y \rangle \leq 0$ for all x in B_r. From Schwarz's inequality, this can happen iff $y = \lambda \hat{x}$, for some $\lambda \geq 0$. □

Let U be an open set in H which contains B_r and suppose $\mathfrak{G}: U \to H$ is a given, not necessarily linear, map.

$\mathfrak{G}$ is said to be a potential operator, or map, if there is a functional $\mathfrak{f}: U \to \mathbb{R}$ such that the Gâteaux derivative $D\mathfrak{f}$ of $\mathfrak{f}$ obeys

(3.3) $$D\mathfrak{f}(x) = \mathfrak{G}(x)$$

for all x in U. When this holds, $\mathfrak{f}$ is called the potential of $\mathfrak{G}$.

One has the following corollary of Theorem 3.1:

COROLLARY. *Suppose $\mathfrak{G}: U \to H$ is a potential map and $\mathfrak{f}: U \to \mathbb{R}$ is a convex, weakly continuous functional which is bounded on B_r and obeys (3.3). Then there is a point $\hat{x}$ in B_r and a number $\lambda \geq 0$ such that*

(3.4) $$\mathfrak{G}(\hat{x}) = \lambda \hat{x}.$$

PROOF. Since $\mathfrak{f}$ is Gâteaux-differentiable on U one has $\partial \mathfrak{f}(x) = \{D\mathfrak{f}(x)\} = \{\mathfrak{G}(x)\}$ for each x in U. The theorem then implies that the maximizers of $\mathfrak{f}$ on B_r are solutions of (3.4).

Let $(\mathfrak{P})$ be the problem of maximizing $\mathfrak{f}$ on B_1, or equivalently, of minimizing f on H. Then if (3.3) holds, the solutions of $(\mathfrak{P})$ are also solutions of (3.4), so they may be regarded as eigenfunctions for the operator $\mathfrak{G}$.

Let
$$(3.5) \qquad \alpha = \inf_{x \in H} f(x) = \sup_{x \in B_1} \mathfrak{f}(x).$$

Here we have normalized the radius to be 1.

Put $\Lambda = I$ and $Y = H$ in (2.9). Then the dual variational principles $(\mathfrak{P}^*)$ or $(\mathfrak{P}^\circledast)$ defined by (2.13) or (2.14) are the same. They are to minimize h on H, where
$$h(y) = \mathfrak{f}^*(y) - \mathfrak{X}_1^*(y) = \mathfrak{f}^*(y) - \|y\|,$$
and we are using the results of Lemma 4.1 in the next section.

Let
$$\beta = \inf_{y \in H} h(y).$$

We will always have that $\alpha = \beta$ (see Theorem 3.3 of [1]). If $(\mathfrak{P})$ and $(\mathfrak{P}^*)$ both have minimizers, they are related by certain relations of Hamiltonian form. Namely, if $\hat{x}$ is a solution of $(\mathfrak{P})$ and $\hat{y}$ is a solution of $(\mathfrak{P}^*)$, then
$$(3.6) \qquad \hat{y} \in \partial \mathfrak{X}_1(\hat{x}) \quad \text{and} \quad \hat{x} \in \partial \mathfrak{f}^*(\hat{y}).$$

It transpires that these dual variational problems provide some new, interesting, and useful variational principles. In the next sections we shall study the dual problems to some well-known extremal problems for eigenvalues.

4. Dual problems to Rayleigh's principle. Rayleigh's principle is a well-known extremal problem for finding the lowest eigenvalues of selfadjoint linear elliptic operators or the largest eigenvalues of selfadjoint compact linear operators or symmetric matrices.

There is a large literature on the principle including Parlett [3], Weinberger [5], and Weinstein and Stenger [6]. Some dual variational principles in the finite-dimensional case have already been discussed in [1, §8].

Let $G: H \to H$ be a compact, selfadjoint, linear operator obeying $\langle Gx, x \rangle \geq 0$ for all x in H.

Let $\mathfrak{f}: H \to \mathbb{R}$ be defined by
$$(4.1) \qquad \mathfrak{f}(x) = \tfrac{1}{2} \langle Gx, x \rangle.$$

Then $\mathfrak{f}$ is a convex, weakly continuous and Gâteaux-differentiable function which is bounded on bounded subsets of H. For each x in H, $D\mathfrak{f}(x) = Gx$.

Define $f_1: H \to \overline{\mathbb{R}}$ by
$$(4.2) \qquad f_1(x) = \mathfrak{X}_1(x) - \tfrac{1}{2} \langle Gx, x \rangle$$

and consider the problem ($\mathfrak{P}_1$) of minimizing f_1 on H. This is equivalent to maximizing $\mathfrak{f}$ on B_1.

It is well known (see Riesz and Sz-Nagy [4, §93]) that $\alpha_1 = \inf_{x \in H} f_1(x) = -\lambda_1/2$ where λ_1 is the largest eigenvalue of G, and this infimum is attained at $\hat{x}$ in B_1 where $\hat{x}$ is an eigenfunction of G corresponding to the eigenvalue λ_1.

More generally, let $v_1, v_2, \ldots, v_n$ be normalized, orthogonal eigenfunctions of G corresponding to the n largest eigenvalues $\lambda_1 \geq \lambda_2 \geq \cdots \geq \lambda_n \geq 0$.

Define $f_{n+1}: H \to \overline{\mathbb{R}}$ by

$$(4.3) \qquad f_{n+1}(x) = \mathfrak{X}_{n+1}(x) - \tfrac{1}{2}\langle Gx, x \rangle$$

where

$$(4.4) \qquad \mathfrak{X}_{n+1}(x) = \begin{cases} 0 & \text{if } \|x\| \leq 1 \text{ and } \langle x, v_k \rangle = 0 \text{ for } 1 \leq k \leq n, \\ +\infty & \text{otherwise.} \end{cases}$$

Then the $(n+1)$st eigenvalue of G is λ_{n+1} where $-\lambda_{n+1}/2 = \inf_{x \in H} f_{n+1}(x)$ and the infimum of f_{n+1} is attained at a point $\hat{x}_{n+1}$ which is a normalized eigenfunction of G corresponding to the eigenvalue λ_{n+1}.

The problem of minimizing f_{n+1} on H will be called ($\mathfrak{P}_{n+1}$).

To describe various dual problems to ($\mathfrak{P}_1$) and ($\mathfrak{P}_{n+1}$) one must compute the polar functionals to $\mathfrak{X}_1$, $\mathfrak{X}_{n+1}$ and $\mathfrak{f}$.

LEMMA 4.1. $\mathfrak{X}_1^*(y) = \|y\|$ and $\mathfrak{X}_{n+1}^*(y) = \|y - P_n y\|$ where $P_n y = \sum_{k=1}^n \langle y, v_k \rangle v_k$ is the projection of y onto the subspace spanned by $\{v_1, v_2, \ldots, v_n\}$.

PROOF. One has $\mathfrak{X}_1^*(y) = \sup_{x \in B_1} \langle x, y \rangle$. From Cauchy–Schwarz the right side is maximized when $x = y/\|y\|$ and hence $\mathfrak{X}_1^*(y) = \|y\|$.

Similarly, $\mathfrak{X}_{n+1}^*(y) = \sup_{x \in C_{n+1}} \langle x, y \rangle$ where $C_{n+1} = \{x \in H: \|x\| \leq 1 \text{ and } \langle x, v_k \rangle = 0 \text{ for } 1 \leq k \leq n\}$. By extending $\{v_1, v_2, \ldots, v_n\}$ to an orthonormal basis of H and using Cauchy–Schwarz as above, one finds the desired result.

LEMMA 4.2. Let $A: H \to H$ be a continuous, selfadjoint, linear operator obeying $\langle Ax, x \rangle \geq 0$ for all x in H. Let $H_0 = \ker A$, $R(A)$ be the range of A and $\mathfrak{f}(x) = \tfrac{1}{2}\langle Ax, x \rangle$, then
 (i) if $y \notin H_0^\perp$ then $\mathfrak{f}^*(y) = +\infty$,
 (ii) if $y \in R(A)$ then $\mathfrak{f}^*(y) = \tfrac{1}{2}\langle A^{-1}y, y \rangle$,
where $A^{-1}: R(A) \to H_0^\perp$ is the (generalized) inverse of A.
When $y \in H_0^\perp - R(A)$, $\mathfrak{f}^*(y)$ may be finite or infinite.

PROOF. Let $y = y_0 + y_1$ be the orthogonal decomposition of y with respect to H_0 and $H_0^\perp$. One has

$$\mathfrak{f}^*(y) = \sup_{x \in H} \left[\langle x, y_0 \rangle + \langle x, y_1 \rangle - \tfrac{1}{2}\langle Gx, x \rangle \right].$$

Suppose $u_0 \, (\neq 0)$ is in H_0 and $\langle u_0, y_0 \rangle \neq 0$. One has $\langle u_0, y_1 \rangle = \tfrac{1}{2}\langle Au_0, u_0 \rangle = 0$ so $\mathfrak{f}^*(y) \geq \alpha \langle u_0, y_0 \rangle$ for all real α. Hence (i) holds.

To prove (ii), let $\hat{x}$ in $H_0^\perp$ be the solution of $Ax = y$. Then $\mathfrak{A}_y(x) = \langle x, y \rangle - \frac{1}{2}\langle Ax, x \rangle$ is maximized at $\hat{x}$ since $\mathfrak{f}$ is convex. Thus $\mathfrak{f}^*(y) = \frac{1}{2}\langle \hat{x}, A\hat{x} \rangle = \frac{1}{2}\langle A^{-1}y, y \rangle$, as required.

Using orthonormal expansions, one can construct specific examples which show $\mathfrak{f}^*(y)$ may be finite or infinite when y is in $H_0^\perp - R(A)$. □

When A is the inverse of an elliptic operator, the functional $\mathfrak{f}^*$ represents a generalized Dirichlet form. In such cases one may identify the effective domain of $\mathfrak{f}^*$ with certain Sobolev spaces. An abstract form of this result follows.

LEMMA 4.3. *Suppose $L: D_L (\subset H) \to H$ is a closed, selfadjoint linear operator whose domain D_L is dense in H. Assume that*

(4.5) $$\langle Lx, x \rangle \geq c\|x\|^2 \quad \text{for all } x \in D_L, \text{ some } c > 0.$$

Define $f: H \to \overline{\mathbb{R}}$ by

$$f(x) = \begin{cases} \frac{1}{2}\langle Lx, x \rangle, & x \in D_L, \\ +\infty, & x \notin D_L. \end{cases}$$

Then $f^*: H \to \overline{\mathbb{R}}$ is $f^*(y) = \frac{1}{2}\langle L^{-1}y, y \rangle$.

PROOF. When L obeys (3.5), one sees that $-\mathfrak{A}_y(x) = f(x) - \langle x, y \rangle$ is a coercive, convex functional on H. Moreover from the Lax–Milgram theorem there is a unique solution $\hat{x} = L^{-1}y$ of $Lx = y$ for each y in H. Hence $\mathfrak{A}_y(x)$ is maximized at $\hat{x}$ and $f^*(y) = \frac{1}{2}\langle L^{-1}y, y \rangle$.

COROLLARY. *Suppose f, L as in the lemma, then the bipolar f^{**} of f is the supremum of the convex, lower semicontinuous functions which are everywhere less than or equal to f.*

PROOF. This follows from Proposition 4.1 of Chapter 1 of [2]. f^{**} is called the Γ-regularization of f.

In particular, one sees that if L corresponds to a linear selfadjoint, elliptic operator on a smooth manifold, or on an open set in R^k with a smooth boundary, then the functional f^* is the quadratic form associated with its inverse operator and f^{**} will be a convex, l.s.c. regularization of f.

A dual principle $(\mathfrak{P}_1^*)$ is obtained as in §3, by putting $\Lambda = I$ and $Y = H$ in (2.9). It is to minimize

(4.6) $$h_1(y) = \mathfrak{f}^*(y) - \|y\|$$

where $\mathfrak{f}^*(y)$ is defined as in Lemma 4.2, with G in place of A.

In particular, if $G = L^{-1}$ where L is an operator obeying the conditions of Lemma 4.3 and with a compact inverse, then

(4.7) $$h_1(y) = \frac{1}{2}\langle Ly, y \rangle - \|y\|$$

for all $y \in D_L$.

Similarly, the dual principles ($\mathfrak{P}^*_{n+1}$) or ($\mathfrak{P}^\otimes_{n+1}$) are to minimize $h_{n+1}\colon H \to \overline{\mathbb{R}}$ where $n \geq 1$ and

(4.8) $$h_{n+1}(y) = \mathfrak{f}^*(y) - \|y - P_n y\|$$

with $\mathfrak{f}^*$ defined as in Lemma 4.2.

There are other dual variational principles to ($\mathfrak{P}_1$) and ($\mathfrak{P}_{n+1}$).

Let $G^{1/2}\colon H \to H$ be the positive square root of G defined by the spectral theorem

(4.9) $$G^{1/2} = \sum_{k=1}^{\infty} \sqrt{\lambda_k}\, E_k.$$

Here $\{E_k\colon k \geq 1\}$ are the spectral projections of G.

Rewrite (4.2) as

(4.10) $$f_1(x) = \mathfrak{X}_1(x) - \tfrac{1}{2}\langle G^{1/2}x, G^{1/2}x\rangle.$$

This may be considered to have the form (2.10) with $Y = H$ and $\Lambda = G^{1/2}$.

Let ($\mathfrak{Q}_1$) be the problem of minimizing (4.10) on H.

From (2.18), the dual principle ($\mathfrak{Q}^*_1$) now is to minimize $g_1\colon H \to \overline{\mathbb{R}}$ defined by

(4.11) $$g_1(y) = \tfrac{1}{2}\|y\|^2 - \|G^{1/2}y\| = \tfrac{1}{2}\|y\|^2 - \langle Gy, y\rangle^{1/2}.$$

To verify this, one uses Lemma 4.2 and the fact that, if $f(x) = \tfrac{1}{2}\|x\|^2$, then $f^*(y) = \tfrac{1}{2}\|y\|^2$.

Similarly, if ($\mathfrak{Q}_{n+1}$) is the problem of minimizing

$$f_{n+1}(x) = \mathfrak{X}_{n+1}(x) - \tfrac{1}{2}\langle G^{1/2}x, G^{1/2}x\rangle$$

on H, the dual principle ($\mathfrak{Q}^*_{n+1}$) is to minimize $g_{n+1}\colon H \to \overline{\mathbb{R}}$ where

(4.12) $$g_{n+1}(y) = \tfrac{1}{2}\|y\|^2 - \|(I - P_n)G^{1/2}y\|$$
$$= \tfrac{1}{2}\|y\|^2 - \langle G(I - P_n)y, (I - P_n)y\rangle^{1/2}$$

since G and P_n commute.

The functionals $\{g_m\colon m \geq 1\}$ defined by (4.11) and (4.12) are defined and finite at each point in H. Thus the dual problems ($\mathfrak{Q}_m$) are unconstrained variational problems. We shall describe some other nice properties of these dual problems in the next section.

The dual variational principles involve the computation of various polar functionals and, from Lemma 4.2, this is easier for these eigenvalue problems if the underlying operator is positive definite. This suggests that instead of looking for the eigenvalues of the compact, selfadjoint linear operator G we look for those of $\mu I + G$ for some $\mu > 0$.

Without loss of generality, we shall take $\mu = 1$ and define $\mathfrak{f}\colon H \to \mathbb{R}$ by

(4.13) $$\mathfrak{f}(x) = \tfrac{1}{2}\langle (I + G)x, x\rangle.$$

Then $\mathfrak{f}$ is uniformly convex on H and
(4.14) $$\mathfrak{f}^*(y) = \tfrac{1}{2}\langle (I+G)^{-1}y, y \rangle.$$
Define
(4.15) $$k_1(x) = \mathfrak{X}_1(x) - \mathfrak{f}(x) \quad \text{for } x \text{ in } H,$$
and let $(\mathfrak{R}_1)$ be the problem of minimizing k_1 on H. Then
(4.16) $$\beta_1 = \inf_{x \in H} k_1(x) = -\tfrac{1}{2}(1 + \lambda_1).$$
Similarly, when $n \geq 1$, let $k_{n+1}(x) = \mathfrak{X}_{n+1}(x) - \mathfrak{f}(x)$ and let $(\mathfrak{R}_{n+1})$ be the problem of minimizing k_{n+1} on H. Then
$$\beta_{n+1} = \inf_{x \in H} k_{n+1}(x) = -\tfrac{1}{2}(1 + \lambda_{n+1}).$$
Put $\Lambda = I$, $Y = H$ in (2.9), then the dual variational principles $(\mathfrak{R}_1^*)$ or $(\mathfrak{R}_1^\circledast)$ are to minimize $j_1 \colon H \to \overline{\mathbb{R}}$ where
(4.17) $$j_1(y) = \tfrac{1}{2}\langle (I+G)^{-1}y, y \rangle - \|y\|.$$
Similarly, $(\mathfrak{R}_{n+1}^*)$ is to minimize $j_{n+1} \colon H \to \overline{\mathbb{R}}$ where
(4.18) $$j_{n+1}(y) = \tfrac{1}{2}\langle (I+G)^{-1}y, y \rangle - \|y - P_n y\|.$$

Again these are functionals which are defined and finite on all of H. In the next sections, we shall study the properties of these dual variational principles.

5. Properties of the dual variational principles. In the preceding section we described some variational principles for the eigenvalues of a linear, compact, selfadjoint operator on a Hilbert space and then described some dual principles. Here we shall analyze these principles.

THEOREM 5.1. *The values of $(\mathfrak{P}_m^*)$ and $(\mathfrak{Q}_m^*)$ are $-\lambda_m/2$. The value of $(\mathfrak{R}_m^*)$ is $-\tfrac{1}{2}(1 + \lambda_m)$, for any $m \geq 1$.*

PROOF. One first notes that the values of $(\mathfrak{P}_m)$ and $(\mathfrak{Q}_m)$ are $-\lambda_m/2$ while that of $(\mathfrak{R}_m)$ is $-\tfrac{1}{2}(1 + \lambda_m)$.

From Theorem 3.3 of (1), it follows that if the dual problems can be derived from a Lagrangian of type II, then the value of the dual problem equals that of the primal. But each of these principles does come from a Lagrangian of the form (2.16), so this theorem holds.

This result says that the minimum values of the functionals h_m, g_m, and k_m defined in the last section are related to the eigenvalues of G. One has much more.

A point $\hat{x}$ in H is a critical point for a functional f on H provided f is Gâteaux differentiable at $\hat{x}$ in H and, moreover, $Df(\hat{x}) = 0$ in H.

LEMMA 5.2. *The functional $g_1 \colon H \to R$ defined by (4.11) is weakly l.s.c. and coercive on H. g_1 is Gâteaux-differentiable on $H_0^\perp = \{y \in H \colon \langle Gy, y \rangle \neq 0\}$ and its derivative is*
(5.1) $$Dg_1(y) = y - \langle Gy, y \rangle^{-1/2} Gy.$$

PROOF. The norm $\|y\|$ is weakly l.s.c. on H while the functional $\langle Gy, y \rangle$ is weakly continuous since G is compact. Hence g_1 is weakly l.s.c.

For all $y \in H$ one has $|\langle Gy, y \rangle| \leq \|G\| \|y\|^2$, so $g_1(y) \geq \frac{1}{2}\|y\|^2 - \frac{1}{2}\|G\|^{1/2}\|y\|$. Thus $\lim_{\|y\| \to \infty} g_1(y) = +\infty$ or g_1 is coercive on H.

Finally, to verify that g_1 is Gâteaux-differentiable at y it suffices to consider the second-term. One has, if $\langle Gy, y \rangle \neq 0$, $h \in H$,

$$t^{-1}\left[\langle G(y+th), y+th \rangle^{1/2} - \langle Gy, y \rangle^{1/2}\right]$$
$$= \frac{\langle Gy, h \rangle}{\langle Gy, y \rangle^{1/2}} + O(t) \quad \text{from Taylor's expansion.}$$

Hence the result.

THEOREM 5.3. *g_1 attains a minimum on H. If $\hat{y}$ minimizes g_1 on H then*
(i) $g_1(\hat{y}) = -\frac{1}{2}\lambda_1$,
(ii) *$\hat{y}$ is an eigenfunction of G corresponding to the eigenvalue λ_1, and*
(iii) $\|\hat{y}\| = \sqrt{\lambda_1}$ *and* $\langle G\hat{y}, \hat{y} \rangle = \lambda_1^2$.

PROOF. Since g_1 is weakly l.s.c. and coercive on H, then g_1 attains a minimum on H, since bounded sets in H are weakly compact.

Let $\hat{y}$ be a minimizer. From Theorem 5.1,
(5.2) $$g_1(\hat{y}) = \inf g_1(y) = -\lambda_1/2,$$
so (i) holds.

One has $0 \in \partial g_1(\hat{y}) = \{Dg_1(\hat{y})\}$, as g_1 is Gâteaux differentiable.

Hence, from (5.1),
(5.3) $$G\hat{y} = \langle G\hat{y}, \hat{y} \rangle^{1/2} \hat{y}.$$

Thus $\hat{y}$ is an eigenfunction of G corresponding to the eigenvalue $\hat{\lambda} = \langle G\hat{y}, \hat{y} \rangle^{1/2}$.

Take the inner product of (5.3) with $\hat{y}$, then $\hat{\lambda}^2 = \langle G\hat{y}, \hat{y} \rangle = \hat{\lambda}\|\hat{y}\|^2$, so $\|\hat{y}\|^2 = \hat{\lambda}$. Substituting into (5.2) one sees that $g_1(\hat{y}) = -\hat{\lambda}/2 = -\lambda_1/2$, so $\hat{\lambda} = \lambda_1$ as required.

COROLLARY. *Suppose $\tilde{y}$ is a critical point of g_1 on H. Then*
(i) *$\tilde{y}$ is an eigenfunction of G corresponding to an eigenvalue $\tilde{\lambda} > 0$, and*
(ii)
(5.4) $$\tilde{\lambda} = \langle G\tilde{y}, \tilde{y} \rangle^{1/2}, \quad \|\tilde{y}\| = \sqrt{\tilde{\lambda}}, \quad \text{and} \quad g(\tilde{y}) = -\tilde{\lambda}/2.$$

PROOF. These properties follow from the definition of a critical point in precisely the same manner as in the proof of the theorem.

Note that g_1 does not have a unique minimizer as if $\hat{y}$ minimizes so does $-\hat{y}$. Moreover, the other critical points of g_1 are precisely certain multiples of the other eigenfunctions of G.

One has similar results for $(\mathfrak{Q}_{n+1}^*)$. One notes from the definition of g_{n+1} that $g_{m+1}(y) \leq g_{n+1}(y)$ whenever $m \leq n$.

Let $H_{n+1} = \{y \in H: \langle y, v_k \rangle = 0 \text{ for } 1 \leq k \leq n\}$. One observes that
$$\inf_{y \in H_{n+1}} g_{n+1}(y) = \inf_{y \in H} g_{n+1}(y),$$

so it suffices to restrict g_{n+1} to H_{n+1}. When this is done one observes that g_{n+1} has the same form as g_1, and one has the following:

THEOREM 5.4. g_{n+1} *attains a minimum on H. If* $\hat{y}_{n+1}$ *minimizes* g_{n+1} *on H, then*
(i) $\hat{y}_{n+1}$ *is an eigenfunction of G corresponding to the eigenvalue* λ_{n+1}, *and*
(ii) $g_{n+1}(\hat{y}_{n+1}) = -\frac{1}{2}\lambda_{n+1}$, $\|\hat{y}_{n+1}\| = \sqrt{\lambda_{n+1}}$, *and* $\langle G\hat{y}_{n+1}, \hat{y}_{n+1} \rangle = \lambda_{n+1}^2$.

If $\tilde{y}$ is any critical point of g_{n+1} on H, then $\tilde{y}$ is an eigenfunction of G on H and the relationships (5.4) hold.

LEMMA 5.5. *The functional* j_1 *defined by* (4.17) *is continuous and coercive on H. It is Gâteaux-differentiable on* $H - \{0\}$ *and*

$$(5.5) \qquad Dj_1(y) = (I + G)^{-1}y - \|y\|^{-1}y.$$

PROOF. One has $\|y\|^2 \leqslant \langle (I + G)y, y \rangle \leqslant (1 + \|G\|)\|y\|^2$ for all y in H, so $\|(I + G)^{-1}\| \leqslant 1$ is finite. Thus each term in the definition of j_1 is continuous and, moreover, $j_1(y) \geqslant \|y\|^2/2(1 + \|G\|) - \|y\|$, so j_1 is coercive on H.

The differentiability follows just as in the proof of Lemma 5.2.

COROLLARY. *Let* $\tilde{y}$ *be a nonzero critical point of* j_1. *Then*
(i) $\tilde{y}$ *is an eigenfunction of G corresponding to an eigenvalue* $\tilde{\lambda}$,
(ii) $\|\tilde{y}\| = 1 + \tilde{\lambda}$ *and* $j_1(\tilde{y}) = -\frac{1}{2}(1 + \tilde{\lambda})$.

PROOF. If $\tilde{y}$ is a nonzero critical point of j_1, then $Dj_1(\tilde{y}) = 0$ or $(I + G)\tilde{y} = \|\tilde{y}\|\tilde{y}$.

Hence $G\tilde{y} = \tilde{\lambda}\tilde{y}$ with $\tilde{\lambda} = \|\tilde{y}\| - 1$.

Substituting this in the expression for j_1 one finds $j_1(\tilde{y}) = \frac{1}{2}(1 + \tilde{\lambda})^{-1}\|\tilde{y}\|^2 - \|\tilde{y}\| = -\frac{1}{2}(1 + \tilde{\lambda})$. □

THEOREM 5.6. j_1 *attains its infimum on H at* $\hat{y} = \pm(1 + \lambda_1)v_1$ *where* v_1 *is a normalized eigenfunction of G corresponding to the largest eigenvalue* λ_1. j_1 *has critical points at* $\pm(1 + \lambda_k)v_k$ *where* $k \geqslant 1$ *and* $\{v_k: k \geqslant 1\}$ *are normalized eigenfunctions of G corresponding to the eigenvalues* $\{\lambda_k: k \geqslant 1\}$.

PROOF. From Theorem 5.1, $\inf_{y \in H} j_1(y) = -\frac{1}{2}(1 + \lambda_1)$. One can verify by direct substitution that, if $\hat{y} = \pm(1 + \lambda_1)v_1$, then $j_1(\hat{y}) = -\frac{1}{2}(1 + \lambda_1)$, hence the result.

Similarly, using Lemma 5.5, one can verify that each of the points $\pm(1 + \lambda_k)v_k$ is a critical point of j_1.

If one defines H_{n+1} as before, the expression (4.18) shows that

$$\inf_{y \in H_{n+1}} j_{n+1}(y) = \inf_{y \in H} j_{n+1}(y),$$

so one may restrict j_{n+1} to H_{n+1}. When this is done, j_{n+1} has the same form as j_1, and one has the following result:

THEOREM 5.7. j_{n+1} *attains its infimum on* H_{n+1} *at* $\hat{y} = \pm(1 + \lambda_{n+1})v_{n+1}$ *where* v_{n+1} *is a normalized eigenfunction of G corresponding to the* $(n + 1)$st *eigenvalue*

λ_{n+1} of G. j_{n+1} has critical points at $\pm(1 + \lambda_k)v_k$ where $k \geq n + 1$ and $\{v_k: k \geq n + 1\}$ are the normalized eigenfunctions of G corresponding to the eigenvalues $\{\lambda_k: k \geq n + 1\}$.

In [1, §8] it is shown that in the finite-dimensional case the critical points of the various functionals are nondegenerate iff the corresponding eigenvalue of G is simple. Also, the Morse index of a nondegenerate critical point equals the number of allowable eigenvalues of G larger than the eigenvalue associated with that critical point.

6. Elliptic eigenvalue problems. Another interesting, and important class of eigenvalue problems arise as the duals of certain elliptic problems.

Let Ω be a bounded, open set in $\mathbb{R}^N$ with a sufficiently smooth boundary $\partial\Omega$ that the Sobolev imbedding theorems hold. Let X be a reflexive Sobolev space of real-valued functions on Ω such that the imbedding $i: X \to L^2(\Omega)$ is compact.

Let $f_1: X \to \mathbb{R}$ be a convex and weakly lower semicontinuous functional and consider

$$(6.1) \qquad f(u) = f_1(u) - \mu\|u\|_2,$$

where $\mu > 0$ and $\|u\|_2 = (\int_\Omega |u(x)|^2 \, dx)^{1/2}$ is the usual norm on $L^2(\Omega)$.

The primal problem ($\mathfrak{P}$) is to minimize f on X. We shall denote the norm on X by $\|u\|$.

THEOREM 6.1. *Suppose f, f_1, X are as above and*

$$(6.2) \qquad \liminf_{\|u\| \to \infty} \frac{f_1(u)}{\|u\|} = +\infty;$$

then f attains a finite minimum on X.

PROOF. The functional $f_2(u) = \mu\|u\|_2$ is weakly continuous on X since the imbedding of X into $L^2(\Omega)$ is compact. Hence, f is weakly l.s.c. on X.

From (6.2), f is coercive on X, so f attains a finite minimum on X as X is reflexive. □

Let $Y = X$ and $\Lambda = I$ in the notation of §2. Then the dual problems ($\mathfrak{P}^*$) or ($\mathfrak{P}^\circledast$) defined by (2.13) or (2.14) is to minimize $g: X^* \to \overline{\mathbb{R}}$ where $g(v) = f_2^*(v) - f_1^*(v)$.

Let the pairing between X and X^* be the usual L^2-inner product $\langle u, v \rangle = \int_\Omega u(x)v(x)\,dx$. Then the dual functional

$$f_2^*(v) = \sup_{u \in X} (\langle u, v \rangle - \mu\|u\|_2)$$

$$= \begin{cases} 0 & \text{if } \|v\|_2 \leq \mu, \\ +\infty & \text{otherwise.} \end{cases}$$

Hence the dual problem ($\mathfrak{P}^*$) is to minimize $-f_1^*$ on the ball $\mathfrak{B}_\mu$ of radius μ in $L^2(\Omega)$. This is equivalent to maximizing f_1^* on $\mathfrak{B}_\mu$ and thus it will be an eigenvalue problem of the type described in §3.

LEMMA 6.2. *Suppose f_1 is Gâteaux-differentiable on X and $\mathfrak{A}(u) = Df_1(u)$ is its derivative. Then every nonzero critical point $\hat{u}$ of f is an eigenfunction of $\mathfrak{A}$ corresponding to an eigenvalue $\hat{\lambda} = \mu \|\hat{u}\|_2^{-1}$. In particular, the minimizers of f on X are eigenfunctions of $\mathfrak{A}$.*

PROOF. One has $Df(u) = \mathfrak{A}(u) - \mu \|u\|_2^{-1} u$ if $u \neq 0$. If $\hat{u}$ ($\neq 0$) is a critical point of f then $Df(\hat{u}) = 0$ or $\mathfrak{A}(\hat{u}) = \hat{\lambda} \hat{u}$ with $\hat{\lambda} = \mu \|\hat{u}\|_2^{-1}$, as claimed.

If $\hat{u} \neq 0$ minimizes f on X, then it must be a critical point of f and hence it is an eigenfunction. When f is minimized at 0 in X, then one has $f_1(u) - f_1(0) \geq \mu \|u\|_2 \geq 0$ for all u in X.

Hence $\langle \mathfrak{A}(0), h \rangle \geq 0$ for all h in X, so $\mathfrak{A}(0) = 0$. Thus $\hat{u} = 0$ is an eigenfunction of $\mathfrak{A}$.

COROLLARY. *Suppose f, $\mathfrak{A}$, X are as above and u_1, u_2 are two nonzero critical points of f corresponding to eigenvalues λ_1, λ_2 of $\mathfrak{A}$. Then $\|u_1\|_2 > \|u_2\|_2$ implies $\lambda_2 > \lambda_1 > 0$. In particular, if $\hat{u}$ is the critical point of f of largest L^2-norm, then $\hat{u}$ is an eigenfunction of $\mathfrak{A}$ of least eigenvalue $\hat{\lambda}$.*

It appears that for nonlinear elliptic eigenvalue problems the appropriate variational principle for finding the eigenvalues is to find the critical points and extrema of (6.1). In this case, however, it is not necessarily the value of the functional f at the critical points that yields the eigenvalues but, instead, it is the norm of the critical points.

In the next section we shall show how functionals of the form (6.1) yield information on the eigenvalues of linear elliptic differential operators, then in §8 we shall apply it to nonlinear elliptic operators.

7. Linear elliptic problems. In this section we shall apply our methods to describe some, apparently new, variational principles for the eigenvalues of linear elliptic operators.

Let $X = W_0^{1,2}(\Omega)$ and

(7.1) $$f_1(u) = \frac{1}{2} \sum_{j,k=1}^{N} \int_\Omega a_{jk}(x) D_j u(x) D_k u(x)\, dx$$

where
  (i) $a_{jk} \in L^\infty(\Omega)$ for $1 \leq j, k \leq N$,
  (ii) $a_{jk}(x) = a_{kj}(x)$ a.e. on Ω for all j, k, and
  (iii) there exists a constant $c > 0$ such that

$$\sum_{j,k=1}^{N} a_{jk}(x) \xi_j \xi_k \geq c |\xi|^2 \quad \text{for all } \xi \text{ in } \mathbb{R}^N.$$

The primal problem ($\mathfrak{P}$) is to minimize

(7.2) $$f(u) = f_1(u) - \mu \|u\|_2$$

on X with f_1 defined by (7.1).

LEMMA 7.1. *The functional f is weakly l.s.c. and coercive on X. It is Gâteaux-differentiable on $X - \{0\}$ and*

$$(7.3) \qquad \langle Df(u), h \rangle = \sum_{j,k=1}^{N} \int_{\Omega} a_{jk} D_j u D_k h \, dx - \mu \|u\|_2^{-1} \langle u, h \rangle$$

for all h in X.

PROOF. From the assumptions (i)–(iii) above one has that f_1 is strictly convex and continuous on X. Hence it is weakly l.s.c.

The imbedding $i: W_0^{1,2}(\Omega) \to L^2(\Omega)$ is compact, so that L^2-norm is weakly continuous on X. Hence f is weakly l.s.c.

(7.3) is verified by a straightforward computation.

From (iii) one observes that

$$f(u) \geq \frac{c}{2} \sum_{j=1}^{n} \int_{\Omega} |D_j u|^2 \, dx - \mu \|u\|_2.$$

Poincaré's inequality says that there is a $c_0 > 0$ such that

$$\int_{\Omega} |\nabla u|^2 \, dx \geq c_0 \int_{\Omega} |u|^2 \, dx \quad \text{for all } u \text{ in } X,$$

so

$$f(u) \geq \frac{c}{4} \int_{\Omega} |\nabla u|^2 \, dx + \frac{cc_0}{4} \int_{\Omega} |u|^2 \, dx - \mu \|u\|_2.$$

Letting $\|u\| \to \infty$ in X one sees that $f(u)/\|u\| \to \infty$ or f is coercive on X.

COROLLARY. *Suppose $\hat{u}$ is a nonzero critical point of f on X. Then $\hat{u}$ is a weak solution of*

$$(7.4) \qquad L\hat{u}(x) = -\sum_{j,k=1}^{N} D_j(a_{jk}(x) D_k \hat{u}(x)) = \nu \hat{u}(x) \quad \text{on } \Omega,$$

$$(7.5) \qquad \hat{u}(x) = 0 \quad \text{on } \partial\Omega,$$

with $\nu = \mu \|\hat{u}\|_2^{-1}$.

PROOF. If $\hat{u}$ is a critical point of f then $\langle Df(\hat{u}), h \rangle = 0$ for all h in X. From Lemma 6.2 and (7.3), this implies that (7.4) holds with $\nu = \mu \|\hat{u}\|_2^{-1}$.

This corollary says essentially that the critical points of f are certain specific eigenfunctions of the elliptic differential operator L.

To obtain further information on ($\mathfrak{P}$), one looks at the dual ($\mathfrak{P}*$).

LEMMA 7.2. *The polar functional $f_1^*: X^* \to \mathbb{R}$, is given by*

$$f_1^*(v) = \tfrac{1}{2} \langle Gv, v \rangle,$$

where $G: X^* \to X$, is the solution map L^{-1} of the linear elliptic equation

$$(7.6) \qquad Lu(x) = -\sum_{j,k=1}^{N} D_j(a_{jk}(x) D_k u(x)) = v(x) \quad \text{in } \Omega,$$

$$(7.5) \qquad u(x) = 0 \quad \text{on } \partial\Omega.$$

PROOF. One has

$$f_1^*(v) = \sup_{u \in X} \int_\Omega \left(uv - \frac{1}{2} \sum_{j,k} a_{jk} D_j u D_k u \right) dx.$$

The expression on the right here is proportional to the usual Dirichlet integral for (7.5)–(7.6). For every $v \in W_0^{1,2}(\Omega)$ this attains a unique maximum at the solution u of (7.5)–(7.6). Let $u = Gv$, then $f_1^*(v) = \frac{1}{2}\langle Gv, v \rangle$, as required.

The dual problem ($\mathfrak{P}^*$) is to minimize $-f_1^*$ on the ball $\mathfrak{B}_\mu$ of radius μ in $L^2(\Omega)$ (see the preceding section). Thus this dual problem is an eigenvalue problem of the type described in §3.

THEOREM 7.3. *The functional f attains a finite minimum on X. The value of ($\mathfrak{P}$) is $-\frac{1}{2}\mu^2 \nu_1^{-1}$ where ν_1 is the least eigenvalue of (7.4)–(7.5). This minimum is attained at $\pm \hat{u}_1$ where $\hat{u}_1$ is a solution of (7.4)–(7.5) corresponding to the eigenvalue ν_1 and with $\|\hat{u}_1\|_2 = \mu \nu_1^{-1}$.*

PROOF. From Lemma 7.1 one has that f attains a finite minimum on X since X is reflexive.

The minimum must either occur at 0 or at a critical point of f since f is Gâteaux-differentiable on $X - \{0\}$. One has $f(0) = 0$.

If $\hat{u}$ is a nonzero critical point of f one has that (7.4)–(7.5) hold, so

$$f(\hat{u}) = -\tfrac{1}{2}\nu\|\hat{u}\|_2^2 \quad \text{from (7.1) and (7.4)}$$
$$= -\tfrac{1}{2}\mu^2 \nu^{-1} < 0.$$

Thus the values of the functional f at the critical points is inversely proportional to the corresponding eigenvalues. In particular, it is minimized at $\nu = \nu_1$ and then $\hat{u} = \pm \hat{u}_1$ with $\|\hat{u}_1\|_2 = \mu \nu_1^{-1}$.

COROLLARY. *Suppose $\hat{u}$ is a critical point of f. Then $\hat{u}$ is an eigenfunction of L corresponding to an eigenvalue $\hat{\nu}$ with $f(\hat{u}) = -\frac{1}{2}\mu^2 \hat{\nu}^{-1}$ and $\|\hat{u}\|_2 = \mu \hat{\nu}^{-1}$.*

These results show that the variational principle ($\mathfrak{P}$) carries the same information as the more familiar Rayleigh's principle. Namely, f is minimized at an eigenfunction corresponding to the least eigenvalue of L and the value of the functional at the minimizers yields the value of the least eigenvalue.

It is natural to ask, by analogy with Rayleigh's principle, if there are similar variational principles for the other eigenvalues. The answer is yes and can be inferred from the results of §4 and by remembering the dual principle ($\mathfrak{P}^*$).

Suppose that $\nu_1, \nu_2, \ldots, \nu_M$ are the M smallest eigenvalues of (7.5)–(7.6) (repeated according to multiplicity) and $v_1, v_2, \ldots, v_M$ are corresponding orthonormal eigenfunctions of L. Let P_M be the projection of X onto the subspace spanned by $\{v_1, v_2, \ldots, v_M\}$.

Define $f_{M+1} \colon X \to \overline{\mathbb{R}}$ by

(7.7) $\qquad f_{M+1}(u) = f_1(u) - \mu\|(I - P_M)u\|_2 \quad \text{for } M \geq 1.$

One has the following analog of Theorem 7.3; its proof is essentially identical to that of (7.3).

THEOREM 7.4. *Suppose $M \geq 1$, then the functional f_{M+1} attains a finite minimum of $-\tfrac{1}{2}\mu^2 v_{M+1}^{-1}$ on X where v_{M+1} is the $(M+1)$st eigenvalue of (7.4)–(7.5). This minimum is attained at $\pm u_{M+1}$ where u_{M+1} is a solution of (7.4)–(7.5) corresponding to the eigenvalue v_{M+1} and with $\|u_{M+1}\| = uv_{M+1}^{-1}$.*

Note that throughout this section we have carried μ as a free parameter. For computational purposes it is taken to be 1 or $\sqrt{2}$.

This example may be generalized in many directions. There is nothing special about second-order elliptic operators or Dirichlet boundary conditions. The crucial fact is the form (7.2) with f_1 being convex and continuous on X and X being compactly imbedded in $L^2(\Omega)$.

8. Nonlinear elliptic eigenvalue problems. The methods and analyses of the preceding sections carry over to many classes of nonlinear elliptic variational problems.

Let $X = W_0^{1,p}(\Omega)$ with $p_0 < p < \infty$ and $f_1: X \to \overline{\mathbb{R}}$ be given by

$$(8.1) \qquad f_1(u) = \int_\Omega F(x, u, Du)\, dx$$

where $F: \Omega \times R^{N+1} \to \mathbb{R}$ is a C^2-function and $p_0 = \max(1, 2N/(N+2))$. Assume

($\mathfrak{F}1$) f_1 is strongly continuous and convex on X,
($\mathfrak{F}2$) $\liminf_{\|u\| \to \infty} (f_1(u)/\|u\|) = +\infty$,
($\mathfrak{F}3$) f_1 is Gâteaux-differentiable on X and

$$\langle Df_1(u), h \rangle = \int_\Omega \left[D_{p_i} F(x, u, Du) D_i h + D_u F(x, u, Du) h \right] dx$$

for all h in X, where $D_{p_i} F$ represents the derivative of F with respect to $D_i u$.

The primal problem ($\mathfrak{P}$) will be to minimize

$$(8.2) \qquad f(u) = f_1(u) - \|u\|_2.$$

THEOREM 8.1. *Assume ($\mathfrak{F}1$)–($\mathfrak{F}3$) hold. Then f attains a finite minimum on X. If this minimum is attained at $\hat{u} \neq 0$, then $\hat{u}$ obeys*

$$(8.3) \qquad \langle Df_1(u), h \rangle = v \langle u, h \rangle$$

for all h in X and $\|\hat{u}\|_2 = v^{-1}$.

PROOF. From ($\mathfrak{F}1$), one has f_1 weakly l.s.c. on X. Since $p > 2N/(N+2)$ one has that the imbedding of X into $L^2(\Omega)$ is compact so $\|u\|_2$ is weakly continuous in X. Hence f is weakly l.s.c. on X. ($\mathfrak{F}2$) now implies that f is coercive on X. Thus f attains a finite minimum on X as X is reflexive.

The functional f is Gâteaux-differentiable on $X - \{0\}$ from ($\mathfrak{F}2$) and the fact that $\|u\|_2$ is differentiable except at 0. Hence (8.3) follows as $Df(u) = Df_1(u) - u/\|u\|_2$.

COROLLARY. *If $\hat{u}$ is a critical point of f on X, then $\hat{u}$ is a (weak) solution of*

(8.4)
$$-D_i\left[D_{p_i}F(x, u, Du)\right] + D_u F(x, u, Du) = \nu u \quad \text{in } \Omega$$
$$\text{subject to } u = 0 \text{ on } \partial\Omega.$$

Moreover, if there is a u in X such that $f(u) < f_1(0)$, then there is a nontrivial solution of (8.4).

PROOF. (8.4) is another version of (8.3). If $f(u) < f_1(0)$ for some u in X, then the minimizer of f will not be 0 and hence there is a nontrivial solution of (8.4). □

The dual problem ($\mathfrak{P}^*$) to ($\mathfrak{P}$) is to minimize $-f_1^*$ on the unit ball B_1 in $L^2(\Omega)$ (from §6 with $\mu = 1$). The dual functional f_1^* may be computed, just as in Lemma 7.2, in terms of solutions of elliptic equations, but it no longer has such a simple form.

REFERENCES

1. G. Auchmuty, *Duality for non-convex variational principles*, J. Differential Equations **50** (1983), 80–145.
2. I. Ekeland and R. Temam, *Analyse convexe et problémes variationelles*, Dunod, Paris, 1974.
3. B. N. Parlett, *The symmetric eigenvalue problem*, Prentice-Hall, Englewood Cliffs, N. J., 1980.
4. F. Riesz and B. Sz.-Nagy, *Functional analysis*, Ungar, New York, 1955.
5. H. F. Weinberger, *Variational methods for eigenvalue approximation*, SIAM, 1974.
6. A. Weinstein and W. Stenger, *Methods of intermediate problems for eigenvalues*, Academic Press, New York 1972.

UNIVERSITY OF HOUSTON

Generalized Elliptic Solutions of the Dirichlet Problem for n-Dimensional Monge–Ampère Equations[1]

ILYA J. BAKELMAN

Abstract. In this paper we are concerned with the Dirichlet problem for n-dimensional Monge–Ampère equations and with related problems of the theory of convex hypersurfaces and functions. The main part of this paper consists of my new theorem devoted to the uniform convergence of convex functions in a closed bounded convex n-domain with τ-parabolic support and to the new existence, uniqueness, and nonuniqueness theorems for elliptic (convex) generalized solutions of the Dirichlet problem for Monge–Ampère equations obtained in the spring and summer of 1983. The paper contains the systematic presentation of the fundamental concepts and facts of the theory of elliptic (convex) generalized solutions of Monge–Ampère equations.

There are two main methods of investigation of the existence theorems of the boundary value problems for Monge–Ampère equations. The first one is the classical continuity method based on a priori estimates of the corresponding Hölder norms for smooth solutions and their derivatives. This method arises from the well-known papers of S. Bernstein [1], H. Weyl [2], H. Lewy [3, 4]. In the last 10 years the significant results were proved by Pogorelov [5], Cheng and Yau [6], Caffarelli, Nirenberg, and Spruck [7], Lions [8], and Krylov [9] for the Dirichlet problem for the n-dimensional Monge–Ampère equations by this method.

The second method is related to the establishment of the existence theorems in the classes of weak and generalized solutions of Monge–Ampère equations. Such solutions arise from the extension of Monge–Ampère equations to the suitable classes of nonsmooth functions, and then we obtain the desired solutions either by the approximation of solutions for some variational problems or by fixed

---

1980 *Mathematics Subject Classification.* Primary 35J60; Secondary 53C45.

[1] This paper was written on the base of my talk at the Summer Institute on Nonlinear Functional Analysis of the AMS, University of California, Berkeley, July 1983. This research was supported by the National Science Foundation Grant MCS 8201106.

© 1986 American Mathematical Society
0082-0717/86 $1.00 + $.25 per page

points theorems for compact and continuous operators. The first outstanding example is the Minkowski Theorem of existence of a convex surface with a prescribed Gaussian curvature in Euclidean space [10]. First Minkowski solved this problem in the class of convex polyhedrons by means of the special variational problem and then obtained the generalized solution of his problem in the general case by the approximation of suitable sequence of convex polyhedrons.

The theory of weak and generalized solutions of the boundary value problems for two-dimensional general elliptic Monge–Ampère equations was developed and constructed by Bakelman [11–18] and Pogorelov [19, 20]. In the same papers they also investigated completely the smoothness of generalized solutions for these Monge–Ampère equations.

The theory of generalized and weak solutions of n-dimensional Monge–Ampère equations based on the papers of Alexandrov [21–23], Bakelman [11, 15, 17, 24–28], Cheng–Yau [6, 29, 30], Pogorelov [5, 31] is developing now in various directions.

In this paper I present my new results relating to necessary and sufficient conditions for the existence of generalized and weak solutions of the Dirichlet problem for n-dimensional Monge–Ampère equations, which were proved in the spring and summer of 1983. I also include a few results of Alexandrov and myself from the papers [11, 17, 23] which can be helpful for the reading of this paper.

1. Monge–Ampère operators and equations.

1.1. *Monge–Ampère operators and equations.* Let G be a bounded domain in the space $R^n = \{x = (x_1, x_2, \ldots, x_n)\}$. Denote by $C^2(G)$ the set of C^2-functions defined in G. The operator

$$H: C^2(G) \to C(G),$$

where

(1.1) $$H(u) = \det(u_{ij})$$

for any $u(x) \in C^2(G)$, is called the *simplest n-dimensional Monge–Ampère operator* or more briefly the *Monge–Ampère operator*. Partial differential equations containing $H(u)$ as the principal term are called the *Monge–Ampère equations*. More precisely, these equations can be described in the following way:

(a) *Classical Monge–Ampère equations* ($n = 2$). Classical Monge–Ampère equations are related to functions with two independent variables x_1 and x_2. These equations have the form

(1.2) $$u_{11}u_{22} - u_{12}^2 = Au_{11} + 2Bu_{12} + Cu_{22} + D$$

where A, B, C, D are given functions of x_1, x_2, u, u_1, u_2. The expression

(1.3) $$\Delta = D + AC - B^2$$

is called the discriminant of the equation (1.2). Let $u(x_1, x_2)$ be a C^2-solution of the equation (1.2); then the identity

$$(u_{11} - C)(u_{22} - A) - (u_{12} + B)^2 = D + AC - B^2$$

yields the ellipticity (hyperbolicity) of the equation (1.2) if and only if $\Delta > 0$ ($\Delta < 0$) for all $(x_1, x_2) \in G, u \in R, (u_1, u_2) \in R^2$.

The theory of the simplest classical Monge–Ampère equations

(1.4) $$u_{11}u_{22} - u_{12}^2 = D, \quad A \equiv B \equiv C \equiv 0$$

is already rich in itself and has various deep applications to global differential geometry, linear and quasilinear PDE and other problems of analysis.

All solutions of elliptic equations (1.4) are necessarily convex or concave functions. All solutions of hyperbolic equations (1.4) have necessarily saddle graphs.

(b) *The n-dimensional simplest Monge–Ampère equations.* These equations have the form

(1.5) $$\det(u_{ij}) = D(x, u, \operatorname{grad} u).$$

Here we consider only elliptic solutions of the equations (1.5). Clearly, they are necessarily convex and concave functions. It is sufficient to consider only convex solutions of the equation (1.5). If the equation (1.5) has convex solutions, then the function D takes only positive values.

(c) *The n-dimensional general Monge–Ampère equations.* First of all the wide class of such equations can be written as

(1.6) $$\det(u_{ij} - A_{ij}(x, u, \operatorname{grad} u)) = D(x, u, \operatorname{grad} u),$$

where A_{ij} form a symmetric matrix optionally.

Narrower classes of Monge–Ampère equations arise from the problems of global differential geometry. For example, the equations

(1.7) $$\det(u_{ij}) = A_{n-1}S_{n-1} + A_{n-2}S_{n-2} + \cdots + A_1 S_1 + A_0$$

form one such class, where $A_0, A_1, \ldots, A_{n-1}$ are given functions of $x, u, \operatorname{grad} u$, and $S_1, S_2, \ldots, S_{n-1}$ are the elementary symmetric functions of the orders $1, 2, \ldots, n-1$ of the principal normal curvatures of the graph of solutions $u(x)$ of the equation (1.7).

1.2. *Normal mapping and R-curvature of convex functions. Weak and generalized solutions for the Monge–Ampère equations* $\det(u_{ij}) = \varphi(x)/R(\operatorname{grad} u)$. The differential equations

(1.8) $$\det(u_{ij}) = \frac{\varphi(x)}{R(\operatorname{grad} u)}$$

arise from geometric problems of the reconstruction of convex hypersurfaces with prescribed Gaussian curvature as a function of the unit normal (Minkowski problem) or as a function of the projection of the variable point of the desired hypersurfaces. The function $R(\operatorname{grad} u) = 1$ corresponds to the Minkowski problem, and the function

$$(1.9) \qquad R(\operatorname{grad} u) = \left(1 + (\operatorname{grad} u)^2\right)^{-(n+2)/2}$$

corresponds to the second problem.

Bakelman [11] established the interlocking necessary and sufficient conditions of the solvability of the Dirichlet problem for the equation (1.8) in the classes of weak and generalized elliptic solutions. The further development was done in [13, 15, 17, 18, 25]. These investigations are based on the concepts of the normal mapping and R-curvature of convex functions introduced in the paper [11]. In this subsection we shall present a survey of these concepts and main results obtained in the papers mentioned above.

Let G be an open bounded convex domain in $R^n = \{x = (x_1, x_2, \ldots, x_n)\}$. The points of $R^{n+1} = R^n \times R$ will be denoted by $(x, z) = (x_1, x_2, \ldots, x_n, z)$. R^n and R^{n+1} will be considered as Euclidean spaces with the canonical metrics.

Let $W^+(G)$ be the set of all convex functions defined in G. Denote by S_u the graph of $u(x) \in W^+(G)$ and introduce the Euclidean n-dimensional space $P^n = \{p = (p_1, p_2, \ldots, p_n)\}$ with the canonical metric. Let α_0 be a supporting hyperplane of S_u, and let

$$z = p_1^0 x_1 + p_2^0 x_2 + \cdots + p_n^0 x_n + q^0$$

be the equation of α_0; then the point $p_0 = (p_1^0, p_2^0, \ldots, p_n^0)$ is called the *normal image* of α_0 and is denoted by $\chi(\alpha_0)$, i.e.

$$(1.10) \qquad \chi(\alpha_0) = \operatorname{grad}\left\{\sum_{i=1}^n p_i^0 x_i + q^0\right\}.$$

We call P^n the gradient space.

Let e by any subset of G. Consider all supporting hyperplanes α of S_u such that the projection of the set $\alpha \cap S_u$ has at least one point in e. Denote by $\chi_u(e) \subset P^n$ the set of all normal images $\chi(\alpha)$ of these hyperplanes α. $\chi_u(e)$ is called *the normal image* of the set e. The mapping χ_u transforming e in $\chi_u(e)$ is called the *normal mapping* (with respect to the convex function $u(x) \in W^+(G)$).

Now we enumerate the properties of the normal mapping:

(1) If the graph of the function $u(x)$ is a convex cone S_u and $x_0 \in G$ is the projection of the vertex of S_u, then $\chi_u(x_0)$ is a convex n-dimensional domain in P^n.

(2) If $u(x) \in C^2(G) \cap W^+(G)$, then the normal mapping χ_u can be considered pointwise and coincides with the tangential mapping $\operatorname{grad} u: G \to P^n$.

(3) If e is a Borel subset of G, then its normal image $\chi_u(e)$ is a Lebesgue measurable subset of P^n.

(4) Let $u_1(x), u_2(x) \in W^+(G)$ and let $u_1(x) \leq u_2(x)$ inside G, and $u_1|_{\partial G} = u_2|_{\partial G}$; then $\chi_{u_1}(G) \supset \chi_{u_2}(G)$.

(5) The supporting hyperplane α of S_u is called singular if $\alpha \cap S_u$ has at least two distinct points.

The equality

(1.11) $$\operatorname{mes}_{P^n} \chi_u\left(\bigcup_\alpha \alpha\right) = 0$$

holds, where $u(x) \in W^+(G)$ and $\bigcup_\alpha \alpha$ is the union of all singular supporting hyperplanes of the graph of the function $u(x)$.

The analogy of this statement for the spherical mapping of convex hypersurfaces is proved by Alexandrov [21]; the scheme of Alexandrov considerations can be used for establishing the equality (1.11) (see [6]).

Let $R(p) > 0$ be a locally summable function in P^n. The set function

(1.12) $$\omega(R, u, e) = \int_{\chi_u(e)} R(p)\, dp$$

is called the *R-curvature of the function* $u(x) \in W^+(G)$.

$\omega(R, u, e)$ is a nonnegative completely additive set function of Borel subsets of G for all convex functions $u(x) \in W^+(G)$. Evidently,

$$\omega(R, u, e) = \int_e R(\operatorname{grad} u) \det(u_{ij})\, dx$$

if $u(x) \in C^2(G) \cap W^+(G)$.

Now we consider the extension of the Dirichlet problem

(1.13) $$\det(u_{ij}) = \varphi(x)/R(\operatorname{grad} u), \qquad u|_{\partial G} = 0$$

to the set $W^+(G)$. The new statement of the Dirichlet problem (1.12) can be formulated in the following way: Find the solutions $u(x) \in W^+(G)$ of the equation

(1.14) $$\omega(R, u, e) = \mu(e)$$

satisfying the boundary condition

(1.15) $$u|_{\partial G} = h(x) \in C(\partial G).$$

The solutions of the boundary value problem (1.14)–(1.15) are called the *weak solutions* for the Dirichlet problem (1.13). Note that any convex function $u(x) \in W^+(G)$ can be obtained as some solution of the equation (1.14) for the corresponding function $\varphi(x)$. The more close relationship between the equations (1.14) and

$$\det(u_{ij}) = \varphi(x)/R(\operatorname{grad} u)$$

can be reached if we additionally assume the absolute continuity of the set function $\mu(e)$, i.e.

(1.16) $$\mu(e) = \int_e \varphi(x)\, dx,$$

where $\varphi(x) \geq 0$ is summable in G and e is any Borel subset of G. If $\mu(e)$ is absolutely continuous, then $\omega(R, u, e)$ is also absolutely continuous for every solution $u(x)$ of the equation (1.14). Moreover, such solutions $u(x)$ satisfy almost everywhere the equation

(1.17) $$\det(u_{ij}) = \varphi(x)/R(\operatorname{grad} u),$$

since du and d^2u exist almost everywhere in G for every convex function $u(x)$.

Weak solutions $u(x)$ of (1.17) having the absolutely continuous R-curvature $\omega(R, u, e)$ are called the *generalized solutions* of the equation (1.17). Conversely, every convex function $u(x) \in W^+(G)$ satisfying (1.17) almost everywhere and having the absolutely continuous R-curvature is a generalized solution of the equation (1.17).

1.3. *The solvability of the Dirichlet problem* (1.13) *in the classes of weak and generalized solutions*. There exist a few obstructions showing that the Dirichlet problem (1.13) cannot be solved for arbitrary functions $\varphi(x)$ and $R(p)$ even if they are strictly positive and continuous. We consider two of them. For the first time they were described by Bakelman in [**13**].

(a) Let

(1.18) $$A(R) = \int_{P^n} R(p)\, dp;$$

the case $A(R) = +\infty$ is not excluded. Since

$$\mu(G) = \omega(R, u, G) = \int_{\chi_u(G)} R(p)\, dp \leq \int_{P^n} R(p)\, dp = A(R)$$

for every solution of the equation (1.14), we obtain the first necessary condition of the solvability of the Dirichlet problem (1.13) and (1.14)-(1.15):

(1.19) (problem (1.13)) $\quad \int_G \varphi(x)\, dx \leq \int_{P^n} R(p)\, dp = A(R),$

(1.20) (problem (1.14)-(1.15)) $\quad \mu(G) \leq \int_{P^n} R(p)\, dp = A(R).$

The inequalities (1.19) and (1.20) describe the first obstruction for generalized and weak solutions of the equation (1.17).

(b) Now consider the Dirichlet problem

(1.21) $$\det(u_{ij}) = K(x)\big(1 + (\operatorname{grad} u)^2\big)^{(n+2)/2},$$

(1.22) $$u|_{\partial G} = kx_1,$$

where G: $\sum_{i=1}^n x_i^2 \leq r^2$ and $0 < \alpha \leq K(x) \leq \beta < +\infty$. We can find the sufficiently small r depending only on β and the number

$$\sigma_n = \int_{P^n} \big(1 + |p|^2\big)^{-(n+2)/2}\, dp < +\infty$$

such that

$$\int_G K(x)\,dx < v_n,$$

where v_n is the volume of the unit n-ball in E^n.

On the other hand, we have the a priori estimate

(1.23) $$\sigma_n \geqslant \int_G K(x)\,d\sigma \geqslant \alpha \cdot \sigma_u$$

where σ_n is the area of the unit hemisphere in E^{n+1} and σ_u is the area of the graph of $u(x)$. Thus the Dirichlet problem (1.21)–(1.22) does not have solutions if the number k is sufficiently large.

Let $u(x) \in W^+(G)$ be a bounded function. Consider the closed convex hull $\overline{\mathrm{Co}}(S_u)$ of the graph of $u(x)$. Let Z be the cylinder with the base ∂G and with generators parallel to the z-axis. Then the closed set

$$M = Z \cap \partial \overline{\mathrm{Co}}(S_u)$$

can be considered as a union of the segments $l(x)$, $x \in \partial G$, and $l(x)$ is orthogonal to G. Evidently, $l(x) = \{(x, h)\}$.

The function $\nu_u(x) = \inf h$ where $x \in \partial G$ and $(x, h) \in l(x)$ is called the *border* of the function $u(x)$.

If the convex function $u(x)$ is defined only inside an open convex bounded domain G, then the border of $u(x)$ is not necessarily continuous.

We consider the Dirichlet problem

(1.24) $$\omega(R, u, e) = \mu(e),$$

(1.25) $$u|_{\partial G} = h(x),$$

in the bounded convex domain G, where $h(x) \in C(\partial G)$, $\mu(e)$ is a given nonnegative completely additive set function, and $u|_{\partial G}$ is the border of a convex function $u(x) \in W^+(G)$.

We denote by $L(\mu, h)$ the set of all solutions $u(x) \in W^+(G)$ of the equation (1.24), whose borders satisfy the condition

(1.26) $$\nu_u(x) \leqslant h(x)$$

for all $x \in \partial G$.

Clearly, if the Dirichlet problem (1.24)–(1.25) has at least one solution, then these solutions belong to the set $L(\mu, h)$.

The convex domain G is called strictly convex if every supporting hyperplane of ∂G has only one common point with ∂G.

Bakelman has proved the following theorem in 1956 (see [11] and also [17, 25]).

THEOREM 1. *If the domain G is bounded and strictly convex and*

(1.27) $$\mu(G) < A(R) = \int_{P^n} R(p)\,dp,$$

then the set $L(\mu, h)$ associated with the problem (1.24)–(1.25) is not empty. Moreover, there exists one and only one solution $u_0(x) \in L(\mu, h)$ such that

(1.28) $$\nu_{u_0}(x) \geq \nu_u(x)$$

for every function $u(x) \in L(\mu, h)$.

Note that the sufficient condition (1.27) of the solvability of the Dirichlet problem (1.24)–(1.25) interlocks with the necessary condition (1.20). Further, generally speaking the boundary condition (1.25) can be satisfied according to the second obstruction (see subsection 1.2) only in some generalized sense. The corresponding definition is given in the statement of Theorem 1.

Let G be an open bounded convex domain in E^n. We denote by G_ε the subset of G consisting of all points $x \in G$ for which the inequality

$$\text{dist}(x, \partial G) > \varepsilon$$

holds. If $\varepsilon > 0$ is sufficiently small, then G_ε is an open subset of G. We shall consider only these small numbers ε.

We denote by $\mathfrak{M}_\varepsilon(G)$ all nonpositive completely additive set functions $\mu(e)$ on the ring of Borel subsets of G, satisfying the additional condition $\mu(G) = \mu(G_\varepsilon)$, i.e. $\mu(G \setminus G_\varepsilon) = 0$. Evidently,

$$\mathfrak{M}_{\varepsilon_1}(G) \subset \mathfrak{M}_{\varepsilon_2}(G)$$

if $\varepsilon_1 > \varepsilon_2 > 0$.

Let $\mu(e)$ be a nonnegative completely additive set function on the ring of Borel subsets of G. We introduce the new set function

(1.29) $$\mu_\varepsilon(G) = \mu(e \cap G_\varepsilon).$$

Evidently $\mu_\varepsilon(e) \subset \mathfrak{M}_\varepsilon(G)$ and $\mu_\varepsilon(e)$ weakly converge to $\mu(e)$ inside G.

We can now state the following theorem (see [17, 25]):

THEOREM 2. *Let G be a bounded open strictly convex domain. Assume that the set function $\mu(e) \in \mathfrak{M}_\varepsilon(G)$ (for sufficiently small $\varepsilon > 0$) and $\mu(G) < A(R)$. Then the Dirichlet problem (1.24)–(1.25) has only one solution $u_\varepsilon(x) \in W^+(G)$ satisfying the boundary condition (1.25) in the classical sense, i.e. $\nu_{u_\varepsilon}(x) = h(x)$ for all $x \in \partial G$.*

Below we will use the following comparison theorem (see [11, 17, 25]):

THEOREM 3. *Let G be an open bounded convex domain in R^n. Let $u_1(x), u_2(x)$ be convex solutions of the Dirichlet problems*

(1.30) $$\omega(R, u_1, e) = \mu_1(e), \quad u_1(x)|_{\partial G} = h(x) \in C(\partial G);$$

(1.31) $$\omega(R, u_2, e) = \mu_2(e), \quad u_2(x)|_{\partial G} = h(x) \in C(\partial G),$$

and let $\mu_1(e) \leq \mu_2(e)$ for every Borel subset e of G. Then, $u_1(x) \geq u_2(x)$ for all $x \in G$.

Now let $\mu(e)$ be any nonnegative completely additive set function on the ring of Borel subsets of G and let $\varepsilon_1 > \varepsilon_2 > 0$ be two sufficiently small arbitrary numbers. Consider two set functions $\mu_{\varepsilon_1}(e) = \mu(e \cap G_{\varepsilon_1})$ and $\mu_{\varepsilon_2}(e) = \mu(e \cap G_{\varepsilon_2})$. Since $G_{\varepsilon_1} \subset G_{\varepsilon_2}$, then

$$\mu_{\varepsilon_1}(e) \leq \mu_{\varepsilon_2}(e)$$

for every Borel subset e of G. Now consider two Dirichlet problems

(1.32,i) $\quad \omega(R, u, e) = \mu_{\varepsilon_i}(e), \quad u|_{\partial G} = h(x) \in C(\partial G), \quad i = 1, 2,$

in a bounded strictly convex domain G and assume that $\mu(G) < A(R)$. Since

$$\mu_{\varepsilon_1}(G) \leq \mu_{\varepsilon_2}(G) \leq \mu(G) < A(R),$$

then from Theorem 2 it follows that there exist the unique solutions $u_{\varepsilon_1}(x) \in W^+(G)$, $u_{\varepsilon_2}(x) \in W^+(G)$ of both problems (1.32, 1 & 2) satisfying the boundary condition in the classical sense.

From Theorem 3 it follows that $u_{\varepsilon_1}(x) \geq u_{\varepsilon_2}(x)$ for all $x \in G$.

According to our assumptions, $R(p)$ is a positive locally summable function in the gradient space P^n. Therefore, the function

$$g_R(\rho) = \int_{P^n} R(p) \, dp$$

is positive strictly increasing and continuous in $[0, +\infty)$. Clearly, $g_R(\rho)$ maps one-to-one $[0, +\infty)$ on $[0, A(R))$. Let $T_R(\tau)$ be the inverse for the function $g_R(\rho)$. Then $T_R(\tau)$ maps one-to-one $[0, A(R))$ on $[0, +\infty)$.

Below we will use the following theorem about estimates of convex functions (see in [17, 26, 27] its proof and various applications to PDE):

THEOREM 4. *Let G be a convex bounded domain in R^n, and let $V(\omega_0)$ be the set of all convex functions $z(x)$ belonging to $W^+(G)$ and satisfying the conditions*
(1) $-\infty < m \leq z|_{\partial G} \leq M < +\infty$;
(2) $\omega(R, z, G) \leq \omega_0 < A(R)$.
Then the inequalities

(1.33) $\quad m - T_R(\omega_0) \operatorname{diam} G \leq z(x) \leq M$

hold for all $x \in G$.

Now consider the Dirichlet problem

(1.34) $\quad \omega(R, u, e) = \mu_\varepsilon(e),$

(1.35) $\quad u|_{\partial G} = h(x) \in C(\partial G),$

where G is a bounded strictly convex domain, $\varepsilon > 0$ is any sufficiently small number, $\mu_\varepsilon(e) = \mu(e \cap G_\varepsilon)$ and $\mu(e)$ is a nonnegative completely additive set function on the ring of Borel subsets of G, satisfying the inequality

(1.36) $\quad \mu(G) < A(R).$

Since $\mu_\varepsilon(G) \leq \mu(G) < A(R)$ for all admissible[2] $\varepsilon > 0$, then there exists the unique solution $u_\varepsilon(x) \in W^+(G)$ of the Dirichlet problem (1.34)–(1.35) satisfying

---

[2] The number $\varepsilon_0 > 0$ is admissible if G_{ε_0} is an open subdomain of a bounded strictly convex domain G. If $\varepsilon_0 > 0$ is an admissible number, then every $\varepsilon \in (0, \varepsilon_0]$ is also admissible.

the boundary condition in the classical sense for all admissible $\varepsilon > 0$. From Theorem 4 it follows that

(1.37) $$m - T_R(\mu(G)) \operatorname{diam} G \leqslant u_\varepsilon(x) \leqslant M$$

where $m = \inf_G h(x)$, $M = \sup_G h(x)$. From (1.37) and (1.32,3) it follows that there exists a convex function

$$u_0(x) = \lim_{\varepsilon \to 0} u_\varepsilon(x).$$

Since $\omega(R, u_\varepsilon, e)$ and $\mu_\varepsilon(e)$ converge weakly inside G correspondingly to $\omega(R, u_0, e)$ and $\mu(e)$, then $u_0(x)$ satisfies the equation $\omega(R, u_0, e) = \mu(e)$ for all Borel subsets e of G.

The convex functions $u_\varepsilon(x)$ converge pointwise to $u_0(x)$ only for all inner points of the open domain G and $\nu_{u_\varepsilon}(x) = h(x)$ for every $x \in \partial G$ and any $\varepsilon > 0$. Therefore,

$$\nu_{u_0}(x) \leqslant h(x)$$

for all $x \in \partial G$. Thus $u_0(x) \in L(\mu, h)$, where the set $L(\mu, h)$ was introduced in Theorem 1. From the construction of $u_0(x)$ it follows that $\nu_u(x) \leqslant \nu_{u_0}(x)$ for every convex function $u(x) \in L(\mu, h)$, and if $\nu_u(x) = \nu_{u_0}(x)$ for all $x \in \partial G$, then $u(x) \equiv u_0(x)$ in G.

Thus Theorem 1 is proved. The detailed proofs of Theorems 2, 3, 4 can be found in Bakelman's papers [**11, 13, 17, 25**]. Note that the proof of Theorem 2 consists of two parts: (a) First we solve the Dirichlet problem for the equation (1.24) specially for convex polyhedrons in the convex polyhedral domains, (b) then consider the special approximation by convex polyhedrons.

The ideas of such investigations arise from the Minkowski papers [**10**] where they were used successfully for the solution of a few important problems of the theory of convex bodies. Further development of these ideas, methods and applications to various problems of differential geometry and partial differential equations was given by Alexandrov [**21, 22**] Pogorelov [**19, 20**], Bakelman [**11, 17**], and Cheng-Yau [**6, 29**].

1.4. *The solvability of the Dirichlet problem* (1.24)–(1.25) *in the classical sense.* Theorem 2 shows that the strict convexity of the bounded domain G and the condition $\mu(G \setminus G_\varepsilon) = 0$ for a nonnegative function $\mu(e) \in \mathfrak{M}_\varepsilon(G)$ (for some admissible $\varepsilon > 0$) provide the classical solvability of the Dirichlet problem (1.24)–(1.25). However, our wish to eliminate the sufficiently overloaded condition $\mu(G \setminus G_\varepsilon) = 0$ by the limiting passage $\varepsilon \to 0$ can bring us to the omission of the boundary condition (1.25) in the classical sense (see Theorem 1). Thus there arises a new concept of generalized satisfaction of the boundary condition (1.25) (see again Theorem 1).

Nevertheless, it turns out to be possible to obtain the necessary and sufficient conditions of the classical solvability of the Dirichlet problem (1.24)–(1.25) in the terms of the asymptotic behavior of the functions $R(p)$ (when $|p| \to \infty$) and of $\mu(e)$ (when $\sup_{x \in e} \operatorname{dist}(x, \partial G) \to 0$) and also of the local properties of a strictly convex domain of G.

In this subsection we present the assumptions and the main results of the solvability of the Dirichlet problem (1.24)–(1.25) in the classical sense, obtained by myself in 1983. The special cases for two-dimensional general strictly convex domains and n-dimensional strictly convex domains with local spherical support were investigated earlier [16, 17, 18, 25]. In 1963–1965 Guberman participated in these investigations.

The basic assumptions.

Assumption (A) *Local parabolic support of the order* $\tau \geq 0$. Let G be an open bounded strictly convex domain in R^n and $\overline{G}$ be the closure of G. Let a_0 be any point of ∂G. Then there exist a supporting $(n - 1)$-plane α of ∂G passing through a_0 and an open ball $U_\rho(a_0)$ with center a_0 and radius $\rho > 0$ such that the convex $(n - 1)$-surface

(1.38) $$\Gamma_\rho(a_0) = \partial G \cap U_\rho(a_0)$$

has one-to-one orthogonal projection $\pi_\alpha \colon \Gamma_\rho(a_0) \to \alpha$. Moreover, ∂G and α have only one common point, a_0, and the unit normal of α directed to the halfspace of R^n, where $\overline{G}$ lies, passes through interior points of G.

All considerations made above also take place for every n-ball $U_{\rho'}(a_0)$ where $0 < \rho' \leq \rho_0$. Denote by $\Pi_\rho(a_0)$ the set $\pi_\alpha(\Gamma_\rho(a_0))$. Let $x_1, x_2, \ldots, x_{n-1}, x_n$ be Cartesian coordinates in R^n introduced in the following way: a_0 is the origin, the axes $x_1, x_2, \ldots, x_{n-1}$ lie in the plane α, and the axis x_n is directed along the interior normal of ∂G at the point a_0. Clearly, the convex $(n - 1)$-surface $\Gamma_\rho(a_0)$ is the graph of some convex function $\psi(x_1, x_2, \ldots, x_{n-1}) \in W^+(\Pi_\rho(a_0))$.

Evidently,

(1.39) $$\psi(0, 0, \ldots, 0) = 0$$

and

(1.40) $$\psi(x_1, x_2, \ldots, x_{n-1}) \geq 0$$

in $\Pi_\rho(a_0) \setminus a_0$. The function $\psi(x_1, x_2, \ldots, x_{n-1})$ is called the *local explicit representation of the closed convex surface* G near the marked point $a_0 \in \partial G$. We will say that ∂G has a *parabolic support of the order* $\tau \geq 0$ *at the point* a_0 if there exist positive numbers $\rho_0 \leq \rho$ and $b(a_0)$ such that

(1.41) $$\psi(x_1, x_2, \ldots, x_{n-1}) \geq b(a_0) \left(\sum_{i=1}^{n-1} x_i^2 \right)^{(2+\tau)/2}$$

for all $(x_1, x_2, \ldots, x_{n-1}) \in \Pi_{\rho_0}(a_0)$.

The equivalent formulation of the last concept is as follows: "The convex $(n - 1)$-surface $\Gamma_{\rho_0}(a_0)$ where $\rho_0 \leq \rho$ can be touched from outside by the $(n - 1)$-dimensional paraboloid

(1.42) $$x_n = b(a_0) \left(\sum_{i=1}^{n-1} x_i^2 \right)^{(2+\tau)/2}$$

of the order $(2 + \tau)/2$ at the point a_0."

We will say that ∂G has a *parabolic support of order not more than* $\tau = \text{const} \geq 0$ if ∂G has a parabolic support of order not more than τ at every point of ∂G.

If $\tau = 0$, then we can replace a $(n-1)$-parabolic segment by some $(n-1)$-spherical segment. The case $\tau = 0$ was investigated in Bakelman's papers [**16, 17, 25**].

Assumption 1. Below in this subsection we suppose that ∂G has a parabolic support of the order not more than $\tau = \text{const} \geq 0$.

Assumption 2. The function $R(p)$ is positive and locally summable in the gradient space P^n and the inequality $R(p) \geq C_0 |p|^{-2k}$ holds for all $p \in P^n$ except $p = (0,0,\ldots,0)$, where $k \geq 0$ and $C_0 > 0$ are constants.

Let $u_m(x)$, $m = 1,2,\ldots$, be the sequence of convex functions defined in an open bounded convex domain G. We call this sequence convergent with the vanishing λ-rate of $\omega(R, u_m, e)$ near ∂G if the following conditions are fulfilled:

(1) The functions $u_m(x)$ converge pointwise in G.

(2) There exists such an n-ball $U_\rho(x_0)$ for every point $x_0 \in \partial G$ that

(1.43) $$\lim_{m \to \infty} \omega(R, u_m, e) \leq C_1 \sup_e (\text{dist}(x, \partial G))^\lambda \text{mes } e$$

where $e \subset U_\rho(x_0) \cap G$ is a Borel subset and $C_1 > 0, \lambda \geq 0$ are constants.

Now we formulate a theorem relating to the convergence of the borders for a pointwise convergent sequence of convex functions inside an open bounded convex domain. The conditions of the same theorem also provide the uniform convergence of the considered sequence of convex functions.

THEOREM 5. *Let* $u_m(x)$, $m = 1, 2, \ldots$, *be the sequence of convex functions defined in an open bounded convex domain* G. *Let the following conditions be fulfilled*:

(1) *The domain* G *satisfies Assumption* 1.

(2) *The convex functions* $u_m(x)$ *form a convergent sequence with the vanishing* λ-*rate of* $\omega(R, u_m, e)$ *near* ∂G *and*

$$u(x) = \lim_{m \to \infty} u_m(x) \in W^+(G).$$

(3) *The function* $R(p)$ *satisfies Assumption* 2.

(4) *The numbers* τ, λ, k *included in Assumptions* 1, 2 *and condition* (2) *satisfy either the inequality*

(1.44a) $$k \leq K \quad (\text{when } k < 1 \text{ or } k \geq n/2)$$

or

(1.44b) $$k < K \quad (\text{when } 1 \leq k < n/2),$$

where $K = (n + \tau + 1)/(\tau + 2) + \lambda/2$ *and* n *is the dimension of the domain* G.

(5) *The borders* $v_{u_m}(x)$ *of convex functions* $u_m(x)$ *are continuous functions on* ∂G *and* $v_{u_m}(x)$ *uniformly converge to some continuous function* $h(x)$ *on* ∂G.

Then $h(x)$ *is the border of the limiting convex function* $u(x)$ *and the convex functions* $u_m(x)$ *uniformly converge to* $u(x)$ *in* G.

We consider the scheme of the proof of Theorem 5. For a detailed proof, see Bakelman [34]. Suppose that the border $\nu_u(x)$ of the limiting convex function $u(x)$ does not coincide with the function $h(x)$ for all $x \in \partial G$. Since the inequality $\nu_u(x) \leq h(x)$ holds for all $x \in \partial G$, then there exists at least one point $x_0 \in \partial G$ such that $\nu_u(x_0) < h(x_0)$.

Now we introduce the special Cartesian coordinates in R^n and R^{n+1} with the origin x_0 (see the text of Assumption (A), subsection 1.4). Thus the axes $x_1, x_2, \ldots, x_{n-1}$ lie in the supporting hyperplane α of ∂G, the axis x_n is orthogonal to α and has points inside G, and finally the z-axis is orthogonal to the hyperplane R^n in the space R^{n+1}.

Let $Q(0, h(0))$ and $\overline{Q} = (0, \nu_u(0))$ be points of the z-axis and let $0 < \delta < 1$ be an arbitrary number. We introduce two new points $Q'(0, h(0) - \delta\Delta h)$ and $Q''(0, \nu_u(0) - \delta\Delta h)$, where $\Delta h = h(0) - \nu_u(0)$. The point Q' lies inside the segment $Q\overline{Q}$ and Q'' lies on the z-axis under the point $\overline{Q}$. Now we consider the hyperplanes

$$\mathcal{B}': z = h(0) - \delta\Delta h - (1/\gamma)x_n,$$
$$\mathcal{B}'': z = \nu_u(0) - \delta\Delta h,$$

where γ is a sufficiently small positive number.

Let $Z = \partial G \times R$ be a cylinder in R^{n+1} with the base ∂G and with the generators parallel to the z-axis. Then Z bounds some convex body K together with the hyperplanes $\mathcal{B}'$ and $\mathcal{B}''$. Evidently the convex hypersurfaces S_{u_m} have nonempty intersections with the convex body K. We denote by Q_m the nearest point of S_{u_m} to the point Q. Now we introduce the sets

$$S_m(K) = S_{u_m} \cap K \quad \text{and} \quad \mathcal{B}'(K) = \mathcal{B}' \cap K.$$

If $\gamma > 0$ is sufficiently small and m is sufficiently large, then

$$S_m(K) \cap \mathcal{B}'' = \varnothing \quad \text{and} \quad S_m(K) \cap Z = \varnothing.$$

Therefore, $Q_m \in S_m(K)$. Let $H_m(K)$ be the projection of $S_m(K)$ on R^n and let V_m be the convex cone with the vertex Q_m and the base $\mathcal{B}'(K)$. Then the normal image of the set $S_m(K)$ covers the normal image of the convex cone V_m. Therefore,

$$\omega(R, V_m) \leq \omega(R, u_m, S_m(K)).$$

Let $H(K)$ be the projection of the convex body K on R^n. Since $H_m(K) \subset H(K)$, then from the last inequality it follows that

$$\omega(R, V_m) \leq \omega(R, u_m, H(K)).$$

If $m \to +\infty$, then the points Q_m converge to the point $\overline{Q}$. Hence, $\lim_{m\to\infty} \omega(R, V_m) = \omega(R, V)$ where V is the convex cone with the vertex $\overline{Q}$ and the base $\mathcal{B}'(K)$. Now we use condition (2) of Assumption 2, i.e. there exists some n-ball $U_\rho(0)$ for which the inequality (1.43) holds. Let $\gamma > 0$ be sufficiently small, then the Borel set

$$H_m(K) \subset U_\rho(0) \cap G.$$

Hence,
$$\lim_{m \to \infty} \omega(R, u_m, H(K)) \leq a \left[\sup_{H(K)} \mathrm{dist}(x, \partial G) \right]^\lambda \mathrm{mes}\, H(K).$$

Evidently,
$$\sup_{H(K)} [\mathrm{dist}(x, \partial G)] = \gamma \Delta h.$$

Let T be the set of points $(x_1, x_2, \ldots, x_n) \in R^n$ satisfying the double inequality
$$b(0) \left[\sum_{i=1}^n x_i^2 \right]^{(2+\tau)/2} \leq x_n \leq \gamma \cdot (\Delta h)$$

(for the definition of $b(0) > 0$, see Assumption 1); then
$$H(K) \subset T \quad \text{and} \quad \mathrm{mes}\, T = d_1 \gamma^{(n+\tau+1)/(\tau+2)}$$

where
$$d_1 = \frac{(2+\tau)^{\mu_{n-1}}}{n+\tau+1} \left(\frac{\Delta h}{b(0)} \right)^{(n+\tau+1)/(2+\tau)} = \mathrm{const} > 0$$

and μ_{n-1} is the volume of the unit $(n-1)$-ball. Thus,

(E.1) $$\omega(R, V) \leq d_2 \gamma^{\lambda + (n+\tau+1)/(2+\tau)},$$

where $d_2 = d_1 (\Delta h)^\lambda a = \mathrm{const} > 0$.

Now we obtain the estimate of $\omega(R, V)$ from below. Our considerations are based on the following

LEMMA. *Let H^n be the solid convex cone in the gradient space P^n, which has the vertex $(0, 0, \ldots, 0, -\delta/\gamma)$ and the base U^{n-1} is the $(n-1)$-dimensional ball*
$$\begin{cases} p_1^2 + p_2^2 + \cdots + p_{n-1}^2 \leq (C')^2 \gamma^{-2/(\tau+2)} \\ p_n = (-C'') \gamma^{-1} \end{cases}$$

and
$$C' = \frac{\tau+2}{\tau+1}(1-\delta)(\Delta h)^{(\tau+1)/(\tau+2)} \gamma^{-2/(\tau+2)}, \qquad C'' = \frac{\tau+2-\delta}{\tau+1}.$$

Then the normal image of the convex cone V contains the set H^n.

The proof of this Lemma is based on the methods of elementary differential geometry (see [34]).

From this Lemma, Assumption 2 and the estimate (E.1) it follows that
$$d_2 \gamma^{\lambda + (n+\tau+1)/(2+\tau)} \geq \int_{H^n} R(p)\, dp \geq C_0 \int_{H^n} \frac{dp}{|p|^{2k}}.$$

But the inequality
$$d_2 \gamma^{\lambda + (n+\tau+1)/(2+\tau)} \geq C_0 \int_{H^n} \frac{dp}{|p|^{2k}}$$

is incompatible with the assumptions

$$k \leqslant K = \frac{n+\tau+1}{\tau+2} + \frac{\lambda}{2} \quad \text{if } k < 1 \text{ or } k \geqslant \frac{n}{2},$$

or

$$k < K = \frac{n+\tau+1}{\tau+2} + \frac{\lambda}{2} \quad \text{if } 1 \leqslant k < \frac{n}{2},$$

if the positive numbers δ and γ are sufficiently small. This completes the proof of Theorem 5.

Now let G be again an open bounded convex domain in R^n and let $\mu(e)$ be any nonnegative completely additive set function of Borel subsets of G.

Assumption 3. There exists an n-ball $U_\rho(x_0)$ for every point $x_0 \in \partial G$ such that

(1.45) $$\mu(e) \leqslant C_2\Big(\sup_e [\text{dist}(x, \partial G)]^\lambda\Big) \text{mes } e,$$

where $C_2 = \text{const} > 0$, $\lambda = \text{const} \geqslant 0$ and e is any Borel subset of $G \cap U_\rho(x_0)$.

If the set function $\mu(e)$ satisfies Assumption 3, then we say, "$\mu(e)$ vanishes with λ-rate near ∂G."

REMARKS. (1) If $\mu(e)$ is absolutely continuous, then $\mu(e) = \int_e \varphi(x)\,dx$, where $\varphi(x) \geqslant 0$ and is summable in G. Therefore, the inequality (1.45) can be replaced by the simpler one

(1.46) $$\varphi(x) \leqslant C_2[\text{dist}(x, \partial G)]^\lambda$$

for all $x \in G \setminus G_\varepsilon$, where $\varepsilon > 0$ is a sufficiently small number.

(2) Let $\mu_\varepsilon(e)$ be the set function constructed above by the formula $\mu_\varepsilon(e) = \mu(e \cap G_\varepsilon)$. Since $\mu_\varepsilon(e) \leqslant \mu(e)$ for all admissible values of $\varepsilon > 0$, then all functions $\mu_\varepsilon(e)$ vanish with λ-rate near ∂G; moreover, the constants C_2 and λ are one and the same for all functions $\mu_\varepsilon(e)$.

THEOREM 6. *We consider the Dirichlet problem*

(1.47) $$\omega(R, u, e) = \mu(e),$$

(1.48) $$u|_{\partial G} = h(x),$$

where the bounded strictly convex domain G satisfies Assumption 1, *the function $R(p)$ satisfies Assumption* 2, *the set function $\mu(e)$ satisfies Assumption* 3, *and $h(x) \in C(\partial G)$. Let*

(1.49) $$\mu(G) < A(R)$$

and

$$k \leqslant K = \frac{n+\tau+1}{\tau+2} + \frac{\lambda}{2} \quad \text{if } k < 1 \text{ or } k \geqslant \frac{n}{2},$$

or

$$k < K = \frac{n+\tau+1}{\tau+2} + \frac{\lambda}{2} \quad \text{if } 1 \leqslant k < \frac{n}{2},$$

where $\tau \geq 0, \lambda \geq 0$ and $k \geq 0$ are numbers participating in Assumptions 1, 2, 3.

Then the Dirichlet problem (1.47)–(1.48) has one and only one solution satisfying the boundary condition (1.48) in the classical sense.

THEOREM 6'. *If all conditions of Theorem 6 are fulfilled and $\mu(e)$ is an absolutely continuous function $\mu(e) = \int_e \varphi(x)\, dx$ with $\varphi(x)$ satisfying the inequality (1.46), then the Dirichlet problem*

(1.50) $$\det(u_{ij}) = \varphi(x)/R(Du),$$

(1.51) $$u|_{\partial G} = h(x) \in C(\partial G)$$

has one and only one generalized solution satisfying the boundary condition in the classical sense.

Theorem 6' follows directly from Theorem 6, because every weak solution of the Dirichlet problem (1.50)–(1.51) is simultaneously a generalized solution of the same boundary value problem.

The proof of Theorem 6 is conducted in the same way as the proof of Theorem 1 (see subsection 1.3). The conditions of Theorem 6 and Remark (2) (see remarks to Assumption 3) permit one to apply Theorem 2 to the Dirichlet problem

(1.52) $$\omega(R, u, e) = \mu_\varepsilon(e),$$

(1.53) $$u|_{\partial G} = h(x).$$

Hence there exists $\varepsilon_0 > 0$ such that the Dirichlet problem (1.52)–(1.53) has the unique solution $u_\varepsilon(x) \in W^+(G)$, $0 < \varepsilon \leq \varepsilon_0$, satisfying the condition (1.53) in the classical sense. Note that this boundary condition does not depend on ε.

The conditions of the same Theorem 6 provide the pointwise convergence of $u_\varepsilon(x)$ for all $x \in G$ to some convex function $u(x) \in W^+(G)$. (We use the same considerations here as in Theorem 1 above.) But the additional assumptions included in the statement of Theorem 6 together with Remark (2) to Assumption 3 allow one to apply the convergence Theorem 5. Hence the border of the convex function $u_\varepsilon(x)$ coincides with the continuous function $h(x)$, and moreover, the convex function $u_\varepsilon(x)$ uniformly converges to $u(x)$ in $\overline{G} = G \cup \partial G$ if ε approaches zero.

Theorems 6 and 6' give the most general conditions of the existence of the inverse operator of the Dirichlet problem for the Monge–Ampère equation

$$\det(u_{ij}) = \varphi(x)/R(\operatorname{grad} u)$$

in the classes of weak and generalized solutions. These theorems will be significantly used for the investigations of more general classes of Monge–Ampère equations in the next sections of this paper. We said above that the particular cases of Theorems 6 and 6' were proved by Bakelman in [11, 17, 25].

2. The Dirichlet problem for elliptic generalized solutions of Monge–Ampère equations $\det(u_{ij}) = f(x, u, \text{grad } u)$.

2.1. Introduction. Bakelman's investigations concerning weak and generalized solutions of the Dirichlet problem for Monge–Ampère equations (see §1) and especially Theorem 1, proved in 1956 [11], have stimulated interest for the Dirichlet problem for the equations

(2.1) $$\det(u_{ij}) = f(x, u, Du)$$

and more general classes of Monge–Ampère equations. We use the notation $Du = \text{grad } u$ here and below in this paper.

In 1958 Alexandrov [23] introduced the concept of weak solutions for the equations (2.1) by means of the extension of the Monge–Ampère operator

$$\frac{1}{f(x, u, Du)} \det(u_{ij})$$

to the set of all convex functions, where the function $f(x, u, p)$ can be estimated from both sides by products $\varphi_1(x)R_1(p)$ and $\varphi_2(x)R_2(p)$. Alexandrov proved that the equation (2.1) has at least one weak solution $u(x) \in W^+(G)$ assuming convexity and boundedness of the domain G and some inequality between the functions f, $\varphi_i(x)$, $R_i(p)$, $i = 1, 2$ (see subsection 2.2). This inequality can be considered as the analogy of the inequality (1.27) of Theorem 1 (see subsection 1.3) for the equation (1.24). Note that the sufficient condition (1.27) of the solvability of the Dirichlet problem for the equation (1.27) interlocks with the necessary condition (1.20) of the solvability of the same problem. The Alexandrov assumption for the more general equation (2.1) is only some sufficient condition providing the existence of at least one weak solution of this equation.

Alexandrov did not investigate the problem when the boundary condition of the Dirichlet problem is fulfilled in any meaning and when the weak solution of this problem is unique.

First he proved the existence of the solution of the Dirichlet problem in the class of convex polyhedrons and then he used the polyhedral approximation for the general case. The open problems about the satisfaction of the boundary condition and uniqueness theorem for n-dimensional Monge–Ampère equations (2.1) were investigated by Bakelman [15, 16, 17, 25], and Cheng and Yau [6, 30].

Bakelman has introduced and realized other methods of the investigations of the Dirichlet problem for the equation (2.1) from 1959 up to the present. His main considerations are based on the reduction of the Dirichlet problem for the equation (2.1) to some operator equation in the Banach spaces and then to applications of global fixed points theorems. The reduction of the equation (2.1) to the operator equation is given by means of the inverse operator for the Dirichlet problem

(2.2) $$\det(z_{ij}) = \varphi(x)/R(Dz),$$

(2.3) $$z|_{\partial G} = h(x) \in C(\partial G).$$

The new most complete and general results about this inverse operator are contained in Theorems 6 and 6' proved by the author in 1983 (see §1 of the present paper). His new theorems of the solvability of the Dirichlet problem for the Monge–Ampère equations (2.1) based on Theorems 6 and 6' will be presented in subsection 2.3.

In 1977 Cheng and Yau [6] investigated the smoothness of generalized solutions for the Monge–Ampère equations

(2.4) $$\det(u_{ij}) = F(x, u).$$

They proved the existence theorems for generalized solutions of the Dirichlet problem for the equations (2.4) which satisfy the boundary condition in the classical sense. They assumed also that

(2.5) $$\frac{\partial}{\partial u} F(x, u) \geq 0$$

for all admissible x and u. This assumption permits them to establish the uniqueness theorem, too.

On the basis of these results Cheng and Yau [6, 29, 35] investigated the smoothness of generalized solutions of equations (2.4) in Hölder spaces $C^{n,a}$ ($n \geq 4$, $0 < a < 1$) and made significant applications to the global problems of differential geometry.

2.2. *Extension of the Monge–Ampère operator $f(x, u, Du)\det(u_{ij})$ and applications to the Monge–Ampère equations.* In this subsection we present a brief survey of the Alexandrov paper [23]. Let G be a convex bounded domain in R^n. Alexandrov writes the Monge–Ampère equation in the form

(2.6) $$f(x, u, Du)\det(u_{ij}) = h(x)$$

and assumes that the functions $h(x)$ and $f(x, u, p)$ satisfy the following conditions:

1. $h(x)$ and $f(x, u, p)$ are nonnegative functions, the first one in G, and the second one in $G \times R \times R^n$. The functions $h(x)$ and $f(x, u, p)$ can also take infinite values.

2. There exists a summable function $f_0(p)$ such that $f(x, u, p) \leq f_0(p)$ for all $(x, u, p) \in H$, where H is a closed bounded subdomain in $G \times R \times R^n$.

3. There exist a number z_0 and a function $f_1(p) \geq 0$ such that $\int_{R^n} f_1(p)\,dp > 0$ and $f(x, u, p) \geq f_1(p)$ for all $x \in G$, $u \leq z_0$, $p \in R^n$.

Note if the function $f(x, u, p) > 0$ for all $x \in G$, $u \in R$, $p \in R^n$, then equation (2.6) can be written as

(2.7) $$\det(u_{ij}) = F(x, u, Du),$$

where

(2.8) $$F(x, u, p) = h(x)/f(x, u, p).$$

Since
$$\frac{h(x)}{f_0(p)} \le F(x, u, p) \le \frac{h(x)}{f_1(p)},$$
where $f_0(p)$ and $f_1(p)$ are functions mentioned in conditions 2, 3, the class of equations introduced by Alexandrov is close to the class of Monge–Ampère equations investigated in §1.

Let $z(x) \in W^+(G)$ and let S_z be the graph of this function. We denote by $(x(p), z(p))$ the point of S_z lying on the supporting hyperplane with normal $\{p, -1\} = (p_1, p_2, \ldots, p_n, -1)$. Evidently, the mapping $p \to (x(p), z(p))$ can be considered as inverse for the normal mapping χ_z. The functions $x(p)$, $z(p)$ take more than one value for some $p \in \chi_z(G)$ if and only if the supporting hyperplane with normal $\{p, -1\}$ touches S_z more than in one point. From the property of the normal mapping it follows that the n-dimensional Lebesgue measure of such points $p \in \chi_z(G)$ is equal to zero (see [22, 6]). Thus the function $f(x(p), z(p), p)$ is definite uniquely for almost all $p \in \chi_z(G)$. The set function

(2.9) $$a(f, z, e) = \int_{x_2(e)} f(x(p), z(p), p) \, dp$$

is defined for all subsets $e \subset G$ and takes only nonnegative values. Evidently, $a(f, z, e) < +\infty$ if e is a closed subset of G.

Let $z(x)$ be any convex function defined in G and $\mu(e)$ be a nonnegative completely additive set function of Borel subsets of G; then the equation

(2.10) $$a(f, z, e) = \mu(e)$$

is the extension of the Monge–Ampère equation (2.6) to the class of all convex functions $W^+(G)$. Every solution of the equation (2.10) is called a *weak solution* for the equation (2.6).

According to condition 3 there exist a number z_0 and a function $f_1(p)$ such that $f(x, z, p) \ge f_1(p)$ (for all $x \in G$ and $z \le z_0$) and
$$\int_{R^n} f_1(p) \, dp > 0.$$

Let
$$V(f) = \sup \int_{R^n} f_1(p) \, dp$$

where sup is taken for all admissible z_0 and for all corresponding functions $f_1(p)$. The case $V(f) = +\infty$ is not excluded.

The main theorem proved by Alexandrov [23] is as follows:

THEOREM 7 (ALEXANDROV). *Let $\mu(e)$ be a completely additive nonnegative set function satisfying the condition*

(2.11) $$\mu(G) < V(f).$$

Then there exists at least one weak solution $z(x) \in W^+(G)$ of equation (2.6).

The proof of this theorem (see [23]) consists of two parts. First the Dirichlet problem in the class of convex polyhedrons was solved and then a weak solution of the equation (2.6) was constructed as the limit of such polyhedral solutions. This method permits one to obtain only the existence of at least one weak solution for the equation (2.6) but does not give any information concerning the satisfaction of the boundary condition for the Dirichlet problem by such solutions. In his papers, Alexandrov did not consider the satisfaction of the boundary Dirichlet condition for weak solutions of the equation (2.6). As we said above this problem was investigated for the n-dimensional Monge–Ampère equations by Bakelman, Cheng and Yau (see subsection 2.1).

2.3. *The Dirichlet problem for the Monge–Ampère equation* $\det(u_{ij}) = f(x, u, Du)$. In this subsection we consider the existence theorem for the Dirichlet problem

(2.12) $$\det(u_{ij}) = f(x, u, Du),$$

(2.13) $$u|_{\partial G} = h(x).$$

We suppose that the bounded convex domain G satisfies Assumption 1 (see subsection 1.4). In this subsection we also suppose that the following conditions with respect to the functions $f(x, u, p)$ and $h(x)$ are fulfilled:

Assumption 3'. (1) $f(x, u, p)$ is continuous in $\overline{G} \times R \times R^n$ and the inequalities

(2.14) $$0 \leq f(x, u, p) \leq \varphi(x)/R(p)$$

hold in the same domain $\overline{G} \times R \times R^n$, where $\varphi(x)$ is nonnegative and locally summable in the gradient space P^n. (Remember that $\overline{G} = G \cup \partial G$ is the closure of G.)

(2) The function $R(p)$ satisfies Assumption 2 (see subsection 1.4).

(3) There exists a sufficient small neighborhood U of ∂G such that the inequality

(2.15) $$\varphi(x) \leq a(\mathrm{dist}(x, \partial G))^\lambda$$

holds for every $x \in U \cap G$, where $\lambda = \mathrm{const} \geq 0$.

We will consider the generalized solutions of the Dirichlet problem (2.12)–(2.13). We call the convex function $u(x) \in W^+(G)$ a *generalized solution of the Dirichlet problem* (2.12)–(2.13) if the border of $u(x)$ coincides with a given continuous function $h(x)$ on ∂G, $u(x)$ satisfies the equation (2.12) almost everywhere in G, and $\omega(R, u, e)$ is an absolutely continuous set function.

Assumption 4, finally, is as follows: $h(x)$ is a continuous function for all $x \in \partial G$.

Now we present our main theorem:

THEOREM 8. *The Dirichlet problem* (2.12)–(2.13) *has at least one generalized solution* $u(x) \in W^+(\overline{G})$ *if the following assumptions are fulfilled*:

(A) *The bounded convex domain G in R^n satisfies Assumption 1, i.e. ∂G has a parabolic support of order not more than* $\tau = \mathrm{const} \geq 0$.

(B) *The function $f(x, u, p)$ satisfies together with functions $\varphi(x)$ and $R(p)$ the conditions* (1), (2), (3) (*see the present subsection* 2.3).

(C) $h(x)$ *is a continuous function on* ∂G.

(D)

(2.16) $$\int_G \varphi(x)\, dx < A(R) = \int_{P^n} R(p)\, dp.$$

(E) *Let* $K = (n + \tau + 1)/(\tau + 2) + \lambda/2$; *then*
$$k \leq K \quad \text{for } k < 1 \text{ and } k \geq n/2$$
and
$$k < K \quad \text{if } 1 \leq k < n/2,$$
where k, λ, τ *are the numbers from Assumptions* 1, 2, 3'.

REMARK. Theorem 8 was proved by Bakelman in 1983. The important particular case $\tau = 0$ of this theorem, i.e., G is strictly convex and has a uniform spherical support in all points of ∂G, was proved in his papers [17, 25] earlier.

PROOF. We denote by $W_h^+(\overline{G})$ the set of all convex functions $u(x)$ satisfying the condition $\nu_u(x) = h(x)$ where $\nu_u(x)$ is the border of $u(x)$. The set $W_h^+(G)$ is not empty, because the Dirichlet problem
$$\det(z_{ij}) = \frac{\varphi(x)}{R(Dz)}, \qquad z|_{\partial G} = h(x)$$
has the generalized solution $z(x) \in W_h^+(\overline{G})$. This follows directly from Theorem 6'.

Evidently, $W_h^+(\overline{G})$ is a convex set in the space $C(\overline{G})$. Let

(2.17) $$F_u(x) = f(x, u, Du) R(Du)$$

for every function $u(x) \in W^+(\overline{G})$. The function $F_u(x)$ is nonpositive in G and

(2.18) $$F_u(x) \leq \varphi(x)$$

almost everywhere in G. If $u(x) \in W^+(\overline{G}) \cap C^1(\overline{G})$, then $F_u(x) \in C(\overline{G})$ and $F_u(x) \in L(G)$, because
$$\int_G F_u(x)\, dx \leq \int_G \varphi(x)\, dx.$$

If $u(x) \in W^+(\overline{G})$, then the same integral inequality can be proved by approximation of functions $u_m(x) \in W^+(\overline{G}) \cap C^1(\overline{G})$. Hence the Dirichlet problem

(2.19) $$\det(z_{ij}) = F_u(x)/R(Dz),$$

(2.20) $$z|_{\partial G} = h(x)$$

has exactly one generalized solution $z(x) \in W_h^+(\overline{G})$, where $u(x) \in W_h^+(\overline{G})$. This statement directly follows from Theorem 6', because all conditions of this theorem are fulfilled.

Thus we have constructed the operator B: $W_h^+(\overline{G}) \to W_h^+(\overline{G})$ and $z(x) = B(u(x))$. Clearly, the fixed points of the operator B are generalized solutions of our initial Dirichlet problem (2.12)–(2.13). The inequality $\int_G \varphi(x)\,dx < A(R)$ yields the estimates

$$\inf h(x) - T_R\left(\int_G \varphi(x)\,dx\right) \operatorname{diam} G \leqslant z(x) = B(u(x)) \leqslant \sup_{\partial G} h(x)$$

for every function $u(x) \in W_h^+(\overline{G})$ (see §1). Thus the set $B(W_h^+(\overline{G}))$ is bounded in $C(\overline{G})$. Therefore, we can take a subsequence $z_{i_m}(x) = B(u_{i_m}(x))$ for each sequence $u_1(x), u_2(x), \ldots, u_m(x), \ldots \in W_h^+(\overline{G})$ converging for all $x \in \overline{G}$. But possibly it converges nonuniformly. From the conditions of Theorem 8 and then from Theorem 5 it follows that the sequence $z_{i_m}(x)$ converges uniformly to some function $z_0(x) \in W_h^+(\overline{G})$. Thus the operator B is compact.

Let the functions $u_m(x) \in W_h^+(\overline{G})$ and let them converge to some function $u_0(x) \in W_h^+(\overline{G})$. Then the set of functions $v_m(x) = B(u_m(x))$ is compact in $W_h^+(\overline{G})$ considered as a subspace of $C(\overline{G})$. This fact is proved in just the same way as the compactness of the set $B(W_h^+(\overline{G}))$. Let $v_0(x) = B(u_0(x))$. We take some uniformly convergent subsequence $v_{m_k}(x)$ in $C(\overline{G})$. Let $\bar{v}(x)$ be the limit of this subsequence. From the conditions of Theorem 8 and then Theorem 5 it follows that $\bar{v}(x) \in W_h^+(\overline{G})$.

Using the properties of the convergent sequence of convex functions and their R-curvatures we obtain that $\bar{v}(x)$ and $v_0(x)$ are generalized solutions for one and the same equation,

$$\det(v_{ij}) = F_{u_0}(x)/R(Dv),$$

and the borders of $\bar{v}(x)$ and $v_0(x)$ coincide. Therefore, these functions coincide in G and the operator B is continuous.

Now all the conditions of Schauder's principle are fulfilled and the Dirichlet problem (2.12)–(2.13) has at least one generalized solution. Theorem 8 is proved.

3. Existence of several different generalized solutions of the Dirichlet problem for the equation $\det(u_{ij}) = f(x, u, Du)$.

3.1. *Introduction.* Theorem 8 provides the existence of at least one generalized solution of the Dirichlet problem

(3.1) $$\det(u_{ij}) = f(x, u, Du),$$

(3.2) $$u|_{\partial G} = h(x) \in C(\partial G),$$

if the function $f(x, u, p)$ satisfies the inequalities

$$0 \leqslant f(x, u, p) \leqslant \varphi(x) R(p)$$

for all $(x, u, p) \in \overline{G} \times R \times P^n$.

In this section we consider the development of Theorem 8 in two directions. The first one is the extension of Theorem 8 to the wider classes of the nonnegative functions $f(x, u, p)$ which can infinitely increase together with $|u| \to +\infty$. The second one is the problem of existence of several different generalized solutions for the Dirichlet problem (3.1)–(3.2). The first theorems in such way were obtained by Bakelman and Krasnoselskiĭ [32] in 1960 (see also the monographs [33, 17]) for two-dimensional Monge–Ampère equations. The improvement and development of these theorems to n-dimensional Monge–Ampère equations were done by Bakelman [25]. In [25] the n-dimensional Monge–Ampère equations in strictly convex domains with the spherical support were considered (i.e. $\tau = 0$ in Assumption 1, §1.4) and the functions $f(x, u, p)$ were estimated from above by $R(p) = 1$. Thus we present the significant improvement of the main results of [25] in the present paper.

Now we formulate lemmas of fixed points of positive operators acting in convex cones of Banach spaces. These lemmas were proved by Krasnoselskii [33], see also [25, 17].

Let B be a Banach space and K be a convex cone in B. We consider only operators $F: K \to K$. Let S be the intersection of K with the sphere $\|x\|_B = \rho$, where ρ is a positive number.

LEMMA A. *Let $F: K \to K$ be a compact and continuous operator in the convex cone K of a Banach space B. If there exists some positive number ρ such that*

$$(3.3) \qquad \|F(x)\|_B \leqslant \|x\|_B$$

for all $x \in S_\rho$ then the operator F has at least one fixed point $x_0 \in K$ and $\|x\|_B \leqslant \rho$.

This lemma follows from the Schauder principle of fixed points in Banach spaces.

LEMMA B. *Let $F: K \to K$ be a compact and continuous operator in the convex cone K of a Banach space B and let ρ_1 and ρ_2 be two different positive numbers. If assumptions $\|F(x)\|_B \leqslant \|x\|_B$ for all $x \in S_{\rho_1}$ and $\|F(x)\|_B \geqslant \|x\|_B$ for all $x \in S_{\rho_2}$ hold, then the operator F has at least one fixed point $x_0 \in K$ such that either*

$$\rho_1 \leqslant \|x_0\|_B \leqslant \rho_2 \quad \text{if } \rho_1 < \rho_2,$$

or

$$\rho_1 \geqslant \|x_0\|_B \geqslant \rho_2 \quad \text{if } \rho_1 > \rho_2.$$

3.2. *Existence Theorem for the Dirichlet problem* (3.4)–(3.5). We consider the Dirichlet problem

$$(3.4) \qquad \det(u_{ij}) = f(x, u, Du),$$

$$(3.5) \qquad u|_{\partial G} = 0.$$

We suppose that the following assumptions hold:

(A) The domain G satisfies Assumption 1 (see subsection 1.4), i.e. G is a bounded convex domain and ∂G has a parabolic support of the order not more than $\tau = \text{const} \geqslant 0$.

(B) We denote by q_G the infimum of distances between parallel supporting hyperplanes of ∂G with opposite outward normals. Let $\varepsilon > 0$ be any number less than $q_G/10$. We introduce the functions

(3.6) $$\varphi(x, \lambda, \varepsilon) = \begin{cases} 1 & \text{if } \operatorname{dist}(x, \partial G) \geq \varepsilon, \\ [\operatorname{dist}(x, \partial G)]^\lambda & \text{if } \operatorname{dist}(x, \partial G) < \varepsilon, \end{cases}$$

and

(3.7) $$R_k(p) = \begin{cases} 1 & \text{if } |p| \leq 1, \\ 1/|p|^{2k} & \text{if } |p| > 1, \end{cases}$$

where $\lambda = \text{const} \geq 0$ and $0 \leq k = \text{const} < n/2$.

Assumption (B) is as follows: The function $f(x, u, p)$ is continuous in $\overline{G} \times R \times P^n$, and the inequalities

(3.8) $$0 \leq f(x, u, p) \leq \frac{(a|u| + b)^\alpha}{R_k(p)} \varphi(x, \lambda, \varepsilon)$$

hold for all $x \in \overline{G}$, $u \leq 0$, $p \in P^n$, where $a = \text{const} \geq 0$, $b = \text{const} \geq 0$, $a^2 + b^2 > 0$, $\alpha = \text{const} \geq 0$.

THEOREM 9. *Let the assumptions (A) (§1.4) and (B) be fulfilled. Then the Dirichlet problem* (3.4)–(3.5) *has at least one convex generalized solution if the inequalities*

(3.9) $$0 \leq \alpha < n - 2k,$$

(3.10) $$k < \frac{n + \tau + 1}{\tau + 2} + \frac{\lambda}{2}$$

hold.

PROOF. We denote by K the set of all convex functions in G, satisfying the condition (3.5) in the classical meaning. The set K is not empty, since the function $z = 0$ belongs to K. Clearly, K is a convex cone in $C(\overline{G})$. Evidently, all convex functions $u(x) \in K$ are nonpositive in G. Let

(3.11) $$F_u(x) = f(x, u(x), Du(x)) R_k(Du(x)),$$

where $u(x)$ is any function from the cone K. From Assumption (B) it follows that $f_u(x)$ is nonnegative and summable in G and

(3.12) $$\int_G F_u(x)\, dx \leq \int_G (a|u(x)| + b)^\alpha dx$$
$$\leq (a\|u\|_{C(\overline{G})} + b)^\alpha \cdot \operatorname{mes} G.$$

We consider the Dirichlet problem

(3.13) $$\det(z_{ij}) = F_u(x)/R_k(Dz), \qquad z|_{\partial G} = 0.$$

Evidently, it is the special case of the Dirichlet problem (1.50)–(1.51), because Assumptions 1, 2, 3 (see subsection 1.4) with respect to functions $F_u(x)$ and $R_k(p)$ hold. Since $k < n/2$, then

(3.14) $$A(R_k) = \int_{p \leq 1} dp + \int_{|p| > 1} \frac{dp}{|p|^{2k}} = +\infty.$$

Therefore,

(3.15) $$\int_G F_u(x)\,dx \leq \left(a\|u(x)\|_{C(\bar{G})} + b\right)^\alpha \operatorname{mes} G < A(R_k) = +\infty.$$

Thus the inequality (1.49) is fulfilled. From the condition (3.10) it follows that

$$k < \frac{n + \tau + 1}{\tau + 2} + \frac{\lambda}{2}.$$

Thus all conditions of Theorem 6' hold. Hence the Dirichlet problem (3.14) has only one generalized solution $z(x) \in K$. Thus the operator $A: K \to K$ is defined such that $z(x) = A(u(x))$, where $z(x)$ is a generalized solution of the Dirichlet problem (3.14). From Theorem 4 (see subsection 1.3) we obtain the estimates

(3.16) $$-T_{R_k}(\omega_u)\operatorname{diam} G \leq z(x) = A(u(x)) \leq 0$$

for all $x \in \bar{G}$, where

(3.17) $$\omega_u = \int_G F_u(x)\,dx$$

and $T_R(\tau)$ is the inverse for the function

$$g_{R_k}(\rho) = \int_{|p| \leq \rho} R_k(p)\,dp = \mu_n\left(1 + \frac{n}{n - 2k}(\rho^{n-2k} - 1)\right)$$

(μ_n is the volume of the unit ball in R^n). Since $0 \leq 2k/n < 1$, $0 < (n - 2k)/n \leq 1$ and $g_{R_k}(+\infty) = A(R_k) = +\infty$, then

(3.18) $$T_{R_k}(\tau) = \left[1 + \frac{n - 2k}{n}\left(\frac{\tau}{\mu_n} - 1\right)\right]^{1/(n-2k)} \leq \left[\left(\frac{\tau}{\mu_n}\right) + 1\right]^{1/(n-2k)}$$

for all $\tau \in [0, +\infty)$.

Thus from (3.15)–(3.18) it follows that

(3.19) $$\|A(u(x))\|_{C(\bar{G})} \leq \left[\frac{\omega_u + \mu_n}{\mu_n}\right]^{1/(n-2k)} \operatorname{diam} G$$

$$\leq \left[\frac{1}{\mu_n}\left(\left(a\|u(x)\|_{C(\bar{G})} + b\right)^\alpha \operatorname{mes} G + \mu_n\right)\right]^{1/(n-2k)}.$$

Therefore the operator $A: K \to K$ maps every bounded subset $Q \subset K$ in a bounded subset $A(Q)$ of K. Using the same considerations as in the proof of Theorem 8 we obtain that the operator $A: K \to K$ is compact and continuous.

Since $\alpha < n - 2k$, $a \geq 0$, $b \geq 0$ and $a^2 + b^2 > 0$, we can find a sufficiently big positive number r_0 such that

(3.20) $$\left[\frac{1}{\mu_n}\left[\left(a\|u(x)\|_{C(\bar{G})} + b\right)^\alpha \operatorname{mes} G + \mu_n\right]\right]^{1/(n-2k)} \operatorname{diam} G < r_0.$$

Then from (3.19) we obtain

$$\|A(u(x))\|_{C(\bar{G})} < \|u(x)\|_{C(\bar{G})}$$

for all $u(x) \in S_{r_0}$ where S_{r_0} is the intersection of the cone K with the sphere $\|x\|_{C(\overline{G})} = r_0$ in $C(\overline{G})$.

Now from Lemma A it follows that the Dirichlet problem (3.4)–(3.5) has at least one generalized solution $z(x) \in K$ and $\|z(x)\|_{C(\overline{G})} \leq r_0$.

Theorem 9 is proved.

REMARKS. (1) Let $\alpha \geq 0$ and $2k \geq 0$ be the orders of the growth of the functions $(a|u| + b)^\alpha$ and

$$\frac{1}{R_k(p)} = \begin{cases} 1 & \text{if } |p| \leq 1; \\ |p|^{2k} & \text{if } |p| > 1. \end{cases}$$

Then the inequalities $0 \leq \alpha + 2k < n$ show the largest values of α and $2k$ in Assumption (B) providing the correctness of Theorem 9.

(2) Theorem 9 is also true if the inequalities (3.8) hold for all $x \in \overline{G}$, $|u| \leq r_0$, $p \in R^n$, where $a = \text{const} \geq 0$, $b = \text{const} \geq 0$, $a^2 + b^2 > 0$, $0 \leq \alpha < n - 2k$, and the positive number r_0 satisfies the inequality (3.20).

3.3. *Existence of several different generalized solutions for the Dirichlet problem* (3.4)–(3.5). Let G be a convex open bounded domain in R^n and let Q be a bounded closed n-dimensional ball inside G and $\text{dist}(Q, \partial G) = \delta > 0$.

LEMMA C. *The inequality*

$$(3.21) \qquad \inf_Q |u(x)| \geq \frac{\delta}{\text{diam } G} \|u(x)\|_{C(\overline{G})}$$

holds for all convex functions $u(x) \in K$, *where* K *is the set of convex functions satisfying the boundary condition* $u|_{\partial G} = 0$ *(evidently* $0 < \delta/\text{diam } G < 1$).

PROOF. The lemma is evident for convex functions $v(x) \in K$, whose graphs are convex cones with the base ∂G. Now let $u(x)$ be any convex function belonging to K. Since the lemma is trivial for the function $u(x) = 0$, we are concerned only with the case $\|u(x)\|_{C(\overline{G})} > 0$. Let $x_0 \in G$ be the point where

$$|u(x_0)| = \|u(x)\|_{C(\overline{G})}.$$

We denote by $v(x)$ the convex function whose graph is the convex cone with the base ∂G and the vertex $(x_0, u(x_0))$. Evidently,

$$(3.22) \qquad \|v(x)\|_{C(\overline{G})} = |u(x_0)| = \|u(x)\|_{C(\overline{G})}$$

and $0 \geq v(x) \geq u(x)$ for all $x \in G$. Now we complete the proof of Lemma C by the following chain of equalities and inequalities:

$$\inf_Q |u(x)| = |u(x^*)| \geq |v(x^*)| \geq \inf_Q |v(x)|$$

$$\geq \frac{\delta}{\text{diam } G} \|v(x)\|_{C(\overline{G})} = \frac{\delta}{\text{diam } G} \|u(x)\|_{C(\overline{G})},$$

where x is the point of Q such that

$$\inf_Q |u(x)| = |u(x^*)|.$$

The lemma is proved.

Now we present the assumptions that we will use below. They are as follows:

(A) The boundary of the open convex bounded domain G has a parabolic support not more than $\tau = \text{const} \geq 0$ (see Assumption 1, subsection 1.4).

(B') The function $f(x, u, p)$ is continuous in $\overline{G} \times R \times P^n$ and the inequalities

$$(3.23) \qquad 0 \leq f(x, u, p) \leq (a|u| + b)^\alpha$$

holds, where $a = \text{const} \geq 0$, $b = \text{const} \geq 0$, $a^2 + b^2 > 0$, $u \leq 0$, $0 \leq \alpha = \text{const} < n$.

Assumption (B') can be considered as the particular case of Assumption (B) (see subsection 3.2) according to $k = 0$. We can use $\lambda = 0$ in our case because the inequality (3.10) is reduced to the inequality $0 < (n + \tau + 1)/(\tau + 2)$ which is correct for all $\tau \in [0, +\infty)$.

We will suppose that assumptions (A) and (B') hold in all our further considerations together with the new assumption (C).

(C) The inequalities

$$(3.24) \qquad \gamma_0 |u|^\beta \leq f(x, u, p) \leq (\gamma_1 |u| + \gamma_2)^\beta$$

hold for all $x \in \overline{G}$, $-u_0 \leq u \leq -u_1$, $p \in P^n$, where $\gamma_0 = \text{const} > 0$, $\beta = \text{const} > n$ and $u_0 = \text{const} > u_1 = \text{const} > 0$, $\gamma_1 = \text{const} > 0$, $\gamma_2 = \text{const} \geq 0$.

We take a convex function $u(x) \in K$ such that

$$(3.25) \qquad u_1 \leq \frac{\delta}{\operatorname{diam} G} \|u\|_{C(\overline{G})} < \|u\|_{C(\overline{G})} \leq u_0,$$

and consider two Dirichlet problems

$$(3.26) \qquad \det(v_{ij}) = \gamma_0 \left(\frac{\delta}{\operatorname{diam} G} \right)^\beta \|u(x)\|_{C(\overline{G})}^\beta,$$

$$(3.27) \qquad v|_{\partial G} = -\frac{\delta}{\operatorname{diam} G} \|u(x)\|_{C(\overline{G})},$$

in the ball Q and

$$(3.28) \qquad \det(z_{ij}) = F_u(x),$$

$$(3.29) \qquad z|_{\partial Q} = 0,$$

in G, where $F_u(x) = f(x, u(x), Du(x))$. From Theorem 6' it follows that there exists only one generalized solution $z(x)$ of the problem (3.28)–(3.29). The function

$$(3.30) \qquad v(x) = \frac{1}{2} B \left(\sum_{i=1}^{n} x_i^2 - r^2 \right) - \frac{\delta}{\operatorname{diam} G} \|u(x)\|_{C(\overline{G})}$$

is the unique generalized solution of the Dirichlet problem (3.28)–(3.29), where $B = \gamma_0^{1/n} (\delta/\operatorname{diam} G)^{\beta/n} \|u(x)\|_{C(\overline{G})}^{\beta/n}$ and r is the radius of the ball Q.

From (3.25)–(3.27) it follows that

$$(3.31) \qquad -\frac{\delta}{\operatorname{diam} G} \|u(x)\|_{C(\overline{G})} \geq v(x) \geq z(x)$$

for all $x \in Q$. Therefore $\|v(x)\|_{C(Q)} \leq \|z(x)\|_{C(\overline{G})}$. From (3.30) we obtain

(3.32) $\quad \|z(x)\|_{C(\overline{G})} \geq \|v(x)\|_{C(Q)}$

$$= \frac{1}{2}\gamma_0^{1/n}\left(\frac{\delta}{\operatorname{diam} G}\right)^{\beta/n} r^2 \|u(x)\|_{C(\overline{G})}^{\beta/n} + \frac{\delta}{\operatorname{diam} G}\|u(x)\|_{C(\overline{G})}.$$

Thus we establish the nontrivial estimate (3.32) for $\|z(x)\|_{C(\overline{G})}$ from below, where $z(x)$ is the solution of the Dirichlet problem (3.28)–(3.29).

Now we prove the existence of two different generalized solutions of the Dirichlet problem

(3.33) $\quad\quad\quad\quad\quad\quad \det(u_{ij}) = f(x, u, Du),$

(3.34) $\quad\quad\quad\quad\quad\quad u|_{\partial G} = 0.$

THEOREM 10. *We assume that the following conditions are fulfilled*:

(1) *Let the domain G satisfy assumption (A) and let $Q \subset G$ be a closed n-ball with radius r and let $\delta = \operatorname{dist}(Q, \partial G) > 0$.*

(2) *The function $f(x, u, p)$ is nonnegative and continuous for all $x \in \overline{G}$, $u \leq 0$, $p \in P^n$, and the inequality*

(3.35) $\quad\quad\quad\quad\quad\quad f(x, u, p) \leq (a|u| + b)^\alpha$

holds for all $x \in G$, $-r_0 \leq u \leq 0$, $p \in P^n$, where $a = \operatorname{const} \geq 0$, $b = \operatorname{const} \geq 0$, $a^2 + b^2 \geq 1$, $0 \leq \alpha < n$, and the number r_0 satisfies the inequality

(3.36) $\quad\quad\quad \left[\frac{(ar_0 + b)}{\mu_n}\right]^{1/n} (\operatorname{mes} G)^{1/n} \operatorname{diam} G < r_0.$

(3) *Let $r_1 > 0$ be a number such that*

(3.38) $\quad \dfrac{\delta}{\operatorname{diam} G} r_1 > r_0 \quad \text{and} \quad r_1 < \dfrac{1}{2}\gamma_0^{1/n}\left(\dfrac{\delta}{\operatorname{diam} G}\right)^{\beta/n} r^2 \left(\dfrac{\delta r_1}{\operatorname{diam} G}\right)^{\beta/n}$

where $\beta = \operatorname{const} > n$.

(4) *The inequality $f(x, u, p) \leq \varphi(u)$ holds for all $x \in \overline{G}$, $u \in (-\infty, 0]$, $p \in P^n$, where $\varphi(u)$ is a strictly positive and continuous function of u. If $r_0 \geq 1$, then we can take $\varphi(u) = (a|u| + b)^\beta$.*

If the conditions (1)–(4) are fulfilled then the Dirichlet problem (3.33)–(3.34) has at least two different convex generalized solutions.

PROOF. First of all we note that the inequalities

$$\gamma_0 |u|^\beta \leq f(x, u, p) \leq (a|u| + b)^\beta$$

hold for all $x \in \overline{G}$, $-r_1 \leq u \leq -(\delta/\operatorname{diam} G)r_1$, $p \in P^n$, where $\beta = \operatorname{const} > n$. This is the consequence from the condition (3).

Now consider the Dirichlet problem (3.28)–(3.29), where $u(x)$ is any function belonging to the convex cone $K \subset C(\overline{G})$. From the conditions of the present theorem it follows that the problem (3.28)–(3.29) has only one generalized solution $z(x)$ and we can consider the operator $A: K \to K$ constructed in the proof of Theorem 9. This operator is compact and continuous. We establish this fact by means of the same methods used in the proof of Theorem 9.

From the condition (2) together with Theorem 9 it follows that operator A has a fixed point $u_1(x) \in K$ such that $\|u_1(x)\|_{C(\overline{G})} \leqslant r_0$.

Now from the conditions (3) and (4), Lemma B, and the inequality (3.32) we obtain the existence of another fixed point $u_2(x) \in K$ such that

$$r_0 < \frac{\delta}{\operatorname{diam} G} r_1 \leqslant \|u_2(x)\|_{C(\overline{G})} \leqslant r_1.$$

Theorem 10 is proved.

The methods used in subsection 3.3 permit one to establish the existence of an infinite number of different generalized solutions if the function $f(x, u, p)$ satisfies the suitable conditions.

References

1. S. Bernstein, *Collected complete works*, Vol. 3, Akad. Sci. SSSR, 1960, 1–439.
2. H. Weyl, *Über die Bestimmung einer geschlossenen konvexen Fläche durch Linien Element*, Vierteljaresschrift der Naturforschenden Gesellschaft in Zurich. **61** (1915), 40–72.
3. H. Lewy, *On differential geometry in the large*. I (*Minkowski problem*), Trans. Amer. Math. Soc. **43** (1938), 258–270.
4. _____, *A priori limitations for solutions of Monge–Ampère equations*. I, II, Trans. Amer. Math. Soc. **37** (1935), **41** (1937).
5. A. V. Pogorelov, *The Minkowski multidimensional problem*, Wiley, New York, 1978.
6. S. Y. Cheng and S. T. Yau, *On the regularity of the Monge–Ampère equation* $\det(u_{ij}) = F(X, u)$, Comm. Pure Appl. Math. **30** (1977), 41–68.
7. L. Caffarelli, L. Nirenberg and J. Spruck, *The Dirichlet problem for nonlinear second order equations*. I, *Monge–Ampère equation*. II, *Complex Monge–Ampère and uniformly elliptic equations* (the part II together with J. J. Kohn), Comm. Pure Appl. Math. (to appear).
8. P. L. Lions, *Sur les equations de Monge–Ampère*. I, Manuscripta Math. **41** (1983), II (to appear).
9. N. V. Krylov, *On degenerate non-linear elliptic equations*, Mat. Sb. **162** (1983), 311–330.
10. H. Minkowski, *Volumen and Oberfläche*, Math. Ann. **57** (1903), 47–495. Gessamelte Abhandlugen II (1911), 230–276.
11. I. Bakelman, *Generalized solutions of the Monge–Ampère equations*, Dokl. Akad. Nauk SSSR **114**, 6 (1957), 1143–1145.
12. _____, *A priori estimates and regularity of generalized solutions of Monge–Ampère equations*, Dokl. Akad. Nauk SSSR **116** (1957), 719–722.
13. _____, *On the theory of Monge–Ampère equations*, Vestnik Leningrad. Univ. Mat. Mekh. Astronom. **13**, 1 (1958), 25–38.
14. _____, *Regularity of solutions of Monge–Ampère equations*, Sci. Notes Leningr. Ped. Inst. **166** (1968), 143–184.
15. _____, *The Dirichlet problem for equations of Monge–Ampère type and their n-dimensional analogues*, Dokl. Akad. Nauk SSSR **126** (1959), 923–926.
16. _____, *Equations with the Monge–Ampère operator*, vol. II, Proc. Fourth All-Union Math. Congress 1961, Leningrad, "Nauka", Moscow, 1964, pp. 469–480.
17. _____, *Geometric methods of solving of elliptic equations*, "Nauka", Moscow, 1965, pp. 1–340.
18. I. Bakelman and I. Guberman, *The Dirichlet problem with the Monge–Ampère operator*, Sibirsk. Mat. Zh. **4** (1963), 1208–1220.
19. A. V. Pogorelov, *Monge–Ampère equations of elliptic type*, Groningen, Noordhoff, 1964.
20. _____, *Extrinsic geometry of convex surfaces*, Transl. Math. Mono., Vol. 35, Amer. Math. Soc., Providence, R. I., 1972.
21. A. D. Alexandrov, *Intrinsic geometry of convex surfaces*, GTTI, Moscow-Leningrad, 1948.
22. _____, *Convex polyhedrons*, GTTI, Moscow-Leningrad, 1950.

23. _____, *Dirichlet problem for the equation* $\det(u_{ij}) = 1$, Vestnik Leningrad. Univ. Mat. Mekh. Astronom. **13** (1958), 3–24.

24. I. Bakelman, *Topological methods in the theory of the Monge–Ampère equations*, Proc. Mathematische Arbeitstagung 1980, Universität Bonn, SFB 40, Theoretische Mathematik, 1980.

25. _____, *The Dirichlet problem for the elliptic Monge–Ampère equations and related problems in the theory of quasi-linear elliptic equations*, Proceedings of a Seminar Monge–Ampère Equations and Related Topics, Firenze, September-October 1980, Istituto Nazionale di Alta Matematica, Roma, 1982, pp. 1–78.

26. _____, *Applications of the Monge–Ampère operators to the Dirichlet problem for quasilinear equations*, Ann. of Math. Stud. **102** (1982), 239–258.

27. _____, *R-curvature, estimates and the stability of solutions of the Dirichlet problem for elliptic equations*, J. Differential Equations **43** (1982), 106–133.

28. _____, *Variational problems and elliptic Monge–Ampère equations*, J. Differential Geom. **19** (1984).

29. S. Y. Cheng and S. T. Yau, *On the regularity of the solution of the n-dimensional Minkowski problem*, Comm. Pure Appl. Math. **19** (1976), 495–516.

30. _____, *The real Monge–Ampère equation and affine flat structures*, vol. 1, Proc. Beijing Sympos. on Differential Geometry and Differential Equations, Science Press Beijing, New York, 1982, pp. 339–370.

31. A. V. Pogorelov, *The Dirichlet problem for the n-dimensional analogue of the Monge–Ampère equations*, Dokl. Akad. Nauk SSSR **12** (1971), 1727–1731.

32. I. Bakelman and M. A. Krasnoselskii, *Non-trivial solutions of the Dirichlet problem with the Monge–Ampère operator*, Dokl. Akad. Nauk SSSR **136** (1961), 161–163.

33. M. A. Krasnoselskii, *Positive solutions of operator equations*, Phismath Publiching Haus, Moscow, 1962, 1–394.

34. I. Bakelman, *The boundary value problems for n-dimensional Monge–Ampère equations*; *n-dimensional plasticity equation*, IHES, Bures-sur-Yvette, (1984), 1–107 (preprint).

35. S. T. Yau, *Survey on partial differential equations in differential geometry*, Ann. of Math. Stud. **102** (1982), 3–73.

TEXAS A & M UNIVERSITY

Existence and Containment of Solutions to Parabolic Systems

PETER W. BATES

Introduction. The positive invariance result of H. Weinberger [8] for weakly coupled parabolic systems has been an extremely useful tool in developing the theory of such systems. H. Amann [1], K. Chueh, C. Conley and J. Smoller [5], J. Bebernes and K. Schmitt [4], and others, have extended some of the results in [8] and used the invariance principle to prove existence results.

In this note we generalize the invariance principle to allow the 'invariant' set to move under some flow. Thus, although there is no positively invariant set there may be a moving set which always contains a solution as it evolves in time. These ideas have been developed to some extent by the author in [2] and C. Reder in [7]. The 'containment principle' is then used to prove existence results similar to those by Bebernes and Schmitt [4]. Finally a Hukuhara–Kneser theorem is stated for these systems.

Basic assumptions and notation. Let $\Omega \subset \mathbf{R}^n$ be a bounded domain with boundary, $\partial\Omega$, of class $C^{2+\alpha}$ for some $\alpha \in (0,1)$. Let $T > 0$ and define
$$\pi_T = \Omega \times (0, T] \quad \text{and} \quad \Gamma_T = (\partial\Omega \times [0, T]) \cup (\Omega \times \{0\}).$$
For a smoothly bounded domain $D \subset \mathbf{R}^n \times \mathbf{R}$, let
$$C^{1,0}(\overline{D}) = \{u \in C(\overline{D}): \partial u/\partial x_i \in C(\overline{D}), 1 \leq i \leq n\}$$
with the norm
$$|u|_1 = |u| + \max_{1 \leq i \leq n} |\partial u/\partial x_i|,$$
where $|\cdot|$ represents the sup norm in $C(\overline{D})$. For nonnegative integers j and k, and numbers $\gamma, \delta \in [0,1)$, let $C^{j+\gamma,k+\delta}(\overline{D})$ denote the space of functions $u(x,t)$ which have, in x, a jth derivative which is Hölder continuous of degree γ, and in

---

1980 *Mathematics Subject Classification.* Primary 35K60, 35B45, 35B05.
Key words and phrases. Invariance, existence, comparison, Hukuhara–Kneser property.

© 1986 American Mathematical Society
0082-0717/86 $1.00 + $.25 per page

t, a kth derivative which is Hölder continuous of degree δ. The space $C^{j+\gamma, k+\delta}$ is a Banach space when equipped with an appropriate norm and if $i \in [0, j + \gamma)$, $m \in [0, k + \delta)$ then the inclusion $C^{j+\gamma, k+\delta} \subset C^{i,m}$ is completely continuous (see Ladyženskaja et al. [6]).

Let $L: C^{2,1} \to C$ be defined by

$$L = \frac{\partial}{\partial t} - \sum_{i,j=1}^{n} a_{ij}(x, t)\frac{\partial^2}{\partial x_i \partial x_j} + \sum_{i=1}^{n} b_i(x, t)\frac{\partial}{\partial x_i}$$

and assume $a_{ij} \in C^{1,\alpha/2}(\bar{\pi}_T)$, $b_i \in C^{\alpha, \alpha/2}(\bar{\pi}_T)$ for $1 \leq i, j \leq n$.

Assume that L is uniformly parabolic, i.e., there exist constants λ, μ with $0 < \lambda < \mu$ such that

$$\lambda |\xi|^2 \leq \sum_{i,j=1}^{n} a_{ij}(x, t)\xi_i \xi_j \leq \mu |\xi|^2$$

for all $\xi \in \mathbf{R}^n$.

For $u = (u_1, \ldots, u_m) \in \mathbf{R}^m$ and $p = (p_1, \ldots, p_n)$, $p_i \in \mathbf{R}^m$, $1 \leq i \leq n$, let $f: \bar{\pi}_T \times \mathbf{R}^m \times \mathbf{R}^{nm} \to \mathbf{R}^m$ be a map such that $f(x, t, u, p)$ is Hölder continuous of degrees α, $\alpha/2$, α, α in x, t, u, p, respectively.

Let $\psi: \Gamma_T \to \mathbf{R}^m$ be the restriction to Γ_T of a function of class $C^{2+\alpha, 1+\alpha/2}(\bar{\pi}_T)$. We shall consider the initial-boundary value problem (IBVP)

(1) $\qquad Lu = f(x, t, u, \nabla u), \qquad (x, t) \in \pi_T,$

(2) $\qquad u(x, t) = \psi(x, t), \qquad (x, t) \in \Gamma_T.$

Under assumptions outlined in the next section comparing f with a vector field associated with a system of ODEs, we shall prove the existence of solutions to (IBVP) and provide pointwise estimates for such solutions through the containment principle mentioned before. Notice that we are not assuming any Lipschitz condition on f.

Let $g: [0, T] \times \mathbf{R}^m \to \mathbf{R}^m$ be continuous and continuously differentiable in its second argument. Consider the system of ODEs

(3) $\qquad \dot{s} = g(t, s), \qquad 0 < t \leq T.$

Let S be a convex open set in $\mathbf{R}^m$. Suppose there is an open set $S_1 \supset \bar{S}$ such that solutions to (3) with initial values in S_1 exist on $[0, T]$ (this is so if S is compact and T is sufficiently small, or if g is globally Lipschitz continuous in s). Let $S(t)$ be the evolution of S under (3), i.e., $S(t) = \{s(t): s \text{ satisfies (3) and } s(0) \in S\}$, $0 \leq t \leq T$. Let $S_T = \{(t, s): s \in \partial S(t), 0 \leq t \leq T\}$. The following is not difficult to prove (see [2]).

LEMMA. *If $S(t)$ is convex for each $t \in [0, T]$ then there exists a field $n: S_T \to \mathbf{R}^m$ such that*

(i) *$n(t, s)$ is an outward unit normal to $S(t)$ at $s \in \partial S(t)$,*

(ii) *if s satisfies (3) and $s(t) \in \partial S(t)$ then $n(\cdot, s(\cdot)): [0, T] \to \mathbf{R}^m$ is continuous.*

Containment and existence. The notation and assumptions of the previous section will be maintained. The first theorem is a version of a result in [2] modified to be used later in proving the existence of solutions to (1) and (2).

THEOREM 1. *Suppose that* $u \in C^{2,1}(\bar{\pi}_T)$ *is a solution to* (1), (2), *that* $\psi(x, t) \in S(t)$ *for* $(x, t) \in \Gamma_T$, *and that* $S(t)$ *is convex for* $0 \leq t \leq T$. *Suppose that* f *and* g *satisfy*

(4) $$n(t, s) \cdot (f(x, t, s, p) - g(t, s)) < 0$$

for all (x, t, s, p) *such that* $(x, t) \in \bar{\pi}_T$, $(t, s) \in S_T$ *and* $p = (p_1, \ldots, p_n)$ *with* $p_i \in \mathbf{R}^m$ *such that* $n(t, s) \cdot p_i = 0$, $1 \leq i \leq n$. *Then* $u(x, t) \in S(t)$ *for all* $(x, t) \in \bar{\pi}_T$.

PROOF. Suppose not, then there exists a point $(x_0, t_0) \in \pi_T$ such that $u(x_0, t_0) \in \partial S(t_0)$, $u(x, t_0) \in \overline{S(t_0)}$ for $x \in \bar{\Omega}$ and $u(x, t) \in S(t)$ for $x \in \bar{\Omega}$, $0 \leq t < t_0$. Let $s(t)$ be the solution to (3) satisfying $s(t_0) = u(x_0, t_0)$, then $s(t) \in \partial S(t)$ for $0 \leq t \leq T$. Define

$$w(x, t) = n(t, s(t)) \cdot (u(x, t) - s(t)),$$

where n is given by the Lemma. By the foregoing remarks, $w \in C^{2,0}(\bar{\pi}_T)$, $w(x_0, t_0) = 0$, $w(x, t_0) \leq 0$ and $w(x, t) < 0$ for $x \in \bar{\Omega}$, $0 \leq t < t$. It follows that, at (x_0, t_0), $\partial w/\partial t$ exists and is nonnegative, $\partial w/\partial x_i = 0$, $(\partial^2 w/\partial x_i \partial x_j)$ is negative semidefinite, and so $Lw \geq 0$. However, since $n \cdot \partial u/\partial x_i = \partial w/\partial x_i = 0$ at (x_0, t_0) we have, by (4),

$$Lw(x_0, t_0) = n(t_0, s(t_0)) \cdot (f(x_0, t_0, s(t_0), \nabla u(x_0, t_0)) - g(t_0, s(t_0))) < 0,$$

a contradiction. This proves the theorem.

REMARKS. 1. If $f(x, t, s, p)$ is locally Lipschitz continuous in s then the theorem can be extended by requiring only $\psi(x, t) \in \overline{S(t)}$ and weak inequality in (4). The conclusion, of course, would become $u(x, t) \in \overline{S(t)}$ for all $(x, t) \in \bar{\pi}_T$ (see [2]).

2. If, instead of being a scalar operator (identity matrix multiplied by the scalar differential operator), $L = \text{diag}(L_i)$ where each scalar operator L_i is uniformly parabolic then some containment results can still be proved. In the case $g \equiv 0$ where $S(t) \equiv S$ this has been done by Chueh, Conley and Smoller in [5] and by Bebernes and Ely in [3] (who treat systems of parabolic functional equations). In that case it is necessary that S be a parallelopiped with edges parallel to the coordinate axes. If $L = \text{diag}(L_i)$ and g is not identically zero then by requiring that $S(t)$ be a parallelopiped with edges parallel to the coordinate axes for all t, one can extend the invariance results of [3] and [5] in the obvious way.

The following existence theorems are based upon those by Bebernes and Schmitt [4] for the case $g \equiv 0$.

THEOREM 2. *Assume that* f, g, ψ, *and* S *are as in Theorem 1. In addition, suppose that* $S([0, T])$ *is bounded and* f *satisfies the growth condition*

$$|f(x, t, u, p)| \leq \phi(|p|)$$

for all $(x, t) \in \bar{\pi}_T$, $u \in \overline{S([0, T])}$, and $p \in \mathbf{R}^{nm}$, where ϕ is nondecreasing, continuous and $r^2/\phi(r) \to \infty$ as $r \to \infty$. Then the IBVP (1), (2) has a classical solution, u, such that $u(x, t) \in S(t)$ for $(x, t) \in \bar{\pi}_T$.

PROOF. We may assume, without loss of generality, that there is a point $s_0 \in S(t)$ for $0 \leq t \leq T$. This is because we may partition $[0, T]$ into finitely many subintervals so that $S(t)$ has the above property on each subinterval. Then solving on successive subintervals, using the final value on one subinterval as the initial data on the next (so preserving the compatibility condition imposed upon ψ) will yield a solution on $[0, T]$.

Now choose $\varepsilon > 0$ so that the ball of radius 2ε about s_0, $B(s_0, 2\varepsilon)$, lies in $S(t)$ for $0 \leq t \leq T$. Let $\hat{g}$ be a function satisfying the same continuity conditions as g but which is identically zero in $[0, T] \times B(s_0, \varepsilon)$ and equal to g on the complement of $[0, T] \times B(s_0, 2\varepsilon)$. Define the Nemytski operators $F, \hat{G}: C^{1,0}(\bar{\pi}_T) \to C(\bar{\pi}_T)$ by

$$(Fu)(x, t) = f(x, t, u(x, t), \nabla u(x, t))$$

and

$$(\hat{G}u)(x, t) = \hat{g}(t, u(x, t)),$$

then F and $\hat{G}$ are continuous and bounded. Also, $F, \hat{G}: C^{1+\alpha,\alpha/2} \to C^{\gamma,\gamma/2}$ are continuous for some $\gamma \in (0, \alpha]$. Define the linear operator K by letting $u = Kv$ be the solution of

$$Lu = v \quad \text{in } \pi_T, \quad u = 0 \quad \text{on } \Gamma_T.$$

We may consider K as a mapping from $C^{\alpha,\alpha/2}$ into $C_0^{2+\alpha,1+\alpha/2}$ (the subscript denoting the zero boundary condition) or as a compact mapping from C into $C^{1+\alpha,\alpha/2} \subset C^{1,0}$ by observing $K: L_q \to W_q^{2,1}$ and using the Sobolev embedding theorem for q sufficiently large (see Ladyzenskaja et al. [6]). It follows that the composite maps $KF, K\hat{G}: C^{1,0} \to C^{1,0}$ are completely continuous. Let $z \in C^{2+\alpha,1+\alpha/2}$ be the solution of

$$Lz = 0 \quad \text{in } \pi_T, \quad z = \psi \quad \text{on } \Gamma_T.$$

Suppose for some $\lambda \in [0, 1]$, there is a solution $u \in C^{1,0}$ of

(5) $$u = \lambda(KFu + z) + (1 - \lambda)(K\hat{G}u + s_0).$$

By the above mapping properties we have immediately $u \in C^{1+\alpha,\alpha/2}$ and so Fu, $\hat{G}u \in C^{\gamma,\gamma/2}$ for some $\gamma \in (0, \alpha]$. Therefore, $u \in C^{2+\gamma,1+\gamma/2}$. By the definition of K and z it follows that u is a classical solution of

(6) $$Lu = \lambda f(x, t, u, \nabla u) + (1 - \lambda)\hat{g}(t, u) \quad \text{in } \pi_T,$$
$$u = \lambda \psi + (1 - \lambda)s_0 \quad \text{on } \Gamma_T.$$

Now, $\lambda f + (1 - \lambda)\hat{g}$ satisfies (4) since $\hat{g} = g$ on S_T. Furthermore, $\lambda \psi + (1 - \lambda)s_0 \in S(t)$ on Γ_T, so Theorem 1 implies that the range of u lies in $S([0, T])$, which is assumed to be bounded. Through the regularity assumptions on the coefficients of L and the continuity and growth conditions on f, the boundedness of u implies the existence of a constant M (independent of λ) such that $|\nabla u| \leq M$ on $\bar{\pi}_T$ (see [6, VII. 6]). The proof will be completed by using the Leray-Schauder degree in $C^{1,0}$ to show that (5) has a solution when $\lambda = 1$. Let $W = \{u \in C^{1,0}: u(x, t) \in S(t), |\nabla u(x, t)| < M + 1 \text{ for } (x, t) \in \bar{\pi}_T\}$, then W is a nonempty bounded open set in $C^{1,0}$. Define $H: [0, 1] \times \overline{W} \to C^{1,0}$ by

$$H(\lambda, u) = u - \lambda(KFu + z) - (1 - \lambda)(K\hat{G}u + s_0).$$

Then H is continuous and $H(\lambda, \cdot)$ is a compact perturbation of the identity in $C^{1,0}$. For each $\lambda \in [0, 1]$, $H(\lambda, u) \neq 0$ for all $u \in \partial W$ for otherwise u would satisfy (5) and so $u(x, t) \in S(t)$ and $|\nabla u(x, t)| \leq M < M + 1$ for $(x, t) \in \bar{\pi}_T$, by the remarks above. Hence, H is a homotopy relative to W and by the homotopy invariance of degree

(7) $$d(I - KF, W, z) = d(I - K\hat{G}, W, s_0).$$

Finally, $\hat{H}: [0, 1] \times \overline{W} \to C^{1,0}$ defined by

$$\hat{H}(\lambda, u) = u - \lambda K\hat{G}u + s_0$$

is a completely continuous homotopy relative to W since the unique solution to

$$Lu = \lambda \hat{g}(t, u) \text{ in } \pi_T, \quad u = s_0 \text{ on } \Gamma_T$$

is $u \equiv s_0 \in W$. Thus, $d(I - K\hat{G}, W, s_0) = d(I, W, s_0) = 1$. This together with (7) completes the proof.

The next result allows a weakening of (4) and the initial boundary data to touch $\partial S(t)$. This is useful for the case $f \equiv g$ when (8), below, is trivially satisfied.

THEOREM 3. *Assume that f, g, ψ and S are as in Theorem 2 except that inequality (4) is replaced by*

(8) $$n(t, s) \cdot (f(x, t, s, p) - g(t, s)) \leq 0$$

on the same (x, t, s, p)-set as before, and allow $\psi(x, t) \in \overline{S(t)}$ on Γ_T. Then the IBVP (1), (2) has a classical solution, u, such that $u(x, t) \in \overline{S(t)}$ for $(x, t) \in \bar{\pi}_T$.

PROOF. Let $s(t)$ be a solution to (3) with $s(0) \in S$, let $\varepsilon > 0$ and let ρ_ε be a smooth function taking values in $[0, 1]$, having support in an ε-neighborhood of S_T and such that $\rho_\varepsilon \equiv 1$ on S_T. Define $\psi_\varepsilon = (1 - \varepsilon)\psi + \varepsilon s$ and

$$f_\varepsilon(x, t, u, p) = f(x, t, u, p) + \varepsilon \rho_\varepsilon(t, u)(s(t) - u).$$

Then, by the convexity of $S(t)$, $\psi_\varepsilon(x, t) \in S(t)$ on Γ_T and f_ε satisfies (4). By applying Theorem 2 we obtain a solution u_ε of

(9) $$u_\varepsilon = K(Fu_\varepsilon + \varepsilon R_\varepsilon u_\varepsilon) + (1 - \varepsilon)z + \varepsilon z_1,$$

where $(R_\varepsilon v)(x, t) = \rho_\varepsilon(t, v(x, t))(s(t) - v(x, t))$, and z_1 is the unique solution of $Lz_1 = 0$ in π_T, $z_1(x, t) = s(t)$ on Γ_T. Furthermore, $u_\varepsilon(x, t) \in S(t)$ in $\bar{\pi}_T$ so $\{R_\varepsilon u_\varepsilon: \varepsilon > 0\}$ is bounded. By the compactness of K we may find a sequence $\varepsilon = \varepsilon_n \to 0$ so that the corresponding sequence of $\{u_\varepsilon\}$ converges, to u say, as $n \to \infty$. Since F is continuous one concludes that $u \in C^{1,0}$ satisfies $u = KFu + z$. As in the proof of Theorem 2, this implies that $u \in C^{2+\gamma, 1+\gamma/2}$ is a solution of (1), (2).

Without any significant changes in the proof of Theorem 7 of [4] one may also prove the following Hukuhara–Kneser property.

THEOREM 4. *Assume the hypotheses of Theorem* 2, *then the solution set of IBVP* (1)–(2) *is a continuum in* $C^{1,0}(\bar{\pi}_T)$.

References

1. H. Amann, *Invariant sets and existence theorems for semi-linear parabolic and elliptic systems*, J. Math. Anal. Appl. **65** (1978), 432–467.
2. P. W. Bates, *Containment for weakly coupled parabolic systems*, 1982 (preprint).
3. J. W. Bebernes and R. Ely, *Existence and invariance for parabolic functional equations*, Nonlinear Anal. **7** (1983), 1225–1235.
4. J. W. Bebernes and K. Schmitt, *Invariant sets and the Hukuhara–Kneser property for systems of parabolic partial differential equations*, Rocky Mountain J. Math. **7** (1977), 557–567.
5. K. Chueh, C. Conley and J. Smoller, *Positively invariant regions for systems of nonlinear diffusion equations*, Indiana Univ. Math. J. **26** (1977), 373–392.
6. O. Ladyženskaja, V. Solonnikov and N. Uralceva, *Linear and quasilinear equations of parabolic type*, Transl. Math. Monographs, vol. 23, Amer. Math. Soc., Providence, R. I., 1968.
7. C. Reder, *Familles de convexes invariantes et equations de diffusion-réaction*, Ann. Inst. Fourier (Grenoble) **32** (1982), 71–103.
8. H. Weinberger, *Invariant sets for weakly coupled parabolic and elliptic systems*, Rend. Mat. Univ. Roma **8** (1975), 295–310.

TEXAS A & M UNIVERSITY

Current address: Brigham Young University

Periodic Hamiltonian Trajectories on Starshaped Manifolds

HENRI BERESTYCKI, JEAN-MICHEL LASRY, GIANNI MANCINI
AND BERNHARD RUF

Introduction. This note, which reports on results from [6, 7], is concerned with the existence of periodic trajectories on a given energy surface for a Hamiltonian system

(1) $$dz/dt = \mathcal{J}H'(z),$$

where $z = z(t)$: $\mathbb{R} \to \mathbb{R}^{2N}$, $H \in C^2(\mathbb{R}^{2N}, \mathbb{R})$ is the Hamiltonian, and $\mathcal{J}$ is the standard skew-symmetric matrix $\mathcal{J} = \begin{pmatrix} 0 & -I \\ I & 0 \end{pmatrix}$ (I is the identity in $\mathbb{R}^N$). Trajectories of (1) remain on energy surfaces $\Sigma = \{u \in \mathbb{R}^{2N}; H(u) = c\}$.

Let us first state some of the known results for this problem (see also the lectures given by I. Ekeland at this conference).

1. P. RABINOWITZ [11]. *Assume that Σ is strictly starshaped and $H'(z) \neq 0$ $\forall z \in \Sigma$. Then (1) has at least one periodic orbit on Σ.*

2. A. WEINSTEIN [12, 13]. *Let $H(0) = 0$, $H'(0) = 0$. If $H''(0)$ is positive definite, then (1) has at least N closed trajectories lying on the energy surface $H = \varepsilon$ for any $\varepsilon > 0$ sufficiently small.*

More recently, I. Ekeland and J.-M. Lasry have obtained the first global existence result for many solutions.

3. I. EKELAND–J.- M. LASRY [9]. *Let H be convex and set $C = \{u \in \mathbb{R}^{2N}; H(u) \leq 1\}$, $\Sigma = \partial C$. If there exists an $\alpha \in \mathbb{R}$ with $1 < \alpha < \sqrt{2}$ and a ball B such that*

(2) $$B \subset C \subset \alpha B,$$

then (1) has at least N closed trajectories on Σ.

1980 *Mathematics Subject Classification.* Primary 58E05, 70H05; Secondary 34C25.

As pointed out by Ekeland and Lasry [9], their result implies the Weinstein theorem only under the additional assumption that $\omega_N < 2\omega_1$, where $\pm i\omega_k$, $1 \leq k \leq N$, are the eigenvalues of $H''(0)$ with $0 < \omega_1 \leq \cdots \leq \omega_N$. Our aim here is to extend this global result in two directions. First, we do not assume that the energy surfaces are convex. Rather, we work with a starshaped Hamiltonian surface. Second, we replace the balls in condition (2) by ellipsoids. We remark that our global result covers the theorem of Weinstein in its full generality. For a more complete review of this type of result, see Berestycki [5].

Let us now precisely state the main result. Let $0 < \omega_1 \leq \cdots \leq \omega_N \in \mathbb{R}$ be given, and set

$$\Omega = \begin{pmatrix} \Omega^1 & 0 \\ 0 & \Omega^1 \end{pmatrix}, \quad \text{where } \Omega^1 = \begin{pmatrix} \omega_1 & & 0 \\ & \ddots & \\ 0 & & \omega_N \end{pmatrix},$$

and define the ellipsoid

$$\mathscr{E} = \left\{ u \in \mathbb{R}^{2N} | (\Omega u, u) \equiv \sum_{i=1}^{N} \frac{\omega_i}{2}(u_i^2 + u_{i+N}^2) \leq 1 \right\},$$

where $u = (u_1, \ldots, u_{2N})$, and $(\cdot, \cdot)$ denotes the scalar product in $\mathbb{R}^{2N}$. We may assume that the energy surface on which we are looking for periodic solutions of (1) is defined by

(3) $$\Sigma = \{ x \in \mathbb{R}^{2N}; H(x) = 1 \}.$$

We assume that $H'(z) \neq 0$ for $z \in \Sigma$,

(4) Σ is a C^2 manifold which is strictly starshaped with respect to the origin and bounds the compact set $\mathscr{R} = \{ x \in \mathbb{R}^{2N}; H(x) \leq 1 \}$,

and

(5) $$\mathscr{E} \subset \mathscr{R} \subset \beta\mathscr{E} \quad \text{for some } \beta > 1.$$

By assumption (4) the tangent plane $T_x\Sigma$ to Σ at a point $x \in \Sigma$ never hits the origin. We may therefore define $\rho > 0$ to be the largest positive real such that

(6) $$T_x\Sigma \cap \mathring{B}_\rho = \varnothing, \quad \forall x \in \Sigma,$$

where $\mathring{B}_\rho = \{ x \in \mathbb{R}^{2N}; |x| < \rho \}$.

THEOREM 1. *Given $\mathscr{E}$, there exists a constant $\delta = \delta(\rho^2, \omega_1, \ldots, \omega_N) > 0$ such that (1) has at least N distinct periodic orbits on any surface Σ satisfying (4)–(6) with $\beta^2 < 1 + \delta$.*

REMARKS. (a) The dependence of δ on ρ^2 and the frequencies $\omega_1, \ldots, \omega_N$ are explicit, but somewhat complicated to write down, and we refer the reader to [6, 7] for the precise statement. We remark, however, that in the simple case $\omega_1 = \cdots = \omega_N$ and Σ convex, one has $\delta = 1$; hence the Ekeland–Lasry theorem is contained in Theorem 1.

(b) Theorem 1 can be sharpened in the following sense: Given $1 \leq p \leq N$, there exist constants δ_p, with $0 < \delta = \delta_N < \delta_{N-1} \leq \cdots \leq \delta_1 = +\infty$, such that if $\beta^2 < 1 + \delta_p$, then (1) has at least p periodic orbits on Σ. Note that for δ_1 ($= +\infty$), condition (5) is satisfied for any strictly starshaped energy surface, and, hence, any such surface carries at least one periodic orbit. Therefore, the theorem of P. Rabinowitz is also contained in Theorem 1. Furthermore, the generalization by A. Ambrosetti–G. Mancini [1] of the Ekeland–Lasry result is contained in Theorem 1 by virtue of the computation of the numbers $\delta_N, \ldots, \delta_1$. (See [6, 7].)

(c) We indicate how the Weinstein theorem can be derived from Theorem 1: We can assume that $H(u) = \frac{1}{2}(\Omega u, u) + R(u)$, $u \in \mathbb{R}^{2N}$, where $R(u) = O(|u|^2)$ for u near 0, and Ω is in canonical form as in Theorem 1. If we set $z = u/\varepsilon$, the energy surface $H(u) = \varepsilon^2$ becomes $H_\varepsilon(z) := \frac{1}{2}(\Omega z, z) + R_\varepsilon(z) = 1$, where $R_\varepsilon(z) = R(\varepsilon z)/\varepsilon^2$. Clearly, for a given $\beta^2 < 1 + \delta$, δ as in Theorem 1, the inclusions

$$\{\tfrac{1}{2}(\Omega z, z) \leq 1/\beta\} \subset \{H_\varepsilon(z) \leq 1\} \subset \beta\{\tfrac{1}{2}(\Omega z, z) < 1/\beta\}$$

hold true for $\varepsilon > 0$ sufficiently small, and $\{H_\varepsilon \equiv 1\}$ carries N closed trajectories by virtue of Theorem 1.

Therefore, our Theorem 1 seems to contain all known results of this type.

Ideas of proof. We first remark that in [7] we present two proofs of Theorem 1, both relying on critical point theory via S^1 action index theories. Here we will give a detailed account of the ideas of one of the proofs and only indicate some of the main features of the second proof. For more details the reader is referred to [7].

1. *Functional framework.* First we give the problem a functional setting. We work in the space $E = H^{1/2}(S^1, \mathbb{R}^{2N})$, the space of 2π-periodic functions from $\mathbb{R}$ to $\mathbb{R}^{2N}$ whose half-derivative is square integrable (this space is obtained either by interpolation between $L^2(S^1, \mathbb{R}^{2N})$ and $H^1(S^1, \mathbb{R}^{2N})$ or by completion of $C^\infty(S^1, \mathbb{R}^{2N})$ with respect to the norm $\|z\|^2 = \sum_{k \in \mathbb{Z}}(1 + |k|)|z_k|^2$, where $z_k \in \mathbb{R}^{2N}$ are the coefficients of the Fourier series expansion $z = \sum_{k \in \mathbb{Z}} z_k e^{ikt}$). The functional we work with is the action integral

$$I(z) = \frac{1}{2\pi} \int_0^{2\pi} (z, -\mathscr{J}\dot{z}),$$

restricted to the manifold

$$G = \left\{ u \in E \Big| \frac{1}{2\pi} \int_0^{2\pi} H(u) = 1 \right\}.$$

Then periodic solutions of (1) on Σ can be obtained via the following

VARIATIONAL PRINCIPLE. *If u is a critical point of $I|G$ such that $\sigma = I(u) > 0$, then $z(t) = u(t/\sigma)$ is a $2\pi\sigma$ periodic solution of (1).*

2. S^1 *symmetry.* The problem of finding periodic solutions of (1) is therefore translated into finding critical points of the functional $I|_G$. We remark that our problem has a natural symmetry: the time shifts. In fact, denoting by T the S^1 action $T_\theta u(t) = u(t + \theta)$, we easily see that I and G are invariant under T; i.e., for all θ, $I(T_\theta u) = I(u)$ and $T_\theta G \subset G$. Therefore it seems natural to call for the S^1

index theories which have recently been developed; see V. Benci [2] and E. Fadell–S. Husseini–P. Rabinowitz [10]. However, one encounters two difficulties in trying to apply these indices directly. For instance, since the functional $I|_G$ is unbounded (from above and below) and the action T has fixed points in E (the constant functions), we cannot use Benci's S^1 index [2]. To handle these problems we introduce a relative index; we remark, however, that another way to overcome this type of difficulties has been developed by Benci [3] in the so-called pseudo-index theories.

3. *Relative index.* Since the precise definition of the relative index is somewhat complicated, we restrict ourselves to just giving a description and its main properties. Let

$$\mathscr{F} = \{A \subset E \setminus \{0\}; A \text{ closed, invariant under } T\}.$$

Furthermore, we decompose $E = E^+ \oplus E^- \oplus E^0$, where

$$E^+ = \text{span}\{\varepsilon_j e^{ikt}; k \in \mathbb{N}, j = 1,\ldots,2N; \varepsilon_j \text{ base vector of } \mathbb{R}^{2N}\},$$

$$E^- = \text{span}\{\varepsilon_j e^{ikt}; -k \in \mathbb{N}, j = 1,\ldots,2N\},$$

$$E^0 = \text{span}\{\varepsilon_j; j = 1,\ldots,2N\} = \mathbb{R}^{2N} \quad (= \text{fixed-point set}).$$

To any $A \in \mathscr{F}$ one can now assign a natural number $\gamma(A/E^+)$, the index of A relative to E^+, with the following properties:

(a) Let $S_k = \{x \in E^- \oplus E^0 \oplus G; G \text{ invariant subspace of } E^+ \text{ with } \dim G = 2k, \|x\| = 1\}$. Then $\gamma(S_k/E^+) = k$.

(b) *Subadditivity.* Let $A, B \in \mathscr{F}$, $B \cap E^0 = \varnothing$. Then $\gamma(A \cup B/E^+) \leq \gamma(A/E^+) + \gamma(B)$, where $\gamma(B)$ denotes the Benci index of B (which is well defined since $B \cap E^0 = \varnothing$).

(c) *Monotonicity.* Let $A, B \in \mathscr{F}$, $B = h(A)$, where

(I) h is equivariant and continuous,

(II) $P_{E^-} h = P_{E^-} + k$; k compact, $P_{E^-} = $ orthogonal projection onto E^-,

(III) $\exists \phi: E^0 \to E^0$ such that $\phi \neq 0$ on $E^0 \setminus \{0\}$ and $\phi(h(u)) = u, \forall u \in E^0$. Then $\gamma(A/E^+) \leq \gamma(B/E^+)$.

(d) Let $A \in \mathscr{F}$, and let $E^+ = F_1 \oplus F_2$ be any decomposition of E^+ into two invariant orthogonal subspaces F_1, F_2: $E^+ = F_1 \oplus F_2$, with $\dim F_1 \leq 2k$. Then $\gamma(A/E^+) \geq k \Rightarrow A \cap F_2 \neq \varnothing$.

4. *Minimax characterization.* Using this index, we now proceed, as usual, to obtain critical values. We define the classes

$$\Gamma_k(G) = \{A \in \mathscr{F}; A \subset G, \gamma(A/E^+) \geq k\}$$

and set

$$C_k = \inf_{A \in \Gamma_k(G)} \max_{x \in A} I(x).$$

Let us note that property (a) indicates that the relative index only "measures" the part of a set lying in E^+; therefore the problem with the fixed point set E^0 disappears. Secondly, we have

$$C_1 = \inf_{\Gamma_1(G)} \max_{A \in \Gamma_1(G)} I \geq \inf_{G \cap E^+} I > 0,$$

since $A \in \Gamma_1(G)$ implies $A \cap E^+ \neq \emptyset$ by property (d); hence, the unboundedness of $I|_G$ does not bother anymore.

We therefore obtain a sequence of values $0 < C_1 \leq C_2 \leq \cdots \leq C_k \leq \cdots$, and, since $I|_G$ satisfies the Palais–Smale condition, one infers by standard "deformation arguments" (using the properties of the relative index) that the values C_k, $k \in \mathbb{N}$, are in fact critical values.

5. *Comparison with linear system.* By the above procedure we find an infinite sequence of critical values. But we note that if u is a critical point, then $u_k(t) = u(kt)$, $k \in \mathbb{N}$, are again critical points with $I(u_k) = kI(u)$. However, these critical points correspond to the same geometric orbit. It is therefore necessary to separate the sequence $(C_k)_{k \in \mathbb{N}}$ into those subsequences which are generated by geometrically different orbits. To do so we compare our problem with a suitable linear problem. We introduce the manifold

$$G_1 = \left\{ \frac{1}{2\pi} \int_0^{2\pi} (\Omega u, u) = 1 \right\}$$

and consider $I|_{G_1}$. Defining, as above, $\alpha_k = \inf \max_{\Gamma_k(G_1)} I$, we obtain again a sequence of critical values; these α_k are the eigenvalues of

$$\mathcal{J}z = \alpha_k \Omega z; \quad \text{i.e.,} \quad \alpha_k = n_i/w_i, \, n_i \in \mathbb{N}.$$

Furthermore, one deduces easily from assumption (5) that

(7) $\qquad \alpha_k \leq C_k \leq \beta^2 \alpha_k, \qquad \forall k \in \mathbb{N}.$

We now demonstrate in a particularly simple situation how the critical values can be distinguished. Assume that, e.g., $\omega_i = \omega_{i+1}$, for some $i \in \{1, \ldots, N\}$. Then $I|_{G_1}$ has a double critical value, say $\alpha_k = \alpha_{k+1} = 1/\omega_i$. By (6) we conclude that $[\alpha_k, \beta^2 \alpha_k]$ contains at least two critical values C_k, C_{k+1} of $I|_G$ with corresponding critical points u_k, u_{k+1}. We claim that if β is chosen appropriately, then u_k and u_{k+1} correspond to geometrically different orbits. Note that we can assume $C_k \neq C_{k+1}$, since otherwise, by the properties of the index,

$$\gamma(\{u; (I|_G)'(u) = 0, I(u) = C_k\}) \geq 2,$$

which yields infinitely many geometrically distinct critical orbits. Now let h_k, $h_{k+1} \in \mathbb{N}$ be such that $\bar{u}_j(t) = u_j(t/h_j)$, $j = k, k+1$, have minimal period 2π. To $\bar{u}_j$ correspond $2\pi I(\bar{u}_j) = 2\pi I(u_j)/h_j$ periodic solutions of (1). Now we note that assumption (6) implies the following lower bound for the period of any closed orbit of (1) on Σ:

(8) $\qquad T_{\min} \geq \pi \rho^2.$

This is a result of C. Croke–A. Weinstein [8]; cf. also the lecture of Lasry at this conference and [7] for an elementary proof and generalizations of this result. Estimate (8) now yields

$$2\pi I(u_j)/h_j \geq T_{\min} \geq \pi \rho^2, \quad j = k, k+1;$$

i.e., $h_j \leq 2I(u_j)/\rho^2 \leq M$, $j = k, k + 1$. Now choose $1 < \beta^2 < (M + 1)/M$ and assume (by way of contradiction) that $\bar{u}_k(t) = \bar{u}_{k+1}(t)$. We then obtain

$$\alpha_k \leq I(u_k) = h_k I(\bar{u}_k) = h_k I(\bar{u}_{k+1}) = \frac{h_k}{h_{k+1}} I(u_{k+1}) \leq \frac{h_k}{h_{k+1}} \beta^2 \alpha_k,$$

and, hence, $h_{k+1}/h_k \leq \beta^2$. But since

$$\frac{M+1}{M} \leq \frac{h_k + 1}{h_k} \leq \frac{h_{k+1}}{h_k} \leq \beta^2 < \frac{M+1}{M},$$

we have a contradiction. Therefore $\bar{u}_k(t) \neq \bar{u}_{k+1}(t)$, i.e., we have found two distinct orbits of (1).

Using this method one also finds in the general situation a number δ, such that $1 < \beta^2 < 1 + \delta$ implies the existence of at least N distinct orbits.

6. *Remarks*. In [6], without imposing any condition of the form (5), we have conjectured that problem (1) has at least N periodic orbits. We refer the reader to the lectures of Ekeland at this conference for *generic* results about the existence of infinitely many closed orbits on any "convex surface" $\Sigma \subset \mathbb{R}^N$.

Remarks on an alternative proof. We indicate here some of the ideas of a second approach to Theorem 1. Here, we use a suitably modified functional in order to apply the methods of convex analysis and the dual variational principle introduced by F. Clarke [14] and Clarke and Ekeland [15]. More precisely, one defines a set of "admissible" functions $\phi \in C^2(\mathbb{R}_+, \mathbb{R}_+)$ with the properties

$\phi(0) = 0$, ϕ strictly concave, $\sup_{s \geq 0} s^2 \phi''(s) < +\infty$,
$\phi'(+\infty) < \phi'(0)$ suitably fixed.

One then finds a $K > 0$ such that $G(z) = \phi \circ H(z) + (K/2)|z|^2$ is strictly convex in $\mathbb{R}^{2N}$ (H can be assumed to have quadratic growth). Then let G^* be the convex conjugate function of G,

$$G^*(\zeta) = \sup_{z \in \mathbb{R}^{2N}} \{(z, \zeta) - G(z)\},$$

and define the functional

$$f(u) = \int_0^{2\pi} \left\{ \frac{1}{2}(\mathscr{J}\dot{u} - Ku, u) + G^*(-\mathscr{J}\dot{u} + Ku) \right\} dt.$$

One shows that critical points of f are (after reparametrization) critical orbits of (1). One notes that the functional f is essentially bounded from below; i.e., there exists a subspace $V \subset E$ of finite codimension such that $f|_V \geq -C > -\infty$. By this fact, the use of the relative index introduced above can be avoided, and, working directly with an S^1 index allowing a finite-dimensional fixed-point space, one derives the following abstract theorem (see also Benci, Capozzi and Fortunato [4] and Fortunato (this conference) for a closely related result that has been obtained independently). This theorem may be useful for other applications.

THEOREM 2. *Let E be a complex Hilbert space with a unitary S^1 representation and finite-dimensional fixed-point space E^0. Let $f \in C^1(E, \mathbb{R})$ be an S^1 invariant functional satisfying the Palais–Smale condition and such that $f(0) = 0$. Assume that there exist invariant subspaces $V, W \subset E$ such that*

(I) $V \subset (E^0)^\perp$,
(II) $f(u) \geq -C > -\infty, \forall u \in V$
(III) $W \supset E^0, \exists \rho > 0$ *such that* $f(u) < 0, \forall u \in W, \|u\| = \rho$,
(IV) $\operatorname{codim}_{\mathbb{C}} V = p < +\infty$, $\dim_{\mathbb{C}} W = m < \infty$,
(V) $(x \in E^0, f(x) \leq 0, f'(x) = 0) \Rightarrow x = 0$.

Then f has at least $m - p$ critical orbits with negative critical levels.

To conclude the proof of Theorem 1, one then shows that the functional f defined above, with $\phi'(+\infty)$, $\phi'(0)$ suitably chosen, yields N distinct critical orbits by virtue of Theorem 2.

References

1. A. Ambrosetti and G. Mancini, *On a theorem by Ekeland and Lasry concerning the number of periodic Hamiltonian trajectories*, J. Differential Equations **43** (1981), 249–256.
2. V. Benci, *A geometrical index for the group S^1 and some applications to the research of periodic solutions of ODE's*, Comm. Pure Appl. Math. **34** (1981), 393–432.
3. _____, *On the critical point theory for indefinite functions in the presence of symmetries*, Trans. Amer. Math. Soc. **274** (1982), 533–572.
4. V. Benci, A. Capozzi and D. Fortunato, *Periodic solutions of Hamiltonian systems of prescribed period*, MRC Tech. Summ. Rep. No. 2508, Madison, WI 53706.
5. H. Berestycki, *Solutions périodiques de systèmes hamiltoniens*, Séminaire Bourbaki, Exposé 603, Février 1983.
6. H. Berestycki, J.-M. Lasry, G. Mancini and B. Ruf, *Sur le nombre des orbites périodiques des équations de Hamilton sur une surface étoilée*, C. R. Acad. Sci. Paris **296** (1983), 15–18.
7. _____, *Existence of multiple periodic orbits on starshaped Hamiltonian surfaces*, Comm. Pure Appl. Math. (to appear).
8. C. Croke and A. Weinstein, *Closed curves on convex hypersurfaces and periods of non linear oscillations*, Invent. Math. **64** (1981), 199–202.
9. I. Ekeland and J.-M. Lasry, *On the number of periodic trajectories for a Hamiltonian flow on a convex energy surface*, Ann. of Math. (2) **112** (1980), 283–319.
10. E. Fadell, S. Husseini and P. H. Rabinowitz, *Borsuk-Ulam theorem for arbitrary S^1 actions and applications*, MRC Tech. Summ. Rep. No. 2301, Madison, WI 53706.
11. P. H. Rabinowitz, *Periodic solutions of Hamiltonian systems*, Comm. Pure Appl. Math. **31** (1978), 141–195.
12. A. Weinstein, *Lagrangian submanifolds and Hamiltonian systems*, Ann. of Math. (2) **98** (1973), 377–410.
13. _____, *Normal modes for non linear Hamiltonian systems*, Invent. Math. **20** (1973), 47–57.
14. F. H. Clarke, *A classical variational principle for periodic Hamiltonian trajectories*, Proc. Amer. Math. Soc. **76** (1979), 186–188.
15. F. H. Clarke and I. Ekeland, *Hamiltonian trajectories having prescribed minimal period*, Comm. Pure Appl. Math. **33** (1980), 103–116.

UNIVERSITÉ PARIS–NORD, FRANCE

UNIVERSITÉ PARIS IX, France

UNIVERSITÀ DI TRIESTE, ITALY

ETH ZÜRICH, SWITZERLAND

Integrability of Nonlinear Differential Equations via Functional Analysis

M. S. BERGER,[1] P. T. CHURCH AND J. G. TIMOURIAN

An important step in studying nonlinear differential equations consists of studying more explicit methods for exhibiting their solutions. Indeed, the attempt to determine nonlinear ordinary differential equations that are "integrable by quadrature" can be traced back to the beginnings of calculus. More recently the "inverse scattering method" has been highly successful in treating a variety of nonlinear partial differential equations involving such evolutionary equations in one space variable as the celebrated Korteweg de Vries equation

$$(1) \qquad u_t + uu_x + u_{xxx} = 0$$

subject to appropriate boundary conditions.

From our point of view, (1) is shown to be "integrable" by showing that (after appropriate coordinate transformations) it is equivalent to its formal linearization (at $u = 0$)

$$u_t + u_{xxx} = 0.$$

In this article we survey, with examples, our recent idea of utilizing functional analysis to extend (for a nonlinear context) the notion of diagonalizability of an operator acting between function spaces. Thus we obtain "global normal forms" for nonlinear boundary value problems. Up to coordinate transformations this notion leads to explicit formulae for the solutions of the equations we study. In particular, we study nonlinear problems that are *not* globally equivalent to any linear problem, an important novel feature of our work.

1. The idea of integrability. Here we define a new type of integrability for nonlinear partial differential equations. In fact, we phrase our definition in terms of nonlinear differentiable mappings acting between Banach spaces. Let A denote

---

1980 *Mathematics Subject Classification.* Primary 58F07, 58G20.
[1] Research supported by NSF and AFOSR grants.

© 1986 American Mathematical Society
0082-0717/86 $1.00 + $.25 per page

a given smooth mapping acting between Banach spaces X_1 and X_2. Then we say A is C^k ($k \geq 0$) *equivalent* to a mapping B acting between Banach spaces Y_1 and Y_2 if there are C^k diffeomorphisms α and β such that the following diagram commutes:

(1)
$$\begin{array}{ccc} X_1 & \xrightarrow{A} & X_2 \\ \alpha \downarrow & & \downarrow \beta \\ Y_1 & \xrightarrow{B} & Y_2 \end{array}$$

This means that the mappings A and B differ by smooth (C^k) coordinate transformations. Thus, for example, in terms of separable Hilbert spaces X_1, X_2 (relative to orthonormal bases) suppose A is written

(3) $$A(x_1, x_2, x_3, \ldots) = (A_1, A_2, A_3, \ldots)$$

with $x = (x_1, x_2, x_3, \ldots)$ and $A_i(x) = P_i A(x)$ as the orthogonal projection P_i of $A(x)$ on the ith basis element in the range. Now we say A is C^k *globally diagonalizable* if A is equivalent to the diagonal map, $D(x)$, defined by

$$D(x_1, x_2, x_3, \ldots) = (b_1(x_1), b_2(x_2), b_3(x_3), \ldots)$$

with $b_i(x_i)$ a smooth function.

More generally A will be called C^k *completely integrable* if A is equivalent to the triangular (nonlinear) operator T

(4) $$T(x_1, x_2, x_3, \ldots) = (T_1(x_1), T_2(x_1, x_2), T_3(x_1, x_2, x_3), \ldots),$$

where the operators $T_i = T_i(x_1, x_2 \cdots x_i)$ are smooth functions of its first i arguments. This last definition implies that (apart from coordinate transformations), an operator equation of the form $Au = g$ can be solved explicitly in terms of the equivalent equation $Tx = \tilde{g}$.

2. The relation to singularity theory. The diagram (2) is a global, infinite-dimensional extension of a similar equivalence in the theory of singularities in differential topology. There is a local analog: Let $x \in X_1$, let U_1 be an open neighborhood of x in X_1, and let U_2 be open in X_2 with $A(U_1) \subset U_2$. If (2) holds with X_i replaced by U_i, then we say that A at x is C^k *locally equivalent* to B at $\alpha(x)$. By defining, as usual, a singularity of the C' Fredholm mapping A to be a point x at which the Fréchet derivative $A'(x)$ is not surjective, we note that the singular points and singular values of A are mapped into singular points and singular values of B by virtue of the diagram (2). Thus the infinite-dimensional singular points and values form local invariants for studying the notion of integrability we have in mind.

We prove that certain nonlinear differential operators A, ordinary or partial, can be globally diagonalized by showing that the "local normal forms" of certain smooth maps in the neighborhood of a singularity can be extended globally to an infinite-dimensional context. In particular, we have extended in [3] the simplest singularities, folds and cusps in the sense of Whitney and Thom, to an infinite-dimensional context as follows.

A. *Infinite-dimensional folds.* For an infinite-dimensional fold we have the following definition and result.

DEFINITION. Let A be a C^k $[k \geq 2]$ Fredholm operator of index 0 acting between the Banach spaces X_1 and X_2. Then A is a *fold* at $\bar{u}$ if
 (i) $\dim \ker A'(\bar{u}) = 1$,
 (ii) for some (and hence for any) nonzero element $e \in \ker A'(\bar{u})$, $A''(\bar{u})(e, e) \notin \text{range } A'(\bar{u})$.

THEOREM 1. *If the C^k $[k \geq 2]$ operator A has a fold at u, then A at u is C^{k-2} locally equivalent to the fold map (a diagonal map) $F: \mathbf{R} \times E \to \mathbf{R} \times E$ defined by $F(t, v) = (t^2, v)$ at $(0,0)$, where E is a real Banach space. The converse holds, provided the equivalence maps are C^2. Moreover, the notion of a fold is invariant under C^k equivalence ($k \geq 2$).*

In the following sections we shall point out some examples of single (globally) diagonalizable operators which have the normal form given in Theorem 1.

B. *Infinite-dimensional cusps.* Let W be the Whitney map $W: \mathbf{R}^2 \to \mathbf{R}^2$ defined by
$$W(t, \lambda) = (t^3 - \lambda t, \lambda); \tag{5}$$
Whitney gave an abstract characterization of W (at $(0,0)$). In [3] we give abstract properties (invariant under local coordinate change) for an operator A at a singular point $\bar{u}$ to be (locally) equivalent to the infinite dimensional cusp map
$$G: \mathbf{R}^2 \times E \to \mathbf{R}^2 \times E, \quad (t, \lambda, v) \to (t^3 - \lambda t, \lambda, v), \tag{6}$$
at $(0, 0, 0)$, where E is any real Banach space. This definition and theorem, given below, parallel those for the fold.

DEFINITION 2. Let A be a C^k $[k \geq 3]$ Fredholm operator of index 0 acting between Banach spaces X_1 and X_2. Then A is an *(intrinsic) cusp* at u if
 (i) $\dim \ker A'(\bar{u}) = 1$;
 (ii) for some (and hence for any) nonzero $e \in \ker A'(\bar{u})$, $A''(\bar{u})(e, e) \in \text{range } A'(\bar{u})$;
 (iii) there is a $w \in X_1$ such that $A''(\bar{u})(e, w) \notin \text{range } A'(\bar{u})$;
 (iv) $A'''(\bar{u})(e, e, e) - 3A''(\bar{u})(e, A'(\bar{u})^{-1}(A''(\bar{u})(e, e))) \notin \text{range } A'(\bar{u})$.

THEOREM 2. *If the C^∞ (resp. C^3) operator A has an (intrinsic) cusp at $\bar{u}$, then A is C^∞ (resp. C^0) locally equivalent to the cusp map G defined above at $(0, 0, 0)$. The converse holds provided that the equivalence maps are C^3. Moreover the notion of an (intrinsic) cusp is invariant under C^k equivalence ($k \geq 3$).*

In [4] Damon considers any infinitesimally stable Fredholm map germ f: $E, 0 \to F, 0$ and map

germ $A: EXY, 0 \to FXY, 0$ with $A(E, Y) = (\overline{A}(e, y), y)$

and $\overline{A}(e, 0) = f(e)$ for $e \in Y$; he proves that A is equivalent to $f \times \text{id}$. The result uses Mather's work, and so requires C^∞, but otherwise applies to more general maps than those in Theorems 1 and 2.

3. Desiderata of a theory of integrability. The goals of our theory of integrability include

(i) (independence of space dimension): The theory should include examples involving differential operators in any number of space dimensions;

(ii) (stability): The methods used in the theory should apply to a perturbation of the integable system ["near" a given integrable one];

(iii) (inclusion of classical examples not integrable by quadrature): The theory should include some examples of classical significance, that are not integrable by quadrature.

We now outline aspects of the results that have been obtained to date to achieve these three goals. We begin with a classical example not integrable by quadrature, but "globally diagonalizable" in our sense.

4. Riccati's equations (the periodic case). We consider the equation

$$u' + u^2 = q(t) \tag{7}$$

where $q(t)$ is an L_2 T-periodic real-valued function. It is well known that (7) is not integrable by quadrature. To approach (7) from our point of view, following the recent thesis of Scovel [5], we define an operator Au acting on T-periodic functions in the Sobolev space $H_1(0, T)$. Then A is a nonlinear operator acting from H_1 to L_2. It is fairly easy to show that A is a C^∞ Fredholm operator of index 0. In fact, by Rellich's lemma, A is a compact perturbation of an invertible linear operator. We can also read off the Fréchet derivatives of A from the expansion $A(u + h) = Au + (h' + 2uh) + h^2$. We now can determine the singularities of A easily from the simple fact that nontrivial T-periodic solutions of $h' + 2uh = 0$ occur exactly when the mean-value $\int_0^T u = 0$ and moreover $h > 0$. Thus dim ker $A'(u) = 1$.

The singular values of A can be determined in this case by direct computation. For a singular point u of S, by setting $h = w^{-2}$ so $u = w'/w$, $A(u) = A(w'/w) = (w'/w)' + (w'/w)^2 = w''/w \equiv q$ (say). Thus q satisfies Hill's equation $w'' - qw = 0$ with $w > 0$ and so, from applying the elementary spectral theory for this situation, w is the positive eigenfunction $\lambda_1(q) = 0$ associated with the lowest eigenvalue of the differential operator $w'' - qw$ subject to periodic boundary conditions; thus

$$A(S) = \{q \in L^2 : \lambda_1(q) = 0\}. \tag{8}$$

Now we interpret the situation geometrically. The set S of singular points of A is a hyperplane through the origin in H_1 orthogonal to the constant function 1. Its image $A(S)$ is a smooth manifold of codimension 1 in L_2. Moreover, locally each singular point of S is a fold. Indeed, by Theorem 1, we compute at a singular point u, from $h \in \text{Ker } A'(u)$ and $h^* \in \text{ker}[A'(u)]^T$

$$\langle A''(u)\{h, h\}, h^* \rangle > 0. \tag{9}$$

Here h and h^* can be computed explicitly as solutions of $h' + 2uh = 0$ and $h^{*\prime} - 2uh^* = 0$. Scovel goes on to explicitly construct the changes of coordinates α and β referred to in §1, based on these explicit formulae on S and $A(S)$ to show that the local folds fit together globally in a smooth fashion. He proves

THEOREM 3 (SCOVEL). *The Riccati operator $Au = u' + u^2$ from $H_1 \to L_2$ is globally C' equivalent to a fold $(t, v) \to (t^2, v)$ by explicit global diffeomorphism α and β as described in §1. Here t is a coordinate measuring the distance of a point from $S \in H_1(0, T)$ and $A(S) \in L_2(0, T)$.*

5. Examples from nonlinear elliptic boundary value problems. In order to fulfill the goal of finding integrable nonlinear systems independent of space dimensions, we turn to boundary value problems for nonlinear elliptic partial differential operators.

EXAMPLE 1. Consider the operator $Au = \Delta u + f(u)$ subject to zero Dirichlet boundary conditions on a bounded domain $\Omega \subset \mathbf{R}^N$, and assume the eigenvalues of Δ on Ω (with the same boundary conditions) are written in ascending order $0 < \lambda_1 < \lambda_2 < \cdots$. We restrict consideration to those C^2 convex functions f with $f''(0) > 0$ and normalization $f(0) = 0$ satisfying the asymptotic conditions

$$(10) \qquad \lambda_2 > \lim_{t \to \infty} \frac{f(t)}{t} > \lambda_1 > \lim_{t \to -\infty} \frac{f(t)}{t} > 0.$$

In [2] we prove that as an operator from $H_1^0(\Omega) \to H_1^0(\Omega)$ (defined by duality), Au can be continuously globally diagonalized with the global normal form F: $(t, v) \to (t^2, v)$. Thus A is a global fold exactly as in the example of Riccati's equation mentioned above.

To achieve this we prove the Cartesian coordinate representation

$$(11) \qquad A[tu_1 + w(t)] = h(t)u_1 + g_1,$$

where u_1 is the positive normalized first eigenvector of Δ, g_1 and $w(t) \perp u_1$. Setting $u(t) = tu_1 + w(t)$, we find by differentiating that at a singular point t_0

$$(12) \qquad \Delta u'(t_0) + f'(u(t_0))u'(t_0) = 0.$$

Since $h'(t_0) = 0$, $u'(t_0)$ is a nontrivial solution of (12) and by the asymptotic conditions (10) we may suppose $u'(t_0) > 0$ in (12). Differentiating again and evaluating at $t = t_0$ we find

$$(13) \qquad A''(u(t))[u'(t)]^2 + A'(u(t_0))u''(t_0) = h''(t_0)u_1.$$

Taking the inner product of both sides with $u'(t_0)$ and using the selfadjointess of $A'(u(t_0))$ to insure

$$\langle A'(u(t_0))u''(t_0), u'(t_0) \rangle = 0$$

we find

$$(14) \qquad \langle A''(u(t_0))(u'(t_0)^2, u'(t_0) \rangle = \int_\Omega f''(u(t_0))[u'(t_0)]^3 > 0.$$

Here we have used the convexity of $f(t)$ and the fact that $f''(0) > 0$.

By virtue of Theorem 1 this result implies that all singular points S of A are folds and moreover a direct analysis of the relation (10) shows that both S and $A(S)$ are manifolds of codimension 1 in H_1 (see [2]). Moreover in [2] we construct diffeomorphisms mapping the singular points [resp., values] of A onto the singular points [resp., values] of the global fold map F. Using these diffeomorphisms we show A is actually globally diagonalizable with normal form the fold map: $(t, v) \to (t^2, v)$ with t a coordinate in the u_1 direction and v orthogonal to u_1. For the details and the added smoothness of the (global) equivalence see [2].

EXAMPLE 2. Consider the nonlinear Dirichlet problem

(15) $$Au \equiv \Delta u + \lambda u - u^3 = \tilde{g}, \quad u/\partial \Omega = 0,$$

defined over a bounded domain Ω in $\mathbf{R}^N$ ($N \leq 3$). This classical formulation gives rise to an operator A_λ on a Sobolev space $H_0^1(\Omega)$ into itself defined by duality as

(16) $$\langle A_\lambda(u), \phi \rangle_H = \int_\Omega \nabla u \nabla \phi - \lambda u \phi + u^3 \phi$$

for $\phi \in C_0^\infty(\Omega)$; set $A(u, \lambda) = (A_\lambda(u), \lambda)$. Let λ_i denote the eigenvalues of the Laplacean on Ω, and order these by magnitude. Then for $\lambda \leq \lambda_1$, an application of Hadamard's theorem [1] shows A_λ is a homeomorphism, but for $\lambda > \lambda_1$ the operator A_λ has a large set of singularities. In fact we can prove for V an open neighborhood of $(0, \lambda_1)$ in $H \times \mathbf{R}$ there are C^∞ diffeomorphism α, β such that that following diagram commutes (for G the operator of §2B):

(17)
$$\begin{array}{ccc} A^{-1}(V) & \xrightarrow{A} & V \\ \alpha \downarrow & & \beta \downarrow \\ \mathbf{R}^2 \times E & \xrightarrow{G} & \mathbf{R}^2 \times E \end{array}$$

We can characterize fold and cusp points of A on $A^{-1}(V)$ as follows: For each u in the singular set $S(A_\lambda)$, $\dim \ker A_\lambda'(u) = 1$ and for nonzero $e \in \ker A_\lambda'(u)$ then (u, λ) is a fold point of A (and A_λ) if and only if $\int_\Omega u e^3 \neq 0$ and a cusp point of A if this integral vanishes. The set of cusp points is a real analytic submanifold Γ of codimension 1 in $S(A)$ and thus codimension two in $A^{-1}(V)$ and the map A: $\Gamma \to A(\Gamma)$ is a real analytic diffeomorphism. Thus for λ near λ_1 and $\tilde{g}$ near 0 (15) has one, two or three solutions. For singular values $\tilde{g}$ (15) has precisely one solution u if u is a cusp, and precisely two solutions u_1, u_2 if u_1 or u_2 is a fold. For $\tilde{g}$ a regular-value of A_λ, the equation has one or three solutions. It has three solutions exactly when $\tilde{g} = A_\lambda(u)$ for u on a ray from 0 such that the segment $[0, u]$ does not meet SA_λ and $\lambda_1 < \lambda < \lambda_2$.

See [4] for a discussion of the number of solutions of a generalization of (15).

EXAMPLE 3. In order to extend the analysis of Example 2 to more general and less explicit nonlinearities we can study (15) via a nonlinear operator

(18) $$A_\lambda(u) = u - \lambda L u + N u$$

mapping an infinite-dimensional real Hilbert space H into itself. Here L is a compact selfadjoint positive linear operator (so that $(Lu, u) \geq 0$ with equality

only if $u = 0$), whose lowest eigenvalue, λ_1, is simple. Let v_1 be the associated unit eigenvector. We assume that N in a C^k operator ($k \geq 3$) such that
   (a) $N^1(u)$ is a nonnegative selfadjoint operator, with $\langle N^1(u)v_1, v_1 \rangle = 0$ implying $u = 0$,
   (b) $N^{(j)}(0) = 0$ for $j = 0, 1, 2$, and
   (c) $\langle N^{(3)}(0)(v, v, v), v \rangle_H > 0$ for any nonzero $v \in H$.

With these restrictions the diagram (17) holds for A, where $A(u, \lambda) = (A_\lambda(u), \lambda)$ and $A_\lambda(u)$ is given in (18). Moreover, we can prove that a singular point $(\bar{u}, \lambda)$ of A near $(0, \lambda_1)$ is a fold or cusp if $\langle N''(\bar{u})\{e_0, e_0\}, e_0 \rangle_H \neq 0$ or equals zero, respectively.

6. Stability. Now we consider the behavior of the globally diagonalizable operators A discussed above under perturbation. For Example 1 in §5, this problem was treated at some length in [2, II]. In that paper we considered both C^2 and C^1 small perturbations of a fixed nonlinear function $f(u)$. For C^2 perturbations, because of the openness of the assumptions on $f(u)$, we find the resulting normal form, the global fold map, is preserved and only the changes of coordinates α and β are slightly deformed. The case of small C^1 perturbations is more subtle. We prove (subject to a suitably restricted C^1 perturbation of $f(u)$) away from a small neighborhood of the singular values of the unperturbed problem, the number of solutions remains unchanged and the solutions of the unperturbed Dirichlet problem are *accurate approximations* to the solutions of the resulting perturbed nonlinear Dirichlet problem.

The perturbation of the Riccati equation follows a similar pattern. Consider the operator
$$\tilde{A}(u) = u' - f(u) \colon H_1(0, T) \to L_2(0, T) \tag{19}$$
mapping T-periodic H_1 functions into L_2. Here we suppose $f(u)$ is a small C^2 perturbation of the Riccati nonlinearity $f(u) = u^2$. The singular points S of $\tilde{A}$ are easily determined to be those T-periodic functions $u(t)$ for which $\int_0^T f'(u(t))\, dt = 0$. In fact, one shows that the resulting map $\tilde{A}$ is C^1 equivalent to a fold, so that $\tilde{A}$ has the same global normal form as the Riccati operator and only the diffeomorphisms α and β change slightly.

References

1. M. S. Berger, *Nonlinearity and functional analysis*, Academic Press, New York, 1977.
2. M. S. Berger and P. T. Church, *Complete integrability and perturbation of a nonlinear Dirichlet problem*. I, II, Indiana Univ. Math. J. **28** (1979), 935–952. ibid. **29** (1980), 715–735, Erratum to I, ibid. **30** (1981), 799.
3. M. S. Berger, P. T. Church and J. G. Timourian, *Cusps and folds in Banach spaces*, I Indiana Univ. Math. J. (to appear).
4. J. Damon, *A theorem of Mather and the local structure of nonlinear Fredholm maps* (preprint).
5. C. Scovel, Ph. D. dissertation, New York Univ., 1983, supervised by H. McKean.

University of Massachusetts

Syracuse University

University of Alberta

Extinction of the Solutions of Some Quasilinear Elliptic Problems of Arbitrary Order

F. BERNIS

1. Introduction. The author proved [2] that all variational solutions of some higher-order semilinear elliptic equations (e.g., (7) with $p = 2$) are "extinct at a finite distance" if the data have compact support. This type of problem has been intensively investigated for second-order equations (see the lecture of J. I. Díaz [6]). We present here a substantial modification of the proof of [2] that applies also to quasilinear equations with strong nonlinearities and to nonconvex singular problems of the calculus of variations. We state these results in their simplest forms.

We always assume that m and n are arbitrary integers ≥ 1, all functions are real-valued, and $1 < p < \infty$. We set

$$(1) \qquad Au = (-1)^m \sum_{|\alpha|=m} D^\alpha \left(|D^\alpha u|^{p-1} \operatorname{sgn} D^\alpha u \right)$$

(for $p = 2$, $A = (-1)^m \Delta^m$, and for $m = 1$, $A = -\Delta_p$),

$$(2) \qquad g \in C(\mathbf{R}), \qquad s \cdot g(s) \geq |s|^r \quad \text{for all } s \in \mathbf{R}$$

(this implies the sign condition $s \cdot g(s) \geq 0$ and $g(0) = 0$, but g is not assumed to be monotone; the nonlinearity is "strong" in the sense that no upper restriction on the growth of g is imposed),

$$(3) \qquad f \in W^{-m,p'}(\Omega), \qquad \operatorname{supp} f \text{ bounded,}$$

$$(4) \qquad |D^j u(x)|^p = \sum_{|\alpha|=j} |D^\alpha u(x)|^p,$$

$$(5) \qquad J_\Omega(u) = \frac{1}{p} \int_\Omega |D^m u|^p \, dx + \frac{1}{r} \int_\Omega |u|^r \, dx - (f, u)_\Omega,$$

where $(\cdot, \cdot)_\Omega$ stands for the duality between $W^{-m,p'}(\Omega)$ and $W_0^{m,p}(\Omega)$.

1980 *Mathematics Subject Classification.* Primary 35J60, 35B99; Secondary 34A40.

THEOREM 1. *Assume (1)–(3) with $\Omega = \mathbf{R}^n$. Assume also $1 < r < p$. Then there exists $u \in W^{m,p}(\mathbf{R}^n)$, with $g(u) \in L^1(\mathbf{R}^n)$, such that u has compact support and is a solution of the equation*

(6) $$Au + g(u) = f$$

in the sense of distributions on $\mathbf{R}^n$. These properties of u imply that $ug(u) \in L^1(\mathbf{R}^n)$.

THEOREM 2. *Assume (3), (4), and $0 < r < p$. Then the problem of minimizing the functional (5) with $\Omega = \mathbf{R}^n$ in the set $W^{m,p}(\mathbf{R}^n) \cap L^r(\mathbf{R}^n)$ has a solution of compact support.*

Theorem 1 implies Theorem 2 for $1 < r < p$ but, of course, not for $0 < r \leq 1$. We state the following corollary.

COROLLARY 1. *Assume (1), (3), and $1 < r < p$. Let u be the unique solution, belonging to $W^{m,p}(\mathbf{R}^n)$, of the equation*

(7) $$Au + |u|^{r-1}\operatorname{sgn} u = f$$

in the sense of distribution on $\mathbf{R}^n$. Then u has compact support.

The strong nonlinearity is dealt with in the results of Brezis and Browder [4]. The key point in the proof below (as in that of [2]) is the application of some *weighted* generalizations of Gagliardo–Nirenberg inequalities (see Appendix I). Antoncev [1] and Díaz and Véron [6, 7] have already applied imbedding-interpolation inequalities in order to prove compactness of the support for very general second-order problems; they use an energy method rather than comparison principles. (The proofs based on comparison principles do not extend to higher-order problems.) We refer to [3] for sharpened results and other developments in dimension 1. Previous work on fourth-order problems in dimension 1 is found in [2] and [3].

REMARK 1. In Theorem 1, $ug(u) \in L^1(\mathbf{R}^n)$, by [4, Theorem 1], and u is unique if g is nondecreasing, by [4, Formula (23)]. In Theorem 2, u is unique if $r \geq 1$ because of strict convexity. (In both cases uniqueness holds without assuming supp u compact.)

2. Proof of Theorem 1. Let G be an open ball such that supp $f \subset G$. The restriction of f to G is also named f. By Brezis–Browder [4, Theorem 7] there exists $u \in W_0^{m,p}(G)$ such that $g(u) \in L^1(G)$, $ug(u) \in L^1(G)$, and u is a solution of (6) in the sense of distributions on G. Therefore, (6) holds in $W^{-m,p'}(G)$ and $g(u) \in W^{-m,p'}(G)$. Thus

$$a(u,v)_G + (g(u), v)_G = (f, v)_G \quad \text{for all } v \in W_0^{m,p}(G),$$

(8) $$a(u,v)_G \equiv \sum_{|\alpha|=m} \int_G |D^\alpha u|^{p-1} \operatorname{sgn} D^\alpha u \cdot D^\alpha v \, dx.$$

Set $x = (x_1 \cdots x_n)$, $y = (x_1 \cdots x_{n-1})$, $t = x_n$. We can trivially suppose that the half-space $\{t > 0\}$ intersects G and does not intersect supp f. For any $t_0 \geq 0$ the function

(9) $$w(x) = \begin{cases} (t - t_0)^m u(x) & \text{if } t > t_0, \\ 0 & \text{if } t \leq t_0, \end{cases} \quad x \in G,$$

belongs to $W_0^{m,p}(G)$. (This is a simple but important remark for the present proof.) Since $wg(u) \geq 0$ (recall that $sg(s) \geq 0$), by [**4**, Theorem 3] we have

$$(g(u), w)_G = \int_{\substack{x \in G \\ t > t_0}} (t - t_0)^m u g(u) \, dx.$$

Setting $v = w$ in (8) we obtain

(10) $$\sum_{|\alpha|=m} \int_{\substack{x \in G \\ t > t_0}} |D^\alpha u|^{p-1} \operatorname{sgn} D^\alpha u \cdot D^\alpha\big((t - t_0)^m u\big) \, dx$$
$$+ \int_{\substack{x \in G \\ t > t_0}} (t - t_0)^m u g(u) \, dx = 0.$$

We compute $D^\alpha((t - t_0)^m u)$. Setting $D_t = \partial/\partial t$, we have, with an obvious notation and for some constants a_{ijm},

(11) $$D^\alpha = D_t^j D_y^\beta, \quad |\beta| = |\alpha| - j, \quad \sum_{|\alpha|=m} = \sum_{j=0}^m \sum_{|\beta|=m-j},$$

(12) $$D^\alpha\big((t - t_0)^m u\big) = (t - t_0)^m D^\alpha u + \sum_{i=1}^j a_{ijm} (t - t_0)^{m-i} D_t^{j-i} D_y^\beta u.$$

We recall (4) and set

(13) $$I_s(t_0) = \int_{\substack{x \in G \\ t > t_0}} (t - t_0)^s |D^m u|^p \, dx + \int_{\substack{x \in G \\ t > t_0}} (t - t_0)^s |u|^r \, dx.$$

We have $|D_t^{j-i} D_y^\beta u(x)| \leq |D^{m-i} u(x)|$ because $|\beta| + j = m$. Then some elementary computations with (2) and (10)–(12) give (we drop $x \in G$ in the integrals)

$$I_m(t_0) \leq C_m \sum_{i=1}^m \int_{t > t_0} (t - t_0)^{m-i} |D^m u|^{p-1} |D^{m-i} u| \, dx$$

$$\leq C_m \sum_{i=1}^m \left(\int_{t > t_0} (t - t_0)^{m-i} |D^m u|^p \, dx \right)^{1/p'}$$

$$\times \left(\int_{t > t_0} (t - t_0)^{m-i} |D^{m-i} u|^p \, dx \right)^{1/p},$$

where we have applied Hölder's inequality. Now we are going to apply to the integrals of the second factor the *weighted interpolation inequalities* of Appendix I, with $j = k = m - i$. Since the zero extension operator maps $W_0^{m,p}(G)$ into $W^{m,p}(\mathbf{R}^n)$ and commutes with D^α, $|\alpha| \le m$, we can apply these inequalities *on half-spaces*, and thus *the constants are independent of G*. So we obtain

$$I_m(t_0) \le C \sum_{i=1}^{m} \left(\int_{t > t_0} (t - t_0)^{m-i} |D^m u|^p \, dx \right)^{1/p' + a_i/p}$$

$$\times \left(\int_{t > t_0} (t - t_0)^{m-i} |u|^r \, dx \right)^{(1 - a_i)/r},$$

(14) $$\frac{1}{p} = \frac{m - i}{n + m - i} + a_i \left(\frac{1}{p} - \frac{m}{n + m - i} \right) + (1 - a_i) \frac{1}{r}.$$

We set $\lambda_i = 1/p' + a_i/p + (1 - a_i)/r$. Applying the inequality $A^a B^b \le C(A + B)^{a+b}$ and explicitly computing λ_i through (14), we obtain

(15) $$I_m(t_0) \le C \sum_{i=1}^{m} I_{m-i}(t_0)^{\lambda_i},$$

(16) $$\lambda_i = 1 + i/(n + m - i + \sigma r) \quad \text{with } \sigma = pm/(p - r),$$

and the constant C of (15) depends only on n, m, p, r. Note that $\sigma > 0$ and $\lambda_i > 1$. Since $I_s' = -sI_{s-1}$, (15) is an ordinary differential inequality. We apply Lemmas 2.III and 3.I (see Appendix II) for $z(t) = I_m(t)$ and $|z^{(m)}(t)| = m! I_0(t)$. The numbers λ_i of (16) satisfy (26). Thus $\mu = \lambda_m > 1 = \lambda_0$. Performing the computations with the exponents, we obtain that supp u is included in the half-space $\{t \le a\}$ and

(17) $$a \le C I_0(0)^{1/(n + \sigma r)} \quad \text{with } \sigma = pm/(p - r),$$

where *the constant C depends only on n, m, p, r*. Again by [4, Theorem 3], $(g(u), u)_G = \int_G u g(u)$. Since the integrals defining $I_0(0)$ are performed on a subset of G, setting $v = u$ in (8) and using (2) and Young's inequality, we obtain

(18) $$I_0(0) \le C\left(p, r, \|f\|_{W^{-m, p'}(\mathbf{R}^n)} \right).$$

If we consider half-spaces orthogonal to the coordinate axes, it is clear that supp u is bounded independently of G. For large G the zero extension of u to $\mathbf{R}^n$ satisfies all requirements of Theorem 1.

3. Proof of Theorem 2. Let G be a ball as before. Let u be a solution of the problem of minimizing J_G in the set $W_0^{m,p}(G)$. (Existence is standard, even for $0 < r < 1$, because G is bounded; an existence proof for unbounded domains can be found in [2].) Proceeding as usual in the calculus of variations, we try to differentiate $J_G(u + \lambda v)$ at $\lambda = 0$. This is not possible for arbitrary $v \in W_0^{m,p}(G)$

if $0 < r \leqslant 1$. But it is trivially possible for $v = u$ and for $v = w$ of (9). The choice $v = w$ gives (10). The choice $v = u$ can be used to obtain (18). Now the proof is completed as before.

4. Further developments.

4.1. *Nonhomogeneous Dirichlet data.* Adapting Brezis–Browder [4, Theorem 7] to the case of nonhomogeneous Dirichlet data, we obtain, from the proof in §2, the following result: Let Ω be an unbounded open set of $\mathbf{R}^n$ with compact boundary. Set $h(s) = \sup_{|s| \leqslant |t|} |g(t)|$. Let $\Phi \in W^{m,p}(\Omega)$ be such that $\Phi h(\Phi) \in L^1(\Omega)$. Assume (1)–(3) and $1 < r < p$. Then there exists u such that $u - \Phi \in W_0^{m,p}(\Omega)$, $g(u) \in L^1(\Omega)$, $Au + g(u) = f$ in $D'(\Omega)$, and u has compact support in $\overline{\Omega}$. These properties of u imply $ug(u) \in L^1(\Omega)$ and $\Phi g(u) \in L^1(\Omega)$.

4.2. *Estimate of the support.* (17) and (18) give an estimate of supp u in terms of the data of the problem. (More details are found in [2].) In the situation of §4.1, (17) is preserved and (18) is to be replaced by

$$I_0(0) \leqslant C\Big(p, r, \|f\|_{W^{-m,p'}(\Omega)}, \|D^m\Phi\|_{L^p(\Omega)}, \|\Phi h(\Phi)\|_{L^1(\Omega)}\Big).$$

4.3. *Intermediate derivatives and variable coefficients.* With the help of Appendix II and [2, §7] we extend Theorem 1 and the proof of §2 to the equation $Bu + g(u) = f$, where

$$Bu = \sum_{|\alpha| \leqslant m} (-1)^{|\alpha|} D^\alpha \big(a_\alpha(x)|D^\alpha u|^{p-1} \operatorname{sgn} D^\alpha u\big),$$

$a_\alpha \in L^\infty(\mathbf{R}^n)$ for all α, $a_\alpha \geqslant 0$ a.e. in $\mathbf{R}^n$ for all α, $a_\alpha \geqslant K > 0$ a.e. in $\mathbf{R}^n$ for $|\alpha| = m$. §4.1 is also extended to this case.

4.4. If $r = p$ the proof in §2 yields a solution decaying exponentially at infinity in the sense of the $W^{m,p}$ norm (also in the uniform sense if $n < mp$). See Lemma 3.II. At the end of the proof we perform a limiting process instead of the zero extension of u.

4.5. The case $r = 1$ of Theorem 2 can be stated in terms of the multivalued equation $f \in Au + \operatorname{sgn} u$. See, e.g., [2].

Appendix I. Some weighted interpolation inequalities. The following lemma is taken from [2, Lemma 8], where derivatives are understood in the sense of distributions. We recall notation (4).

LEMMA 1. *Set* $H = \{x \in \mathbf{R}^n : x_n > 0\}$, $t \equiv x_n$. *Let* m, j, k *be integers,* $m \geqslant 1$, $0 \leqslant j < m$, $k \geqslant 0$. *Let* $1 \leqslant p < \infty$ *and* $1 \leqslant r \leqslant p$. *Then*

$$\left(\int_H t^k |D^j u|^p \, dx\right)^{1/p} \leqslant C \left(\int_H t^k |D^m u|^p \, dx\right)^{a/p} \left(\int_H t^k |u|^r \, dx\right)^{(1-a)/r}$$

if the Lebesgue integrals of the right side exist, where a is given by

$$\frac{1}{p} = \frac{j}{n+k} + a\left(\frac{1}{p} - \frac{m}{n+k}\right) + (1-a)\frac{1}{r},$$

and the constant C depends only on n, m, j, p, r, k.

The lemma is easily extended to the case $0 < r < 1$, assuming that $\int_H t^k |u|^p \, dx$ is finite, using the method of [2, Lemma 9].

Appendix II. An ordinary differential inequality. We present precise global bounds for nonoscillatory solutions of the differential inequality

$$|z(t)|^{\lambda_0} \leq B \sum_{i=1}^{m} |z^{(i)}(t)|^{\lambda_i} \quad \text{for all } t \geq 0, \tag{19}$$

where $B, \lambda_0, \ldots, \lambda_m$ are arbitrary positive numbers. We associate to (19) a new set of exponents μ_i defined as follows:

$$\frac{1}{\mu} = \frac{1}{\lambda_0} + \max_{1 \leq i \leq m} \frac{m}{i}\left(\frac{1}{\lambda_i} - \frac{1}{\lambda_0}\right), \tag{20}$$

$$\frac{1}{\mu_i} = \frac{i}{m}\frac{1}{\mu} + \frac{m-i}{m}\frac{1}{\lambda_0}, \quad i = 0, 1, \ldots, m. \tag{21}$$

Therefore $\mu_m = \mu \leq \lambda_m$, $\mu_0 = \lambda_0$, all $\mu_i > 0$. The exponent μ is the greatest number such that the μ_i defined by (21) satisfy

$$\mu_i \leq \lambda_i \quad \text{for } i = 0, 1, \ldots, m. \tag{22}$$

The definition of the μ_i is guided by dimensional analysis (see Remark 2.1). We also note that

$$\lambda_0 \lesseqgtr \mu \quad \text{if} \quad \lambda_0 \lesseqgtr \min\{\lambda_1, \ldots, \lambda_m\}, \quad \text{respectively.} \tag{23}$$

LEMMA 2. *I. Assume* (1) $z \in C^m(\overline{\mathbf{R}}_+) \cap L^\infty(\mathbf{R}_+)$, $m \geq 1$; (2) $z^{(m)}$ *is monotone in* $\overline{\mathbf{R}}_+$; *and* (3) z *satisfies the differential inequality* (19). *Then*

$$|z(t)|^{\lambda_0} \leq K |z^{(m)}(t)|^{\mu} \quad \text{for all } t \geq 0, \tag{24}$$

where μ is given by (20) (*thus* (23) *holds*), *and the constant K depends only on $m, \lambda_0, \lambda_1, \ldots, \lambda_m, B, |z^{(m)}(0)|$, and $|z(0)|$.*

II. The constant K can be chosen to be independent of $|z(0)|$ if, in addition,

$$\frac{m-i}{m}\lambda_i < \lambda_0 \quad \text{for } i = 1, \ldots, m-1. \tag{25}$$

III. The constant K depends only on $m, \lambda_0, \lambda_m, B$ if, in addition to I, we have $\mu_i = \lambda_i$ for $i = 1, \ldots, m$ (in particular, $\mu = \lambda_m$); i.e., if

$$\frac{1}{\lambda_i} = \frac{i}{m}\frac{1}{\lambda_m} + \frac{m-i}{m}\frac{1}{\lambda_0} \quad \text{for } i = 1, \ldots, m-1. \tag{26}$$

This lemma is completed by

LEMMA 3. *Assume hypotheses* (1) *and* (2) *of Lemma 2. Assume that z satisfies* (24), K, λ_0 *and μ now being arbitrary positive numbers.*

I. If $\lambda_0 < \mu$, then z has compact support $[0, a]$ and $a \leq C|z^{(m)}(0)|^{1/(\tau - m)}$.

II. If $\lambda_0 = \mu$, then, for $0 \leq i \leq m - 1$, $|z^{(i)}(t)| \leq C_i e^{-Ct}$ for all $t \geq 0$.

III. If $\lambda_0 > \mu$, then, for $0 \leq i \leq m - 1$ and for all $t \geq 0$,

$$|z^{(i)}(t)| \leq C_i \left(|z^{(m)}(0)|^{1/(\tau - m)} + Ct\right)^{\tau - i}.$$

C and C_i are positive constants depending only on m, λ_0, μ, K and m, i, λ_0, μ, K, respectively, and τ is defined by

$$\tau = m\mu/(\mu - \lambda_0). \tag{27}$$

REMARK 2. I. The numbers μ_i of (21) have the property that all terms $|z^{(i)}(t)|^{\mu_i}$ have the same dimensionality, i.e., that there exists τ such that $(\tau - i)\mu_i = \tau\lambda_0$ for all i (τ is given by (27)). The dimensionality exponent of $z^{(i)}(t)$ is $\tau - i$. Note that $\tau > m$ if $\lambda_0 < \mu$, and $\tau < 0$ if $\lambda_0 > \mu$.

II. Lemma 2 is partly preserved if we replace μ by any smaller positive number. The choice (20) of μ gives the optimal information about the behaviour of z at infinity.

III. Lemma 3.III implies that $z^{(m)}(t) = O(t^{\tau-m})$ as $t \to \infty$. See [3, §7.3] if $z^{(m-1)}$ (rather than $z^{(m)}$) is assumed to be monotone.

PROOF OF LEMMA 2. The hypotheses imply that, for $0 \le i \le m$, $z^{(i)}(t) \to 0$ as $t \to \infty$ and $z^{(i)}(t)$ is monotone with constant sign in $\overline{\mathbf{R}}_+$. (If not, z would be unbounded.) Therefore,

$$|z^{(i)}(t)| = \|z^{(i)}\|_{L^\infty(t,\infty)}.$$

Thus, by L^∞ interpolation (for half-lines),

$$|z^{(i)}(t)| \le C|z^{(m)}(t)|^{i/m}|z(t)|^{(m-i)/m}. \tag{28}$$

From (22) and (28),

$$|z^{(i)}(t)|^{\lambda_i} \le |z^{(i)}(0)|^{\lambda_i - \mu_i}|z^{(i)}(t)|^{\mu_i} \le K_i|z^{(i)}(t)|^{\mu_i}, \tag{29}$$

where K_i depends only on m, i, $\lambda_0, \lambda_1, \ldots, \lambda_m$, $|z^{(m)}(0)|$, and $|z(0)|$.

$$|z^{(i)}(t)|^{\mu_i} \le (C/\varepsilon)|z^{(m)}(t)|^{(i/m)\mu_i}\varepsilon|z(t)|^{((m-i)/m)\mu_i} \tag{30}$$
$$\le C\big((1/\varepsilon)^{q_i}|z^{(m)}(t)|^\mu + \varepsilon^{q'_i}|z(t)|^{\lambda_0}\big),$$

where we have used Young's inequality and (21), setting

$$\frac{1}{q_i} = \frac{i}{m}\frac{\mu_i}{\mu}, \qquad \frac{1}{q'_i} = \frac{m-i}{m}\frac{\mu_i}{\lambda_0}.$$

Inserting (30) and (29) in (19) and choosing ε small enough (depending on m, B, and the λ_i, K_i), we obtain Lemma 2.I. III is clear from (29). Setting $t = 0$ in (19) and (28), we obtain

$$|z(0)|^{\lambda_0} \le C \sum_{i=1}^m |z^{(m)}(0)|^{(i/m)\lambda_i}|z(0)|^{((m-i)/m)\lambda_i},$$

which implies (e.g., by Young's inequality) that $|z(0)|$ can be bounded in terms of $|z^{(m)}(0)|$ if (25) holds. This proves II.

PROOF OF LEMMA 3. From (28), (24), and (27),

$$|z^{(m-1)}(t)| \le C|z^{(m)}(t)|^{(m-1)/m + \mu/m\lambda_0} = C|z^{(m)}(t)|^{(\tau-m+1)/(\tau-m)}.$$

Integrating this first order differential inequality (for $z^{(m-1)} \neq 0$), we obtain Lemma 3 for $i = m - 1$. The proof is completed by successive integrations between t and ∞. (Recall that $z^{(i)}(t) \to 0$ as $t \to \infty$, and $\tau < 0$ in case III.)

REMARK 3. Kiguradze (see the survey [9]) has studied the asymptotic behaviour of *nonoscillatory* solutions of general classes of *higher-order*, ordinary differential equations. In particular, the compactness of the support of z in Lemma 3.I is already proved in [8, Theorem 4] (see also [9, Theorem 2.10]) with the help of comparison principles. (Kiguradze says "singular solution of the first kind".) For systems of equations see Čanturia [5, Theorem 2]. In [3] we prove that Lemma 3.I, II, III still hold for some classes of *oscillatory* solutions.

REFERENCES

1. S. N. Antoncev, *On the localization of solutions of nonlinear degenerate elliptic and parabolic equations*, Soviet Math. Dokl. **24** (1981), 420–424.

2. F. Bernis, *Compactness of the support for some nonlinear elliptic problems of arbitrary order in dimension n*, Comm. Partial Differential Equations **9**(3) (1984), 271–312.

3. _____, *Asymptotic rates of decay for some nonlinear ordinary differential equations and variational problems of arbitrary order*, Ann. Fac. Sci. Toulouse **6** (1984), 121–151.

4. H. Brezis and F. E. Browder, *Some properties of higher order Sobolev spaces*, J. Math. Pures Appl. (9) **61** (1982), 245–259.

5. T. A. Čanturia, *On singular solutions of nonlinear systems of ordinary differential equations*, Differential Equations (M. Farkas, ed.), Colloq. Math. Soc. János Bolyai, Vol. 15, North-Holland, Amsterdam, 1977, pp. 107–109.

6. J. I. Díaz, *Elliptic and parabolic quasilinear equations giving rise to a free boundary: the boundary of support of the solutions*, These Proceedings.

7. J. I. Díaz and L. Véron, *Compacité du support des solutions d'équations quasilinéaires elliptiques ou paraboliques*, C. R. Acad. Sci. Paris Sér. A **297** (1983), 149–152.

8. I. T. Kiguradze, *The problem of oscillation of solutions of nonlinear differential equations*, Differential Equations **1** (1965), 773–782.

9. _____, *On the oscillatory and monotone solutions of ordinary differential equations*, Arch. Math. (Brno) **14** (1978), 21–44.

UNIVERSIDAD POLITÉCNICA, SPAIN

On a System of Degenerate Diffusion Equations

M. BERTSCH, M. E. GURTIN, D. HILHORST AND L. A. PELETIER

1. Introduction. In this lecture we discuss the system of equations

(1.1a) $\quad u_t = (u(u+v)_x)_x, \quad x \in (-1,1), t > 0,$

(1.1b) $\quad v_t = k(v(u+v)_x)_x, \quad x \in (-1,1), t > 0,$

in which $k \geq 0$.

This system originated in a model of Gurtin and MacCamy [8] and Gurtin and Pipkin [9] describing the dispersal of two interacting biological populations with respective densities u and v. According to this model, u and v satisfy the conservation laws

(1.2) $\quad u_t + \operatorname{div} J^u = 0, \quad v_t + \operatorname{div} J^v = 0$

in which the fluxes J^u and J^v are related to u and v by the equations

(1.3) $\quad J^u = -k_1 u \operatorname{grad}(u+v), \quad J^v = -k_2 v \operatorname{grad}(u+v), \quad x \in \mathbb{R}^N,$

where $k_1, k_2 \geq 0$. Substituting (1.3) into (1.2) yields (1.1) if we scale t by a factor $k_1 (\neq 0)$, set $k = k_2/k_1$ and $N = 1$.

We consider the cases $k = 0$ and $k > 0$ separately. If $k = 0$, (1.1b) implies that $v = v(x)$, and thus we are dealing with a *mobile* population u in the presence of a *sedentary* population v. In discussing this case we freely quote from [4, 5]. In this context we must also mention a recent paper by Alt [1]. We only make some preliminary observations about the case $k > 0$. In particular, we show that if initially the populations are *segregated*—i.e., they occupy different parts of the habitat—they will remain segregated for all later times if $u_0 + v_0 \geq m > 0$ in $[-1, 1]$ and uniqueness of solutions of (1.1a) and (1.1b) is assumed.

Finally, we observe that if $k = 1$, (1.1a) and (1.1b) can be added to yield a single equation

$$P_t = (PP_x)_x, \quad P = u + v.$$

1980 *Mathematics Subject Classification.* Primary 35B35, 35K65.

© 1986 American Mathematical Society
0082-0717/86 $1.00 + $.25 per page

Having obtained P, we can obtain u and v from the first-order equations
$$u_t = (uP_x)_x, \qquad v_t = (vP_x)_x.$$
This problem is related to one recently considered by Busenberg and Ianelli [6].

2. The case $k = 0$. We write $\Omega = (-1, 1)$, $\mathbb{R}^+ = (0, \infty)$, $Q = \Omega \times \mathbb{R}^+$, and for any $T > 0$, $Q_T = \Omega \times (0, T]$. Postulating no flux boundary conditions, we consider the problem

(I) $$\begin{aligned} u_t &= (u(u+v)_x)_x, & (x,t) &\in Q_T, \\ u(u+v)_x &= 0, & (x,t) &\in \partial\Omega \times \mathbb{R}^+, \\ u(x,0) &= u_0(x), & x &\in \overline{\Omega}, \end{aligned}$$

in which we make the following assumptions about v and u_0:

A1. v is Lipschitz continuous on $\overline{\Omega}$.

A2. There exists a constant M such that
$$\frac{v'(x) - v'(y)}{x - y} \geq -M \quad \text{for almost all } x, y \in \overline{\Omega}, x \neq y.$$

A3. $u_0 \in C(\overline{\Omega})$, $u_0 \geq 0$.

If $v(x) \equiv \text{constant}$ in $\overline{\Omega}$, (1.1a) becomes

(2.1) $$u_t = \tfrac{1}{2}(u^2)_{xx}.$$

Thus, we essentially obtain the porous medium equation, about which many results are known [7, 10]. In particular, if $\text{supp } u_0 \subset \Omega$, it need not have a classical solution. Therefore, it is necessary to define solutions of Problem (I) in a weak sense.

DEFINITION. *A (weak) solution u of Problem* (I) *on $[0, T]$ is a bounded continuous function defined on $\overline{Q}_T$ with the property*

(2.2) $$\int_\Omega u(t)\psi(t) = \int_\Omega u_0 \psi(0) + \int_0^t \int_\Omega \left(\frac{1}{2} u^2 \psi_{xx} + u\psi_t - uv_x \psi_x \right)$$

for any $t \in (0, T]$ and $\psi \in C^{2,1}(\overline{Q}_T)$ such that $\psi \geq 0$ in $\overline{Q}_T$ and $\psi_x = 0$ on $\partial\Omega \times (0, T]$.

A (weak) subsolution (supersolution) is defined similarly, but with equality in (2.2) replaced by $(\leq) \geq$.

THEOREM 1 [5]. *Let v and u_0 satisfy, respectively,* A1 *and* A3. *Then Problem* (I) *has a solution in $\overline{Q}_T$. If, in addition, v satisfies* A2, *the solution is unique.*

When $v(x) \equiv \text{constant}$ and u satisfies (2.1), it is known that for each $u_0 \geq 0$ ($\not\equiv 0$) there exists a time $T > 0$ such that
$$u(x, t) > 0 \quad \text{for all } x \in \overline{\Omega}, t > T;$$
i.e., the mobile species eventually occupies the entire habitat $\overline{\Omega}$ [2]. This raises the question as to whether this is so if $v(x) \not\equiv \text{constant}$. In other words,

> *can a sedentary population (v) prevent the mobile one (u) from spreading throughout the entire habitat?*

To answer this question we shall consider functions v with support inside Ω, and u_0 with support to one side of the support of v.

Similarly, if $v(x) \equiv$ constant, the solution u of Problem (I) has the property [2]
$$u_0(x_0) > 0 \Rightarrow u(x_0, t) > 0 \quad \forall t \geq 0,$$
and one can ask whether this property continues to hold if $v(x) \not\equiv$ constant. In biological terms:

> *can a sedentary population (v) ever remove the mobile one (u) completely from some part of the habitat?*

We show, by means of an example, that this is possible.

Let $\mathscr{E}$ denote the set of equilibrium solutions of Problem (I). Then $q \in \mathscr{E}$ if (i) $q \geq 0$, (ii) $q + v$ is absolutely continuous on $\overline{\Omega}$, and (iii) $q(q + v)' = 0$ in $\overline{\Omega}$.

Let $\hat{v} = \max_{\overline{\Omega}} v$. Then we write $\mathscr{E} = \mathscr{E}_1 \cup \mathscr{E}_2$, where
$$\mathscr{E}_1 = \{q \in \mathscr{E}: q + v > \hat{v} \text{ on } \overline{\Omega}\}, \quad \mathscr{E}_2 = \{q \in \mathscr{E}: q + v \leq \hat{v} \text{ on } \overline{\Omega}\}.$$
To see this, let $q \in \mathscr{E}$ and let $\mathscr{P}$ be a connected component of the set $\{x \in \overline{\Omega}: q(x) + v(x) > \hat{v}\}$. Since $q + v \in C(\overline{\Omega})$, $\mathscr{P}$ is open. Clearly, $q > 0$ on $\mathscr{P}$, whence
$$q(x) + v(x) \equiv c > \hat{v} \quad \text{in } \mathscr{P}.$$
Thus, again by the continuity of $q + v$, $\mathscr{P} = \overline{\mathscr{P}}$ and $\mathscr{P}$ is closed. It follows that either $\mathscr{P} = \overline{\Omega}$ and $q \in \mathscr{E}_1$, or $\mathscr{P} = \emptyset$ and $q \in \mathscr{E}_2$. Thus, $\mathscr{E}_1$ is a one-parameter family of functions and $\mathscr{E}_2$, in Figure 1, a two-parameter family.

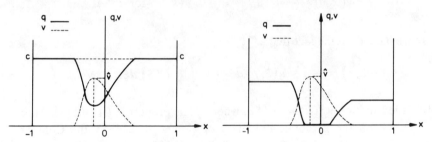

FIGURE 1. The sets $\mathscr{E}_1$ and $\mathscr{E}_2$.

Suppose supp $v = [a, b] \subset \Omega$ and write $\Omega_L = (-1, a)$, $\Omega_R = (b, 1)$. Set $\hat{u}_0 = \max_{\overline{\Omega}} u_0$, $\hat{v} = \max_{\overline{\Omega}} v$, and finally, let $u = u(t, u_0)$ denote the solution of Problem (I).

The next two theorems give an answer to the question of whether the sedentary population can act as an effective barrier against the mobile population.

THEOREM 2. *Suppose* supp $u_0 \subset \overline{\Omega}_L$. *Then*
$$\hat{u}_0 \leq \hat{v} \Rightarrow \text{supp } u(t, u_0) \cap \overline{\Omega}_R = \emptyset, \quad \forall t \geq 0.$$

Theorem 2 is proved by means of the *Comparison Principle*.

LEMMA 1. *Let $\underline{u}(t)$ be a subsolution and $\overline{u}(t)$ a supersolution of Problem (I). Then*
$$\underline{u}(0) \leq \overline{u}(0) \Rightarrow \underline{u}(t) \leq \overline{u}(t), \quad \forall t \geq 0.$$

Let

(2.3) $$\hat{q}(x) = \begin{cases} \hat{u}_0 & \text{when } x \in \overline{\Omega}_L, \\ [\hat{u}_0 - v(x)]_+ & \text{when } x \in (a, \hat{x}), \\ 0 & \text{when } x \in [\hat{x}, 1], \end{cases}$$

where $[f]_+ = \max\{0, f\}$ and $\hat{x} = \sup\{x > a : v < \hat{v} \text{ on } (a, x]\}$. Then $\hat{q} \in \mathscr{E}_2$, $u_0 \leq \hat{q}$, and, hence, by Lemma 1, $u(t, u_0) \leq \hat{q}$ for all $t \geq 0$. Thus $u(t, u_0)$ is in $\overline{\Omega}_L$ for all $t \geq 0$.

THEOREM 3. *Suppose there exists a $q \in \mathscr{E}$ such that*
(i) $\operatorname{supp} q \cap \Omega_R = \emptyset$,
(ii) $\int_x^1 u_0 \leq \int_x^1 q$, $\forall x \in \overline{\Omega}$.
Then $\operatorname{supp} u(t, u_0) \cap \overline{\Omega}_R = \emptyset$ *for all* $t \geq 0$.

PROOF (SKETCH). Define the function $w(x, t) = -\int_x^1 u(\xi, t) \, d\xi$. Then w is a solution of the problem

(2.4) $$\begin{aligned} w_t &= \tfrac{1}{2}(w_x^2)_x + v_x w_x && \text{in } Q, \\ w(-1, t) &= -\int_{-1}^1 u_0, \; w(1, t) = 0 && \text{for } t > 0, \\ w(x, 0) &= -\int_x^1 u_0 && \text{for } x \in \overline{\Omega}, \end{aligned}$$

where we have used the fact that

$$w(-1, t) = -\int_{-1}^1 u(\xi, t) \, d\xi = -\int_{-1}^1 u_0(\xi) \, d\xi, \quad \forall t \geq 0.$$

Define the function $w^*(x) = -\int_x^1 q$, where q has properties (i) and (ii). Then, by (ii), $w \geq w^*$ on the parabolic boundary of Q, and, hence, by the Comparison Principle for equation (2.4), $w \geq w^*$ in $\overline{Q}$.

By (i), $w^* = 0$ in $\overline{\Omega}_R \times \mathbb{R}^+$, whence $w \geq 0$ in $\overline{\Omega}_R \times \mathbb{R}^+$. But by definition, $w \leq 0$ in $\overline{Q}$. Hence, $w = 0$ in $\overline{\Omega}_R \times \mathbb{R}^+$, which implies that $u = 0$ in $\overline{\Omega}_R \times \mathbb{R}^+$.

What the distribution u will eventually look like can often be determined with the help of the following theorem.

THEOREM 4 [4, 5]. *Let v and u_0 satisfy, respectively, A1, A2, and A3, and let $u(t, u_0)$ be the solution of Problem (I). Then there exists an equilibrium solution q such that $u(t, u_0) \to q$ in $C(\overline{\Omega})$ as $t \to \infty$. Moreover, $\int_\Omega q = \int_\Omega u_0$.*

The proof is based on two ingredients:
(i) A contraction property of the solution: if u_{01} and u_{02} are two initial functions satisfying A3, then

(2.5) $$\|u(t, u_{01}) - u(t, u_{02})\|_{L^1(\Omega)} \leq \|u_{01} - u_{02}\|_{L^1(\Omega)};$$

and if, in addition, u_{01} and u_{02} *intertwine*—i.e., if there exists an interval $I \subset \Omega$ on which $u_{0i} > 0$, $i = 1, 2$, and $u_{01} - u_{02}$ changes sign—then the inequality in (2.5) is *strict*. (ii) A Lyapunov function argument, using the function $V: L^1(\Omega) \to \mathbb{R}$ defined by $V(w) = \int_\Omega |w - q^*|$, where q^* is an appropriately chosen element in $\mathscr{E}$.

Suppose v and u_0 are as in Theorem 2, $\hat{u}_0 \leq \hat{v}$, and v has the property that for each $c \in [0, \hat{v})$,

(2.6) $\qquad \{x \in \Omega: v(x) > c\}$ is a connected set.

Then the asymptotic density profile $q = \lim_{t \to \infty} u(t, u_0)$ can be uniquely determined. By Lemma 1, $u(t, u_0) \leq \hat{q}$ for all $t \geq 0$, where $\hat{q}$ is defined by (2.3). Hence, $q \leq \hat{q}$, which implies that $q \in \mathscr{E}_2$. In view of (2.6), q is now uniquely determined by the condition $\int_\Omega q = \int_\Omega u_0$ of Theorem 4.

We conclude this section with a remark about the second question: Can a sedentary population—according to this model—drive a mobile one out of part of a habitat? In the following example we show that this can indeed happen.

Set $v(x) = 1 + x$. Then (1.1a) becomes

(2.7) $\qquad u_t = \tfrac{1}{2}(u^2)_{xx} + u_x,$

or, in terms of the variables $\xi = x + t$, $\tau = t$, and $\bar{u} = \tfrac{1}{2}u$,

$$\bar{u}_\tau = (\bar{u}^2)_{\xi\xi},$$

which is the porous medium equation. Remembering the Barenblatt–Pattle (exact) solution of this equation [3], we conclude that the function

(2.8) $\quad u(x, t; a) = \tfrac{1}{6}(t + 1)^{-1/3}[a^2 - \eta^2]_+, \quad \eta = (x + t)/(t + 1)^{1/3},$

in which $a \in \mathbb{R}^+$, is an exact solution of (2.7) and thus of Problem (I) on $[0, T]$, provided $a \in (0, 1)$, u_0 is chosen appropriately, and T is chosen so small that $\operatorname{supp} u(\cdot, t; a) \subset [-1, 1]$ for $0 \leq t \leq T$.

It follows from (2.8) that

$$\operatorname{supp} u(\cdot, t; a) = [\zeta_1(t), \zeta_2(t)], \quad \text{where } \zeta_i(t) = -t + (-1)^i a(t + 1)^{1/3}.$$

Hence, $[\zeta_1(t), \zeta_2(t)] \subset [-1, 1]$ if $0 \leq t \leq T < 1$ and $a \leq 2^{-1/3}(1 - T)$.

Note that if $a < T(1 + 2^{1/3})$, there exists a time $T_0 \in (0, T)$ such that

$$\operatorname{supp} u(t, u_0) \cap \operatorname{supp} u_0 = \varnothing \quad \text{for } T_0 < t \leq T.$$

Thus at time T_0 the mobile individuals have been driven out of their original territory.

3. The case $k > 0$. Postulating no flux boundary conditions again, we now consider the problem

(II) $\qquad \begin{aligned} u_t &= (u(u + v)_x)_x & \text{in } Q_T, \\ v_t &= k(v(u + v)_x)_x & \text{in } Q_T, \\ u(u + v)_x &= 0, v(u + v)_x = 0 & \text{on } \partial\Omega \times (0, T], \\ u(x, 0) &= u_0(x), v(x, 0) = v_0(x) & \text{in } \overline{\Omega}, \end{aligned}$

in which u_0 and v_0 are nonnegative and Lipschitz continuous in $\overline{\Omega}$.

DEFINITION. *A weak solution of Problem* (II) *on* $[0, T]$ *is a pair of functions* (u, v) *with the properties*

(i) $u, v \in L^\infty(Q_T), u(t), v(t) \in L^\infty(\Omega)$ for all $t \in (0, T]$,
(ii) $u(u + v)_x, v(u + v)_x \in L^2(Q_t)$ for all $t \in (0, T]$,
(iii)
$$\int_\Omega u(t)\psi(t) - \int_\Omega u_0\psi(0) = \int_0^t \int_\Omega \{u\psi_t - u(u+v)_x\psi_x\},$$
$$\int_\Omega v(t)\psi(t) - \int_\Omega v_0\psi(0) = \int_0^t \int_\Omega \{v\psi_t - kv(u+v)_x\psi_x\}$$

for all $t \in (0, T]$ and all $\psi \in C^{1,1}(\overline{Q}_T)$.

REMARK 1. Suppose (u, v) is a solution of Problem (II). Then by setting $\psi = 1$, we can deduce from (iii) the conservation laws

(3.1) $\quad\displaystyle\int_\Omega u(t) = \int_\Omega u_0, \quad \int_\Omega v(t) = \int_\Omega v_0 \quad$ for all $t \in [0, T]$.

REMARK 2. Suppose $(\bar{u}, \bar{v})$ is an equilibrium solution of Problem (II). Then
$$\bar{u}(\bar{u} + \bar{v})' = 0, \quad \bar{v}(\bar{u} + \bar{v})' = 0 \quad \text{in } \overline{\Omega}$$
in view of the boundary conditions. Hence, $\{(\bar{u} + \bar{v})^2\} = 0$ in $\overline{\Omega}$ or

(3.2) $\quad\quad\quad\quad\quad\quad\quad \bar{u}(x) + \bar{v}(x) \equiv \text{constant}.$

Let $f: \overline{\Omega} \to \mathbb{R}$. Then we write $I(f) = \text{int supp } f$. We say that the populations u and v are *segregated* at time t if $I(u(t)) \cap I(v(t)) = \emptyset$.

In this section we make the observation that if the two populations are initially segregated, they may—according to this model—remain segregated for all later times and never mix.

Suppose that u_0 and v_0 are segregated, and that $x_1 \in I(u_0), x_2 \in I(v_0) \Rightarrow x_1 < x_2$. Write

(3.3) $\quad\quad\quad\quad\quad\quad\quad U = \displaystyle\int_\Omega u_0 \quad\text{and}\quad V = \int_\Omega v_0.$

Suppose that (u, v) is a segregated solution of Problem (II) on $[0, T]$; that is, $u(t)$ and $v(t)$ are segregated for all $t \in [0, T]$. Then the function $w: \overline{Q}_T \to \mathbb{R}$ defined by

$$w(x, t) = \begin{cases} \int_{-1}^x u(\xi, t)\, d\xi & \text{if } 0 \leq w < U, \\ U + V - \int_x^1 v(\xi, t)\, d\xi & \text{if } U < w \leq U + V, \end{cases}$$

is (formally) a solution of the problem

(III)
(3.4a) $\quad c(w)_t = (w_x^2)_x \quad\quad\quad\quad\quad$ in Q_T,
(3.4b) $\quad w(-1, t) = 0, w(+1, t) = U + V \quad$ for $t \in (0, T]$,
(3.4c) $\quad w(x, 0) = \displaystyle\int_{-1}^x (u_0 + v_0) \quad\quad\quad$ for $x \in \overline{\Omega}$,

where $c: \mathbb{R} \to \mathbb{R}$ is defined by

$$c(s) = \begin{cases} 2(s - U), & s \leq U, \\ (2/k)(s - U), & s > U, \end{cases}$$

and U, V are given by (3.3).

In the following proposition we list a few properties of the solution to Problem (III). Their proofs will appear elsewhere.

PROPOSITION. (a) *There exists a unique function* $w \in C^{0+1}(\overline{Q}_T)$, *such that* $w_x \in C(\overline{Q}_T)$, *which satisfies* (3.4) *in some weak sense.*

(b) $w(\cdot, t) \to \overline{w}$ *exponentially in* $C^1(\overline{\Omega})$ *as* $t \to \infty$, *where*

$$\overline{w}(x) = \frac{1}{2}(U + V)(1 + x).$$

(c) *If* $u_0 + v_0 \geq m > 0$ *in* $\overline{\Omega}$, *the function* $\zeta(t) = \{x \in \Omega : w(x, t) = U\}$ *is well defined for each* $t \in [0, T]$ *and* $\zeta \in C^{1/2}([0, T])$.

If $u_0 + v_0 \geq m > 0$ in $\overline{\Omega}$, it can be shown that the pair of functions (u, v) defined by

$$u(x, t) = \begin{cases} w_x(x, t) & \text{if } w \leq U, \\ 0 & \text{if } w > U, \end{cases}$$

$$v(x, t) = \begin{cases} 0 & \text{if } w \leq U, \\ w_x(x, t) & \text{if } w > U, \end{cases}$$

is a segregated solution of Problem (III) on $[0, T]$.

ACKNOWLEDGEMENT. The authors are grateful to Professor J. L. Vazquez for suggesting Theorem 3.

REFERENCES

1. W. Alt, *Degenerate diffusion equations with drift functionals modelling aggregation* (preprint).
2. D. G. Aronson and L. A. Peletier, *Large time behavior of solutions of the porous medium equation*, J. Differential Equations **39** (1981), 378–412.
3. G. I. Barenblatt, *On some nonstationary motions of fluid and gas in porous media*, Prikl. Mat. Mekh. **17** (1953), 637–732.
4. M. Bertsch, M. E. Gurtin, D. Hilhorst and L. A. Peletier, *On interacting populations that disperse to avoid crowding: the effect of a sedentary colony*, J. Math. Biol. **19** (1984), 1–12.
5. M. Bertsch and D. Hilhorst, *A density dependent diffusion equation in population dynamics: stabilization to equilibrium*, SIAM J. Math. Anal. (to appear).
6. S. Busenberg and M. Lanelli, *A degenerate nonlinear diffusion problem in age-structured population dynamics*, Harvey Mudd College Tech. Rep., 1983.
7. A. Friedman, *Variational principles and free-boundary problems*, Wiley, New York, 1982.
8. M. E. Gurtin and R. C. MacCamy, *On the diffusion of biological populations*, Math. Biosci. **33** (1977), 35–49.
9. M. E. Gurtin and A. C. Pipkin, *On interacting populations that disperse to avoid crowding*, Quart. Appl. Math. **42** (1984), 87–94.
10. L. A. Peletier, *The porous media equation*, Application of Nonlinear Analysis in the Physical Sciences (H. Amann, N. Bazley and K. Kirchgassner, eds.), Pitman, 1981, pp. 229–241.

UNIVERSITY OF LEIDEN, THE NETHERLANDS (Current address of M. Bertsch and L. A. Peletier)

CARNEGIE–MELLON UNIVERSITY (Current address of M. E. Gurtin)

UNIVERSITÉ PARIS–SUD, FRANCE (Current address of D. Hilhorst)

Pointwise Continuity for a Weak Solution of a Parabolic Obstacle Problem

MARCO BIROLI

The results in this paper were obtained jointly with U. Mosco.

1. Notations. Let Ω be an open set in R^N, $N \geq 3$, and $Q = \Omega \times (0, T)$. For sets in R^{N+1} we use the following notion of capacity: For a compact set E contained in an open parabolic cylinder C, define

$$\operatorname*{cap}_C(E) = \operatorname{Inf}\left[\int_C |D_x w|^2 \, dx \, dt; \, w \in D(C), \, w = 1 \text{ in a neighborhood of } E\right].$$

We have so defined a Choquet capacity [6]. For arbitrary sets (with closure in C) we denote the *external* capacity by cap_C. We now give some easily proved properties of this notion of capacity.

(1) Let $C = \omega \times (T_0, T_1)$. Then

$$\operatorname*{cap}_C(E) = \int_{T_0}^{T_1} \operatorname*{cap}_\omega(E_t) \, dt,$$

where $\operatorname{cap}_\omega$ is the usual elliptic capacity, and E_t is the section of E at the instant t.

(2) Let C_1 be a parabolic cylinder whose closure is contained in C (C_1 open). If $w \in L^2(C_1)$ and $D_x w \in L^2(C_1)$, then w is quasi-continuous in C_1; moreover, if $w_n \to w$ in $L^2(C_1)$ and $D_x w_n \to D_x w$ in $L^2(C_1)$, then $w_n \to w$ quasi-uniformly in C_1.

Now fix $z_0 = (x_0, t_0)$ and set

$$B(r, x_0) = \{x; |x - x_0| < r\},$$

$$Q(r, z_0) = \{(x, t); |x - x_0| < r, |t - t_0| < r^2\},$$

$$Q_\theta^-(r, z_0) = \{(x, t); |x - x_0| < (1 - 6\theta)^{1/2} r^2, t_0 - r^2 < t < t_0 - 6\theta r^2\}.$$

1980 *Mathematics Subject Classification.* Primary 49A29; Secondary 35K55, 35R35, 35B65.

© 1986 American Mathematical Society
0082-0717/86 $1.00 + $.25 per page

Let ψ be a function defined quasi-everywhere and set

$$E(\psi, z_0, \eta, r) = \{z \in Q_\theta^-(r, z_0); \psi(z) \geq \operatorname{Sup} \psi(z) - \eta; |t - t_0| < r^2, |x - x_0| < r\}, \quad \eta > 0,$$

$$\Delta_\theta(\psi, z_0, \eta, r) = \Delta_\theta(\eta, r) = \underset{Q(2r, z_0)}{\operatorname{cap}} E(\psi, z_0, \eta, r),$$

$$\delta_\theta(\eta, r) = \Delta_\theta(\eta, r)/\sigma_N r^N,$$

where $\sigma_N = \operatorname{cap}_{Q(2, z_0)} Q(1, z_0)$. The *Wiener modulus* of ψ is defined by

$$\omega_\theta(r, R) = \operatorname{Inf}\left\{\omega \geq 0; \omega \exp \int_r^R \delta_\theta(\omega, \rho) \frac{d\rho}{\rho} \geq 1\right\}.$$

We let G^{z_0} be the Green function of the heat operator, and

$$G_\rho^{z_0} = G^{z_0} * \alpha_\rho \quad \left(\alpha \in D(Q_1), \int \alpha\, dx\, dt = 1, \alpha_\rho = \rho^{-(N+1)} \alpha(\rho^2 t, \rho x)\right)$$

is the regularized Green function.

2. Results. The continuity of the solution of a parabolic obstacle problem has been studied for regular continuous obstacles in [**2, 3**] for linear and (some) nonlinear cases and in [**12**] for obstacles that are Hölder continuous in time and one-sided Hölder continuous in space. In this paper we also consider the case of irregular obstacles and give pointwise estimates on the modulus of continuity. We observe that the techniques used here are derived from the analogous elliptic problem solved by J. Frehse and U. Mosco [**7**]–[**11**], and from the proof of a Wiener estimate at boundary points for the solution of a parabolic equation [**4, 5**]; moreover, in this last paper a different notion of capacity is used.

Let ψ be a function bounded above, which we assume to be quasi-everywhere defined in Q (for the above-defined notion of capacity); a given function $u \in L^2(0, T; H^1(\Omega)) \cap L^\infty(0, T; L^2(\Omega))$ is a *local solution* of the parabolic obstacle problem relative to the heat operator and ψ if

$$\int_0^t \int_\Omega \{D_t v(v - u)\phi + D_x u D_x(\phi(v - u))\}\, dx\, dt + \frac{1}{2}\int_0^t \int_\Omega D_t \phi(v - u)^2 dx\, dt \geq 0,$$

$$\forall \phi \in C^\infty(Q), \phi|_\Sigma = 0 \ (\Sigma = \partial\Omega \times (0, T)), \phi(x, 0) = 0,$$

$$\forall v \in H^1(0, T; L^2(\Omega)) \cap L^2(0, T; H^1(\Omega)),$$

$$v \geq \psi \text{ quasi-everywhere in } \operatorname{supp}(\phi) \subset \Omega \times (0, t).$$

REMARK 1. Using the test function $v = M + \lambda\mu$, $\mu \in D(Q)$, $M = \operatorname{Sup}_Q \psi$, we obtain, as $\lambda \to \infty$, that u is a supersolution of the heat equation.

We obtain the following results:

THEOREM 1. *Let u be a local solution of the parabolic obstacle problem relative to the heat operator and ψ; then for $z_0 \in Q$ we have*

(2.1.) $$M(r) \leq K\left\{M(R)\omega_\theta(r, R)^\beta + \omega_\theta(r, R) \wedge \underset{Q(R, z_0)}{\operatorname{osc}} \psi\right\},$$

where $\beta \in (0,1)$, $R \leq R_0$ (R_0 suitable), $\theta \in (0, \theta_0)$, and

$$M(r) = \left(\int_{Q(r, z_0)} |D_x u|^2 G^{z_0} \, dx \, dt \right)^{1/2} + \operatorname*{osc}_{Q(r, z_0)} u.$$

THEOREM 2. *Let the assumptions of Theorem 1 hold; then*

$$\operatorname*{osc}_{Q(r, z_0)} u \leq K\left(R^\gamma + \operatorname*{osc}_{Q(R, z_0)} \psi \right) \omega_\theta^\gamma(r, R) + \omega_\theta(r, R) \wedge \operatorname*{osc}_{Q(R; z_0)} \psi,$$

where $\gamma \in (0, 1)$.

Such an estimate has been obtained, for the first time, for the elliptic case in [10, 11].

COROLLARY 1. *If*

$$(2.2) \qquad \lim_{r \to 0} \omega_\theta(r, R) = 0,$$

u is continuous at z_0; moreover, if

$$(2.3) \qquad \omega_\theta(r, R) \leq K(r/R)^\nu, \quad \nu \in (0, 1),$$

then u is Hölder continuous at z_0.

COROLLARY 2. *If ψ is continuous (Hölder continuous) at z_0, then u is continuous (Hölder continuous) at z_0.*

REMARK 2. We observe that if $\delta_\theta(\eta, r)$ is decreasing in θ, then $\omega_\theta(r, R)$ is increasing in θ, and if (2.2) ((2.3)) holds for θ_0, then it also holds for $\theta \in (0, \theta_0)$.

REMARK 3. Condition (2.2) holds if

$$(2.4) \qquad \lim_{r \to 0} \int_r^1 \delta_\theta(\eta, \rho) \frac{d\rho}{\rho} = +\infty, \quad \forall \eta > 0.$$

Condition (2.3) holds if

$$(2.5) \qquad \liminf_{r \to 0} |\log r|^{-1} \int_r^1 \delta_\theta(\eta, \rho) \frac{d\rho}{\rho} \geq K > 0.$$

REMARK 4. The results given here also hold for general linear parabolic operators and can also be proved in the nonlinear case with quadratic growth in the spatial gradient.

3. Sketch of the proofs. The proof of Theorem 2 can be obtained by Theorem 1 and an iteration method, as in [10, 11]. The main tool in the proof of Theorem 1 is a Poincaré inequality involving only the spatial gradient, which is given for supersolution of the heat operator.

PROPOSITION 1. *Let w be a supersolution of the heat operator in $Q(2; 0)$ such that*

$$\|w^+ \phi(t_2)\|^2_{L^2(B(1;0))} - \|w^+ \phi(t_1)\|^2_{L^2(B(1;0))} \leq C \int_{t_1}^{t_2} \int_\Omega D_x w^+ D_x \phi \phi w^+ \, dx \, dt,$$

$$t_1 \leq t_2, t_1, t_2 \in (-1, 1), \phi \in D\big(B\big((1-\chi)^{1/2}; 0\big)\big), u \in L^\infty(Q(2, 0));$$

then

$$\int_{Q_{(1-\chi)^{1/2}}} |w^+|^2 \, dx \, dt \leq \frac{K}{\delta^+} \int_{Q_1} |D_x w^+|^2 \, dx \, dt,$$

$$\int_{Q_{(1-\chi)^{1/2}}} |w^-|^2 \, dx \, dt \leq \frac{K}{\delta^-} \int_{Q_1} |D_x w^-|^2 \, dx \, dt,$$

where δ^+ and δ^- are the normalized capacities of the sets

$$N^+ = \left\{ z \in (-1, -1+\chi) \times B\big((1-\chi)^{1/2}; 0\big); w^+(z) = 0 \right\},$$

$$N^- = \left\{ z \in (-1, -1+\chi) \times B\big((1-\chi)^{1/2}; 0\big); w^-(z) = 0 \right\},$$

and K depends on χ.

The result of Proposition 1 is obtained by the compactness Lemma 4.4 of [5] and by a proof ab absurdo.

Now let $d \geq \psi$ in $Q(R, z_0)$ q.e. and let $\bar{z} \in Q(\theta^{1/2}R; z_0)$. We choose, in the variational inequality, $v = d$ and $\phi = \eta^2 \tau^2 G^{\bar{z}}$, where $\eta \in D(B(R, x_0))$ is such that $\eta = 1$ in $B(R/8; \bar{x})$, $\eta = 0$ outside $B(R/4; \bar{x})$, $0 \leq \eta \leq 1$, and $\tau \in D(t_0 - r^2, t_0 + r^2)$ is such that $\tau = 1$ for $t \geq \bar{t} - 3R^2$, $\tau = 0$ for $t \leq \bar{t} - 5R^2$, and $0 \leq \tau \leq 1$. We can suppose $|D_x \eta| \leq C/R$, $|D_{xx} \eta| \leq C/R^2$, $|D_t \tau| \leq C/\theta R^2$. By standard computations we obtain

$$(3.1) \quad \int_{Q(R, z_0)} |D_x u|^2 G_\rho^{\bar{z}} \eta^2 \tau^2 \, dx \, dt + \left(|u - d|^2 \eta^2 \tau^2 * \alpha_\rho \bar{z} \right)(\bar{z})$$

$$\leq C_1 R^{-2} \int_{\bar{t} - 3\theta R^2}^{\bar{t}} \int_{B(R/4; \bar{x}) - B(R/8; \bar{x})} |u - d|^2 G_\rho^{\bar{z}} \, dx \, dt$$

$$+ C_2(1 + \theta^{-1}) \int_{\bar{t} - 5\theta R^2}^{\bar{t} - 3\theta R^2} \int_{B(R/4; \bar{x})} |u - d|^2 G_\rho^{\bar{z}} \, dx \, dt.$$

Taking the limit as $\rho \to 0$ and using the estimates on the Green function [1], we have

$$(3.2) \quad \int_{Q(R, z_0)} |D_x u|^2 G^{\bar{z}} \eta^2 \tau^2 \, dx \, dt + |u - d|^2(\bar{z})$$

$$\leq C_3 \exp\left(\frac{-C_4}{\theta}\right) \theta^{-N/2} \sup_{(\bar{t} - 3\theta R^2, \bar{t}) \times B(R/4; \bar{x})} |u - d|^2$$

$$+ C_5 \theta^{-(N+2)/2} R^{-(N+2)} \int_{\bar{t} - 5\theta R^2}^{\bar{t} - 3\theta R^2} \int_{B(R/4; \bar{x})} |u - d|^2 \, dx \, dt;$$

then, taking the supremum for $\bar{z}$ in $Q(\theta^{1/2}R; z_0)$,

$$(3.3) \quad \int_{t_0-\theta R^2}^{t_0} \int_{B(\theta^{1/2}R; x_0)} |D_x u|^2 G^{z_0} \, dx \, dt + \sup_{Q(\theta^{1/2}R; z_0)} |u - d|^2$$

$$\leq 2C_3 \exp\left(\frac{-C_4}{\theta}\right) \theta^{-N/2} \sup_{Q(R; z_0)} |u - d|^2$$

$$+ 2C_5 \theta^{-(N+2)/2} R^{-(N+2)} \int_{t_0-6\theta R^2}^{t_0-2\theta R^2} \int_{B(R/2; x_0)} |u - d|^2 \, dx \, dt.$$

Now let $\hat{d}$ be such that

$$\text{cap}\big(z \in E(\psi, z_0, \eta, R), (u(z) - \hat{d})^+ = 0\big) \geq \Delta_\theta(\eta, R)/4,$$

$$\text{cap}\big(z \in E(\psi, z_0, \eta, R), (u(z) - \hat{d})^- = 0\big) \geq \Delta_\theta(\eta, R)/4,$$

and $d = \hat{d}_+ \eta$. We observe that

$$\sup_{Q(\theta^{1/2}R, z_0)} (u(z) - d)^2 \geq 4^{-1} \left(\operatorname*{osc}_{Q(\theta^{1/2}R, z_0)} (u(z) - d)\right)^2 = 4^{-1} \left(\operatorname*{osc}_{Q(\theta^{1/2}R; z_0)} u\right)^2,$$

$$\sup_{Q(R; z_0)} (u(z) - \hat{d})^2 \leq \left(\operatorname*{osc}_{Q(R; z_0)} (u(z) - \hat{d})\right)^2 = \left(\operatorname*{osc}_{Q(R; z_0)} u\right)^2.$$

From (3.3), using the Poincaré inequality and the estimates on the Green function [1], we have

$$(3.4) \quad \int_{t_0-\theta R^2}^{t_0} \int_{B(\theta^{1/2}R; x_0)} |D_x u|^2 G^{z_0} \, dx \, dt + \left(\operatorname*{osc}_{Q(\theta^{1/2}R; z_0)} u\right)^2$$

$$\leq C_6 C_7(\theta) \left(\operatorname*{osc}_{Q(R; z_0)} u\right)^2$$

$$+ C_8(\theta)(C_9(\theta)\delta_\theta(\eta, R))^{-1} \int_{t_0-R^2}^{t_0-\theta R^2} \int_{B(R; x_0)} |D_x u|^2 G^{z_0} \, dx \, dt$$

$$+ C_{10}(\theta)\eta,$$

where

$$C_7(\theta) = \exp(-C_4/\theta), \quad C_9(\theta) = \exp(-C_{11}/\theta)\theta,$$

and $C_8(\theta)$, $C_{10}(\theta)$ are decreasing in θ.

Now, using a hole-filling method in time as in [5] and the integration lemma of [9], we have

$$M(r) \leq KM(R)\exp\left(-\beta \int_r^R \delta_\theta(\eta, \rho) \frac{d\rho}{\rho}\right) + C_{12}\eta$$

for $\theta \in (0, \theta_0)$ and β, C_{12} depending on θ. Choosing $\eta = \omega_\theta(r, R) + \varepsilon$, $\varepsilon \geq 0$, and taking into account that

$$\frac{r}{R} \leq \omega_\theta(r, R) \leq \frac{r}{R} V \underset{Q(R; z_0)}{\operatorname{osc}} \psi,$$

we easily obtain, as $\varepsilon \to 0$, Theorem 1; Theorem 2 follows by an iteration technique as in [10, 11]. The results of Corollaries 1 and 2 can be easily obtained from the estimate in Theorem 2.

References

1. D. G. Aronson, *Non-negative solutions of linear parabolic equations*, Ann. Scuola Norm. Sup. Pisa Cl. **22** (1968), 607–694.
2. M. Biroli, *Hölder regularity for parabolic obstacle problem*, Boll. Un. Mat. Ital. B(6) **1** (1982), 1079–1088.
3. _____, *Existence d'une solution hölderienne pour des inéquations paraboliques non linéaires avec obstacle*, C. R. Acad. Sci. Paris **296** (1983), 7–9.
4. M. Biroli and U. Mosco, *Estimations ponctuelles sur le bord du module de continuité des solutions faibles des équations paraboliques*, C. R. Acad. Sci. Paris **296** (1983),
5. _____, *Wiener estimates at boundary points for parabolic equations*, Sonderforschungbereich 72, Universität Bonn, 1984 (preprint).
6. G. Choquet, *Lectures on analysis*, Vol. I, Benjamin, 1969.
7. J. Frehse and U. Mosco, *Irregular obstacles and quasi-variational inequalities of the stochastic impulse control*, Ann. Scuola Norm. Sup. Pisa Cl. Sci(4) **9** (1982), 105–157.
8. _____, *Sur la continuité ponctuelle des solutions faibles du problème d' obstacle*, C. R. Acad. Sci. Paris **295** (1982), 571–574.
9. _____, *Wiener obstacles* (Sem. Nonlinear Partial Differential Equations, College de France), Res. Notes in Math., Pitman, 1984.
10. U. Mosco, *Module de Wiener pour un problème d' obstacle*, C. R. Acad. Sci. Paris (in press).
11. _____, Lecture at the Summer Research Institute, UCLA, July 1983.
12. M. Struwe and M. A. Vivaldi, Personal communication.

Politecnico di Milano, Italy

A Note on Iteration for Pseudoparabolic Equations of Fissured Media Type

MICHAEL BÖHM

Abstract. The nonlinear pseudoparabolic fissured-media equation is considered, and a linear iterative approximation scheme is formulated. Various a priori estimates for the solutions of this scheme are shown to converge to the solution of the fissured-media equation.

0. Introduction. Diffusion in fissured media [1, 4, 16], modeling of the flow of second-order fluids in hydrodynamics [8, 10], and heat conduction involving two temperatures [4, 17] lead to the following nonlinear pseudoparabolic problem:

Let $G \subset R^N$, $N = 2, 3$, be a bounded domain, $\Gamma := \partial G$ and smooth, $S := [0, T]$ a finite (time-)interval, $Q_T := S \times G$. We are looking for a function $u = u(t, x)$, $t \in S$, $x \in G$ such that

$$(0.1) \quad u'(t) - m \cdot \operatorname{div}(k_1(u) \cdot \operatorname{grad} u') - \operatorname{div}(k_2(u) \cdot \operatorname{grad} \alpha(u))$$
$$= f_2 - m \cdot \operatorname{div}(k_1(u) \cdot \operatorname{grad} f_1),$$

$$(0.2) \quad u(0, x) = u_0(x), \quad x \in G,$$

$$(0.3) \quad u/\Gamma = 0$$

($u' := \partial u/\partial t$, grad := gradient with respect to $x \in G$, k_i and α denote real-valued functions to be specified later, $f_i = f_i(t, x)$, $i = 1, 2$, $m = $ const. > 0).

For the sake of simplicity we assume $k := k_1 = k_2$, $f := f_1 = f_2$. Note that most of what follows can be extended (more general second-order operators instead of the div-operator above, other function spaces, variational inequalities).

In this note we formulate an iteration procedure (P_n) which approximates (0.1)–(0.3). (P_n) is based on a reformulation of (0.1)–(0.3) as an ordinary differential equation. Problems of type (0.1)–(0.3) have been considered in [1]–[4],

---

1980 *Mathematics Subject Classification.* Primary 35K70, 58D25, 65M99, 35K55; Secondary 76S05, 76A05

[11]; e.g., [1]–[3] present, among other things, several variants of existence theorems, sufficient conditions for obtaining uniqueness, maximum principles, and (for special cases) a discussion of nonregularity aspects.

1. Technical preliminaries.

1.1. *Function spaces.* Let $|\cdot|$ denote the Euclidean norm in R^N, $p, r \in [1, \infty]$, $s \in [-1, 1]$, $\lambda \in [0, 1]$. $L^p(G)$ denotes the usual space of all (equivalence classes of) p-Bochner integrable functions, $|\cdot|_p$ is the usual integral norm in $L^p(G)$ if $p < \infty$, the ess sup-norm if $p = \infty$.

$(\,,\,)$ denotes the scalar product in $L^2(G)$ and $L^2(G)^N$, resp.

$W^{s,p}(G)$ is the Sobolev space of all functions whose s-th (fractional) derivatives are in $L^p(G)$. $W^{s,p}(G)$ is normed by the usual norm $\|\cdot\|_{s,p}$ in these spaces (cf. [13, 14]), $H^1(G) := W^{1,2}(G)$.

By $W_0^{s,p}(G)$ we mean the subspace of all functions in $W^{s,p}(G)$ having zero trace on ∂G. This space is also normed by $\|\cdot\|_{s,p}$, $H_0^1(G) := W_0^{1,2}(G)$ (cf. [13, 14]). Furthermore, set $H^{-1}(G) := (H_0^1(G))^*$.

$C^{0,\lambda}(\overline{G})$ is the space of all Hölder-continuous (with exponent λ) functions defined on $\overline{G}$. $|\cdot|_{C^{0,\lambda}}$ denotes the usual Hölder norm (cf. [13]).

$C(\overline{G})$ is the set of all continuous functions defined on $\overline{G}$. $|\cdot|_C$ denotes the sup-norm on $C(\overline{G})$.

Suppose $\{V, \|\,\|\}$ is a Banach space. $L^r(S, V) := L^r(0, T; V)$ is the space of all (equivalence classes of) Bochner-measurable functions $u: S \to V$ having finite norm

$$|u|_{L^r(S,V)} := \begin{cases} \operatorname*{ess\,sup}_{s \in S} \|u(s)\| & \text{if } r = \infty, \\ \left(\int_0^T \|u(s)\|^r ds \right)^{1/r} & \text{if } r < \infty. \end{cases}$$

$C(S, V) := C(0, T; V)$ is the set of all continuous $u: S \to V$ normed by

$$|u|_{C(S,V)} := \max_{t \in S} \|u(t)\|.$$

$W^{1,r}(S, V) := W^{1,r}(0, T; V)$ is the space of all $u \in L^r(S, V)$ whose distributional derivative u' belongs to $L^r(S, V)$. This space is normed by

$$\|u\|_{W^{1,r}(S,V)} := |u|_{L^r(S,V)} + |u'|_{L^r(S,V)}.$$

For a discussion of these spaces see [14, 15].

The symbol $(\,,\,)$ is also used for duality pairing between $H_0^1(G)$ and $H^{-1}(G)$.

1.2. *Assumptions on the coefficients.* Let $k \in L^\infty(R)$ s.t. there are constants k_0, k_1 with

(1.1) $0 < k_0 \leq k(r) \leq k_1$ a.a. $r \in R$;

(1.2) $k \in C^{0,1}(R)$ with Lipschitz constant L_k;

(1.3) $\alpha \in C^{0,1}(R)$ with Lipschitz constant L_α, $\alpha(0) = 0$.

REMARK. In most of the cases below where (1.2) is used, it would suffice to require Hölder continuity of k. The condition $\alpha(0) = 0$ is for pure technical convenience and could be omitted. Furthermore, working with strictly positive u_0,

u_{0n}, f, f_n and monotone $\alpha(\cdot)$ (which implies in some cases that the solution u (u_n) of (0.1)–(0.3) ((P$_n$)) is also strictly positive) opens the door to weakening the boundedness conditions imposed in (1.1) on k.

1.3. *Operator formulation for* (0.1)–(0.3). Let $u \in L^1(G)$ and formally set

(1.4) $$A_u := -\operatorname{div}(k(u) \cdot \operatorname{grad}(\cdot)),$$

(1.5) $$B_u := (I + m \cdot A_u)^{-1}.$$

To make these definitions more precise we give

LEMMA 1.1. (i) *If k is as in* (1.1), $p \in [1, \infty)$ *for* $N = 2$ *and* $p \in [1, 2N/(N-2)]$ *for* $N > 2$, *then*

$$B_u: H^{-1}(G) \to H_0^1(G) \hookrightarrow L^p(G) \text{ is continuous.}$$

(ii) *If k is as in* (1.1), $p \in (1, 2N/(N+2)]$, *then*

$$B_u: L^p(G) \to W_0^{1,p}(G) \text{ is continuous.}$$

(iii) *Let k be as in* (1.1), $p > N/2$. *Then there is a* $\bar{\lambda} \in (0,1)$ *s.t.*

$$B_u: L^p(G) \to C^{0,\bar{\lambda}}(\overline{G}) \text{ is continuous.}$$

(iv) *Let k be as in* (1.1), $q \in [1, N/(N-1))$. *Then*

$$B_u: L^1(G) \to W_0^{1,q}(G) \text{ is continuous.}$$

(v) *Let k be as in* (1.1) *and* (1.2) (*actually, Hölder continuity of k is sufficient*), $u \in C^{0,\lambda}(\overline{G})$ *for some* $\lambda \in (0,1)$, $p > 1$. *Then*

$$B_u: L^p(G) \to W_0^{1,p}(G) \text{ is continuous.}$$

PROOF. (i)–(iv)—cf. [6, 7]; (v)—cf. [19].

REMARK. The continuity of the linear operator B_u implies, in the usual way, estimates for $B_u v$ in terms of v. Furthermore, note that $\bar{\lambda}$ in (iv) depends only on mk_0, k_1, p, N, and G, resp. The Lipschitz constant of B_u in (v) depends only on the modulus of continuity of $k(u(\cdot))$, on mk_0, k_1, p, N, and G, resp.

Now set

(1.6) $$A_u^m := (1/m)(I - mB_u).$$

(A_u^m is the Yoshida approximation of A_u.)

A short calculation shows that (0.1)–(0.3) can be reformulated as

(P) $\quad u \in W^{1,1}(S, L^p(G)), \quad u' + A_u^m \alpha(u) = f, \quad u(0) = u_0,$

if $u_0 \in L^p(G), f \in L^1(S, L^p(G))$. (P) has been considered in [1]–[3].

2. Iteration. Let k and α satisfy (1.1) and (1.3), resp. We consider the iteration scheme

$$u_1 := u_0.$$

(P$_n$) Given u_n, u_{0n+1}, f_n we are looking for u_{n+1} s.t.

$$u'_{n+1}(t) = f_n(t) - A_{u_n(t)}^m \alpha(u_n(t)), \quad u_{n+1}(0) = u_{0n+1}.$$

To show the convergence of the sequence $\{u_n\}$, defined by (P_n), to the solution u of problem (P), we proceed in three steps: First we make the meaning of the equation in (P_n) a little more precise (Lemma 2.1); then in Lemma 2.2 we verify some boundedness properties of $\{u_n\}$; and, finally, we prove some variants of $u_n \to u$.

LEMMA 2.1. *Let* $s \in [0,1]$, $p, r \in [1, \infty]$. *The solution of* (P_n) *is formally given by*

$$(2.1) \qquad u_{n+1}(t) = u_{0n+1} + \int_0^t f_n(s)\,ds - \int_0^t A^m_{u_n(s)}\alpha(u_n(s))\,ds,$$

and we have the following:

(i) *If* $u_{0n+1} \in L^p(G)$, $f_n, u_n \in L^1(S, L^p(G))$, *and* k *and* α *are as in* (1.1) *and* (1.3), *resp., then*

$$u_{n+1} \in W^{1,1}(S, L^p(G)).$$

If, in addition, $f_n \in L^r(S, L^p(G))$, *then* $u_{n+1} \in W^{1,r}(S, L^p(G))$.

(ii) *Let* k *and* α *be as in* (i). *Then there is a* $\bar{\lambda} \in (0, 1)$ *s.t. if* $u_{0n+1} \in C^{0,\bar{\lambda}}(\bar{G})$, $f_n, u_n \in L^1(S, C^{0,\bar{\lambda}}(\bar{G}))$, *then*

$$u_{n+1} \in W^{1,1}(S, C^{0,\bar{\lambda}}(\bar{G})).$$

If, in addition, $f_n \in L^r(S, C^{0,\bar{\lambda}}(\bar{G}))$, *then* $u_{n+1} \in W^{1,r}(S, C^{0,\bar{\lambda}}(\bar{G}))$.

(iii) *Let* k *and* α *be as in* (i), *and let* k *satisfy a Hölder condition*, $\lambda \in (0,1)$. *If* $u_{0n+1} \in C^{0,\lambda}(\bar{G})$, $f_n, u_n \in L^1(S, C^{0,\lambda}(\bar{G}))$, *then*

$$u_{n+1} \in W^{1,1}(S, C^{0,\lambda}(\bar{G})).$$

If $f_n \in L^r(S, C^{1,\lambda}(\bar{G}))$, *then* $u_{n+1} \in W^{1,r}(S, C^{0,\lambda}(\bar{G}))$.

(iv) *Let* k *and* α *fulfill the same assumptions as in* (iii), $p > N/2$, $u_{0n+1} \in W_0^{1,p}(G)$, $u_n, f_n \in L^1(S, W_0^{1,p}(G))$. *Then*

$$u_{n+1} \in W^{1,1}(S, W_0^{1,p}(G)).$$

$f_n \in L^r(S, W_0^{1,p}(G))$ *implies* $u_{n+1} \in W^{1,r}(S, W_0^{1,p}(G))$.

PROOF. (2.1) is obviously a representation for a solution of (P_n). (i)–(iv) follow from Lemma 2.1 and corresponding measurability considerations along the lines of [12].

REMARK. It is easily checked that in all the above cases $f_n \in C(S,\dots)$ implies $u_{n+1} \in C^1(S,\dots)$ (cf. [2, 3]). To list some boundedness properties of $\{u_n\}$ we have

LEMMA 2.2. (i) *Let* $p \in [1, \infty]$, $a_{1n} := |u_{0n+1}|_p + |f_n|_{L^1(S, L^p(G))} \leq$ const. *(independent of n)*, k *and* α *as in* (1.1), (1.3), *resp. Then there is a constant* b_1 *(independent of n) s.t.* $|u_n|_{C(S, L^p(G))} \leq b_1$.

(ii) *Let* $\bar{\lambda} \in (0,1)$ *be sufficiently small (cf. Lemma 1.1(iii))*, $a_{2n} := |u_{0n}|_{C^{0,\bar{\lambda}}} + |f_n|_{L^1(S, C^{0,\bar{\lambda}}(\bar{G}))} <$ const. *(independent of n)*, k *and* α *as in* (i). *Then there is a constant* b_2 *(independent of n) s.t.* $|u_n|_{C(S, C^{0,\bar{\lambda}}(\bar{G}))} \leq b_2$.

(iii) *If* $p > N$, $a_{3n} := \|u_{0n+1}\|_{1,p} + \|f_n\|_{L^1(S, W^{1,p}(G))} <$ const. *(independent of n)*, *and if* k *and* α *satisfy* (1.1)–(1.3), *then there is a constant* b_3 *(independent of n) s.t.* $|u_n|_{C(S, W^{1,p}(G))} < b_3$.

(iv) *Let k and α satisfy* (1.1) *and* (1.3), *resp.*, $a_{4n} := \|u_{0n+1}\|_{1,2} + \|f_n\|_{L^1(S, H_0^1(G))}$ < *const.* (*independent of n*). *Then there is a constant b_4 s.t.* $|u_n|_{C(S, H_0^1(G))} < b_4$.

(i') ((ii')–(iv')) ($W^{1,r}(S, \ldots)$-*boundedness*). *Let* $r \in [1, \infty]$, *$p, k, \alpha, u_{0n+1}, f_n$ be as in* (i) ((ii)–(iv), *resp.*) *and suppose* $|f_n|_{L^r(S, L^p(G))} \leq$ *const.* (*independent of n*) ($\|f_n\|_{L^r(S, C^{0,\lambda}(\overline{G}))}$, $\|f_n\|_{L^r(S, W^{1,p}(G))}$, $\|f_n\|_{L^r(S, H_0^1(G))} <$ *const.* (*independent of n*), *resp.*). *Then* $\|u_n\|_{W^{1,r}(S, L^p(G))} <$ *const.* (*independent of n*) ($\|u_n\|_{W^{1,r}(S, C^{0,\lambda}(\overline{G}))}$, $\|u_n\|_{W^{1,r}(S, W^{1,p}(G))}$, $\|u_n\|_{W^{1,r}(S, H_0^1(G))} <$ *const.*, *resp.*).

PROOF. One obtains, after taking the L^p-norm on both sides of (2.1) for $t \in S$,

$$|u_{n+1}(t)|_p \leq a_{1n} + \int_0^t |A_{u_n(s)}^m \alpha(u_n(s))|_p \, ds.$$

By regularity theory for elliptic equations (cf. [5; 4; 2, Lemma 3.1], Lemma 1.1), there is a constant $c_1 = c_1(G, m, N, p, k_0, k_1, \alpha)$ s.t. $|A_{u_n}^m \alpha(u_n)|_p \leq c_1 |u_n|_p$. Therefore

(2.2) $$|u_{n+1}(t)|_p \leq a_{1n} + c_1 \int_0^t |u_n(s)|_p \, ds.$$

Setting

$$b_{1n} := \exp\left(\frac{1}{2} c_1 T\right)\left(2^{-n} |u_{0n+1}|_p + \sum_{k=0}^n 2^{-k} a_{1n-k}\right),$$

we get from our assumptions that $b_1 := \sup_n b_{1n} < \infty$ and, from (2.2), by iteration,

(2.3) $$|u_{n+1}|_{C(S, L^p(G))} \leq b_{1n} \leq b_1,$$

which proves (i).

To see (ii), note that $u_n \in C^{0,\lambda}(\overline{G})$ implies $\alpha(u_n) \in C^{0,\lambda}(\overline{G})$, and thus $B_{u_n}\alpha(u_n) \in C^{0,\lambda}(\overline{G})$ (cf. [6], Lemma 1.1), and there is a constant $c_2 := c_2(G, N, m, k_0, k_1, \overline{\lambda}, \alpha)$ such that

$$|A_{u_n}^m \alpha(u_n)|_{C^{0,\lambda}} \leq c_2 |u_n|_{C^{0,\lambda}};$$

therefore

$$|u_{n+1}(t)|_{C^{0,\lambda}} \leq a_{2n} + c_2 \int_0^t |u_n(s)|_{C^{0,\lambda}} \, ds,$$

which implies (in total analogy to (i)) the existence of a constant b_2 s.t.

(2.4) $$|u_{n+1}|_{C(S, C^{0,\lambda}(\overline{G}))} \leq b_2 < \infty.$$

To verify (iii) remember that, by the imbedding theorems, $W^{1,p}(G) \hookrightarrow C^{0,a}(\overline{G})$ ($a := 1 - N/p$). Therefore there is a modulus of continuity for $k(u_n(t, \cdot))$ which depends neither on t nor on n. Hence, the $W^{1,p}$-regularity theory for elliptic operators (cf. [19], Lemma 1.1) applies to the present situation—i.e., there is a constant $c_3 = c_3(G, N, p, k_0, k_1, m, L_k)$ with

$$\|B_{u_n}\alpha(u_n)\|_{1,p} \leq c_3 |u_n|_p.$$

Taking the $W^{1,p}$-norm on both sides of (2.1), one arrives at

$$\|u_{n+1}\|_{1,p} \leq a_{3n} + c_3 \int_0^t \|u_n(s)\|_{1,p} \, ds,$$

and thus (as in (2.2))

(2.5) $$|u_{n+1}|_{C(S,W^{1,p}(G))} \leq b_3 < \infty.$$

(iii) is proved.

To show (iv) take the H^1-norm on both sides of (2.1), observe that $u_n \in H^1(G)$ implies $\alpha(u_n) \in H^1(G) \hookrightarrow H^{-1}(G)$, and therefore, with $c_4 = c_4(G, N, k_0, k_1, m)$ = const.,

$$\|B_{u_n}\alpha(u_n)\|_{1,2} \leq c_4\|u_n\|_{1,2},$$

which finally—as in the preceding cases—results in

(2.6) $$|u_n|_{C(S,H^1(G))} \leq b_4 = \text{const.} \quad (\text{independent of } n).$$

To obtain the corresponding results for the time derivatives, differentiate (2.1) with respect to t and apply (i)–(iv).

The next lemma lists some sufficient conditions for which the solutions u_n of (P_n) converge to the solution u of (P). What we need is unique solvability of (P) and some kind of integrability of u. Both are discussed in [1]–[3]; the main requirements there are (1.1)–(1.3), $W^{1,q}$-regularity of k, and $W^{1,p}$-regularity of u_0 and f.

LEMMA 2.3. *Suppose k and α satisfy (1.1)–(1.3).*

(i) *Let the assumptions of Lemma 2.2(i) be fulfilled, $p \in [1, 2N/(N-2)]$ if $N > 2$, $p \in [1, \infty)$ if $N = 2$, $u \in L^1(S, L^{\bar{p}}(G)) \cap C(S, L^p(G))$ with $\bar{p} := 2p/(p-1)$, $u_{0n} \to u_0$ in $L^p(G)$, $f_n \to f$ in $L^1(S, L^p(G))$. Then $u_n \to u$ in $C(S, L^p(G))$.*

(ii) *Suppose the assumptions of Lemma 2.2(iv) are satisfied, $p, \bar{p}, u$ as in (i), and $u \in C(S, H_0^1(G))$, $u_{0n} \to u_0$ in $H_0^1(G)$, $f_n \to f$ in $L^1(S, H_0^1(G))$. Then $u_n \to u$ in $C(S, H_0^1(G))$.*

(i') *Let $r \in [1, \infty]$ and let the assumptions of (i) be fulfilled. If, in addition, $f_n \to f$ in $L^r(S, L^p(G))$, then $u_n \to u$ in $W^{1,r}(S, L^p(G))$.*

(ii') *Adding the condition $f_n \to f$ in $L^r(S, H_0^1(G))$, $r \in [1, \infty]$ to (ii), one obtains $u_n \to u$ in $W^{1,r}(S, H_0^1(G))$.*

(iii) *Let $p > N$, $r \in [1, \infty]$, and let, for $q > p$, $u \in L^1(S, W^{1,q}(G)) \cap C(S, W^{1,p}(G))$, $\|u_{0n}\|_{1,q} + \|f_n\|_{L^1(S,W^{1,q}(G))} <$ const. (independent of n), $u_{0n} \to u_0$ in $W^{1,p}(G)$, $f_n \to f$ in $L^1(S, W^{1,p}(G))$. Then $u_n \to u$ in $C(S, W^{1,p}(G))$.*

PROOF. Set $w_n := u_n - u$, $w_{0n} := u_{0n} - u_0$, $\bar{f}_n := f_n - f$. After subtraction of the equations for $t \in S$, (P) and (P_n) yield

(2.7) $$w'_{n+1}(t) = \bar{f}_n(t) + A^m_{u_n(t)}\alpha(u_n(t)) - A^m_{u(t)}\alpha(u(t)),$$

and after taking the L^p-norm and integrating with respect to t, they yield

$$|w_{n+1}(t)|_p \leq a_n + b_n(t) + c_n(t) + d_n(t),$$

with
$$a_n := |w_{0n+1}|_p,$$
$$b_n(t) := \int_0^t |\bar{f}_n(s)|_p \, ds,$$
$$c_n(t) := \int_0^t \left| A^m_{u_n(s)}(\alpha(u_n(s)) - \alpha(u(s))) \right|_p ds,$$
$$d_n(t) := \int_0^t \left| \left(A^m_{u_n(s)} - A^m_{u(s)} \right) \alpha(u(s)) \right|_p ds.$$

(1.2) and the L^p-estimates for elliptic operators (cf. also Lemma 1.1) imply that there is a constant $c_5 = c_5(G, N, p, m, k_0, k_1, \alpha)$ such that
$$c_n(t) \leqslant c_5 \int_0^t |w_n(s)|_p \, ds.$$

To get an estimate for $d_n(t)$, note that with
(2.8) $$v := B_u \alpha(u), \quad v_n := B_{u_n} \alpha(u),$$
(2.9) $$d_n(t) \leqslant \int_0^t \frac{1}{m} |\alpha(u_n) - \alpha(u)|_p + \frac{1}{m} |v_n - v|_p \, ds.$$

$y_n := v_n - v$ solves the problem
(2.10) $$y_n + mA_{u_n} y_n = g_n$$
with $g_n := m(A_u - A_{u_n})\alpha(u)$.

The assumption $p \geqslant 2$ implies $g_n \in H^{-1}(G)$, so that, by the imbedding theorems, $H^1(G) \hookrightarrow L^p(G)$, and known estimates for weak solutions of elliptic problems of type (2.10) (cf. [5], e.g.), there is a constant $c = c(G, N, m, k_0, k_1)$ such that
$$|y_n|_p \leqslant c\|y_n\|_{1,2} \leqslant c\|g_n\|_{-1,2}.$$

(1.2) (Lipschitz continuity of k) and Hölder's inequality imply
$$\|g_n\|_{-1,2} \leqslant c|w_n|_p \|u\|_{1,q} \quad (q \text{ is defined in (iii)}).$$

Thus
(2.11) $$|y_n|_p + \|y_n\|_{1,2} \leqslant c|w_n|_p \|u\|_{1,q},$$
and with constant $c = c(G, N, p, m, k_0, k_1, \alpha)$,
$$d_n(t) \leqslant c \int_0^t |w_n|_p + \|u\|_{1,q} |w_n|_p \, ds.$$

Summarizing, we obtain
$$|w_{n+1}(t)|_p \leqslant a_n + b_n(t) + c \int_0^t \left(1 + \|u\|_{1,q} \right) |w_n(s)|_p \, ds.$$

Iteration of this estimate finally yields (cf. [9])
$$|w_{n+1}|_{C(S, L^p(G))}$$
$$\leqslant \exp\left(2c \left(1 + \|u\|_{L^1(S, W^{1,q}(G))} \right) \right) \left(2^{1-n} |w_1|_{C(S, L^p(G))} + \sum_{k=0}^n 2^{-k}(a_{n-k} + b_{n-k}(T)) \right).$$

A short calculation shows that the right side tends to zero as $n \to \infty$ (remember that $a_n + b_n(T) \to 0$ as $n \to \infty$ and $\sum_{k=0}^{\infty} 2^{-k}$ is finite!). (i) is proved.

To verify (ii), integrate (2.7) from 0 to t and take the H_0^1-norm on both sides. Now, reasoning as in (i), we obtain, by (2.10), (2.11),

$$\|w_{n+1}(t)\|_{1,2} \leqslant \|w_{0n+1}\|_{1,2} + \int_0^t \|\bar{f}_n(s)\|_{1,2}\, ds$$
$$+ c\int_0^t \left(1 + \|u\|_{1,q}\right)\|w_n\|_{1,2}\, ds.$$

This implies, as in (i), $w_n \to 0$ in $C(S, H_0^1(G))$ as $n \to \infty$.

To get the corresponding results (i') and (ii'), resp., take in (2.7) the $L^p(G)$-norm (H_0^1-norm, resp.) to arrive at

$$\left|w'_{n+1}\right|_{L^r(S, L^p(G))} \leqslant \left|\bar{f}_n\right|_{L^r(S, L^p(G))} + \left|A_{u_n}^m \alpha(u_n) - A_u^m \alpha(u)\right|_{L^r(S, L^p(G))}.$$

(i) implies the convergence to zero of the right side, which shows (i'). (ii') follows similarly.

Because $q > p > N$, there is an $a \in (0,1)$ s.t. $p^{-1} = (1-a)q^{-1} + 2a$. Thus $W^{1,p}(G) = (W^{1,q}(G), W^{1,2}(G))_{a,p}$ ((,)$_{a,p}$ denotes the (a, p)-interpolation space obtained by the real method; cf. [**14**]); therefore

$$\left|w_n\right|_{C(S, W^{1,p}(G))} \leqslant c\left|w_n\right|_{C(S, W^{1,q}(G))}^{1-a}\left|w_n\right|_{C(S, W^{1,2}(G))}^{a}.$$

In view of Lemma 2.2(iii), the first expression on the right side remains bounded as $n \to \infty$. By (ii) the second goes to zero as $n \to \infty$. This proves (iii).

REMARK. Using the preceding two lemmas, we easily obtain additional convergence results in other spaces. Furthermore, estimates such as those in (2.12) allow us to find upper bounds for the rate of convergence of the sequence $\{u_n\}$.

REFERENCES

1. M. Böhm and R. E. Showalter, *Diffusion in fissured media* SIAM J. Math. Anal., April 1985.

2. _____, *On a non-linear diffusion equation* SIAM J. Math. Anal. (to appear).

3. M. Böhm, *Existence-uniqueness results for pseudoparabolic equations*, LCDS-report 83-15.

4. E. DiBenedetto and R. E. Showalter, *A pseudoparabolic variational inequality and the Stefan problem*, Nonlinear Anal., **6** (1982), 279–291.

5. O. A. Ladyzhenskaya and N. N. Ural'tseva, *Linear and quasi-linear elliptic equations*, "Nauka", Moscow 1964. (Russian)

6. G. Stampacchia, *Le problème de Dirichlet pour les équations elliptiques du second ordre à coefficients discontinuous*, Ann. Inst. Fourier (Grenoble) **15** (1965), 189–258.

7. D. Kinderlehrer and G. Stampacchia, *An introduction to variational inequalities and their applications*, Academic Press, New York, 1980.

8. T. Ting, *Certain non-steady flows of second-order fluids*, Arch. Rational Mech. Anal. **14** (1963), 1–26.

9. M. Böhm, *On a nonlinear evolution equation*, Seminarberichte, 35. Sektion Mathematik, Humboldt Universität, Berlin, 1981.

10. R. Huilgol, *On a second-order fluid of differential type*, Internat. J. Nonlinear Mech. **3** (1968), 471–482.

11. R. W. Carroll and R. E. Showalter, *Singular and degenerate Cauchy problems*, Academic Press, New York, 1976.

12. R. E. Showalter, *Existence and representation theorems for a semilinear Sobolev equation in Banach spaces*, SIAM J. Math. Anal. **3** (1972), 527–543.

13. R. A. Adams, *Sobolev spaces*, Academic Press, New York, 1975.
14. J. Bergh and J. Löfström, *Interpolation spaces. An introduction*, Grundlehren Math. Wiss., no. 223, Springer-Verlag, 1976.
15. H. Gajewski, K. Gröger and K. Zacharias, *Nichtlineare Operatorgleichungen und Operatordifferentialgleichungen*, Akademie-Verlag, Berlin, 1974.
16. G. Barenblatt, I. Zheltov and I. Kochina, *Basic concepts in the theory of seepage of homogeneous liquids in fissured rocks*, J. Appl. Math. Mech. **24** (1960).
17. P. Chen and M. Gurtin, *On a theory of heat conduction involving two temperatures*, Z. Angew. Math. Phys. **19** (1968).
18. E. DiBenedetto and M. Pierre, *On the maximum principle for pseudoparabolic equations*, Indiana Univ. Math. J. **30** (1981), 821–854.
19. Ch. J. Simader, *On Dirichlet's boundary value problem*, Lecture Notes in Math., vol. 268, Springer, Berlin and New York, 1973.

HUMBOLDT-UNIVERSITÄT ZU BERLIN, SEKTION MATHEMATIK, GERMAN DEMOCRATIC REPUBLIC

Homogenization of Two-Phase Flow Equations

ALAIN BOURGEAT

Introduction. We use the equations of incompressible two phase flows developed by G. Chavent [1]. Rock permeability and porosity are rapidly oscillating functions of space variables because of the heterogenities in media. By using tools of homogenization [2]–[4], essentially weak convergence, we obtain "homogenized" equations which define flow behaviors in a homogeneous medium "globally" equivalent to the original.

Our results apply to all kinds of heterogeneous media, such as fractures, strata, etc., so long as they are assumed to be periodically spaced.

The physical problem. Let $\Omega \subset \mathbb{R}^n$ be a porous medium with boundary $\Gamma = \Gamma_1 \cup \Gamma_2 \cup \Gamma_3$. Γ_1 is the injection boundary, Γ_2 the production boundary, and Γ_3 the impermeable boundary. (See the Figure.)

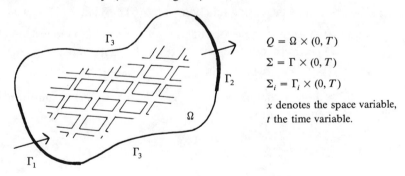

$Q = \Omega \times (0, T)$

$\Sigma = \Gamma \times (0, T)$

$\Sigma_i = \Gamma_i \times (0, T)$

x denotes the space variable,
t the time variable.

Within the bounded domain Ω, there is a network of uniformly spaced heterogeneities. Then rock porosity and rock permeability are assumed periodic functions of x. ε is the ratio of the period scale to the domain scale.

We assume the two fluids and the rocks are incompressible and the gravity effects negligible.

1980 *Mathematics Subject Classification.* Primary 65C20, 65P05; Secondary 35B40, 35K60.

© 1986 American Mathematical Society
0082-0717/86 $1.00 + $.25 per page

Equations. Let S be the "reduced" wetting phase saturation and P the "global" pressure [1]. This "global" pressure produces a mean oil + water filtration velocity vector q. The rock porosity Φ and the rock permeability tensor Ψ depend on ε, and, consequently, the flow equations in this heterogeneous medium have solutions $(P_\varepsilon, S_\varepsilon)$ depending on ε.

From the model of [1] we have a system of coupled quasilinear partial differential equations:

The global pressure equation (Darcy's Law) is

(1)
$$-q^\varepsilon(x;t) = d(S_\varepsilon)\Psi^\varepsilon(x) \nabla P_\varepsilon(x;t),$$
$$\nabla \cdot q^\varepsilon = 0 \quad \text{in } Q,$$
$$q^\varepsilon \cdot \nu|_{\Sigma_1} = q_1(x;t); \quad P_\varepsilon(x;t)|_{\Sigma_2} = 0; \quad q^\varepsilon \cdot \nu|_{\Sigma_3} = 0.$$

The wetting phase saturation equation (mass conservation law) is

(2)
$$\Phi_\varepsilon(x)\frac{\partial S_\varepsilon}{\partial t} - \nabla \cdot (b(S_\varepsilon)q^\varepsilon(x;t) + \Psi^\varepsilon(x)\nabla\alpha(S_\varepsilon)) = 0,$$
$$S_\varepsilon(x;t)|_{\Sigma_1 \cup \Sigma_2} = 0; \quad \Psi^\varepsilon \nabla\alpha(S_\varepsilon) \cdot \nu|_{\Sigma_3} = 0,$$
$$S_\varepsilon(x;0) = s_0(x) \quad \text{in } \Omega.$$

$\alpha(S)$ is a monotone increasing function. In the immiscible case its derivatives vanish at $S = 0$ and $S = 1$, in which case the diffusion term in (2) is degenerated. Let a be the primitive of α, i.e.,

$$\alpha(S) = \int_0^S a(s)\, ds,$$

$b(S)$ is a monotone increasing function with $b(0) = 0$ and $b(1) = +1$.

(3) a, b, and d are assumed to be continuous and bounded functions from $\mathbb{R}$ into $\mathbb{R}$. $d(S)$ is strictly positive, with $d(S) \geq d > 0$ in Q.

(4) $\Phi_\varepsilon \in L^\infty(\Omega)$, $\Psi_{ij}^\varepsilon \in L^\infty(\Omega)$ for all $i, j = 1, \ldots, n$ independently of ε.

(5) In Ω: $\exists C > 0$ such that $\Phi_\varepsilon(x) \geq C$, and $\Psi^\varepsilon(x) \nabla u(x)$ is uniformly elliptic independently of ε.

Define
$$V = \{v \in H^1(\Omega), v|_{\Gamma_1 \cup \Gamma_2} = 0\}, \quad H = L^2(\Omega),$$
$$W = \{w \in H^1(\Omega), w|_{\Gamma_2} = 0\}.$$

The nonlinear variational version of equations (1) and (2) is

(6)
$$\int_\Omega q^\varepsilon \cdot \nabla w\, dx = \int_{\Gamma_1} wq_1\, d\sigma \quad \text{for all } w \in W \text{ a.e. on } (0, T),$$
$$P_\varepsilon(t) \in W \text{ a.e. on } (0, T).$$

(7)
$$\Phi_\varepsilon\left(\frac{\partial S_\varepsilon}{\partial t}, v\right) + \int_\Omega b(S_\varepsilon)q^\varepsilon \cdot \nabla v\, dx + \int_\Omega \Psi^\varepsilon \nabla\alpha(S_\varepsilon) \cdot \nabla v\, dx = 0$$
$$\text{for all } v \in V \text{ a.e. on } (0, T),$$
$$S_\varepsilon(0) = s_0, \quad S_\varepsilon(t) \in V \text{ a.e. on } (0, T).$$

$(\cdot, \cdot)$ denotes the duality product of V with its dual V' space.

Let β be the primitive of $(a)^{1/2}$, i.e., $\beta(S) = \int_0^S (a(s))^{1/2} \, ds$.
We require a result from [1].

THEOREM 1. *Assuming that*
(8) $s \in L^2(\Omega)$, $q_1 \in L^2(\Sigma_1)$, $0 \leq s_0(x) \leq 1$ *a.e. in* Ω, *and*
(9) β^{-1} *is a Hölder function of order* $\theta \in (0, 1)$,
then, for all fixed ε, system (6), (7) has at least a solution $(P_\varepsilon, S_\varepsilon)$ such that

(10)
(i) $S_\varepsilon \in L^\infty(Q)$, $0 \leq S_\varepsilon(x; t) \leq 1$ *a.e. in* Q,
(ii) $\beta(S_\varepsilon) \in L^2(0, T; V)$; $\partial S_\varepsilon / \partial t \in L^2(0, T; V')$,
(iii) $P_\varepsilon \in L^2(0, T; W)$.

Note. Furthermore, if the capillary diffusion term is nondegenerate (miscible flows), then

(11) $$S_\varepsilon \in L^2(0, T; V) \cap L^\infty(0, T; H).$$

Note. This theorem does not allow uniqueness in the n-dimensional case. Uniqueness holds only in some cases: Either the domain is one-dimensional, or d is a constant function (i.e., we can decouple (1) and (2) [1, 5]), or the space dimension is two and the equations are nondegenerated [6].

A priori estimates.

LEMMA 1. *There exists a subsequence of $(S_\varepsilon, P_\varepsilon)$, still denoted $(S_\varepsilon, P_\varepsilon)$, such that*

(12)
(i) $\beta_\varepsilon = \beta(S\varepsilon) \rightharpoonup \beta$ *in* $L^2(0, T; V)$ *weakly*,
(ii) $P_\varepsilon \rightharpoonup P$ *in* $L^2(0, T; W)$ *weakly*,
(iii) $\gamma_\varepsilon = b(S_\varepsilon)q_\varepsilon$, $\mu_\varepsilon = b(S_\varepsilon)d(S_\varepsilon)\nabla P_\varepsilon$, $\lambda_\varepsilon = \nabla \alpha(S_\varepsilon)$, *and*
(iv) $\nu_\varepsilon = d(S_\varepsilon)\nabla P_\varepsilon$ *tend to* $\gamma, \mu, \lambda, \nu$ *in* $[L^2(Q)]^n$ *weakly*,
(v) $\chi_\varepsilon = \Phi_\varepsilon(\partial S_\varepsilon/\partial t) \rightharpoonup \chi$ *and* $\Delta_\varepsilon = \nabla \cdot \gamma_\varepsilon \rightharpoonup \Delta$ *in* $L^2(0, T; V')$ *weakly.*

PROOF. (a) Setting $v = S_\varepsilon$ and $w = \delta(S_\varepsilon) = \int_0^{S_\varepsilon} b(s) \, ds$ in (6)–(7), we get

(13)
(i) $\int_\Omega q^\varepsilon \cdot b(S_\varepsilon)\nabla S_\varepsilon \, dx = 0$,
(ii) $\frac{1}{2} \frac{d}{dt} \left| (\Phi_\varepsilon)^{1/2} S_\varepsilon \right|_H^2 + \int_\Omega \Psi^\varepsilon \nabla \beta(S_\varepsilon) \cdot \nabla \beta(S_\varepsilon) \, dx = 0$.

Integrating equation (13)(ii) over $(0, T)$, one obtains $\beta(S_\varepsilon)$ bounded independently of ε, and, hence, (12)(i) follows.

(b) Setting $w = P_\varepsilon$ and integrating equation (6) over $(0, T)$ gives (12)(ii).
(c) By hypotheses (3)–(4) we get

$$a_S(v) = \int_\Omega \Psi^\varepsilon \nabla \beta(S_\varepsilon) \cdot (a(S_\varepsilon))^{1/2} \nabla v \, dx \leq K_1 \|\beta(S_\varepsilon)\|_V \|v\|_V,$$

and

$$b_P(v) = \int_\Omega b(S_\varepsilon) q \cdot \nabla v \, dx \leq K_2 \|P_\varepsilon\|_W \|v\|_V.$$

Equation (7) and the above inequalities give (12)(iv).

(d) (12)(iii) follows from (3), (10)(i), and (12)(i)–(ii).

Note. In the case of miscible fluids, $a(S_\varepsilon) \geq C > 0$, and, hence, (13)(ii) gives $S_\varepsilon \rightharpoonup S$ in $L^2(0, T; V)$ weakly. And, because $\partial S_\varepsilon/\partial t \rightharpoonup \partial S/\partial t$ in $L^2(0, T; V')$ weakly, we have $S_\varepsilon \to S$ in $L^2(Q)$ strongly. But when $a(S_\varepsilon)$ is only assumed nonnegative (immiscible fluids), the proof of strong convergence is more difficult.

LEMMA 2. *Assuming only* $a(S_\varepsilon) \geq 0$, *we get*

(14)
  (i) $S_\varepsilon \to S$ in $L^2(Q)$ *strongly*,
  (ii) $\chi = [\tilde{\Phi}](\partial S/\partial t)$.

PROOF OF (i). By (12)(i), we get

(15) $\beta(S_\varepsilon)$ is bounded in $L^2(0, T; W^{s,2}(\Omega))$ for all $s \in (0, 1)$.

Since $\beta^{-1}(\beta(S_\varepsilon)) = S_\varepsilon$, hypothesis (9) yields

(16) $\|S_\varepsilon\|^r_{L^r(0,T;W^{\theta s,r}(\Omega))} \leq K_3 \|\beta(S_\varepsilon)\|^2_{L^2(0,T;W^{s,2}(\Omega))} \|\beta^{-1}\|^r_{\text{Höl }\theta}$, with $r = 2/\theta$.

That is, S_ε is bounded in $L^r(0, T; W^{\theta s, r}(\Omega))$ independently of ε. Because the injection of $W^{\theta s, r}(\Omega)$ in $L^r(\Omega)$ is compact and the injection of $W^{\theta s, r}(\Omega)$ in V' is continuous, we get from (15) and (16) that

$$S_\varepsilon \to S \text{ in } L^r(Q) \text{ strongly.}$$

Since r is greater than 2, (14)(i) follows.

PROOF OF (ii). By (14)(i) we get

$$\Phi_\varepsilon S_\varepsilon \rightharpoonup [\tilde{\Phi}] S \text{ in } L^2(Q) \text{ weakly,}$$

where $[\tilde{\Phi}]$ is the average of Φ_ε over one period. ($[\tilde{\ }\,]$ denotes the averaging operator.) By a change of variables ($x = x$, $y = x/\varepsilon$), the period in the x becomes the period Y, and $\Phi_\varepsilon(x)$ becomes $\Phi(y) = \Phi(x/\varepsilon)$. Then, by definition,

$$[\tilde{\Phi}] = \frac{1}{\text{meas}(Y)} \int_Y \Phi(y)\, dy \quad (\text{see } [2, 3]).$$

The above result, (12)(iv), and the fact that $\Phi_\varepsilon(x)(\partial S_\varepsilon/\partial t) = \partial/\partial t(\Phi_\varepsilon S_\varepsilon)$ gives us (14)(ii).

LEMMA 3.

(17) $\qquad \lambda = \nabla\alpha(S), \quad \mu = b(S)d(S)\nabla P, \quad \nu = d(S)\nabla P.$

By (3), (10)(i), (14)(i), and the Lebesgue theorem, one sees that $\alpha(S_\varepsilon)$, $b(S_\varepsilon)$, $d(S_\varepsilon)$ tend to $\alpha(S)$, $b(S)$, $d(S)$ in $L^2(Q)$ strongly. Hence, by (12)(ii), we get (17).

Limits as $\varepsilon \to 0$. To continue, we need the same additional hypotheses used to get uniqueness. So for the remainder of the paper we will assume $a(S) \geq C > 0$ and the following hypotheses:

either

(H1) Ω is one-dimensional,

or

(H2) $d(S) = cte\ (\equiv 1,\ \text{for example}).$

Under either of these hypotheses we have

THEOREM 2. *P is the solution of*

(i) $\quad q = \Psi^{\#} \nabla P; \qquad \nabla \cdot q = 0 \quad \text{in } Q,$
$\quad q \cdot \nu|_{\Sigma_1} = q_1; \qquad P|_{\Sigma_2} = 0; \quad q \cdot \nu|_{\Sigma_3} = 0,$

with

(18) (ii) $\quad \Psi^{\#} = 1/\left[\widetilde{1/\Psi}\right]$ *under hypothesis* (H1).

And under hypothesis (H2), $\Psi^{\#}$ *is a tensor given by*

(iii) $\quad \Psi^{\#}_{il} = \left[\Psi_{il} + \Psi_{ij}\dfrac{\partial w^l}{\partial y_j}\right]\;$ *where $w^l(y)$ is the*

$\qquad\qquad$ *Y-periodic solution of* $-\dfrac{\partial}{\partial y_i}\left[\Psi_{ij}\dfrac{\partial w^l}{\partial y_j}\right] = \dfrac{\partial}{\partial y_i}[\Psi_{il}].$

PROOF. Let $\phi \in C_0^{\infty}(0, T)$ and define

$$\xi_\varepsilon(\phi) = \int_0^T P_\varepsilon \phi(t)\, dt, \qquad \theta^\varepsilon(\phi) = \int_0^T q^\varepsilon(x; t)\phi(t)\, dt.$$

By (12)(iii), (12)(ii), and (4) we have

$\qquad \theta^\varepsilon \to \theta \;\text{in}\; (L^2(\Omega))^n \;\text{weakly} \qquad \text{and} \qquad \xi_\varepsilon \to \xi \;\text{in}\; W \;\text{weakly}.$

(a) Assume (H1) holds. Then (1) gives $\theta^\varepsilon(\phi) = cte$—i.e., bounded in $H^1(\Omega)$ independently of ε. Hence,

(19) $\qquad\qquad \theta^\varepsilon \to \theta \;\text{in}\; L^2(\Omega) \;\text{strongly},$

and

(20) $\qquad \displaystyle\int_0^T d(S_\varepsilon)\dfrac{\partial P_\varepsilon}{\partial x}\phi(t)\, dt = \dfrac{\theta^\varepsilon}{\Psi^\varepsilon} \to \left[\dfrac{\tilde{1}}{\Psi}\right]\theta \;\text{in}\; L^2(\Omega) \;\text{weakly}.$

By (17) we get

$$\int_0^T d(S_\varepsilon)\dfrac{\partial P_\varepsilon}{\partial x}\phi(t)\, dt \to \int_0^T d(S)\dfrac{\partial P}{\partial x}\phi(t)\, dt \quad \text{in}\; L^2(\Omega)\; \text{weakly}.$$

Hence, by (20)

$$\theta(\phi) = \left[\dfrac{\tilde{1}}{\Psi}\right]\int_0^T d(S)\dfrac{\partial P}{\partial x}\phi(t)\, dt \quad \text{for all}\; \phi \in C_0^\infty(0, T).$$

That is, θ^ε converges to $\Psi^{\#} d(S)(\partial P/\partial x)$ in $\mathscr{D}'(0, T; L^2(\Omega))$, so the proof is complete.

(b) Assume (H2) holds. Then $\theta^\varepsilon(\phi) = \Psi^\varepsilon \nabla \xi_\varepsilon$.

Equation (1) gives $\nabla \cdot (\Psi^\varepsilon \nabla \xi_\varepsilon) = 0$. This is a standard elliptic problem of homogenization [2], and the limit is $\nabla \cdot \Psi^{\#} \nabla \xi = 0$, with $\Psi^{\#}$ as in (18)(iii).

LEMMA 4. *For all $\phi \in C_0^\infty$, define*

$$u_\varepsilon = \int_0^T \alpha(S_\varepsilon)\phi(t)\, dt,$$

$$v_\varepsilon = \int_0^T \nabla\alpha(S_\varepsilon)\phi(t)\, dt = \nabla u_\varepsilon,$$

$$w_\varepsilon = \int_0^T b(S_\varepsilon) q^\varepsilon \phi(t)\, dt.$$

Then

(21) $$[\nabla \cdot \Psi^\varepsilon \nabla u_\varepsilon] = \nabla \cdot w_\varepsilon + g_\varepsilon,$$

and

(22)
(i) $\quad u_\varepsilon \to u = \int_0^T \alpha(S)\phi(t)\, dt \quad \text{in } L^2(\Omega),$

(ii) $\quad v_\varepsilon \to v = \int_0^T \nabla\alpha(S)\phi(t)\, dt = \nabla u \quad \text{in } L^2(\Omega) \text{ weakly},$

(iii) $\quad g_\varepsilon \to g = [\tilde{\Phi}]\int_0^T S\frac{d\phi}{dt}\, dt \quad \text{in } V'.$

Moreover, with either (H1) *or* (H2), *and assuming q^ε bounded in $L^\infty(\Omega)$,*

(iv) $\quad \nabla \cdot w_\varepsilon \to \nabla \cdot w = \nabla \cdot \int_0^T b(S)q\phi(t)\, dt \quad \text{in } V'.$

PROOF. (a)

$$\int_0^T \nabla \cdot \Psi^\varepsilon(x) \nabla\alpha(S_\varepsilon)\phi(t)\, dt = \nabla \cdot (\Psi^\varepsilon \nabla u_\varepsilon),$$

$$\int_0^T \nabla \cdot b(S_\varepsilon) q^\varepsilon(x; t)\phi(t)\, dt = \nabla \cdot w_\varepsilon.$$

Let

$$g_\varepsilon = -\int_0^T \Phi_\varepsilon(x) \frac{\partial S_\varepsilon}{\partial t}\phi(t)\, dt.$$

Multiplying (2) by $\phi(t)$ and integrating over $(0, T)$ gives (21).

(b) (14)(i), (10)(i), and the Lebesgue theorem give (22)(i). (17) gives (22)(ii). (14)(ii) and (12)(iv) give (22)(iii).

(c) Assume either (H1) or (H2) holds. By (12)(ii) and (18)(i) we get $q_\varepsilon \to q = \Psi^\# \nabla P$ in $L^2(Q)$ weakly. By the Lebesgue theorem $b(S)q^\varepsilon \to b(S)q$ in $L^2(Q)$ weakly. In (12)(iv), $\Delta = \nabla \cdot b(S)q$, and then (22)(iv) is verified, with (1) and by compactness.

THEOREM 3. *S is the solution of*

(23)
$$[\tilde{\Phi}]\frac{\partial S}{\partial t} - \nabla \cdot (b(S)q) + \Psi^{\#}\nabla\alpha(S) = 0 \quad in\ Q,$$
$$S(x;t)\big|_{\Sigma_1 \cup \Sigma_2} = 0, \quad \Psi^{\#}\nabla S \cdot \nu\big|_{\Sigma_3} = 0,$$
$$S(x;0) = s_0(x) \quad in\ \Omega.$$

PROOF. Equation (21) with (22)(i), (iii), and (iv) is a standard linear elliptic equation. Its limit as $\varepsilon \to 0$ is given by

$$\nabla \cdot (\Psi^{\#}\nabla u) = \nabla \cdot w + g \quad \text{for all } \phi \in C_0^{\infty}(0, T).$$

Hence, we obtain (23) as a limit in $\mathscr{D}'(0, T; L^2(\Omega))$.

Conclusion. These results were also obtained by using asymptotic expansions [7]. Asymptotic expansion is only a heuristic way to compute limits. It cannot be used to get reasonable convergence results. Our results are consistent with the various averaging methods classically used in reservoir simulators. They may be of practical use to disconnect the mesh size from the heterogeneities size in a wide class of reservoir simulations.

REFERENCES

1. G. Chavent, *A new formulation of diphasic incompressible flows in porous media*, Lecture Notes in Math., vol. 503, Springer-Verlag, Berlin, 1976, pp. 258–270.

2. A. Bensoussan, J. L. Lions and G. Papanicolaou, *Asymptotic analysis for periodic structure*, North-Holland, Amsterdam, 1978.

3. E. Sanchez–Palencia, *Non-homogeneous media and vibration theory*, Lecture Notes in Physics, vol. 127, Springer-Verlag, Berlin, 1980.

4. B. Dacorogna, *Weak continuity and weak lower semi continuity of non linear functionals*, Lecture Notes in Math., vol. 922, Springer-Verlag, Berlin, 1982.

5. G. Gagneux, *Déplacement de fluides non-miscibles incompressibles dans un cylindre poreux*, J. Mécanique **19** (1980), 295–325.

6. S. N. Kruzkov and S. M. Sukorjanskii, *Boundary value problems for systems of equations of two-phase porous flow type; statement of the problems, questions of solvability, justification of approximate methods*, Math. USSR-Sb. **33** (1977).

7. A. Bourgeat, *Homogenized behavior of two-phase flows in naturally fractured reservoirs with uniform fractures distribution*, Computer Methods in Mechanics and Engineering, North-Holland, Amsterdam, 1984.

INSTITUT NATIONAL DES SCIENCES APPLIQUÉES DE LYON, FRANCE

Some Variational Problems with Lack of Compactness

HAIM BREZIS

0. Introduction. In these lectures I propose to describe various nonlinear elliptic equations (or systems) which admit a variational structure. Their common feature is that the standard variational techniques (minimization, critical point theory) do not apply in a straightforward way because of *lack of compactness* or *lack of the Palais–Smale condition* (which is a form of compactness). Such a difficulty seems to be inherent to a number of problems in geometry and physics in view of their invariance under some transformations (dilations, etc. ...). Technically, the lack of compactness arises because the Sobolev inequality—or some isoperimetric inequality—occurs with its *limiting exponent*.

Most of the results below have been obtained in collaboration, either with L. Nirenberg or with J.-M. Coron. I start with a simple, but very instructive example.

PROBLEM (I). Let $\Omega \subset \mathbf{R}^N$ be a (smooth) bounded domain with $N \geq 3$. We look for a function $u \colon \overline{\Omega} \to \mathbf{R}$ satisfying

$$\text{(I)} \quad \begin{cases} -\Delta u = u^p + \lambda u & \text{on } \Omega, \\ u > 0 & \text{on } \Omega, \\ u = 0 & \text{on } \partial\Omega, \end{cases}$$

where $1 < p \leq (N+2)/(N-2)$ and $\lambda \in \mathbf{R}$. The *subcritical* case, i.e., $1 < p < (N+2)/(N-2)$, has been extensively studied—see e.g. the review article by P. L. Lions [42]. It is easy to establish

PROPOSITION 0. *Suppose* $1 < p < (N+2)/(N-2)$. *Then, for each* $\lambda \in (-\infty, \lambda_1)$ *there is a solution of* (I). *There is no solution of* (I) *for* $\lambda \geq \lambda_1$.

Here, λ_1 denotes the first eigenvalue of $-\Delta$ with Dirichlet boundary condition.

1980 *Mathematics Subject Classification.* Primary 35J60, 49A45, 58E20.

© 1986 American Mathematical Society
0082-0717/86 $1.00 + $.25 per page

PROOF. Set

(1) $$S_\lambda = \inf_{\substack{u \in H_0^1(\Omega) \\ \|u\|_{p+1}=1}} \left\{ \int |\nabla u|^2 - \lambda \int u^2 \right\}.$$

Since $p + 1 < 2N/(N - 2)$ it follows that $H_0^1(\Omega) \subset L^{p+1}(\Omega)$ with *compact injection*. As a consequence the infimum in (1) is achieved. Indeed let (u_j) be a minimizing sequence, so that

(2) $$u_j \in H_0^1, \quad \|u_j\|_{p+1} = 1,$$
$$\int |\nabla u_j|^2 - \lambda u_j^2 = S_\lambda + o(1).$$

Thus u_j is bounded in H_0^1 and we may assume that $u_j \to u$ weakly in H_0^1 and *strongly* in L^{p+1}. Passing to the limit in (2) we obtain

$$\int |\nabla u|^2 - \lambda \int u^2 \leq S_\lambda \quad \text{and} \quad \|u\|_{p+1} = 1.$$

Hence u is a minimizer for (1). Moreover, we may assume that $u \geq 0$ in Ω—otherwise we replace u by $|u|$. Writing the Euler equation for (1), we obtain

$$-\Delta u - \lambda u = \mu u^p \quad \text{on } \Omega$$

for some Lagrange multiplier μ—and, in fact, $\mu = S_\lambda$ (just multiply the equation by u). If $\lambda < \lambda_1$, then $S_\lambda > 0$ and we may "stretch out" μ by playing on the difference of homogeneities. Finally, the strong maximum principle implies that $u > 0$ on Ω. In order to see that (1) has no solution when $\lambda \geq \lambda_1$ it suffices to multiply (I) by ϕ_1, the eigenfunction of $-\Delta$ corresponding to λ_1.

REMARK 1. Under some appropriate restrictions (either on Ω or on p) one can assert that there is a continuous *branch* of solutions (λ, u) of (I) which bifurcates from $(\lambda_1, 0)$ and which covers the interval $(-\infty, \lambda_1)$; see Figure 1.

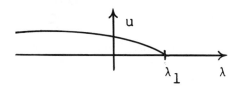

FIGURE 1

This follows from the abstract results of P. Rabinowitz [48] combined with a priori estimates for the solutions of (I) (see H. Brezis–R. Turner [17], D. de Figueiredo–P. L. Lions–R. Nussbaum [26]). The question of *uniqueness* is essentially open even if Ω is a ball—in which case all solutions of (I) are radial by the results of Gidas-Ni-Nirenberg [28]. When Ω is a ball the solution is known to be unique in various special cases: $\lambda = 0$ (via a simple scaling argument), or $\lambda \geq 0$ and $p \leq N/(N - 2)$ (see Ni–Nussbaum [44]).

We now turn to the *limiting case* $p = (N + 2)/(N - 2)$. This case is especially interesting because it resembles the *Yamabe problem* in differential geometry: Find a function u satisfying

$$-4\frac{(N-1)}{(N-2)}\Delta u = u^{(N+2)/(N-2)} - R(x)u \quad \text{on } M,$$

$$u > 0 \qquad\qquad\qquad\qquad\qquad \text{on } M.$$

Here M is an N-dimensional Riemannian manifold and $R(x)$ is the scalar curvature—for further details we refer to Th. Aubin [4, 5, 6], J. Kazdan [37, 72] and to R. Schoen [64] for the complete solution of Yamabe's conjecture. *The argument used in the proof of Proposition 0 fails in the limiting case. Indeed $u_j \to u$ weakly in H_0^1 and weakly in L^{p+1}. However, u_j need not converge strongly in L^{p+1}.* Passing to the limit in (2), we find only

$$\int |\nabla u|^2 - \lambda \int u^2 \leq J_\lambda \quad \text{and} \quad \|u\|_{p+1} \leq 1,$$

since *the unit sphere in L^{p+1} is not closed for the weak H^1 topology. Thus the weak limit u of a minimizing sequence u need not be a minimizer. This is the fundamental difficulty in all the variational problems with lack of compactness.*

In fact, we shall see in §1 that if $\lambda \leq 0$ the infimum in (1) is never achieved. When Ω is starshaped this follows easily from:

THEOREM (POHOŽAEV [47]). *Suppose Ω is starshaped and $p = (N + 2)/(N - 2)$. Then there is no solution of* (I) *with $\lambda \leq 0$.*

Pohožaev's Theorem is a direct sequence of *Pohožaev's identity*: If

(3) $$\begin{cases} -\Delta u = g(u) & \text{on } \Omega, \\ u = 0 & \text{on } \partial\Omega, \end{cases}$$

and g is a continuous function, then

(4) $$\left(1 - \frac{N}{2}\right)\int_\Omega g(u)u + N\int_\Omega G(u) = \frac{1}{2}\int_{\partial\Omega}(x\cdot n)\left(\frac{\partial u}{\partial n}\right)^2,$$

where $G(u) = \int_0^u g(s)\,ds$ and n is the outward normal (the proof of Pohožaev's identity is elementary: multiply (3) by u and by $\sum x_i \partial u/\partial x_i$).

In the special case where $g(u) = u^p + \lambda u$, Pohožaev's identity reduces to

(5) $$\lambda \int_\Omega u^2 = \int_{\partial\Omega}(x\cdot n)\left(\frac{\partial u}{\partial n}\right)^2$$

and therefore $\lambda > 0$ if Ω is starshaped (since $x \cdot n > 0$ a.e. on $\partial\Omega$).

Pohožaev's interesting nonexistence result had a disastrous impact: most authors carefully avoided the "taboo" limiting case by invoking Pohožaev as an excuse! It turns out that *in the limiting case the lower-order terms—which are irrelevant in the subcritical case—play here a very important role* and they "help" to establish the existence by "lowering the infimum".

The main results from [6] concerning (I) are the following.

THEOREM 1. *Let Ω be any bounded domain in $\mathbf{R}^N$ with $N \geq 4$, and let $p = (N + 2)/(N - 2)$. Then, for every $\lambda \in (0, \lambda_1)$ there is a solution of* (I). *Moreover the infimum in* (1) *is achieved.*

The case of dimension three is much more difficult and we have a satisfactory result only when Ω is a ball:

THEOREM 2. *Let Ω be a ball in $\mathbf{R}^3$ and let $p = 5$. Then for every $\lambda \in (\lambda_1/4, \lambda_1)$ there is a solution of* (I); *moreover, the infimum in* (1) *is achieved. There is no solution of* (I) *for $\lambda \notin (\lambda_1/4, \lambda_1)$.*

The striking difference between the cases $N \geq 4$ and $N = 3$ remains a mystery. In §1 I will describe some of the ingredients which enter in the proofs of Theorems 1 and 2. The best constant S for the Sobolev inequality plays an important role. The proof of existence involves two independent steps:

Step 1. Show that if $S_\lambda < S$, then the infimum in (1) is achieved.

Step 2. Show that indeed $S_\lambda < S$ under some appropriate restrictions on λ or on Ω.

More precisely, we prove that *below the level S some form of compactness is restored*. This approach, which has been introduced by Th. Aubin in [4], is also used by R. Schoen [64]. As we shall see, Step 1 is straightforward, while Step 2 involves a heavy technical machinery. Alternative proofs have been obtained by Atkinson–Peletier [66] and McLeod–Norbury [67].

PROBLEM (II). Let $\Omega \subset \mathbf{R}^N$ be a (smooth) bounded domain with $N \geq 3$. We look for a function $u: \overline{\Omega} \to \mathbf{R}$ satisfying

(II) $\quad \begin{cases} -\Delta u = u^p + \mu u^q & \text{on } \Omega, \\ u > 0 & \text{on } \Omega, \\ u = 0 & \text{on } \partial\Omega, \end{cases}$

where $p = (N + 2)/(N - 2)$, $1 < q < p$ and $\mu \in \mathbf{R}$. Pohožaev's identity applied to a solution of (II) says that

(6) $\quad \mu\left(1 - \frac{N}{2} + \frac{N}{q+1}\right)\int_\Omega u^{q+1} = \frac{1}{2}\int_{\partial\Omega} (x \cdot n)\left(\frac{\partial u}{\partial n}\right)^2$

and therefore $\mu > 0$ when Ω is starshaped. The main results from [16] concerning (II) are the following.

THEOREM 3. *Let Ω be any bounded domain in $\mathbf{R}^N$ with $N \geq 4$. Then for every $\mu > 0$ there is a solution of* (II).

Again, dimension three is much more delicate:

THEOREM 4. *Let $\Omega \subset \mathbf{R}^3$ be any bounded domain and let $p = 5$. We distinguish two cases*:

 (i) *if $3 < q < 5$, then for every $\mu > 0$ there is a solution of* (II),

 (ii) *if $1 < q \leq 3$, then for every μ large enough there is a solution of* (II).

Moreover, (II) *has no solution for $\mu > 0$ small when Ω is strictly starshaped.*

Problem (II) cannot be solved by a simple minimizing argument; one has to use more sophisticated variational tools. The solutions of (II) correspond to nonzero critical points of the functional

$$F(u) = \frac{1}{2}\int_\Omega |\nabla u|^2 - \frac{1}{p+1}\int_\Omega (u^+)^{p+1} - \frac{\mu}{q+1}\int_\Omega (u^+)^{q+1}$$

on the space $H_0^1(\Omega)$. All the assumptions of the Mountain Pass Lemma of Ambrosetti and Rabinowitz (see [1] and Lemma 7 below) are satisfied, *except for the Palais–Smale* (PS) *condition*. Usually—when the (PS) condition is satisfied—one can assert that the constant

$$c = \text{Inf Sup } F$$

(Inf Sup over appropriate sets) is a *critical value* of F. Here, however, one cannot conclude that c is a critical value because the (PS) condition fails. Again, our approach for existence involves essentially two steps:

Step 1. Show that if $c < (1/N)S^{N/2}$ then c is indeed a critical value. More precisely we show that *below the "magic" level* $(1/N)S^{N/2}$ *the* (PS) *condition is restored*.

Step 2. Show that indeed $c < (1/N)S^{N/2}$ under some appropriate conditions (on μ, Ω, q).

PROBLEM (III) (RELLICH'S CONJECTURE). Let $\Gamma \subset \mathbf{R}^3$ be a Jordan curve with $\Gamma \subset B_R$—a ball of radius R. Let $H > 0$ be a fixed constant. We look for a surface Σ of constant mean curvature H spanned by Γ (i.e., $\partial\Sigma = \Gamma$).

For example, if Γ is a *circle of radius* R and $HR < 1$, there are two such surfaces: the small spherical "bubble" Σ_1 and the large spherical "bubble" Σ_2 (see Figure 2).

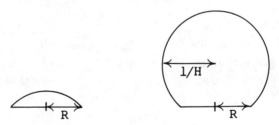

FIGURE 2

Incidentally, it is not known whether these are the only solutions (see Remark 11).

If $HR = 1$ there is one solution (the hemisphere) and if $HR > 1$ there is no solution (this intuitive fact has been proved rigorously by E. Heinz [31]). A conjecture attributed to Rellich—and publicized more recently by S. Hildebrandt and L. Nirenberg—asserts that a similar result holds for any (smooth) Jordan curve Γ. By a solution Σ we mean a generalized surface parametrized on the unit disk with possible self-intersections, multiple coverings, etc. More precisely, let

$$\Omega = \{(x, y) \in \mathbf{R}^2, x^2 + y^2 < 1\}.$$

We look for a surface Σ of the form $\Sigma = u(\Omega)$ where $u: \overline{\Omega} \to \mathbf{R}^3$ satisfies the system

$$(\text{III}_\text{P}) \quad \begin{cases} \Delta u = 2Hu_x \wedge u_y & \text{on } \Omega, \\ u_x^2 - u_y^2 = u_x \cdot u_y = 0 & \text{on } \Omega, \\ u(\partial \Omega) = \Gamma. \end{cases}$$

The last two conditions are called the *Plateau conditions*. We may also consider the same system with the usual *Dirichlet condition*:

$$(\text{III}_\text{D}) \quad \begin{cases} \Delta u = 2H_x \wedge u_y & \text{on } \Omega, \\ u = \gamma & \text{on } \partial\Omega, \end{cases}$$

where $\gamma: \partial\Omega \to \mathbf{R}^3$ is a given smooth map such that $\gamma(\partial\Omega) \subset B_R$. The system (III_D) has no simple geometric interpretation; however, it is easier to handle than (III_P) and it involves essentially the same difficulties as (III_P).

The main results of [9] concerning (III) are the following.

THEOREM 5. *Suppose $HR < 1$; then there exist at least two solutions of* (III_P).

THEOREM 6. *Suppose $HR < 1$ and γ is not a constant; then there exist at least two solutions of* (III_D).

Here are the principal ideas which enter in the proof of Theorem 6. A first solution $\underline{u}$ (the counterpart of the small bubble) of (III_D) had been obtained by S. Hildebrandt [32] (following some earlier work of E. Heinz [30]). It plays the same role as the zero solution in Problems (I) and (II). We look for a second solution $\bar{u}$ of the form $\bar{u} = \underline{u} - v$ which leads to the following system for v:

$$(7) \quad \begin{cases} \mathscr{L}v \equiv -\Delta v + 2H\underline{u}_x \wedge v_y + 2Hv_x \wedge \underline{u}_y = 2Hv_x \wedge v_y & \text{on } \Omega, \\ v \not\equiv 0 & \text{on } \Omega, \\ v = 0 & \text{on } \partial\Omega. \end{cases}$$

Problem (7) has a variational structure which resembles (I) because of its *lack of compactness*. I will explain in §3 how to adapt some of the methods used for Problem (I). We shall see that *below some magic level—related to the best constant for an isoperimetric inequality—a kind of compactness is restored*.

PROBLEM (IV) (LARGE HARMONIC MAPS). Let $\Omega = \{(x, y) \in \mathbf{R}^2; x^2 + y^2 < 1\}$ and let S^2 be the unit sphere in $\mathbf{R}^3$. We look for a mapping $u: \overline{\Omega} \to \mathbf{R}^3$ satisfying

$$(\text{IV}) \quad \begin{cases} -\Delta u = u|\nabla u|^2 & \text{on } \Omega, \\ u \in S^2 & \text{on } \Omega, \\ u = \gamma & \text{on } \partial\Omega, \end{cases}$$

where $\gamma: \partial\Omega \to S^2$ is given.

Problem (IV) has a variational structure; namely, the solutions of (IV) correspond to the critical points of the functional

$$E(u) = \int |\nabla u|^2,$$

subject to the constraint
$$u \in \mathscr{E} = \{u \in H^1(\Omega; S^2); u = \gamma \text{ on } \partial\Omega\}.$$

The solutions of (IV) are *harmonic maps* with the prescribed boundary condition $u = \gamma$ on $\partial\Omega$. It is clear that there is some $\underline{u} \in \mathscr{E}$ which is an *absolute minimum* of E on $\mathscr{E}$, that is,
$$E(\underline{u}) \leq E(v) \qquad \forall v \in \mathscr{E}.$$
In [27], Giaquinta and Hildebrandt have raised the question whether (IV) admits other solutions besides $\underline{u}$. They had observed that for some special γ's (such that $\gamma(\partial\Omega)$ is a circle) there are indeed at least two solutions of (IV). The main result of [10] concerning (IV) is the following.

THEOREM 7. *Suppose γ is not a constant. Then there is some $\bar{u} \neq \underline{u}$ which is a solution of* (IV).

The method—which will be presented in §4—is the following. The set $\mathscr{E}$ is not connected and we split $\mathscr{E}$ into its connected components $\mathscr{E} = \bigcup_{k \in Z} \mathscr{E}_k$. This is done using degree theory. Then we try to minimize E on each $\mathscr{E}_k$. The main difficulty is that the sets $\mathscr{E}_k$ are not closed for the weak H^1 topology (the degree is continuous for the strong H^1 topology but not for the weak H^1 topology). Here again the same methodology applies:

Step 1. We prove that if $\text{Inf}_{\mathscr{E}_k} E$ is less than some "magic" level, then $\text{Inf}_{\mathscr{E}_k} E$ is achieved (and even some form of compactness is restored).

Step 2. We prove that indeed $\text{Inf}_{\mathscr{E}_k} E$ is less than the magic level for at least two different k's.

Finally I should mention that there are many other problems where a similar lack of compactness also occurs. Here are some of them:

Best constants in Hardy–Littlewood, Sobolev, traces and related inequalities, see Th. Aubin [3], Talenti [57], E. Lieb [40], P. L. Lions [43], P. Cherrier [19].

Best constants in inequalities involving analytic functions, see Jacobs [34].

Variants of the Yamabe problem and questions involving the scalar curvature, see Bourguignon [8], Jerison-Lee [35], P. L. Lions [43], A. Bahri–J.-M. Coron [65].

Problems related to the Yang–Mills equations, see K. Uhlenbeck [61], C. Taubes [59], S. Sedlacek [50].

Of course a lack of compactness also occurs when a problem is posed in all of $\mathbf{R}^N$ and is translation invariant. This leads to other kinds of difficulties (not related to the critical exponents but rather to the noncompactness of $\mathbf{R}^N$), see e.g. P. L. Lions [42] for a detailed analysis and an extensive bibliography; see also Brezis–Lieb [14] for some vector field equations and C. Taubes [58, 60] for the Yang–Mills–Higgs equations on $\mathbf{R}^3$.

1. Problem (I). Sobolev's inequality asserts that there is some constant $\alpha > 0$ such that
$$\|\nabla\phi\|_2^2 \geq \alpha\|\phi\|_{2^*}^2 \qquad \forall \phi \in H_0^1(\Omega) = \{\phi \in L^{2^*}(\Omega), \nabla\phi \in L^2, \phi = 0 \text{ on } \partial\Omega\},$$

where $2^* = 2N/(N-2)$; *the best Sobolev constant is by definition*

$$S = \inf_{\substack{\phi \in H_0^1(\Omega) \\ \|\phi\|_{2^*} = 1}} \|\nabla \phi\|_2^2. \tag{8}$$

We recall some known facts about S:

(a) S is independent of Ω and depends only on N, since the ratio $\|\nabla\phi\|_2 / \|\phi\|_{2^*}$ is invariant under dilations.

(b) When $\Omega = \mathbf{R}^N$ the infimum in (8) is achieved by the function

$$U(x) = \frac{C}{(1 + |x|^2)^{(N-2)/2}}$$

or, after scaling, by any of the functions

$$U_\varepsilon(x) = \frac{C_\varepsilon}{(\varepsilon + |x|^2)^{(N-2)/2}}, \quad \varepsilon > 0, \tag{9}$$

where C and C_ε are some normalization constants. In addition, the functions $U_\varepsilon(x - x_0)$ $(x_0 \in \mathbf{R}^N)$ are the *only* functions for which the infimum in (8) is achieved. More precisely, a minimizer U of (8) (in $\mathbf{R}^N$) must satisfy the Euler equation

$$-\Delta U = SU^p \quad \text{in } \mathbf{R}^n \text{ with } \quad p = (N+2)/(N-2) \tag{10}$$
$$\text{and} \quad U > 0 \quad \text{in } \mathbf{R}^n, \, U \in L^{2^*}, \, \nabla U \in L^2.$$

However, all the solutions of (10) are known to be of the form (9).

For these facts we refer to Th. Aubin [3], G. Talenti [57], E. Lieb [40], P. L. Lions [43], Gidas–Ni–Nirenberg [29].

(c) As a direct consequence of (a) and (b) we find that the best constant in (8) is *never achieved* except when $\Omega = \mathbf{R}^N$.

The following quantity plays an important role in the solution of Problem (I):

$$S_\lambda = \inf_{\substack{u \in H_0^1(\Omega) \\ \|u\|_{2^*} = 1}} \left\{ \int |\nabla u|^2 - \lambda u^2 \right\}. \tag{11}$$

Our first lemma provides a fundamental estimate for S_λ:

LEMMA 1. *Assume $N \geq 4$, then*

$$S_\lambda < S \quad \forall \lambda > 0, \tag{12}$$

$$S_\lambda = S \quad \forall \lambda \leq 0. \tag{13}$$

In other words, the graph of the function $\lambda \mapsto S_\lambda$ looks as in Figure 3.

FIGURE 3

The proof of Lemma 1 is rather technical and we refer to [**16**] for details. The idea is the following. If the infimum in (8) were achieved by some function ϕ we could use it as a testing function in (11) and the conclusion would be trivial. Since it is not achieved, we try instead

$$u_\varepsilon(x) = \zeta(x) U_\varepsilon(x) \qquad (\text{assuming } 0 \in \Omega),$$

where $\zeta \in \mathscr{D}(\Omega)$ is any fixed function such that $\zeta \equiv 1$ near 0, so that as $\varepsilon \to 0$ the function $U_\varepsilon(x)$ "concentrates" near $x = 0$ where ζ has "no influence". We set

$$Q_\lambda(u) = \frac{\int |\nabla u|^2 - \lambda u^2}{\|u\|_{2^*}^2}.$$

A careful expansion as $\varepsilon \to 0$ shows that

$$Q_\lambda(u_\varepsilon) = \begin{cases} S + O(\varepsilon^{(N-2)/2}) - \lambda K\varepsilon & \text{if } N \geq 5, \\ S + O(\varepsilon) - \lambda K\varepsilon |\log \varepsilon| & \text{if } N = 4, \end{cases}$$

for some constant K, which depends only on N. The conclusion of Lemma 1 follows easily. If we try the same argument when $N = 3$ we find

$$Q_\lambda(u_\varepsilon) = S + A\varepsilon^{1/2} - B\lambda\varepsilon^{1/2} + O(\varepsilon),$$

where $A > 0$, $B > 0$ are constants (depending on ζ). We may only conclude that

$$S_\lambda \leq S \qquad \forall \lambda \in \mathbf{R},$$
$$S_\lambda = S \qquad \forall \lambda \leq 0.$$

We postpone for a little while the analysis of the case $N = 3$, which is rather delicate.

Theorem 1 follows easily from Lemma 1 and our next lemma:

LEMMA 2. *Assume $N \geq 3$ and let (u_j) be a minimizing sequence for (11). Then for some subsequence still denoted by (u_j), we have either*
  (a) $u_j \to u$ *strongly in H_0^1 and u is a minimizer for (11), or*
  (b) $u_j \to 0$ *weakly in H_0^1.*
Moreover, if $S_\lambda < S$, then case (a) *occurs.*

The argument is rather simple and I will describe two different proofs.

FIRST PROOF. By definition of (u_j) we have

(14) $$\int |\nabla u_j|^2 - \lambda \int u_j^2 = S_\lambda + o(1),$$

(15) $$\int |u_j|^{p+1} = 1 \qquad (p = (N+2)/(N-2)).$$

It follows that (u_j) is bounded in H_0^1, and we may always assume that $u_j \to u$ weakly in H_0^1; we write

$$u_j = u + v_j$$

with $v_j \to 0$ weakly in H_0^1 and in L^{p+1}, $v_j \to 0$ strongly in L^q, $q < 2^*$ and a.e.

We deduce from (14) that

(16) $$\int |\nabla u|^2 + \int |\nabla v_j|^2 - \lambda \int u^2 = S_\lambda + o(1).$$

On the other hand, we may use the following:

LEMMA 3 (BREZIS–LIEB [13]). *Suppose $u \in L^{p+1}$, (v_j) is bounded in L^{p+1} and $v_j \to 0$ a.e. Then*

(17) $$\int |u + v_j|^{p+1} = \int |u|^{p+1} + \int |v_j|^{p+1} + o(1).$$

Combining (15) and (17), we obtain

(18) $$1 = \int |u|^{p+1} + \int |v_j|^{p+1} + o(1),$$

and by convexity

(19) $$1 \leq \left[\int |u|^{p+1} + \int |v_j|^{p+1} \right]^{2/(p+1)} + o(1) \leq \|u\|_{p+1}^2 + \|v_j\|_{p+1}^2 + o(1).$$

Using (16) and (19) we find

(20) $$\int |\nabla u|^2 - \lambda \int u^2 + \int |\nabla v_j|^2 \leq S_\lambda \left[\|u\|_{p+1}^2 + \|v_j\|_{p+1}^2 \right] + o(1).$$

However, we have

(21) $$\int |\nabla u|^2 + - \lambda \int u^2 \geq S_\lambda \|u\|_{p+1}^2$$

and

(22) $$S_\lambda \|v_j\|_{p+1}^2 \leq \frac{S_\lambda}{S} \int |\nabla v_j|^2 \leq \int |\nabla v_j|^2$$

(by definition of S_λ, S and since $S_\lambda \leq S$). We deduce from (20), (21), (22) that several *equalities* hold, namely

(23) $$\begin{aligned} \int |\nabla u|^2 - \lambda \int u^2 &= S_\lambda \|u\|_{p+1}^2, \\ \int |\nabla v_j|^2 &= \frac{S_\lambda}{S} \int |\nabla v_j|^2 + o(1), \end{aligned}$$

(24) $$\left[\int |u|^{p+1} + \int |v_j|^{p+1} \right]^{2/(p+1)} = \|u\|_{p+1}^2 + \|v_j\|_{p+1}^2 + o(1).$$

If $S_\lambda < S$, from (23) we obtain $\int |\nabla v_j|^2 = o(1)$, and thus $u_j \to u$ strongly in H_0^1; hence u is a minimizer.

Therefore we may assume that $S_\lambda = S$ and also $u \neq 0$ (otherwise (b) holds). We set

$$t_j = \|v_j\|_{p+1}^{p+1} / \|u\|_{p+1}^{p+1}$$

so that, by (24),
$$(1 + t_j)^{2/(p+1)} = 1 + t_j^{2/(p+1)} + o(1).$$
This implies that $t_j = o(1)$ since the function $1 + t^{2/(p+1)} - (1 + t)^{2/(p+1)}$ is increasing for $t \in [0, \infty)$. Hence we conclude again that $u_j \to u$ strongly in H_0^1.

SECOND PROOF. By definition of S_λ we have

(25) $\quad \int |\nabla(u_j + \phi)|^2 - \lambda \int (u_j + \phi)^2 \geq S_\lambda \|u_j + \phi\|_{p+1}^2 \quad \forall \phi \in H_0^1.$

On the other hand, we have by convexity
$$|u_j + \phi|^{p+1} - |u_j|^{p+1} \geq (p+1)|u_j|^{p-1} u_j \phi,$$
and therefore

(26) $\quad \|u_j + \phi\|_{p+1}^2 \geq \left[\left(1 + (p+1) \int |u_j|^{p-1} u_j \phi\right)^+\right]^{2/(p+1)}.$

Combining (25) and (26) and passing to the limit, we obtain
$$S_\lambda + 2\int \nabla u \nabla \phi + \int |\nabla \phi|^2 - 2\lambda \int u\phi - \lambda \int \phi^2$$
$$\geq S_\lambda \left[\left(1 + (p+1) \int |u|^{p-1} u\phi\right)^+\right]^{2/(p+1)}.$$

Replacing t by $t\phi$, we find, as $t \to 0$,
$$\int \nabla u \nabla \phi - \lambda \int u\phi = S_\lambda \int |u|^{p-1} u\phi \quad \forall \phi \in H_0^1;$$
that is,
$$-\Delta u - \lambda u = S_\lambda |u|^{p-1} u,$$
and, in particular,

(27) $\quad \int |\nabla u|^2 - \lambda \int u^2 = S_\lambda \|u\|_{p+1}^{p+1}.$

Combining (21) and (27), we obtain that either $\|u\|_{p+1} \geq 1$ or $\|u\|_{p+1} = 0$. In the first case we must have $\|u\|_{p+1} = 1$ (since $\|u\|_{p+1} \leq 1$); it follows that $u_j \to u$ strongly in H_0^1 since $\int |\nabla u_j|^2 = S_\lambda + \lambda \int u^2 + o(1) = \int |\nabla u|^2 + o(1)$ (by (27)). Finally, it is easy to see that $S_\lambda < S$ implies that case (a) occurs. Indeed, we have
$$\int |\nabla u_j|^2 - \lambda \int u_j^2 = S_\lambda + o(1)$$
$$\geq S\|u_j\|_{p+1}^2 - \lambda \int u^2 + o(1) = S - \lambda \int u^2 + o(1),$$
and therefore
$$\lambda \int u^2 \geq S - S_\lambda > 0.$$

REMARK 2. A variant of the first proof was originally due to E. Lieb—who uses a similar device in [40]; it was later simplified by the introduction of Lemma 3. The strong convergence was originally pointed out by F. Browder, at an earlier stage when the argument was less transparent. The second proof is inspired by a device used in [14].

REMARK 3. Suppose (u_j) is a minimizing sequence for (11) and $u_j \to 0$ weakly in H_0^1. Then much more can be said about (u_j): there exist a sequence $\varepsilon_j \to 0$ ($\varepsilon_j > 0$) and a bounded sequence (a_j) in $\mathbf{R}^N$ such that

$$\left\| u_j - \varepsilon_j^{-(N-2)/2} \omega\left(\frac{\cdot - a_j}{\varepsilon_j}\right) \right\|_{H^1} \to 0$$

where $\omega(x) = C/(1+|x|^2)^{(N-2)/2}$ and $\|\omega\|_{p+1} = 1$.

In other words, (u_j) "concentrates" near a point where it resembles an extremal function (for the Sobolev inequality) with a high peak. Indeed, in view of Lemma 2, we have $S_\lambda = S$ and $\int |\nabla u_j|^2 = S + o(1)$, $\|u_j\|_{p+1} = 1$. Therefore (u_j) is a *minimizing sequence for the Sobolev inequality*, and the conclusion follows from the results of P. L. Lions [43].

We now turn to the analysis of the case $N = 3$.

LEMMA 4. *Suppose $\Omega \subset \mathbf{R}^3$ is strictly starshaped (i.e. $x \cdot n \geq \alpha > 0 \ \forall x \in \partial\Omega$). Then, there is some $\underline{\lambda} > 0$ depending on Ω such that (I) has no solution for $\lambda \leq \underline{\lambda}$. When Ω is a ball we may take $\underline{\lambda} = \lambda_1/4$.*

SKETCH OF THE PROOF. Suppose u is a solution of (I) with $\lambda \geq 0$. By Pohožaev's identity (5) we have

$$\lambda \int_\Omega u^2 = \frac{1}{2} \int_{\partial\Omega} (x \cdot n)\left(\frac{\partial u}{\partial n}\right)^2 \geq \frac{\alpha}{2} \int_{\partial\Omega} \left(\frac{\partial u}{\partial n}\right)^2 \geq b\left(\int_{\partial\Omega} \frac{\partial u}{\partial n}\right)^2$$

$$= b\left(\int_\Omega \Delta u\right)^2 = b\left(\int_\Omega |\Delta u|\right)^2 \geq c \int_\Omega u^2.$$

Here we have used the fact that $-\Delta u = u^5 + \lambda u \geq 0$ and also that $(\Delta)^{-1}$ is a bounded operator from L^1 into L^2.

When Ω is a ball there is no solution of (I) for $\lambda \leq \lambda_1/4$. The argument is somewhat tricky and I refer to [16] for the details. Here is the main idea. Suppose u is a solution of (I); we know from Gidas-Ni-Nirenberg [28] that u must be radial and we write $u(r)$ with $0 < r < 1$ (assuming Ω is the unit ball). First one proves that

$$(28) \quad \int_0^1 u^2\left(\lambda\psi' + \frac{1}{4}\psi'''\right)r^2 dr = \frac{2}{3}\int_0^1 u^6(r\psi - r^2\psi')\, dr + \frac{1}{2}|u'(1)|^2\psi(1)$$

for every smooth function such that $\psi(0) = 0$. Identity (28) is a sharpening of Pohožaev's identity—which corresponds to the special choice $\psi(r) = r$. Next, we assume by contradiction that $0 < \lambda \leq \pi^2/4$ (we recall that $\lambda_1 = \pi^2$ for the unit ball). In (28) we choose $\psi(r) = \sin(\sqrt{4\lambda}\, r)$, so that $\lambda\psi' + \frac{1}{4}\psi''' \equiv 0$, while $r\psi - r^2\psi' > 0$ on $(0,1)$ and $\psi(1) \geq 0$, which is absurd.

The following lemma allows us to conclude the proof of Theorem 2.

LEMMA 5. *Suppose $\Omega \subset \mathbf{R}^3$ is any bounded domain. Then there is some $\lambda^* \in (0, \lambda_1)$, depending on Ω, such that*

$$S_\lambda = S \quad \text{for } \lambda \leqslant \lambda^*,$$
$$S_\lambda < S \quad \text{for } \lambda > \lambda^*.$$

When Ω is a ball we have $\lambda^* = \lambda_1/4$.

In other words, the graph of the function $\lambda \mapsto S_\lambda$ looks as in Figure 4. Note the difference between the cases $N = 3$ and $N \geqslant 4$.

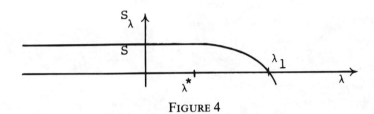

FIGURE 4

REMARK 4. It follows from Lemma 5 and Lemma 2 that for $\lambda > \lambda^*$, the infimum in (11) is achieved and it provides (after stretching) a solution of (I). On the other hand, if $\lambda < \lambda^*$, then the infimum for S_λ (i.e. (11)) is *not* achieved. Indeed, suppose by contradiction that it is achieved by some u_0. Let μ be such that $\mu > \lambda$. Then $S_\mu < S_\lambda = S$, since we may use u_0 as a testing function when computing S_μ. This is absurd since $S_\mu = S$ for $\mu \leqslant \lambda^*$.

We do not know if there can be solutions of (I) when Ω is starshaped and $\lambda < \lambda^*$; obviously, if there are any they would correspond to critical points of the functional $\int |\nabla u|^2 - \lambda \int u^2$ on the sphere $\|u\|_{p+1} = 1$—but they would not be minima. When Ω is a ball we know (see Lemma 4) that there is no solution of (I) for $\lambda \leqslant \lambda^* = \lambda_1/4$. On the other hand, when Ω is an annulus, there are solutions of (I) for *all* $\lambda \in (-\infty, \lambda_1)$.

Incidentally, we do not know whether for some domains Ω the infimum in (11) is achieved when $\lambda = \lambda^*$; this is an interesting open problem (if Ω is a ball it is not achieved since there is no solution of (I) when $\lambda = \lambda^*$.

SKETCH OF THE PROOF OF LEMMA 5. We start with the case where Ω is the unit ball. If we try the same argument as in the proof of Lemma 1 we find, as $\varepsilon \to 0$,

$$Q_\lambda(u_\varepsilon) = S + A\varepsilon^{1/2} - B\lambda\varepsilon^{1/2} + O(\varepsilon),$$

where A and B depend on ζ. We conclude that $S_\lambda < S$ for $\lambda > A/B$. Therefore one should try to minimize the ratio A/B. The optimal choice of ζ corresponds to $\zeta(x) = \cos(\pi|x|/2)$, i.e.,

$$u_\varepsilon(x) = \cos(\pi|x|/2)(\varepsilon + |x|^2)^{1/2}.$$

A careful expansion as $\varepsilon \to 0$ (see [16]) shows that
$$Q_\lambda(u_\varepsilon) = S + \left(\tfrac{1}{4}\pi^2 - \lambda\right)K\varepsilon^{1/2} + O(\varepsilon)$$
where $K > 0$ is a constant. Therefore we see that $S_\lambda < S$ for $\lambda > \pi^2/4$.

On the other hand, we must have $S_\lambda = S$ for $\lambda \leqslant \pi^2/4$. Otherwise—if $S_\lambda < S$ —we would deduce from Lemma 2 that there is a solution of (I); this is impossible, by Lemma 4.

We turn now to the case of a *general bounded domain* Ω. Let $\tilde{\Omega}$ be a ball with $\Omega \subset \tilde{\Omega}$. Given $u \in H_0^1(\Omega)$, we extend it by 0 outside Ω. The previous analysis shows that
$$\int_{\tilde{\Omega}} |\nabla u|^2 \geqslant S\|u\|_{L^6(\tilde{\Omega})}^2 + \frac{\lambda_1(\tilde{\Omega})}{4}\|u\|_{L^2(\tilde{\Omega})}^2$$
and hence there is some $\delta > 0$ such that
$$\int_{\Omega} |\nabla u|^2 \geqslant S\|u\|_{L^6(\Omega)}^2 + \delta\|u\|_{L^2(\Omega)}^2 \qquad \forall u \in H_0^1(\Omega).$$

(Alternatively, one could also use symmetrization as in [16].) This implies that $S_\delta \geqslant S$; but on the other hand $S_\lambda \leqslant S$ $\forall \lambda \in \mathbf{R}$ (see the proof of Lemma 1). Therefore we conclude that $S_\delta = S$.

ADDITIONAL PROPERTIES. OPEN PROBLEMS.

1. *Uniqueness*. It is not known—even when Ω is a ball—whether the solution of (I) is unique, except for λ near λ_1, where the standard bifurcation analysis gives uniqueness.

2. *Regularity. Further estimates*. So far, we were concerned only with the existence of weak H^1 solutions of (I). It can be shown that H^1 solutions of (I) are smooth. This is a consequence of the following.

LEMMA 6 (BREZIS-KATO [12]). *Suppose* $a(x) \in L^{N/2}(\Omega)$ *and* $u \in H_0^1(\Omega)$ *satisfies*
$$-\Delta u = au \quad \text{in } \Omega.$$
Then $u \in L^q(\Omega)$ *for all* $q < \infty$.

In the case of Problem (I) we use Lemma 6 with $a = u^{p-1} + \lambda \in L^{N/2}$ (since $(p-1)(N/2) = 2N/(N-2)$). Therefore, we conclude that $\Delta u \in L^t(\Omega)$ for all $t < \infty$, and thus $u \in W^{2,t}(\Omega)$ for all $t < \infty$, etc.... *It is, however, surprising to find out that there is no estimate of the type*
$$\|u\|_\infty \leqslant F(\|u\|_{H^1}, |\lambda|)$$
for the solutions of (I), where F is bounded on bounded sets. To see this, suppose $N \geqslant 4$ and let u_λ be a solution of (I) with $\lambda \in (0, \lambda_1)$ corresponding to a minimum v_λ of (11). More precisely, we have
$$\int |\nabla v_\lambda|^2 - \lambda \int v_\lambda^2 = S_\lambda, \qquad \|v_\lambda\|_{p+1} = 1$$

and some multiple of v_λ, namely $u_\lambda = S_\lambda^{1/(p-1)} v_\lambda$, satisfies (I). It follows that

$$\|\nabla u_\lambda\|_2^2 \leq S_\lambda^{2/(p-1)} \left(S_\lambda + \lambda \int v_\lambda^2 \right)$$

and, in particular, u_λ remains bounded in H_0^1 as $\lambda \to 0$.

On the other hand, we claim that

$$\lim_{\lambda \downarrow 0} \|u_\lambda\|_q = \infty \quad \text{for any } q > 2^*.$$

Indeed suppose that $\|u_{\lambda_n}\|_q \leq C$ for some sequence $\lambda_n \to 0$ and some $q > 2^*$. It follows from (I) that (u_{λ_n}) is bounded in $W^{2,q/p}$ and therefore (u_{λ_n}) is relatively compact in H_0^1 since $p/q - 1/N < 1/2$. Thus (v_{λ_n}) is also relatively compact in H_0^1 and we may assume that $v_{\lambda_n} \to v$ strongly in H_0^1 with

$$\int |\nabla v|^2 = S, \quad \|v\|_{p+1} = 1;$$

but that is absurd.

3. *Continuous branches of solutions.* It follows from the abstract bifurcation theory (see P. Rabinowitz [48]) that there is a continuous branch $\mathscr{C}$ of solutions (λ, u) of (I) which emanates from $(\lambda_1, 0)$ and which goes to infinity in $\mathbf{R} \times L^\infty$. It would be interesting to prove, for example, that if $N \geq 4$, then $\mathscr{C}$ covers at least the interval $(0, \lambda_1)$. *This is related to the open problem of finding L^∞ estimates for all solutions of (I) when $\lambda > 0$ stays bounded away from 0.* Even H^1 estimates are not known—but as we have pointed out above, they would not imply L^∞ estimates.

Presumably, the bifurcation diagrams, when Ω is a ball, look as in Figure 5.

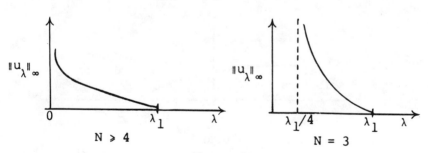

FIGURE 5

4. *The case of an annulus.* Suppose Ω is an annulus, say

$$\Omega = \{ x \in \mathbf{R}^N;\ a < |x| < b \} \quad \text{with } N \geq 4.$$

We claim that for every $\lambda > 0$, small enough, Problem (I) admits both radial and nonradial solutions. Indeed, set

(29) $$\Sigma_\lambda = \inf_{\substack{u \in H_0^1(\Omega) \\ u \text{ radial} \\ \|u\|_{p+1} = 1}} \left\{ \int |\nabla u|^2 - \lambda \int u^2 \right\}.$$

Since the *injection* $H_0^1 \subset L^{p+1}$ *is compact for radial functions* we see that the infimum in (29) is achieved for all $\lambda \in \mathbf{R}$ and—after stretching—it provides a solution of (I) for all $\lambda \in (-\infty, \lambda_1)$. The functions $\lambda \mapsto \Sigma_\lambda$ and $\lambda \mapsto S_\lambda$ are continuous and, moreover, $S_0 < \Sigma_0$ (since the best Sobolev constant is not achieved). It folows that $S_\lambda < \Sigma_\lambda$ for $\lambda > 0$ small enough. *We deduce that the infimum in* (11) *is achieved by some nonradial function.* We do not know whether these nonradial solutions occur by "secondary bifurcation" from the "branch" of radial solution. Also, it is plausible that there is some constant $\lambda_c \in (0, \lambda_1)$ such that

$$\Sigma_\lambda = S_\lambda \quad \text{for } \lambda \geqslant \lambda_c,$$
$$\Sigma_\lambda > S_\lambda \quad \text{for } \lambda < \lambda_c,$$

and that a secondary bifurcation occurs at λ_c. One may guess the diagrams in Figure 6. Other results concerning *symmetry breaking* for positive solutions of semilinear elliptic equations have been obtained by C. Coffman [20] and J. Smoller–A. Wasserman [52].

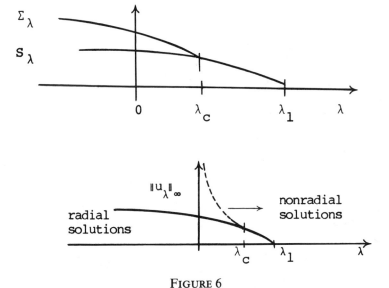

FIGURE 6

5. *Domains with topology.* Suppose Ω is not contractible in itself to a point. It is a very interesting open problem to determine whether there exist solutions of (I) for all $\lambda \in (-\infty, \lambda_1)$; but some exciting partial results have been obtained recently by J.-M. Coron [22] and J.-M. Coron–A. Bahri [69].

More generally, it is tempting to conjecture that if Ω is not contractible to a point the problem

$$\begin{cases} -\Delta u = u^p + a(x)u & \text{on } \Omega, \\ u > 0 & \text{on } \Omega, \\ u = 0 & \text{on } \partial\Omega, \end{cases}$$

admits a solution under the single assumption $\int |\nabla \phi|^2 - a(x)\phi^2 \geq \delta \int |\nabla \phi|^2$, $\forall \phi \in H_0^1$, with $\delta > 0$.

6. *Solutions with variable sign.* Some results concerning the problem

$$\begin{cases} -\Delta u = |u|^{p-1}u + \lambda u & \text{on } \Omega, \\ u \neq 0 & \text{on } \Omega, \\ u = 0 & \text{on } \partial \Omega, \end{cases}$$

have been obtained recently by Cerami–Fortunato–Struwe [18], Struwe [56], Capozzi–Fortunato–Palmieri [68] and D. Zhang [70].

7. *Improved Sobolev inequalities with best constants.* Let $\Omega \subset \mathbf{R}^3$ be (any) bounded domain. The conclusion of Lemma 5 may be read as follows. There is a (best) constant $\lambda^* > 0$ depending on Ω, such that

$$\|\nabla u\|_2^2 \geq S\|u\|_6^2 + \lambda^* \|u\|_2^2 \quad \forall u \in H_0^1(\Omega).$$

Suppose now that $\Omega \subset \mathbf{R}^N$ with $N \geq 4$. In view of Lemma 1 there is *no* inequality of the type

$$\|\nabla u\|_2^2 \geq S\|u\|_{2^*}^2 + \delta \|u\|_2^2 \quad \forall u \in H_0^1(\Omega),$$

with $\delta > 0$. However one can still exhibit a "bonus term", namely we have

$$\|\nabla u\|_2^2 \geq S\|u\|_{2^*}^2 + \delta [u]_{N/(N-2)}^2 \quad \forall \in H_0^1(\Omega)$$

for some $\delta > 0$, where $[\ ]_q$ denotes the weak (Marcinkiewicz) q-norm; see [15]. We recall that in $\mathbf{R}^N$ the Sobolev inequality

$$\|\nabla u\|_2^2 \geq S\|u\|_{2^*}^2$$

is strict except if u belongs to the set $\mathscr{E}$ of extremal functions, i.e., u is of the form $u(x) = CU_\varepsilon(x - x_0)$. In particular, the inequality is strict if u has its support in a bounded domain. It is an interesting question to estimate from below the quantity

$$\|\nabla u\|_2^2 - S\|u\|_{2^*}^2$$

by an expression which involves the "distance from u to $\mathscr{E}$".

That question recalls the Bonnesen-type isoperimetric inequality (see Osserman [46]); it provides an estimate from below for $L^2 - 4\pi A$, where A is the area of a convex set in $\mathbf{R}^2$, bounded by a curve of length L.

I should also mention that an improvement of the Hardy–Littlewood inequality (with the best constant) for functions with support in a fixed ball has been obtained and used by Daubechies–Lieb [24].

8. *Equations with variable coefficients.* It is fairly easy to obtain results comparable to Theorem 1 for the problem

$$\begin{cases} -\Delta u = u^p + a(x)u & \text{on } \Omega, \\ u > 0 & \text{on } \Omega, \\ u = 0 & \text{on } \partial \Omega. \end{cases}$$

However, the problem

$$\begin{cases} -\Delta u = a(x)u^p + \lambda u & \text{on } \Omega, \\ u > 0 & \text{on } \Omega, \\ u = 0 & \text{on } \partial\Omega, \end{cases}$$

with $a(x)$ smooth on $\overline{\Omega}$ and $a(x) \geq a > 0$ seems more delicate and largely open (except for some special cases, see e.g. A. Bahri–J.-M. Coron [65] and Escobar–Schoen [71]).

2. Problem (II). The main abstract tool which we shall use is the following variant of the Mountain Pass Lemma of Ambrosetti–Rabinowitz [1]:

LEMMA 7. *Let F be a C^1 function on a Banach space E. Assume there are constants $\rho > 0$ and $R > 0$ such that*

(30) $\qquad F(u) \geq \rho$ *for all $u \in E$ with $\|u\| = R$*

and

(31) $\qquad F(0) = 0$ *and $F(v_0) \leq 0$ for some $v_0 \in E$ with $\|v_0\| > R$.*

Set

$$M = \{ p \in C([0,1]; E); \ p(0) = 0, \ p(1) = v_0 \}$$

and

(32) $\qquad c = \inf_{p \in M} \max_{t \in [0,1]} F(p(t)) \geq \rho.$

Then there exists a sequence (u_j) in E such that $F(u_j) \to c$ and $F'(u_j) \to 0$ in E^*.

REMARK 5. I emphasize that the (PS) condition is not among the assumptions of Lemma 7. When the (PS) condition is assumed we may conclude that c is a critical value, i.e., there exists some $u \in E$ such that $F(u) = c$ and $F'(u) = 0$. This is the standard version of the Ambrosetti–Rabinowitz Lemma. Indeed, the (PS) condition says that (u_j) is relatively compact whenever $F(u_j)$ is bounded and $F'(u_j) \to 0$.

Recently J. P. Aubin and I. Ekeland [2] have found a very elegant proof of the Mountain Pass Lemma. Their proof fits very well to our situation and I will sketch it briefly. It relies on *Ekeland's minimization principle*.

LEMMA 8. (*See* [25] *or* [2].) *Let M be a complete metric space and let $f: M \to (-\infty, +\infty]$ be a l.s.c. function. Assume $c = \text{Inf}_M f > -\infty$. Then for every $\varepsilon > 0$ there exists a $p_\varepsilon \in M$ such that*

(33) $\qquad c \leq f(p_\varepsilon) \leq c + \varepsilon$

and

(34) $\qquad f(p) - f(p_\varepsilon) + \varepsilon d(p, p_\varepsilon) \geq 0 \qquad \forall p \in M.$

SKETCH OF THE PROOF OF LEMMA 7. Let $M = \{ p \in C([0,1]; E); \ p(0) = 0, \ p(1) = v_0 \}$ with its usual sup norm. We consider the mapping $f: M \to \mathbf{R}$ defined by

$$f(p) = \max_{t \in [0,1]} F(p(t))$$

so that
$$c = \inf_{p \in M} f(p) \geq \rho.$$

We deduce from Lemma 8 that there is a $p_\varepsilon \in M$ satisfying (33) and (34); in other words $p_\varepsilon([0,1])$ is almost an "optimal path" joining 0 to v_0. A careful analysis of (33) and (34) (see [2]) shows that there exists $t_\varepsilon \in [0,1]$ such that
$$f(p) = c \leq F(p_\varepsilon(t_\varepsilon)) = \max_{t \in [0,1]} F(p_\varepsilon(t)) \leq c + \varepsilon$$
and
$$\|F'(p_\varepsilon(t_\varepsilon))\| < \varepsilon.$$

The conclusion of Lemma 7 follows easily by choosing $\varepsilon_j = 1/j$ and $u_j = p_{\varepsilon_j}(t_{\varepsilon_j})$.

In order to prove Theorems 3 and 4 we introduce on the space $E = H_0^1(\Omega)$ the C^1 functional

(35) $\quad F(u) = \frac{1}{2}\int |\nabla u|^2 - \frac{1}{p+1}\int (u^+)^{p+1} - \frac{\mu}{q+1}\int (u^+)^{q+1}.$

In general F does not satisfy the (PS) condition (see Remark 6 below). However, the following lemma *makes up for the lack of the* (PS) *condition*.

LEMMA 9. *Suppose $\mu \geq 0$ and let (u_j) be a sequence in H_0^1 such that*

(36) $\quad F(u_j) \to c$ *and* $F'(u_j) \to 0$ *in* H^{-1},

with

(37) $\quad c < (1/N)S^{N/2}.$

Then (u_j) is relatively compact in H_0^1.

REMARK 6. In view of Lemma 9 we can say that the "(PS)$_c$ condition" holds when c is less than the critical level $(1/N)S^{N/2}$. It is very easy to see that the (PS)$_c$ condition does not hold at the level $c = (1/N)S^{N/2}$. It suffices to choose a sequence of the form $u_\varepsilon(x) = S^{1/(p-1)}\zeta(x)SU_\varepsilon(x)$, so that $F(u_\varepsilon) \to (1/N)S^{N/2}$, $F'(u_\varepsilon) \to 0$ in H^{-1} as $\varepsilon \to 0$; moreover $u_\varepsilon \rightharpoonup 0$ weakly in H_0^1 but u_ε does not converge to 0 strongly in H_0^1.

PROOF OF LEMMA 9. We rewrite (36) as

(38) $\quad \frac{1}{2}\int |\nabla u_j|^2 - \frac{1}{p+1}\int (u_j^+)^{p+1} - \frac{\mu}{q+1}\int (u_j^+)^{q+1} = c + o(1)$

and

(39) $\quad -\Delta u_j = (u_j^+)^p + \mu(u_j^+)^q + \varepsilon_j,$

with $\varepsilon_j \to 0$ in H^{-1}. Multiplying (39) by u_j, we find

(40) $\quad \int |\nabla u_j|^2 = \int (u_j^+)^{p+1} + \mu\int (u_j^+)^{q+1} + \langle \varepsilon_j, u_j \rangle.$

Combining (38) and (40), one shows easily that (u_j) is bounded in H_0^1. We may therefore assume that $u_j \rightharpoonup u$ weakly in H_0^1 and u satisfies
$$-\Delta u = (u^+)^p + \mu(u^+)^q \quad \text{in } \Omega.$$

Hence we also have

(41) $$\int |\nabla u|^2 = \int (u^+)^{p+1} + \mu \int (u^+)^{q+1}.$$

We write
$$u_j = u + v_j.$$

A variant of Lemma 3 (see [13]) leads to

(42) $$\int (u_j^+)^{p+1} = \int (u^+)^{p+1} + \int (v_j^+)^{p+1} + o(1).$$

Combining (38) with (42) and (40) with (42) we obtain

(43) $$F(u) + \frac{1}{2}\int |\nabla v_j|^2 - \frac{1}{p+1}\int (v_j^+)^{p+1} = c + o(1)$$

and
(44)
$$\int |\nabla u|^2 + \int |\nabla v_j|^2 = \int (u^+)^{p+1} + \int (v_j^+)^{p+1} + \mu\int (u^+)^{q+1} + o(1).$$

Then, using (41), we deduce that

(45) $$\int |\nabla v_j|^2 = \int (v_j^+)^{p+1} + o(1).$$

We may therefore assume that
$$\int |\nabla v_j|^2 \to k, \qquad \int (v_j^+)^{p+1} \to k.$$

By Sobolev's inequality, we have
$$\int |\nabla v_j|^2 \geq S \|v_j^+\|_{p+1}^2,$$

and at the limit we have $k \geq S k^{2/(p+1)}$. It follows that *either* $k = 0$ *or* $k \geq S^{N/2}$. We shall now see that $k \geq S^{N/2}$ is excluded—which implies that $k = 0$, i.e., $u_j \to u$ strongly in H_0^1 and the proof of Lemma 9 will be concluded. Suppose $k \geq S^{N/2}$. Passing to the limit in (43) we obtain
$$F(u) + \frac{1}{N}k = c,$$

and with assumption (37) we find

(46) $$F(u) < 0.$$

On the other hand we have, by definition of F and in view of (41),
$$F(u) = \frac{1}{2}\int |\nabla u|^2 - \frac{1}{p+1}\int (u^+)^{p+1} - \frac{\mu}{q+1}\int (u^+)^{q+1}$$
$$= \frac{1}{N}\int (u^+)^{p+1} + \mu\left(\frac{1}{2} - \frac{1}{q+1}\right)\int (u^+)^{q+1} \geq 0,$$

a contradiction of (46).

SKETCH OF THE PROOF OF THEOREM 3. The functional F defined by (35) obviously satisfies (30) if $R > 0$ is small enough. Also, there are plenty of v_0's satisfying (31): it suffices to pick any $w_0 \in H_0^1$, $w_0 \neq 0$ and then $v_0 = kw_0$ satisfies (31) when k is large enough. In general, when the (PS) condition holds, the choice of v_0 is irrelevant. Here, however the situation is different since c *defined by* (31) *depends* on v_0 and we would like to achieve $c < (1/N)S^{N/2}$. The choice of v_0 becomes a delicate matter. Clearly, it suffices to find a $w_0 \in H_0^1$ such that $w_0 \geq 0$, $w_0 \neq 0$ and

$$(47) \qquad \sup_{t \geq 0} F(tw_0) < \frac{1}{N} S^{N/2}.$$

Indeed we may then take $v_0 = kw_0$ with $k > 0$ so large that $k\|w_0\| > R$ and $F(kw_0) < 0$; also we have

$$c \leq \sup_{t \in [0,1]} F(tv_0) \leq \sup_{t \geq 0} F(tw_0) < \frac{1}{N} S^{N/2}.$$

It is not clear that one can find a w_0 satisfying (47). For example, if $\mu = 0$ an easy computation shows that

$$\sup_{t \geq 0} F(tw_0) = \frac{1}{N}\left(\frac{\|\nabla w_0\|_2^2}{\|w_0\|_{p+1}^2}\right)^{N/2} > \frac{1}{N} S^{N/2}.$$

This computation suggests, however, that the term μu^q "helps" to lower c and that w_0 should be chosen close to the extremal function for the Sobolev inequality. Therefore we set, as in the proof of Lemma 1,

$$w_\varepsilon(x) = \zeta(x) U_\varepsilon(x).$$

A careful expansion as $\varepsilon \to 0$ (see [16]) shows that

$$\sup_{t \geq 0} F(tw_\varepsilon) \leq \frac{1}{N} S^{N/2} + O(\varepsilon^{(N-2)/2}) - \mu K \varepsilon^\alpha,$$

where $K > 0$ is a constant and $\alpha = ((N+2) - q(N-2))/4 < (N-2)/2$. The conclusion follows by choosing $\varepsilon > 0$ small.

SKETCH OF THE PROOF OF THEOREM 4. When $N = 3$ the argument above leads to

$$\sup_{t \geq 0} F(tw_\varepsilon) \leq \frac{1}{3} S^{3/2} + O(\varepsilon^{1/2}) - \mu K \varepsilon^\alpha$$

for some constants $K > 0$ and α such that

$$\begin{aligned}&0 < \alpha < 1/2 && \text{if } 3 < q < 5, \\ &\alpha = 1/2 && \text{if } q = 3, \\ &\alpha \geq 1/2 && \text{if } 1 < q < 3.\end{aligned}$$

This concludes the proof of Theorem 4 for $3 < q < 5$.

When $1 < q \leq 3$ the argument is different. We fix *any* $w_0 \geq 0$, $w_0 \not\equiv 0$. In order to emphasize the μ dependence I write now F_μ instead of F. I claim that

(48) $$\sup_{t \geq 0} F_\mu(tw_0) \to 0 \quad \text{as } \mu \to \infty.$$

Indeed, the sup in (48) is achieved for some $t = t_\mu$ such that

$$t_\mu \int |\nabla w_0|^2 - t_\mu^p \int w_0^{p+1} - \mu t_\mu^q \int w_0^{q+1} = 0$$

and thus

$$t_\mu \leq \left[\frac{1}{\mu} \frac{\int |\nabla w_0|^2}{\int w_0^{q+1}} \right]^{1/(q-1)}.$$

Hence $\sup_{t \geq 0} F_\mu(tw_0) \leq \frac{1}{2} t_\mu^2 \int |\nabla w_0|^2 \to 0$ as $\mu \to \infty$. It follows that for μ large enough, condition (47) is satisfied and (II) has a solution. We establish now that (II) has no solution when Ω is strictly starshaped, $1 < q \leq 3$ and $\mu > 0$ is small. Indeed, suppose u is a solution of (II). Pohožaev's identity (6) leads to

$$\mu \int_\Omega u^{q+1} \geq a \int_{\partial\Omega} \left(\frac{\partial u}{\partial n}\right)^2 \geq b \left(\int_{\partial\Omega} \frac{\partial u}{\partial n}\right)^2 = b \left(\int_\Omega \Delta u\right)^2 = b \left(\int_\Omega |\Delta u|\right)^2.$$

Next, we use the fact that Δ^{-1} is a bounded operator from L^1 into L_w^3 (weak L^3) and that $|\Delta u| \geq u^5$. Hence we find

(49) $$\mu \int_\Omega u^{q+1} \geq c[u]_3^2,$$

(50) $$\mu \int_\Omega u^{q+1} \geq c\|u\|_5^{10}.$$

By Hölder and interpolation we have

(51) $$\|u\|_{q+1} \leq C\|u\|_r \leq C[u]_3^a \|u\|_5^{1-a}$$

with $r = 6(q+1)/(q+3)$, $a/3 + (1-a)/5 = 1/r$, i.e., $a = (9-q)/r(q+1)$ (note that $r \geq q+1$ and $3 < r \leq 5$ since $1 < q \leq 3$). Combining (49), (50), and (51), we find

$$\|u\|_{q+1} \leq C \left(\mu \int u^{q+1} \right)^{a/2 + (1-a)/10}.$$

But $a/2 - (1-a)/10 = 1/(q+1)$ and thus we conclude that $\mu \geq \mu_0 > 0$.

ADDITIONAL PROPERTIES. OPEN PROBLEMS.

1. Instead of (II) we now consider the more general problem

(52) $$\begin{cases} -\Delta u = \lambda u^p + f(x, u) & \text{on } \Omega, \\ u > 0 & \text{on } \Omega, \\ u = 0 & \text{on } \partial\Omega, \end{cases}$$

where $\lambda > 0$, $p = (N + 2)/(N - 2)$, $f: \Omega \times \mathbf{R}_+ \to \mathbf{R}_+$ and $f(x, 0) = 0$. Using the same method as above one can prove (see [**16**]) that (52) has a solution under the following assumptions.

(53) $\quad\quad\quad\quad f(x, u) = o(u^p) \quad \text{as } u \to +\infty;$

(54) $\quad \begin{cases} -\Delta - f_u(x, 0) \text{ is positive, i.e.,} \\ \int |\nabla \phi|^2 - f_u(x, 0)\phi^2 \geq \delta \int |\nabla \phi|^2 \quad \text{with } \delta > 0, \forall \phi \in H_0^1; \end{cases}$

(55) $\quad\quad\quad f(x, u) \geq \delta u^q \quad \text{with } \delta > 0, q \geq 1, \forall x \in \Omega, \forall u \geq 0.$

When $N \geq 5$ it suffices to assume, instead of (55),

(55') $\quad\quad\quad f(x_0, u_0) > 0$ for some $x_0 \in \Omega$ and some $u_0 > 0.$

This result may be used in order to solve the problem: Find u satisfying

(56) $\quad \begin{cases} -\Delta u = \lambda(1 + u)^p & \text{on } \Omega, \\ u > 0 & \text{on } \Omega, \\ u = 0 & \text{on } \partial\Omega, \end{cases}$

where $\lambda > 0$ and $p = (N + 2)/(N - 2)$.

Then there exists a constant $\bar{\lambda} > 0$ such that problem (56) admits:
(a) at least two solutions for every $\lambda \in (0, \bar{\lambda})$,
(b) one solution for $\lambda = \bar{\lambda}$,
(c) no solution for $\lambda > \bar{\lambda}$.

In other words, $\bar{\lambda}$ is a turning point, and we have the diagram in Figure 7.

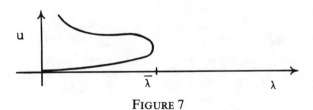

FIGURE 7

Indeed it is already known (see [**38, 23**]) that there is a constant $\bar{\lambda}$ such that problem (56) admits:
(a) a minimal solution $\underline{u}$ for every $\lambda \in (0, \bar{\lambda})$ and that $-\Delta - \lambda p(1 + \underline{u}(x))^{p-1}$ is positive,
(b) one solution for $\lambda = \bar{\lambda}$,
(c) no solution for $\lambda > \bar{\lambda}$.

We look for a second solution of (56) of the form
$$u = \underline{u} + v$$
so that (56) becomes

(57) $\quad \begin{cases} -\Delta u = \lambda(1 + \underline{u} + v)^p - \lambda(1 + \underline{u})^p & \text{on } \Omega, \\ v > 0 & \text{on } \Omega, \\ v = 0 & \text{on } \partial\Omega. \end{cases}$

Finally, we write
$$\lambda(1 + \underline{u} + v)^p - \lambda(1 + \underline{u})^p \equiv \lambda v^p + f(x, v)$$
and it is easy to check that (53), (54), and (55) hold.

2. There are still many unsolved problems[1] concerning Problem (II) even when Ω is a *ball*, $N = 3$ and $p = 5$. Numerical computations due to O. Bristeau (at INRIA) suggest that:

(i) *When $q = 3$* there is some μ_0 such that
  (a) for $\mu > \mu_0$ there is a unique solution of (II),
  (b) for $\mu \leq \mu_0$ there is no solution of (II)
[when Ω is the unit ball, it seems that $\mu_0 = 8/\sqrt{3}\,\pi$];

(ii) *when $1 < q < 3$* there is some μ_0 such that
  (a) for $\mu > \mu_0$ there are two solutions of (II),
  (b) for $\mu = \mu_0$ there is one solution of (II),
  (c) for $\mu < \mu_0$ there is no solution of (II).

3. Problem (III) (Rellich's conjecture). I will explain in detail how to handle the Dirichlet problem (III$_D$) and then I will say a few words about the Plateau problem (III$_P$).

The existence of a first solution $\underline{u}$ of (III$_D$)—the "*small solution*— is due to Hildebrandt [32, 33]. His argument is the following. The solutions of (III$_D$) are the critical points of the functional

$$\Phi(u) = \int |\nabla u|^2 + \frac{4H}{3} \int u \cdot u_x \wedge u_y$$

subject to the constraint

$$u \in H^1_\gamma = \{ u \in H^1(\Omega; \mathbf{R}^3); \ u = \gamma \ \text{on} \ \partial\Omega \}.$$

Unfortunately, $\inf_{u \in H^1_\gamma} \Phi(u) = -\infty$. Therefore, we consider instead the "variational inequality"

(58) $$A = \inf_{\substack{u \in H^1_\gamma \\ \|u\|_\infty \leq R'}} \Phi(u),$$

where R' is a fixed constant with $R < R' < 1/H$. It is easy to see that the infimum in (58) is achieved. Indeed, let (u^j) be a minimizing sequence. Since

$$|a \wedge b| \leq |a||b| \leq \tfrac{1}{2}(|a|^2 + |b|^2)$$

and since $HR'/3 < \tfrac{1}{2}$ it follows that (u^j) is bounded in H^1. We may assume that $u^j \rightharpoonup \underline{u}$ weakly in H^1 and $u^j \to \underline{u}$ a.e. Writing

$$u^j = \underline{u} + v^j,$$

we obtain

(59) $$\int |\nabla \underline{u}|^2 + \int |\nabla v^j|^2 + \frac{4H}{3} \int u^j \cdot \left(\underline{u}_x + v^j_x \right) \wedge \left(\underline{u}_y + v^j_y \right) = A + o(1).$$

---

[1] Very recently Atkinson and Peletier [66] have partially answered the problem described below.

By dominated convergence we have

$$u^j \wedge \underline{u}_x \to \underline{u} \wedge \underline{u}_x \quad \text{and} \quad u^j \wedge \underline{u}_y \to \underline{u} \wedge \underline{u}_y \quad \text{strongly in } L^2,$$

and thus

$$\int u^j \cdot \underline{u}_x \wedge v_y^j = o(1), \quad \int u^j \cdot v_x^j \wedge \underline{u}_y = o(1).$$

Since

$$\frac{4H}{3}\left|\int u^j \cdot (v_x^j \wedge v_y^j)\right| \leq \frac{2}{3}\int |\nabla v^j|^2,$$

we deduce from (59) that $\underline{u}$ is a minimizer for (58).

Next, one shows (using $HR' < 1$) that $-\Delta|\underline{u}|^2 \leq 0$ on Ω and the maximum principle implies that

$$\sup_{\Omega} |\underline{u}| \leq \sup_{\partial\Omega} |\underline{u}| \leq R.$$

Therefore $\underline{u}$ satisfies $\Phi'(\underline{u}) = 0$ (and not just the variational inequality), that is, $\underline{u}$ is a solution of (III_D).

Using the fact that the second variation of Φ is ≥ 0 we obtain

$$\int |\nabla \phi|^2 + 4H\int \underline{u} \cdot \phi_x \wedge \phi_y \geq 0 \quad \forall \phi \in H_0^1.$$

A more delicate argument (see [9]) shows that $\underline{u}$ is a strict minimum, i.e., there is a constant δ such that

(60) $$\int |\nabla \phi|^2 + 4H\int \underline{u} \cdot \phi_x \wedge \phi_y \geq \delta \int |\nabla \phi|^2 \quad \forall \phi \in H_0^1.$$

Then, we look for a second solution $\bar{u}$ of (III_D) of the form

$$\bar{u} = \underline{u} - v$$

and we are led to the system

(7) $$\begin{cases} \mathscr{L}v \equiv -\Delta v + 2H(\underline{u}_x \wedge v_y + v_x \wedge \underline{u}_y) = 2Hv_x \wedge v_y & \text{on } \Omega, \\ v \neq 0 & \text{on } \Omega, \\ v = 0 & \text{on } \partial\Omega, \end{cases}$$

which has again a variational structure.

The linear operator $\mathscr{L}$ is selfadjoint and is the Fréchet derivative of the functional $\frac{1}{2}(\mathscr{L}v, v)$ with

$$(\mathscr{L}v, v) = \int |\nabla v|^2 + 4H\int \underline{u} \cdot v_x \wedge v_y,$$

while the nonlinear term $v_x \wedge v_y$ is the Fréchet derivative of the volume functional $\frac{1}{3}Q(v)$ where

(61) $$Q(v) = \int v \cdot v_x \wedge v_y.$$

In view of the analogy between (7) and (I) (a linear selfadjoint operator versus a nonlinear potential operator which is homogeneous) it is tempting to try the following program. Consider

(62) $$J = \inf_{\substack{v \in H_0^1 \\ Q(v)=1}} (\mathscr{L}v, v)$$

and suppose that the Inf in (62) is achieved by some v^0. Then we have

$$\mathscr{L}v^0 = \mu v_x^0 \wedge v_y^0$$

where μ is a Lagrange multiplier—and clearly $\mu = J$. Thus $v = (J/2H)v^0$ satisfies (7).

There are several difficulties with this program:

1. It is not clear that $Q(v)$, defined by (61), makes sense for all $v \in H_0^1$, since $v \notin L^\infty$ in general. But, as we shall see, it is indeed possible to extend Q by continuity to all of H_0^1.

2. The function Q, which is continuous for the strong H_0^1 topology, is *not* continuous for the weak H^1 topology and the set $\{v \in H_0^1; Q(v) = 1\}$ is *not* closed for the weak H_0^1 topology. This generates the same difficulty as in Problem (I) and we overcome it with the same strategy. The inequality

$$|Q(v)|^{2/3} \leq C \int |\nabla v|^2 \quad \forall v \in H_0^1$$

takes the place of the Sobolev inequality, and we have a best constant

$$S = \inf_{\substack{v \in H_0^1 \\ Q(v)=1}} \int |\nabla v|^2.$$

Then we split the argument into two parts:

Step 1. Show that if $J < S$, then the infimum in (62) is achieved,

Step 2. Show that indeed $J < S$: the additional term $4H \int \underline{u} \cdot v_x \wedge v_y$ "helps" to lower the infimum.

Here are some of the technical details. We start with a very useful estimate (see Lemma A.1 in [9]):

LEMMA 10. *Let* $\phi, \psi \in H^1(\Omega; \mathbf{R})$ *and let* v *be the solution of*

$$\begin{cases} \Delta v = \phi_x \psi_y - \psi_x \phi_y & \text{on } \Omega, \\ v = 0 & \text{on } \partial\Omega. \end{cases}$$

Then $v \in L^\infty \cap H^1$ *and*

(63) $$\|v\|_\infty \leq C \|\nabla \phi\|_2 \|\nabla \psi\|_2,$$

(64) $$\|\nabla v\|_2 \leq C \|\nabla \phi\|_2 \|\nabla \psi\|_2.$$

REMARK 7. The conclusion of Lemma 10 is somewhat surprising. Indeed the function $f = \phi_x \psi_y - \psi_x \phi_y$ lies in L^1 and *not better*! Therefore u could have a logarithmic singularity; however, this does not happen, because the special form of f produces some cancellations. Also note that (64) is a direct consequence of

(63) since
$$\int |\nabla v|^2 \leq \|v\|_\infty \int |\phi_x \psi_y - \psi_x \phi_y| \leq \|v\|_\infty \|\nabla \phi\|_2 \|\nabla \psi\|_2.$$

LEMMA 11. *There is a constant C such that*

(65) $\quad \left| \int u \cdot \phi_x \wedge \phi_y \right| \leq C \|\nabla u\|_2 \|\nabla \phi\|_2^2 \quad \forall u \in \mathcal{D}(\Omega; \mathbf{R}^3), \forall \phi \in H^1(\Omega; \mathbf{R}^3).$

PROOF. Indeed, we introduce the solution v of the problem
$$\begin{cases} \Delta v = \phi_x \wedge \phi_y & \text{on } \Omega, \\ v = 0 & \text{on } \partial\Omega; \end{cases}$$
we write
$$\int u \cdot \phi_x \wedge \phi_y = \int u \Delta v = -\int \nabla u \cdot \nabla v$$
and we use (64).

We set
$$R(u, v) = \int u \cdot v_x \wedge v_y \quad \text{for } u \in \mathcal{D}(\Omega; \mathbf{R}^3), v \in H^1(\Omega; \mathbf{R}^3)$$
and extend R by continuity to $H_0^1 \times H^1$ with the help of Lemma 11. Clearly we have
$$|R(u, v)| \leq C \|\nabla u\|_2 \|\nabla v\|_2^2 \quad \forall u \in H_0^1, \quad \forall v \in H^1$$
and
$$|R(u, v + w) - R(u, v) - R(u, w)| \leq C \|\nabla u\|_2 \|\nabla v\|_2 \|\nabla w\|_2$$
$$\forall u \in H_0^1, \quad \forall v, w \in H^1.$$

We set
$$Q(u) = R(u, u) \quad \text{for } u \in H_0^1,$$
and obviously Q is continuous on H_0^1 for the strong H_0^1 topology. However, one should be careful with the meaning of R and Q. If $u \in H_0^1$, the function $u \cdot u_x \wedge u_y$ need not be integrable in the Lebesgue sense and $Q(u)$ is a kind of *improper integral* of $u \cdot u_x \wedge u_y$.

An easy integration by parts shows that

(66) $\quad Q(u + v) = Q(u) + Q(v) + 3R(u, v) + 3R(v, u) \quad \forall u, v \in H_0^1.$

Our next two lemmas describe some useful properties of R and Q under weak limits.

LEMMA 12. *Suppose (v^j) is a sequence in H_0^1 such that $v^j \to 0$ weakly in H_0^1. Then*

(67) $\quad R(u, v^j) \to 0 \quad \forall u \in H_0^1$

and also

(68) $\quad \int u \cdot v_x^j \wedge v_y^j \to 0 \quad \forall u \in H^1 \cap L^\infty.$

REMARK 8. The conclusion of Lemma 12 is rather surprising. Indeed v_x^j and v_y^j converge to 0 weakly in L^2; therefore their products are dangerous!

PROOF. We establish for example (67); for the proof of (68) see [9]. Given $\varepsilon > 0$ there is some $\tilde{u} \in \mathscr{D}(\Omega; \mathbf{R}^3)$ such that
$$\|u - \tilde{u}\|_{H^1} < \varepsilon.$$
We write
$$\int \tilde{u} \cdot v_x^j \wedge v_y^j = \frac{1}{2} \int \tilde{u} \cdot \left(v^j \wedge v_y^j\right)_x + \tilde{u} \cdot \left(v_x^j \wedge v^j\right)_y$$
$$= \frac{1}{2} \int v^j \cdot \tilde{u}_x \wedge v_y^j + v^j \cdot v_x^j \wedge \tilde{u}_y \to 0$$
since $v^j \to 0$ strongly in L^2. On the other hand we have
$$\left|R(u, v^j) - R(\tilde{u}, v^j)\right| \leq C\|\nabla u - \nabla \tilde{u}\|_2 \|\nabla v^j\|_2^2 \leq C\varepsilon \|\nabla v^j\|_2^2$$
and therefore $\limsup_{j \to \infty} |R(u, v^j)| \leq C\varepsilon$, $\forall \varepsilon > 0$, i.e. $\lim_{j \to \infty} R(u, v^j) = 0$.

As a direct consequence of (66) and (67) we have

LEMMA 13. *Suppose (v^j) is a sequence in H_0^1 such that $v^j \rightharpoonup$ weakly in H_0^1. Then we have for every $u \in H_0^1$*
$$Q(u + v^j) = Q(u) + Q(v^j) + o(1).$$

REMARK 9. Note the analogy between Lemma 3 and Lemma 13.

We deduce from (65) the following important inequality:

(69) $$|Q(v)|^{2/3} \leq C\|\nabla v\|_2^2 \quad \forall v \in H_0^1,$$

which has some *striking similarities with the Sobolev inequality*. In particular, let us consider the best constant for (69) which is, by definition,

(70) $$S = \inf_{\substack{v \in H_0^1(\Omega) \\ Q(v) = 1}} \int_\Omega |\nabla v|^2,$$

so that we have

(71) $$|Q(v)|^{2/3} \leq (1/S)\|\nabla v\|_2^2 \quad \forall v \in H_0^1$$

[this S has nothing to do with the best Sobolev constant of §1].

We now describe some properties of S:

(a) Because of scale invariance, S is unchanged if we replace the unit disk Ω by any domain in $\mathbf{R}^2$;

(b) In *all of* $\mathbf{R}^2$ the infimum (70) is achieved provided we replace H_0^1 by $\{v \in L^\infty; \nabla v \in L^2 \text{ and } v \to 0 \text{ at infinity}\}$, and one *extremal function* is
$$U(x, y) = \frac{C}{1 + x^2 + y^2} \begin{pmatrix} x \\ y \\ 1 \end{pmatrix},$$
which is essentially a stereographic projection from $\mathbf{R}^2$ onto S^2. The *exact value of S* is

(72) $$S = (32\pi)^{1/3}.$$

Any dilate of U, namely

$$U_\varepsilon(x, y) = \frac{C_\varepsilon}{(\varepsilon^2 + x^2 + y^2)} \begin{pmatrix} x \\ y \\ \varepsilon \end{pmatrix}, \qquad \varepsilon > 0,$$

is also an extremal function for (70).

Incidentally, *all the extremal functions for* (70) *are known.* More precisely, any extremal function for (70) must satisfy the equation

(73) $\qquad -\Delta U = SU_x \wedge U_y \quad \text{in } \mathbf{R}^2$

and all the solutions of (73) with $\int |\nabla U|^2 < \infty$ are described in [11].

(c) The best constant in (70) is *never achieved in any bounded domain* Ω. This can be proved in two different ways:

Method 1. Suppose U is an extremal function for (70) and extend U by 0 outside Ω. Then U is an extremal function for (70) in $\mathbf{R}^2$. However, they are all known and satisfy $\operatorname{Supp} U = \mathbf{R}^2$ (see [11])—a contradiction.

Method 2. We can always assume that Ω *is a disk* (otherwise fix a large disk containing Ω and extend U by 0 outside Ω). We must have

(74) $\qquad \begin{cases} -\Delta U = SU_x \wedge U_y & \text{on } \Omega, \\ U = 0 & \text{on } \partial\Omega, \end{cases}$

and also $U \not\equiv 0$ since $Q(U) = 1$. Then we obtain a contradiction with

LEMMA 14 (WENTE [63]). *Suppose U satisfies* (74). *Then we have* $U \equiv 0$.

REMARK 10. Wente's Lemma may be viewed as the "counterpart" of Pohožaev's Theorem for Problem (III). It also shows that the assumption $\gamma \not\equiv C$ in Theorem 6 is essential. When $\gamma \equiv C$, the only solution of (III$_D$) is $u \equiv C$.

(d) *Inequality* (71) *is also related to the classical isoperimetric inequality*

(75) $\qquad V^{2/3} \leq \frac{1}{(36\pi)^{1/3}} A,$

where V denotes the volume of a set in $\mathbf{R}^3$ bounded by a surface of area A, and equality holds only for spheres.

Suppose $\phi \in H_0^1(\Omega; \mathbf{R}^3)$; then $\phi(\Omega)$ is a closed surface enclosing a set of (algebraic) volume

$$V = \tfrac{1}{3} Q(\phi).$$

On the other hand, its area is given by

$$A = \int |\phi_x \wedge \phi_j|.$$

We deduce from (75) that

(76) $\qquad |Q(\phi)|^{2/3} \leq \frac{1}{(4\pi)^{1/3}} \int |\phi_x \wedge \phi_y|,$

which implies (71) with $S = (32\pi)^{1/3}$ since $|\phi_x \wedge \phi_y| \leq \frac{1}{2}|\nabla\phi|^2$. Equality in (76) holds whenever $\phi(\Omega)$ is a sphere. However, equality in (71) holds only if $\phi(\Omega)$ is a sphere *and* $|\phi_x \wedge \phi_y| = \frac{1}{2}|\nabla\phi|^2$, i.e., $\phi_x^2 - \phi_y^2 = \phi_x \cdot \phi_y = 0$. But this is impossible —except if $\phi(\Omega)$ is a point—since there is no conformal map from the disk onto a sphere.

We may now sketch the proof of Theorem 6. It is divided into two steps.

STEP 1. Let J be defined by (62) and assume

(77) $$J < S.$$

Then the infimum in (62) is achieved. More precisely, every minimizing sequence for (62) is relatively compact in H_0^1.

PROOF. The argument resembles the proof of Lemma 2. Let (v^j) be a minimizing sequence for (62), i.e.,

(78) $$(\mathscr{L}v^j, v^j) = J + o(1),$$

(79) $$Q(v^j) = 1.$$

It follows from (60) that (v^j) is bounded in H_0^1 and we may assume that $v^j \rightharpoonup v$ weakly in H_0^1. We write

$$v^j = v + w^j$$

and we deduce from Lemma 13 that

$$1 = Q(v^j) = Q(v) + Q(w^j) + o(1).$$

Thus we have

(80) $$1 \leq |Q(v)|^{2/3} + |Q(w^j)|^{2/3} + o(1).$$

On the other hand, we obtain from (68) that

(81) $$(\mathscr{L}v^j, v^j) = (\mathscr{L}v, v) + \int |\nabla w^j|^2 + o(1).$$

Combining (78), (80), and (81) we find

$$(\mathscr{L}v, v) + \int |\nabla w^j|^2 \leq J\left[|Q(v)|^{2/3} + |Q(w^j)|^{2/3}\right] + o(1).$$

However, we know that

$$(\mathscr{L}v, v) \geq J|Q(v)|^{2/3}$$

and

$$\int |\nabla w^j|^2 \geq S|Q(w^j)|^{2/3}.$$

Hence we conclude that $\int |\nabla w^j|^2 = o(1)$; therefore $v^j \to v$ *strongly* in H_0^1 and v is a minimizer for (62).

STEP 2. We claim that $J < S$. The argument resembles the proof of Lemma 1. Here, one uses the fact that $\gamma \not\equiv C$, which implies $\underline{u} \not\equiv C$. Therefore there is some point $(x_0, y_0) \in \Omega$ where $\nabla \underline{u}(x_0, y_0) \neq 0$. We choose an orthonormal basis

$\{i, j, k\}$ of $\mathbf{R}^3$ such that $\alpha = u_x \cdot i + u_y \cdot j < 0$ at (x_0, y_0). We consider the ratio

$$R(v) = \frac{(\mathscr{L}v, v)}{|Q(v)|^{2/3}}$$

and the function v_ε defined in the basis $\{i, j, k\}$ by

$$v_\varepsilon(x, y) = \phi(x, y)U_\varepsilon(x - x_0, y - y_0),$$

where $\phi \in \mathscr{D}(\Omega)$ is any fixed function with $\phi \equiv 1$ near (x_0, y_0). A careful expansion as $\varepsilon \to 0$ shows that

$$R(v_\varepsilon) = S + \alpha S H \varepsilon + O(\varepsilon^2 |\log \varepsilon|)$$

and the conclusion follows, since $\alpha < 0$.

The proof of Theorem 5 is somewhat technical and I will only describe its main steps. Given any $\gamma\colon \partial\Omega \to \mathbf{R}^3$ with $\gamma(\partial\Omega) \subset B_R$ and $HR < 1$ we have obtained a small solution $\underline{u}_\gamma$ and a large solution $\bar{u}^\gamma$ of the Dirichlet problem (III$_D$).

An easy computation shows that

$$\Phi(\bar{u}^\gamma) = \Phi(\underline{u}_\gamma) + \frac{1}{12H^2}J_\gamma^3,$$

where

$$J_\gamma = \inf_{\substack{v \in H_0^1 \\ Q(v)=1}} \left\{ \int |\nabla v|^2 + 4H \int \underline{u}_\gamma \cdot v_x \wedge v_y \right\}.$$

[We recall that $\Phi(u) \equiv \int |\nabla u|^2 + (4H/3)\int u \cdot u_x \wedge u_y$.]

One can prove that the small solution $\underline{u}_\gamma$ is *unique* (see [9]), and therefore the expression

$$A(\gamma) \equiv \Phi(\bar{u}^\gamma)$$

is defined without ambiguity even though $\bar{u}^\gamma$ need not be unique.

Next, one "slides" γ along the given Jordan curve Γ by introducing the family

$$\mathscr{C} = \{\gamma\colon \partial\Omega \to \mathbf{R}^3;\ \gamma(\partial\Omega) = \Gamma \text{ and } \gamma \text{ is "nondecreasing"}\}.$$

One proves (see [9]) that $\inf_{\gamma \in \mathscr{C}} A(\gamma)$ is achieved by some γ^0 and that $\bar{u}^{\gamma^0}$ is a "large" solution of the Plateau problem (III$_P$).

REMARK 11. It is a very interesting open problem to determine whether, or under what conditions (on γ, resp. Γ), there exist *other* solutions of (III) besides $\underline{u}$ and $\bar{u}$.

REMARK 12. It would also be interesting to show that given γ (resp. Γ) there is some *critical constant* $H^* > 0$ such that Problem (III$_D$) (resp. (III$_P$)) admits
   (a) at least two solutions for every $H \in (0, H^*)$,
   (b) one solution for $H = H^*$,
   (c) no solution for $H > H^*$.

REMARK 13. Many works have been devoted to the question of surfaces of constant mean curvature spanned by a given curve Γ, see e.g. [30, 31, 51, 53, 54, 55, 62] and their references.

4. Problem (IV) (large harmonic maps). We recall that we look for critical points of the functional

$$E(u) = \int |\nabla u|^2$$

subject to the constraint

$$u \in \mathscr{E} = \{u \in H^1(\Omega; S^2); u = \gamma \text{ on } \partial\Omega\}.$$

Clearly the absolute minimum $\underline{u}$ of E on $\mathscr{E}$ exists, i.e., $\underline{u} \in \mathscr{E}$ and $E(\underline{u}) \leq E(v)$ $\forall v \in \mathscr{E}$. First we split $\mathscr{E}$ into its (connected) components $\mathscr{E} = \bigcup_{k \in \mathbf{Z}} \mathscr{E}_k$ using degree theory. Since we deal with H^1 *maps in two dimensions they need not be continuous and their degree is not defined in a standard way.* However, according to Schoen–Uhlenbeck [49], we know that H^1 maps from S^2 into S^2 do have a well-defined degree.

Given two elements $u_1, u_2 \in \mathscr{E}$ we "transport" them on S^2 and "glue" them into a single map ϕ from S^2 into S^2: we write $S^2 = S_N^2 \cup S_S^2$, the northern hemisphere and the southern hemisphere, and we identify S_N^2, S_S^2, and Ω by stereographic projection; note that u_1, u_2 glue well on the equator since $u_1 = u_2 = \gamma$ on $\partial\Omega$. Hence $\phi \in H^1(S^2, S^2)$ and it makes sense to talk about deg ϕ. There is even an analytic expression for deg ϕ (see [45]), namely,

$$\deg \phi = (1/4\pi)[Q(u_1) - Q(u_2)],$$

where $Q(u) = \int_\Omega u \cdot u_x \wedge u_y$.

For each $k \in \mathbf{Z}$ we set

$$\mathscr{E}_k = \{u \in \mathscr{E}; (1/4\pi)[Q(u) - Q(\underline{u})] = k\}$$

($\underline{u}$ does not play any special role—we could fix instead any element in $\mathscr{E}$). Clearly $\mathscr{E} = \bigcup_{k \in \mathbf{Z}} \mathscr{E}_k$ and each $\mathscr{E}_k$ is both open and closed for the strong H^1 topology; moreover, $\mathscr{E}_k \neq \emptyset$ $\forall k$ and $\underline{u} \in \mathscr{E}_0$.

In order to find critical points of E in $\mathscr{E}$ it is tempting to consider, for $k \neq 0$,

(82) $$J_k = \underset{\mathscr{E}_k}{\text{Inf }} E.$$

However, there is a difficulty with this program. Indeed, suppose (u^j) is a minimizing sequence for (82) so that (u^j) is bounded in H^1 and we may assume that $u^j \rightharpoonup u$ weakly in H^1. However, $\mathscr{E}_k$ is *not closed for the weak H^1 topology* since $Q(u)$ is not continuous for the weak H^1 topology. Hence it may well happen that $u \notin \mathscr{E}_k$. This does indeed happen each time $\text{Inf}_{\mathscr{E}_k} E$ is not achieved —a frequent situation (see Remark 15). We use again the same strategy as in Problems (I) and (III) and we divide the argument in two parts.

Step 1. We assume that for some $k \in \mathbf{Z}$

(83) $$J_k < J_0 + 8\pi.$$

Then the infimum in (82) is achieved.

PROOF. Let (u^j) be a minimizing sequence for (82), so that

(84) $$\int |\nabla u^j|^2 = J_k + o(1),$$

(85) $$Q(u^j) - Q(\underline{u}) = 4\pi k.$$

We may assume that

$$u^j \rightharpoonup \bar{u} \text{ weakly in } H^1,$$
$$u^j \to \bar{u} \text{ strongly in } L^2 \text{ and a.e.}$$

Clearly we have by lower semicontinuity

$$\int |\nabla \bar{u}|^2 \leq J_k,$$

and the main difficulty is to show that $\bar{u} \in \mathscr{E}_k$, i.e.,

(86) $$Q(\bar{u}) - Q(\underline{u}) = 4\pi k.$$

As always, we write

$$u^j = \bar{u} + v^j$$

so that $v^j \to 0$ weakly in H_0^1, and by (84) we have

(87) $$\int |\nabla \bar{u}|^2 + \int |\nabla v^j|^2 = J_k + o(1).$$

On the other hand, we have

$$Q(u^j) = \int u^j \cdot (\bar{u}_x + v_x^j) \wedge (\bar{u}_y + v_y^j)$$
$$= \int \bar{u} \cdot \bar{u}_x \wedge \bar{u}_y + \int u^j \cdot v_x^j \wedge v_y^j + o(1),$$

since products of the form $u^j \wedge \bar{u}_x$ and $u^j \wedge \bar{u}_y$ converge strongly in L^2 by dominated convergence. Thus we obtain

(88) $$|Q(u^j) - Q(\bar{u})| \leq \frac{1}{2} \int |\nabla v^j|^2 + o(1).$$

Combining (85), (86), (88), and (87), we find

$$|Q(\underline{u}) - Q(\bar{u}) + 4\pi k| \leq \frac{1}{2} \int |\nabla v^j|^2 + o(1)$$
$$= \frac{1}{2} \left[J_k - \int |\nabla \bar{u}|^2 \right] + o(1).$$

Using assumption (83), we conclude that

$$|Q(\underline{u}) - Q(\bar{u}) + 4\pi k| < \frac{1}{2} \left[J_0 - \int |\nabla \bar{u}|^2 \right] + 4\pi \leq 4\pi,$$

since J_0 is the absolute minimum of E on $\mathscr{E}$. Finally, we note that

$$(1/4\pi)[Q(\underline{u}) - Q(\bar{u})] \in \mathbf{Z}$$

and thus $Q(\underline{u}) - Q(\bar{u}) + 4\pi k = 0$, so that $\bar{u} \in \mathscr{E}_k$.

Step 2. We claim that indeed $J_k < J_0 + 8\pi$ *either for* $k = 1$ *or* $k = -1$.
SKETCH OF THE PROOF. It suffices to find a function $v \in \mathscr{E}$ such that

(89) $$|Q(v) - Q(\underline{u})| = 4\pi$$

and

(90) $$\int |\nabla v|^2 < \int |\nabla \underline{u}|^2 + 8\pi.$$

The construction of v is explicit and technical. Again it involves a *blow-up* near a point $(x_0, y_0) \in \Omega$ where $\nabla \underline{u}(x_0, y_0) \neq 0$ (the assumption that γ is nonconstant enters here). Set

$$D_\varepsilon = \{(x, y) \in \mathbf{R}^2; (x - x_0)^2 + (y - y_0)^2 < \varepsilon^2\}.$$

One chooses some $v^\varepsilon \in \mathscr{E}$ such that

(a) $$v^\varepsilon = \underline{u} \text{ on } \Omega \setminus D_{2\varepsilon},$$

(b) $$v^\varepsilon(x, y) = \frac{2\lambda}{\lambda^2 + r^2} \begin{pmatrix} x - x_0 \\ y - y_0 \\ -\lambda \end{pmatrix} + \begin{pmatrix} 0 \\ 0 \\ 1 \end{pmatrix} \text{ in } D_\varepsilon,$$

where $\lambda = c\varepsilon^2$, $r^2 = (x - x_0)^2 + (y - y_0)^2$, and c is a constant to be fixed. A careful expansion as $\varepsilon \to 0$ shows that

$$(1/4\pi)|Q(v^\varepsilon) - Q(\underline{u})| = 1$$

and

$$\int |\nabla v^\varepsilon|^2 = \int |\nabla \underline{u}|^2 + 8\pi - \alpha\varepsilon^2 + o(\varepsilon^2)$$

with $\alpha > 0$ (for an appropriate choice of c).

REMARK 14. When $\gamma \equiv C$ is a constant, it is known (see Lemaire [39]) that $u \equiv C$ is the only solution of Problem (IV). This result may be viewed as the "counterpart" of Pohožaev's Theorem and Wente's Lemma (Lemma 14).

REMARK 15. We consider now the special case where

(91) $$\gamma(x, y) = \begin{pmatrix} Rx \\ Ry \\ \sqrt{1 - R^2} \end{pmatrix} \text{ with } 0 < R < 1.$$

Then there are two *explicit* solutions of (IV), namely,

$$\underline{u}(x, y) = \frac{2\lambda}{\lambda^2 + r^2} \begin{pmatrix} x \\ y \\ \lambda \end{pmatrix} + \begin{pmatrix} 0 \\ 0 \\ -1 \end{pmatrix}$$

and

$$\bar{u}(x, y) = \frac{2\mu}{\mu^2 + r^2} \begin{pmatrix} x \\ y \\ -\mu \end{pmatrix} + \begin{pmatrix} 0 \\ 0 \\ 1 \end{pmatrix},$$

where $r^2 = x^2 + y^2$, while $\lambda = 1/R + ((1/R^2) - 1)^{1/2}$ and $\mu = 1/R - ((1/R^2) - 1)^{1/2}$. It is known (see [10]) that:

(a) $\text{Min}_{\mathscr{E}} E$ is achieved only at $\underline{u}$,
(b) $\bar{u} \in \mathscr{E}_{-1}$ and $\text{Min}_{\mathscr{E}_{-1}} E$ is achieved only at $\bar{u}$,
(c) $\text{Inf}_{\mathscr{E}_k} E$ is *not* achieved except when $k = 0$ and $k = -1$.

However, *there could possibly be other critical points (nonminimal solutions) on* $\mathscr{E}_k$. *This is an interesting open problem, even for general* γ—*not just the special* γ *given by* (91).

REMARK 16. A result comparable to Theorem 7 has been obtained independently by Jost [36].

REMARK 17. Benci and Coron have recently considered an extension of Problem (IV): Find a mapping $u: \Omega \to \mathbf{R}^{n+1}$ (Ω is still the unit disk in $\mathbf{R}^2$) such that

(92) $$\begin{cases} -\Delta u = u|\nabla u|^2 & \text{on } \Omega, \\ u \in S^n & \text{on } \Omega, \\ u = \gamma & \text{on } \partial\Omega. \end{cases}$$

They prove (see [7] and [21]) that if γ is nonconstant there exist at least two solutions of (92). The argument is quite different from the one described here since $\mathscr{E}$ is connected when $n \geq 3$.

REFERENCES

1. A. Ambrosetti and P. Rabinowitz, *Dual variational methods in critical point theory and applications*, J. Funct. Anal. **14** (1973), 349–381.

2. J. P. Aubin and I. Ekeland, *Applied nonlinear analysis*, Wiley, New York, 1984.

3. Th. Aubin, *Problèmes isopérimétriques et espaces de Sobolev*, J. Differential Geom. **11** (1976), 573–598.

4. _____, *Equations différentielles nonlinéaires et problème de Yamabe concernant la courbure scalaire*, J. Math. Pures Appl. **55** (1976), 269–293.

5. _____, *Best constants in the Sobolev imbedding Theorem: The Yamabe problem*, Seminar on Differential Geometry (S. T. Yau, ed.), Princeton Univ. Press, Princeton, N. J., 1982.

6. _____, *Nonlinear analysis on manifolds, Monge–Ampère equations*, Springer, Berlin and New York, 1982.

7. V. Benci and J.-M. Coron, *The Dirichlet problem for harmonic maps from the disk into the euclidean n-sphere*, Ann. Inst. H. Poincaré Anal. Non Linéaire **2** (1985), 119–142.

8. J.-P. Bourguignon, *Sur une condition d'intégrabilité d'origine géométrique pour une famille d'équations aux dérivées partielles nonlinéaires sur la sphère*, Séminaire Goulaouic–Meyer–Schwartz, 1981/1982, Exp. No. XVI, 13 pp., École Polytech., Palaiseau, 1982; see also a detailed paper by Bourguignon and Ezin (to appear).

9. H. Brezis and J.-M. Coron, *Multiple solutions of H-systems and Rellich's conjecture*, Comm. Pure Appl. Math. **37** (1984), 149–187 [announced in C. R. Acad. Sci. Paris **295** (1982), 615–618].

10. _____, *Large solutions for harmonic maps in two dimensions*, Comm. Math. Phys. **92** (1983), 203–215.

11. _____, *Convergence of solutions of H-systems, or How to blow bubbles*, C. R. Acad. Sci. Paris **298** (1984), 389–392 and Arch. Rational Mech. Anal. **89** (1985), 21–56.

12. H. Brezis and T. Kato, *Remarks on the Schrödinger operator with singular complex potential*, J. Math. Pures Appl. **58** (1979), 137–151.

13. H. Brezis and E. Lieb, *A relation between pointwise convergence of functions and convergence of integrals*, Proc. Amer. Math. Soc. **88** (1983), 486–490.

14. _____, *Minimum action solutions of some vector field equations*, Comm. Math. Phys. **96** (1984), 97–113.

15. _____, *Sobolev inequalities with remainder terms*, J. Funct. Anal. **62** (1985), 73–86.

16. H. Brezis and L. Nirenberg, *Positive solutions of nonlinear elliptic equations involving critical Sobolev exponents*, Comm. Pure Appl. Math. **36** (1983), 437–477.

17. H. Brezis and R. E. L. Turner, *On a class of superlinear elliptic problems*, Comm. Partial Differential Equations **2** (1977), 601–614.

18. G. Cerami, D. Fortunato and M. Struwe, *Bifurcation and multiplicity results for nonlinear elliptic problems involving critical Sobolev exponents*, Ann. Inst. Poincaré Anal. Non Linéaire **1** (1984), 341–350.

19. P. Cherrier, *Problèmes de Neumann non linéaires sur les varétés riemanniennes*, J. Funct. Anal. **57** (1984), 154–206.

20. C. V. Coffman, *A nonlinear boundary value problem with many positive solutions* (to appear).

21. J.-M. Coron, *Harmonic maps from the disk into the euclidean n-sphere*, in these Proceedings.

22. _____, *Topologie et cas limite des injections de Sobolev*, C. R. Acad. Sci. Paris **299** (1984), 209–212.

23. M. Crandall and P. Rabinowitz, *Some continuation and variational method for positive solutions of nonlinear elliptic eigenvalue problems*, Arch. Rational Mech. Anal. **58** (1975), 207–218.

24. I. Daubechies and E. Lieb, *One-election relativistic molecules with Coulomb interaction*, Comm. Math. Phys. **90** (1983), 497–510.

25. I. Ekeland, *On the variational principle*, J. Math. Anal. Appl. **47** (1974), 324–353.

26. D. de Figueiredo, P. L. Lions and R. Nussbaum, *A priori estimates for positive solutions of semilinear elliptic equations*, J. Math. Pures Appl. **61** (1982), 41–63.

27. M. Giaquinta and S. Hildebrandt, *A priori estimates for harmonic mappings*, J. Reine Angew. Math. **336** (1982), 124–164.

28. B. Gidas, W. M. Ni and L. Nirenberg, *Symmetry and related properties via the maximum principle*, Comm. Math. Phys. **68** (1979), 209–243.

29. _____, *Symmetry of positive solutions of nonlinear elliptic equations in* $\mathbf{R}^n$, Mathematical Analysis and Applications (L. Nachbin, ed.) Academic Press, New York, 1981, pp. 370–401.

30. E. Heinz, *Über die Existenz einer Fläche konstanter mittlerer Krümmung bei vorgegebner Berandung*, Math. Ann. **127** (1954), 258–287.

31. _____, *On the nonexistence of a surface of constant mean curvature with finite area and prescribed rectifiable boundary*, Arch. Rational Math. Anal. **35** (1969), 249–252.

32. S. Hildebrandt, *On the Plateau problem for surfaces of constant mean curvature*, Comm. Pure Appl. Math. **23** (1970), 97–114.

33. _____, *Nonlinear elliptic systems and harmonic mappings* Proc. Beijing Sympos. Differential Geometry and Differential Equations, (Beijing, 1982), Science Press and Gordon & Breach.

34. S. Jacobs, *An isoperimetric inequality for functions analytic in multiply connected domains*, Report Mittag-Leffler Inst. (1970).

35. D. Jerison and J. Lee, *A subelliptic, nonlinear eigenvalue problem and scalar curvature on CR manifolds*, in Microlocal Analysis (Baouendi, Beals and Rothschild, eds.) Amer. Math. Soc., Providence, R. I., 1984, pp. 57–63.

36. J. Jost, *The Dirichlet problem for harmonic maps from a surface with boundary onto a 2-sphere with nonconstant boundary values*, J. Differential Geom. (to appear).

37. J. Kazdan, *Gaussian and scalar curvature, an update*, Seminar on Differential Geometry (S. T. Yau, ed.) Princeton Univ. Press, Princeton, N. J., 1982.

38. J. Keener and H. Keller, *Positive solutions of convex nonlinear eigenvalue problems*, J. Differential Equations **16** (1974), 103–125.

39. L. Lemaire, *Applications harmoniques de surfaces riemanniennes*, J. Differential Geom. **13** (1978), 51–78.

40. E. Lieb, *Sharp constant in the Hardy–Littlewood–Sobolev and related inequalities*, Ann. Math. **118** (1983), 349–374.

41. P. L. Lions, *On the existence of positive solutions of semilinear elliptic equations*, SIAM Rev. **24** (1982), 441–467.

42. _____, *The concentration-compactness principle in the calculus of variations. The locally compact case.* Parts I and II, Ann. Inst. H. Poincaré Anal. Non Linéaire **1** (1984), 109-145 and 223-284.

43. _____, *The concentration-compactness principle in the calculus of variations. The limit case.* Parts I and II (to appear) [announced in C. R. Acad. Sci. Paris **296** (1983), 645-648].

44. W. M. Ni and R. Nussbaum, *Uniqueness and nonuniqueness for positive radial solutions of* $\Delta u + f(u,r) = 0$ (to appear).

45. L. Nirenberg, *Topics in nonlinear functional analysis*, N. Y. U. Lecture Notes (1973-74).

46. R. Osserman, *Bonnesen-style isoperimetric inequalities*, Amer. Math. Monthly **86** (1979), 1-29.

47. S. Pohožaev, *Eigenfunctions of the equation* $\Delta u = \lambda f(u) = 0$, Soviet Math. Dokl. **6** (1965), 1408-1411 (translated from Dokl. Akad. Nauk SSSR **165** (1965), 33-36).

48. P. Rabinowitz, *Some global results for nonlinear eigenvalue problems*, J. Funct. Anal. **7** (1971), 487-513.

49. R. Schoen and K. Uhlenbeck, *Boundary regularity and the Dirichlet problem for harmonic maps*, J. Differential Geom. **18** (1983), 253-268.

50. S. Sedlacek, *A direct method for minimizing the Yang-Mills functional over 4-manifolds*, Comm. Math. Phys. **86** (1982), 515-527.

51. J. Serrin, *On surfaces of constant mean curvature which span a given space curve*, Math. Z. **112** (1969), 77-88.

52. J. Smoller and A. Wasserman, *Symmetry breaking for positive solutions of semilinear elliptic equations* (to appear).

53. K. Steffen, *Flächen konstanter mittlerer krümmung mit vorgegebenem volumen öder flächeninhalt*, Arch. Rational Mech. Anal. **49** (1972), 99-128.

54. _____, *On the nonuniqueness of surfaces with prescribed mean curvature spanning a given contour* (to appear).

55. M. Struwe, *Nonuniqueness in the Plateau problem for surfaces of constant mean curvature* (to appear).

56. _____, *A global existence result for elliptic boundary value problems involving limiting nonlinearities* (to appear).

57. G. Talenti, *Best constant in Sobolev inequality*, Ann. Mat. Pura Appl. **110** (1976), 353-372.

58. C. Taubes, *The existence of a non-minimal solution to the* SU(2) *Yang-Mills-Higgs equations on* $\mathbf{R}^3$. Part I, Comm. Math. Phys. **86** (1982), 257-298; Part II, Comm. Math. Phys. **86** (1982), 299-320.

59. _____, *Path connected Yang-Mills moduli spaces*, J. Differential Geom. (to appear).

60. _____, *Min-max theory for the Yang-Mills-Higgs equations* (to appear).

61. K. Uhlenbeck, *Variational problems for gauge fields* Seminar on Differential Geometry (S. T. Yau, ed.) Princeton Univ. Press, Princeton, N. J., 1982.

62. H. Wente, *An existence theorem for surfaces of constant mean curvature*, J. Math. Anal. Appl. **26** (1969), 318-344.

63. _____, *The differential equation* $\Delta x = 2H(x_u \wedge x_v)$ *with vanishing boundary values*, Proc. Amer. Math. Soc. **50** (1975), 131-137.

64. R. Schoen, *Conformal deformation of a Riemannian metric to constant scalar curvature*, J. Differential Geometry (to appear).

65. A. Bahri and J. M. Coron, *Critical points at infinity in the Yamabe equation and the Kazdan-Warner problem*, C. R. Acad. Sci. Paris (to appear).

66. F. Atkinson and L. A. Peletier, *Emden-Fowler equations involving critical exponents*, J. Nonlinear Anal. (to appear).

67. J. B. McLeod and J. Norbury (in preparation).

68. A. Capozzi, D. Fortunato and G. Palmieri, *An existence result for nonlinear elliptic problems involving critical Sobolev exponents* (to appear).

69. J.-M. Coron and A. Bahri, *Sur une équation elliptique non linéaire avec l'exposant de Sobolev critique*, C. R. Acad. Sci. Paris (to appear).

70. D. Zhang, *Multiplicity results for the probelm* $-\Delta u = u|u|^{4/(n-2)} + \lambda u$ (to appear).

71. J. Escobar and R. Schoen, *Conformal metrics with prescribed scalar curvature* (to appear).

72. J. Kazdan, *Prescribing the curvature of a Riemannian manifold*, CBMS Reg. Conf. Ser. in Math., Amer. Math. Soc., Providence, R.I., 1985.

UNIVERSITÉ PARIS VI

Degree Theory for Nonlinear Mappings

FELIX E. BROWDER

Introduction. For half a century (since the celebrated paper of Leray and Schauder in 1934), the *degree of mapping* has been one of the principal tools in nonlinear functional analysis applied to obtain existence and multiplicity results for solutions of nonlinear partial differential and functional equations. It is our purpose in the present paper to present some of the recent results obtained by the writer concerning degree theory for various classes of nonlinear mappings of monotone type from a reflexive Banach space X to its conjugate space X^*, as part of a larger projected program of studying the mapping degree for general classes of mappings with domains contained in one Banach space X and ranges in another Y.

What is a theory of degree of mapping? From our present point of view, which is motivated by analytical rather than topological considerations, such a theory consists of an *algebraic count* of the number of solutions of the equation $f(x) = y_0$, where in the simplest situation f is a mapping (in many cases, continuous) of cl(G), the closure of an open subset G of the domain space X, into the image space Y, while y_0 is a point of the target space Y which is always assumed disjoint from the image under f of the boundary of G (i.e., $y_0 \notin f(\text{bdry}(G))$). The degree theories we shall consider in the present discussion are *classical*, i.e. the value of the degree function for any such count is an ordinary integer. The integer may be positive or negative, and we recall that in the simplest finite-dimensional situations, positive counts correspond to solutions at which the mapping f is orientation-preserving, while negative counts correspond to solutions at which f is not orientation-preserving. In our general situation, we distinguish such *classical* degree theories from the broader category of *generalized* degree theories in which the value of the degree function may fall in $\mathbf{Z}_2$, the

---

1980 *Mathematics Subject Classification* (1985 *Revision*). Primary 35J60, 47H05, 47H10, 58C30;

integers modulo 2, or into a more general abelian group obtained as a compactification of the integers, the nonstandard integers, cobordism groups, infinite cohomology classes, or any other abstract construction which has been applied to describe degree functions with noninteger values.

The basic prescription that we propose to delimit a degree theory for a particular class of mappings is very simple in outline. It consists of the following:

(1) We are given a class F of mappings f, each of which is defined on the closure of an open subset G of the space X, with f mapping cl(G) into the space Y. The corresponding class of open sets G is implicitly described by giving the class F and may consist of all open sets, or of a more restricted subclass such as the bounded open sets of X.

(2) For a given subset of points y_0 in Y and for a mapping f in the class F just described, we propose to give an integer

$$d(f, G, y_0)$$

called the degree of the mapping f over G with respect to y_0. In general, we presume this degree function will only be defined if $y_0 \notin f(\text{bdry}(G))$. We may impose still further conditions upon the target points y_0, but only by explicit stipulation.

(3) We assume that we are also given a class H of *homotopies* in the class F, where an element h of H is a one-parameter family

$$h = \{f_t: 0 \leq t \leq 1\}$$

with each f_t a mapping of the class F, and all the f_t having a common domain cl(G) independent of t.

Within this framework, we impose the basic restrictions that form the definition of a degree theory as it will be understood in the subsequent discussion.

A. **Normalization.** *If $d(f, G, y_0)$ is well defined and different from 0, then y_0 lies in $f(G)$.*

There exists a normalizing mapping f_0 from X to Y such that for any permissible open set G in X, $f_0|_{\text{cl}(G)}$ lies in the class F, while for any permissible target point y_0 with respect to this latter mapping

$$d(f_0|_{\text{cl}(G)}, G, y_0) = +1, \quad \text{if } y_0 \in f_0(G).$$

B. **Additivity on domain.** *Let f be a mapping in the class F defined on cl(G), y_0 a permissible target point for f in Y. Suppose that G_1 and G_2 are two disjoint open subsets of G such that if $f_1 = f|_{\text{cl}(G_1)}$ and $f_2 = f|_{\text{cl}(G_2)}$, then f_1 and f_2 lie in the class F. Suppose moreover that*

$$y_0 \notin f(\text{cl}(G) \setminus (G_1 \cup G_2)).$$

Then y_0 is a permissible target point for f_1 and f_2, and we have

$$d(f, G, y_0) = d(f_1, G_1, y_0) + d(f_2, G_2, y_0).$$

C. **Invariance of the degree under permissible homotopies.** *Let $h = \{f_t: 0 \leq t \leq 1\}$ be an element of the class H of permissible homotopies, with cl(G) being the common domain of the mappings f_t. Let $\{y_t: 0 \leq t \leq 1\}$ be a continuous curve in Y*

such that for each t, y_t is a permissible target point for f_t (and in particular, $y_t \notin f_t(\mathrm{brdy}(G))$). Then

$$d(f_t, G, y_t)$$

is independent of t in $[0,1]$.

The conditions A, B, and C represent the most basic properties of such classical degree functions as the Brouwer degree for maps of $\mathbf{R}^n$ into $\mathbf{R}^n$ or the Leray–Schauder degree in infinite-dimensional Banach spaces. To give our stipulations and restrictions a more concrete flavor, let us consider them explicitly in these two cases.

For the Brouwer degree function, X and Y are finite-dimensional with the same dimension, $X = \mathbf{R}^n = Y$ for some given n. F is the class of continuous mappings f of $\mathrm{cl}(G)$ into $\mathbf{R}^n$, where G is a bounded open subset of $\mathbf{R}^n$. For any such mapping f and for any y_0 such that $y_0 \notin f(\mathrm{bdry}(G))$, the Brouwer degree $d(f, G, y_0)$ is a well-defined integer. The class of homotopies H is given by $h = \{f_t: 0 \leqslant t \leqslant 1\}$ where h is a continuous homotopy in the usual sense of maps f_t of $\mathrm{cl}(G)$, with G a fixed bounded open subset of $\mathbf{R}^n$. The conditions A, B, and C of the degree definition are then satisfied by the basic properties of the Brouwer degree. For normalization, we choose the normalizing mapping f_0 to be the identity map of $\mathbf{R}^n$ onto $\mathbf{R}^n$.

For the Leray–Schauder degree function, $X = Y$ is an infinite-dimensional Banach space, G is restricted to the class of bounded open subsets of X, while F is the class of maps f of $\mathrm{cl}(G)$ into X which are continuous and such that $(I - f)(\mathrm{cl}(G))$ is relatively compact in X. (Here I is the identity map of X, and the latter condition is usually written in the form $f = I - g$, where g is a compact map of $\mathrm{cl}(G)$ into X.) The normalizing mapping f_0 is the identity mapping I. The crucial definition is that of the class H of permissible homotopies, where $h = \{f_t: 0 \leqslant t \leqslant 1\}$ lies in H if h is a continuous homotopy of maps f_t of $\mathrm{cl}(G)$ into X such that there exists a compact subset K of X with $(I - f_t)(\mathrm{cl}(G)) \subset K$ for all t in $[0, 1]$. A target point y_0 is permissible if $y_0 \notin f(\mathrm{bdry}(G))$, and the Leray–Schauder degree function $d(f, G, y_0)$ satisfies the conditions A, B, and C with these stipulations.

We note explicitly that in this very classical special case, not all continuous homotopies through the class F are permissible. It is a crucial part of the definition of a degree function on a given class of mappings F to give an appropriate specification for the class H of permissible homotopies, making the class H as large as possible for results on the existence of a degree function and as small as possible for results on the uniqueness of the degree.

One very important justification of our definition of the degree function in terms of the conditions A, B, and C is the result established in 1972 by Amann–Weiss and by Fuhrer that the Brouwer and Leray–Schauder degree functions are unique in their respective contexts if A, B, and C hold. It is an important heuristic principle to which we hold that the possibility of extending degree functions to broader classes of nonlinear mappings as classical degree functions is closely related to the possibility of showing the uniqueness of this extension if the conditions A, B, and C are to hold. Moreover, such uniqueness results are important in their own right to show that the degree functions obtained on these extended classes do not depend upon the particular extension or approximation method used to define them.

The existence of a degree function for a given class of mappings F with a nontrivial class of homotopies H has immediate nontrivial consequences with respect to existence theorems for nonlinear equations involving maps in that class. The simplest illustration of this principle is the kind of argument by which one deduces the Brouwer and Schauder fixed point theorems from the existence of the Brouwer and Leray-Schauder degree functions, respectively. Suppose that the class F under consideration is *convex*, i.e. for any two maps f_0 and f_1 in F with the same domain $\text{cl}(G)$, the convex linear combination $f_t = (1 - t)f_0 + tf_1$ also lies in F for $0 \leq t \leq 1$. Suppose that H includes all such affine homotopies. Suppose finally that $X = Y$, that f_0 is the identity mapping I, and that G is the unit ball in X. Let g be a self-mapping of $\text{cl}(G)$ into itself such that $f = I - g$ belongs to F. Then g has a fixed point. The proof applies the permissible homotopy $f_t = (1 - t)I + tf$, which by hypothesis belongs to H. By the normalization principle, $d(f_0, G, 0) = +1$. If $0 \notin f_t(\text{bdry}(G))$ for all t in $[0, 1]$, then the homotopy invariance principle C implies that $d(f, G, 0) = +1$ and hence that f has a zero in G, i.e. g has a fixed point in G. However, if for x on $\text{bdry}(G)$ we have $f_t(x) = 0$, t must lie in the open interval $(0, 1)$ and $x = tg(x)$ with $t < 1$. Thus, $\|x\| = 1 \leq t\|g(x)\| < 1$, which is a contradiction.

Thus, if we are considering a class of mappings F in an infinite-dimensional Banach space X such that f in F maps $\text{cl}(G)$ into X where, for the corresponding map $g = I - f$, g does not always have a fixed point if G is a ball in X and g is a self-map of $\text{cl}(G)$, no degree function can exist. In particular, this is the case when F is the class of all continuous mappings for X of infinite dimension by a well-known result of Dugundji. The basic task of finding degree theories more general than the Leray-Schauder degree is therefore in part a problem of characterizing suitable classes of nonlinear mappings which have enough structure and are sufficiently narrowed down from the broad class of all continuous mappings so that a degree theory can exist.

One possible candidate at first sight might seem to be the class of mappings f from an infinite-dimensional Banach space X to another Y with f a proper C^1 Fredholm mapping of index zero. As was observed in [9], it follows that if $X = Y$ is a Hilbert space, for example, no classical degree theory exists for this class (though a degree theory modulo two does exist). This follows immediately from the uniqueness of the Leray-Schauder degree together with the connectedness of the general linear group on X. If we replace F by a more restricted subclass F_0, however, with suitable restrictions on the linear Fredholm mappings which are the differentials of f, we can define a degree function on F_0 or even on $F_0 + C$, where C is the class of compact mappings. For example, F_0 could be taken as the class of Fredholm mappings whose differentials df_x lie in the sum of a convex class of invertible linear operators W and the class of compact linear operators C_0. Then each f in F_0 can be written in the form $f(x) = S(x, x)$ where $S(x, u)$ is a homeomorphism in x on a neighborhood of u, and depends compactly on u. Moreover, the family of maps S involved also lies in a convex class of such representations, and a degree theory for f can be derived from the Leray–Schauder theory (as in [3]). In the case in which F_0 is even further restricted to consist of a single linear Fredholm operator, this definition coincides with the coincidence degree of Mawhin.

Our principal focus of attention in the present paper will be mappings f from cl(G) in X with values in X^*, the conjugate space of X. Such mappings are generated for example in the study of nonlinear elliptic partial differential operators in divergence form

$$A(u) = \sum_{|\alpha| \leq m} (-1)^{|\alpha|} D^\alpha A_\alpha(x, u, \ldots, D^m u),$$

where u is taken from a Sobolev space $W_0^{m,p}(\Omega)$ for an open subset Ω of $\mathbf{R}^n$. We let $X = W_0^{m,p}(\Omega)$, and its conjugate space X^* is the space of distributions $W^{-m,p'}(\Omega)$ with $p' = p(p-1)^{-1}$. If the coefficient functions A_α satisfy the Carathéodory condition and a growth condition in u and its derivatives of power not greater than $(p - 1)$, A gives a continuous, bounded mapping of X into X^* (bounded in the usual sense that A maps bounded sets of X into bounded sets of X^*). If, in addition, Ω is bounded, A satisfies the Leray–Lions ellipticity condition, and the coerciveness condition

$$\sum_{|\alpha| \leq m} A_\alpha(x, \xi) \xi_\alpha \geq c_0 |\xi|^p - c_1,$$

then the operator A from X to X^* falls within the following class of nonlinear mappings:

$(S)_+$: *A mapping f: cl(G) $\to X^*$, where G is a bounded open subset of X, is said to lie in the class $(S)_+$ if for any weakly convergent sequence $\{u_j\}$ in cl(G) with weak limit u in X for which $\overline{\lim} \langle f(u_j), u_j - u \rangle \leq 0$, it follows that u_j converges strongly to u and $f(u_j)$ converges weakly to $f(u)$.*

The class $(S)_+$ is closed under compact perturbations, i.e. if f belongs to $(S)_+$ and g is a compact map, then $f + g$ also lies in $(S)_+$. In particular, adding any lower-order terms to A keeps us in the class $(S)_+$. Under more general and much weaker hypotheses on A, including abandonment of the hypothesis that Ω is bounded, the corresponding nonlinear mapping of X into X^* falls into the broader class of pseudo-monotone maps of X into X^*, given by the following definition:

(p.m.) *Let X be a reflexive Banach space, f a mapping of X into X^*. Then f is said to be pseudo-monotone if for any weakly convergent sequence $\{u_j\}$ with weak limit u for which $\overline{\lim} \langle f(u_j), u_j - u \rangle \leq 0$, we have $f(u_j)$ converging weakly to $f(u)$ and $\langle f(u_j), u_j \rangle$ converging to $\langle f(u), u \rangle$.*

The class (p.m.) is not closed under compact perturbations. While each map f in $(S)_+$ is proper on each closed bounded subset of a reflexive space X, this is no longer the case for pseudo-monotone maps f. However, it is true that each pseudo-monotone mapping f is the uniform limit on a bounded set G in X of a sequence of mappings $\{f_k\}$ in the class $(S)_+$. If the space X is locally uniformly convex, as well as the space X^* (as is the case in our application), an example of a mapping of class $(S)_+$ is the duality mapping J of X into X^* given by

$$J(u) = \left\{ w \mid w \in X^*, \|w\| = \|u\|, \langle w, u \rangle = \|u\|^2 \right\}.$$

We use this mapping J as the normalizing mapping for degree functions from X to X^*. If X is merely reflexive, we use the duality mapping corresponding to an equivalent norm having the desired properties.

To construct a degree theory for mappings in the class $(S)_+$, we need an appropriate class H of homotopies. This is provided by the following extension of the $(S)_+$ concept:

DEFINITION 1. Let X be a reflexive Banach space, G a bounded open subset of X, $h = \{f_t: 0 \leq t \leq 1\}$ a one-parameter family of mappings of $\mathrm{cl}(G)$ into X^*. Then h is said to lie in the class $H_{(S)_+}$ if the following condition holds:

For any sequence $\{u_j\}$ in $\mathrm{cl}(G)$ converging weakly in X to u, and any sequence $\{t_j\}$ in $[0,1]$ converging to t, for which

$$\varlimsup \langle f_{t_j}(u_j), u_j - u \rangle \leq 0,$$

we have u_j converging strongly to u, $f_{t_j}(u_j)$ converging weakly to $f_t(u)$.

THEOREM 1. *Let X be a reflexive Banach space, F the class of mappings f of type $(S)_+$ from $\mathrm{cl}(G)$ to X^* with G any bounded open subset of X. Let H be the class of homotopies $h = \{f_t: 0 \leq t \leq 1\}$ of class $(S)_+$. Then there exists one and only one degree function on F corresponding to homotopy invariance with respect to H and normalized by $f_0 = J$, the duality mapping of X into X^* corresponding to an equivalent norm on X with respect to which both X and X^* are locally uniformly convex.*

The uniqueness result remains valid if H is replaced by the smaller class of affine homotopies.

If we examine the special case in which the context of Theorem 1 of maps from X to X^* intersects the classical context of mappings from X to X in the Leray–Schauder theory, the common case is that in which $X = X^*$ is a Hilbert space H. Here the duality mapping J coincides with the identity mapping I of H, which lies in the class $(S)_+$. As we have already remarked, the class $(S)_+$ is invariant under compact perturbations. Hence every mapping in the Leray–Schauder class, $f = I - g$ with g compact, must lie in the class $(S)_+$. Moreover, every permissible homotopy in the Leray–Schauder class is a homotopy of class $(S)_+$. Hence, when X is a Hilbert space, the degree theory for mappings of type $(S)_+$ is an extension, and in fact a very broad extension, of the classical Leray–Schauder degree theory.

The method of proof of Theorem 1 is based upon Galerkin approximations of f on finite-dimensional subspaces of X. We sketch it very formally here, referring for the detailed argument to the exposition given in [9]. If X_γ is a finite-dimensional subspace of X, we may assume without loss of generality that $0 \in G$, and we set $G_\gamma = G \cap X_\gamma$, and let ζ_γ be the injection map of X_γ into X. The map of $\mathrm{cl}(G_\gamma)$ into X_γ^* given by

$$f_\gamma(x) = \zeta_\gamma^* f(x), \qquad x \in \mathrm{cl}(G_\gamma),$$

is called the Galerkin approximant of f with respect to X. From the fact that f lies in the class $(S)_+$, it follows that if $y_0 \notin f(\mathrm{bdry}(G))$, then for all subspaces X_γ containing some fixed X_0 of finite dimension, the corresponding target point $\zeta_\gamma^*(y_0)$ does not lie in $f_\gamma(\mathrm{bdry}(G_\gamma))$. Hence, the Brouwer degree function yields the approximating degree $d(f_\gamma, G_\gamma, \zeta_\gamma^*(y_0))$. Furthermore, we may show that all these degrees stabilize for X_γ containing another finite-dimensional subspace X_1. Similarly, we may begin with a homotopy $\{f_t\}$ of class $(S)_+$ and show that for X_γ containing another finite-dimensional subspace X_2, the corresponding homotopy $\{f_{t,\gamma}\}$ is a permissible homotopy on G_γ with respect to the curve $\{y_{t,\gamma}\}$.

This schema of proof as described for Galerkin approximations for this particular class may be translated formally for approximation processes of any kind, uniform approximations in the case of Leray–Schauder theory, Galerkin approximations in the above case, and still other kinds of approximation schemes in the cases considered later in this paper. We wish to construct a degree function over a given class of mappings F corresponding to a given class H of homotopies. The approximation process produces an ordered family of approximating mappings f_γ lying in some class for which a degree function already exists. If y_0 is a permissible target point for f, it must be shown that a suitable approximant y_γ is a permissible target point for the approximating mapping f_γ. Applying the degree theory already constructed, we obtain the approximating degrees $d(f_\gamma, G_\gamma, y_\gamma)$. To obtain a classical degree function, this family must be shown to *stabilize*, i.e. past some point in the approximating sequence, all the values $d(f_\gamma, G_\gamma, y_\gamma)$ must be the same. Similar arguments are necessary for the homotopy and its approximations.

Let us apply this schema to the treatment of the class of pseudo-monotone maps f. In this case, Galerkin approximation does not yield a degree theory and another mode of approximation is necessary. As we have implicitly noted above, f can be approximated by $f_\varepsilon = f + \varepsilon J$ for small $\varepsilon > 0$, where J is the duality mapping which lies in the class $(S)_+$ and, for each $\varepsilon > 0$, f_ε lies in the class $(S)_+$. Thus, on each bounded set, f is uniformly approximated by f_ε since

$$\|f(x) - f_\varepsilon(x)\| \leq \varepsilon \|x\| \leq \varepsilon M.$$

To obtain a degree function for f by using the degree function on the class $(S)_+$, we must modify the definition of the degree function given above to an extended or weak degree function in the following form:

The degree $d(f, G, y_0)$ is defined when $y_0 \notin \mathrm{cl}(f(\mathrm{bdry}(G)))$.

The normalization condition A is replaced by the weaker condition A':

If $d(f, G, y_0) \neq 0$, then $y_0 \in \text{cl}(f(G))$.

In the additivity-on-domain condition, the hypothesis becomes

$$y_0 \notin \text{cl}(f(\text{cl}(G) \setminus (G_1 \cup G_2))).$$

In the homotopy invariance condition, the hypothesis becomes: There exists $\delta > 0$ such that $\text{dist}(y_t, f_t(\text{bdry}(G))) > \delta$ for all t in $[0, 1]$.

THEOREM 2. *There exists one and only one degree function on the class F of pseudo-monotone mappings on bounded open sets G of X with values in X* in the extended sense just described, where the class H of permissible homotopies is the class of pseudo-monotone homotopies in the Definition given below. For each map f in this class and for each $y_0 \notin \text{cl}(f(\text{bdry}(G)))$, $d(f, G, y_0))$ is the common value of $d(f_\varepsilon, G, y_0)$ for all $\varepsilon > 0$ and sufficiently small. If G is convex and if $d(f, G, y_0) \neq 0$, the weakened normalization condition still implies that $y_0 \in f(G)$, but not necessarily for more general open sets G.*

DEFINITION 2. *Let X be a reflexive Banach space, $\{f_t: 0 \leq t \leq 1\}$ a one-parameter family of mappings of X into X*. Then this family is said to be a pseudo-monotone homotopy if for any sequence $\{u_j\}$ converging weakly to u in X and for any sequence $\{t_j\}$ converging to t in $[0, 1]$ for which*

$$\overline{\lim} \langle f_{t_j}(u_j), u_j - u \rangle \leq 0,$$

we have $f_{t_j}(u_j)$ converging weakly to $f_t(u)$, and $\langle f_{t_j}(u_j), u_j \rangle$ converging to $\langle f_t(u), u \rangle$.

A good deal of the argument of the proof of Theorem 2 goes over to the more general case of any mapping f which can be uniformly approximated by maps in the class $(S)_+$ on bounded sets of X. Indeed, suppose that $\text{dist}(y_0, f(\text{bdry}(G))) = \delta > 0$, and suppose that on $\text{bdry}(G)$, $\|f(x) - f_\varepsilon(x)\| < \delta$ for a suitable approximant f_ε. For any two such approximants f_ε and f_β, it follows that the one-parameter family $f_t = (1 - t)f_\varepsilon + tf_\beta$ is a homotopy of class $(S)_+$ and that on the boundary of G, $f_t(x) \neq y_0$. By the homotopy invariance property of the degree function on the class $(S)_+$, it follows that $d(f_\varepsilon, G, y_0) = d(f_\beta, G, y_0)$ and the approximating degrees stabilize. The weakened degree properties follow automatically. The only part of the construction that fails to extend in a simple way is the description of the class of permissible homotopies, which must be transformed in this case into the class of homotopies which can be uniformly approximated by homotopies in the class $(S)_+$.

THEOREM 3. *Let X be a reflexive Banach space, G a bounded open subset of X invariant under the involution $x \to -x$. Let f be a mapping of class $(S)_+$ which maps $\text{cl}(G)$ into X* such that, for x in the boundary of G, $f(-x) = -f(x)$. Suppose moreover that $0 \notin f(\text{bdry}(G))$. Then $d(f, G, 0)$ is an odd integer.*

The same conclusion holds if f is merely pseudo-monotone, provided that one assumes that $\text{dist}(0, f(\text{bdry}(G))) > 0$.

Proof of Theorem 3. Suppose first that f is of class $(S)_+$. If f is odd on the boundary of G, each Galerkin approximant of f is odd on the boundary of G_γ. By the classical Borsuk-Ulam property, once $d(f_\gamma, G_\gamma, 0)$ is well defined, it must be an odd integer. Hence $d(f, G, 0)$ is an odd integer.

If f is merely pseudo-monotone and if $d(f, G, 0)$ is well defined, then $d(f, G, 0) = d(f_\varepsilon, G, 0)$ where $f_\varepsilon = f + \varepsilon J$ for $\varepsilon > 0$. Since J is odd, f_ε is odd on the boundary of G. By the preceding argument $d(f_\varepsilon, G, 0)$ is an odd integer. Hence $d(f, G, 0)$ is an odd integer. q.e.d.

In the following sections, we shall consider the extension of the degree function to other classes of nonlinear mappings from X to X^*. In §1, we describe the extension to maps of the form $T + f$ with T maximal monotone, f bounded and pseudo-monotone. In §2, we treat the case of maps from $W_0^{m,p}(\Omega)$ to its conjugate space of the form $f + N_g$, where f is a pseudo-monotone map and N_g is the Niemytskiĭ operator corresponding to a nonlinear function $g(x, r)$ satisfying merely a sign condition. In §3, we treat a class of nonlinear maps of the form $L + f$, where f is once more a pseudo-monotone bounded mapping of a Hilbert space H into itself and L is a linear operator of a class introduced by Brezis-Nirenberg [2] in the study of periodic solutions of nonlinear wave equations, namely L has closed range with $N(L) = N(L^*)$ and L has a compact inverse on $R(L)$.

1. Perturbations of maximal monotone operators. An important extension of the theory of nonlinear mappings of monotone type from a reflexive Banach space X to its conjugate space X^* involves the concept of a (possibly multivalued) maximal monotone operator from a domain in X with values in X^*.

DEFINITION 3. *Let X be a Banach space, T a mapping from X into 2^{X^*}. Then*:

(1) *T is said to be monotone if for any pair of elements $[u, w]$, $[v, z]$ of the graph of T, $G(T)$, we have*

$$\langle w - z, u - v \rangle \geq 0.$$

(2) *T is said to be maximal monotone if it is monotone and maximal in the sense of inclusion of graphs among monotone mappings from X to 2^{X^*}.*

This last part of the definition is couched in terms of multivalued mappings and in terms of set-theoretic categories. It might be rephrased in slightly more concrete terms by saying:

(2′) T is maximal monotone if it is monotone and if for any pair $[u_0, w_0]$ in $X \times X^*$, we have

$$\langle w_0 - w, u_0 - u \rangle \geq 0$$

for all w in $T(u)$ for any u, then $w_0 \in T(u_0)$.

Some important examples of maximal monotone mappings are the following:

(a) If f is a continuous (or even hemicontinuous) monotone mapping from X to X^* (in the single-valued sense), then f is maximal monotone.

(b) If T is maximal monotone from X to 2^{X^*}, then T^{-1} is maximal monotone from X^* to 2^X.

(c) If G is a closed convex subset in X, and if for each u in G,
$$T(u) = \{w \mid w \in X^*, \langle w, u \rangle \geq \langle w, v \rangle \text{ for all } v \text{ in } G\},$$
then T (extended to have no values outside of G) is maximal monotone from X to 2^{X^*}.

(d) Let f be a proper lower-semi-continuous function from the Banach space X to the reals together with $\{+\infty\}$. The subgradient ∂f is defined by
$$(\partial f)(u) = \{w \mid w \in X^*; \text{ for all } v, f(v) \geq f(u) + \langle w, v - u \rangle\}.$$
then ∂f is maximal monotone mapping from X to 2^{X^*}.

(e) If L is a densely defined linear operator from the reflexive Banach space X to X^* with L monotone, then L is maximal monotone if and only if L^* is monotone.

The list of examples corresponds to a number of domains in which the theory of maximal monotone operators and their perturbations finds various applications. Suppose for example that we consider the generalized Hammerstein operator
$$I + f \circ g,$$
where g is a continuous monotone mapping from the Banach space X to its conjugate space X^*, while f is a continuous monotone mapping from X^* back to X. (In the classical case, f is linear.) If we wish to solve
$$u + f(g(u)) = h,$$
h given and u unknown in the space X, the equation is equivalent to the multivalued relation
$$g(u) \in f^{-1}(h - u),$$
or to the relation involving the maximal monotone operator $T(u) = -f^{-1}(h - u)$,
$$0 \in (T + g)(u).$$

Changing variables and setting $v = g(u)$, we obtain another, equivalent equation
$$u = h - f(v) \in g^{-1}(v),$$
i.e.
$$h \in (f + g^{-1})(v),$$
involving the maximal monotone mapping g^{-1}. Similarly, the equation
$$f(u) + g(u) = 0$$
is equivalent to the multivalued relation
$$0 \in g^{-1}(v) + (-f^{-1}(-v)).$$

The variational inequality
$$\langle f(u), u - v \rangle \geq 0, \quad v \in G,$$

for an element u of G is equivalent to the multivalued relation

$$f(u) \in T(u),$$

where T is the maximal monotone mapping of example (c).

A fundamental theorem which is a criterion for maximal monotonicity is the following:

PROPOSITION 1. *Let X be a reflexive Banach space, J a duality mapping of X into X^* corresponding to an equivalent norm in which X and X^* are strictly convex. Let T be a monotone mapping of X into 2^{X^*}. Then T is maximal monotone if and only if $(T + J)$ has all of X^* as its range. In that case, $(T + \varepsilon J)^{-1}$ is a well-defined single-valued demicontinuous monotone mapping from X^* to X for $\varepsilon > 0$.*

It is our purpose in this part of the discussion to develop the theory of the mapping degree for (possibly multivalued) mappings of the form $T + f$, where T is maximal monotone from X to 2^{X^*} and f is a bounded mapping of class $(S)_+$ from the closure of a bounded open subset G of X into X^*. The concept of the degree of mapping for multivalued mappings is formally the same as for the single-valued case, provided that one replaces equations of the type $y = f(x)$ by the corresponding inclusion relation $y \in T(x)$. Thus the condition $y_0 \notin f(\text{bdry}(G))$ must be replaced by the corresponding condition $y_0 \notin (T + f)(\text{bdry}(G))$, where the latter is the union of the values assumed by $(T + f)$ at all points of the set $\text{bdry}(G)$.

DEFINITION 4. *Let $\{T_t: 0 \leq t \leq 1\}$ be a one-parameter family of maximal monotone operators from X to 2^{X^*}. Then the homotopy is said to be permissible if for each v in X^*, the mapping $t \to (T + J)^{-1}(v)$ is continuous from $[0, 1]$ to the strong topology of X.*

THEOREM 4. *Let X be a reflexive Banach space, F the class of (possibly multivalued) mappings of the form $T + f$, where T is a maximal monotone mapping from X to 2^{X^*} and f is a bounded mapping of class $(S)_+$ from $\text{cl}(G)$ to X^*, with G a bounded open subset of X. Let y_0 be a point of Y which is not in $(T + f)(\text{bdry}(G))$. Then there exists one and only one degree function for the class F with normalizing mapping J and with the class of permissible homotopies of the form*

$$h = \{T_t + f_t: 0 \leq t \leq 1\},$$

where T_t is a permissible homotopy of maximal monotone mappings in the sense of the above Definition and $\{f_t\}$ is a bounded homotopy of class $(S)_+$.

The detailed proof of Theorem 4 is given in [9]. We shall merely sketch it here. The basic idea is to reduce the degree theory for the present class to the already constructed degree theory for mappings of type $(S)_+$ by approximating the maximal monotone mapping T by a family of bounded monotone mappings, its Yosida approximants. We may assume without loss of generality (by translating in each variable by a constant element) that $[0, 0]$ lies in $G(T)$. Then for each $\varepsilon > 0$, the Yosida approximant T_ε of T is a bounded, continuous, single-valued

monotone mapping of X into X^* given by
$$T_\varepsilon = (T^{-1} + \varepsilon J^{-1})^{-1}.$$
Similarly, if we are given a homotopy $\{T_t\}$ of maximal monotone operators, one obtains its Yosida approximation $\{T_{t,\varepsilon}\}$.

Suppose that $y_0 \notin (T + f)(\text{bdry}(G))$. One shows that for $\varepsilon > 0$ and sufficiently small, $y_0 \notin (T_\varepsilon + f)(\text{bdry}(G))$. However, the mapping $T_\varepsilon + f$ lies in the class $(S)_+$ since the class $(S)_+$ is a cone and the convex linear combination of an element of $(S)_+$ and a monotone operator lies in $(S)_+$.

Thus, for $\varepsilon > 0$, we can use the degree theory for the class $(S)_+$ and obtain the approximating degree $d(T_\varepsilon + f, G, y_0)$. The concluding (and more difficult) step of the proof of the existence of the degree function is to show that these degree approximations stabilize for ε sufficiently small, and that the corresponding approximation for the homotopy is valid.

A variant of Theorem 4 which is very useful for parabolic problems concerns the class of mappings of the form $L + f$, where L is a linear maximal monotone mapping and f is a bounded L-pseudo-monotone mapping, where the pseudo-monotonicity condition is imposed only upon sequences for which the sequence $\{Lu_j\}$ is bounded. This discussion will appear elsewhere.

We remark finally that using the technique of passing from pseudo-monotone mappings by uniform approximation to mappings of class $(S)_+$, one obtains a generalization of Theorem 4 to the class of maps of the form $T + f$ with T maximal monotone and f bounded and pseudo-monotone. In this case, as in the earlier treatment, the degree function satisfies the conditions in the weakened form.

2. Strongly nonlinear perturbations. Let us turn to another extension of degree theory but in a more concrete context of operators of a particular analytic form in Sobolev spaces. This study is directed toward the theory of nonlinear elliptic operators in divergence form with strongly nonlinear zeroth-order perturbations:
$$A(u) + g(x, u),$$
where
$$A(u) = \sum_{|\alpha| \leq m} (-1)^{|\alpha|} D^\alpha A_\alpha(x, u, \ldots, D^m u)$$
satisfies the polynomial growth and ellipticity conditions that make it a pseudo-monotone operator or an operator of class $(S)_+$ from the Sobolev space $W_0^{m,p}(\Omega)$ to its conjugate space $W^{-m,p}(\Omega)$, while the strongly nonlinear zeroth-order term $g(x, u)$ satisfies no a priori growth condition but only a sign condition of the following form.

DEFINITION 5. *Let Ω be an open subset of $\mathbf{R}^n$, g a real-valued function on $\Omega \times \mathbf{R}^1$. Then g is said to satisfy the sign condition if the following properties hold*:

(1) $g(x, r)$ *is continuous in r for fixed x in Ω, measurable in x for fixed r. For each integer $s > 0$, there exists a function h_s in $L^1_{\text{loc}}(\Omega)$ such that for $|r| \leq s$,*
$$|g(x, r)| \leq h_s(x).$$

(2) *For all values of x and r,*

$$g(x,r)r \geq 0.$$

A consequence of these conditions is the fact that $g(x,0) = 0$ for all x.

For any strongly nonlinear function g which satisfies the conditions of the above definition we may form the Niemytskiĭ (i.e. evaluation) operator corresponding to g with domain in a prescribed Sobolev space $W_0^{m,p}(\Omega)$ for a given p with $1 < p < \infty$, and with values in its conjugate space $W^{-m,p'}(\Omega)$. Formally, the Niemytskiĭ operator N_g acts by the prescription

$$N_g(u)(x) = g(x, u(x)), \quad x \in \Omega,$$

and it is defined with domain

$$D(N_g) = \{u | u \in W_0^{m,p}(\Omega), N_g(u) \in W^{-m,p'}(\Omega) \cap L^1_{\text{loc}}(\Omega)\}.$$

The meaning of this definition is, of course, that one tests the measurable function $N_g(u)$ as to whether it belongs to $L^1_{\text{loc}}(\Omega)$, and if it does, since it is then a distribution on the open set Ω, as to whether it lies in the particular space of distributions $W^{-m,p'}(\Omega)$.

We shall consider mappings of the form $f + N_g$ acting from the subset $D(N_g)$ of $X = W_0^{m,p}(\Omega)$ to $X^* = W^{-m,p'}(\Omega)$, where, for simplicity, we shall assume for the first part of our discussion that f is a mapping of class $(S)_+$ from $\text{cl}(G)$, the closure of an open subset G of X, to the space X^*. We propose to show the existence and uniqueness of a degree function for $f + N_g$ using the properties of the degree function on the class $(S)_+$.

To do this, we shall apply still another approximation method to the mappings we propose to study by using the following truncation procedure on the strong nonlinearity $g(x, r)$:

DEFINITION 6. Let $\{\Omega_k\}$ be a sequence of relatively compact open subsets of Ω whose union is the whole of Ω, with $\Omega_k \subset \Omega_{k+1}$ for each k. Let ξ_k be the characteristic function of Ω_k. Consider the truncation of the function g at level k, i.e.

$$g^{(k)}(x,r) = \begin{cases} g(x,r), & \text{if } |g(x,r)| \leq k, \\ k \, \text{sign}(g(x,r)) & \text{if } |g(x,r)| \geq k. \end{cases}$$

We let the kth approximant N_k of the Niemytskiĭ operator N_g be the map of X into X^* given by

$$N_k(u)(x) = \xi_k(x) g^{(k)}(x, u(x)).$$

Each such operator N_k is a compact mapping of X into $X^* \cap L^\infty_c(\Omega)$.

THEOREM 5. *Suppose that f is a bounded mapping of class $(S)_+$ of $\text{cl}(G)$ into X^*, where G is a bounded open subset of $X = W_0^{m,p}(\Omega)$, N_g the Niemytskiĭ operator from X to X^* corresponding to a strongly nonlinear function g satisfying the sign* condition in the sense of the above Definition. Suppose that y_0 is a target point in X^* such that $y_0 \notin (f + N_g)(\text{bdry}(G))$. Then:

(i) *For k sufficiently large, $y_0 \notin (f + N_k)(\text{bdry}(G))$.* Since $(f + N_k)$ lies in the class $(S)_+$, $d(f + N_k, G, y_0)$ is well defined by the degree theory for mappings of class $(S)_+$.

(ii) *For k sufficiently large, $d(f + N_k, G, y_0)$ is independent of k.* We define this ultimate common value to be $d(f + N_g, G, y_0)$.

The proof of Theorem 5, as well as of the other results about Niemytskiĭ operators for strongly nonlinear functions satisfying the sign condition, rests upon the approximation theorems for elements of Sobolev spaces due to L. I. Hedberg [14]. One result which follows from Hedberg's theory (including the extension in collaboration with T. Wolff [15]) is the following.

Given an element u in the Sobolev space $W_0^{m,p}(\Omega)$, there exists a sequence $\{w_j\}$ of measurable functions on Ω with compact support such that $0 \leq w_j \leq 1$ on Ω while $u_j = w_j u$ lies in $W_0^{m,p}(\Omega) \cap L_c^\infty(\Omega)$ and the sequence $\{u_j\}$ converges to u in $W_0^{m,p}(\Omega)$.

A direct consequence of this approximation theorem is the following theorem on the action of distributions in $W^{-m,p'}(\Omega)$ given in Brezis-Browder [1], which will serve as the principal tool of our arguments in the present section.

PROPOSITION 2. *Let u be an element of $W_0^{m,p}(\Omega)$, T an element of $W^{-m,p'}(\Omega) \cap L_{\text{loc}}^1(\Omega)$ such that $T(x)u(x) \geq -h(x)$, where h is a summable function on Ω. Let $\langle T, u \rangle$ be the distribution action of T on u. Then $T(x)u(x)$ is summable on Ω and*

$$\langle T, u \rangle = \int_\Omega T(x)u(x)\, dx.$$

Proof of Theorem 5: Proof of (i). Suppose the assertion of (i) is false. Then we may find a sequence $\{u_k\}$ of bdry(G) corresponding to an infinite sequence of values of k going to infinity such that

$$f(u_k) + N_k(u_k) = y_0.$$

We identity this subsequence with the original sequence and reason to a contradiction from our assumption. We may assume that the sequence $\{u_k\}$ is weakly convergent in X to an element u, and since $f(\text{cl}(G))$ is bounded, that $\{f(u_k)\}$ converges weakly in X^* to an element w of X^*. Using the compactness of the Sobolev imbedding on compact subdomains of Ω, we may assume as well that $u_k(x)$ converges a.e. on Ω to $u(x)$. It follows by the construction of the approximant N_k that $N_k(u_k)(x)$ converges a.e. to $N_g(u)(x)$ on Ω.

On the other hand,

$$\int_\Omega N_k(u_k)(x)u_k(x)\, dx = \langle y_0 - f(u_k), u_k \rangle \leq M$$

for all k, while the integrand is always nonnegative and converges to $N_g(u)(x)u(x)$. Applying Fatou's Lemma, we see that $N_g(u)u$ is a summable function and that

$$\int_\Omega N_g(u) u \, dx \leq \varliminf_k \langle y_0 - f(u_k), u_k \rangle.$$

From the local bounds on the nonlinear function $g(x, r)$ for bounded r, it follows easily that $N_g(u)$ is locally summable on Ω. If we apply Proposition 2, it follows that if $N_g(u) \in W^{-m,p'}(\Omega) \cap L^1_{\text{loc}}(\Omega)$, then

$$\langle N_g(u), u \rangle = \int_\Omega N_g(u)(x) u(x) \, dx.$$

From the approximating equation, we see that $N_k(u_k) = y_0 - f(u_k)$ converges weakly in $W^{-m,p'}(\Omega)$ to $y_0 - w$. We have already remarked that $N_k(u_k)$ converges a.e. in Ω to $N_g(u)$, while

$$\int_\Omega N_k(u_k) u_k \, dx \leq M.$$

It follows from the boundedness of these integrals, together with the local bounds on $g(x, r)$ for r bounded, that we can apply the Vitali equi-integrability criterion on compact subsets of Ω and conclude that $N_k(u_k)$ converges to $N_g(u)$ in $L^1(\Omega')$ for each relatively compact open subset Ω' of Ω, i.e., $N_k(u_k)$ converges to $N_g(u)$ in $L^1_{\text{loc}}(\Omega)$. Since we have two modes of convergence for $N_k(u_k)$, both implying convergence in the distribution sense, their limits must coincide as distributions, i.e. $N_g(u) = y_0 - w$, and therefore $N_g(u) \in W^{-m,p'}(\Omega)$.

Finally, we consider

$$\langle f(u_k), u_k - u \rangle = \langle y_0 - N_k(u_k), u_k \rangle - \langle f(u_k), u \rangle.$$

Since $f(u_k)$ converges weakly to w in X^*, it follows that $\langle f(u_k), u \rangle \to \langle w, u \rangle$. Moreover,

$$\langle y_0, u_k \rangle \to \langle y_0, u \rangle,$$

while

$$\varlimsup \{ -\langle N_k(u_k), u_k \rangle \} = \varlimsup \left\{ -\int_\Omega N_k(u_k)(x) u_k(x) \, dx \right\}$$

$$\leq -\int N_g(u)(x) u(x) \, dx.$$

Therefore

$$\varlimsup \langle f(u_k), u_k - u \rangle \leq \langle y_0 - w, u \rangle - \int N_g(u)(x) u(x) \, dx$$

$$\leq \langle y_0 - w, u \rangle - \langle N_g(u), u \rangle = 0,$$

since $N_g(u) = y_0 - w$. We now apply the properties of f as a mapping of the class $(S)_+$, and we find that u_k converges strongly to u in $W_0^{m,p}(\Omega)$ and u is therefore an element of the closed set $\text{bdry}(G)$ in X to which all the u_k belong. Moreover, $f(u_k)$ converges weakly to $f(u)$, and hence $w = f(u)$. Thus, $N_g(u) = y_0 - w = y_0 - f(u)$, i.e. u is a solution on $\text{bdry}(G)$ of the equation $f(u) + N_g(u) = y_0$, which is excluded by assumption. Hence, the proof of (i) is complete.

Proof of (ii). Suppose the assertion of (ii) were false. Then we could find an infinite sequence of pairs of integers k and $j > k$ such that during the affine homotopy from $f + N_k$ to $f + N_j$, the target point y_0 would fall in the image of bdry(G). (Indeed, if this were not the case we should have $d(f + N_k, G, y_0) = d(f + N_j, G, y_0)$ for all $j > k$ past some point by the homotopy invariance property of the degree function on affine homotopies in the class $(S)_+$). In other words, we should have elements u_k of the boundary of G and real numbers s_k in the interval $[0, 1]$ such that

$$f(u_k) + (1 - s_k)N_k(u_k) + s_k N_j(u_k) = y_0.$$

We proceed as in the proof of (i) by extracting an infinite subsequence (which we again for convenience of notation identify with the original sequence) such that u_k converges weakly in X to u, $f(u_k)$ converges weakly in X^* to w, and u_k converges a.e. on Ω to u. Let

$$z_k = (1 - s_k)N_k(u_k) + s_k N_j(u_k).$$

We note that since the truncation process never changes the sign of the functions involved, $z_k(x)u_k(x) \geqslant 0$. Moreover, as k and hence j go to infinity, $z_k(x)$ converges a.e. to $N_g(u)(x)$ and $z_k(x)u_k(x)$ converges a.e. to $N_g(u)(x)u(x)$. Since

$$z_k(x)u_k(x)\,dx = \langle y_0 - f(u_k), u_k \rangle \leqslant M,$$

it follows once more by Fatou's Lemma that $N_g(u)u$ is summable on Ω and that

$$\int_\Omega N_g(u)u\,dx \leqslant \underline{\lim} \langle y_0 - f(u_k), u_k \rangle.$$

Paralleling the preceding argument, we see that $z_k = y_0 - f(u_k)$ converges weakly in X^* to $y_0 - w$, while by the equi-integrability of the sequence $\{z_k\}$ on relatively compact subdomains Ω' of Ω and since z_k converges a.e. on Ω to $N_g(u)$, it follows that z_k converges strongly in $L^1_{\text{loc}}(\Omega)$ to $N_g(u)$. Hence $N_g(u) \in W^{-m,p'}(\Omega) \cap L^1_{\text{loc}}(\Omega)$, and $N_g(u) = y_0 - w$.

Applying our basic proposition, we know that

$$\int N_g(u)(x)u(x)\,dx = \langle N_g(u), u \rangle = \langle y_0 - w, u \rangle.$$

On the other hand, we have

$$\overline{\lim} \langle f(u_k), u_k - u \rangle = \lim\{-N_k(u_k)u_k\,dx\} + \langle y_0 - w, u \rangle$$

$$\leqslant -\int_\Omega N_g(u)u\,dx + \langle y_0 - w, u \rangle = 0.$$

Since f lies in the class $(S)_+$, u_k must converge strongly to u, $f(u_k)$ converges weakly to $f(u)$ in X^*, u lies in bdry(G), and $w = f(u)$. Thus

$$f(u) + N_g(u) = y_0,$$

which is a contradiction, proving (ii). q.e.d.

From the definition of the extended degree function $d(f + N_g, G, y_0)$, the first two basic properties of the degree function A and B follow by a mechanical argument. The homotopy invariance property C demands a detailed prescription of the class of homotopies which are to be taken as permissible.

DEFINITION 7. *Let $\{g_t(x, r)\}$ be a family of functions parametrized by $[0, 1]$, each of which satisfies the sign condition as above. Such a family will be said to be a permissible homotopy if $g_t(x, r)$ is jointly continuous in t and r for fixed x in Ω, and if the inequality*

$$|g_t(x, r)| \leq h_s(x), \quad |r| \leq s,$$

holds uniformly for all t in $[0, 1]$ for fixed functions h_s in $L^1_{\text{loc}}(\Omega)$.

DEFINITION 8. *H, the class of permissible homotopies of maps of the form $f + N_g$ with g satisfying the sign condition, will consist of all homotopies $h = \{f_t + N_{g_t}\}$ with $\{f_t\}$ a bounded homotopy of class $(S)_+$ and $\{g_t\}$ a permissible homotopy of functions satisfying the sign condition as described in the preceding Definition.*

THEOREM 6. *Let $h = \{f_t + N_{g_t}\}$ be a homotopy in the class H, and let $\{y_t: 0 \leq t \leq 1\}$ be a continuous curve in X^* such that for all t, $y_t \notin (f_t + N_{g_t})(\text{bdry}(G))$. Then*

$$d(f_t + N_{g_t}, G, y_t) \text{ is independent of t in } [0, 1].$$

Proof of Theorem 6. Let $N_{t,k}$ be the kth approximant to the Niemytskiĭ operator N_{g_t}. If the conclusion of Theorem 6 were false, then for a sequence of integers k going to infinity, the homotopy $\{f_t + N_{t,k}\}$ in the class $(S)_+$ would have y_0-points on the boundary of G. Hence we could find a sequence $\{u_k\}$ on the boundary of G such that for each k, there exists t_k in $[0, 1]$ such that

$$f_{t_k}(u_k) + N_{t_k, k}(u_k) = y_{t_k}.$$

We may assume by passing to an infinite subsequence (which we denote by the original index k) that u_k converges weakly to u in X, $f(u_k)$ converges weakly to w in X^*, t_k converges to t in $[0, 1]$, and y_{t_k} converges strongly to y_t in X^*. Moreover, we may assume also that u_k converges to u a.e. in Ω, and hence that $N_{t_k, k}(u_k)$ converges to $N_{g_t}(u)$ a.e. in Ω. If $z_k = N_{t_k, k}(u_k)$, we know by the definition of the truncation processes and the assumption that each g_t satisfies the sign condition that $z_k(x)u_k(x) \geq 0$, and that $z_k(x)u_k(u)$ converges a.e. to $N_{g_t}(u)u$. Hence we may apply Fatou's Lemma, and observe that

$$\int_\Omega N_{g_t}(u)u \leq \varliminf \int_\Omega z_k(x)u_k(x)\,dx \leq \varliminf \langle y_{t_k} - f_{t_k}(u_k), u_k \rangle \leq M.$$

Hence $N_{g_t}(u)u$ lies in $L^1(\Omega)$.

Let us now establish explicitly the fact that the sequence $\{z_k\}$ is equi-integrable on Ω', where Ω' is any relatively compact open subset of Ω. Let $R > 0$ be given.

Then for any measurable subset A of Ω',

$$\int_A |z_k(x)|\,dx \leq \left\{\int_{A \cap \{u_k(x) \geq R\}} + \int_{A \cap \{u_k(x) < R\}}\right\} |z_k(x)|\,dx$$

$$\leq R^{-1} \int_\Omega z_k(x) u_k(x)\,dx + \int_A |h_R(x)|\,dx,$$

where h_R is the bound function for $g_t(x, r)$ in the definition of a permissible homotopy. Given $\varepsilon > 0$, we first choose R so large that

$$R^{-1} M < \varepsilon/2,$$

where M is the bound for the integrals $\int z_k(x) u_k(x)\,dx$. Then having fixed R, we choose meas(A) so small that $\int_A h_R(x)\,dx < \varepsilon/2$. It then follows that $\int_A |z_k(x)|\,dx < \varepsilon$.

It follows from this equi-integrability and the convergence of z_k to $N_{g_t}(u)$ a.e. that z_k converges to N_{g_t} strongly in $L^1_{\text{loc}}(\Omega)$, and in particular that $N_{g_t}(u)$ lies in $L^1_{\text{loc}}(\Omega)$. On the other hand, $z_k = y_{t_k} - f(u_k)$ converges to $y_t - w$ weakly in X^*. It follows that the two distribution limits for z_k must coincide, and hence $N_{g_t}(u) = y_t - w$ lies in $L^1_{\text{loc}}(\Omega) \cap X^*$.

We now seek to apply the hypothesis that $\{f_t\}$ is a homotopy of class $(S)_+$. We note that

$$\langle f_{t_k}(u_k), u_k - u \rangle = \langle y_{t_k}, u_k - u \rangle + \langle z_k, u \rangle - \int z_k(x) u_k(x)\,dx.$$

Hence

$$\overline{\lim} \langle f_{t_k}(u_k), u_k - u \rangle \leq \langle y_t - w, u \rangle - \int N_{g_t}(u) u\,dx = 0,$$

as we see by applying the proposition for pairing of distributions. Hence u_k converges strongly to u, $f_{t_k}(u_k)$ converges weakly to $f_t(u)$, and $w = f_t(u)$. Since all the u_k lie on the closed subset bdry(G) of X, so does u. Hence u is a solution of the equation $f_t(u) + N_{g_t}(u) = y_t$ on the boundary of G, which contradicts the hypothesis of Theorem 6. q.e.d.

As a consequence of Theorem 6 and the preceding arguments, we have a degree function on the class F of maps of the form $\{f + N_g\}$ with f of class $(S)_+$ and bounded, g satisfying the sign condition, with the homotopy invariant under the class of homotopies H.

THEOREM 7. *There is one and only one degree function on the class F with the homotopies H, where F consists of sums of bounded maps f of class $(S)_+$ and of Niemytskiĭ operators N_g for g satisfying the sign condition, and H is the class of homotopies described in Definition 8.*

Proof of Theorem 7. We have already constructed such a degree function $d(f + N_g, G, y_0)$. Suppose there were another $d_1(f + N_g, G, y_0)$. To show that the two coincide, we shall show that $d_1(f + N_g, G, y_0) = \lim_k d(f + N_k, G, y_0)$, where the degrees in the limit are understood in the sense of the unique degree theory on the class $(S)_+$. By assumption, the new degree theory d_1 has the invariance property under some class of homotopies H_0 which we need only assume includes all affine homotopies. We now consider the homotopy

$$f + (1-t)N_g - tN_k.$$

If it were not true that $d_1(f + N_g, G, y_0)$ were equal to $d_1(f + N_k, G, y_0) = d(f + N_k, G, y_0)$, it would follow that during the homotopy just constructed, there would have to be a y_0-point on the boundary of G for some value of t in $[0, 1]$. We are assuming of course that y_0 does not lie in $(f + N_g)(\text{bdry}(G))$ so that the degrees written above are well defined.

Suppose then that we could find a sequence of k such that $d_1(f + N_g, G, y_0) \neq d(f + N_k, G, y_0)$. Then for this sequence, we could find another sequence $\{u_k\}$ on the boundary of G and a sequence $\{t_k\}$ in $[0, 1]$ such that

$$f(u_k) + (1-t_k)N_g(u_k) + t_k N_k(u_k) = y_0.$$

Let $z_k = (1-t_k)N_g(u_k) + t_k N_k(u_k)$. It follows as before that $z_k(x)u_k(x) \geq 0$. We may assume without loss of generality, as in the preceding proofs, that u_k converges to u weakly in X, $f(u_k)$ converges weakly to w in X^*, u_k converges to u a.e. in Ω, $z_k(x)$ converges to $N_g(u)(x)$ a.e. in Ω, the sequence $\{z_k\}$ is equi-integrable on every relatively compact open subset of Ω and hence converges strongly to $N_g(u)$ in $L^1_{\text{loc}}(\Omega)$, while

$$\int N_g(u)u \, dx \leq \underline{\lim} \int z_k(x)u_k(x) \, dx \leq M.$$

Hence $N_g(u)$ lies in $L^1_{\text{loc}}(\Omega)$ and $N_g(u) = \lim z_k = y_0 - w$, so that

$$\langle N_g(u), u \rangle = \langle y_0 - w, u \rangle.$$

Finally,

$$\overline{\lim} \langle f(u_k), u_k - u \rangle \leq \langle y_0 - w, u \rangle - \int N_g(u)u \, dx = 0,$$

so that by the $(S)_+$-property of f, u_k converges strongly to u in X and $f(u_k)$ converges weakly to $f(u)$ in X^*. Thus u lies in $\text{bdry}(G)$, $f(u) = w$, and $f(u) + N_g(u) = y_0$, which is a contradiction. q.e.d.

The extension of the degree function in the weakened sense to the broader class $f + N_g$, where f is pseudo-monotone and bounded instead of being of class $(S)_+$, is obtained by approximating f by the family of $f_\varepsilon = f + \varepsilon J$ of maps in the class $(S)_+$. The existence and uniqueness arguments for such an extension follow along lines similar to those which one applies when N_g is missing.

3. Periodic solutions of nonlinear wave equations.

In the study of nonlinear wave equations, Brezis and Nirenberg [2] introduced a class of nonlinear mappings in a Hilbert space $\tilde{H}$ of an interesting type which give an operator-theoretic representation of the problem of finding periodic solutions of nonlinear wave equations in one space dimension.

This class of operators consists of $L + f$, where L is a closed densely defined linear operator in $\tilde{H}$ satisfying certain conditions stated below and f is a bounded monotone continuous mapping from $\tilde{H}$ to $\tilde{H}$. One can extend this class by letting f be a bounded pseudo-monotone continuous mapping and obtain a degree theory in the weakened sense for such mappings on degree theories for maps of the form $L + f$, with $f_\varepsilon + f + \varepsilon J$ lying in the class $(S)_+$.

DEFINITION 9. *L is a suitable linear operator for the present problem if it satisfies the following three conditions.*

(1) $H_0 = R(L)$ *is a closed subspace of $\tilde{H}$.*
(2) $N(L) = N(L^*)$, *and hence H_0 is invariant under L.*
(3) *If L_0 is the restriction of L to H_0, then L_0^{-1} mapping H_0 into H_0 is a compact linear operator.*

The properties of L are modeled on the one-dimensional wave operator $Lu = u_{xx} - u_{tt}$ defined for functions u on a rectangle R with sides parallel to the x and t axes and with the side lengths having a rational ratio, where u is assumed to lie in $L^2(R)$ and have periodic behavior in both x and t with periods corresponding to the side lengths of R. The spectrum of L is pure point spectrum but the null space of L is of infinite dimension. Moreover, not only is L not a Fredholm operator, but its spectrum goes to infinity in both directions.

We assume that f maps $\text{cl}(G)$ into $\tilde{H}$, where G is a bounded open subset of $\tilde{H}$, that f is continuous with $f(\text{cl}(G))$ bounded in $\tilde{H}$, and that f lies in the class $(S)_+$. We shall sketch a proof of the construction of an appropriate degree function for $(L + f)$ using a Galerkin-like procedure studied by Mawhin [16], who did not obtain a stabilization of the values of the approximating degrees. Our results include such a stabilization and are developed in full detail in [12].

The approximation process is as follows.

DEFINITION 10. *Let $\tilde{H}$ be a Hilbert space, L a closed densely defined linear operator satisfying the three conditions of Definition 9. Let $\{H_k\}$ be an increasing sequence of finite-dimensional subspaces of $N(L)$ whose union is dense in $N(L)$. Let W_k be the direct sum of H_0 and H_k, P_k the orthogonal projection of $\tilde{H}$ on W_k, L_k the restriction of L to $D(L) \cap W_k$, $G_k = G \cap W_k$, and f_k the restriction of the mapping $P_k f$ to $\text{cl}(G_k)$. Then L_k is a closed densely defined linear mapping in W_k, f_k maps $\text{cl}(G_k)$ into W_k. The (generalized Galerkin) approximant of $(L + f)$ with respect to W_k is the mapping $L_k + f_k$.*

PROPOSITION 3. *Suppose that L satisfies conditions (1), (2), and (3) of Definition 9 and that f mapping $\text{cl}(G)$ into $\tilde{H}$ has bounded image and is of class $(S)_+$. Let y_0 be a point in $\tilde{H}$ which does not lie in $(L + f)(\text{bdry}(G))$, and let $y_k = P_k y_0$. Then: for k sufficiently large, y_k does not lie in $(L_k + f_k)(\text{bdry}(G_k))$.*

Proof. Suppose the assertion of the proposition were false. Then corresponding to an infinite sequence of k, we could find a sequence $\{u_k\}$ in the boundary of G, $u_k \in \text{bdry}(G_k)$, such that

$$L_k(u_k) + P_k f(u_k) = y_k.$$

Since G and $f(\mathrm{cl}(G))$ are both bounded, we may assume that u_k converges weakly to some u in $\tilde{H}$ and $f(u_k)$ converges weakly to some w in $\tilde{H}$.

Let Q_k be the orthogonal projection of $\tilde{H}$ on H_k, P_0 the orthogonal projection of $\tilde{H}$ on H_0. Then $P_k = P_0 + Q_k$, and if we set $v_k = P_0 u_k$, $z_k = Q_k u_k$, we have
$$u_k = v_k + z_k, \qquad L_k u_k = L_0 v_k.$$

Since
$$L_0 v_k = L_k u_k = -P_k f(u_k)$$
is bounded, and L_0^{-1} is assumed to be compact, the sequence $\{v_k\}$ is relatively compact in the strong topology on H_0. On the other hand, $v_k = P_0 u_k$ converges weakly to $P_0 u = v$ in H. Hence v_k converges strongly to v.

Since the sequence $\{W_k\}$ is increasing with dense union in $\tilde{H}$, it follows that P_k converges to I strongly on $\tilde{H}$. It follows that $f_k(u_k) - f(u_k)$ converges weakly to 0 in H since
$$(f_k(u_k) - f(u_k), h) = (f(u_k), (P_k - I)h) \to 0, \qquad h \in \tilde{H}.$$
Therefore,
$$(f(u_k), u_k - u) - (P_k f(u_k), u_k - u) = (P_k f(u_k) - f(u_k), u) \to 0$$
as $k \to \infty$. On the other hand,
$$(P_k f(u_k), u_k - u) = -(L_0 v_k, u_k - u) = -(L_0 v_k, v_k - v) \to 0,$$
since $L_0 v_k$ is bounded in norm and $v_k - v$ converges strongly to 0. Thus
$$\lim (f(u_k), u_k - u) = 0$$
and, since f is of class $(S)_+$, u_k converges strongly to u and $f(u_k)$ converges weakly to $f(u)$, so that $f_k(u_k)$ converges weakly to $f(u)$. Taking weak limits on the terms of the equation $L_k u_k + f_k(u_k) = y_k$, we find that
$$Lu + f(u) = y,$$
with u being the strong limit of points in $\mathrm{bdry}(G)$ also lying in $\mathrm{bdry}(G)$. This contradicts our initial assumption. q.e.d.

Using the result of the proposition, we can try to obtain a degree function for $L + f$ by using degree functions on the approximants $L_k + f_k$ in the spaces W_k. The simplification involved in passing to the approximations consists in replacing the linear operator L by the new linear operators L_k, each of which is a Fredholm operator. We choose δ with $0 < \delta < 1$, and write the equation
$$L_k(u_k) + f_k(u_k) = y_k$$
in the equivalent form
$$(L_k + \delta Q_k) u_k + (f_k - \delta Q_k)(u_k) = y_k.$$
This suggests setting
$$d(L_k + f_k, G_k, y_k) = d_{\mathrm{LS}}(I + C_k, U_k, y_k),$$
where
$$C_k = (f_k - \delta Q_k)(L_0 + \delta Q_k)^{-1}$$

is compact, and the open set
$$U_k = (L_0 + \delta Q_k)^{-1}(G).$$
The notation d_{LS} denotes the Leray–Schauder degree, but the right-hand side is not well defined without further consideration since U_k is in general not bounded. If we can show that all solutions h of the equation $(I + C_k)(h) = y_k$ are bounded, we could replace U_k by any open subset which is bounded and contains all such solutions h, and get a well-defined and unambiguous value for the degree. Suppose then that h lies in U_k and $(I + C_k)(h) = y_k$. Then if $u = (L_k + \delta Q_k)(h)$, u is a solution of $(L_k + f_k)(u) = y_k$ with u_k in G. Such elements u are bounded since G is bounded. Moreover, $L_0(P_0 h) = P_0 u$ is bounded so that $P_0 h$ is bounded. Hence, since $Q_k u = \delta Q_k h$ implies that $Q_k h$ is bounded, it follows tht $h = P_0 h + Q_k h$ is uniformly bounded for all such solutions.

THEOREM 8. *Under the hypothesis of Proposition 3, the sequence of degrees $d(L_j + f_j, G_j, y_j)$ stabilizes.*

The proof of Theorem 8 depends upon the following proposition, which extends Proposition 11 of [9] and is established in detail in [12].

PROPOSITION 4. *Suppose that for $j > k$, $d(L_j + f_j, G_j, 0) \neq d(L_k + f_k, G_k, 0)$. Then there exists u_j in $\text{bdry}(G_j) \cap D(L_j)$ such that*
$$(L_j u_j + f_j(u_j), u_j) \leq 0,$$
while for all z in W_k,
$$(L_j u_j + f_j(u_j), z) = 0.$$

Proof of Theorem 8. We may assume without loss of generality that $y_0 = 0$. Suppose that the sequence of approximating degrees does not stabilize. Then we may find a sequence $\{u_j\}$, as in Proposition 4, of elements of $\text{bdry}(G) \cap D(L)$ such that
$$(Lu_j + f(u_j), u_j) \leq 0$$
while for all $j > k$, and all z in W_k,
$$(Lu_j + f(u_j), z) = 0$$
We may assume that u_j converges weakly to u in $\tilde{H}$ and that $f(u_j)$ converges weakly to w. Since H_0 is contained in each W_k, it follows that for z in H_0, if we set $v_j = P_0 u_j$,
$$(Lv_j, z) = (Lu_j, z) = -(f(u_j), z) \to -(w, z).$$
Thus Lv_j converges weakly to $-w$, while v_j converges weakly to $P_0 u$. Since L_0^{-1} is compact on H_0, it follows that v_j converges strongly to v while $Lv_j = L_0 v_j$ converges weakly to Lv.

Let h be any element of $\tilde{H}$. Then
$$(Lv_j + f(u_j), h) = (Lv_j + f(u_j), P_k h) + (Lv_j + f(u_j), (h - P_k h)).$$

For $j > k$, the first summand vanishes. We can make the second summand arbitrarily small for all j by making k sufficiently large, since Lv_j and $f(u_j)$ are both bounded. Hence if we first choose k large and then $j > k$, we can make $|(Lv_j + f(u_j), h)| < \varepsilon$ for a given $\varepsilon > 0$. Thus $Lv_j + f(u_j)$ converges weakly to 0, i.e. $f(u_j)$ converges weakly to $-Lv = -LP_0u = -Lu$.

Finally,

$$\left(f(u_j), u_j - u\right) \leq -\left(Lv_j, v_j\right) - \left(f(u_j), P_k u\right) + \left(f(u_j), P_k u - u\right).$$

If we choose k large, we can make $|(f(u_j), P_k u - u)| < \varepsilon$ for all j. For j greater than this k, $(f(u_j), P_k u) = -(Lv_j, P_k u)$. The sum of the first two terms on the right side of the inequality becomes

$$\left(Lv_j, P_k u - v_j\right) = \left(Lv_j, v - v_j\right),$$

which goes to zero as j goes to infinity since Lv_j is bounded and $v - v_j$ goes to 0 strongly. Hence

$$\overline{\lim} \left(f(u_j), u_j - u\right) \leq 0.$$

Since f is of class $(S)_+$, u_j converges strongly to u and $f(u_j)$ converges weakly to $f(u)$. Since all u_k lie in bdry(G), u must lie in bdry(G). Moreover $f(u_j)$ converges simultaneously to $f(u)$ and $-Lu$. Hence $f(u) = -Lu$, i.e. $Lu + f(u) = 0$ for u on the boundary of G. This is a contradiction. q.e.d.

Bibliography

1. H. Brezis and F. E. Browder, *Some properties of higher order Sobolev spaces*, J. Math. Pures Appl. **61** (1982), 245–259.

2. H. Brezis and L. Nirenberg, *Characterizations of the ranges of some nonlinear operators and applications to boundary value problems*, Ann. Scuola Norm. Sup. Pisa **5** (1978), 255–326.

3. F. E. Browder, *Nonlinear equations of evolution and nonlinear operators in Banach spaces*, Nonlinear Functional Analysis, Proc. Sympos. Pure Math., vol. **18**, part 2, American Math. Soc., Providence, R.I., 1975.

4. _____, *Degree of mapping for nonlinear operators of monotone type*, Proc. Nat. Acad. Sci. U.S.A. **80** (1983), 1771–1773.

5. _____, *Degree of mapping for nonlinear operators of monotone type: Densely defined mappings*, Proc. Nat. Acad. Sci. U.S.A. **80** (1983), 2405–2407.

6. _____, *Degree of mapping for nonlinear operators of monotone type: Strongly nonlinear mappings*, Proc. Nat. Acad. Sci. U.S.A. **80** (1983), 2408–2409.

7. _____, *L'unicité du degré topologique pour des applications de type monotone*, C. R. Acad. Sci. Paris Ser. I Math. **296** (1983), 145–148.

8. _____, *The degree of mapping and its generalizations*, Contemp. Math. **21** (1983), 15–39.

9. _____, *Fixed point theory and nonlinear problems*, Bull. Amer. Math. Soc. **9** (1983), 1–39 (also in Mathematical Heritage of Henri Poincaré, Proc. Sympos. Pure Math., vol. **39**, part 2, 1984, pp. 49–87).

10. _____, *The theory of degree of mapping for nonlinear mappings of monotone type* (to appear in Brezis-Lions Seminar, vol. VI, Pitman, 1985).

11. _____, *Uniqueness of the mapping degree for elliptic operators with strong zeroth-order nonlinearities* (to appear in *Aspects of mathematics and its applications*, Elsevier Science Publishers, 1986).

12. _____, *Degree theories for general classes of nonlinear mappings* (to appear in *Proc. NATO Conf. on nonlinear problems and fixed point theorems*, May 1985).

13. F. E. Browder and W. V. Petryshyn, *Approximation methods and the generalized topological degree for nonlinear mappings in Banach spaces*, J. Funct. Anal. **3** (1969), 217–245.

14. L. I. Hedberg, *Spectral synthesis in Sobolev spaces and uniqueness of solutions of the Dirichlet problem*, Acta Math. **147** (1981), 237–264.

15. L. I. Hedberg and T. H. Wolff, *Thin sets in nonlinear potential theory*, Ann. Inst. Fourier (Grenoble) **33** (1983), 161–187.

16. J. Mawhin, *Nonlinear functional analysis and periodic solutions of semilinear wave equations*, Nonlinear Phenomena in Mathematical Sciences (V. Lakshmikantham, ed.), Academic Press, New York, 1982, pp. 671–681.

17. J. R. L. Webb, *Boundary value problems for strongly nonlinear elliptic equations*, J. London Math. Soc. (2) **21** (1980), 123–132.

UNIVERSITY OF CHICAGO

Construction of Periodic Solutions of Periodic Contractive Evolution Systems from Bounded Solutions

RONALD E. BRUCK[1]

Abstract. Given an approximate solution of a periodic contractive evolution system, we show how to construct an approximately-periodic approximate solution by averaging the values of the given approximate solution. We also give a necessary and sufficient condition for a solution to be weakly asymptotic to a periodic solution.

0. Introduction. Let E be a real Banach space. In this paper we consider variations on a problem which includes as perhaps its most interesting application the following problem: Suppose $u\colon \mathbf{R}^+ \to E$ is a *bounded* function which satisfies

(0.1) $$du/dt + A(t)u(t) \ni f(t)$$

for a.e. $t \geqslant 0$, where each $A(t)$ is accretive, f is a "suitable" forcing function, and the functions $t \to A(t)$, $t \to f(t)$ are periodic with the same period $p > 0$. How can we construct a p-periodic solution ω of

(0.2) $$d\omega/dt + A(t)\omega(t) \ni f(t) \quad \text{a.e. } t \in \mathbf{R}$$

from the solution u?

We actually phrase the problem more generally to permit discrete and discontinuous evolution systems, and relax the notion of a "solution" of (0.1) to that of an "approximate" solution (to take into account, e.g., roundoff and discretization errors).

By a *contractive evolution system* (abbreviated as CES) on E we mean a family

$$\mathscr{U} = \{U(t, s)\colon D(s) \subset E \to D(t) \mid -\infty < s \leqslant t < +\infty\}$$

1980 *Mathematics Subject Classification.* Primary 47H09, 47H20.
[1] Partially supported by NSF Grant MCS 81-02806.

of nonexpansive mappings $U(t, s)$ with domain $D(s)$ contained in E, which satisfy the transition relations

(0.3) $\qquad U(t, t) = \mathrm{id}_{D(t)} \quad \text{for all } t \text{ in } \mathbf{R}$,

(0.4) $\qquad U(t, s)U(s, r) = U(t, r) \quad \text{for all } r \leqslant s \leqslant t$.

(A mapping is said to be *nonexpansive* if it has Lipschitz constant 1 in the usual metric, e.g. if it is an isometry.) We say the CES $\mathscr{U}$ is *periodic* with period p provided $p > 0$ and

(0.5) $\qquad U(t + p, s + p) = U(t, s) \quad \text{for all } s \leqslant t$.

The periodicity of $\mathscr{U}$ thus entails the periodicity of its domains: $D(s + p) = D(s)$.

One usually visualizes CES's as arising as the family of transition operators for initial value problems based on (0.1), extended (via continuity) to the closure of the set of possible starting values for which (0.1) has a solution. The periodicity of $A(t)$ and $f(t)$ then leads very naturally to the definition (0.5). But we emphasize that we have placed no continuity assumptions on $t \to U(t, s)x$ or on $s \to U(t, s)x$. Consider, for example, a bi-infinite sequence $\{T_n\}_{n=-\infty}^{+\infty}$ of nonexpansive mappings $T_n \colon D_n \to D_{n+1}$, and define $D(s) = D_m$, $m \in \mathbf{Z}$ and $m - 1 < s \leqslant m$, and $U(t, s) = T_n \cdots T_m$, where $m = \inf[s, t[\cap \mathbf{Z}$, $n = \sup[s, t[\cap \mathbf{Z}$ when such integers exist. Otherwise $U(t, s) = $ identity on $D(s)$. Periodicity in this instance corresponds exactly to periodicity of the sequence $\{T_n\}$.

By a *trajectory* of the CES $\mathscr{U}$ we mean a function $u \colon \mathbf{R} \to E$ which satisfies

(0.6) $\qquad u(t) = U(t, s)u(s) \quad \text{for all } s \leqslant t$.

We also introduce the notions of *semitrajectory*, *asymptotic semitrajectory*, and *ε-approximate semitrajectory* for $\mathscr{U}$ as follows: Let $a \in \mathbf{R}$ and suppose $u \colon [a, +\infty[\to E$ is a function. Then u is:

a *semitrajectory* of $\mathscr{U}$ if u satisfies (0.6) for all $a \leqslant s \leqslant t < \infty$;

an *asymptotic semitrajectory* of $\mathscr{U}$ provided

$$\lim_{s \to \infty} \inf_{x \in D(s)} \sup_{t \geqslant s} \|u(t) - U(t, s)x\| = 0;$$

an *ε-approximate semitrajectory* of $\mathscr{U}$ provided $u(t) \in D(t)$ for all t and

(0.7) $\qquad \|u(t) - U(t, s)u(s)\| \leqslant \varepsilon \quad \text{for all } a \leqslant s \leqslant t \leqslant s + 1$.

The last definition is tailored to account for the fact that numerically computed solutions always involve error, e.g., roundoff, "fuzzy" data, or discretization error, and that the error can be controlled only on bounded intervals. The stipulation that $s \in [t, t + 1]$ is a normalization: it easily follows from nonexpansiveness that if u satisfies (0.7), then u also satisfies

(0.8) $\qquad \|u(t) - U(t, s)u(s)\| \leqslant (R + 1)\varepsilon \quad \text{for all } a \leqslant s \leqslant t \leqslant s + R$.

Throughout the rest of this paper, E always denotes a *uniformly convex* Banach space, and (except in Theorem 0.4) $\mathscr{U}$ denotes a p-periodic CES which satisfies the additional conditions

(0.9) $\qquad \text{each domain } D(t) \text{ is closed convex,}$

and
(0.10) $\{\text{diam } D(t)\}$ is bounded, uniformly in t.

The main results of this paper may then be stated as follows.

THEOREM 0.1 (CONSTRUCTIBILITY). *For each $\varepsilon > 0$ there exists $N \in \mathbf{Z}^+$ and $\delta > 0$ with the following properties: Whenever u is a δ-approximate solution of $\mathscr{U}$ and $n \geq N$, then the function σ_n defined by*

$$\sigma_n(t) = \frac{1}{n} \sum_{k=0}^{n-1} u(t + kp)$$

is an ε-approximate solution of $\mathscr{U}$, and satisfies

$$\|\sigma_n(t) - \sigma_n(t + p)\| \leq \varepsilon \text{ for all } t \geq 0.$$

To state the next theorem we need Lorentz's definition [1] of *almost-convergence*. A sequence $\{x_n\}$ in a Banach space E is said to be *weakly almost-convergent* to a point x provided

$$\text{w-}\lim_{n \to \infty} \frac{1}{n} \sum_{i=1}^{n} x_{i+k} = x$$

uniformly in $k \geq 0$.

THEOREM 0.2 (MEAN BEHAVIOR). *Suppose in addition to the standing hypotheses that E has Fréchet-differentiable norm. If $u: \mathbf{R}^+ \to E$ is a bounded asymptotic semitrajectory of $\mathscr{U}$, then for each t in $\mathbf{R}$ the sequence $\{u(t + np)\}$ is weakly almost-convergent to a point $\omega(t)$ of $D(t)$, and $\omega(t)$ defines a p-periodic trajectory of $\mathscr{U}$. Moreover, if u is Pettis integrable on bounded intervals, then*

$$\frac{1}{t} \int_c^{c+t} u(\tau) \, d\tau \to \frac{1}{p} \int_0^p \omega(\tau) \, d\tau$$

as $t \to \infty$, uniformly in $c \geq 0$.

THEOREM 0.3 (PERIODIC FORCING). *If $u: \mathbf{R}^+ \to E$ is a bounded, **uniformly continuous** asymptotic semitrajectory of $\mathscr{U}$, then a necessary and sufficient condition that there exist a p-periodic trajectory ω of $\mathscr{U}$ such that*

$$u(t) - \omega(t) \to 0 \quad \text{as } t \to \infty$$

is that $u(t + p) - u(t) \to 0$ as $t \to \infty$.

THEOREM 0.4 (ASYMPTOTIC SUPERPOSITION). *Suppose $\mathscr{U}$ is a CES—not necessarily periodic—on a uniformly convex Banach space E, and that (0.9) and (0.10) are satisfied. If u_1 and u_2 are asymptotic semitrajectories of $\mathscr{U}$, so is $cu_1 + (1 - c)u_2$ for any $0 \leq c \leq 1$.*

1. Known results in the discrete case. The proofs of Theorems 0.1–0.4 rely very heavily on the asymptotic behavior of the iterates of a *single* nonexpansive mapping. This section is devoted to a review of known results for such mappings; the proofs can be found in Bruck [2, 3]. For a survey of the broader issue of asymptotic behavior in general, and a list of further references, see Bruck [4].

THEOREM A. *Suppose E is uniformly convex. Then for each $R > 0$ there exists a continuous, convex, strictly increasing function $\gamma: \mathbf{R}^+ \to \mathbf{R}^+$ such that whenever C is a convex subset of E with diameter $\leq R$, then for any nonexpansive $T: C \to E$ and any convex combination $\sum \lambda_i x_i$ of elements of C,*

(1.1) $\quad \gamma\left(\left\|\sum \lambda_i T x_i - T\left(\sum \lambda_i x_i\right)\right\|\right) \leq \max_{i,j} \left\{\|x_i - x_j\| - \|Tx_i - Tx_j\|\right\}.$

In principle, γ is constructible. In Hilbert space, for example, it is a multiple of t^2. An immediate and useful consequence of Theorem A is what might be called the "approximate convexity" of approximate fixed-point sets of nonexpansive mappings. We denote by $F_\varepsilon(T)$ the set of points x such that $\|x - Tx\| \leq \varepsilon$ (the "ε-approximate" fixed points of T).

THEOREM B. *Suppose E is uniformly convex. Then for each $R > 0$ and each $\varepsilon > 0$ there exists $\delta > 0$ such that, for all closed convex subsets C of E with $\operatorname{diam} C \leq R$ and all nonexpansive mappings $T: C \to E$,*

$$\operatorname{clco} F_\delta(T) \subset F_\varepsilon(T).$$

Theorem B is easily proved directly from Theorem A (in fact, more easily than by its original proof in [3]). Another result proved in [3] seems not to be directly derivable from Theorem A:

THEOREM C. *Suppose E is uniformly convex. Then for every $\varepsilon > 0$ and $R > 0$ there exists $\delta > 0$ and $N \in \mathbf{Z}^+$ such that for every closed convex subset C of E with $\operatorname{diam} C \leq R$, every nonexpansive $T: C \to E$, and every sequence $\{x_n\}$ in C which satisfies $\|x_{n+1} - Tx_n\| \leq \delta$ there holds*

$$(x_1 + \cdots + x_n)/n \in F_\varepsilon(T) \quad \text{for all } n \geq N.$$

The final external result we shall require is the nonlinear mean ergodic theorem. For a proof see Baillon [5], Bruck [2], or Reich [6].

THEOREM D. *Suppose E is uniformly convex and has Fréchet-differentiable norm. Then for any bounded closed convex subset C of E, any nonexpansive $T: C \to C$, and any x in C, the sequence $\{T^n(x)\}$ is weakly almost-convergent to a fixed point of T.*

2. Proofs. We begin with Theorem 0.2, since its proof is in many respects the simplest.

We first prove the theorem when u is a semitrajectory. Without loss of generality we may take $u(t) = U(t, 0)u(0)$ for $t \geq 0$. We shall denote $u(t + np)$ by $u_n(t)$. It follows from the periodicity of $\mathscr{U}$ that $U(t + np, t) = U(t + p, t)^n$, and hence that

$$u_n(t) = U(t + p, t)^n u(t) = T^n u(t)$$

for the nonexpansive mapping $T = U(t + p, t): D(t) \to D(t)$. Since E is uniformly convex and $D(t)$ is a bounded closed convex subset of E, Theorem D shows that $\{u_n(t)\}$ is weakly $(C, 1)$-convergent to a fixed point of T. Define $\omega(t)$

to be the limit. Clearly ω is defined on all of $\mathbf{R}$ and has period p. The crux of the proof is that ω really is a *trajectory* of $\mathscr{U}$, and to prove this we use a device of Michel Lapidus [7] which is simpler and more direct than our original proof.

Fix $s \leq t$. We must prove $\omega(t) = U(t, s)\omega(s)$ for all $t \geq s$. Since $\mathscr{U}$ satisfies (0.3), (0.4), it suffices to prove this when $s \leq t \leq s + p$. For example, when $s + p \leq t < s + 2p$, we have

$$\omega(t) = U(t, s + p)\omega(s + p) \quad \text{and} \quad \omega(s + p) = U(s + p, s)\omega(s),$$

from which the desired conclusion follows.

Consider the product space $E \times E$ as the Hilbert direct sum of its factors, and the map $Q: D(s) \times D(t) \to D(s) \times D(t)$ defined by

$$Q[x, y] = [U(s + p, t)y, U(t, s)x].$$

$E \times E$ is uniformly convex, $D(s) \times D(t)$ is a bounded closed convex subset, and Q is obviously nonexpansive. Moreover, for the semitrajectory u we have

$$Q[u(s), u(t)] = [u(s + p), u(t)],$$
$$Q^2[u(s), u(t)] = [u(s + p), u(t + p)].$$

More generally,

$$Q^{2n}[u(s), u(t)] = [u_n(s), u_n(t)],$$
$$Q^{2n+1}[u(s), u(t)] = [u_{n+1}(s), u_n(t)].$$

Since $\{u_n(s)\}$ and $\{u_n(t)\}$ are weakly almost-convergent to $\omega(s)$ and $\omega(t)$, respectively, we find that $\{Q^n[u(s), u(t)]\}$ is weakly almost-convergent to $[\omega(s), \omega(t)]$, *which is a fixed point of Q.* Equating $Q[\omega(s), \omega(t)]$ to $[\omega(s), \omega(t)]$, we find $\omega(t) = U(t, s)\omega(s)$, as claimed.

In the general case, when u is an asymptotic semi trajectory, let $\{k(n)\}$ be a sequence of positive integers and let v be a weak subsequential limit of $\{n^{-1}\sum_{i=1}^{n} u_{i+k(n)}(t)\}$. Given $\varepsilon > 0$ we can choose $s_\varepsilon > 0$ and $x_\varepsilon \in D(s_\varepsilon)$ so that for $u^*(t) = U(t, s_\varepsilon)x_\varepsilon$ $(t \geq s_\varepsilon)$ we have $\|u(t) - u^*(t)\| \leq \varepsilon$ for all $t \geq s_\varepsilon$. Thus for any t in $\mathbf{R}$,

(2.1) $$\|u_n(t) - u_n^*(t)\| \leq \varepsilon$$

for n sufficiently large. By the earlier case, $\{u_n^*(t)\}$ is weakly almost-convergent to some p-periodic trajectory ω_ε. But (2.1) implies that

(2.2) $$\|v - \omega_\varepsilon(t)\| \leq \varepsilon.$$

This implies, first, that the set of weak subsequential limits of $\{n^{-1}\sum_{i=1}^{n} u_{i+k(n)}(t)\}$ has diameter $\leq 2\varepsilon$ for *any* $\varepsilon > 0$—so the sequence must be weakly convergent. This proves $\{u_n(t)\}$ is weakly almost-convergent. If its limit is denoted by $\omega(t)$, then (2.2) also implies that $\omega_\varepsilon(t) \to \omega(t)$ strongly as $\varepsilon \to 0$. Since ω_ε is a semitrajectory and each $U(t, s)$ is continuous, it follows that ω is a trajectory. Its periodicity is obvious.

To prove the second half of Theorem 0.2, suppose u is Pettis integrable. Note that for positive integers k, n we have

(2.3)
$$(np)^{-1} \int_{kp}^{kp+np} u(t)\, dt = (np)^{-1} \sum_{i=0}^{n-1} \int_0^p u_{k+i}(t)\, dt$$
$$= \frac{1}{p} \int_0^p \sigma_{k,n}(t)\, dt,$$

where $\sigma_{k,n}(t) = n^{-1} \sum_{i=0}^{n-1} u_{k+i}(t)$. As $n \to \infty$, $\sigma_{k,n}(t)$ converges weakly to $\omega(t)$ (uniformly in k). Applying the dominated convergence theorem to (2.3) we find that ω is also Pettis integrable, and

$$(np)^{-1} \int_{kp}^{kp+np} u(t)\, dt \to \frac{1}{p} \int_0^p \omega(t)\, dt$$

as $n \to \infty$, uniformly in $k \geq 0$. The final conclusion of the theorem is deduced from the trivial observation that

$$\lim_{t \to \infty} \frac{1}{t} \int_c^{c+t} u(s)\, ds - (np)^{-1} \int_{kp}^{kp+np} u(s)\, ds = 0$$

uniformly in $c \geq 0$, for $n = [t/p]$, $k = [c/p]$.

PROOF OF THEOREM 0.1. Obviously

$$\|\sigma_n(t) - \sigma_n(t+p)\| = \|u(t) - u(t+np)\|/n \leq (\operatorname{diam} D(t))/n$$

can be made $< \varepsilon$ for sufficiently large n. The critical assertion is that σ_n is an ε-approximate solution if δ is sufficiently small.

As in the proof of Theorem 0.2, $E \times E$ is uniformly convex, and the sets $D(s) \times D(t)$ have diameters uniformly bounded in s, t. For any $s \leq t \leq s + p$, the map

$$Q[x, y] = [U(s+p, t)y, U(t, s)x]$$

is nonexpansive and, by Theorem D, we can choose $\delta > 0$ and $N \in \mathbf{Z}^+$ such that

$$\|[x_{n+1}, y_{n+1}] - Q[x_n, y_n]\| \leq \delta \quad \text{for all } n$$

implies that, for $n \geq N$,

$$\frac{1}{n} \sum_{i=1}^n [x_i, y_i] \in F_\varepsilon(Q).$$

Thus if u satisfies

$$\|u(t) - U(t, s)u(s)\| \leq \delta \quad \text{for all } 0 \leq s \leq t \leq s + p$$

(note the use of (0.8)) we have, alternately,

$$\|[u_{n+1}(s), u_n(t)] - Q[u_n(s), u_n(t)]\| \leq \delta,$$
$$\|[u_{n+1}(s), u_{n+1}(t)] - Q[u_{n+1}(s), u_n(t)]\| \leq \delta,$$

and in particular

$$\|\sigma_n(t) - U(t, s)\sigma_n(s)\| \leq \varepsilon$$

for $n \geq N$.

PROOF OF THEOREM 0.3. By Theorem 0.2 the sequence $\{u_n(t)\}$ is weakly almost-convergent to a p-perodic trajectory ω (where, as usual, $u_n(t) = u(t + np)$). If $u(t) - \omega(t) \to 0$ as $t \to \infty$, then obviously $u(t + p) - u(t) \to 0$.

Conversely, suppose $u(t + p) - u(t) \to 0$. Then $u_{n+1}(t) - u_n(t) \to 0$ weakly as $n \to \infty$ for each t. But this is a Tauberian condition for almost-convergence, and implies that $u_n(t) \to \omega(t)$ for all $t \in \mathbf{R}$.

Let $t_n \to \infty$: we must show $u(t_n) - \omega(t_n) \to 0$. Since the sequence is bounded, it suffices to show that the only possible weak subsequential limit is 0. To this end, suppose $u(t_n) - \omega(t_n) \to f$. Write $t_n = k_n p + r_n$, where $k_n \in \mathbf{Z}$ and $0 \leqslant r_n < p$. Passing to a subsequence, we may assume $\{r_n\}$ converges to some $r \in [0, p]$. By the hypothesized uniform continuity of u,

$$u(t_n) - u_{k(n)}(r) \to 0 \quad \text{as } n \to \infty,$$

hence $u(t_n) \to \omega(r)$. ω is also continuous and has period p, hence $\omega(t_n) = \omega(r_n) \to \omega(r)$. Subtracting, we have $f = 0$, proving the theorem.

PROOF OF THEOREM 0.4. Let us first note that if u_1 and u_2 are asymptotic semitrajectories of $\mathscr{U}$, then the $\lim_{t \to \infty} \|u_1(t) - u_2(t)\|$ exists. Indeed, let $\varepsilon > 0$. Then there exists $s_\varepsilon > 0$ and x_1, x_2 such that

$$\|u_i(t) - U(t, s)x_i\| \leqslant \varepsilon \quad \text{for all } t \geqslant s_\varepsilon.$$

Thus for $t \geqslant s \geqslant s_\varepsilon$, we have

$$\|u_1(t) - u_2(t)\| \leqslant \|u_1(t) - U(t, s_\varepsilon)x_1\| + \|U(t, s_\varepsilon)x_1 - U(t, s_\varepsilon)x_2\|$$
$$+ \|U(t, s_\varepsilon)x_2 - u_2(t)\|$$
$$\leqslant 2\varepsilon + \|U(s, s_\varepsilon)x_1 - U(s, s_\varepsilon)x_2\|$$
$$\leqslant 4\varepsilon + \|u_1(s) - u_2(s)\|.$$

Taking, first, the lim sup as $t \to \infty$ and then the lim inf as $s \to \infty$, the assertion is proved.

The hypothesis (0.10) guarantees the existence of a continuous, strictly increasing, convex function $\gamma: \mathbf{R}^+ \to \mathbf{R}^+$ such that (1.1) is satisfied for every mapping $T = U(t, s)$. In particular, taking $v(t) = cu_1(t) + (1 - c)u_2(t)$ (for $c \in [0, 1]$), we find that for $t \geqslant s \geqslant s_\varepsilon$,

$$\gamma(\|v(t) - U(t, s)[cU(s, s_\varepsilon)x_1 + (1 - c)U(s, s_\varepsilon)x_2]\|)$$
$$\leqslant \|U(s, s_\varepsilon)x_1 - U(s, s_\varepsilon)x_2\| - \|U(t, s_\varepsilon)x_1 - U(t, s_\varepsilon)x_2\|$$
$$\leqslant \|u_1(s) - u_2(s)\| - \|u_1(t) - u_2(t)\| + 2\varepsilon.$$

Since, as we have remarked, $\{\|u_1(t) - u_2(t)\|\}$ converges as $t \to \infty$, we see that once s is large enough,

$$\|v(t) - U(t, s)[cU(s, s_\varepsilon)x_1 + (1 - c)U(s, s_\varepsilon)x_2]\| \leqslant \gamma^{-1}(3\varepsilon).$$

Thus v is also an asymptotic semitrajectory of $\mathscr{U}$. Q.E.D.

3. Periodic forcing of "convexe mobile". In this section we apply the previous results to J. J. Moreau's "convexe mobile". Throughout, E will be a Hilbert space.

Recall that the *indicator function* I_K of a closed convex subset K of E is defined by

$$I_K(x) = \begin{cases} +\infty & \text{if } x \notin K, \\ 0 & \text{if } x \in K, \end{cases}$$

and that the *subdifferential* ∂I_K is given by

$$\partial I_K(x) = \begin{cases} \varnothing & \text{if } x \notin K, \\ \{w \in E | (w, x-y) \geq 0 \text{ for all } y \text{ in } K\} & \text{if } x \in K. \end{cases}$$

That is, $\partial I_K(x)$ is the cone of outward normals to K at x (if $x \in K$).

Moreau [8, 9] introduced the following evolution problem: Let $\{K(t) | t \in \mathbf{R}^+\}$ be a family of closed convex subsets of E, and consider the initial value problem

$$du/dt + \partial I_{K(t)}(u) \ni 0 \quad \text{for a.e. } t \geq 0,$$
$$u(0) = x \in K(0).$$

We shall instead consider the more general "forced" problem:

(3.1)
$$du/dt + \partial I_{K(t)}(u) \ni f(t) \quad \text{for a.e. } t \geq s,$$
$$u(s) = x \in K(s).$$

THEOREM 3.1. *Suppose $\{K(t)\}$ has period p, each $K(t)$ is a bounded closed convex set with nonempty interior, and every boundary point of every $K(t)$ is strongly exposed. Suppose the map $t \to K(t)$ is absolutely continuous in the Hausdorff metric and $f \in L^2_{\text{loc}}(\mathbf{R}, E)$. Finally, suppose that for every s in $\mathbf{R}$ and every x in $K(s)$ the initial value problem (3.1) has a unique strong solution on $[s, +\infty[$.*

Then for every strong solution u of (3.1) there exists a p-periodic strong solution ω of

(3.2) $$d\omega/dt + \partial I_{K(t)}(\omega(t)) \ni f(t) \quad \text{for a.e. } t \in \mathbf{R}$$

such that $u(t) - \omega(t) \to 0$ strongly as $t \to \infty$.

PROOF. We define a CES $\mathscr{U}$ as follows: $D(s) = K(s)$ and $U(t, s)x$ is found by solving (3.1) for an absolutely continuous function v on $[s, t]$, and putting $U(t, s)x = v(t)$. Standard monotonicity results show that $\mathscr{U}$ is a p-periodic CES.

Consider the given strong solution u of (3.1), which we may assume to be defined on $\mathbf{R}^+$, and put $u_n(t) = u(t + np)$. By Theorem 0.2, $\{u_n(t)\}$ is weakly almost-convergent to a p-periodic trajectory $\omega(t)$ of $\mathscr{U}$—and by the existence hypothesis of the present theorem, ω is a *strong* solution of (3.2). We distinguish two cases.

Case I. There exists t_0 such that $\omega(t_0) \in \text{bdry } K(t_0)$. Since by hypothesis every point of the boundary is strongly exposed, there must exist $w \in E$ such that

(3.3) $$(w, \omega(t_0) - y) \geq 0 \quad \text{for all } y \text{ in } K(t_0),$$

and

(3.4) if $\{y_n\} \subset K(t_0)$ satisfies $\lim_n(w, \omega(t_0) - y_n) = 0$, then $y_n \to \omega(t_0)$ strongly.

But the Cèsaro means of $\{u_n(t_0)\} \subset K(t_0)$ are weakly convergent to $\omega(t_0)$; thus (3.3) and (3.4) assure the existence of a *subsequence* of $\{u_n(t_0)\}$ which converges strongly to $\omega(t_0)$.

But the function $t \to \|u(t) - \omega(t)\|$ is nonincreasing, since each $U(t, s)$ is nonexpansive and since a subsequence of
$$\|u(t_0 + np) - \omega(t_0)\| = \|u(t_0 + np) - \omega(t_0 + np)\|$$
converges to 0. Therefore $u(t) - \omega(t) \to 0$ strongly as $t \to \infty$.

Case II. $\omega(t) \notin \text{bdry } K(t)$ for all t. Since ω is continuous and $t \to K(t)$ is absolutely continuous in the Hausdorff metric, $t \to \text{dist}(\omega(t), \text{bdry } K(t))$ is continuous and periodic and always positive. So there exists $\varepsilon > 0$ such that
$$B_\varepsilon(\omega(t)) \subset K(t) \quad \text{for all } t \text{ in } \mathbf{R},$$
where $B_\varepsilon(x)$ denotes the closed ball of radius ε centered at x. Thus for any h in E with $\|h\| < \varepsilon$, $\omega(t) + h \in \text{int } K(t)$ and therefore $\partial I_{K(t)}(\omega(t)) = \partial I_{K(t)}(\omega(t) + h) = \{0\}$. It follows that $\omega(t) + h$ is also a strong solution of (3.2) for any $\|h\| < \varepsilon$.

In particular, $\|\omega(t) + h - u(t)\|$ is a nonincreasing function of t. For $s \leq t$ we have
$$\|\omega(t) + h - u(t)\| \leq \|\omega(s) + h - u(s)\|$$
which in Hilbert space implies
$$2\varepsilon \|\{\omega(t) - u(t)\} - \{\omega(s) - u(s)\}\| \leq \|\omega(s) - u(s)\|^2 - \|\omega(t) - u(t)\|^2.$$
This proves $\{\omega(t) - u(t)\}$ is strongly Cauchy as $t \to \infty$, and since its mean value is 0 (by the second half of Theorem 0.2), $\omega(t) - u(t) \to 0$, as claimed.

References

1. G. G. Lorentz, *A contribution to the theory of divergent sequences*, Acta Math. **80** (1948), 167–190.
2. R. E. Bruck, *A simple proof of the mean ergodic theorem for nonlinear contractions in Banach spaces*, Israel J. Math. **32** (1979), 107–116.
3. _____, *On the convex approximation property and the asymptotic behavior of nonlinear contractions in Banach spaces*, Israel J. Math. **38** (1981), 304–314.
4. _____, *Asymptotic behavior of nonexpansive mappings*, Contemporary Math., vol. 18, Amer. Math. Soc., Providence, R. I., 1983, pp. 1–47.
5. J. B. Baillon, *Comportement asymptotique des itérés de contractions non linéaires dans les espaces L^p*, C. R. Acad. Sci. Paris Sér. A–B **286** (1978), A157–A159.
6. S. Reich, *Nonlinear ergodic theory in Banach spaces*, Report ANL-79-69, Argonne Nat. Lab., 1979.
7. M. Lapidus, Personal communication.
8. J. J. Moreau, *Problème d'évolution associé a un convexe mobile d'un espace hilbertien*, C. R. Acad. Sci. Paris Sér. A–B **276** (1973), 791–794.
9. _____, *Evolution problem associated with a moving convex set in a Hilbert space*, J. Differential Equations **26** (1977), 347–374.

UNIVERSITY OF SOUTHERN CALIFORNIA

An Abstract Critical Point Theorem for Strongly Indefinite Functionals[1]

A. CAPOZZI AND D. FORTUNATO

0. Introduction. Let E be a real Hilbert space, and let $L: E \to E$ be a continuous selfadjoint operator such that

(0.1) 0 does not belong to the essential spectrum of L.

Let $A: E \to E$ be a (nonlinear) operator such that

(0.2) A is compact, $A(0) = 0$, and there exists $\psi \in C^1(E, \mathbb{R})$ such that $\nabla \psi = A$ and $\psi(0) = 0$.

We will be concerned with the following problem:

(0.3) Find $u \in E$, $u \neq 0$ s.t. $Lu = A(u)$.

Observe that the solutions of (0.3) are the critical points of the functional

(0.4) $$f(u) = \tfrac{1}{2}(Lu|u)_E - \psi(u), \qquad u \in E.$$

Let $\{P_\lambda: \lambda \in \mathbb{R}\}$ be the spectral decomposition of L; we set

(0.5) $$P_- = \int_{-\infty}^{0^-} dP_\lambda, \qquad P_+ = \int_{0^+}^{+\infty} dP_\lambda,$$
$$E_- = P_-(E), \qquad E_+ = P_+(E), \qquad E_0 = \operatorname{Ker} L.$$

If E_+ (resp. E_-) is finite dimensional, then, by (0.2), it is easy to see that the functional (0.4) is bounded from above (resp. from below) modulo a compact perturbation. We say that such a functional is semidefinite. Examples of semidefinite functionals occur in the study of elliptic nonlinear boundary value problems. Multiplicity results for these problems have been obtained by using variants of Ljusternik–Schnirelmann theory (cf., e.g., [10, 1, 15, 4] and their references).

---

[1]Work supported by Ministero della Pubblica Istruzione, Italia (fondi 40% e 60%).
1980 *Mathematics Subject Classification.* Primary 58E05, 58F05; Secondary 34C15.

If $\dim E_+ = \dim E_- = +\infty$, then f is unbounded both from above and from below, and this indefiniteness cannot be removed by a compact perturbation. In this case we say that (0.4) is strongly indefinite.

Strongly indefinite functionals occur in the study of periodic solutions of Hamiltonian systems. In fact, consider the Hamiltonian system of $2n$ ordinary differential equations

(0.6) $$-J\dot{z} = H_z(z),$$

where $H \in C^1(\mathbb{R}^{2n}, \mathbb{R})$, $z = (p, q) \in \mathbb{R}^{2n}$, and J is the symplectic matrix in $\mathbb{R}^{2n}$, i.e.,

$$J = \begin{bmatrix} 0 & -I \\ I & 0 \end{bmatrix},$$

I being the identity matrix in $\mathbb{R}^n$. If $T > 0$ it is easy to see that the T-periodic solutions of (0.6) are the critical points of the "action" functional

(0.7) $$f(z) = \frac{1}{2} \int_0^{2\pi} (-J\dot{z}/z)_{\mathbb{R}^{2n}} dt - \omega \int_0^{2\pi} H(z)\, dt,$$

where $\omega = T/2\pi$. If there exist positive constants K_1, K_2, β such that

(0.8) $$|H_z(z)| \leq K_1 + K_2|z|^\beta,$$

then it can be shown that f is continuously Fréchet differentiable on the space $W^{1/2}(S^1, \mathbb{R}^{2n})$ of $2n$-tuples of 2π-periodic functions which possess square-integrable derivatives of "fractional order $1/2$" (cf. [8, 6]). Moreover, the spectrum of the operator $z \to Lz = -J\dot{z}$, with periodic conditions, contains infinitely many positive and negative eigenvalues, then (0.7) is strongly indefinite.

Here we state some results concerning the study of critical points of strongly indefinite functionals in the case in which the functional is invariant under the action of a suitable compact Lie group (cf. Theorem 1.1).

1. The abstract result. Suppose that f, defined in (0.4), is invariant under the action of the group $S^1 = \mathbb{R}/2\pi\mathbb{Z}$. This symmetry property of f permits the use of a topological invariant, defined in [14] and [2], in order to obtain a lower bound to the number of nontrivial critical points of (0.4). Such a number, roughly speaking, represents the number of eigenvalues, counted with their multiplicities, which "interact" with the nonlinear operator A.

Before stating the theorem, we need some definitions. Let $r: S^1 \to U(E)$ be a representation of S^1 on the group of unitary linear transformations on E. We still denote $r(S^1)$ by S^1, and we say that a functional f on E is S^1-*invariant* if $f \circ g = f$ for any $g \in S^1$. A subset $A \subset E$ is called S^1-*invariant* if $g(A) \subset A$ for any $g \in S^1$. Finally, we set Fix $S^1 = \{u \in E | g(u) = u \text{ for any } g \in S^1\}$, and if $u \in E$ the *orbit* of u is the set $\{g(u): g \in S^1\}$.

The following theorem can be proved (cf. [6]):

THEOREM 1.1. *Suppose that a unitary representation of the group S^1 acts on the real Hilbert space E. Let f be the functional defined by (0.4) with L and ψ satisfying* *assumptions* (0.1) *and* (0.2). *Moreover, suppose that*

(f_1) *given* $C \in]0, +\infty[$, *every sequence* $\{u_k\}$ *s.t.* $f(u_k) \to C$ *and* $\|\nabla f(u_k)\| \cdot \|u_k\| \to 0$, *possesses a bounded subsequence;*

(f_2) *f is S^1-invariant;*

(f_3) *there exist two closed linear subspaces* $V, W \subset E$ *and positive constants* C_0, C_∞, ρ *such that*

  (i) *V and W are S^1-invariant,*
  (ii) $\dim(V \cap W) < +\infty, \operatorname{codim}(V + W) < +\infty$,
  (iii) *Fix* $S^1 \subset V$ *or Fix* $S^1 \subset W$,
  (iv) $f(u) \geq C_0$ *for any* $u \in V, \|u\| = \rho$,
  (v) $f(u) < C_\infty$ *for any* $u \in W$,
  (vi) $f(u) < C_0$ *for any* $u \in$ *Fix* S^1 *s.t.* $\nabla f(u) = 0$.

Under the above assumptions f possesses at least

$$\tfrac{1}{2}(\dim(V \cap W) - \operatorname{codim}(V + W))$$

orbits of critical points, whose critical values are in $[C_0, C_\infty]$.

REMARK 1.2. An analogous version of Theorem 1.1 can be obtained if f is even, i.e., invariant under the action of the group Z_2 (cf. [6]).

REMARK 1.3. Theorem 1.1 generalizes Theorem 4.1 of [3] in two points. Assumption (f_1) is weaker than assumption (f_2) in [3]. Moreover, in [3] (f_3.iii) is replaced by the stronger assumption Fix $S^1 \subset W$.

REMARK 1.4. If in (f_3) we also assume that $\dim V < +\infty$, then Theorem 1.1 becomes a variant of an abstract theorem contained in [9]. If f is even and we also assume, in (f_3), that $\dim W < +\infty$ (resp. $\dim V < +\infty$), Theorem 1.1 becomes a variant of some results contained in [1] and [4] (resp. [10]).

Observe that if f is strongly indefinite, both V and W are, in general, infinite dimensional.

REMARK 1.5. If ψ is strongly convex, then the study of the critical points of a strongly indefinite functional can be reduced to the study of a semidefinite functional by using a duality method introduced by Clarke and Ekeland (cf. [11]–[13]).

REMARK 1.6. Theorem 1.1 can be applied as well if ψ "grows more than quadratically" as if it "grows less than quadratically" at infinity. The choices of the right spaces V and W depend on the particular situation under study.

2. An application. Theorem 1.1 has been applied to the study of nonconstant periodic solutions of prescribed period T of system (0.6) in the cases when the Hamiltonian function $H(z)$ is asymptotically quadratic (cf. [6, 7]) and of the type (cf. [5, 6]):

$$H(z) = H(p, q) = \sum_{i,j} a_{ij}(q) p_i p_j + V(q),$$

where $\{a_{ij}(q)\}$ is a positive definite matrix and

$$V(q)/|q|^2 \to \infty \quad \text{as } |q| \to \infty.$$

In this case the different behaviour at infinity of the variables p and q causes technical difficulties in the choice of the right spaces V and W of assumption (f_3).

Here we want to give an easy application of Theorem 1.1 to the study of periodic solutions of (0.6). In order to avoid technical difficulties we shall assume

(H_o) $H(z)/|z|^2 \to \infty$ as $|z| \to \infty$.

The following theorem holds:

THEOREM 2.1. *Consider the Hamiltonian system* (0.6), *with H satisfying* (0.8) *and* (H_0). *Moreover, suppose that there exist positive constants C_1, C_2, α such that*

(H_1) $\frac{1}{2}(H_z(z)|z) - H(z) \geq C_1|z|^\alpha - C_2$ *for any* $z \in \mathbb{R}^{2n}$,

(H_2) $H(z) \geq 0$, $H(0) = 0$, $H_z(0) = 0$.

Then for any prescribed period T, (0.6) *possesses infinitely many nonconstant T-periodic solutions which are geometrically distinct.*

REMARK 2.2. Theorem 2.1 is a variant of a result of Rabinowitz (cf. [16]), in which assumptions (H_0), (H_1) are replaced by

(H^*) $(H_z(z)|z) \geq pH(z) > 0$, $|z| > R^*$,

where $p > 2$ and $R^* > 0$. Observe that by (H^*) it follows that $H(z) \geq |z|^p$ for $|z| > R^*$. On the other hand, the "superquadraticity" conditions (H_0)–(H_1) cover cases which are not covered by (H^*). For example, the function

$$H(z) = |z|^2 \log(1 + |z|^2)$$

satisfies (H_0)–(H_1) but not (H^*).

SKETCH OF THE PROOF OF THEOREM 2.1. (I) Let $T > 0$ and set

$$\psi(z) = \int_0^{2\pi} \omega H(z)\, dt, \quad z \in W^{1/2}(S^1, \mathbb{R}^{2n}) = E, \quad Lz = -J\dot{z},$$

with $\omega = T/2\pi$. By previous considerations and standard arguments, we have that the functional f, defined by (0.7), satisfies (0.1) and (0.2).

If we consider the unitary representation of S^1 on $W^{1/2}(S^1, \mathbb{R}^{2n})$ given by the time translations (i.e., if $g \in S^1$, $r(g)$ is the unitary map in $W^{1/2}(S^1, \mathbb{R}^{2n})$ defined by $z(t) \mapsto z(t + g)$), then, since H does not depend on t, we have that f is S^1-invariant and (f_2) is satisfied.

(II) By using (H_1) it can be verified that (f_1) is satisfied.

(III) We denote by λ_j (resp. E_j), $j \in \mathbb{Z}$, the eigenvalues (resp. the eigenspaces) of L. We set

$$E_k^+ = \overline{\bigoplus_{j \geq k} E_j}, \quad E_k^- = \overline{\bigoplus_{j \leq k} E_j},$$

where the closures are taken in $W^{1/2}(S^1, \mathbb{R}^{2n})$. It can be shown that there exist $j \in \mathbb{Z}$ and $R > 0$ such that

$$f(z) \geq C_0 > 0 \quad \text{for any } z \in E_j^+, \|z\| = R.$$

Now, if $m \in \mathbb{N}$, we set $V = E_j^+$, $W = E_{j+m}^-$; then it is easy to see that V and W satisfy assumption (f_3). Hence, we can conclude that (0.6) possesses at least

$$\tfrac{1}{2}[\dim(V \cap W) - \mathrm{codim}(V + W)] = \tfrac{1}{2}\dim(V \cap W) \geq m$$

nonconstant T-periodic solutions which are geometrically distinct. Since m is arbitrary, the conclusion follows.

References

1. A. Ambrosetti and P. H. Rabinowitz, *Dual variational methods in critical point theory and applications*, J. Funct. Anal. **14** (1973), 349–381.
2. V. Benci, *A geometrical index for the group S^1 and some applications to the study of periodic solutions of ordinary differential equations*, Comm. Pure Appl. Math. **34** (1981), 393–432.
3. _____, *On the critical point theory for indefinite functionals in the presence of symmetries*, Trans. Amer. Math. Soc. **274** (1982), 533–572.
4. V. Benci, P. Bartolo and D. Fortunato, *Abstract critical point theorems and applications to some nonlinear problems with strong resonance at infinity*, Nonlinear Anal. **9** (1983), 981–1012.
5. V. Benci, A. Capozzi and D. Fortunato, *Periodic solutions of a class of Hamiltonian systems*, Lecture Notes in Math., vol. 964, Springer-Verlag, 1982, pp. 86–94.
6. _____, *Periodic solutions of Hamiltonian systems with a prescribed period*, Univ. of Wisconsin, Math. Res. Center, Tech. Summ. Rep. 2508.
7. _____, *On asymptotically quadratic Hamiltonian systems*, Nonlinear Anal. **8** (1983), 929–931.
8. V. Benci and P. H. Rabinowitz, *Critical point theorems for indefinite functionals*, Invent. Math. **52** (1979), 336–352.
9. H. Berestycki, J. M. Lasry, G. Mancini and B. Ruf, *Existence of multiple periodic orbits on star-shaped Hamiltonian surface*, Comm. Pure Appl. Math. (to appear).
10. B. C. Clark, *A variant of Ljusternik–Schnirelmann theory*, Indiana Univ. Math. J. **22** (1972), 65–74.
11. F. H. Clarke, *Periodic solutions to Hamiltonian inclusion*, J. Differential Equations **40** (1981), 1–6.
12. F. H. Clarke and I. Ekeland, *Hamiltonian trajectories having prescribed minimal period*, Comm. Pure Appl. Math. **33** (1980), 103–116.
13. I. Ekeland, *Periodic solutions of Hamiltonian equations and a theorem of P. Rabinowitz*, J. Differential Equations **34** (1979), 523–534.
14. E. R. Fadell and P. H. Rabinowitz, *General cohomological index theories for Lie group actions with an application to bifurcation questions for Hamiltonian systems*, Invent. Math. **45** (1978), 139–174.
15. P. H. Rabinowitz, *Variational methods for nonlinear eigenvalue problems* (CIME, Varenna, 1974), Edizione Cremonese, Rome, 1974, pp. 141–195.
16. _____, *Periodic solutions of Hamiltonian systems*, Comm. Pure Appl. Math. **31** (1978), 157–184.
17. _____, *Periodic solutions of Hamiltonian systems: a survey*, SIAM J. Math. Anal. **13** (1982), 343–352.

UNIVERSITÀ DEGLI STUDI DI BARI, ITALY

Uniqueness of Positive Solutions for a Sublinear Dirichlet Problem

ALFONSO CASTRO

Abstract. Uniqueness of positive solutions for a class of sublinear elliptic boundary value problems is proved. Our method concentrates on estimates near the boundary.

1. Introduction. Here we study the uniqueness of positive solutions of

$$(1.1)_\lambda \qquad -\Delta u = \lambda f(u) \quad \text{for } x \in \Omega, \qquad u = 0 \quad \text{for } x \in \partial\Omega,$$

where $\lambda > 0$, Ω is a smooth bounded region in R^n and f is a function of class C^1 such that

(1.2) There exist $0 < c < d < \infty$ such that $f(u) > 0$ if $0 \leq u < d$, $f'(u) \leq 0$ on (c, d), and $f(u) < 0$ if $u > d$.

In addition we assume that

(1.3) There exists a subgroup X of rotations of R^n such that Ω is invariant under X and, for every x in a neighborhood of $\partial\Omega$, $\{\theta(x); \theta \in X\}$ is an $(n-2)$-dimensional sphere.

Clearly every smooth region in R^2, every ball in R^n, the region between two balls in R^n, tori in R^3, etc., satisfy (1.3).

This paper extends results of [5] and [9] in that we allow more general domains and we do not assume $f'(d) < 0$. E. Dancer has recently announced results that overlap with ours. In [4] a similar problem was studied with a different type of nonlinearity. For applications see [3, 5, 8]. Our main result is

THEOREM 1.1. *Under the above assumptions, there exists λ_0 such that if $\lambda > \lambda_0$, then problem $(1.1)_\lambda$ has a unique positive solution.*

1980 *Mathematics Subject Classification.* Primary 35J65; Secondary 47H12.
Key words and phrases. Nonlinear Dirichlet problem, unique positive solution, sub-super solution.

We prove this theorem by obtaining estimates near the boundary. We do this first in the ball. Most of our estimates are independent of (1.3). For this reason we suspect Theorem 1.1 to hold assuming (1.2) only.

2. Estimates in a ball. Throughout this section Ω is a ball in R^n. Without loss of generality we can assume Ω to be centered at the origin. Let ρ denote the radius of Ω.

LEMMA 2.1. *For each $N > 0$ there exists $M(N) \equiv M > 0$ and $\lambda(N)$ such that if $\lambda \geq \lambda(N)$, $|x| \leq \rho - N\lambda^\delta$, $\delta \in (-\frac{1}{2}, 0)$, and u is a solution to $(1.1)_\lambda$, then $f(u(x)) \leq M\lambda^{-(1+2\delta)}$. Moreover, $M(N) = O(N^{-2})$.*

PROOF. Since $f(0) > 0$, without loss of generality we can assume that $f(c) \neq f(y)$ if $y \neq c$. Let G denote the Green's function of $-\Delta$ on Ω with Dirichlet boundary condition. Because $f(u) < 0$ if $u > d$, by the maximum principle we have $u(x) \leq d$ for any $x \in \Omega$. In particular we obtain

(2.1) $\quad d \geq u(0)$

$$= \lambda \int_\Omega G(0, x) f(u(x))\, dx$$

$$\geq \lambda \min\{f(u(x)); |x| > \rho - N\lambda^\delta\} c_n \left(\int G(0, x) |x|^{n-1} d(|x|) \right)$$

$$\geq K\lambda \min\{f(u(x)); |x| > \rho - N\lambda^\delta\} N^2 \lambda^{2\delta},$$

where the integral on the second line is over the interval $[\rho - N\lambda^\delta, \rho]$ and K is a constant depending only on n for λ large enough. Hence there exists x_0 such that $|x_0| \geq \rho - N\lambda^\delta$ and $f(u(x_0)) \leq M\lambda^{-(1+2\delta)}$, where $M = d/KN^2$. Now we choose $\lambda(\delta)$ large enough so that $M\lambda^{-(1+2\delta)} \leq f(c)$ and K depend only on n if $\lambda \geq \lambda(\delta)$. Since the solutions to $(1.1)_\lambda$ are radially symmetric and radially decreasing (see [6]) we have

(2.2) $\qquad\qquad f(u(x)) \leq f(u(x_0)) \leq M\lambda^{-(1+2\delta)}$

for $|x| \leq \rho - N\lambda^\delta \leq |x_0|$, and the lemma is proved.

REMARK 2.2. Double-checking the above proof we see that the statement also holds for $\delta = 0$ if $N < \rho d_n$, with d_n depending only on n. Also from the proof of Lemma 2.1 follows that if $N \geq N_0 \equiv (d/Kf(c))^{1/2}$, then $u(x) \geq c$ for $|x| \leq \rho - N^{-1/2}$.

Now we obtain estimates on the normal derivative of u at $\partial\Omega$.

LEMMA 2.3. *There exists a constant $c_1 > 0$, independent of λ, such that if u is a solution of $(1.1)_\lambda$, then for $|x| = \rho$*

(2.3) $\qquad\qquad u_\eta(x) \leq -c_1 \lambda^{1/2},$

where u_η denotes the outward unit normal derivative of u.

PROOF. In spherical coordinates $(1.1)_\lambda$ is equivalent to

$$-u''(r) - ((n-1)/r) u'(r) = \lambda f(u(r)), \qquad 0 \leq r \leq \rho, \quad u(\rho) = 0.$$

Multiplying by u' we obtain a first order linear equation in $(u')^2$. Integrating this equation we have

$$u'(r) = -\left(2\lambda\left(\int_0^r \frac{F(u(t))2(n-1)t^{2n-3}}{r^{2(n-1)}} dt - F(u(r))\right)\right)^{1/2},$$

where $F(r) = \int_0^r f(s)\,ds$. Taking N as in Remark 2.2 and using that F is nondecreasing we have

(2.4) $$u'(\rho) \leq -\left(2\lambda \int F(c)2(n-1)\left(\frac{\rho}{2}\right)^{2n-3} \rho^{2(n-1)}\right)^{1/2}$$

where the integral is over the interval $[0, \rho/2]$, and this proves the lemma.

3. Estimates in a general region. Since $\partial\Omega$ is a smooth manifold, the function assigning to each $x \in \partial\Omega$ the outward unit normal to Ω at x is a smooth function. We will denote this function by η. Also it is well known (see [7]) that there exists $\rho > 0$ such that if $x, y \in \partial\Omega$, $x \neq y$, and $s \in (-\rho, 0)$, then $x + s\eta(x) \neq y + s\eta(y)$. In addition, ρ can be chosen so that if $x \in \Omega$ and $\min\{|x - y|;\ y \in \partial\Omega\} = s < \rho$, then $x = y - s\eta(y)$ for some $y \in \partial\Omega$. Actually y is the point of minimal distance from x to $\partial\Omega$. For $s \in (0, \rho)$ we will denote by $\Omega_s = \{x \in \Omega;\ d(x, \partial\Omega) \geq s\}$. Clearly Ω_s is also a smooth region. Furthermore, ρ can be chosen so that for each $x \in \partial\Omega$ there exists a closed ball $B_x \subset \Omega$ of radius $\rho/2$ such that $B_x \cap \partial\Omega = \{x\}$.

Since every solution of $(1.1)_\lambda$ in Ω is a supersolution for $(1.1)_\lambda$ in B_x, the assertions of Lemma 2.1, Remark 2.2 and Lemma 2.3 hold in Ω. It follows from the maximum principle that $(1.1)_\lambda$ has a minimal and a maximal solution (see [3]). Let them be u and v respectively. The dependence of u and v on λ will be clear from the context. Let $w = v - u \geq 0$. We set $a_j = \max\{\int_{\partial\Omega_s} -w_\eta;\ s \in [0, j/\lambda]\}$ for each positive integer $j \in [0, N_0\lambda^{1/2} + 1]$, where N_0 is as in Remark 2.2.

LEMMA 3.1. *There exists a constant K_1 such that if λ is large enough, then*

(3.1) $$a_j \leq K_1 a_0.$$

PROOF. Suppose $a_j > a_{j-1}$. Hence there exists $s' \in [(j-1)/\lambda, j/\lambda]$ with $a_j = \int_{\partial\Omega_{s'}} -w_\eta$. Since w satisfies

(3.2) $$-\Delta w = \lambda(f(v) - f(u)) \quad \text{in } \Omega, \quad w = 0 \quad \text{on } \partial\Omega,$$

integrating over $\Omega_{(j-1)/\lambda} - \Omega_{s'}$, we have

$$a_j = \int_{\partial\Omega_{s'}} -w_\eta$$

$$= \int_{\partial\Omega_{(j-1)/\lambda}} -w_\eta - \lambda \int_{\Omega_{(j-1)/\lambda} - \Omega_{s'}} (f(v) - f(u))$$

(3.3) $$\leq a_{j-1} + \lambda K \iint w\, dt\, dy$$

(3.4) $$\leq a_{j-1} + K(ja_j/\lambda),$$

where in (3.3) the variable t ranges over $[(j-1)/\lambda, s']$, y ranges over $\partial\Omega$ and K is a constant depending on the maximum of the derivative of f and the Jacobian of the transformation that identifies each point in $\Omega_{(j-1)/\lambda} - \Omega_{s'}$ with its projection onto $\partial\Omega$ and its distance to $\partial\Omega$. Certainly K is independent of λ. From (3.4) and the assumption that $a_j > a_{j-1}$ we see that

$$(3.5) \qquad a_j \leqslant (1-(Kj)/\lambda)^{-1} a_{j-1}.$$

Therefore we have

$$(3.6) \qquad a_j \leqslant \left(1-\left(KN_0\lambda^{1/2}+K\right)/\lambda\right)^{-N_0\lambda^{1/2}-1} a_0.$$

Since $j < N_0\lambda^{1/2} + 1$ the factor of a_0 in (3.6) is bounded independently of λ, which proves the lemma.

Let now z be any positive solution to $(1.1)_\lambda$. By Stokes' theorem we have

$$(3.7) \qquad \int_{\Omega-\Omega_s} \nabla z \cdot \nabla(z_\eta) - \int_{\partial\Omega} (z_\eta)^2 + \int_{\partial\Omega_s} (z_\eta)^2 = \lambda \int_{\Omega-\Omega_s} f(z) z_\eta \, dx.$$

Also

$$(3.8) \qquad \left(|\nabla z|^2\right)_\eta = 2\sum c_j(x) z_{x_i} z_{x_i x_j},$$

where the indices i, j range from 1 to n, and $c_j(x)$ denotes the jth component of the outward unit normal at the point in $\partial\Omega$ closest to x. On the other hand

$$(3.9) \qquad \nabla z \cdot \nabla(z_\eta) = \frac{1}{2}\left(|\nabla z|^2\right)_\eta + \sum_{i,j=1}^{n} (c_j)_{x_i} z_{x_i} z_{x_j}.$$

Since on $\partial\Omega$ we have $|\nabla z|^2 = (z_\eta)^2$, from (3.7)–(3.9) we conclude

$$(3.10) \qquad -\int_{\partial\Omega} \frac{(z_\eta)^2}{2} = -\int_{\Omega-\Omega_s} \sum_{i,j} (c_j)_{x_i} z_{x_i} z_{x_j}$$
$$+ \int_{\partial\Omega_s} \frac{|\nabla z|^2}{2} - (z_\eta)^2 + \lambda \int_{\Omega-\Omega_s} f(z) z.$$

Applying (3.10) to u and v and subtracting we have

$$(3.11) \qquad \frac{1}{2}\int_{\partial\Omega} \left((v_\eta)^2 - (u_\eta)^2\right) = \lambda \int_{\partial\Omega_s} (F(v) - F(u)) J$$
$$+ \lambda \int_{\Omega-\Omega_s} (F(v) - F(u)) J_\eta$$
$$+ \int_{\partial\Omega_s} \left(\frac{|\nabla u|^2 - |\nabla v|^2}{2} - (u_\eta)^2 + (v_\eta)^2\right)$$
$$+ \sum_{i,j} \int_{\Omega-\Omega_s} \left((c_j)_{x_i} v_{x_j} v_{x_i} - (c_j)_{x_i} u_{x_j} u_{x_i}\right),$$

where J is the Jacobian of the transformation that identifies each point in $\Omega - \Omega_s$ with its projection on $\partial\Omega$ and its distance to $\partial\Omega$.

Let e be a regular value of u and v with $e \in (f^{-1}((M+1)/\lambda), f^{-1}(M/\lambda))$, with M such that $f(u(x)) \leq M/\lambda$ and $f(v(x)) \leq M/\lambda$ if $d(x, \partial\Omega) \geq \rho/2$ (see Remark 2.2). Let $\Omega_u = \{x \in \Omega; u(x) > e\} \supset \{x \in \Omega; d(x, \partial\Omega) \geq \rho/2\}$. Similarly we define Ω_v. Hence

$$(3.12) \quad \int_{\Omega_u} \nabla u \cdot \nabla u + \int_{\Omega_v} \nabla v \cdot \nabla v = \lambda \int_{\Omega_u} f(u)(u-e) + \lambda \int_{\Omega_v} f(v)(v-e).$$

Since $\lambda f(u)$ and $\lambda f(v)$ are bounded on $\Omega_{\rho/2}$ (see Remark 2.2), from (3.12) we see that there exists $t \in (\rho/2, \rho)$ such that

$$(3.13) \quad \int_{\partial\Omega_s} \left(|\nabla u|^2 + |\nabla v|^2\right) \leq K_2,$$

with K_2 independent of λ.

In order to estimate the second term in the right side of (3.11) we observe that, from Lemma 2.1 and Remark 2.2, there exists a constant M_1 such that if $d(x, \partial\Omega) \geq N_0^{-1/4}$, then $f(u(x)) \leq M_1^{-1/2}$. Therefore

$$(3.14) \quad \begin{aligned} &\lambda \int_{\Omega - \Omega_t} (F(v) - F(u)) J_\eta \\ &= \lambda \int_{\Omega - \Omega_{N_0\lambda^{-1/2}}} (F(v) - F(u)) J_\eta + \lambda \int f(y)(v-u) J_\eta \\ &\quad + \lambda \int_{\Omega_{N_0\lambda^{-1/4}} - \Omega_t} f(y)(v-u) J_\eta \\ &\leq K_2 \lambda^{1/2} + M_2 \int r\, dr\, dq, \end{aligned}$$

where the third and last integrals are over $\Omega_{N_0\lambda^{-1/2}} - \Omega_{N_0\lambda^{-1/4}}$, K_2 is a constant that depends on J and the maximum value of f, and M_2 is a constant that depends on M_1 and the maximum value of f. Since the last term in (3.14) is of order $\lambda^{1/2}$, as λ tends to infinity, we have

$$(3.15) \quad \lambda \int_{\Omega - \Omega_t} (F(v) - F(u)) J_\eta = O(\lambda^{1/2}).$$

Adding and subtracting $\sum_{i,j} (c_j)_{x_i} v_{x_j} u_{x_i}$ to the last term in (3.11) we obtain

$$(3.16) \quad \sum_{i,j} \int (c_j)_{x_i} (v_{x_j} v_{x_i} - u_{x_j} u_{x_i}) = O\left(\left(\int \left(|\nabla u|^2 + |\nabla v|^2\right)\right)^{1/2} \left(\int |\nabla w|^2\right)^{1/2}\right),$$

where the integrals are over $\Omega - \Omega_t$.

Since $f(u(x)) = O(1/\lambda)$ for every $x \in \partial\Omega_t$, replacing this, (3.13), (3.15) and (3.16) in (3.11) for $s = t$ we obtain

$$(3.17) \quad \int_{\partial\Omega} \left((v_\eta)^2 - (u_\eta)^2\right) = O\left(\lambda^{1/2} + \left(\int \left(|\nabla u|^2 + |\nabla v|^2\right)\right)^{1/2} \left(\int |\nabla w|^2\right)^{1/2}\right),$$

where the last two integrals are over $\Omega - \Omega_t$.

By Lemma 2.3 there exists a constant K_3 such that if z is any solution of $(1.1)_\lambda$ and λ is large enough, then $f(u(x)) \leq K_3 \lambda^{-1-\delta}$ if $\delta \in (-\frac{1}{2}, -\frac{1}{4})$ and $d(x, \partial\Omega) \in [N_0\lambda^{-1/2}, N_0\lambda^{-1/4}]$. In addition, $f(u(x)) \leq K_3\lambda^{-1/2}$ if $d(x, \partial\Omega) \geq N_0\lambda^{-1/4}$ and $f(z(x)) \leq K_3\lambda^{-1/2}$ if $d(x, \partial\Omega) \geq t$. Therefore we have

$$(3.18) \quad \int_{\partial\Omega} ((v_\eta)^2 - (u_\eta)^2) = O\left(\lambda^{1/2}\left(1 + \lambda^{-1/4}\left(\int_\Omega |\nabla w|^2\right)^{1/2}\right)\right),$$

where we have used that $\int_\Omega |\nabla(u+v)|^2 = O(\lambda^{1/2})$ (see Lemma 5.1).

Since estimate (2.3) holds also in a general region, we infer from (3.18) that

$$(3.19) \quad \int_{\partial\Omega} (u-v)_\eta = O\left(1 + \lambda^{-1/4}\left(\int_{\Omega-\Omega_t} |\nabla w|^2\right)^{1/2}\right).$$

Since $f(v(x)) - f(u(x)) \leq 0$ for $d(x, \partial\Omega) > N_0\lambda^{-1/2}$, integrating (3.2) over Ω_s we have $\int_{\partial\Omega_s} w_n \geq 0$ if $s \geq N_0\lambda^{-1/2}$. Therefore by Gronwall's inequality we have

$$(3.20) \quad \int_{\partial\Omega_s} w = O\left(\max\left\{\int_{\partial\Omega_r} w; r \in [0, N_0^{-1/2}]\right\}\right).$$

Again using Gronwall's inequality and (3.20) we have

$$(3.21) \quad \int_{\partial\Omega_s} w = O\left(\lambda^{-1/2}\left(1 + \lambda^{-1/4}\left(\int_\Omega |\nabla w|^2\right)^{1/2}\right)\right).$$

Replacing (3.21) in (3.14) as well as in the first term in the right side of (3.11) we have

$$(3.22) \quad \int_{\partial\Omega} -w_\eta = O\left(\lambda^{-1/2} + \lambda^{-1/4}\int_\Omega |\nabla w|^2\right).$$

Now multiplying (3.2) by w and using that $\int_{\partial\Omega_s} w = O(\lambda^{-1/2}\int_{\partial\Omega} - w_\eta)$ we have

$$(3.23) \quad \int_\Omega |\nabla w|^2 \leq \lambda \int_{\Omega - \Omega_{N_0\lambda^{-1/2}}} (f(v) - f(u))w = O\left(\int_{\partial\Omega} |w_\eta|\right),$$

where we have used that $w \geq 0$, and that $(f(v) - f(u)) \leq 0$ on $\Omega_{N_0\lambda^{-1/2}}$. Combining (3.22) and (3.23) we conclude

$$(3.24) \quad \int_{\partial\Omega} |w_\eta| = O(\lambda^{-1/2}).$$

4. Proof of Theorem 1.1. First we observe that if $T \in X$, then $u(T(x)) = u(x)$ and $v(T(x)) = v(x)$ for all $x \in \Omega$. In fact, since $-\Delta$ is invariant under rotations if z is a solution to $(1.1)_\lambda$, then so is $z_1(x) \equiv z(T(x))$. Hence if $u(y) \neq u(T(y))$ for some y, say $u(y) < u(T(y))$, then considering $u_1(x) \equiv u(T^{-1}(y))$ we have not only that u_1 is a solution of $(1.1)_\lambda$ but also we have $u_1(T(y)) = u(y) < u(T(y))$, which contradicts that u is the minimal solution of $(1.1)_\lambda$. A similar argument shows that $v(x) = v(T(x))$ for all $x \in \Omega$, all $T \in X$. Therefore we also have

$$(4.1) \quad w(x) = w(T(x)) \quad \text{for all } (x, T) \in \Omega \times X.$$

Let W denote the solution to $-\Delta W = 1$ in Ω, $W = 0$ on $\partial\Omega$. By standard a priori estimates (see [2]), all derivatives of W are bounded in $L_\infty(\Omega)$. In particular there exists a constant K such that
$$(4.2) \qquad W(x) \leq Kd(x, \partial\Omega) \quad \text{for all } x \in \Omega.$$
Multiplying (3.2) by W and integrating by parts we have
$$(4.3) \quad \int_\Omega w \leq \lambda \int_\Omega W(f(v) - f(u)) \leq O(\lambda^{+1/2}) \int_{\Omega - \Omega_{N_0\lambda^{-1/2}}} (f(v) - f(u))$$
where we have used that, by Lemma 2.3, $\int_{\partial\Omega_s}(f(v) - f(u)) = O(s)\int_{\partial\Omega} w_\eta$. Hence from (4.3) we obtain
$$(4.4) \quad \int_\Omega |f(v) - f(u)|^p \leq b^{p-1}K\left(\int_\Omega w\right) \leq b^{p-1}\lambda^{-1/2}\left(\int_{\partial\Omega}|w_\eta|\right)L,$$
where K and L are constants independent of λ, and $b = b(\lambda)$ denotes the $L_\infty(\Omega)$ of w. From (4.4) we infer
$$(4.5) \qquad |w|_{2,p} = O\left(b^{(p-1)/p}\lambda^{(2p-1)/2p}\left(\int_{\partial\Omega}|w_\eta|\right)^{1/p}\right),$$
where $|\ |_{2,p}$ denotes the norm in the Sobolev space $H^{2,p}(\Omega)$. Since $H^{2,p}$ can be continuously imbedded in $C^{0,(2p-n)/p}$ (see [1]), from (4.5) follows that there exists a constant K such that if
$$(4.6) \qquad |y| \leq Kb^{1/(2p-n)}\lambda^{(1-2p)/(4p-2n)}\left(\int_{\partial\Omega}|w_\eta|\right)^{-1/(2p-n)},$$
then
$$(4.7) \qquad |w(x+y)| \geq b/2,$$
where x denotes an element of Ω such that $w(x) = \max\{w(r); r \in \Omega\}$.

By the maximum principle x cannot lie in $\Omega_{N_0\lambda^{-1/2}}$; therefore by hypothesis (1.3) the n-dimensional Lebesgue measure of $S = \{T(y); T \in X, |x-y| \leq \tau\}$ is bounded below by $c_3\tau^2$, where c_3 is a constant independent of λ and τ is the right side of (4.6). Thus
$$(4.8) \qquad \int_S w \geq c_3\tau^2 b/2.$$
On the other hand, $\int_{\partial\Omega_s} w = O(\lambda^{-1/2}\int_{\partial\Omega}|w_\eta|)$ for every $s \in (0, \rho)$. Hence
$$(4.9) \qquad \tau b = O\left(\lambda^{-1/2}\int_{\partial\Omega}|w_\eta|\right).$$
Going back to the definition of τ (see (4.6)), by (4.9) we have
$$(4.10) \qquad b = O\left(\lambda^{(n-1)/(4p-2n+2)}\int_{\partial\Omega}|w_\eta|\right).$$
Now let t be as in (3.14). Since $\max\{f(u(x)); x \in \partial\Omega_t\} = O(\lambda^{-1})$ by Lemma 2.1, we have
$$(4.11) \qquad \int_{\partial\Omega_t}(F(v(x)) - F(u(x))) = O\left(\lambda^{-1/2}\int_{\partial\Omega}|w_\eta|\right).$$

In obtaining (4.11) we have used that $\int_{\partial\Omega_r} w = O(\lambda^{-1/2}\int_{\partial\Omega}|w_\eta|)$ for any $r \in (0, \rho)$. This also implies (see (3.14))

$$(4.12) \quad \lambda\int_{\Omega-\Omega_t}(F(v)-F(u))J_\eta = O\left(\int_{\partial\Omega}|w_\eta|\right).$$

Since

$$(4.13) \quad \int_{\Omega-\Omega_\rho}||\nabla u|^2 - |\nabla v|^2| \leq \left(\int_\Omega |\nabla w|^2\right)^{1/2}\left(\int_\Omega |\nabla(u+v)|^2\right)^{1/2}$$

$$= O\left(\lambda^{1/4}\left(\int_\Omega |\nabla w|^2\right)^{1/2}\right),$$

we see that t in (3.14)–(3.16) can be chosen so that

$$(4.14) \quad \int_{\partial\Omega_t}\left(\frac{|\nabla u|^2 - |\nabla v|^2}{2} - (u_\eta)^2 + (v_\eta)^2\right) = O\left(\lambda^{1/4}\left(\int_\Omega |\nabla w|^2\right)^{1/2}\right).$$

Therefore, since $|(u_\eta + v_\eta)(x)| \geq c_1\lambda^{1/2}$ for every $x \in \partial\Omega$, replacing this, (3.16), (4.11)–(4.14) in (3.11) for $s = t$ we have

$$(4.15) \quad \int_{\partial\Omega}|w_\eta| = O\left(\lambda^{-1/4}\left(\int_\Omega |\nabla w|^2\right)^{1/2}\right)$$

$$= O\left(\lambda^{-1/4+1}\left(\int_\Omega (f(v)-f(u))w\right)^{1/4}\right).$$

Finally, replacing (4.10) in (4.15) we obtain

$$\int_{\partial\Omega}|w_\eta| = O\left(\lambda^{-1/2+1}b\left(\int_{\Omega-\Omega_{N_0\lambda^{1/2}}}(f(y)-f(u))\right)^{1/2}\right)$$

$$= O(\lambda^{-1/4+((n-1)/(8p-4n+2))})\left(\int_{\partial\Omega}|w_\eta|\right).$$

The latter equation shows that, by choosing p large enough, there exists λ_0 such that if $\lambda > \lambda_0$, then $\int_{\partial\Omega}|w_\eta| = 0$. Therefore by (4.10) we have $w = 0$, which proves the theorem.

5. Appendix. The purpose of this appendix is to prove an estimate used in deriving (3.18), namely

LEMMA 5.1. *There exists a constant K, independent of λ, such that if z is a solution to $(1.1)_\lambda$, then*

$$(5.1) \quad \int_\Omega |\nabla z|^2 \leq K\lambda^{1/2}.$$

PROOF. Let N_0 be as in Remark 2.2. Hence multiplying $(1.1)_\lambda$ by z and integrating by parts we have

$$(5.2) \quad \int_\Omega |\nabla z|^2 = \lambda \left(\int_\Omega f(z)z \right) = \lambda \left(\int_{\Omega_{N_0 \lambda^{-1/2}}} f(z)z + \int_{\Omega - \Omega_{N_0 \lambda^{-1/2}}} f(z)z \right)$$

$$\leqslant \lambda \left(\int_{\Omega_{N_0 \lambda^{-1/2}} - \Omega_{N_0 \lambda^{-1/4}}} + \int_{\Omega_{N_0 \lambda^{-1/4}}} f(z)z + O(\lambda^{-1/2}) \right),$$

where we have used that f and z are bounded and that the measure of $\Omega - \Omega_{N_0 \lambda^{-1/2}}$ is of order $\lambda^{-1/2}$. From Lemma 2.1 we have $f(z(x)) = O(\lambda^{-1/2})$ for $d(x, \partial\Omega) \geqslant N_0 \lambda^{-1/2}$, and that $f(z(x)) = O(\lambda^{-1} d(x, \partial\Omega)^{-2})$ for $d(x, \partial\Omega) \in [N_0 \lambda^{-1/2}, N_0 \lambda^{-1/4}]$. Replacing these in (5.2) we infer

$$(5.3) \quad \int |\nabla z|^2 = \lambda O\left(\lambda^{-1} \int_{\lambda^{-1/2}}^{\lambda^{-1/4}} s\, ds + \lambda^{-1/2} \right) = O(\lambda^{1/2}),$$

and this proves the lemma.

ACKNOWLEDGEMENT. The author wishes to thank Professor R. Shivaji for helpful comments.

REFERENCES

1. R. A. Adams, *Sobolev spaces*, Academic Press, New York, 1975.
2. S. Agmon, A. Douglis and L. Nirenberg, *Estimates near the boundary of solutions of elliptic partial differential equations satisfying general boundary conditions*. I, Comm. Pure Appl. Math. **12** (1959), 623–727.
3. K. J. Brown, M. M. Ibrahim and R. Shivaji, *S-shaped bifurcation curves*, Nonlinear Anal. **5** (1981), 475–486.
4. A. Castro and R. Shivaji, *Uniqueness of positive solutions for a class of elliptic boundary value problems*, Proc. Royal Soc. Edin. 98A, (1984), 267–269.
5. E. N. Dancer, *On the structure of solutions of an equation arising in catalysis theory*, J. Differential Equations **37** (1980), 404–437.
6. B. Gidas, Wei Ming Ni and L. Nirenberg, *Symmetry and related properties via the maximum principle*, Comm. Math. Phys. **68** (1979), 209–243.
7. V. Guillemin and A. Pollack, *Differential topology*, Prentice Hall, N. J., 1974.
8. J. P. Kernevez, G. Joly, M. C. Duban, B. Bunow and D. Thomas, *Hysteresis, oscillations and pattern formation in realistic immobilized enzyme systems*, J. Math. Biol. **7** (1979), 41–56.
9. R. Shivaji, *Uniqueness results for a class of positone problems*, Nonlinear Anal. **37** (1983), 223–230.

SOUTHWEST TEXAS STATE UNIVERSITY

Applications of Homology Theory to Some Problems in Differential Equations

CHANG KUNG-CHING

In this paper, a unified homology approach is used to carry out the following three results: (1) A generalization of a theorem due to Amann and Zehnder [**AZ1**] as well as an improvement due to the author [**Ch1**] to the infinite-dimensional case; (2) The existence of nontrivial solutions of an asymptotically linear equation with resonance at ∞, which can be considered as an improvement of a theorem due to Landesman, Lazer, and Rabinowitz [**Ra1**]; (3) An extension as well as a simpler proof of a result due to Conley and Zehnder [**CZ1**], which solves an Arnold's conjecture on the number of fixed points of a measure preserving mapping on a torus.

Let H be a real Hilbert space and let A be a bounded linear symmetric operator on H which splits the space H into $H_+ \oplus H_{-1} \oplus H_0$, according to its spectral decomposition. We denote $P_\pm$ and P_0 as the orthogonal projections onto the positive/negative spectrum space $H_\pm$, and the kernel of A, H_0, respectively. The following hypotheses are made:

(H_1) The restriction $A|_{H_\pm}$ is invertible, i.e. $A|_{H_\pm}$ has bounded inverse on $H_\pm$.
(H_2) $\gamma := \dim(H_- \oplus H_0) < +\infty$.
(H_3) $g \in C^2(H, \mathbf{R}^1)$ has a bounded and compact gradient $dg(x)$ for $x \in H$. If further, $\dim(H_0) > 0$, we assume that

$$g(P_0 x) \to -\infty \quad \text{as } \|P_0 x\| \to +\infty.$$

We are concerned with the number of critical points of the functional

(1) $$f(x) = \tfrac{1}{2}(Ax \cdot x) + g(x).$$

1980 *Mathematics Subject Classification.* Primary 58E05; Secondary 58F05, 35J65.

© 1986 American Mathematical Society
0082-0717/86 $1.00 + $.25 per page

Namely, we have

THEOREM 1. *Under the assumptions* (H_1), (H_2), *and* (H_3), *suppose that* $\{p_i\}_1^k$ *are given critical points of f with*

$$\text{ind } d^2f(p_i) > \gamma, \quad i = 1, 2, \ldots, k.$$

Then f has at least $k + 1$ critical points.

This is an existence theorem related to the asymptotically linear operator equation with resonance at ∞: $Ax + dg(x) = \theta$.

A C^2 function f is said to be asymptotically quadratic if there is a symmetric bounded linear operator A_∞ such that $\|dg(x)\| = o(\|x\|)$ as $\|x\| \to \infty$, where $g(x) = f(x) - \frac{1}{2}(A_\infty x, x)$. In case A_∞ is invertible, we have

THEOREM 2. *Suppose that f is an asymptotically quadratic function with asymptotic operator A_∞ satisfying the assumptions (H_1), (H_2) with $H_0 = \{\theta\}$. Suppose that $g(x) = f(x) - \frac{1}{2}(A_\infty x, x)$ has a compact gradient. Assume that $\{p_i\}_1^k$ are given critical points of f with*

$$\text{ind } d^2f(p_i) > \gamma, \quad i = 1, 2, \ldots, k.$$

Then f has at least $k + 1$ critical points.

One more theorem, which is related to a Conley–Zehnder theorem, reads as

THEOREM 3. *Suppose that A is a bounded linear symmetric operator defined on a Hilbert space satisfying (H_1) and (H_2), with $\ker(A) = \{\theta\}$. Assume that V^n is a C^2-compact manifold without boundary, and that $g \in C^2(H \times V^n, \mathbf{R}^1)$ with a bounded and compact differential vector field $dg(x, v)$. Then the functional*

$$(2) \qquad f(x, v) = \tfrac{1}{2}(Ax, x) + g(x, v)$$

has at least cuplength $(V^n) + 1$ critical points. If further, f is a Morse function, then f has at least $\sum_{i=0}^n \beta_i(V^n)$ critical points, where $\beta_i(V^n)$ is the ith Betti number of V^n, $i = 0, 1, 2, \ldots, n$.

As a corollary of Theorem 3, we obtain an estimate of the number of periodic orbits of the following Hamiltonian systems with periodic Hamiltonians.

THEOREM 4. *Suppose that $H \in C^2(\mathbf{R}^1 \times \mathbf{R}^{2n}, \mathbf{R}^1)$ is 2π-periodic in all variables. Then the Hamiltonian system*

$$(3) \qquad \dot{z} = JH'_z(t, z),$$

where $z \in C^1(S^1, \mathbf{R}^{2n})$ and

$$J = \begin{pmatrix} 0 & -I_n \\ I_n & 0 \end{pmatrix},$$

possesses at least $2n + 1$ 2π-periodic orbits. If further, all of the 2π-periodic solutions have no Floquet multiplier equal to 1, then (3) has at least 2^{2n}-periodic orbits.

This is the theorem which gives an answer to a problem raised by Arnold [**Ar1**].

As applications to other partial differential equation problems, we start with a resonance problem

(4) $-\Delta u = \hat{\lambda} u - \phi(x, u)$, $x \in \Omega$, $u|_{\partial\Omega} = 0$,

where Ω is a bounded open domain in $\mathbf{R}^n$ with smooth boundary $\partial\Omega$, $\hat{\lambda} \in \sigma(-\Delta)$ is an eigenvalue of $-\Delta$ under Dirichlet condition, and $\phi \in C^1(\overline{\Omega} \times \mathbf{R}^1, \mathbf{R}^1)$.

THEOREM 5. *Suppose that*
(1) $\phi(x, u)$ *is bounded, i.e.* $\exists$ *constant m such that* $|\phi(x, u)| \leq m$, *and*
(2) *the Landesman–Lazer condition holds, i.e.*

$$\int_\Omega \Phi\left(x, \sum_{i=1}^N t_i \psi_i(x)\right) dx \to -\infty \quad \text{as} \quad \sum_{i=1}^N |t_i|^2 \to \infty,$$

where $\Phi(x, t) = \int_0^t \phi(x, s)\, ds$ *and* $\operatorname{span}\{\psi_1, \ldots, \psi_N\} = \ker(-\Delta - \hat{\lambda} I)$.
Then:
(1) *The BVP* (4) *has a solution.*
(2) *If further* $\phi(x, 0) = 0$ *and* $\phi'_u(x, 0) > \lambda_+ - \hat{\lambda}$, *where* $\hat{\lambda} < \lambda_+$ *are consecutive eigenvalues in* $\sigma(-\Delta)$, *then the BVP* (4) *has a nontrivial solution.*

Similar results for periodic solution problems (PSP in abbreviation) of a nonlinear wave equation as well as of a Hamiltonian system are also valid.

Finally, we turn to the multiple solution problem

(5) $-\Delta u = \phi(x, u)$, $x \in \Omega$, $u|_{\partial\Omega} = 0$.

THEOREM 6. *Suppose that equation* (5) *has two pairs of strictly sub- and super-solutions* $\underline{\varphi}_1 < \overline{\varphi}_1$, $\underline{\varphi}_2 < \overline{\varphi}_2$, *with disjoint order intervals* $[\underline{\varphi}_1, \overline{\varphi}_1]$, $[\underline{\varphi}_2, \overline{\varphi}_2]$. *Assume that* $\hat{\lambda} = \lim_{u \to \infty}(\phi(x, u)/u)$ *exists and that* $\lambda_2 < \hat{\lambda} \notin \sigma(-\Delta)$. *Then the BVP* (5) *has at least four solutions.*

1. Proofs of Theorems 1, 2, and 3. Let $\mathfrak{M}$ be a C^2-Riemannian–Hilbert manifold and suppose $f \in C^2(\mathfrak{M}, \mathbf{R}^1)$ satisfies the Palais-Smale (P.S.) condition. The following basic facts from Morse theory are presumed (cf. Chang [**Ch3**]).

For an isolated critical point p of f, we attach a sequence of groups, which is called the critical group:

$$C_q(f, p) := H_q(f_c \cap U_p, (f_c \setminus \{p\}) \cap U_p; Q), \qquad q = 0, 1, 2, \ldots,$$

where U_p is a neighborhood of p such that p is the unique critical point of f in U_p, $c = f(p)$, f_c denotes the level set $\{x \in \mathfrak{M} | f(x) \leq c\}$, and Q stands for the rational field.

The critical groups are well defined, i.e. they do not depend on the special choice of the neighborhood U_p.

Suppose that f has only isolated critical values, each of which corresponds to a finite number of critical points, say

$$\cdots < c_{-2} < c_{-1} < c_0 < c_1 < c_2 < \cdots$$

with $K_{c_i} := f^{-1}(c_i) \cap K = \{z_1^i, \ldots, z_{m_i}^i\}$, $i = 0, \pm 1, \pm 2, \ldots$, where K is the critical set of f.

For each interval $(a, b) \subset \mathbf{R}^1$, where a, b are not critical values of f, we define

$$m_q := m_q(a, b) = \sum_{a<c_i<b} \sum_{j=1}^{m_i} \operatorname{rank} c_q(f, z_j^i)$$

and

$$\beta_q := \beta_q(a, b) = \operatorname{rank} H_q(f_b, f_a; Q), \quad q = 0, 1, 2, \ldots.$$

The following Morse inequalities hold:

$$m_p \geqslant \beta_0,$$
$$m_1 - m_0 \geqslant \beta_1 - \beta_0$$
$$\cdots \qquad \cdots$$
$$m_q - m_{q-1} + \cdots + (-1)^q m_0 \geqslant \beta_q - \beta_{q-1} + \cdots + (-1)^q \beta_0$$
$$\cdots \qquad \cdots$$

If the function f is nondegenerate, i.e. all critical points of f are nondegenerate, then the number m_q has a geometrical interpretation: the number of critical points with prescribed Morse index q, $q = 0, 1, 2, \ldots$.

The foregoing Morse inequalities are valid also in cases where $\mathfrak{M}$ is a manifold with boundary $\partial \mathfrak{M}$, under the condition that the negative gradient flow of f points inward to $\mathfrak{M}$ at each point of $\partial \mathfrak{M}$.

After these preparations, we turn to the proofs.

PROOF OF THEOREM 1. 1° It is easily verified that the function f satisfies the Palais-Smale condition on H. In fact, for $\{x_n\}_1^\infty \subset H$, $df(x_n) \to \theta$ implies that $\forall \varepsilon > 0$, $\exists N = N(\varepsilon)$ such that for $n > N$,

$$\left|(Ax_n, x_n^\pm) + (dg(x_n), x_n^\pm)\right| \leqslant \varepsilon \|x_n^\pm\|,$$

where $x_n^\pm = P_\pm x$, $P_\pm$ is the orthogonal projection onto $H_\pm$. This implies that $\{\|x_n^\pm\|\}$, and then $\{(Ax_n, x_n)\}$ is bounded. Hence

(6) $\quad |g(P_0 x_n)| \leqslant |g(x_n) - g(P_0 x_n)| + |g(x_n)| \leqslant m\|x_n^+ + x_n^-\| + |g(x_n)|,$

where $m = \sup\{\|dg(x)\| | x \in H\}$. Suppose that along the sequence $\{x_n\}$, $f(x_n)$ is bounded. Then $|g(x_n)|$ must be bounded. Therefore $|g(P_0 x_n)|$, and then $\{\|P_0 x_n\|\}$, is bounded. Because of the compactness of dg and the finite dimensional condition on H_0, we have a convergent subsequence of $\{x_n\}_1^\infty$. This is the P.S. condition.

2° Denote $\varepsilon_\pm = \inf\{\|Ax_\pm\| \mid \|x_\pm\| = 1\}$, which is positive, and $R_+ = (m + 1)/\varepsilon_+$. Let $\mathfrak{M} = (H_+ \cap B_{R_+}) \oplus H_- \oplus H_0$. From

$$(df(x), x_+) = (Ax_+, x_+) - (dg(x), x_+) \geqslant \varepsilon_+ \|x_+\|^2 - m\|x_+\|$$

we know that f has no critical point outside $\mathfrak{M}$, and that $-df(x)$ points inward to $\mathfrak{M}$ on each point of $\partial \mathfrak{M}$.

Now we apply the critical point theory on the manifold $\mathfrak{M}$. Noticing that

$$-\tfrac{1}{2}\|A\|\,\|x_-\|^2 - m(\|x_-\| + R_+) + g(P_0 x)$$
$$\leqslant f(x) \leqslant \tfrac{1}{2}\|A\|R_+^2 - \tfrac{1}{2}\varepsilon_-\|x_-\|^2 + m(\|x_-\| + R_+) + g(P_0 x)$$

we obtain

$$f(x) \to -\infty \Leftrightarrow \|x_- + P_0 x\| \to \infty \quad \text{uniformly in } x_+,$$

i.e. $\forall T > 0,\ \exists a_1 < a_2 < -T,\ R_1 > R_2 > 0$, such that

$$(H_+ \cap B_{R_+}) \oplus ((H_- \oplus H_0) \setminus B_{R_1}) \subset f_{a_1}$$
$$\subset (H_+ \cap B_{R_+}) \oplus ((H_- \oplus H_0) \setminus B_{R_2}) \subset f_{a_2}.$$

3° Suppose that the conclusion of this theorem is not true, i.e. f has only critical points $\{p_i\}_1^k$; we choose $T > 0$ large enough such that $\{p_i\}_1^k \not\subset f_{-T}$. A strong deformation retract $\tau_1\colon f_{a_2} \to f_{a_1}$ can then be defined by the negative gradient flow. There is another strong deformation retract τ_2:

$$(H_+ \cap B_{R_+}) \oplus ((H_- \oplus H_0) \setminus B_{R_2}) \to (H_+ \cap B_{R_+}) \oplus ((H_- \oplus H_0) \setminus B_{R_1}),$$

defined by $\tau_2 = \eta(1, \cdot)$, where $(x_0 = P_0 x)$

$$\eta(t; x_+, x_- + x_0) = \begin{cases} (x_+, x_- + x_0), & \|x_- + x_0\| \geqslant R_1, \\ \left(x_+, \dfrac{x_- + x_0}{\|x_- + x_0\|}(tR_1 + (1-t)\|x_- + x_0\|)\right), & \|x_- + x_0\| \leqslant R_1. \end{cases}$$

Combining these two strong deformation retracts, $\tau = \tau_2 \circ \tau_1$, we obtain a strong deformation retract $\tau\colon f_{a_2} \to (H_+ \cap B_{R_+}) \oplus ((H_- \oplus H_0) \setminus B_{R_1})$. Then we have

$$H_q(\mathfrak{M}, f_{a_2}) \cong H_q\big(\mathfrak{M}, (H_+ \cap B_{R_+}) \oplus ((H_- \oplus H_0) \setminus B_{R_1})\big)$$
$$\cong H_q\big(H_- \oplus H_0, (H_- \oplus H_0) \setminus B_{R_1}\big)$$
$$\cong H_q\big((H_- \oplus H_0) \cap B_{R_1}, \partial((H_- \oplus H_0) \setminus B_{R_1})\big)$$
$$\cong \begin{cases} Q, & q = \gamma, \\ 0, & q \neq \gamma. \end{cases}$$

However, we have assumed ind $d^2 f(p_i) > \gamma$, according to the shifting theorem due to Gromoll–Meyer (for instance, cf. [**Ch3**]) we see

$$c_q(f, p_i) = 0 \quad \text{as } q \leqslant \gamma.$$

Now a contradiction follows from the γth Morse inequality.

REMARK 1. A special case where H_+ is of finite dimension and $k = 1$ was proved in Liu [**Li1**], by a homotopy argument.

PROOF OF THEOREM 2. The only difference between this proof and the proof of Theorem 1 is a weaker assumption on the asymptotics. The following lemma is applied to reduce Theorem 2 to Theorem 1.

LEMMA 1. *Suppose that $f \in C^2(H, \mathbf{R}^1)$ is asymptotically quadratic with an asymptotic operator A_∞. Then there are positive numbers $R_1 < R_2$, and a function $\rho \in C^\infty(\mathbf{R}^1_+, \mathbf{R}^1)$ satisfying*

$$\rho(t) = \begin{cases} 1, & 0 \leq t \leq R_1, \\ 0, & t > R_2, \end{cases}$$

such that the function $\tilde{f}(x) = \frac{1}{2}(A_\infty x, x) + \rho(\|x\|)g(x)$ satisfies

(7) $$\|d\tilde{f}(x)\| \geq 1 \quad \forall x \in B_{R_2} \setminus B_{R_1}.$$

PROOF. The function ρ is constructed in [Ch1] with a slight modification. According to Lemma 1, we see that
  (i) $\tilde{f}$ and f have the same critical set,
  (ii) $\tilde{f}$ has the same kind of form as in Theorem 1, except for the compactness of $d(\rho(\|x\|) \cdot g(x))$.

However, the compactness of the gradient vector field was used only in verifying the P.S. condition. Now suppose $d\tilde{f}(x_n) \to \theta \cdot$. Then we have $\{x_n\}_1^\infty \subset B_{R_1}$ except finite points, provided by (7) and the invertibility of A_∞. Then $\tilde{f}$ satisfies the P.S. condition and accordingly, the theorem is proved.

REMARK 2. Theorem 2 is an extension of a result due to Chang [Ch1], which implies the following corollary due to Amann and Zehnder [AZ1].

COROLLARY 1. *Suppose that $f \in C^2(\mathbf{R}^n, \mathbf{R}^1)$ is an asymptotically quadratic function with an invertible asymptotic matrix A_∞. If θ is a critical point of f with the condition*

$$\operatorname{ind}(A_\infty) \notin [m_-^0, m_-^0 + m_0^0],$$

where $m_-^0 = \operatorname{ind}(d^2 f(\theta))$ and $m_0^0 = \dim \ker d^2 f(\theta)$, then f has at least a nontrivial critical point.

Turning to Theorem 3, first we need a

DEFINITION. Let (X, Y) be a pair of topological spaces with $Y \subset X$. Two nontrivial homology classes $[z_1], [z_2] \in H_*(X, Y)$ satisfy the relation $[z_1] < [z_2]$ if there is an $\omega \in H^*(X)$ with $\dim \omega > 0$, such that $[z_1] = [z_2] \cap \omega$, where $\cap$ is the cap product.

Second, we need a lemma which is a modification of a result mentioned in Bott [Bo1].

LEMMA 2. *Suppose that M is a Finsler C^2 manifold, and that $f \in C^1(M, \mathbf{R}^1)$, satisfying the Palais-Smale condition. If f has only isolated critical points, if there are $[z_1], [z_2] \in H_*(f_b, f_a)$ with $[z_1] < [z_2]$, and if $a < b$ are not critical values, let*

$$c_i = \inf_{\tilde{z}_i \in [z_i]} \sup_{x \in |\tilde{z}_i|} f(x), \quad i = 1, 2,$$

where $|\tilde{z}|$ denotes the support of the singular chain $\tilde{z}$. Then c_1, c_2 are critical values of f with $a < c_1 < c_2 \leq b$.

PROOF. According to the minimax principle, c_1, c_2 are critical values with $a < c_1 \leq c_2 \leq b$. It remains to prove $c_1 < c_2$. Suppose not, i.e. $c = c_1 = c_2$, $\forall \varepsilon > 0$ we have $\tilde{z}_2 \in [z_2]$ with $|\tilde{z}_2| \subset f_{c+\varepsilon}$. Since $K_c = K \cap f^{-1}(c)$ consists of isolated critical points, we may choose two contractible neighborhoods $N \subset N'$ of K_c. Because of $\dim \omega > 0$, we can choose $\tilde{\omega} \in \omega$ such that $\tilde{\omega}$ is a cochain with support in $f_b \setminus N'$. Subdividing $\tilde{z}_2$: $\tilde{z}_2' + \tilde{z}_2''$ such that $|\tilde{z}_2'| \subset N'$ and $|\tilde{z}_2''| \subset f_b \setminus N$, we have

$$|\tilde{z}_1| \subset f_{c+\varepsilon} \setminus N$$

provided by $\tilde{z}_1 = \tilde{z}_2 \cap \tilde{\omega} = \tilde{z}_2'' \cap \tilde{\omega}$. According to the deformation lemma, there is a homeomorphism $\eta: f_{c+\varepsilon} \setminus N \to f_{c-\varepsilon}$. This implies $\eta|\tilde{z}_1| \subset f_{c-\varepsilon}$. But $\eta(\tilde{z}_1) \in |z_1|$, so we arrive at a contradiction.

PROOF OF THEOREM 3. Similarly to the proof of Theorem 1, the function f satisfies the P.S. condition on $H \times V^n$. Also similarly, we see that there is $B_{R_+} > 0$ such that f has no critical point outside $\mathfrak{M} := (H_+ \cap B_{R_+}) \times H_- \times V^n$, and that $f(x, v) \to -\infty$ as $\|x\| \to \infty$ in $\mathfrak{M}$, uniformly in $v \in V^n$. Hence, for $-a$ large enough, there is $R_- > 0$ such that

$$H_*(\mathfrak{M}, f_a) \cong H_*(\mathfrak{M}, (H_+ \cap B_{R_+}) \times (H_- \setminus B_{R_-}) \times V^n)$$
$$\cong H_*((H_- \cap B_{R_-}) \times V^n, \partial(H_- \cap B_{R_-}) \times V^n)$$
$$\cong H_*(H_- \cap B_{R_-}, \partial(H_- \cap B_{R_-})) \otimes H_*(V^n).$$

Therefore,

(8) $$H_q(\mathfrak{M}, f_a) \cong H_{q-\gamma}(V^n),$$

and similarly $H^*(\mathfrak{M}) \cong H^*(V^n)$.

First, suppose that f is a Morse function. Then f has at least

$$\sum_{q=0}^{\infty} \operatorname{rank} H_q(\mathfrak{M}, f_a) = \sum_{j=0}^{n} \operatorname{rank} H_j(V^n) = \sum_{j=0}^{n} \beta_j(V^n)$$

critical points, provided by the Morse inequalities.

Second, for general f, suppose that cuplength $(V^n) = l$, then $\exists \omega_1, \ldots, \omega_l \in H^*(\mathfrak{M})$, such that $\omega_1 \cup \cdots \cup \omega_l \neq 0$. Accordingly, we have $[z_1] \in H_*(\mathfrak{M}, f_a)$ such that the duality

$$[[z_1], \omega_1 \cup \cdots \cup \omega_l] \neq 0.$$

Let $[z_{j+1}] = [z_j] \cap \omega_j, j = 1, 2, \ldots, l$, which give $l + 1$ nontrivial singular chains in $H_*(\mathfrak{M}, f_a)$. According to Lemma 2, we obtain $l + 1$ distinct critical values.

2. Proofs of Theorems 4, 5, and 6. All these theorems are applications of the theorems in the previous section to concrete equations.

PROOF OF THEOREM 5. Define a functional

$$J(u) = \int_\Omega \left[\frac{(\nabla u)^2}{2} - \frac{\lambda u^2}{2} - \Phi(x, u) \right] dx$$

on the Hilbert space $H_0^1(\Omega)$. (J may not be C^2 in $H_0^1(\Omega)$, but is C^2 in $C_0^1(\overline{\Omega})$, the subspace of $C^1(\overline{\Omega})$, with null boundary condition, Theorem 1 is valid in this case, as remarked in Chang [**Ch3**].) Define

$$A = \text{id} - \lambda(-\Delta)^{-1} \quad \text{and} \quad g(u) = \int_\Omega \Phi(x, u(x))\, dx.$$

Obviously all the assumptions (H_1), (H_2), and (H_3) are satisfied. The first conclusion follows directly from Theorem 1 with $k = 0$. As to the second conclusion, applying an argument in [**AZ1**], we see

$$\text{ind } d^2 J(\theta) \geq \dim \bigoplus_{\lambda \leq \lambda_+} \ker(\text{id} - \lambda(-\Delta)^{-1}) > \dim \bigoplus_{\lambda \leq \hat\lambda} \ker(\text{id} - \lambda(-\Delta)^{-1}) \stackrel{\Delta}{=} \gamma.$$

Again, by Theorem 1, we obtain the conclusion.

REMARK 3. The conclusion (1) is a Landesman–Lazer type result (cf. Rabinowitz [**Ra1**]).

REMARK 4. Combining with a saddle point reduction due to Amann (cf. [**AZ1**]), the conclusion (2) can be extended as follows:

(2′) If $\phi(x, 0) = 0$, $\phi'_u(x, 0) < \lambda_- - \hat\lambda$ where $\lambda_- < \hat\lambda$ are consecutive eigenvalues in $\sigma(-\Delta)$, while the assumption (2) is replaced by

$$\int_\Omega \Phi\left(x, \sum_{i=1}^N t_i \psi_i(x)\right) dx \to +\infty \quad \text{as} \quad \sum_{i=1}^N (t_i)^2 \to \infty,$$

then the BVP (4) has a nontrivial solution if $|\phi'_u(x, u)|$ is bounded.

REMARK 5. A similar result for PSP of the nonlinear wave equation has been obtained in Liu [**Li1**].

PROOF OF THEOREM 6. At this moment, we assume that J satisfies the P.S. condition. According to a theorem in Chang [**Ch2**], there are two local minima of the functional

$$J(u) = \int_\Omega \left[\frac{(\nabla u)^2}{2} - \Phi(x, u(x))\right] dx, \quad u \in C_0^1(\overline{\Omega}).$$

Suppose $u_i \in [\underline{\varphi}_i, \overline{\varphi}_i]$, $i = 1, 2$, where [,] denotes the ordered interval in the function space $C_0^1(\overline{\Omega})$. It follows that

$$C_q(J, u_i) = \begin{cases} Q, & q = 0, \\ 0, & q \neq 0, \end{cases} \quad i = 1, 2.$$

An extended Mountain Pass Theorem can then be applied to obtain one more solution u_3 (cf. Chang [**Ch2**]). According to the homological characterization of the mountain pass point due to Tian [**Ti1**] and Hofer [**Ho1**], we may choose u_3 such that

$$C_q(J, u_3) = \begin{cases} Q, & q = 1, \\ 0, & q \neq 1. \end{cases}$$

However, the argument used in Theorem 3 shows that

$$H_q(\mathfrak{M}, J_a) \cong 0 \quad \text{for } q = 0, 1$$

if $-a > 0$ is large enough. If there were no other solutions of the BVP (5), then a contradiction would occur, provided by the first Morse inequality.

Returning to the verification of the P.S. condition, we only want to point out that J is an asymptotically quadratic function with invertible asymptotics.

REMARK 6. One easily extends Theorem 6 to cases where $\hat{\lambda} \in \sigma(-\Delta)$, under the Landesman–Lazer condition.

PROOF OF THEOREM 4. This proof is now a direct consequence of Theorem 3. Since the procedure is basically the same as in Conley and Zehnder [**CZ1**], we confine ourselves to sketch an outline of the proof.

First, since $H(t, z)$ is periodic, it follows that $\|H_{zz}(t, z)\|$ is bounded, say by m. Using a saddle point reduction, the functional

$$f(z) = \int_{S^1} \left\{ \tfrac{1}{2} \langle \dot{z}, Jz \rangle - H(t, z(t)) \right\} dt,$$

where $z \in H^1(S^1, \mathbf{R}^{2n})$ is reduced to a C^2-function $a(x, v)$ defined on a finite-dimensional space $\mathbf{R}^{4n[m]} \times T^{2n}$, according to the Fourier series expansion, where

$$\mathbf{R}^{4n[m]} \cong \bigoplus_{1 \leq |j| \leq [m]} M(j),$$

$M(j) = \operatorname{span}\{\cos 2\pi jt e_k + \sin 2\pi jt J e_k \mid k = 1, 2, \ldots, 2n\}$, $j = 0, 1, 2, \ldots$ and $\{e_k\}_1^{2n}$ is the basis of $\mathbf{R}^{2n}$. More precisely, $a(x, v)$ has the form

$$a(x, v) = \tfrac{1}{2}(Ax, x) + g(x, v),$$

where A is the diagonal matrix preserving $M(j)$ as invariant subspaces with eigenvalues $j, j = \pm 1, \pm 2, \ldots, \pm[m]$, and $g(x, v) \in C^2(\mathbf{R}^{4n[m]} \times T^{2n}; \mathbf{R}^1)$, having bounded compact gradient vector field.

Since cuplength $(T^{2n}) = 2n$ and

$$\operatorname{rank} H_q(T^{2n}) = C_q^{2n}, \quad q = 0, 1, \ldots, 2n,$$

we obtain the conclusion via Theorem 3 directly.

REMARK 7. Theorem 4 has many different proofs following that of Conley and Zehnder, cf. A. Weinstein [**We1**], Hofer, Berestycki, and Lasry.

REFERENCES

[AZ1] H. Amann and E. Zehnder, *Nontrivial solutions for a class of nonresonance problems and applications to nonlinear differential equations*, Ann. Scuola Norm. Sup. Pisa Cl. Ser. (4) **7** (1980), 539–603.

[Ar1] V. I. Arnold, *Mathematical methods of classical mechanics*, Springer, 1978.

[Bo1] R. Bott, *Lectures on Morse theory, old and new*, Bull. Amer. Math. Soc. (N.S.) **7** (1982), 331–358.

[Ch1] K. C. Chang, *Solutions of asymptotically linear operator equations via Morse theory*, Comm. Pure Appl. Math. **34** (1981), 693–712.

[Ch2] _____, *Variational method and sub- and super-solutions*, Scientia Sinica **26** (1983), 1256–1265.

[Ch3] _____, *Morse theory and its applications to PDE*, Séminaire Mathématiques Superieures, Univ. de Montreal (to appear).

[CZ1] C. C. Conley and E. Zehnder, *The Birkhoff-Lewis fixed point theorem and a conjecture of V. I. Arnold*, preprint.

[**Ho1**] H. Hofer, *A note on the topological degree at a critical point of mountain pass-type*, Proc. Amer. Math. Soc. (to appear).

[**Li1**] G. Q. Liu, Thesis, Beijing, 1983.

[**Ra1**] P. Rabinowitz, *Some minimax theorems and applications to nonlinear PDE*, Nonlinear Analysis, Academic Press, New York, 1978, pp. 161–177.

[**Ti1**] G. Tian, *On the mountain pass theorem*, Chinese Bulletin of Science **14** (1983), 833–835.

[**We1**] A. Weinstein, C^0 *perturbation theorems for symplectic fixed points and Lagrangian intersections*, preprint.

PEKING UNIVERSITY, PEOPLE'S REPUBLIC OF CHINA

On the Continuation Method and the Method of Monotone Iterations

PHILIPPE CLÉMENT

Abstract. We consider the following abstract problem. Let (E, P) be an ordered Banach space with cone P having a nonempty interior P°. Let $\lambda_1, \lambda_2 \in R$, $\lambda_1 < \lambda_2$, $a, b \in P$, such that $b - a \in P^\circ$. Let the operator $K: [\lambda_1, \lambda_2] \times [a, b] \to [a, b]$ be compact, strongly increasing with respect to the second variable for fixed $\lambda \in (\lambda_1, \lambda_2)$, and strictly increasing with respect to the first variable for fixed $u \in [a, b]$. Moreover, assume that a is the only fixed point of $K(\lambda_1, \cdot)$ and that b is the only fixed point of $K(\lambda_2, \cdot)$. Consider the equation
$$(*) \qquad u = K(\lambda, u).$$
Under the above assumptions we prove that any closed connected subset of solutions of $(*)$ in $[\lambda_1, \lambda_2] \times [a, b]$, which meets (λ_1, a) and (λ_2, b), contains the maximal and the minimal solutions of $(*)$, which are obtained by monotone iterations. Such a subset of solutions is shown to exist. Applications to a semilinear elliptic eigenvalue problem are studied.

1. Introduction. Let (E, P) be an ordered Banach space (see [1, p. 627]). For $a, b \in E$, $a < b$, $[a, b]$ denotes the order-interval $\{u \in E | a \leq u \leq b\}$.

Let $K: [a, b] \to [a, b]$ be a compact mapping: i.e., K is continuous, and the range of K is relatively compact in $[a, b]$.

Since $[a, b]$ is closed and convex in E, it is a consequence of Schauder's Theorem that K possesses at least one fixed point in $[a, b]$. If K is also increasing, i.e., $u < v$ implies $K(u) \leq K(v)$, then the existence of a minimal (resp. maximal) fixed point of K, which we denote by $\check{u}$ (resp. $\hat{u}$), is easily established by an iteration procedure [1, p. 639].

$$\check{u} = \lim_{n \to \infty} K^{(n)}(a), \qquad \hat{u} = \lim_{n \to \infty} K^{(n)}(b).$$

If we also assume that (E, P) is normal [1, p. 627] and that P has a nonempty interior P°, then $[a, b]$ is a bounded set of E, with nonempty interior $[a, b]^\circ$, provided that $a \ll b$, i.e., $b - a \in P^\circ$.

1980 *Mathematics Subject Classification.* Primary 47H07; Secondary 47H10, 35J65.

© 1986 American Mathematical Society
0082-0717/86 $1.00 + $.25 per page

The Leray–Schauder degree of $I - K$ relative to $[a, b]°$, $d(I - K, [a, b])$ (see for example [8]), is then well defined whenever K has no fixed point on the boundary of $[a, b]$—e.g., when K maps $[a, b]$ into its interior. Note that this implies

(1.0) $\qquad a \ll K(a) \quad \text{and} \quad K(b) \ll b.$

Conversely, if K satisfies (1.0), then a sufficient condition for K to map $[a, b]$ into its interior is that K is strongly increasing, i.e., $u < v$ implies $K(u) \ll K(v)$. We assume K is strongly increasing and satisfies (1.0). $d(I - K, [a, b]°)$ is easily computed by considering the compact homotopy

(1.1) $\qquad H(t, u) := u - (1 - t)c - tK(u), \quad t \in [0, 1], u \in [a, b],$

$$\text{with } c \in [a, b]°.$$

Then
$$d(I - K, [a, b]°) = d(H(1, \cdot), [a, b°]) = d(H(t, \cdot), [a, b°])$$
$$= d(H(0, \cdot), [a, b°]) = 1,$$

$t \in [0, 1]$, by noting that the solutions (t, u) of

(1.2) $\qquad H(t, u) = 0, \quad t \in [0, 1], u \in [a, b],$

satisfy $u \in [a, b]°$.

Since $d(H, (t, \cdot), [a, b]°)$ is constant $\neq 0$ for $t \in [0, 1]$, it follows from [3; 8, Corollaire 10, p. V–16], that there exists a subset C of solutions of (1.2), which is connected in $[0, 1] \times [a, b]$, equipped with the product topology, and meets $(0, c)$, and at least one point $(1, \bar{u})$, where $\bar{u}$ is a fixed point of K. A natural question arises: namely, which fixed points $\bar{u}$ can be "reached by the homotopy"? or, more precisely, which fixed points $\bar{u}$ of K belong to the component of $(0, c)$ in $[0, 1] \times [a, b]$? In particular, do $(1, \check{u})$, $(1, \hat{u})$ belong to C?

In §2 we prove that if $a \leqslant c \leqslant K(c) < \check{u}$, then the component C of $(0, c)$ in $[0, 1] \times [a, b]$ meets $[1, \check{u}]$, and that $(t, u) \in C$, $0 < t < 1$, implies $u \ll \check{u}$. Similarly, one could consider the homotopy $\tilde{H}(t, u) = u - (1 - t)K(u) - tc, t \in [0, 1]$, $u \in [a, b]$. Then, provided $\check{u} < K(c) \leqslant c \leqslant b$, $\tilde{C}$, the component of $(1, c)$, meets $(0, u)$, and $(t, u) \in \tilde{C}, 0 < t < 1$, implies $\hat{u} \ll u$.

If $\check{u} < \hat{u}$ and there exist $u_1, u_2 \in [a, b]$ satisfying

(1.3) $\qquad \check{u} < u_1 < u_2 < \hat{u}, \quad K(u_1) < u_1, u_2 < K(u_2),$

then Amann [2] proved that there exists a third fixed point $\bar{u}$, such that $\check{u} \ll \bar{u} \ll \hat{u}$, satisfying $\bar{u} \not\leqslant u_1$ and $u_2 \not\leqslant \bar{u}$. See [1, Theorem 14.2, p. 666]. Consider the homotopy

(1.4) $\qquad H(t, u) = \begin{cases} u - (1 - 2t)a - 2tK(u), & t \in [0, \frac{1}{2}], \\ u - 2(1 - t)K(u) - (2t - 1)b, & t \in [\frac{1}{2}, 1], \end{cases}$

$u \in [a, b]$, and define

$$S := \{ (t, u) \in [0, 1] \times [a, b] | H(t, u) = 0 \}.$$

Then if C_1 is the component of $(0, a)$ in S, we know, by the preceding, that C_1 contains $(\frac{1}{2}, \check{u})$; similarly, C_2, the component of $(1, b)$ in S, contains $(\frac{1}{2}, \hat{u})$; we prove in §2 that there exist a connected set C_3 in S, which meets $(\frac{1}{2}, \check{u})$ $(\frac{1}{2}, \hat{u})$, and at least a third point $(\frac{1}{2}, \bar{u})$, where $\bar{u}$ is a fixed point of K satisfying $\bar{u} \not\leqslant u_1$, $u_2 \not\leqslant \bar{u}$.

These results are special cases of Theorem 2.1, where a general homotopy

(1.5) $$u = K(\lambda, u), \quad \lambda \in [\lambda_1, \lambda_2], u \in [a, b],$$

is considered. There, K is a compact mapping which is strongly increasing "in u" for $\lambda \in (\lambda_1, \lambda_2)$ and strictly increasing "in λ" for $(\lambda, u) \in S$. Then if a (resp. b) is the only fixed point of $K(\lambda_1, \cdot)$ (resp. $K(\lambda_2, \cdot)$), we prove that C, the component of (λ_1, a) in $S := \{(\lambda_1, \lambda_2) \in [\lambda_1, \lambda_2] \times [a, b] | \ u = K(\lambda, u)\}$, meets (λ_2, b) and contains all maximal and minimal solutions of (1.5) for $\lambda \in (\lambda_1, \lambda_2)$. Thus Theorem 2.1 relates the solutions of (1.5), obtained by applying the continuation method of Leray–Schauder–Rabinowitz [8], and the solutions of (1.5), obtained by monotone iterations [1].

In §3 we give an application of the results of §2 to a semilinear elliptic problem

$$-\Delta u = \lambda g(\cdot, u) \quad \text{in } \Omega \subset R^N, \quad u = 0 \quad \text{in } \partial\Omega,$$

where we refine some results of [4, 5].

2. The main result. Throughout this section, (E, P) denotes an ordered Banach space with cone P having a nonempty interior P° (we do not assume (E, P) is normal), $a, b \in E$ such that $a \ll b$, and $\lambda_1, \lambda_2 \in R$ such that $\lambda_1 < \lambda_2$.

$K: [\lambda_1, \lambda_2] \times [a, b] \to [a, b]$ is continuous and has a relatively compact range in $[a, b]$ (where $[\lambda_1, \lambda_2] \times [a, b]$ is equipped with the product topology). S denotes the set of solutions of

(2.1) $$u = K(\lambda, u), \quad (\lambda, u) \in [\lambda_1, \lambda_2] \times [a, b].$$

THEOREM 2.1. *Let K, defined as above, satisfy the following assumptions*:

(i) *For each $\lambda \in (\lambda_1, \lambda_2)$, $K(\lambda, \cdot)$ is strongly increasing.*

(ii) *For each $(\lambda, u) \in S$, and $\mu_1 < \lambda < \mu_2$, $K(\mu_1, u) < K(\lambda, u) < K(\mu_2, u)$.*

(iii)$_{a,b}$ *a (resp. b) is the only fixed point of $K(\lambda_1, \cdot)$ (resp. $K(\lambda_2, \cdot)$).*
Then

(1) *C, the component of (λ_1, a) in S, meets (λ_2, b).*

(2) *Any closed connected set D in S which meets (λ_1, a) and (λ_2, b) contains all maximal $\hat{u}(\lambda)$ and minimal $\check{u}(\lambda)$ fixed points of $K(\lambda, \cdot)$, $\lambda \in (\lambda_1, \lambda_2)$. Moreover, for each $\lambda \in (\lambda_1, \lambda_2)$,*

(2.2) $$\check{u}(\lambda) = \sup_{\lambda_1 < \mu < \lambda} \check{u}(\mu) = \lim_{\mu \uparrow \lambda} \check{u}(\mu),$$

(2.3) $$\hat{u}(\lambda) = \inf_{\lambda < \mu < \lambda_2} \hat{u}(\mu) = \lim_{\mu \downarrow \lambda} \hat{u}(\mu).$$

(3) *If for some $\bar{\lambda} \in (\lambda_1, \lambda_2)$, $\check{u}(\bar{\lambda}) < \hat{u}(\bar{\lambda})$, and if there exist u_1, u_2 satisfying*

(2.4) $$\check{u}(\bar{\lambda}) < u_1 < u_2 < \hat{u}(\bar{\lambda}),$$

(2.5) $$u_1 > K(\bar{\lambda}, u_1), \quad u_2 < K(\bar{\lambda}, u_2),$$

then any closed connected set D in S which meets $[\lambda_1, a]$ and $[\lambda_2, b]$ contains a point $[\bar\lambda, \bar u]$ where $\bar u \not< u_1, u_2 \not< \bar u$.

Proof of Theorem 2.1.

LEMMA 1. *For $\lambda_1 \leq \mu_1 < \mu_2 \leq \lambda_2$ let A be a closed connected subset of $S \cap ([\mu_1, \mu_2] \times [a, b])$ which meets some $(\mu_1, u_1) \in S$ and $(\mu_2, u_2) \in S$. Let A_0 denote the component of $A \cap ([\mu_1, \mu_2] \times [a, b])$ which meets (μ_1, u_1). Then $\sup_{(\mu, u) \in A_0} \mu = \mu_2$.*

PROOF. Assume to the contrary that $\sup_{(\mu, u) \in A_0} \mu < \mu_2$. Then there exists $\lambda_0 \in (\mu_1, \mu_2)$ such that $A_0 \cap (\{\lambda_0\} \times [a, b]) = \varnothing$. Set $B := A \cap (\{\lambda_0\} \times [a, b])$. $B \neq \varnothing$ since $\mathrm{Proj}_{[\mu_1, \mu_2]} A = [\mu_1, \mu_2]$. If we define $F = A \cap ([\mu_1, \lambda_0] \times [a, b])$, then F is a compact metric space, and A_0, B are closed disjoint subsets of F. There exists no connected subset of F which meets both A_0 and B, otherwise $A_0 \supseteq G$, by using the maximality of A_0, and $A_0 \cap B \neq \varnothing$, a contradiction. Thus, by a lemma of point-set topology (see for instance [8, Lemma 1.9]), there are closed disjoint subsets of F, F_1 and F_2, such that $A_0 \subseteq F_1$, $B \subseteq F_2$, and $F_1 \cup F_2 = F$. Then define $F_3 = \{(\mu, u) \in A | \mu \geq \lambda_0\}$. $F_2 \cup F_3$ is closed, $F_1 \cap (F_2 \cup F_3) = \varnothing$, and $A = F_1 \cup (F_2 \cup F_3)$, contradicting the connectedness of A. Thus $\sup_{(\mu, u) \in A_0} \mu = \mu_2$.

Next we prove

LEMMA 2. *For $\lambda_1 \leq \mu_1 < \mu_2 \leq \lambda_2$ let A be a closed connected subset of $S \cap ([\mu_1, \mu_2] \times [a, b])$ which meets some $(\mu_1, u_1) \in S$ and $(\mu_2, u_2) \in S$. If $u_1 \leq \check u(\mu_2)$, then $(\mu_2, \check u(\mu_2)) \in A$. Similarly, if $u_2 \geq \hat u(\mu_1)$, then $(\mu_1, \hat u(\mu_1)) \in A$.*

PROOF. Define A_0 as in Lemma 1, and define $B = \{(\mu, u) \in A_0 | u \leq \check u(\mu_2)\}$. B is not empty and closed in A_0. We claim that B is open: that is, for each $(\mu, u) \in B$, $u \ll \check u(\mu_2)$. First we prove $u < \check u(\mu_2)$. Otherwise, $u = \check u(\mu_2)$ and (by (ii))

$$u = K(\mu, u) = K(\mu, \check u(\mu_2)) < K(\mu_2, \check u(\mu_2)) = \check u(\mu_2),$$

a contradiction. From (i) we get

$$u = K(\mu, u) \ll K(\mu, \check u(\mu_2)) < K(\mu_2, \check u(\mu_2)) = \check u(\mu_2).$$

Thus B is open. Since A_0 is connected, $B = A_0$. By using Lemma 1 and the compactness of K, we get a sequence $(\lambda_n, u_n) \in A_0$ such that $\sup_{n \to \infty} \lambda_n = \lim_{n \to \infty} \lambda_n = \mu_2$ and $\lim_{n \to \infty} u_n = u$. Since K is continuous, $u = K(\mu_2, u)$. Moreover, by the preceding, $u \leq \check u(\mu_2)$. From the minimality of $\check u(\mu_2)$ we have $u = \check u(\mu_2)$ and $(\mu_2, \check u(\mu_2)) \in A$.

Next we prove assertion (1). From (iii)$_{a,b}$ and the compactness of K it follows that $\lim_{\lambda \downarrow \lambda_1} \hat u(\lambda) = a$ and $\lim_{\lambda \uparrow \lambda_2} \check u(\lambda) = b$. Thus, since $b - a \in P^\circ$, there are $\alpha \in (0, \tfrac{1}{2})$ and $\varepsilon \in (0, \lambda_2 - \lambda_1)$ such that $\hat u(\lambda) < a + \alpha(b - a)$ for $\lambda \in (\lambda_1, \lambda_1 + \varepsilon)$ and $\check u(\lambda) > b - \alpha(b - a)$ for $\lambda \in (\lambda_2 - \varepsilon, \lambda_2)$. Hence, $\hat u(\lambda) \ll \check u(\mu)$ for $\lambda \in (\lambda_1, \lambda_1 + \varepsilon)$ and $\mu \in (\lambda_2 - \varepsilon, \lambda_2)$. We claim that, for each $\lambda \in (\lambda_1, \lambda_1 + \varepsilon)$ and

$\mu \in (\lambda_2 - \varepsilon, \lambda_2)$, there is a maximal connected set $C_{\lambda,\mu}$ in $S \cap ([\lambda, \mu] \times [a, b])$ which meets $(\lambda, \hat{u}(\lambda))$ and $(\mu, \check{u}(\mu))$. For $t \in [\lambda, \mu]$, define $O_t := [K(t, a), K(t, b)]^\circ$. Note that $K(t, a) \ll \check{u}(t) \leq \hat{u}(t) \ll K(t, b)$ for $t \in [\lambda, \mu]$. Then $O := \bigcup_{t \in [\lambda,\mu]} \{t\} \times O_t$ is an open subset of $[\lambda, \mu] \times [a, b]$ containing no solution of (2.1) on its boundary (as a subset of $[\lambda, \mu] \times [a, b]$). We know that $d(I - K(t_1, \cdot), O_t) = 1$, $t \in [\lambda, \mu]$. By [3; 8, Corollaire 10, v–16] there exists a component $C_{\lambda,\mu}$ of $S \cap ([\lambda, \mu] \times [a, b])$ which meets $\{\lambda\} \times S_\lambda$ and $\{\mu\} \times S_\mu$, where $S_t := \{u \in [a, b] | (t, u) \in S\}$, $t \in [\lambda_1, \lambda_2]$. It follows from Lemma 2 that $(\lambda, \hat{u}(\lambda))$ and $(\mu, \check{u}(\mu))$ belong to $C_{\lambda,\mu}$. We conclude the proof of assertion (i) by considering the connected set $\bigcup_{\lambda \in (\lambda_1, \lambda_1 + \varepsilon); \mu \in (\lambda_2 - \varepsilon, \lambda_2)} C_{\lambda,\mu}$. Its closure meets (λ_1, a) and (λ_2, b). Hence, the component of S which contains (λ_1, a) meets (λ_2, b).

Next we prove assertion (2). For $\lambda \in (\lambda_1, \lambda_2)$ let D_λ denote the component of $D \cap ([\lambda_1, \lambda) \times [a, b])$ which contains (λ_1, a). Then by Lemma 1, $\sup_{(\mu, u) \in D_\lambda} \mu = \lambda$. As above we can choose a sequence $(\lambda_n, u_n) \in D_\lambda$ such that $\lambda_n \to \lambda$ and $u_n \to \check{u}(\lambda)$. Thus $(\lambda, \check{u}(\lambda)) \in D$. By considering $D = C$, the component of S which contains (λ_1, a) we can choose $u_n = \check{u}(\lambda_n)$. Thus $\check{u}(\lambda) = \sup_{\mu < \lambda} \check{u}(\mu) = \lim_{\mu \uparrow \lambda} \check{u}(\mu)$.

For the proof of assertion (3) we refer the reader to [6].

3. An example. We consider the nonlinear eigenvalue problem

(P) $\qquad -\Delta u = \lambda g(\cdot, u) \quad \text{in } \Omega, \qquad u = 0 \quad \text{on } \Gamma = \partial\Omega,$

where Ω is a bounded domain of R^N with smooth boundary Γ.

$g: \overline{\Omega} \times R \to R$ is continuous, and g_u exists and is continuous. A solution of (P) is a pair $(\lambda, \mu) \in R \times W^{2,P}(\Omega)$ with $p > N$ satisfying (P). Let $\bar{u}$ be a positive, superharmonic, bounded, lower-semicontinuous function on Ω such that $g(x, \bar{u}(x)) = 0$ a.e. in Ω and such that $g(x, u) > 0$ for $0 \leq u \leq \bar{u}(x)$, $x \in \Omega$. It was shown in [4, 5] that if S denotes the set of solutions of (P) in $R^+ \times W^{2,P}$ equipped with the $R \times C^1$ topology and if C is the component of S containing $(0, 0)$, then C satisfies

(1) $(\lambda, u) \in C \setminus (0, 0)$ implies u is positive, superharmonic, and $u(x) < \bar{u}(x)$, $x \in \Omega$.

(2) for every $\lambda > 0$, C has a minimal solution $\bar{u}(\lambda)$;

(3) $\lim_{\lambda \to \infty} |\bar{u}(\lambda) - \bar{u}|_{L^P} = 0$, $P < \infty$.

For $\lambda > 0$ we say that $\check{u}(\lambda)$ is the minimal (resp. maximal) solution of (P) in $[0, \bar{u}]$ if $(\lambda, \check{u}(\lambda))$ (resp. $(\lambda, \hat{u}(\lambda))$) is a solution of (P), and for any solution (λ, u) satisfying $0 \leq u(x) \leq \bar{u}(x)$, $x \in \Omega$, $u(x) \geq \check{u}(\lambda)(x)$, $x \in \Omega$ (resp. $u(x) \leq \hat{u}(\lambda)(x)$, $x \in \Omega$).

The aim of this section is to prove

THEOREM 3.1. *C, defined above, contains, for each $\lambda > 0$, the minimal and the maximal solutions in $[0, \bar{u}]$.*

PROOF. (a) C contains the minimal solution $\check{u}(\lambda)$ for each $\lambda > 0$. Let $\bar{\lambda} > 0$ and $w(\bar{\lambda}) \geq 0$ be such that

(3.1) $\qquad w(\bar{\lambda}) + \lambda g_u(x, u) \geq 0 \quad \text{for } \lambda \in [0, \bar{\lambda}],$

and for $0 \leq u \leq \bar{u}(x), x \in \Omega$. Then we rewrite (P) as

(P') $\qquad \begin{aligned} -\Delta u + w(\bar{\lambda})u &= w(\bar{\lambda})u + \lambda g(\cdot, u) \quad \text{in } \Omega, \\ u &= 0 \quad \text{on } \Gamma. \end{aligned}$

(P') is then equivalent to

(P'') $\qquad u = K(\lambda, u), \quad u \in E := \{ v \in C^1(\bar{\Omega}) | v = 0 \text{ on } \Gamma \},$

equipped with the C^1 norm, and

(3.2) $\qquad K(\lambda, u)(x) = \int_\Omega G_w(x, y)(x(\bar{\lambda})u(y) + \lambda g(y, u)) \, dy,$

where $G_w(\cdot, \cdot)$ denotes the Green function relative to $-\Delta u + w(\bar{\lambda})u$ on Ω with Dirichlet boundary conditions.

Note that by (3.2), K is defined on $R \times L^\infty(\Omega)$ and takes its values in E. In E we introduce the cone

$$P := \{ u \in E | u(x) \geq 0, x \in \Omega \}$$

of positive solutions; it is standard that P has a nonempty interior P°. Next we define $\tilde{u}(\lambda) := K(\bar{\lambda}, \bar{u})$. $\tilde{u}(\lambda)$ satisfies

$$\begin{aligned} -\Delta u(\bar{\lambda}) + w(\bar{\lambda})\bar{u}(\lambda) &= w(\bar{\lambda})\bar{u} \quad \text{in } \Omega, \\ u(\bar{\lambda}) &= 0 \quad \text{on } \Gamma. \end{aligned}$$

Then $u(\bar{\lambda}) \in P^\circ$ and $\tilde{u}(\lambda) < \bar{u}$ in Ω. Our choice of $w(\bar{\lambda})$ then implies $K(\bar{\lambda}, \tilde{u}(\bar{\lambda})) < \tilde{u}(\bar{\lambda})$. Since $K(\bar{\lambda}, \cdot)$ is increasing in u, $K(\bar{\lambda}, \cdot): [0, \tilde{u}(\bar{\lambda})] \to [0, \tilde{u}(\bar{\lambda})]$, and thus (P) has a minimal solution $\check{u}(\bar{\lambda})$. Note that $\check{u}(\bar{\lambda}) \in P^\circ$. By our choice of $w(\bar{\lambda})$, $u \to K(\lambda, u)$ is strongly increasing for $\lambda \in (0, \bar{\lambda})$, increasing for $\lambda = 0$, and $K(\lambda, 0) > 0$ for $\lambda \in (0, \bar{\lambda})$ and $K(\lambda, \check{u}(\bar{\lambda})) < \check{u}(\bar{\lambda})$ for $\lambda \in (0, \bar{\lambda})$. Moreover $K(\cdot, u)$ is strictly increasing in λ for each $u \in [0, u(\bar{\lambda})]$. Thus, $K: [0, \bar{\lambda}] \times [0, \check{u}(\bar{\lambda})] \to [0, \check{u}(\bar{\lambda})]$ satisfies the assumptions of Theorem 2.1. There exists a connected set D of solutions in $R \times C^1$ which meets $(0, 0)$ and $(\bar{\lambda}, \check{u}(\bar{\lambda}))$. Since $D \subset C$, C contains $(\bar{\lambda}, \check{u}(\bar{\lambda}))$ and, obviously, $\check{u}(\bar{\lambda}) = \bar{u}(\bar{\lambda})$.

(b) C contains the maximal solution $\check{u}(\lambda)$ in $[0, \bar{u}]$ for each $\lambda > 0$. Let $\bar{\lambda} > 0$ and $w(\bar{\lambda}) \geq 0$ be chosen as in (a). We set $S_{\bar{\lambda}} := \{ u \in [0, \bar{u}] | (\bar{\lambda}, u) \text{ is a solution of (P)} \}$. Define K as in (a); we know that $u \in S_{\bar{\lambda}}$ implies $u < K(\bar{\lambda}, \bar{u}) = \tilde{u}(\bar{\lambda})$, and $u > 0$. Thus $S_{\bar{\lambda}} \in [0, \tilde{u}(\bar{\lambda})]^\circ = [0, K(\bar{\lambda}, \bar{u})]^\circ$. Define $\lambda_n := \bar{\lambda} + n, n \in N$. Then, since $\check{u}(\lambda_n) < \check{u}(\lambda_{n+1}), n \in N$ (easily verified), we have $K(\bar{\lambda}, \check{u}(\lambda_n)) < K(\bar{\lambda}, \check{u}(\lambda_{n+1}))$. Moreover, $K(\bar{\lambda}, \bar{u}) = \lim_{n \to \infty} K(\bar{\lambda}, \check{u}(\lambda_n))$ follows from statement (3) before Theorem 3.1. We claim that $[0, K(\bar{\lambda}, \bar{u})]^\circ = \bigcup_{n=1}^\infty [0, K(\bar{\lambda}, \check{u}(\lambda_n))]^\circ$. Indeed, let $v \in [0, K(\bar{\lambda}, \bar{u})]^\circ$. By definition, there is

$\alpha > 0$ such that $K(\bar{\lambda}, \bar{u}) - v \geq \alpha e$, where e is an element of P°. Moreover, $K(\bar{\lambda}, \bar{u}) = \lim_{n\to\infty} K(\bar{\lambda}, \check{u}(\lambda_n))$ in C^1 implies the existence of a sequence $\{\beta_n\}$ with $\lim_{n\to\infty} \beta_n = 0$ such that

$$K(\bar{\lambda}, \bar{u}) - K(\bar{\lambda}, \check{u}(\lambda_n)) \leq \beta_n e, \quad n \in N.$$

Thus, $K(\bar{\lambda}, \check{u}(\lambda_n)) - v \geq (\alpha - \beta_n)e$, $n \in N$, and there are $N \in \mathbb{N}$ and $c > 0$ such that

$$K(\bar{\lambda}, \check{u}(\lambda_N)) - v \geq ce.$$

Thus, $v \in [0, K(\bar{\lambda}, \check{u}(\lambda_N))]^\circ \subset \bigcup_{n=1}^\infty [0, K(\bar{\lambda}, \check{u}(\lambda_n))]^\circ$. Hence, $S_{\bar{\lambda}} \subset \bigcup_{n=1}^\infty [0, K(\bar{\lambda}, \check{u}(\lambda_n))]^\circ$. Next we observe that $S_{\bar{\lambda}}$ is compact in C^1. Hence, there is $m \in N$ such that

$$S_{\bar{\lambda}} \subset \bigcup_{n=1}^m [0, K(\bar{\lambda}, \check{u}(\lambda_n))]^\circ \subset [0, K(\bar{\lambda}, \check{u}(\lambda_m))].$$

Note that $\bar{v} := K(\bar{\lambda}, \check{u}(\lambda_m))$ satisfies

$$-\Delta \bar{v} + w(\bar{\lambda})\bar{v} = w(\bar{\lambda})\check{u}(\lambda_m) + \bar{\lambda}g(\cdot, \check{u}(\lambda_m)) \quad \text{in } \Omega,$$
$$\bar{v} = 0 \quad \text{on } \Gamma,$$

and $\check{u}(\lambda_m)$ satisfies

$$-\Delta \check{u}(\lambda_m) + w(\bar{\lambda})\check{u}(\lambda_m) = w(\bar{\lambda})\check{u}(\lambda_m) + \lambda_m g(\cdot, \check{u}(\lambda_m)) \quad \text{in } \Omega,$$
$$\check{u}(\lambda_m) = 0 \quad \text{on } \Gamma.$$

Since $\lambda_m > \bar{\lambda}$ and $g(\cdot, \check{u}(\lambda_m)) > 0$, we have

$$-\Delta(\bar{v} - \check{u}(\lambda_m)) + w(\bar{\lambda})(\bar{v} - \check{u}(\lambda_m)) \leq 0 \quad \text{in } \Omega,$$
$$\bar{v} - \check{u}(\lambda_m) = 0 \quad \text{on } \Gamma.$$

Thus, $\bar{v} < \check{u}(\lambda_m)$ and $S_{\bar{\lambda}} \subset [0, \check{u}(\lambda_m)]$. Next, choosing $w(\lambda_m) \geq 0$ such that $w(\lambda_m) + \lambda g_u(x, u) > 0$ for $\lambda \in [0, \lambda_m]$, $0 \leq u \leq \bar{u}(x)$, $x \in \Omega$, one defines K as in (a) and verifies that with this choice of w, K satisfies the assumptions of Theorem 2.1 on $[0, \lambda_m] \times [0, \check{u}(\lambda_m)]$. Then there is a connected set D of solutions of (P) in $R \times C^1$ which contains the maximal solution $\hat{u}(\bar{\lambda})$ in $[0, \check{u}(\lambda_m)]$. But since $S_{\bar{\lambda}} \subset [0, \check{u}(\lambda_m)]$, $\hat{u}(\bar{\lambda})$ is the maximal solution in $[0, \bar{u}]$. Since $D \subset C$, C contains the maximal solution $\hat{u}(\lambda)$ of (P) in $[0, \bar{u}]$. This completes the proof of Theorem 3.1.

REMARK 1. It is also a consequence of the proof that if we denote by C_λ the component of solutions of (P) in $[0, \lambda) \times C^1$ which contains $(0, 0)$, then $C = \bigcup_{\lambda > 0} C_\lambda$.

REMARK 2. In the "bifurcation case", i.e., when g satisfies $g(x, 0) = 0$, $x \in \Omega$, but $g_u(x, 0) > 0$, $x \in \Omega$, a similar analysis shows that $C = \bigcup_{\lambda > \lambda_1} C_\lambda$ when C is the component of positive solutions "emanating" from $(\lambda_1, 0)$, the bifurcation point. Then for $\lambda > \lambda_1$, C contains all maximal solutions in $[0, \bar{u}]$. Note that in this case the minimal solution in $[0, \bar{u}]$ is 0, but it is shown in [4, 5] that C possesses a minimal solution for each $\lambda > \lambda_1$.

REMARK 3. By looking at the proofs it appears that the conclusions of Theorem 2.1. remain true if we replace assumption (i) by "for each $\lambda \in (\lambda_1, \lambda_2)$, $K(\lambda, \circ)$ is increasing ($u \leqslant v$ implies $K(\lambda, u) \leqslant K(\lambda, v)$)", and assumption (ii) by "there is $n \in \mathbb{N}$ such that for each $(\lambda, u) \in S$, and $\mu_1 < \lambda < \mu_2$, $K^{(n)}(\mu_1, u) \ll K^{(n)}(\lambda, u) \ll K^{(n)}(\mu_2, u)$". This version is actually stronger than Theorem 2.1 (here $n = 2$).

References

1. H. Amann, *Fixed point equations and nonlinear eigenvalue problems in ordered Banach spaces*, SIAM Rev. **18** (1976), 620–709.

2. _____, *On the number of solutions of nonlinear equations in ordered Banach spaces*, J. Funct. Anal. **11** (1972), 348–384.

3. F. E. Browder, *On continuity of fixed points under deformations of continuous mappings*, Summa Brasil. Math. **4** (1960), 253–293.

4. Ph. Clément and L. A. Peletier, *Sur les solutions superharmoniques de problèmes aux valeurs propres elliptiques*, C.R. Acad. Sci. Paris **293** (1981).

5. _____, *On positive superharmonic solutions to semi-linear elliptic eigenvalue problems*, J. Math. Anal. Appl. **100** (1984), 561–582.

6. Ph. Clément, *Some remarks on the continuation method of Leray–Schauder–Rabinowitz and the method of monotone iterations*, MRC Tech. Summary Rep. #2454, Univ. of Wisconsin, Madison, 1982.

7. P. H. Rabinowitz, *Some aspects of nonlinear eigenvalue problems*, Rocky Mountain J. Math. **3** (1973), 161–202.

8. _____, *Théorie du degré topologique et applications à des problèmes aux limites non linéaires*, Université Paris VI et CNRS, 1975, (*notes rédigés par H. Berestycki*).

9. _____, *Pairs of positive solutions of nonlinear elliptic partial differential equations*, Indiana Univ. Math. J. **23** (1973), 173–186.

TECHNISCHE HOGESCHOOL DELFT, THE NETHERLANDS

Existence Theorems for Superlinear Elliptic Dirichlet Problems in Exterior Domains

CHARLES V. COFFMAN[1] AND MOSHE M. MARCUS

1. Introduction. We are concerned with the existence of positive solutions to the Dirichlet problem

(1.1) $\qquad \Delta u - u + f(u) = 0 \quad \text{in } \Omega, \qquad u = 0 \quad \text{on } \partial\Omega,$

on an exterior domain Ω in $\mathbb{R}^3$ and where f is a "superlinear" nonlinearity which vanishes to higher than first order at $u = 0$. Whatever the domain Ω, we shall, throughout the paper, always understand a solution to (1.1) to be a function belonging to $W_0^{1,2}(\Omega)$. When Ω is a bounded domain, this existence question is substantially simpler than in the case of unbounded domains because of the compactness of the imbedding $W_0^{1,2}(\Omega) \subset L_p(\Omega)$, $2 \le p < 6$. In the case $\Omega = \mathbb{R}^3$, where (1.1) with various specific nonlinearities arises in a variety of physical problems [5, 11, 14, 17], the first rigorous proof of the existence theorem was given by Nehari [13] for $f(u) = |u|^k$, $1 < k < 5$. Subsequently, this result was extended to more general f (allowed also to depend on $|x|$) by Ryder [15]. Since the appearance of the papers of Nehari and Ryder, and particularly in recent years, there has been considerable interest in the problem (1.1) in $\mathbb{R}^N$. We make no attempt to include a complete bibliography but refer the reader to [1] where the most general results, as well as an extensive bibliography, can be found.

Nehari and Ryder obtained radially symmetric solutions by ordinary differential equation methods. An understanding of the role that is played by radial symmetry is to be found in the observation [2, 9, 16] that the subspace of $W^{1,2}(\mathbb{R}^3)$ that consists of radially symmetric functions imbeds compactly in $L_p(\mathbb{R}^3)$, $2 < p < 6$. Given this fact one can formulate an existence proof that does not depend on reduction of the problem to an ordinary differential equation

---

1980 *Mathematics Subject Classification.* Primary 35J65; Secondary 35J20.
[1] This research was supported in part by NSF Grant MCS 80-02851.

© 1986 American Mathematical Society
0082-0717/86 $1.00 + $.25 per page

[16]. One can also prove the existence of positive solutions to (1.1) when Ω is the exterior of a ball [4]. These considerations give rise to the following questions: Does (1.1) have nonradially symmetric solutions in $\mathbb{R}^3$? Does (1.1) have nontrivial solutions in nonradially-symmetric unbounded domains Ω, specifically in exterior domains? At least so far as positive solutions are concerned, the first question is answered negatively in [7]. Negative results in connection with the second question are given in [4] where it is shown that, if there is a vector that has its projection along the exterior normal to $\partial\Omega$ nonnegative at every point of $\partial\Omega$, then (1.1) does not admit nontrivial solutions. This leaves open, however, the question of existence for exterior domains.

Here we give two existence theorems for positive solutions on nonradially-symmetric exterior domains. The first result concerns the case where Ω has a substantial degree of symmetry, specifically, where there exists a group G of orthogonal transformations of $\mathbb{R}^3$ *having only the origin as a common fixed point and leaving Ω invariant*. When this is the case, then, provided $\partial\Omega$ has sufficiently small diameter (1.1), with suitably restricted f, will admit a positive solution. Secondly, we prove the existence of a positive solution to problem (1.1) in the case where Ω is nearly spherical, i.e. where $\partial\Omega$ is contained in a sufficiently thin ring about $|x| = R$. This result holds for all positive R outside an exceptional set which is at most countable.

We have limited our attention here to the case of dimension 3 because, in some respects, the presentation is simpler in this case. However, the main results—with some obvious modifications—remain valid for domains in $\mathbb{R}^N$ with $N \geq 2$.

We remark that there appears to be some similarity between the methods used in the proof of our Theorem 1 and the "concentration compactness" method of P.-L. Lions [10].

Finally, we wish to express our gratitude to Professor E. Dancer of the University of New England and Professor John Toland of the University of Bath for several very helpful suggestions in connection with this work.

2. Preliminaries. In what follows, Ω will denote a domain in $\mathbb{R}^3$, mostly, but not always an exterior domain (i.e., the exterior of a compact domain). We shall always assume that $\partial\Omega$ is sufficiently smooth so that the problem

$$-\Delta u + u = h \quad \text{in } \Omega, \quad u = 0 \quad \text{on } \partial\Omega$$

has a Green's function in the usual sense.

Concerning the nonlinear term f in (1.1) we shall make the following assumptions:

(2.1) $\quad f \in C^1(\mathbb{R}), \quad f(0) = 0 \quad \text{and} \quad f(u) > 0 \quad \text{for } u > 0.$

(2.2) $\quad uf(u) > (2 + \varepsilon)F(u) \quad \text{for some } \varepsilon > 0 \text{ and all } u > 0,$

where F is defined by

(2.3) $$F(u) = \int_0^u f(t)\, dt.$$

Condition (2.2) is a condition of "superlinearity". It implies that the function $u \to u^{-2-\varepsilon}F(u)$ is monotone increasing on $(0, \infty)$. From this fact we deduce

(2.4) $\quad f'(0) = 0 \quad$ and $\quad f(u) > (2 + \varepsilon)F(1)u^{1+\varepsilon} \quad$ for $u > 1$.

Let $\alpha > 0$ be such that $f(\alpha)/\alpha = 1$ and $f(u)/u < 1$ for $u \in (0, \alpha)$. Setting $\tilde{f}(v) = f(\alpha v)/\alpha$ we observe that (1.1) is equivalent to
$$\Delta v - v + \tilde{f}(v) = 0 \quad \text{(with } v = u/\alpha\text{)}.$$
Therefore, without loss of generality, we shall assume

(2.5) $\quad\quad\quad f(1) = 1 \quad$ and $\quad f(u) < u \quad$ for $u \in (0, 1)$.

In some cases we shall require a condition of superlinearity stronger than (2.2), namely

(2.6) $\quad\quad uf'(u) > (1 + \varepsilon)f(u) \quad$ for some $\varepsilon > 0$ and all $u > 0$.

(The condition assumed by Ryder in [15] reduces to (2.6) when $f \in C^1$.) Condition (2.6) implies that the function $u \to u^{-1-\varepsilon}f(u)$ is monotone increasing in $(0, \infty)$ and, consequently (in view of (2.5)),

(2.7) $\quad\quad f(u) < u^{1+\varepsilon} \quad$ for $u < 1 \quad$ and $\quad f(u) > u^{1+\varepsilon} \quad$ for $u > 1$.

In what follows we shall also have occasion to assume that f satisfies a polynomial growth condition of the form

(2.8) $\quad\quad\quad\quad f(u) \leq au + bu^\sigma, \quad u > 0,$

where $1 < \sigma < 5$. In view of (2.5) we may and shall assume that $0 < a < 1$.

Our notation for function spaces will follow that of [8]. In what follows, when $\Omega \subset \Omega'$ we shall regard $W_0^{1,2}(\Omega)$ as a subspace of $W_0^{1,2}(\Omega')$ using the natural isometric immersion.

Let Ω be a domain in $\mathbb{R}^3$ and let $\mathscr{K}$ be a cone of functions in $W_0^{1,2}(\Omega)$ containing a nonzero element. (Here we mean a "cone" in the widest sense; the only condition required: $u \in \mathscr{K} \Rightarrow \alpha u \in \mathscr{K} \,\forall \alpha > 0$.) Associated with (1.1), we consider the variational problem (see Nehari [12])

(2.9) $\quad\quad\quad \text{minimize } H(u) := \int_\Omega \left[\frac{1}{2}uf(u) - F(u)\right] dx,$

subject to

(2.10)
(a) $u \in \mathscr{K} \setminus \{0\}$,

(b) $\int_\Omega \left(|\nabla u|^2 + u^2\right) dx \leq \int_\Omega uf(u)\, dx.$

From (2.8),
$$\int_\Omega uf(u)\, dx \leq a\|u\|^2_{W^{1,2}(\Omega)} + \text{const.}\|u\|^{\sigma+1}_{W^{1,2}(\Omega)}, \quad \forall u \in W^{1,2}(\Omega).$$
Hence, there exists $\delta_0 > 0$ such that $\|u\|_{W^{1,2}(\Omega)} > \delta_0$ whenever u is in $W^{1,2}(\Omega)$ and satisfies (2.10)(b). In view of (2.4) it is clear that the set of functions u satisfying (2.10) is not empty. Further, if u satisfies (2.10) then (by (2.2))

(2.11) $\quad\quad 2H(u) \geq \varepsilon(2 + \varepsilon)^{-1} \int_\Omega uf(u)\, dx \geq \varepsilon(2 + \varepsilon)^{-1} \|u\|^2_{W^{1,2}(\Omega)}.$

Therefore, the infimum of H subject to (2.10) is positive.

Next we shall say that $\mathscr{K}$ is an *acceptable* cone with respect to (1.1) if it satisfies the following additional conditions:

(i) If $u \in \mathscr{K}$ and v is the solution of

(2.12) $\qquad -\Delta v + v = f(u)$ in Ω, $\qquad v = 0$ on $\partial \Omega$,

then $v \in \mathscr{K}$.

(ii) $\mathscr{K}$ is closed, and bounded subsets of $\mathscr{K}$ are precompact in $L_p(\Omega)$, $2 < p < 6$.

PROPOSITION 2.1. *Let $\mathscr{K}$ be an acceptable cone with respect to (1.1). Suppose that f satisfies (2.1), (2.6) and (2.8). Then the variational problem (2.9), (2.10) is solvable in $\mathscr{K}$. A solution of the variational problem is also a solution of (1.1).*

The main ingredient of the proof is the following observation: If $u \in \mathscr{K}$ and u and v are related by (2.12), then one can choose $\alpha > 0$ so that, setting $w = \alpha v$,

$$\int_\Omega \left(|\nabla w|^2 + w^2 \right) dx = \int_\Omega w f(w) \, dx.$$

With w defined in this manner, one can show that $H(w) \leq H(u)$ with equality if and only if u is a solution to (1.1) (see [**12**]).

We mention two important examples of acceptable cones.

EXAMPLE 1. If Ω is a bounded domain with Lipschitz boundary, then $\mathscr{K} = W_0^{1,2}(\Omega)$ is an acceptable cone. In this case we shall denote the minimum in problem (2.9), (2.10) by $h(\Omega)$.

EXAMPLE 2. If Ω is the exterior of a ball

$$\Omega = \left\{ x \in \mathbb{R}^3 : |x| > \rho \right\} \qquad (\rho > 0),$$

then the cone $\mathscr{K}$ of nonnegative, radially-symmetric functions in $W_0^{1,2}(\Omega)$ is an acceptable cone. (The same is true if $\Omega = \mathbb{R}^3$.) In this case we shall denote the minimum in problem (2.9), (2.10) by h_ρ. (When $\Omega = \mathbb{R}^3$, the minimum will be denoted by h_0.)

In one of our main results we shall be concerned with regions Ω which are invariant with respect to a group G of orthogonal transformations in $\mathbb{R}^3$. If G is such a group, we denote

$$\nu(G) = \inf_{x \neq 0} \operatorname{card}\{ g(x) : g \in G \}.$$

We shall be interested in those groups G for which $\nu(G) > 1$. A finite group G will have this property if and only if it is of one of the following two types (see [**3**]):

(1) G contains a rotation through an angle $2\pi/n$ (n an integer) about an axis l and at least one of the following two transformations: the reflection with respect to the plane through the origin, orthogonal to l; a rotation through the angle π about some axis in the above-mentioned plane.

(2) G is the symmetry group of one of the regular polyhedra. (In this case reflexions may be included, but it is not necessary.)

If G is a group of the type (1) then $\nu(G) = 2$. If G is a group of the type (2) then either $\nu(G) = 4$ or $\nu(G) = 6$ or $\nu(G) = 12$ according as G is the tetrahedral, octahedral or icosahedral group. (The value of $\nu(G)$ is not affected by the presence or absence of reflexions in this case.)

When Ω is a domain invariant with respect to a group G we shall denote by $\mathcal{K}(\Omega, G)$ the cone of nonnegative, G-invariant functions in $W_0^{1,2}(\Omega)$. If Ω is unbounded this cone in general will not be an acceptable cone, since the compactness condition will not be satisfied in general. However, as we have shown, the infimum in problem (2.9), (2.10) will be positive. This infimum will be denoted by $h(\Omega, G)$.

3. Statement of main results. In our first result we consider symmetrical domains.

THEOREM 1. *Let f satisfy conditions (2.1), (2.6), (2.5), (2.8). Suppose that Ω is an exterior domain, invariant with respect to a group of orthogonal transformations G such that $\nu(G) \geqslant 2$. If*

(3.1) $$h(\Omega, G) < \nu(G)h_0,$$

then (1.1) has a positive G-invariant solution. Further, given f and G there exists a positive number ρ (which depends only on f and $\nu(G)$) such that (3.1) is satisfied whenever

$$\partial\Omega \subset \{x: |x| \leqslant \rho\}. \quad \square$$

In our second result we consider nearly spherical domains.

THEOREM 2. *Let f satisfy conditions (2.1), (2.5), (2.6). Assume, in addition, that f is real analytic on $[0, \infty)$. Then there exists a set $E \subset (0, \infty)$, which is at most countable, such that for every $R \in (0, \infty) \setminus E$ there exists $\delta = \delta(R) > 0$ with the following property: If Ω is an exterior domain and*

(3.2) $$\partial\Omega \subset \{\alpha: R - \delta < |x| < R + \delta\},$$

then (1.1) has a positive solution. $\square$

Note that the analyticity assumption on f implies that f has an analytic continuation in a neighborhood of $[0, \infty)$ in the complex plane. This analytic continuation will still be denoted by f.

The results stated above (with some obvious modifications) remain valid for $\Omega \subset \mathbb{R}^N$ with $N \geqslant 2$. Also the main arguments in their proofs are unaffected by the dimension. However, some technical details are simpler in $\mathbb{R}^3$. For this reason we confine our presentation in this note to the case $\mathbb{R}^3$.

The analysis of the various possibilities for the group of transformations G which was described in the previous section is, of course, specific to $\mathbb{R}^3$. In general, in the case of odd dimension, there is a finite bound on $\nu(G)$ over all possible groups G,[2] just as in the case of $\mathbb{R}^3$. Consequently, in this case, for a

---

[2] Assuming radial symmetry, hence $\nu(G) = \infty$, to be excluded.

given f, Theorem 1 imposes an absolute limitation on the size of $\partial\Omega$ in the nonspherically symmetric case. On the other hand, in the case of even dimension, there are groups G with arbitrarily large $\nu(G)$.

4. Symmetric domains, proof of Theorem 1. We obtain a solution to the exterior problem (1.1) as a limit of solutions to approximating problems on bounded domains. Accordingly, let Ω be an exterior domain with

(4.1) $$\partial\Omega \subset \{x: |x| \leq \rho\},$$

let $\{R_n\}$ be an unbounded increasing sequence of real numbers

$$\rho < R_1 < R_2 < \cdots < R_n < R_{n+1} < \cdots, \quad \lim_{n\to\infty} R_n = \infty,$$

and let

$$\Omega_n = \{x: x \in \Omega, |x| < R_n\}.$$

We now consider the problem (2.9), (2.10), with Ω replaced by Ω_n, for each $n = 1, 2, \ldots$. Let G be a group of orthogonal transformations (possibly the trivial group) that leaves Ω, hence each Ω_n, invariant. Let

$$\mathcal{K}_n = \{u \in W_0^{1,2}(\Omega_n): u(x) \geq 0 \text{ a.e. in } \Omega, u \text{ is } G \text{ invariant}\}.$$

It is clear that each $\mathcal{K}_n$ is an acceptable cone, thus, by Proposition 2.1, there exists for each n a solution u_n to (1.1) on Ω_n which minimizes $H(u)$ in $\mathcal{K}_n$ subject to (2.10)(b). Clearly then,

$$H(u_1) > H(u_2) > \cdots > H(u_n) > \cdots$$

and thus, from (2.11), it follows that $\{u_n\}$ is a bounded sequence in $W_0^{1,2}(\Omega)$. Any function that is a weak limit in $W_0^{1,2}(\Omega)$ of $\{u_n\}$ or a subsequence thereof is a solution to (1.1). Thus, unless $\{u_n\}$ converges weakly to 0 in $W_0^{1,2}(\Omega)$, there exists a nontrivial solution to (1.1). If the full sequence converges weakly to 0 we shall say that the sequence *collapses*. We now examine the case of a collapsing sequence.

(i) *If the sequence $\{u_n\}$ collapses then $u_n \to 0$ uniformly in each compact subset of* Ω.

This follows from standard results on regularity.

Let c be a number in $(0, 1)$, to be kept fixed in the following discussion. Let ξ be a point on the unit sphere in $\mathbb{R}^3$. We shall say that ξ is a *distinguished direction* if there exists a subsequence $\{u_{n_k}\}$ of $\{u_n\}$ and a sequence of points $x_k \in \Omega_{n_k}$ ($k = 1, 2, \ldots$) such that

$$u_{n_k}(x_k) \geq c \quad \text{and} \quad x_k/|x_k| \to \xi.$$

The set of distinguished points will be denoted by $\mathcal{D} = \mathcal{D}(\{u_n\})$. The set $\mathcal{D}$ is independent of c but we shall not use this fact. In view of our assumptions on f, the set $\mathcal{D}$ is not empty, because each solution u_n must be larger than 1 at some points of Ω_n.

Denote

$$\chi(\{u_n\}) = \inf \operatorname{card} \mathcal{D}(\{u_{n_k}\})$$

where the infimum is taken over all subsequences $\{u_{n_k}\}$. Since u_n is G-invariant, we have
(4.2) $$\nu(G) \leq \chi(\{u_n\}).$$

MAIN LEMMA. *Let the sequence $\{u_n\}$ be constructed as above. If the sequence collapses then $\chi(\{u_n\})$ is finite and*
(4.3) $$\liminf_{n \to \infty} H(u_n) \geq \chi(\{u_n\})h_0.$$

PROOF. Let $\xi \in \mathbb{R}^3$ with $|\xi| = 1$ be a distinguished direction for $\{u_n\}$. There exists then a subsequence $\{u_{n_k}\}$ of $\{u_n\}$ and a sequence $\{x_k\}$ in $\mathbb{R}^3$ such that
(4.4) $$u_{n_k}(x_k) \geq c, \quad k = 1, 2, \ldots, \quad \lim_{k \to \infty} \frac{x_k}{|x_k|} = \xi,$$
and we can assume, as we shall, that
$$\frac{\partial u_{n_k}}{\partial r}(x_k) \geq 0, \quad k = 1, 2, \ldots;$$
here $r = |x|$. It follows from a result of [6] that $\partial u_n/\partial r < 0$ for $|x| \geq (\rho + R_n)/2$, thus we have, in view of (i),
(4.5) $$\lim_{k \to \infty} |x_k| = \infty, \quad |x_k| \leq \frac{\rho + R_{n_k}}{2}.$$

We now put $v_k(x) = u_{n_k}(x + x_k)$ so that
(4.6) $$v_k(0) \geq c, \quad k = 1, 2, \ldots.$$
The sequence $\{v_k\}$ is bounded in $W^{1,2}(\mathbb{R}^3)$, we can assume therefore that $\{v_k\}$ is weakly convergent in $W^{1,2}(\mathbb{R}^3)$. Let the weak limit be v_0. In view of (4.5), v_0 is a solution to (1.1) in $\mathbb{R}^3$. It follows from the regularity of the v_k and (4.6) that (the smooth representative of) v_0 satisfies $v_0(0) \geq c$, and v_0 is thus not the trivial solution. Consequently, equality holds in (2.10)(b) when $u = v_0$, $\Omega = \mathbb{R}^3$, and hence, we have
$$\int_{\mathbb{R}^3} \left[\frac{1}{2}v_0 f(v_0) - F(v_0)\right] dx \geq h_0.$$
If $\delta > 0$ is given then we can pick a ball B centered at zero such that
$$\int_B \left[\frac{1}{2}v_0 f(v_0) - F(v_0)\right] dx \geq h_0 - \delta.$$
Suppose ρ_1 is the radius of B. Then for k sufficiently large, we have
$$\int_{|x - x_k| \leq \rho_1} \left[\frac{1}{2}u_{n_k} f(u_{n_k}) - F(u_{n_k})\right] dx \geq h_0 - 2\delta.$$
If $\chi(\{u_n\}) > 1$, so that card $\mathscr{D}(\{u_{n_k}\}) > 1$, let ξ' be an element of $\mathscr{D}(\{u_{n_k}\})$ different from ξ. Repeat the previous argument with ξ and $\{u_n\}$ replaced by ξ' and $\{u_{n_k}\}$. If χ is an integer, $\chi \leq \chi(\{u_n\})$, then repeating this process χ times we finally obtain a subsequence $\{u_{n'}\}$ and sequences $\{x_n^i\}$, $i = 1, \ldots, \chi$, and $\rho_1 = \rho_1(\delta)$ such that
$$\lim_{n \to \infty} \left(\frac{x_n^i}{|x_n^i|}\right) = \xi_i,$$

where $\xi_1, \ldots, \xi_\chi$ are distinct distinguished directions and

(4.7) $$\liminf_{n \to \infty} \int_{|x - x_n^i| \leq \rho_1} \left[\frac{1}{2} u_{n'} f(u_{n'}) - F(u_{n'})\right] dx \geq h_0 - 2\delta$$

for $i = 1, \ldots, \chi$. Obviously, we have

$$\lim_{n \to \infty} |x_n^i - x_n^j| = \infty \quad \text{for } i \neq j,$$

thus (4.7) implies

$$\liminf_{n \to \infty} H(u_{n'}) \geq \chi(h_0 - 2\delta).$$

Since δ can be taken arbitrarily small, the conclusion readily follows, and the proof of the main lemma is complete.

We now complete the proof of Theorem 1. Let Ω, G be as in the statement of that theorem. For the sequence $\{u_n\}$ constructed above we have, in the notation of §3, $H(u_n) = h(\Omega_n, G)$, moreover, one easily sees that

$$\lim_{n \to \infty} h(\Omega_n, G) = h(\Omega, G).$$

Thus if (3.1) holds, then it follows from the main lemma and (4.2) that $\{u_n\}$ cannot collapse and hence (1.1) must have a positive solution.

To see that (3.1) indeed holds when $\mathbb{R}^3 \setminus \Omega$ is sufficiently small, we note first that when (4.1) holds then, for any G,

(4.8) $$h(\Omega, G) \leq h_\rho$$

(for the definition of h_ρ see Example 2 in §2). The final step is to observe that

$$\lim_{\rho \downarrow 0} h_\rho = h_0.$$

5. Proof of Theorem 2.

Suppose that u is a solution to the problem (1.1) on the exterior domain $\Omega_R = \{x: |x| > R\}$. It is fairly obvious that by perturbation methods one can obtain a solution to (1.1) on the exterior domain Ω, where δ is small and (3.2) holds, provided the eigenvalue problem

(5.1) $$\Delta w - w + \mu f'(u) w = 0 \quad \text{in } |x| > R,$$

(5.2) $$w = 0, \quad |x| = R$$

does not admit $\mu = 1$ as an eigenvalue. Accordingly, most of this section is devoted to an analysis of the spectrum of the problem (5.1), (5.2) for R, u as indicated. The details are lengthy and somewhat tedious so only an outline shall be given here.

We shall assume that f satisfies (2.1), (2.5), and (2.6). No polynomial growth law is assumed but in the proof of Theorem 2 we assume f to be analytic.

We consider positive radially-symmetric solutions to the problem (1.1) where Ω is the exterior of a ball of radius R. These solutions can be obtained of course as solutions to the boundary value problem

(5.3) $$u'' + (2/r)u' - u + f(u) = 0, \quad r > R,$$

(5.4) $$u(R) = 0, \quad \int_R^\infty (u'^2 + u^2) r^2 \, dr < \infty.$$

A nontrivial solution to (5.3), (5.4) will be called a *variational solution* if it minimizes

$$\int_R^\infty \left[\frac{1}{2} uf(u) - F(u)\right] r^2 \, dr$$

subject to

$$0 < \int_R^\infty (u'^2 + u^2) r^2 \, dr \le \int_R^\infty uf(u) r^2 \, dr,$$

and will be called a *local variational solution* if it is a local solution to this variational problem (i.e., local with respect to the norm

$$\|u\| = \left(\int_R^\infty (u'^2 + u^2) r^2 \, dr\right)^{1/2}.$$

PROPOSITION 5.1. *The problem* (5.3), (5.4) *has a variational solution for every* $R > 0$.

Functions that satisfy (5.3) and the second condition of (5.4) have the asymptotic behavior

(5.5) $\qquad u(r) \sim ar^{-1} e^{-r} \quad \text{as } r \to \infty.$

On the other hand, a solution $u = u(r, a)$ to (5.3) that satisfies (5.5) also satisfies the integral equation

(5.6) $\qquad u(r, a) = ar^{-1} e^{-r} + r^{-1} \int_r^\infty \sinh(r - \rho) f(u(\rho, a)) \rho \, d\rho.$

PROPOSITION 5.2. *The integral equation* (5.6) *is uniquely solvable (on any interval* (R, ∞), $R > 0$) *and the solution satisfies* (5.3).

Let $a > 0$ and let $u = u(r, a)$ be the solution to (5.6) on $(0, \infty)$. We define $R(a)$ to be the infimum of all positive numbers ρ such that $u(\rho, a) > 0$ in (ρ, ∞).

PROPOSITION 5.3. *There exists an* $a_0 \ge 0$ *such that*

$$R(a) > 0, \qquad a > a_0, \qquad \lim_{a \downarrow a_0} R(a) = 0.$$

$R(a)$ *is a* C^1-*function of* a *on* (a_0, ∞) *and*

(5.7) $\qquad \lim_{a \to \infty} R(a) = \infty.$

For $R = R(a)$, $u(r, a)$ *is a positive solution to* (5.3), (5.4). *For all sufficiently large* R *every solution of* (5.3), (5.4) *is of the form* $u = u(r, a)$ *with* $a \in (a_0, \infty)$. *Finally, if* (5.3) *has a positive solution* u *on* $(0, \infty)$ *that satisfies*

$$\int_0^\infty (u'^2 + u^2) r^2 \, dr < \infty$$

(*in particular, if* (2.8) *holds, with* $1 < \sigma < 5$), *then* $u = u(r, a_0)$ *is such a solution*.

We shall refer to the family $\{u(r, a): a \in (a_0, \infty)\}$ as the *principal branch* of solutions to (5.3), (5.4).

Given a solution u belonging to the principal branch we next consider the eigenvalue problem (5.1), (5.2), $u = u(r, a)$, $R = R(a)$. The eigenvalues of (5.1), (5.2) are the eigenvalues of the family of problems

$$(5.8_n) \qquad v'' + \frac{2}{r}v' - v - \frac{n(n+1)}{r^2}v + \mu v f'(u) = 0, \qquad r > R,$$

$$(5.9) \qquad v(R) = 0, \qquad \int_R^\infty (v'^2 + v^2) r^2 \, dr < \infty,$$

$n = 0, 1, \ldots$. For $u = u(r, a)$, $R = R(a)$ the kth eigenvalue of (5.8_n), (5.9) will be denoted by $\mu_{k,n}(a)$.

PROPOSITION 5.4. *The eigenvalues $\mu_{k,n}(a)$ depend continuously on a on (a_0, ∞) for each fixed pair (k, n); this dependence is analytic if f is analytic as in Theorem 2. We have, for $a \in (a_0, \infty)$,*

$$(5.10) \quad \mu_{k',n}(a) > \mu_{k,n}(a), \quad k' > k, \qquad \mu_{k,n'}(a) > \mu_{k,n}(a), \quad n' > n,$$

and

$$(5.11) \qquad \lim_{k \to \infty} \mu_{k,n}(a) = \infty, \qquad \lim_{n \to \infty} \mu_{k,n}(a) = \infty.$$

For every n,

$$(5.12) \qquad \varlimsup_{a \to \infty} \mu_{1,n}(a) \leq \frac{1}{1+\varepsilon},$$

where ε is the constant in (2.6). For all $a \in (a_0, \infty)$,

$$(5.13) \qquad \mu_{1,1}(a) < 1,$$

while

$$(5.14) \qquad \mu_{k,n}(a) > 1 \quad \text{for } k > 1 \text{ and } n > 0,$$

and, if $u = u(r, a)$ is a local variational solution, then

$$(5.15) \qquad u_{k,0}(a) \geq 1 \quad \text{for } k \geq 1.$$

The inequality (5.12) follows from the Rayleigh quotient characterization of $\mu_{1,n}(a)$, (2.6), (5.7), and the relation

$$\int_R^\infty (u'^2 + u^2) r^2 \, dr = \int_R^\infty u f(u) r^2 \, dr.$$

The assertion concerning (5.15) is a standard variational principle. Inequalities (5.13) and (5.14) follow by observing that $\partial u/\partial r$ satisfies (5.8_1), with $\mu = 1$, and the second, but not the first condition of (5.9). Thus $\mu = 1$ is never an eigenvalue of (5.8_1), (5.9) for u on the principal branch; (5.13) now follows from (5.12) and the continuity of the eigenvalues. Since the principal branch contains variational solutions, (5.14) follows similarly in view of (5.15) and the monotonicity relations (5.10).

We shall say that a_0 is *n-critical* if $n \geq 1$ and $\mu_{1,n}(a_0) = 1$ or if $n = 0$ and $\mu_{k,0}(a_0) = 1$ for some $k \geq 2$. It is not difficult to show that a_0 is a 0-critical point if and only if $dR(a_0)/da = 0$. Since $R(a)$ is analytic the set of 0-critical points is at most countable.

Proposition 5.4 has the following

COROLLARY. *When f satisfies the hypothesis of Theorem 2 then* (5.1), (5.2) *admits the eigenvalue* 1 *for at most a countable set of radii.*

In view of (5.14), the only way that (5.1), (5.2) can admit $\mu = 1$ as an eigenvalue is if $R = R(a)$ where a is *n*-critical. As we mentioned before, the set of 0-critical points is at most countable. Since $\mu_{1,n}$ is an analytic function of a and (by (5.12)) $\mu_{1,n}$ is not identically equal to one, the set $\{a: \mu_{1,n}(a) = 1, n = 1, 2, \ldots\}$ is at most countable. This proves our assertion.

Suppose now that R, Ω_R, u are as in the first paragraph of this section and 1 is not an eigenvalue of (5.1), (5.2). Let Ω denote a region which satisfies (3.2) for some $\delta > 0$ to be determined. Let G, G_1 denote the Green's functions for the Dirichlet problem for $-\Delta u + u = h$ on Ω_R and Ω, respectively. It will be convenient here to assume that G, G_1 are defined as extended-real-valued functions, on $\mathbf{R}^3 \times \mathbf{R}^3$ with, e.g.

$$G_1(x, y) = G_1(y, x) = 0 \quad \text{for } y \in \mathbf{R}^3, x \notin \overline{\Omega}.$$

We have

$$u(\cdot) = \int_{\mathbf{R}^3} G(\cdot, y) f(u(y)) \, dy,$$

and we seek a solution w to

(5.16) $$w(\cdot) = \int_{\mathbf{R}^3} G_1(\cdot, y) f(w(y)) \, dy.$$

We put $g = G_1 - G$, $v = w - u$, and upon substituting into (5.16), we obtain

(5.17)
$$v(\cdot) - \int_{\mathbf{R}^3} G(\cdot, y) v(y) f'(u(y)) \, dy$$
$$= \int_{\mathbf{R}^3} G(\cdot, y) \mathscr{H}(y, v(y)) \, dy + \int_{\mathbf{R}^3} g(\cdot, y) f(w(y)) \, dy,$$

where

$$f(w(x)) = f(u(x)) + v(x) f'(u(x)) + \mathscr{H}(x, v(x))$$

so that

$$\mathscr{H}(x, v) = O(|v|^2) \quad \text{as } v \to 0$$

uniformly in $\mathbf{R}^3$. Finally, it can be shown that if $\nu > 0$ is given and $0 \leq \alpha < 1$, then there exists $\delta = \delta(\nu, \alpha)$ such that

$$\sup_{x \in \mathbf{R}^3} e^{\alpha |x|} \int_{\mathbf{R}^3} |g(x, y)| e^{-\alpha |y|} \, dy \leq \nu$$

when (3.2) holds.

The completion of the analysis of (5.16) is now routine and requires only the choice of an appropriate function space. One possible choice is the space of functions $v(x)$ that are continuous on $\mathbf{R}^3$ with

$$\lim_{|x|\to\infty} e^{|x|/2}|v(x)| = 0,$$

normed by

$$\|v\| = \sup_x e^{|x|/2}|v(x)|.$$

We readily conclude that (5.14), hence (1.1) has a positive solution provided (3.2) holds with $\delta > 0$ sufficiently small.

References

1. H. Berestycki and P.-L. Lions, *Nonlinear scalar field equations*, I. *Existence of a ground state*, II. *Existence of infinitely many solutions*, Arch. Rational Mech. Anal. **82** (1983), 313–375.

2. C. V. Coffman, *Uniqueness of the ground state solution for $\Delta u - u + u^3 = 0$ and a variational characterization of other solutions*, Arch. Rational Mech. Anal. **46** (1972), 81–95.

3. H. S. M. Coxeter and W. O. J. Moser, *Generators and relations for discrete groups*, Springer-Verlag, 1972.

4. M. J. Esteban and P.-L. Lions, *Existence and non-existence results for semilinear elliptic problems in unbounded domains*, Proc. Roy. Soc. Edinburgh Sect. A **93** (1982), 1–14.

5. R. Finkelstein, R. LeLevier and M. Ruderman, *Non-linear spinor fields*, Phys. Rev. **83** (1951), 326.

6. B. Gidas, W.-M. Ni and L. Nirenberg, *Symmetry and related properties via the maximum principle*, Comm. Math. Phys. **68** (1979), 209–243.

7. _____, *Symmetry of positive solutions of nonlinear elliptic equations in R^n*, Mathematical Analysis and Applications, Part A, Adv. in Math. Suppl. Studies, Vol. 7A, Academic Press, New York, 1981.

8. A. Kufner, O. John and S. Fucik, *Function spaces*, Noordhoff, Leyden, 1977.

9. P.-L. Lions, *Symétrie et compacité dans les espaces de Sobolev*, J. Funct. Anal. **49** (1982), 315–334.

10. _____, *Principe de concentration-compacité en calcul des variations*, C. R. Acad. Sci. Paris Sér. I Math. **294** (1982), 261–264.

11. N. V. Mitskevich, *The scalar field of a stationary nucleon in a non-linear theory*, Soviet Phys. JETP **2** (1956), 197.

12. Z. Nehari, *On a class of nonlinear second-order differential equations*, Trans. Amer. Math. Soc. **95** (1960), 101–123.

13. _____, *On a nonlinear differential equation arising in nuclear physics*, Proc. Roy. Irish Acad. Sect. A **62** (1963), 117–135.

14. N. Rosen and H. B. Rosenstock, *The forces between particles in a non-linear field theory*, Phys. Rev. **85** (1952), 257.

15. G. H. Ryder, *Boundary value problems for a class of nonlinear differential equations*, Pacific J. Math. **22** (1967), 477–503.

16. W. A. Strauss, *Existence of solitary waves in higher dimensions*, Comm. Math. Phys. **55** (1977), 149–162.

17. J. L. Synge, *On a certain non-linear differential equation*, Proc. Roy. Irish Acad. Sect. A **62** (1961), 17–42.

CARNEGIE-MELLON UNIVERSITY

TECHNION-ISRAEL INSTITUTE OF TECHNOLOGY

A Global Fixed Point Theorem for Symplectic Maps and Subharmonic Solutions of Hamiltonian Equations on Tori

C. CONLEY[1] AND E. ZEHNDER

1. Introduction and results. In this paper we describe some results presented at the AMS Summer Institute on Nonlinear Functional Analysis and Applications in Berkeley.

The results belong to the circle of old questions in celestial mechanics related to the so-called Poincaré–Birkhoff fixed point theorem. In his search for periodic solutions in the restricted three body problem, H. Poincaré constructed a symplectic section map of an annulus A on the energy surface. This annulus is bounded by the so-called direct and retrograde periodic orbits. It lead him in 1912 to the formulation of the following theorem [21]:

Every area preserving homeomorphism of an annulus $A := S^1 \times [a, b]$ rotating the two boundaries in opposite directions possesses at least 2 fixed points in the interior.

Using strictly two-dimensional arguments, G. Birkhoff succeeded in 1913 in proving this statement in an ingenious way [4, 5]. The standard topological fixed point techniques are not applicable and in fact the statement is wrong if the area preserving character is dropped. As pointed out by V. Arnold in his book [1], the theorem could be derived, at least in the differentiable case, from the existence of fixed points for area preserving maps on a two-dimensional torus T^2.

The theorem gave rise to the topological fixed point theory of Lefschetz, which, however, is applicable neither to the annulus nor to the torus, because their Euler characteristic is zero. If f is a map on a compact surface F_g of genus g which is

---

[1] Deceased November 20, 1984.
1980 *Mathematics Subject Classification.* Primary 58F22, 58F05, 58C30; Secondary 58E05, 54H25, 70F15, 70H15.

homotopic to the identity mapping then its Lefschetz-number is $L(f) = L(\text{id}) = \chi(F_g) = 2 - 2g$. Hence if $g \neq 1$, then $L(f) \neq 0$ and f has at least one fixed point.

In particular every continuous map of S^2 homotopic to the identity map possesses at least one fixed point. The map may have only one fixed point, as the translation $z \to z + 1$ on the Riemann sphere shows. It is, however, a striking fact that f possesses at least two fixed points, if it preserves, in addition, a regular measure (see [20] and [24]). The proof is again based on a strictly two-dimensional argument: If p^* is the fixed point of f guaranteed by the Lefschetz theory, then $f|S^2 \setminus p^* = g$ can be identified with a homeomorphism of the plane $\mathbf{R}^2$. If g had no fixed point, then by Brouwer's translation theorem there would be an open set D with the property that $g^j(D) = f^j(D)$ are mutually disjoint for all $j > 0$. Consequently, since f preserves the measure μ,

$$\sum_{j=0}^{n} \mu(f^j(D)) = (n+1)\mu(D) \leq \mu(S^2)$$

for every n and hence $\mu(D) = 0$, contradicting the assumption that $\mu(D) \neq 0$ for an open set D. This proves the claim that there must be a second fixed point. In particular an area preserving diffeomorphism of S^2 possesses at least two fixed points, i.e. at least as many fixed points as a function on S^2 has critical points.

In contrast to S^2, the Lefschetz fixed point theory is not adequate to find fixed points of maps on the torus $T^2 = \mathbf{R}^2/\mathbf{Z}^2$ whose Euler characteristic vanishes. There are indeed many maps on T^2 without fixed points. For example, on the covering space $\mathbf{R}^2$ of T^2, the area preserving translation map

$$X = x + c_1, \qquad Y = y + c_2$$

has no fixed point on T^2 if $c = (c_1, c_2) \notin \mathbf{Z}^2$. The class of mappings on T^2 has therefore to be restricted, if they necessarily should possess fixed points.

In the following we consider measure preserving diffeomorphisms ψ of T^2 which are homologous to the identity map, hence, on $\mathbf{R}^2$ are given by

$$\psi : \begin{aligned} X &= x + p(x, y), \\ Y &= y + q(x, y) \end{aligned}$$

with two periodic functions p and q and require that ψ preserves the center of mass, hence excluding in particular the above translation map. Summarizing, we require that the diffeomorphism of T^2 satisfy

(i) ψ is homologous to the identity,
(ii) $dX \wedge dY = dx \wedge dy$,
(iii) $\int_{T^2} p = 0 = \int_{T^2} q$.

V. Arnold conjectured in [1] and in [2] that such a diffeomorphism ψ possesses as many fixed points as a function on T^2 has critical points.

CONJECTURE BY V. ARNOLD. *Every diffeomorphism on T^2 satisfying* (i)–(iii) *possesses at least 3 fixed points. Moreover, if all the fixed points are nondegenerate, then ψ possesses at least four fixed points.*

Here a fixed point p of ψ is called nondegenerate if 1 is not an eigenvalue of $d\psi(p)$. As a side remark we first observe that there are maps even close to the identity map having 3 fixed points only. To see this we take a smooth function G on T^2 with precisely 3 critical points, for example $G(x, y) = \sin \pi x \cdot \sin \pi y \cdot \sin \pi(x + y)$. The map $\psi: (x, y) \to (X, Y)$ of T^2, implicitly defined by

$$X = x + \varepsilon \frac{\partial G}{\partial Y}(x, Y), \quad y = Y + \varepsilon \frac{\partial G}{\partial x}(x, Y)$$

is for $\varepsilon > 0$ sufficiently small close to the identity, satisfies (i)–(iii), and has 3 fixed points, namely the critical points of G on T^2.

We point out that the conjecture is global. Under the additional assumption that ψ is close to the identity map in the C^1-topology it is in fact easy to prove, since under this restriction there is a direct and one-to-one correspondence between the fixed points of ψ and the critical points of a function on T^2. Indeed, following H. Poincaré (Les Methodes Nouvelles de la Mécanique Céleste, tome III, Gauthiers-Villars, Paris, 1899, p. 214) one can consider the one-form on T^2:

$$(X - x)(dY + dy) - (Y - y)(dX + dx) = dS(x, y),$$

which due to the assumptions (i)–(iii) is an exact form on T^2, i.e. the function S is a function on the torus. Obviously, the critical points of S are in one-to-one correspondence to the fixed points of ψ provided the two one-forms $(dY + dy)$ and $(dX + dx)$ are linearly independent. But this is the case if and only if (-1) is not an eigenvalue of the Jacobian matrix $d\psi(p)$, as one readily verifies. Since a function on T^2 possesses at least 3 critical points, the conjecture follows under the above additional smallness assumption.

The idea of relating fixed points of symplectic maps to critical points of a related function on the manifold in question is being used quite frequently in order to establish existence results. It was extended by A. Weinstein [26], who uses it, for example, to show that a symplectic diffeomorphism of a compact and simply connected manifold M possesses at least as many fixed points as a function on M has critical points, provided however the map is sufficiently C^1-close to the identity map on M. For more general results using this generating function technique and for references we refer to J. Moser [19]. So far, however, this method has lead to perturbation results only.

The proof of the Arnold conjecture [10] uses quite a different idea. The fixed points will also be found as critical points of a function, which, however, is not defined on T^2 but on a high dimensional manifold $T^2 \times \mathbf{R}^{2N}$ and which originates in a variational problem for forced oscillations of a Hamiltonian equation on T^2. This has the advantage that one avoids the difficulty of the "eigenvalues -1". Instead of looking for fixed points of the map ψ we rather search for periodic orbits of a Hamiltonian equation.

It can be shown that the assumptions (i)–(iii) on the map ψ on T^2 are equivalent to the fact that ψ is the time 1 map of a flow $\psi = \phi^1$, where ϕ^1 is the flow of a time-dependent Hamiltonian vector field on T^2, i.e. satisfies

$$\frac{d}{dt}\phi^t(x) = J\nabla h(t, \phi^t(x)) \quad \text{and} \quad \phi^0(x) = x,$$

where $x \in \mathbf{R}^2$ and where $h = h(t, x)$ is periodic in all its variables of period 1. The proof of this equivalence is strictly two-dimensional and we refer to [10]. As usual, the matrix J stands for the symplectic structure

$$J = \begin{pmatrix} 0 & 1 \\ -1 & 0 \end{pmatrix}.$$

Now a periodic solution of the Hamiltonian equation on T^2 having period 1 gives rise to a required fixed point of the map ψ. This way the problem of finding fixed points is reduced to the problem of finding periodic solutions of a Hamiltonian vector field. More generally, we now consider Hamiltonian vector fields on $T^{2n} = \mathbf{R}^{2n}/\mathbf{Z}^{2n}$ for any $n \geqslant 1$:

(1) $$\dot{x} = J\nabla h(t, x), \quad (t, x) \in \mathbf{R} \times \mathbf{R}^{2n},$$

with $h \in C^2(\mathbf{R} \times \mathbf{R}^{2n})$ being periodic in all its variables of period 1. The conjecture by Arnold is then a consequence of the following existence statement for forced oscillations.

THEOREM 1 [10]. *Every Hamiltonian vector field* (1) *on* T^{2n} *possesses at least* $(2n + 1)$ *periodic solutions of period* 1. *Moreover, if all the* 1-*periodic solutions are nondegenerate, then there are at least* 2^{2n} *such solutions.*

Here a periodic solution is called nondegenerate if none of its Floquet multipliers is equal to 1. This condition requires effectively that the forced oscillations are isolated among such solutions. We observe that the periodic solutions found by the theorem belong to the same homotopy class of loops on T^{2n}, in fact they are all contractible loops. Other periodic solutions may not exist. This is in contrast to the closed geodesics on T^{2n}, where one finds very easily a closed geodesic in every homotopy class of loops. In this connection one should recall that in the Morse theory of closed geodesics each cohomology class in the loopspace of the manifold is represented in the index of some critical point. It turns out that in the Morse theory for forced oscillations this is not the case, only the cohomology of the manifold itself has to be represented.

Postponing the discussion of forced oscillations on general symplectic manifolds to the last section we now turn to the problem of subharmonic solutions of the Hamiltonian vector field (1) on T^{2n}. The periodic solutions of Theorem 1 are obviously also periodic solutions of period T for every integer $T > 1$, i.e. $x(t + T) = x(t)$. The period $T > 1$ however is not minimal. It is our next aim to find solutions having integers $T > 1$ as their minimal periods. These solutions are called subharmonic solutions. To find such solutions further assumptions on the vector field are required as the example $h \equiv 0$ shows.

We first recall that there is an abundance of subharmonic solutions locally in every neighborhood of a 1-periodic solution $x(t)$, provided however this solution is of "general elliptic type" and provided the Hamiltonian system is sufficiently smooth. To be more precise, let $p = x(0)$ be the corresponding fixed point $p = \psi(p)$ for the time 1 map $\psi = \phi^1$ of the flow of the Hamiltonian system, the periodic points of ψ nearby correspond to the subharmonic solutions near $x(t)$. Assume $p = 0 \in \mathbf{R}^{2n}$ and assume it is an elliptic fixed point of the exact symplectic map ψ, requiring that all the Floquet multipliers of $x(t)$ are on the unit circle. They can be ordered in pairs $(\lambda_j, \lambda_j^{-1})$ with $1 \leq j \leq n$. Excluding the lower order resonances by requiring that

$$\lambda_1^{j_1}\lambda_2^{j_2} \cdots \lambda_n^{j_n} \neq 1 \quad \text{for } j \in \mathbf{Z}^n, 0 < |j| \leq 4,$$

the map ψ can be put, by an analytic symplectic diffeomorphism into the following so-called Birkhoff normal form. With the symplectic coordinates $z_j = (x_j, y_j), 1 \leq j \leq n$:

$$z_j \to \begin{pmatrix} \cos\theta_j & -\sin\theta_j \\ \sin\theta_j & \cos\theta_j \end{pmatrix} z_j + O_4(z),$$

where $\theta_j = \alpha_j + \sum_{s=1}^n \beta_{js}|z_s|^2$, $1 \leq j \leq n$. The matrix $\beta = (\beta_{js})$ is a symplectic invariant, and we require that $\det \beta \neq 0$, hence postulating effectively that the map ψ is, close to 0, nonlinear. Assume now $\psi \in C^r$ with $r \geq \max\{4n, 5\}$ and define the neighborhood B_δ of the fixed point 0 by $B_\delta = \{z \in \mathbf{R}^{2n} | |z_j|^2 \leq \delta^2, 1 \leq j \leq n\}$. If $P_\delta \subset B_\delta$ denotes the closure of the set of periodic orbits of ψ which are contained in B_δ, then the Lebesgue measure $m(B_\delta)$ is positive. More precisely it can be proved that

$$m(B_\delta \setminus P_\delta) = O(\sqrt{\delta})m(B_\delta).$$

The proof of this estimate is based on the fact that the invariant tori guaranteed by the so-called KAM-theory are in the closure of the set of periodic orbits (see [12]).

We point out again that this local existence statement for subharmonic solutions near the 1-periodic solutions requires an excessive amount of smoothness of the system and a strong nonlinearity condition near the periodic solution. The global existence proof for the 1-periodic solutions of the theorem being based on Morse theory arguments gives no such information about the behaviour nearby. We shall therefore formulate different assumptions, namely assumptions on the linearized systems along the 1-periodic solutions, which guarantee the existence of infinitely many subharmonic solutions, which are not necessarily near the 1-periodic solutions.

THEOREM 2. *Assume that no Floquet multiplier of the 1-periodic solutions of the system* (1) *on* T^{2n}, $n \geq 2$, *is a root of unity. Then there is an integer* $T_0 > 1$ *such that*:

(i) *For every prime* $T \geq T_0$ *there is a periodic solution of* (1) *having minimal period T.*

(ii) *If for such a T all the T-periodic solutions are nondegenerate, then there are at least 2 subharmonic solutions with minimal period T. Moreover, if, in addition, for all the 1-periodic solutions* $x = x(t)$

$$\lim_{l \to \infty} \frac{1}{l} j(x^l) = \Delta(x) \neq 0,$$

then there are at least 2^{2n} subharmonic solutions having minimal period T. All these solutions are contractible loops on T^{2n}.

The integer $j(x^l)$ in the statement denotes the winding number of the periodic solution x^l, which is the l-times iterated solution $x = x(t)$. It will be defined in §3. We should mention that there are many recent results on subharmonic solutions of general Hamiltonian vector fields on $\mathbf{R}^{2n}$ under various strong nonlinearity asssumptions of the system at infinity; the solutions are established by intricate mini-max arguments for which we refer to P. Rabinowitz.

In the following we shall first recall the proof of Theorem 1 from [10], which is based on a classical variational principle, and which gives a new look at the Ljusternik–Schnirelman theory. Using parts of the proof we then prove in §3 the second theorem by means of a Morse theory for forced oscillations. This theory is somewhat different from the familiar Morse theory for the closed geodesic problem, since the variational problem is not coercive.

2. A classical variational principle and a different view on Ljusternik–Schnirelman theory. The forced oscillations claimed in Theorem 1 will be found by means of a classical variational principle for which the periodic orbits are the critical points. It is well known that in general the flow of a Hamiltonian vector field is very complicated, since solutions of quite different behaviour over an infinite interval of time are mixed. There is, however, a variational principle which picks out precisely the periodic solutions among all the solutions, thereby avoiding the complexity of the flow. Consider the contractible loops on T^{2n} which are, in the covering space $\mathbf{R}^{2n}$ described by the periodic functions $t \to x(t) \in \mathbf{R}^{2n}$ with $x(0) = x(1)$. The action functional on the periodic functions is then defined to be

$$(2) \qquad f(x) = \int_0^1 \left\{ \tfrac{1}{2} \langle \dot{x}, Jx \rangle - h(t, x(t)) \right\} dt.$$

On D this functional belongs to C^2, where $D = \{ x \in H^1([0,1], \mathbf{R}^{2n}) | x(0) = x(1) \}$. We claim that the critical points of f are the required 1-periodic solutions of (1). In fact for the derivative one finds

$$(3) \qquad df(x)y = \int_0^1 \langle -J\dot{x} - \nabla h(t, x(t)), y \rangle \, dt,$$

so that the gradient with respect to $L_2 = L_2((0,1); \mathbf{R}^{2n})$ is given by

$$(4) \qquad \nabla f(x) = -J\dot{x} - \nabla h(t, x).$$

Since $J^2 = -1$ one concludes indeed that $\nabla f(x) = 0$ if and only if x satisfies the Hamiltonian equation (1), so that it remains to find critical points of f. Here one is confronted with the difficulty that f is bounded neither from below nor from above. All the critical points have infinite dimensional stable and unstable invariant manifolds so that Morse theory cannot be applied directly. This can be seen from the derivative of $\nabla f\colon D \to L_2$ at x^*:

$$d\nabla f(x^*)x = -J\dot{x} - h_{xx}(t, x^*)x. \tag{5}$$

This linear operator is on $D \subset L_2$ into L_2, selfadjoint, and has compact resolvent; its discrete spectrum is bounded neither from below nor from above. This is in sharp contrast to the energy functional for closed geodesics on a Riemannian manifold, which is bounded from below. There are, of course, other variational principles for periodic solutions of Hamiltonian systems which are coercive, but they require conditions on the Hamiltonian, as for example convexity, not satisfied in our case; we refer to I. Ekeland [14] and the references therein, also to his article in these proceedings. That in the indefinite case the variational approach can be used effectively for existence proofs was first demonstrated by P. Rabinowitz [22] and subsequently used by many authors. It turns out that in our special case the analytical difficulties are only minor due to the fact that the Hamiltonian function h is periodic.

In view of the periodicity of h the function f above satisfies $f(x + j) = f(x)$ for every loop x and for every constant loop $j \in \mathbf{Z}^{2n}$. Splitting $x = [x] + \xi$ into its meanvalue $[x] \in \mathbf{R}^{2n}$ over a period and into the remainder we can identify the meanvalues with points on the torus $T^{2n} = \mathbf{R}^{2n}/\mathbf{Z}^{2n}$ and can view the function f as a function on $T^{2n} \times E$:

$$f\colon T^{2n} \times E \to \mathbf{R},$$

where E is the linear space of periodic functions having meanvalue zero. In case $h = 0$ one sees that the torus $T^{2n} \times \{0\}$ is a hyperbolic invariant manifold of the gradient flow $dx/ds = -\nabla f(x)$ consisting of the rest points. Since h and its derivatives are bounded, the dominant term in f is in fact its quadratic part. This will now be used in order to reduce the problem of finding critical points of f on the infinite dimensional space to the problem of finding the critical points of a related function g on the finite dimensional submanifold $T^{2n} \times \mathbf{R}^{2N}$ for some large N depending on the C^2-size of h. This will be done by a standard but global Lyapunov–Schmidt reduction (see [10]).

Let P be the orthogonal projection onto the eigenspaces of the eigenvalues contained in the interval $\{\lambda \in \mathbf{R}|\, |\lambda| \leq \tau\}$ of the operator $-J\dot{x}$, which is a selfadjoint operator on $D \subset L_2$ into L_2. Denote the splitting $L_2 = PL_2 + P^\perp L_2$ by $L_2 = Z + Y$. Let $x = z + y$, then $\nabla f(x) = 0$ is equivalent to

$$P\nabla f(z + y) = 0, \qquad P^\perp \nabla f(z + y) = 0.$$

For fixed $z \in Z$ the second equation is easily solved for $y = v(z) \in Y$ by the contraction principle in L_2 if τ is sufficiently large. By the implicit function theorem one then concludes that $z \to v(z)$ is a C^1 map from Z into $Y_1 = Y \cap D$. With $u(z) = z + v(z)$ the first equation becomes

$$\nabla g(z) = 0 \quad \text{with } g(z) = f(u(z)).$$

Since $\nabla g(z) = (\nabla f)(u(z)) = -Jz - P\nabla h(t, u(z))$ one concludes that $\nabla g: Z \to Z$ is a C^1-map. To find the critical points of g one now studies the gradient flow

(6) $$\frac{d}{ds}z = -\nabla g(z) \quad \text{on } z \in T^{2n} \times \mathbf{R}^{2N}$$

which, more explicitly, with $z = [z] + \xi \in T^{2n} \times \mathbf{R}^{2N}$ is of the form

$$\frac{d}{ds}[z] = 0 + v_0(z), \quad \frac{d}{ds}\xi = A\xi + v_1(z),$$

with a differentiable and uniformly bounded vector field $v = (v_0, v_1)$. Moreover, there is an invariant splitting $\xi = (\xi_+, \xi_-)$ for the linear map A such that $A\xi = (A_+\xi_+, A_-\xi_-)$ satisfies $(A_+\xi_+, \xi_+) \geq 2\pi|\xi_+|^2$ and $(A_-\xi_-, \xi_-) \leq -2\pi|\xi_-|^2$. From this one finds a constant $K > 0$ satisfying $(d/ds)|\xi_+|^2 \geq 1$ if $|\xi_+| \geq K$ and $(d/ds)|\xi_-|^2 \leq -1$ if $|\xi_-| \geq K$. In order to interpret these estimates geometrically we define the compact sets:

(7) $$B = T^{2n} \times D_1 \times D_2, \quad B^- = T^{2n} \times \partial D_1 \times D_2,$$

with the discs $D_1 = \{\xi_+ \in \mathbf{R}^n \mid |\xi_+| \leq K\}$ and $D_2 = \{\xi_- \in \mathbf{R}^N \mid |\xi_-| \leq K\}$. One now sees that the flow leaves the set B through the boundary part B^- of ∂B in forward time. The set B^- is called the strict exit set of B. The pair (B, B^-) is an example of an isolating block with exit set B^- in the sense of [13]. It is clear from the estimates that the set of bounded solutions of the gradient flow, hence in particular the rest points, are contained in the interior of B.

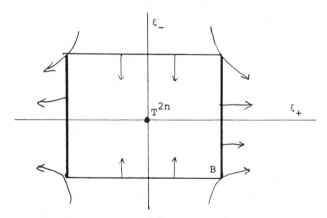

There are now several ways to find the rest points in B. They can be found for example by applying the Ljusternik–Schnirelman minimax theory and the Morse theory to the gradient flow in (B, B^-), recalling that B^- is the exit set (see also K.-C. Chang [27]).

In order to apply Morse theory we assume that there are only finitely many rest points z_j, $1 \leq j \leq k$ in B. Then the algebraic invariants of the pair (B, B^-), represented by the Poincaré polynomial $p(t, (B, B^-))$, are related to the local algebraic invariants $p(t, z_j)$ of the isolated rest points z_j by the Morse inequalities:

$$\text{(8)} \qquad \sum_{j=1}^{k} p(t, z_j) = p(t, (B, B^-)) + (1 + t)q(t),$$

where q is a polynomial having nonnegative integer coefficients. Since $H^*(B, B^-) \cong H^*(T^{2n}) \times H^*(S^N)$ we conclude that

$$\text{(9)} \qquad P(t) = p(t, (B, B^-)) = \sum_{j=0}^{2n} \binom{2n}{j} t^{j+N}.$$

In the special case where the critical points are nondegenerate it is well known that $p(t, z_j) = t^{m_j}$, with m_j being the Morse index of the critical point, so that in view of (9) the Morse inequalities become

$$\text{(10)} \qquad \sum_{j=1}^{k} t^{m_j} = \sum_{s=0}^{2n} \binom{2n}{j} t^{s+N} + (1 + t)q(t).$$

In this case we find in particular the number of critical points $k \geq 2^{2n}$, which is equal to the sum of the Betti numbers of the torus T^{2n}. In order to finish the second part of Theorem 1 we have to show that a critical point z_j is nondegenerate if and only if no Floquet multiplier of the corresponding periodic solutions $x_j = u(z_j)$ is equal to one. This follows from the following technical lemma (see also [9]).

LEMMA 1. *Let $x = u(z)$. Then the following statements are equivalent.*
(i) 1 *is not a Floquet multiplier of the loop x.*
(ii) $d(\nabla f)(u(z)): D \to L_2$ *is a bijection.*
(iii) $d(\nabla g)(z): Z \to Z$ *is a bijection.*

PROOF. In order to prove (i) ⇔ (ii) we have, in view of (5), to show that the kernel of $d(\nabla f)(x)$ is zero, but this is Floquet theory. In order to prove (ii) ⇔ (iii) we express $d(\nabla f)(x): D \to L_2$ in matrix-notation with respect to the reduction splitting $Z \times Y_1 \to Z \times Y$, where $Y_1 = Y \cap D$:

$$d(\nabla f) = \begin{pmatrix} \frac{\partial}{\partial z}(\nabla_z f) & \frac{\partial}{\partial y}(\nabla_z f) \\ \frac{\partial}{\partial z}(\nabla_y f) & \frac{\partial}{\partial y}(\nabla_y f) \end{pmatrix}.$$

One shows readily that at $x = u(z)$ the linear map $(\partial/\partial y)(\nabla_y f): Y_1 \to Y$ is a bijection. We therefore can define the isomorphism I, at $x = u(z)$, from D into itself by

$$I = \begin{pmatrix} 1 & 0 \\ \gamma & 1 \end{pmatrix},$$

with $\gamma = -\{(\partial/\partial y)(\nabla_y f)\}^{-1} \cdot (\partial/\partial z)(\nabla_y f)$. Then one finds at $x = u(z)$

$$d(\nabla f) \cdot I = \begin{pmatrix} d(\nabla g) & * \\ 0 & \frac{\partial}{\partial y}(\nabla_y f) \end{pmatrix},$$

and the lemma follows. □

In order to prove the first part of Theorem 1 one could apply the Ljusternik–Schnirelman theory to the gradient flow in (B, B^-). We proceed differently and describe a more direct and more flexible approach to determine the structure of the set of bounded solutions of the gradient flow in B. The method is also applicable to topological flows which are not necessarily gradient-like, on spaces which are not manifolds. We first claim that in fact every topological flow defined in a neighborhood of B contains an invariant set S in B whose topology is inherited from that of the surrounding B, provided only it behaves at the boundary of B like the gradient flow in the above picture. This will follow from a general result due to V. Benci, which describes conditions about the way the exit set B^- sits in B, which allow conclusions about the topology of S from that of B. We only formulate a special case.

PROPOSITION 1 (V. BENCI [15]). *Consider a continuous flow on a locally compact metric space in a neighborhood of a compact pair* (B, B^-) *which is assumed to be an isolating block having exit set* B^-. *Let* S *be the maximal invariant set contained in* B, *i.e.* $S := \{\gamma \in B | \gamma \cdot \mathbf{R} \in B\}$. *Assume*

(a) *there are classes* $\alpha_j \in H^*(B) \setminus 1$, $1 \leqslant j \leqslant m$, *such that* $\alpha := \alpha_1 \cup \alpha_2 \cup \cdots \cup \alpha_m \neq 0$;

(b) *there is a class* $\beta \in H^*(B, B^-)$ *such that* $\beta \cup \alpha \neq 0$ *in* $H^*(B, B^-)$.

Then $l(S) \geqslant m + 1$; *recall that in case* $B^- = \varnothing$, *where* S *is an attractor, one has* $H^*(S) = H^*(B)$.

Here $l(X)$ stands for the cup long of a compact space X. If $H^*(X)$ is the Alexander cohomology it is defined as

$$l(X) = 1 + \sup\{k \in \mathbf{N} | \text{there are classes}$$
$$\alpha_1, \ldots, \alpha_k \in H^*(X) \setminus 1 \text{ with } \alpha_1 \cup \cdots \cup \alpha_k \neq 0\},$$

and $l(X) = 1$ if no such class exists. The above proposition can be viewed as an extension of the Ljusternik–Schnirelman theory to the case of topological flows. In fact, in the special case of a gradient flow on a manifold one concludes using duality from the conditions (a) and (b) on (B, B^-) a chain of length $m + 1$ of subordinated homology classes and therefore at least $m + 1$ rest points of the flow in B by using mini-max over the homology classes.

PROOF. Define $A^- = \{\gamma \in B | \gamma \cdot \mathbf{R}^- \subset B\}$ to be the set in B which stays in B for all negative times. From $S = \bigcap_{t<0} A^- \cdot t$ we conclude by the continuity of the Alexander cohomology that $H^*(S) = H^*(A^-)$. Similarly it can be shown that

$H^*(B, B^- \cup A^-) = 0$. Consider now the diagram:

$$\begin{array}{ccccc} H^*(B, B^-) \times & H^*(B, A^-) & \stackrel{\cup}{\to} & H^*(B, B^- \cup A^-) \cong 0 \\ \downarrow \text{id} & \downarrow j^* & & \downarrow j^* \\ H^*(B, B^-) \times & H^*(B) & \stackrel{\cup}{\to} & H^*(B, B^-) \\ & \downarrow i^* & & \\ & H^*(A^-) \cong H^*(S) & & \end{array}$$

Assume $i^*\alpha = 0$, with $\alpha = \alpha_1 \cup \cdots \cup \alpha_m$. By the exactness of the short sequence there is an $\hat{\alpha} \in H^*(B, A^-)$ such that $\alpha = j^*(\hat{\alpha})$. Take $\beta \in H^*(B, B^-)$ as in assumption (b) of the Proposition. From the commutativity of the diagram it then follows that $\beta \cup \alpha = j^*(\beta \cup \hat{\alpha}) = j^*(0) = 0$, contradicting (b). Hence $i^*\alpha \neq 0$ and therefore $l(A^-) = l(S) \leq m + 1$. □

We apply the proposition to the special block (B, B^-) given by (7). Here $H^*(B) \cong H^*(T^{2n})$ and $H^*(B, B^-) \cong H^*(T^{2n}) \times H^*(S^N)$; and hence there are classes $\alpha_s \in H^1(B)$ with $\alpha := \alpha_1 \cup \alpha_2 \cup \cdots \cup \alpha_{2n} \neq 0$ being the volume form on T^{2n}. If β is the image of the volume form on S^N, then $\beta \cup \alpha \neq 0$ in $H^*(B, B^-)$ and the assumptions of the proposition are met. It follows that

(11) $$l(S) \leq 2n + 1 = l(B) = l(T^{2n}).$$

In order to count the rest points of the flow in S we recall the concept of a Morse decomposition.

DEFINITION. *Let S be a compact invariant set of a continuous flow. A Morse decomposition of S is a finite collection $\{M_p\}_{p \in P}$ of disjoint compact and invariant subsets of S which can be ordered, say $\{M_1, M_2, \ldots, M_k\}$, so that the following property holds true. If*

$$\gamma \in S \setminus \bigcup_{p \in P} M_p,$$

then there is a pair of indices $i < j$ such that the positive $(t \to +\infty)$ and the negative $(t \to -\infty)$ limit sets $\omega(\gamma)$ and $\omega^*(\gamma)$ of γ satisfy $\omega(\gamma) \subset M_i$ and $\omega^*(\gamma) \subset M_j$.

From this definition one deduces immediately

PROPOSITION 2 [10]. $l(S) \leq \sum_{p \in P} l(M_p)$.

PROOF. It is sufficient to consider a decomposition of S into two sets $\{M_1, M_2\}$. From the definition we conclude two compact subsets of S satisfying $S_1 \cup S_2 = S$ and $M_1 = \bigcap_{t > 0} S_1 \cdot t$ and $M_2 = \bigcap_{t < 0} S_2 \cdot t$. By the continuity of the Alexander cohomology $H^*(S_j) = H^*(M_j)$, $j = 1, 2$, and it remains to prove $l(S) \leq l(S_1) + l(S_2)$, which is however well known (see also [10]). □

We finally use the gradient structure of the flow in B. Consider a compact invariant set S of a continuous flow, which is on S gradient-like, i.e. for which there is a continuous real valued function g on S which is strictly decreasing on nonconstant orbits. If there are only finitely many rest points in S, they constitute a Morse decomposition of S. An ordering of the rest points is simply an ordering of their values under the function g. Since in this case the sets M_p are points and hence have cuplong equal to 1 we deduce from Proposition 2 that S contains at least $l(S)$ rest points. If we apply this remark in particular to the set S of bounded solutions of the gradient flow $dz/ds = -\nabla g(z)$ we conclude, in view of estimate (11), that g possesses at least $2n + 1$ critical points z_j. The corresponding loops $x_j = u(z_j)$ are the required forced oscillations of Theorem 1.

3. Morse theory for subharmonic solutions on T^{2n}, $n \geq 2$, and proof of Theorem 2. If $T > 0$ is an integer, then a T-periodic solution of (1), which is a contractible loop on T^{2n}, is a critical point of the functional

$$(12) \qquad f(x) = \int_0^T \left\{ \frac{1}{2} \langle \dot{x}, Jx \rangle - h(t, x(t)) \right\} dt,$$

defined on the T-periodic functions $x: [0, T] \to \mathbf{R}^{2n}$. As in the previous section the critical points of f are in one-to-one correspondence with the rest points of the gradient flow

$$(13) \qquad \frac{d}{ds} z = -\nabla g(z), \quad \text{with } z \in T^{2n} \times \mathbf{R}^{2N},$$

on the finite dimensional submanifold, where N depends on the chosen period T. The rest points will be found by means of Morse theory in (B, B^-). It will be crucial that, beyond the statement of Lemma 1, the Morse index of a nondegenerate critical point z of g is related to the winding number of the corresponding periodic solution $x = u(z)$ of (1). This winding number is, similarly to the Maslov index, intrinsically defined in the phase space by means of the linearized flow along the solution x. This additional structure of the Morse index will allow us, by means of an iterated index formula, to separate in the function space those critical points belonging to T-periodic solutions with minimal period T from those belonging to T-times iterated 1-periodic solutions.

In order to recall from [11] the winding number we take a periodic solution $x(t) = x(t + T)$ of (1) having period T. The linearized equation along $x(t)$ is then given by $\dot{y} = Jh_{xx}(t, x(t))y$. Setting $S(t) = h_{xx}(t, x(t))$ we have $S(t) = S(t + T)$ and symmetric. The resolvent $X(t) \in \mathscr{L}(\mathbf{R}^{2n})$ satisfies

$$(14) \qquad \dot{X}(t) = JS(t) \cdot X(t) \quad \text{and} \quad X(0) = 1.$$

The eigenvalues of the Monodromy matrix $\Omega = X(T)$, T being the period, are the Floquet multipliers of $x(t)$. The matrix $X(t)$, $t \in \mathbf{R}$, belongs to the group of symplectic matrices $\mathrm{Sp}(n, \mathbf{R})$ which we abbreviate by $W := \mathrm{Sp}(n, \mathbf{R})$. We shall call the T-periodic solution $x(t)$ nondegenerate if 1 is not an eigenvalue of $\Omega = X(T)$.

Set $W^* = \{ M \in W \mid 1 \notin \sigma(M) \}$. For $M \in W$ there is a unique polar decomposition $M = PO$ with P and O belonging to W and where P is positive symmetric and O is orthogonal. Hence in particular

$$O = \begin{pmatrix} u_1 & -u_2 \\ u_2 & u_1 \end{pmatrix} \quad \text{with } \bar{u} := u_1 + iu_2 \in U(n),$$

$U(n)$ being the group of unitary matrices in $\mathbf{C}^n$. Since $|\det \bar{u}| = 1$ there is a homomorphism $0 \to \bar{u} \to \det \bar{u} \in S^1$. For an arc $\gamma(t)$ in W, $t_0 \le t \le t_1$, there is an associated arc $\bar{u}(t) \in U(n)$ and we can pick a continuous function $\Delta(t) \in \mathbf{R}$ with $\det \bar{u}(t) = \exp(i\Delta(t))$. Then $\Delta(\gamma) := \Delta(t_1) - \Delta(t_0)$ depends only on γ. If $\gamma(t_0) = \gamma(t_1)$, then $\Delta(\gamma) = 2\pi m$ for $m \in \mathbf{Z}$ and the loop γ is contractible in W if and only if $\Delta(\gamma) = 0$.

To the above periodic solution $x(t)$ with resolvent $X(t) = \gamma(t)$, $0 \le t \le T$, we associate $\Delta(\gamma) = \Delta(x)$. Assume now x to be nondegenerate. Then $\gamma(0) = \text{id}$ and $\Omega = \gamma(T) \in W^*$. The set W^* has two components each of which is simply connected relative to W. One component contains the matrix $Y_+ = -\text{id}$ with degree $(+1)$. The other component contains the matrix

$$Y_- = \begin{pmatrix} 2 & & \\ & -I & \\ & & 1/2 \\ & & & -I \end{pmatrix}$$

having degree (-1), where I stands for the identity matrix in $(n-1)$ dimensions. We can continue the arc γ by an arc $\hat{\gamma}$ in W^* from $\Omega = X(T)$ to either Y_+ or Y_-. $\Delta(\hat{\gamma})$ depends only on Ω and we shall write $\Delta(\hat{\gamma}) = r(\Omega)$. It moreover can be shown that $0 \le |r(\Omega)| < \pi n$ and $r(\Omega) = 0$ if Ω is a hyperbolic matrix. Now $\Delta(\gamma \cup \hat{\gamma}) = \Delta(\gamma) + \Delta(\hat{\gamma})$ is an integer multiple of π and we define the winding number of the periodic solution $x(t)$ to be the integer

$$j(x) = \pi^{-1}(\Delta(\gamma) + r(\Omega)) \in \mathbf{Z},$$

where $\Delta(\gamma) \in \mathbf{R}$ and $|r(\Omega)| < \pi n$. If now $x(t) = x(t + T)$ is a T-periodic solution we denote its l-times iterated loop $x(t) = x(t + lT)$ by x^l, it has period lT. If it is nondegenerate, it has the winding number

$$j(x^l) = \pi^{-1}(\Delta(\gamma^l) + r(\Omega^l)),$$

where $\gamma^l(t) = X(t)$, $0 \le t \le lT$, and $\Omega^l = X(lT)$. The following iterated index formula then holds true.

PROPOSITION 3. *Assume x^l is nondegenerate. Then*

$$j(x^l) = \pi^{-1}(l\Delta(\gamma) + r(\Omega^l))$$

with $0 \le |r(\Omega^l)| < \pi n$. Moreover, if Ω is hyperbolic, then $j(x^l) = l \cdot j(x)$.

PROOF. We have to prove $\Delta(\gamma^l) = l\Delta(\gamma)$. Recall that $X(t + lT) = X(t)\Omega^l$ for $0 \le t \le T$. Now $\Delta(\gamma^l) = \Delta(\gamma_1) + \Delta(\gamma_2) + \cdots + \Delta(\gamma_l)$, where $\gamma_k(t) = X(t)\Omega^{k-1}$ for $0 \le t \le T$. With the polar decompositions $\Omega^{k-1} = e^A O$ and $X(t) = e^{A(t)}O(t)$ we define the continuous deformation δ_s of γ_k by $\delta_s(t) = e^{sA(t)} \cdot O(t) \cdot e^{sA} \cdot O$, with $0 \le s \le 1$ and $0 \le t \le T$. From the diagram

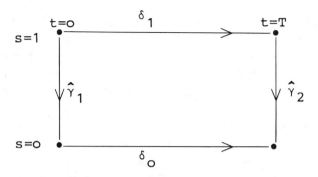

with $\delta_1(t) = \gamma_k(t)$ and $\delta_0(t) = O(t) \cdot O$ for $0 \leq t \leq T$ we conclude

$$\Delta(\delta_1 \cup \hat{\gamma}_2 \cup (-\delta_0) \cup (-\hat{\gamma}_1)) = 0,$$

as the loop is contractible in W. Since obviously $\Delta(\hat{\gamma}_1) = \Delta(\hat{\gamma}_2) = 0$ we conclude $\Delta(\delta_1) = \Delta(\delta_0) = \Delta(O(t) \cdot O)$ which is equal to $\Delta(O(t)) = \Delta(\gamma)$ so that $\Delta(\gamma_k) = \Delta(\gamma)$ and the claim follows. □

The relation between the winding number of a nondegenerate periodic orbit and the Morse index of its corresponding rest point of the gradient flow is given by the following lemma, which is proved in [11] by a deformation argument.

PROPOSITION 4. *Let $x = u(z)$ be a nondegenerate T-periodic orbit of (1) with winding number j. Then the corresponding critical point z of g in $T^{2n} \times \mathbf{R}^{2N}$ is nondegenerate and has Morse index $m = N + n - j$.*

In order to prove Theorem 2 we assume that all the 1-periodic orbits of (1) are nondegenerate. There are then only finitely many of them, $x_1, x_2, \ldots, x_m$, and we denote their winding numbers by $j(x_k) = \pi^{-1}(\Delta(x_k) + r(\Omega_k))$. Assume now that no Floquet multiplier is a root of unity. Then all the iterates x_k^l of these 1-periodic solutions are nondegenerate as well. From Proposition 3 we conclude that either $|j(x_k^l)| \to +\infty$ as $l \to +\infty$ (in case $\Delta(x_k) \neq 0$) or $|j(x_s^l)| < n$ for all $l \geq 1$ (in case $\Delta(x_s) = 0$). Therefore there is an integer T_0 such that for every $l \geq T_0$ and for all 1-periodic solutions we have $|j(x_k^l)| > n$ if $\Delta(x_k) \neq 0$ and $|j(x_s^l)| < n$ if $\Delta(x_s) = 0$. Now let $T \geq N_0$ be prime and assume there are only finitely many T-periodic solutions. Their corresponding rest points in S, which is contained in (B, B^-), constitute a Morse decomposition of S. By Proposition 4 we find for the Morse equation, in view of (9):

$$\sum_{k=1}^{m} t^{N+n-j_k} + \sum_{s=1}^{r} p(t, z_s) = \sum_{j=0}^{2n} \binom{2n}{j} t^{N+j} + (1+t)q(t),$$

where the first sum is the contribution of the T-times iterated 1-periodic solutions having indices $j_k = j(x_k^T)$ with $1 \leq k \leq m$. Since $|j_k| \neq n$ for every k, these iterated periodic solutions do not represent any cohomology of the torus T^{2n} on the right side in dimension zero and in dimension $2n$. Therefore the second term representing the contribution of the periodic solutions having minimal period T does not vanish, and we conclude at least one critical point z_s corresponding to a periodic solution having T as minimal period. If the T-periodic solutions are nondegenerate, then $p(t, z_s) = t^{m_s}$ and therefore we must have $r \geqslant 2$ in order to represent the lowest and highest cohomology of T^{2n}, which means that there are at least two critical points corresponding to periodic solutions with minimal period T. Finally, if, in addition, $\Delta(x_k) \neq 0$ for all $1 \leqslant k \leqslant m$, then $|j_k| > n$ for all k and therefore $r \geqslant 2^{2n}$ as claimed in Theorem 2.

4. Concluding remarks. (a) The proof of Arnold's conjecture outlined above makes heavy use of the global coordinate system available on T^{2n} on which, moreover, the symplectic structure is constant. For this reason the action functional for $h = 0$ dominates the gradient flow, so that an isolating block for the set of bounded solutions is immediately found for every Hamiltonian function. On an arbitrary symplectic compact manifold M such coordinates do exist only locally but the action functional for $h = 0$ still dominates in a neighborhood of the constant loops in the loopspace if the Hamiltonian vector field is sufficiently C^0-small. This way A. Weinstein [25] succeeded to establish for C^0-*small* Hamiltonian vector fields the existence of at least cuplong (M) forced oscillations on an arbitrary compact symplectic manifold.

We point out that in the special case $n = 1$, the statement of Theorem 1 for T^{2n} holds true for every symplectic structure. This follows in fact from J. Moser's theorem [17] that two volume forms are equivalent (by means of a diffeomorphism) if they have the same total volume. For $n > 1$ the classification of symplectic forms according to equivalence is not understood as far as we know, not even for the torus T^{2n}.

More generally, on every compact symplectic manifold M one might hope to find at least as many forced oscillations for exact Hamiltonian vector fields depending periodically on time, as a function on M has critical points. For some special cases proofs have been given recently, namely for $M = CP^n$ with its standard symplectic structure by B. Fortune and A. Weinstein [16] and for two-dimensional surfaces and certain Kähler manifolds by A. Floer [31] and J. Sikorav [23].

(b) We already pointed out that the Morse theory of forced oscillations for a general Hamiltonian system on a symplectic manifold is different from the problem of closed geodesics on a Riemannian manifold, which are the critical points of a coercive variational principle. Still one could ask whether, similar to the approach of broken geodesics (see for example G. Birkhoff [7]), there is a different variational approach on a finite dimensional manifold *directly* for which the forced oscillations are the critical points, by using that locally and piecewise the loops satisfy the Hamiltonian equation. For the case of the torus M. Chaperon [28] constructed indeed such a variational principle.

(c) Finally it should be pointed out that the above approach which turns a problem of symplectic geometry into a variational problem on an infinite dimensional function space is applicable also to global intersection problems of

Lagrangian manifolds. We do not describe these results here, but refer to M. Chaperon [8] and [9] and H. Hofer [29]; see also the recent beautiful work by M. Gromov [30].

This work is supported by the Stiftung Volkswagenwerk.

References

1. V. I. Arnold, *Mathematical methods of classical mechanics* (Appendix 9), Springer, 1978.
2. _____, "Fixed points of symplectic diffeomorphisms", problem XX in *Problems of present-day mathematics*, Proc. Sympos. Pure Math., Vol. 28, Amer. Math. Soc., Providence, R. I., 1976, p. 66.
3. A. Banyaga, *Sur la structure du groupe des difféomorphisms qui préservent une forme symplectique*, Comment. Math. Helv. **53** (1978), 174–227.
4. G. D. Birkhoff, *Proof of Poincaré's geometric theorem*, Trans. Amer. Math. Soc. **14** (1913), 14–22.
5. _____, *An extension of Poincaré's last geometric theorem*, Acta Math. **47** (1925) 297–311.
6. _____, *The restricted problem of three bodies*, Rend. Circ. Mat. Palermo **39** (1915), 265–334.
7. _____, *Dynamical systems with two degrees of freedom*, Trans. Amer. Math. Soc. **18** (1917), 199–300.
8. M. Chaperon, *Quelques questions de géométrie symplectique*, Séminaire Bourbaki 1982/83, n^0 610; Astérisque **105–106** (1983), 231–249.
9. M. Chaperon and E. Zehnder, *Quelques résultats globeaux en géometrie symplectique*, vol. III: *Travaux en cours*, Acte des Journées de Lyon 1983, Hermann, 1984.
10. C. Conley and E. Zehnder, *The Birkhoff–Lewis fixed point theorem and a conjecture of V. I. Arnold*, Invent. Math. **73** (1983), 33–49.
11. _____, *Morse type index theory for flows and periodic solutions of Hamiltonian equations*, Comm. Pure Appl. Math. **37** (1984), 207–253.
12. _____, *An index-theory for periodic solutions of a Hamiltonian system*, Geometric Dynamics (Rio 1981), Lecture Notes in Math., vol. 1007, Springer, pp. 122–145.
13. C. Conley, *Isolated invariant sets and the Morse index*, CBMS Regional Conf. Ser. in Math., vol. 38, Amer. Math. Soc., Providence, R.I., 1978.
14. I. Ekeland, *Une théorie de Morse pour les systèmes hamiltoniens convexes*, Ann. Inst. H. Poincaré, Anal. Non Linéaire **1** (1984), 19–78.
15. V. Benci, MRC report (to appear).
16. B. Fortune and A. Weinstein, *A symplectic fixed point theorem for complex projective spaces*, Bull. A.M.S. **12** (1985), 128–130.
17. J. Moser, *On the volume elements on a manifold*, Trans. Amer. Math. Soc. **120** (1965), 286–294.
18. _____, *Periodic orbits near an equilibrium and a theorem by Alan Weinstein*, Comm. Pure Appl. Math. **29** (1976), 727–747.
19. _____, *A fixed point theorem in symplectic geometry*, Acta Math. **141** (1978), 17–34.
20. N. Nikishin, *Fixed points of diffeomorphisms on the twosphere that preserve area*, Funkts. Anal. i Prilozhen **8** (1974), 84–85. (Russian)
21. H. Poincaré, *Sur un théorème de géométrie*, Rend. Circ. Mat. Palermo **33** (1912), 375–407.
22. P. Rabinowitz, *Periodic solutions of Hamiltonian systems*, Comm. Pure Appl. Math. **31** (1978), 157–184.
23. J. C. Sikorav, *Points fixes d'un symplectomorphisme homologue à l'identité*, preprint, Orsay, 1984.
24. C. P. Simon, *A bound for the fixed point index of an area-preserving map with applications to mechanics*, Invent. Math. **26** (1974), 187–200.
25. A. Weinstein, C^0-*perturbation theorems for symplectic fixed points and Lagrangian intersections*, Lecture Notes of the AMS Summer Institute on Nonlinear Functional Analysis and Applications, Berkeley, 1983.
26. _____, *Lectures on symplectic manifolds*, CBMS Regional Conf. Ser. in Math., vol. 29, Amer. Math. Soc., Providence, R.I., 1977.
27. Kung-Ching Chang, *Applications of homology theory to some problems in differential equations*, MSRI 064-83, Berkeley, 1983.

28. M. Chaperon, *Une idée du type "géodésique brisée" pour les systèmes hamiltoniens*, C. R. Acad. Sci. Paris **298**, no. 13 (1984), 293–296.

29. H. Hofer, *Lagrangian embedding and critical point theory*, Bath, 1984 (preprint).

30. M. Gromov, *Pseudoholomorphic curves in symplectic manifolds*, IHES, Bûres-Sûr-Yvette, 1985 (preprint).

31. A. Floer, *Proof of the Arnold conjecture for surfaces and generalizations for certain Kähler manifolds*, RUB Bochum, 1984 (preprint).

UNIVERSITY OF WISCONSIN, MADISON AND UNIVERSITÄT BOCHUM, FEDERAL REPUBLIC OF GERMANY

Harmonic Maps from the Disk into the Euclidean N-Sphere

JEAN-MICHEL CORON

I. Introduction. This article is about some joint work [BrC2, BeC] with H. Brézis and V. Benci. The problem is the following: Let

$$\Omega = \{(x, y) \in \mathbf{R}^2 \mid x^2 + y^2 < 1\},$$

$$S^N = \{v \in \mathbf{R}^{N+1} \mid |v| = 1\},$$

and γ be a function from $\partial\Omega$ into S^N; we seek functions u from $\overline{\Omega}$ into S^N such that

(1) $\qquad -\Delta u = u|\nabla u|^2 \quad \text{in } \Omega,$

(2) $\qquad u = \gamma \quad \text{on } \partial\Omega.$

Let $\varepsilon = \{u \in H^1(\Omega; \mathbf{R}^{n+1}) \mid u = \gamma \text{ on } \partial\Omega \text{ and } |u| = 1 \text{ a.e.}\}$. If $N = 2$, we assume that ε is nonempty (this is the case if, for example, $\gamma \in C^1(\partial\Omega; S^2)$). If $N \geqslant 3$, we assume that $\gamma \in C^{2,\delta}(\partial\Omega; S^N)$ for some $\delta \in (0,1)$, which means that $\gamma \in C^2(\partial\Omega; S^N)$ and that the second derivatives of γ are Hölder continuous with exponent δ. Then ε is also nonempty.

We have the following theorem.

THEOREM. *If γ is not a constant, then there exist at least two solutions of (1)–(2) which are in ε.*

REMARKS. When $N = 2$, the theorem has been obtained independently by Jost [J] and Brézis-Coron [BrC2]. The case $N \geqslant 3$ was later proved by Benci-Coron [BeC].

2. The existence of at least one solution of (1)–(2) in ε is trivial (and remains true even if γ is a constant). To see this, let

$$E(u) = \int_\Omega |\nabla u|^2, \quad u \in \varepsilon.$$

1980 *Mathematics Subject Classification.* Primary 58E20; Secondary 35J65.

Clearly, there exists some $\underline{u}$ in ε such that

$$E(\underline{u}) = \operatorname*{Inf}_{\varepsilon} E = m, \tag{3}$$

and it is easy to prove that $\underline{u}$ is a solution of (1)–(2).

3. If u is a continuous solution of (1)–(2) in ε, then it is well known (see [**LU, HW, Wi**]) that $u \in C^\infty(\Omega; S^2)$. Moreover, thanks to Morrey's regularity theory, $\underline{u} \in C^\infty(\Omega; S^2)$. Our second solution is also in $C^\infty(\Omega; S^2)$.

4. When γ is a constant, L. Lemaire [**Le**] has proved that if u is a continuous solution of (1)–(2) in ε, then u is also a constant. Therefore, if continuous solutions are sought, the assumption that γ is not a constant is essential to our theorem.

5. When $N = 2$, then the theorem remains true if S^2 is replaced by any Riemannian surface homeomorphic to S^2 (see [**BrC2**]).

II. Sketch of the proof for $N = 2$. See [**BrC2**] for the details. We begin with the remark that ε is not connected. In fact, let

$$Q(u) = \frac{1}{4\pi} \int_\Omega u \cdot u_x \wedge u_y \quad \text{for } u \in \varepsilon.$$

Then we have the following:

LEMMA 1. *If u_1 and u_2 are in ε, then $Q(u_1) - Q(u_2) \in \mathbf{Z}$.*

PROOF. (See [**BrC2**]). In the proof the crucial point is a density theorem ($C^\infty(S^2; S^2)$ is dense in $H^1(S^2; S^2)$), which is due to R. Schoen and K. Uhlenbeck [**ScU**]. Let $\varepsilon_k = \{u \in \varepsilon \mid Q(u) - Q(\underline{u}) = k\}$. It is easy to prove that ε_k is nonempty. It is natural to try to minimize E in ε_k, but there is a major difficulty due to the fact that ε_k is not weakly closed for the H^1-topology. In fact, one can find examples (see [**BrC2**]) where the infinimum of E on ε_k is achieved if and only if $k \in \{0, 1\}$. Similar difficulties also occur in [**A; BrC1; BN; J; Lb; Ls; SaU; S; T**] and [**We**].

We are going to prove that at least one of the two infima, $\operatorname{Inf}_{\varepsilon_{-1}} E$ or $\operatorname{Inf}_{\varepsilon_1} E$, is achieved. The key (inspired by [**BrC1**]) to this is the following lemma.

LEMMA 2. $\exists v \in \varepsilon$ *such that*

$$v \in \varepsilon_1 \cup \varepsilon_{-1}, \tag{4}$$

$$E(v) < E(\underline{u}) + 8\pi. \tag{5}$$

REMARK. When γ is constant Lemma 2 is false.

PROOF OF THE THEOREM. (See [**BrC2**]). Let us assume that, for example $v \in \varepsilon_1$. We are going to prove that $\operatorname{Inf}_{\varepsilon_1} E$ is achieved. Let $u^n \in \varepsilon_1$ be such that

$$E(u^n) = \operatorname*{Inf}_{\varepsilon_1} E + O(1).$$

Clearly, u^n is bounded in $H^1(\Omega; \mathbf{R}^3)$, and therefore we may assume that u^n tends weakly to some $\bar{u}$ in $H^1(\Omega; \mathbf{R}^3)$. We have

$$E(\bar{u}) \leq \underline{\lim} E(u^n).$$

Hence, we need only prove that $\bar{u} \in \varepsilon_1$. We argue by contradiction. Assume that $Q(\bar{u}) \neq 1$. Therefore, $|Q(u^n) - Q(\bar{u})| \geq 1$; for example,

(6) $$Q(u^n) \geq Q(u) + 1.$$

Let

$$F(u^n) = E(u^n) - 8\pi Q(u^n) = \int |\nabla u^n|^2 - 2u^n \cdot (u_x^n \wedge u_y^n).$$

(If instead of (6) we have $Q(u^n) \leq Q(\bar{u}) - 1$, we consider $F(u^n) = E(u^n) + 8\pi Q(u^n)$.) Since the map $\mathbf{R}^3 \times \mathbf{R}^3 \to \mathbf{R}$, $(p, q) \to |p|^2 + |q|^2 - 2a \cdot (p_x \wedge p_y)$ is convex if $|a| \leq 1$, we have $F(u) \leq \lim F(u^n)$. Hence, (with (6))

$$\mathrm{Inf}_\varepsilon E \geq E(\bar{u}) + 8\pi, \quad \text{but} \quad E(\bar{u}) \geq E(\underline{u}).$$

Therefore,

$$\mathrm{Inf}\, E \geq E(\underline{u}) + 8\pi,$$

but (see (5))

$$\mathrm{Inf}\, E \leq E(v) < E(\underline{u}) + 8\pi,$$

a contradiction.

III. Sketch of the proof for $N \geq 3$. (See [BeC] for the details.) In this case we use the Ljusternik–Schnirelmann topological method; since ε is not a manifold, following J. Sacks–K. Uhlenbeck [SaU] we study an approximate problem, i.e., the critical points of the functional

$$E_\alpha(u) = \int_\Omega \left(1 + |\nabla u|^2\right)^\alpha - 1 \text{ with } u \in \varepsilon^\alpha \text{ and } \alpha > 1,$$

where

$$\varepsilon^\alpha = \{ u \in W^{1,2\alpha}(\Omega; \mathbf{R}^{N+1}) | u = \gamma \text{ on } \partial\Omega \text{ and } |u| = 1 \},$$

$$\Sigma_\alpha = \{ \sigma | \sigma \in C^0(S^{n-2}; \varepsilon^\alpha), \sigma \text{ is not homotopic to a constant} \}.$$

Σ_α satisfies the Palais–Smale condition and $\Sigma_\alpha \neq \emptyset$. Let, for $\alpha > 1$,

$$C_\alpha = \inf_{\sigma \in \Sigma_{2\alpha}} \mathrm{Max}_{S \in S^{n-2}} E_\alpha(\sigma(S)).$$

c_α is a critical value of E_α. Clearly, c_α is nondecreasing in α. Let $c = \lim_{\alpha \to 1^+} c_\alpha$. We have two cases.

Case 1: $c > m$.

Let u be such that $E'_\alpha(u_\alpha) = 0$ and $E_\alpha(u_\alpha) = c_\alpha$. Obviously u_α is bounded in H^1, and therefore we can extract a subsequence u_{α_n} which converges weakly in H^1 to some u; one can prove that u satisfies (1)–(2) (see [SaU] or [BeC]), and the crucial point is to prove that $u \neq \underline{u}$. For the proof of $u \neq \underline{u}$, as for $N = 2$, the key point is the strict inequality $c < m + 8\pi$. Using this inequality, a theorem of E. Calabi [C], arguments involved in Sacks-Uhlenbeck [SaU], and classical Morrey estimates [M] (see also [N]), we prove the strong convergence of u_{α_n} to u. Hence, $E(u) = c > m = E(\underline{u})$, which proves that $u \neq \underline{u}$.

Case 2: $c = m$.

We prove that in this case problem (1)-(2) has infinitely many solutions in ε. The crucial point is

LEMMA 3. *For every w in ε there is $\theta > 0$ such that* $\{u \in \varepsilon | \|w - u\|_{H_1} \leq \theta\} \cap \varepsilon^\alpha$ *is contractible to a point in ε^α for every $\alpha \in [1, 2]$.*

For the proof of Lemma 3 we use an argument involved in Schoen–Uhlenbeck [ScU]. For a proof of Lemma 3 and the end of the proof, see [BeC].

REFERENCES

[A] Th. Aubin, *Equations différentielles non linéaires et problème de Yamabe concernant la courbure scalaire*, J. Math. Pures Appl. **55** (1976), 269–296.

[BeC] V. Benci and J. M. Coron, *Dirichlet problem for harmonic maps from the disk into the Euclidean n-sphere*, Ann. Inst. H. Poincaré. Anal. Non Linéaire (1985).

[BrC1] H. Brézis and J. M. Coron, *Multiple solutions of H-systems and Rellich's conjecture*, Comm. Pure Appl. Math. **37** (1984), 149–187.

[BrC2] _____, *Large solutions for harmonic maps in two dimensions*, Comm. Math. Phys. **92** (1983) 203, 215.

[BN] H. Brézis and L. Nirenberg, *Positive solutions of nonlinear elliptic equations involving critical Sobolev exponents*, Comm. Pure Appl. Math. **36** (1983), 437–477.

[C] E. Calabi, *Minimal immersions of surfaces in Euclidean spheres*, J. Differential Geom. **1** (1967), 111–125.

[W] S. Hildebrandt and K. O. Widman, *Some regularity results for quasilinear elliptic systems of second order*, Math. Z. **142** (1975), 67–86.

[J] J. Jost, *The Dirichlet problem for harmonic maps from a surface with boundary onto a 2-sphere with non-constant boundary values*, J. Differential Geom. (to appear).

[LU] O. A. Ladyženskaya and N. N. Ural'ceva, *Linear and quasilinear elliptic equations*, 2nd Russian ed., "Nauka" Moscow, 1973.

[Le] L. Lemaire, *Application harmonique de surfaces riemanniennes*, J. Differential Geom. **13** (1978), 51–78.

[Lb] E. Lieb, *Sharp constants in the Hardy-Littlewood-Sobolev and related inequalities*, Ann. of Math. (2) **118** (1983), 349–374.

[Ls] P. L. Lions, *The concentration compactness principle in the calculus of variations, the limit case*, Riv. Iberoamericana (to appear). Announced in C. R. Acad. Sci. Paris **296** (1983), 643–648.

[M] C. B. Morrey, *Multiple integrals in the calculus of variations*, Springer-Verlag, 1966.

[N] L. Nirenberg, *On nonlinear elliptic partial differential equations and Hölder continuity*, Comm. Pure Appl. Math. **6** (1953), 103–156.

[SaU] J. Sacks and K. Uhlenbeck, *The existence of minimal immersions of 2-spheres*, Ann. of Math. (2) **113** (1981), 1–24.

[ScU] R. Schoen and K. Uhlenbeck, *Boundary regularity and the Dirichlet problem for harmonic maps*, J. Differential Geom. **18** (1983), 253–268.

[S] M. Struwe, *Nonuniqueness in the Plateau problem for surfaces of constant mean curvature*,

[T] C. Taubes, *The existence of a non-minimal solution to the SU(2) Yang-Mills-Higgs equations on R^3*, Part I Comm. Math. Phys. **86** (1982), 257–298, Part II Comm. Math. Phys. **86** (1982), 299–320.

[We] H. Wente, *The Dirichlet problem with a volume constraint*, Manuscripta Math. **11** (1974), 141–157.

[Wi] M. Wiegner, *A priori Schranken für Lösungen gewisser elliptischer systeme*, Math. Z. **147** (1976), 21–28.

L'ECOLE POLYTECHNIQUE, CENTRE DE MATHÉMATIQUE, FRANCE

Nonlinear Semigroups and Evolution Governed by Accretive Operators[1]

MICHAEL G. CRANDALL

Abstract. This is a review paper which outlines the main points of the theory of nonlinear semigroups and evolution governed by accretive operators. The subject is now rather mature, so most of the principal ideas and results are not new. However, the presentation here is organized differently from that in other sources and does touch upon recent results. An attempt has been made to make this paper a pleasant route to a certain view of the subject.

In this review article we will outline some of the main points of the theory of nonlinear semigroups and evolution governed by accretive operators. As the subject has achieved a certain maturity, most of the principal ideas and results are not new. However, the current presentation is somewhat different from others in its style and choice of topics, and we have tried to make it pleasant reading for newcomers to the subject. It does touch upon some recent results, and we hope it will be of interest.

The material is organized into eight sections. §1 contains preliminaries and introduces the subject via a "generation" theory point of view. Here one finds elementary notions about semigroups, their generators, strong solutions and accretive operators. §2 introduces the notion of a "mild solution" of an abstract initial-value problem, a notion which allows a certain unity and ease of expression in the following discussion which would otherwise be severely hampered by a lack of regular solutions. Mild solutions are, roughly, uniform limits of solutions of suitable difference approximations of the problem under consideration. §3 presents the basic convergence results, which state that if suitable difference ap-

---

1980 *Mathematics Subject Classification*. Primary 47H20, 47H06 47H15; Secondary 47H09, 35-02, 39-02.

Key words and phrases. Nonlinear evolution equation, nonlinear semigroups, generator, accretive operators, mild solutions, difference approximations.

[1] Technical Summary Report #2724, August 1984; sponsored by the United States Army under Contract No. DAAG29-80-C-0041 and in part by the National Science Foundation under Grant No. MCS-8002946.

proximations can be solved, then their solutions will converge. These results lie at the heart of the theory and provide an ample supply of mild solutions.

§4 is off the usual track and presents something a bit more novel. Here, in a model case, a relationship between Kato's theory of quasilinear evolution and the results of §3 is exhibited.

§5 is also organized in an unusual way. Here we return to the generation question and explain some of the highlights as well as a couple of open problems. These considerations are used to introduce more subtle conditions under which it can be proved that there are solvable difference approximations and to point out recent remarks by Kobayashi on the question of necessary conditions. This section is not referred to in what follows it.

In §6, at last, we concede to the conventional and discuss the regularity of our mild solutions. Here the standard conditions guaranteeing the differentiability of mild solutions and the pointwise satisfaction of the equations are given. This is also a natural place to describe the inequalities which Benilan proved uniquely identify a mild solution when it exists.

§7 briefly describes the most useful auxiliary results of the theory. These results concern the continuous dependence of solutions on the equation, representation of solutions and compactness criteria of various sorts.

Few of the results stated here are proved, although some description of the line of argument is given from time to time. Similarly, we have omitted all references in the text proper, as comments as to who did what interrupt the flow and do not serve a browsing reader well. We partially correct this in §8 where further comments are made on the material of the previous sections and (incomplete) references are given. Here we also attempt to refer to some of the current activity in this area of which we are aware, but the field has become too vast to attempt any sort of completeness in describing either the old or the new in an article of moderate length. For example, we have not attempted to discuss questions of asymptotic behaviour, an area which has enjoyed a great deal of relatively recent activity, in the text (but we do give some references in §8). An even more profound omission concerns applications, which we at first thought to approach somehow. However, this idea was abandoned owing to our inability to come up with a satisfactory scheme that was not inferior to suggesting the reader refer to Crandall [26], Evans [43] and Barbu [3] (in that order). This will not yield an up-to-date view of the situation, but it will provide some simple examples and then an accurate impression of the nature and range of applications. More recent references are given in §8.

It should be mentioned that this review is affected by work the author has done with Benilan and Pazy on the small book [13], which is in a state of perpetual preparation. The author is also indebted to S. Oharu, A. Pazy and S. Reich for their comments on this paper.

1. Preliminaries. The origins of this subject lie in questions posed by its pioneers about the "generation" of semigroups of transformations. We adopt this point of view as a pedagogical device, although a more "applied" attitude holds sway at the moment. Thus we begin by defining the class of semigroups under consideration and observing properties which their "generators" might be expected to have. This leads us naturally to the class of accretive operators.

Let X be a real Banach space with the norm $\| \ \|$. The norm of the dual space X^* of X will also be denoted by $\| \ \|$. If C is a subset of X, a *semigroup on C* will mean a collection $\{S(t): 0 \leq t\}$ of self-maps of C with the properties (i) below:

(i) $S(0)x = x$ and $S(t)S(\tau)x = S(t + \tau)x$ when $0 \leq t, \tau$ and $x \in C$.

Note that the value of $S(t)$ at $x \in C$ is written $S(t)x$ even if $S(t)$ is not a linear function. A semigroup S on C is a *continuous* semigroup if

(ii) The mapping $[0, \infty) \times C \ni (t, x) \to S(t)x \in C$ is continuous when C carries the norm topology of x.

We are mainly interested in the situation in which the continuity of $S(t)x$ in the "state variable" x is special. A continuous semigroup S on C which satisfies

(iii) $\|S(t)x - S(t)y\| \leq \|x - y\|$ for $0 \leq t$ and $x, y \in C$,

is said to be *nonexpansive* or a *semigroup of contractions*. More generally, if there is a number ω such that

(iv) $\|S(t)x - S(t)y\| \leq e^{\omega t}\|x - y\|$ for $0 \leq t$ and $x, y \in C$,

then we say that S is a *quasicontractive* (of type ω) semigroup on C. Of course, if either (iii) or (iv) hold, then S is continuous as soon as $t \to S(t)x$ is continuous for each $x \in C$.

The notion of a semigroup is an abstraction of the notion of a uniquely solvable initial-value problem of the form

(IVP)
$$du/dt + Au = 0,$$
$$u(0) = x,$$

where A is a (nonlinear) operator $A: D(A) \subseteq X \to X$. If for each $x \in C$ (IVP) has a unique solution (in some sense!) $u(t)$ on $[0, \infty)$, then putting $S(t)x = u(t)$ should define a semigroup on C. Further properties of the semigroup should correspond to further properties of the operator A and the notion of solution involved, and ways of relating semigroups and initial-value problems (or related objects) will be referred to here as "generation theory".

The most obvious way to attempt to associate an initial-value problem (IVP) with a semigroup S on C is to compute the operator

$$A_S x = \lim_{t \downarrow 0} \frac{x - S(t)x}{t}$$

whose domain $D(A_S)$ is the set of $x \in C$ such that the limit exists, and then hope "solving" (IVP) with $A = A_S$ will return S. The operator $-A_S$ is called the *infinitesimal generator* of S. Let us see how the quasicontractive property (iv) of a semigroup S would be reflected in its infinitesimal generator. If S satisfies (iv), then, for $0 < \lambda, t$,

$$\left\| x - \hat{x} + \lambda \left(\frac{(x - S(t)x)}{t} - \frac{(\hat{x} - S(t)\hat{x})}{t} \right) \right\|$$

$$\geq \left(1 + \frac{\lambda}{t}\right)\|x - \hat{x}\| - \frac{\lambda}{t}\|S(t)x - S(t)\hat{x}\|$$

$$\geq \left(1 + \frac{\lambda(1 - e^{\omega t})}{t}\right)\|x - \hat{x}\|$$

so if $x, \hat{x}$ are in $D(A_S)$ we may pass to the limit to find that

$$\|x - \hat{x} + \lambda(A_S x - A_S \hat{x})\| \geq (1 - \lambda\omega)\|x - \hat{x}\| \quad \text{for } x, \hat{x} \in D(A_S).$$

We will refer to this property by saying that $A_S + \omega I$ is accretive. More precisely, if A is an operator then $A + \omega I$ *is accretive* if

(1.1) $\quad \|x - \hat{x} + \lambda(Ax - A\hat{x})\| \geq (1 - \lambda\omega)\|x - \hat{x}\| \quad \text{for } x, \hat{x} \in D(A)$

and $\lambda \geq 0$.

In the special case that $A + 0I$ is accretive we simply say that A is accretive. (There is a little subtlety here, and we leave it to the reader to check that $A + \omega I$ is accretive if and only if $A + \omega I$ is accretive.) It follows from the above remarks that if S is a semigroup of contractions, then A_S is accretive.

If $z = x - \hat{x}$ and $w = Ax - A\hat{x}$ and $\omega = 0$, then (1.1) can be written as $0 \leq [z, w]_\lambda$ for $\lambda > 0$ where

(1.2) $\qquad\qquad [z, w]_\lambda = (\|z + \lambda w\| - \|z\|)/\lambda$

defines $[z, w]_\lambda$ for $\lambda \neq 0$. Since $\lambda \to \|z + \lambda w\|$ is a convex function of λ we may define

(1.3) $\qquad\qquad [z, w] = \lim_{\lambda \downarrow 0} [z, w]_\lambda = \inf_{\lambda > 0} [z, w]$

and then observe that an operator $A: D(A) \subseteq X \to X$ is accretive if and only if

(1.4) $\qquad\qquad 0 \leq [x - \hat{x}, Ax - A\hat{x}] \quad \text{for } x, \hat{x} \in D(A).$

Let us list some properties of the bracket [,] before continuing. One further concept, namely that of the "duality map" $J: X \to X^*$ is required. It is given by

(1.5) $\qquad J(x) = \{x^* \in X^*: x^*(x) = \|x\| \text{ and } \|x^*\| \leq 1\}$

where $x^*(x)$ denotes the value of $x^* \in X^*$ at $x \in X$. For example, $J(0)$ is the closed unit ball in X^*.

PROPOSITION 1. *Let $x, y, z \in X$ and $\alpha, \beta \in \mathbf{R}$. We have*
(i) $\quad$ [,]: $X \times X \to \mathbf{R}$ *is upper-semicontinuous,*
(ii) $\quad [\alpha x, \beta y] = |\beta|[x, y]$ *if $\alpha\beta > 0$,*
(iii) $\quad [x, \alpha x + y] = \alpha\|x\| + [x, y]$,
(iv) $\quad |[x, y]| \leq \|y\|$ *and* $[0, y] = \|y\|$,
(v) $\quad -[x, -y] \leq [x, y]$,
(vi) $\quad [x, y + z] \leq [x, y] + [x, z]$,
(vii) $\quad |[x, y] - [x, z]| \leq \|y - z\|$,
(viii) $\quad [x, y] = \max\{x^*(y): x^* \in J(x)\}$.

Let us consider still another way to say that $A + \omega I$ is accretive. If we put $z = x + \lambda Ax$ and $\hat{z} = \hat{x} + \lambda A\hat{x}$ in (1.1), then formally, $x = (I + \lambda A)^{-1}z$ and $\hat{x} = (I + \lambda A)^{-1}\hat{z}$, and (1.1) may be reformulated as

$$(1.6) \quad \|(I + \lambda A)^{-1}z - (I + \lambda A)^{-1}\hat{z}\| \leq (1 - \lambda\omega)^{-1}\|z - \hat{z}\|$$

$$\text{for } z, \hat{z} \in R(I + \lambda A),$$

from which we see that $(I + \lambda A)^{-1}$ is indeed a function with $(1 - \lambda\omega)^{-1}$ as a Lipschitz constant if $\lambda\omega < 1$. Just as f^{-1} need not be a function if f is, A need not be a function in order that (1.6) hold, and we will continue the discussion in the multivalued generality that this suggests.

More precisely, we will call a mapping $A\colon X \to 2^X$ (the subsets of X) an "operator" in X. Functions with domain and range in X are identified with the corresponding single-valued operators, where a single-valued operator A is one whose values Ax are either singletons or the empty set. The effective domain $D(A)$ of an operator A is

$$D(A) = \{x \in X\colon Ax \text{ is not empty}\}.$$

If A is a single-valued operator (or the corresponding function) and $x \in D(A)$, we will use Ax, depending on the context, to denote either the singleton set or its corresponding element. If A and B are operators and $\lambda \in \mathbf{R}$ then we form new operators A^{-1}, λA and $A + B$ in the expected ways. For example,

$$A^{-1}x = \{y \in X\colon y \in Ax\}.$$

We formulate the notion of accretiveness for operators. The equivalence of the four conditions given is clear from the above.

DEFINITION 1. If A is an operator in X and $\omega \in \mathbf{R}$, then $A + \omega I$ is accretive (or, for short, $A \in \mathbf{A}(\omega)$) if the following equivalent conditions hold:
  (i) $(1 - \lambda\omega)\|x - \hat{x}\| \leq \|x - \hat{x} + \lambda(y - \hat{y})\|$ for $y \in Ax, \hat{y} \in A\hat{x}$, and $\lambda \geq 0$.
  (ii) $[x - \hat{x}, y - \hat{y}] \geq -\omega\|x - \hat{x}\|$ for $y \in Ax$ and $\hat{y} \in A\hat{x}$.
  (iii) If $y \in Ax$ and $y \in Ax$, then there is an $x^* \in J(x - \hat{x})$ such that

$$x^*(y - \hat{y}) \geq -\omega\|x - \hat{x}\|.$$

  (iv) If $\lambda > 0$ and $\lambda\omega < 1$, then $(I + \lambda A)^{-1}$ is single-valued and has $(1 - \lambda\omega)^{-1}$ as a Lipschitz constant.

In practice, it is usually (ii) which is used to verify accretiveness. We complete this section by recalling the notion of a strong solution of the inclusion

$$(\text{DE})_f \qquad u'(t) + Au(t) \ni f(t)$$

in which A is an operator in X and $f\colon [0, T] \to X$ is Bochner integrable with respect to Lebesgue measure (that is, $f \in L^1(0, T\colon X)$). The space $W^{1,1}(0, T\colon X)$ consists of those functions u which have the form

$$(1.7) \qquad u(t) = u(0) + \int_0^t h(s)\, ds$$

for some $h \in L^1(0, T: X)$. It is well known that $W^{1,1}(0, T: X)$ consists of exactly those absolutely continuous functions $u: [0, T] \to X$ which are differentiable a.e. on $[0, T]$ and that, when (1.7) holds with $h \in L^1(0, T: X)$, then $u'(t) = h(t)$ a.e. Moreover, if X is reflexive, then every absolutely continuous $u: [0, T] \to X$ belongs to $W^{1,1}(0, T: X)$, while there are spaces X and absolutely continuous functions u which are nowhere differentiable.

DEFINITION 2. A strong solution of $(DE)_f$ on $[0, T]$ is a $u \in W^{1,1}(0, T: X)$ such that $f(t) - u'(t) \in Au(t)$ almost everywhere on $[0, T]$.

As a sample exercise in the concepts we have introduced so far, let us prove that if $A + \omega I$ is accretive, then strong solutions of $(DE)_f$ are determined by their initial values. More precisely,

PROPOSITION 2. *Let* $f, \hat{f} \in L^1(0, T: X)$, $A \in \mathbf{A}(\omega)$ *and* $u, \hat{u}$ *be strong solutions of* $u' + Au \ni f$, $\hat{u}' + A\hat{u} \ni \hat{f}$, *respectively, on* $[0, T]$. *Then*

(1.8)
$$\|u(t) - \hat{u}(t)\| \leq e^{\omega t}\|u(0) - \hat{u}(0)\|$$
$$+ \int_0^t e^{\omega(t-s)}[u(s) - \hat{u}(s), f(s) - \hat{f}(s)]\, ds$$
$$\leq e^{\omega t}\|u(0) - \hat{u}(0)\| + \int_0^t e^{\omega(t-s)}\|f(s) - \hat{f}(s)\|\, ds.$$

PROOF. Let $f: [0, T] \to X$ be differentiable from the right at $s \in [0, T)$. Then for $h > 0$,

$$\frac{\|f(s+h)\| - \|f(s)\|}{h} = \frac{\|f(s) + hf_R'(s)\| - \|f(s)\|}{h} + o(1)$$
$$= [f(s), f_R'(s)]_h + o(1)$$

where $f_R'(s)$ denotes the right derivative of f at s. Upon letting $h \downarrow 0$ we find that $\|f(t)\|$ has a right derivative at $t = s$ and

(1.9) $$D_R\|f(t)\|\bigg|_{t=s} = [f(s), f_R'(s)],$$

where D_R denotes the right derivative. Similarly,

(1.10) $$D_L\|f(t)\|\bigg|_{t=s} = -[f(s), -f_L'(s)],$$

and we conclude that if both f and $\|f(t)\|$ are differentiable at $t = s$, then

(1.11) $$(d/dt)\|f(t)\|\bigg|_{t=s} = [f(s), f'(s)] = -[f(s), -f'(s)].$$

If u and $\hat{u}$ are absolutely continuous, then so is $g(t) = \|u(t) - \hat{u}(t)\|$, which is therefore differentiable a.e. If u and $\hat{u}$ are as in the proposition, then $u, \hat{u}$ and g are all differentiable at almost every t, and, by the above, for such values of t we have

$$(d/dt)\|u(t) - \hat{u}(t)\| = -[u(t) - \hat{u}(t), \hat{u}'(t) - u'(t)]$$
$$= -[u(t) - \hat{u}(t), (f(t) - u'(t)) - (\hat{f}(t) - \hat{u}'(t)) + (\hat{f}(t) - f(t))].$$

Since u and $\hat{u}$ are strong solutions of their respective equations, we have $f(t) - u'(T) \in Au(t)$ and $\hat{f}(t) - \hat{u}'(t) \in A\hat{u}(t)$ a.e. At such points t, by Proposition 1(vi) and Definition 1(ii),

$$[u(t) - u(\hat{t}), (f(t) - u'(t)) - (\hat{f}(t) - \hat{u}'(t)) + (\hat{f}(t) - f(t))]$$
$$\geq [u(t) - \hat{u}(t), (f(t) - u'(T)) - (\hat{f}(t) - \hat{u}'(t))]$$
$$\quad - [u(t) - \hat{u}(t), f(t) - \hat{f}(t)]$$
$$\geq -\omega \|u(t) - u(t)\| - [u(t) - \hat{u}(t), f(t) - \hat{f}(t)].$$

We conclude that $g(t) = \|u(t) - \hat{u}(t)\|$ satisfies

$$g'(t) \leq \omega g(t) + [u(t) - \hat{u}(t), f(t) - \hat{f}(t)]$$

and the integration of this elementary inequality yields the first inequality of (1.8). The first inequality of (1.8) comes from Proposition 1(iv).

In particular, if the assumptions of Proposition 2 hold and $f = \hat{f}$, $u(0) = \hat{u}(0)$, then $u = \hat{u}$ and strong solutions of the initial-value problem are unique. Even more, they depend continuously on initial data and the forcing term f according to the estimate (1.8). If $f = \hat{f}$, the estimate of (1.8) amounts to the same quasicontractive estimate from which we motivated the condition that "$A + \omega I$" is accretive, and the circle is complete.

However, it is an unpleasant but very interesting fact that we cannot think only of strong solutions. Indeed, we will have to travel rather far from this notion to accommodate the full range of phenomena in this subject. In particular, there is an example of a quasicontractive semigroup S on $X = C([0,1])$ such that $S(t)x$ is not differentiable for any $0 \leq t$ or $x \in X$ and there are natural accretive operators A arising in partial differential equations for which the problem (DE)$_f$ with $f = 0$ has no strong solutions on $[0, \infty)$, a phenomenon which corresponds to the development of "shocks"—that is, discontinuities develop in the solution or its derivatives in such a way as to render it subtle to well-pose the corresponding problem. We take up another, broader, notion of "solution" in the next section.

2. Mild solutions. Let us motivate the notion to be introduced the following way. The existence question for the classical Cauchy problem

(CP)
$$(du/dt)(t) + Au(t) = f(t),$$
$$u(0) = x,$$

where A is a continuous function in some $\mathbf{R}^n = X$ is often approached via the method of "Euler lines". Now, this just amounts to solving (which is a trivial matter when it is possible—see below) the explicit difference approximation

(2.1)
$$(u_\lambda(i\lambda) - u_\lambda((i-1)\lambda))/\lambda + Au_\lambda((i-1)\lambda) = f((i-1)\lambda),$$
$$u_\lambda(0) = x,$$

for the nodal values $u_\lambda(i\lambda)$ of the approximation u_λ, interpolating them linearly, and then studying the (subsequential) convergence of u_λ to a solution u. Of course, if linear interpolation produces approximate solutions u_λ which converge uniformly to a continuously differentiable solution, so will piecewise constant interpolation. Now the scheme (2.1) is often not a good one if A is a partial differential operator, since the formal solution of (2.1) is, in the case $f = 0$,

$$u_\lambda(i\lambda) = (I - \lambda A)u_\lambda((i - 1)\lambda) = (I - \lambda A)^i u_\lambda(0) = (I - \lambda A)^i x$$

and one is applying high powers $(I - \lambda A)^i$ of a differential operator to a fixed x, and this is a bad idea in general. On the other hand, if we replace (2.1) by

(2.2) $$(u_\lambda(i\lambda) - u_\lambda((i - 1)\lambda))/\lambda + Au_\lambda(i\lambda) = f(i\lambda)$$

and $f = 0$, we formally have

$$u_\lambda(i\lambda) = (I + \lambda A)^{-1} u_\lambda((i - 1)\lambda)$$

or

(2.3) $$u_\lambda(i\lambda) = J_\lambda^i x$$

where

(2.4) $$J_\lambda = (I + \lambda A)^{-1}.$$

Now, provided the inverses have an adequately large domain, this procedure is well defined and proceeds by applying powers of the inverses, which have a better chance of behaving well. The method (2.2) is called "implicit Euler". One is naturally led to consider the idea of accretive operators when approaching things from this point of view, since the method (2.3) has a better chance of success if the operators J_λ^i are equicontinuous for $i\lambda$ bounded. About the only way to guarantee something like this is to ask each factor J_λ to have a Lipschitz constant K_λ, and then J_λ^i has K_λ^i as a Lipschitz constant. If $K_\lambda = (1 - \lambda\omega)^{-1}$, then $K_\lambda^i \to e^{\omega t}$ as $i\lambda \to t$ and $i \to \infty$. Moreover, one is naturally led to place conditions on the domain of the *resolvents* J_λ. The best thing is to have them defined everywhere (at least for small λ).

DEFINITION 3. Let $A \in \mathbf{A}(\omega)$. Then $A + \omega I$ is *m-accretive* if $R(I + \lambda A) = X$ for $\lambda > 0$ and $\lambda\omega < 1$.

The condition of *m*-accretivity is enjoyed by many important operators in applications. The simplest example of an A such that $A + \omega I$ is *m*-accretive is a Lipschitz continuous function A on all of X which has ω as a Lipschitz constant. Also, every continuous function A on X which is accretive is *m*-accretive (although this is not so easy to prove). However, examples from differential equations will not be continuous and will typically have small domains.

With this prologue, we will adopt below a notion of solution which refers directly to the approximation method used to prove the existence of solutions, the implicit Euler method. There is ample evidence that this is a reasonable thing to do. It provides a great unity to the spectrum of examples developed over the last decade or so and we can speak of "mild solutions" of very different problems. On the abstract side, we find it a completely adequate notion for discussing $(DE)_f$ and the basic existence theory and generation theory involving accretive operators which has been developed. It has shortcomings when one attempts to deal with generalizations of (DE)$_f$, which we call a "quasiautonomous equation", to more general time dependencies, e.g. $u' + A(t)u = 0$, as will be mentioned later.

We now define the notion. For a while, there will be no restrictions of accretiveness imposed and the discussion is completely general in this respect. In order to accommodate the natural generality $f \in L^1(0, T: X)$ and other results discussed later, we will need more general approximations than (2.3)—in particular, we will want to refer to variable meshes rather than the constant step size λ above.

Let $f \in L^1(0, T: X)$ and $\varepsilon > 0$. An ε-*discretization* on $[0, T]$ of $u' + Au \ni f$ on $[0, T]$ consists of a partition $\{0 = t_0 \leqslant t_1 \leqslant t_2 \leqslant \cdots \leqslant t_N\}$ of the interval $[0, t_N]$ and a finite sequence $\{f_1, f_2, \ldots, f_n\} \subseteq X$ such that

(2.5)
(a) $t_i - t_{i-1} < \varepsilon$ for $i = 1, \ldots, N$ and $T - \varepsilon < t_N \leqslant T$, and

(b) $\sum_{i=1}^{N} \int_{t_{i-1}}^{t_i} \|f(s) - f_i\| \, ds < \varepsilon.$

The inequalities (2.5)(a) require that the step sizes of the partition not exceed ε and the endpoint $t_N \leqslant T$ not miss T by more than ε. The inequality (b) requires that the forcing term f be approximated within ε in $L^1(0, t_N: X)$ by the piecewise constant function whose value on $(t_{i-1}, t_i]$ is f_i. These data do not refer to the operator A occurring in the equation. We will indicate it by writing $\mathbf{D}_A(0 = t_0, t_1, \ldots, t_N: f_1, \ldots, f_N)$ for the discretization.

A *solution* of a discretization $\mathbf{D}_A(0 = t_0, t_1, \ldots, t_N: f_1, \ldots, f_N)$ is a piecewise constant function $v: [0, t_N] \to X$ whose values v_i on $(t_{i-1}, t_i]$ satisfy

(2.6) $\qquad (v_i - v_{i-1})/(t_i - t_{i-1}) + Av_i \ni f_i \quad \text{for } i = 1, \ldots, N.$

The value $v_0 = v(0)$ is not otherwise restricted. An ε-*approximate solution* of an inclusion $u' + Au \ni f$ is a solution v of an ε-discretization $\mathbf{D}_A(0 = t_0, \ldots, t_N: f_1, \ldots, f_N)$.

DEFINITION 4. Let A be an operator in X, $T > 0$, and $f \in L^1(0, T: X)$. Then a mild solution of $u' + Au \ni f$ on $[0, T]$ is a $u \in C[0, T: X]$ (the continuous functions from $[0, T]$ into X) with the property that for each $\varepsilon > 0$ there is an ε-approximate solution v of $u' + Au \ni f$ on $[0, T]$ such that $\|v(t) - u(t)\| \leqslant \varepsilon$ for t in the domain of v.

Less formally, mild solutions are the continuous uniform limits of approximate solutions. We reiterate that, while this notion is just broad enough to provide an adequate extension of the class of strong solutions for our purposes, mild solutions represent a very considerable generalization of the strong notion.

Let us mention that while we have treated the endpoints 0 and T of the interval $[0, T]$ asymmetrically in our definition of ε-discretizations and ε-approximate solutions, this difference disappears at the level of mild solutions (although this requires an argument). That is, the class of mild solutions is unchanged if one requires only $0 \leqslant t_0 \leqslant \varepsilon$ in the definition of an ε-discretization and makes the obvious subsequent modifications. We do not know if the class of mild solutions is left unchanged in general if we require both $t_0 = 0$ and $t_N = T$. Without further ado, if $a < b$ and $f \in L^1(a, b: X)$, then one defines strong solutions on $[a, b]$, ε-discretizations of $u' + Au \ni f$ on $[a, b]$ and mild solutions on $[a, b]$ in the obvious way. If $\mathbf{Q}$ is an arbitrary interval and $f \in L^1_{\text{loc}}(\mathbf{Q}: X)$, a mild (strong) solution of $u' + Au \ni f$ on $\mathbf{Q}$ is a $u \in C[\mathbf{Q}: X]$ which is a mild (respectively, strong) solution on every compact subinterval of $\mathbf{Q}$ (and this is consistent with the original definitions, as must be checked).

The next result collects a variety of properties of mild solutions of differential equations.

PROPOSITION 3. PROPERTIES OF MILD SOLUTIONS. *Let A be an operator in X, $\mathbf{Q}$ an interval in $\mathbf{R}$, and $f \in L^1_{\text{loc}}(\mathbf{Q}: X)$.*

(i) *If u is a strong solution of $u' + Au \ni f$ on $\mathbf{Q}$, then u is a mild solution of $u' + Au \ni f$ on $\mathbf{Q}$.*

(ii) *If u is a mild solution of $u' + Au \ni f$ on $\mathbf{Q}$, then $u(\mathbf{Q}) \subseteq \overline{D(A)}$ (the closure of $D(A)$).*

(iii) *Let $\mathbf{Q} = [0, T]$ and u be a mild solution of $u' + Au \ni f$ on $[0, T]$. If $D(A)$ is closed and A is single-valued and continuous, then*

$$u(t) = u(0) - \int_0^t Au(s)\, ds + \int_0^t f(s)\, ds \quad \text{for } 0 \leq t \leq T.$$

In particular, u is a strong solution, and it is a classical solution if f is continuous.

(iv) CONTINUATION PROPERTY. *Let $\mathbf{Q} = \mathbf{Q}_1 \cup \mathbf{Q}_2$ where $\mathbf{Q}_i$ is a subinterval of $\mathbf{Q}$. If $u \in C(\mathbf{Q}: X)$ is a mild solution of $u' + Au \ni f$ on each $\mathbf{Q}_i$, then it is a mild solution on $\mathbf{Q}$.*

(v) CLOSURE PROPERTY. *Let $f_n \in L^1_{\text{loc}}(\mathbf{Q}: X)$, $u_n \in C(\mathbf{Q}: X)$, and u_n be a mild solution of $u'_n + Au_n \ni f_n$ for $n = 1, 2, \ldots$. If f_n converges to f in L^1 and u_n converges to u uniformly on compact subsets of $\mathbf{Q}$ as $n \to \infty$, then u is a mild solution of $u' + Au \ni f$ on $\mathbf{Q}$.*

(vi) TRANSLATION PROPERTY. *Let u be a mild solution of $u' + Au \ni f$ on $\mathbf{Q}$ and $h \in \mathbf{R}$. Then $v(t) = u(t + h)$ is a mild solution of $v' + Av \ni g$ on $\mathbf{Q} - h$ where $g(t) = f(t + h)$.*

(vii) PERTURBATION PROPERTY. *Let p be a continuous mapping of $\overline{D(A)}$ into X. Then u is a mild solution of $u' + Au + p(u) \ni f$ on $\mathbf{Q}$ if and only if $u \in C(\mathbf{Q}: X)$, u has its values in $\overline{D(A)}$, and u is a mild solution of $u' + Au \ni g$ on $\mathbf{Q}$, where $g(t) = f(t) - p(u(t))$.*

As the reader can see, many of the usual properties we associate with the solutions of ordinary differential equations remain true in the context of mild solutions, a comforting fact. One of the things we can do with the notion of a mild solution is define a semigroup from an arbitrary operator A. Indeed, given an operator A in X, put

$$D_A = \{ x \in X : u' + Au \ni 0,\ u(0) = x,\ \text{has a unique mild solution on } [0, \infty) \}$$

and define $S_A(t): D_A \to X$ by

(2.7) $\quad S_A(t)x = u(t) \quad$ where u is the mild solution of $u' + Au \ni 0$, $u(0) = x$.

PROPOSITION 4. *Let A be an operator in X. Then S_A is a semigroup on D_A.*

We will call S_A the "semigroup generated by $-A$".

3. Convergence of approximate solutions. In this section we outline the main facts concerning the existence of mild solutions of the initial-value problem

$$(\text{IVP})_{x,p} \qquad \begin{aligned} du/dt + Au &\ni f(t), \\ u(0) &= x, \end{aligned}$$

nwhere $A + \omega I$ is accretive and $f \in L^1(0, T: X)$. Consider a discretization $\mathbf{D}_A(0 = t_0, t_1, \ldots, t_N: f_1, \ldots, f_N)$ of $u' + Au \ni f$. The nodal values $v_i = v(t_i)$ of a solution v of this discretization satisfy

(3.1) $\qquad (v_i - v_{i-1})/(t_i - t_{i-1}) + Av_i \ni f_i \quad$ for $i = 1, \ldots, N$

or, equivalently,

(3.2) $\qquad v_i = J_{\delta_i}(v_{i-1} + \delta_i f_i) \quad$ for $i = 1, \ldots, N$,

where $\delta_i = t_i - t_{i-1}$ and $J_\lambda = (I + \lambda A)^{-1}$, and we are assuming that J_λ is a function in the range of λ where we use it. If $A + \omega I$ is accretive, then this is the case in (3.2) provided that $\delta_i \omega < 1$ for $i = 1, \ldots, N$. If $A + \omega I$ is also m-accretive, then the domain of J_λ is all of X for $\lambda \omega < 1$, and (3.2) uniquely determines the v_i for any choice of the discretization $\mathbf{D}_A(0 = t_0, \ldots, t_N: f_1, \ldots, f_N)$ of small mesh and initial-value v_0 of v. Thus for a large and interesting class of operators A, every discretization of small mesh of $u' + Au \ni f$ is uniquely solvable once the initial-value of the solution is specified. This adds interest to the next theorem. In the theorem, an ε-approximate solution of $(\text{IVP})_{x,f}$ on $[0, T]$ means an ε-approximate solution of $u' + Au \ni f$ on $[0, T]$ which further satisfies

(3.3) $\qquad \|v(0) - x\| < \varepsilon.$

THEOREM 1. (CONVERGENCE) *Let $A + \omega I$ be accretive, $x \in \overline{D(A)}$ (the closure of $D(A)$), and $f \in L^1(0, T: X)$. For each $\varepsilon > 0$ let $(\text{IVP})_{x,f}$ have an ε-approximate solution on $[0, T]$. Then the problem $(\text{IVP})_{x,f}$ has a unique mild solution u. Moreover, there is a function $\kappa(\varepsilon)$ such that $\kappa(0+) = 0$ and if v is an ε-approximate solution of $(\text{IVP})_{x,f}$, then*

$$\|u(t) - v(T)\| \leq \kappa(\varepsilon) \quad \text{for } 0 \leq t \leq T - \varepsilon.$$

The convergence theorem asserts that if ε-approximate solutions exist for each ε, then they converge as $\varepsilon \to 0$ to a mild solution u, and the difference between u and any ε-approximate solution is estimable in terms of ε. The estimate κ will depend on T, ω, the behaviour of A near x (or simply $\|y\|$ for $y \in Ax$ if $x \in D(A)$) and the modulus of continuity of f in L^1 with respect to translations. In particular, if $A + \omega I$ is m-accretive, and $f \in L^1(0, T: X)$, then $(\text{IVP})_{x,f}$ has a unique mild solution for every $x \in \overline{D(A)}$.

We will not give the details of the proof of the Convergence Theorem, but we will outline a simple and interesting attack which gives much more information than claimed so far.

First step. Let v be a solution of a discretization $\mathbf{D}_A(0 = s_0, s_1, \ldots, s_M: f_1, \ldots, f_M)$ and w be a solution of a discretization $\mathbf{D}_A(0 = t_0, t_1, \ldots, t_N: g_1, \ldots, g_N)$ with nodal values v_i and w_j, respectively. Put $a_{i,j} = \|v_i - w_j\|$, $\gamma_i = s_i - s_{i-1}$, $\delta_j = t_j - t_{j-1}$. Then

$$(3.4) \quad \left(1 - \omega \frac{\gamma_i \delta_j}{\gamma_i + \delta_j}\right) a_{i,j} \leq \frac{\delta_j}{\gamma_i + \delta_j} a_{i-1,j} + \frac{\gamma_i}{\gamma_i + \delta_j} a_{i,j-1} + \frac{\gamma_i \delta_j}{\gamma_i + \delta_j} \|f_i - g_j\|$$

for $1 \leq i \leq M$, $1 \leq j \leq N$. Moreover, for any $x \in D(A)$ and $y \in Ax$,

$$a_{i,0} \leq \alpha_{i,1} \|v_0 - x\| + \|w_0 - x\| + \sum_{k=1}^{i} \alpha_{i,k} \gamma_k (\|f_k\| + \|y\|) \quad \text{for } 0 \leq i \leq M,$$

(3.5) and

$$a_{0,j} \leq \beta_{j,1} \|w_0 - x\| + \|v_0 - x\| + \sum_{k=1}^{j} \beta_{j,k} \delta_k (\|g_k\| + \|y\|) \quad \text{for } 0 \leq j \leq N,$$

where

$$(3.6) \quad \alpha_{i,k} = \prod_{m=k}^{i} (1 - \omega \gamma_m)^{-1} \quad \text{and} \quad \beta_{j,k} = \prod_{m=k}^{j} (1 - \omega \delta_m)^{-1}.$$

The inequalities (3.4) and (3.5) are elementary consequences of the assumptions. The proof of the Convergence Theorem then reduces to estimating solutions of (3.4) and (3.5). One way to recognize the behaviour of solutions of these inequalities is outlined next.

Second step. Consider real-valued functions $\psi(s, \tau)$, $\varphi(s, \tau)$ of two real variables s, τ which are related by the differential equation
(3.7)
$$\psi_s(s, \tau) + \psi_\tau(s, \tau) - \omega \psi(s, \tau) = \varphi(s, \tau) \quad \text{for } 0 \leq s \leq T \text{ and } 0 \leq \tau \leq T.$$

Let us introduce a grid

$$(3.8) \quad \Delta = \{(s_i, t_j): 0 = s_0 \leq \cdots \leq s_M \leq T; 0 = t_0 \leq \cdots \leq t_N \leq T\}$$

and approximate (3.7) on this grid by the difference relations
(3.9)
$$\frac{\Psi_{i,j} - \Psi_{i-1,j}}{\gamma_i} + \frac{\Psi_{i,j} - \Psi_{i,j-1}}{\delta_j} - \omega \Psi_{i,j} = \Phi_{i,j} \quad \text{for } i = 1, \ldots, M, \; j = 1, \ldots, N,$$

where $\gamma_i = s_i - s_{i-1}$ and $\delta_j = t_j - t_{j-1}$. When (3.9) is solved for $\Psi_{i,j}$ (given the values $\Psi_{i,j}$ for $i = 0$ or $j = 0$) we might hope $|\Psi_{i,j} - \psi(s_i, \tau_j)|$ is small when it is small on the "boundary" $i = 0$ or $j = 0$, the mesh $\max\{\gamma_i, \delta_j\}$ is small, and the values $\Phi_{i,j}$ approximate Φ on $(s_{i-1}, s_i] \times (\tau_{j-1}, \tau_j]$ in some good way. Now (3.9) becomes, upon rearranging,

$$(3.10) \quad \left(1 - \omega \frac{\gamma_i \delta_j}{\gamma_i + \delta_j}\right) \Psi_{i,j} = \frac{\delta_j}{\gamma_i + \delta_j} \Psi_{i-1,j} + \frac{\gamma_i}{\gamma_i + \delta_j} \Psi_{i,j-1} + \frac{\gamma_i \delta_j}{\gamma_i + \delta_j} \Phi_{i,j}$$

and the relationship with (3.5) becomes apparent. The terms corresponding to $\Phi_{i,j}$ in (3.4) are $\|f_i - g_j\|$ where the f_i and g_j are nodal values of piecewise constant approximations of functions $f, g \in L^1(0, T: X)$. It is not exactly clear how such $\Phi_{i,j}$ approximate this φ, but they do so in the sense required to apply the convergence result described next.

Let $b \in C([-T, S])$. The integration of (3.6) subject to the boundary condition $\psi(s, \tau) = b(s - \tau)$ if $s = 0$ or $\tau = 0$ leads to the following formulae for $\psi = G(b, \varphi)$:

(3.11)
$$G(b, \varphi) = \begin{cases} e^{\omega \tau} b(s - \tau) + \int_0^\tau e^{\omega(\tau - \alpha)} \varphi(s - \tau + \alpha, \alpha) \, d\alpha & \text{for } 0 \leq \tau \leq s \leq T, \\ e^{\omega s} b(s - \tau) + \int_0^s e^{\omega(s - \alpha)} \varphi(\alpha, \tau - s + \alpha) \, d\alpha & \text{for } 0 \leq s \leq \tau \leq T. \end{cases}$$

In what follows φ is such that $G(b, \varphi)$ is well defined and continuous.

We want to discuss a solution operator for the corresponding discrete problem and begin by introducing the norm in which approximation of φ will be required. To this end, if $\Omega = [0, S] \times [0, T]$ and $\varphi: [0, S] \times [0, T] \to \mathbf{R}$, put

(3.12)
$$\|\varphi\|_\Omega = \inf\{\|f\|_{L^1(0, S)} + \|g\|_{L^1(0, T)}: |\varphi(s, \tau)| \leq |f(s)| + |g(\tau)| \text{ a.e. on } \Omega\},$$

where it is understood that the inf of the empty set is $-\infty$. Next, if Δ is the grid (3.8), put $\Omega(\Delta) = [0, s_M] \times [0, t_N]$. Let

$$B: [-t_N, s_M] \to \mathbf{R} \quad \text{and} \quad \Phi: \Omega(\Delta) \to \mathbf{R}$$

be piecewise constant on Δ, i.e. there are constants $B_{i,j}$, $\Phi_{i,j}$ such that $B(0) = B_{0,0}$ and $B(r) = B_{i,j}$ for $i = 0$ and $-t_j \leq r < -t_{j-1}$ or $j = 0$ and $s_{i-1} < r \leq s_i$, and

$$\Phi(s, \tau) = \Phi_{i,j} \quad \text{for } (s, \tau) \in (s_{i-1}, s_i] \times (\tau_{j-1}, \tau_j].$$

If the mesh $m(\Delta) = \max\{\gamma_i, \delta_j\}$ of Δ satisfies $m(\Delta)\omega < 1$, the equations (3.9) are obviously uniquely solvable for $\Psi_{i,j}$ given that

(3.13) $\qquad\qquad\qquad \Psi_{i,j} = B_{i,j} \quad \text{for } i = 0 \text{ or } j = 0.$

Let $\Psi = H_\Delta(B, \Phi)$ denote the piecewise constant function on Δ obtained by solving (3.9) subject to (3.13).

THEOREM 2. *Let $b \in C([-T, S])$ and $\varphi: [0, T] \times [0, T] \to \mathbf{R}$. Then*

$$\|G(b, \varphi) - H_\Delta(B, \Phi)\|_{L^\infty(\Omega(\Delta))} \to 0$$

as $m(\Delta) + \|b - B\|_{L^\infty(-T_N, S_M)} + \|\varphi - \Phi\|_{\Omega(\Delta)} \to 0.$

There is a lot packed into this result, and we have formulated it in a sort of sneaky way. In particular, no assumptions were made on Φ in the theorem. This was possible, because the result asserts nothing unless we can approximate Φ with piecewise constant functions in the norms (3.12), and not every function can be so approximated. In particular, not every bounded and measurable function on $[0, S] \times [0, T]$ can be so approximated. However, functions of the form $\|f(s) - g(\tau)\|$ with integrable f and g can.

$g(\tau)\|$ with integrable f and g can.

Let us sketch the application of Theorem 2 to Theorem 1. Let $f, g \in L^1(0, T: X)$, x_0, $\hat{x}_0 \in \overline{D(A)}$, v be an ε-approximate solution of

(3.14) $\qquad u' + Au \ni f, \qquad u(0) = x_0,$

and w be an ε-approximate solution of $u' + Au \ni g$, $u(0) = \hat{x}_0$, and the discretizations solved by v and w be the ones in the first step. The piecewise constant function B on $[-t_N, s_M]$, whose values $B_{i,j}$ for $i = 0$ or $j = 0$ are the right-hand sides of (3.5), tends, as $\varepsilon \to 0$, uniformly to the function

(3.15)
$$b(s) = e^{\omega s}\|x_0 - x\| + \int_0^s e^{\omega(s-\alpha)}(\|f(\alpha)\| + \|y\|) \, d\alpha + \|x - \hat{x}_0\|$$
$$\text{for } 0 \leqslant s \leqslant T,$$
$$b(-\tau) = e^{\omega \tau}\|\hat{x}_0 - x\| + \int_0^\tau e^{\omega(\tau-\alpha)}(\|g(\alpha)\| + \|y\|) \, d\alpha + \|x - x_0\|$$
$$\text{for } 0 \leqslant \tau \leqslant T.$$

To prove this one uses the fact that the functions whose nodal values are the f_i and g_j differ from f and g in L^1 by at most ε and elementary estimates. Moreover, if $\varphi(s, \tau) = \|f(s) - g(\tau)\|$ and Φ is the piecewise constant function on Δ given by $\Phi_{i,j} = \|f_i - g_j\|$, then

(3.16) $\qquad \|\varphi - \Phi\|_{\Omega(\Delta)} \leqslant 2\varepsilon.$

Finally, since the $a_{i,j}$ satisfy the inequalities (3.5) and (3.6) they may be estimated above by the solution $\Psi_{i,j} = H(B, \Phi)_{i,j}$ of the corresponding equalities. Recalling the meaning of the $a_{i,j}$ and Theorem 2 we conclude that for any $\eta > 0$

(3.17) $\qquad \|v(s) - w(\tau)\| \leqslant G(b, \varphi)(s, \tau) + \eta$

as soon as ε is small enough.

We use (3.17) in three situations. If $f = g$, $x_0 = \hat{x}_0$, and $s = \tau = t$, we compute $G(b, \varphi)(t, t) = 2\|x_0 - x\|$ and conclude that

$$\|v(t) - w(t)\| \leqslant 2\|x_0 - x\| + \eta,$$

as soon as ε is small. Since $x_0 \in \overline{D(A)}$ and $x \in D(A)$ is arbitrary, this verifies the Cauchy criterion for the net of ε-approximate solutions of (3.14). Let u be the limit of the ε-approximate solutions of (3.14) as $\varepsilon \to 0$. Now we take the limit in (3.17) with $s = t + h$, $f = g$ and $x_0 = \hat{x}_0$ to conclude that

(3.18)
$$\|u(t + h) - u(t)\| \leqslant G(b, \varphi)(t + h, t)$$
$$= e^{\omega t}\left((e^{\omega h} + 1)\|x_0 - x\| + \int_0^h e^{\omega(h-\alpha)}(\|f(\alpha)\| + \|y\|) \, d\alpha\right)$$
$$+ \int_0^t e^{\omega(t-\alpha)}\|f(\alpha + h) - f(\alpha)\| \, d\alpha,$$

for every $y \in Ax$ and $x \in D(A)$. It follows easily that u is continuous. In a similar way, (3.15) in the general case implies that if u and $\hat{u}$ are mild solutions of problems $u' + Au \ni f$ and $\hat{u}' + A\hat{u} \ni \hat{f}$, then

(3.19) $\quad \|u(t) - \hat{u}(t)\| \leqslant e^{\omega t}\|u(0) - \hat{u}(0)\| + \int_0^t e^{\omega(t-\alpha)}\|f(\alpha) - \hat{f}(\alpha)\| \, d\alpha.$

(The change in notation was made because we ran out of suitable letters.) The inequality (3.19) reproduces the extreme inequalities of (1.8) for mild solutions. Given the convergence, Theorem 2 also quickly implies the validity of the next proposition.

PROPOSITION 5. *Let $A + \omega I$ be accretive, $f, \hat{f} \in L^1(0, T: X)$ and $u, \hat{u}$ be mild solutions of $u' + Au \ni f$ and $\hat{u}' + A\hat{u} \ni \hat{f}$, respectively. Then*

$$(3.20) \quad \|u(t) - \hat{u}(t)\| \leq e^{\omega(t-s)}\|u(s) - \hat{u}(s)\|$$
$$+ \int_s^t e^{\omega(t-\tau)}[u(\tau) - \hat{u}(\tau), f(\tau) - \hat{f}(\tau)]\, d\tau$$

for $0 \leq s \leq t \leq T$.

The convergence theorem does not address the question of when approximate solutions exist. Let us point out a couple of simple situations when this is not a problem. Recall that if $A + \omega I$ is m-accretive, then every discretization $\mathbf{D}_A(0 = t_0, t_1, \ldots, t_N: f_1, \ldots, f_N)$ of small mesh is uniquely solvable when an initial value is specified. We summarize the situation as regards the case in which $A + \omega I$ is m-accretive.

THEOREM 3. *Let $A + \omega I$ be m-accretive, $x \in \overline{D(A)}$ and $f \in L^1(0, T: X)$. Then $u' + Au \ni f$, $u(0) = x$ has a unique mild solution on $[0, T]$. Moreover, if u and $\hat{u}$ are, respectively, mild solutions of $u' + Au \ni f$ and $\hat{u}' + A\hat{u} \ni \hat{f}$ on $[0, T]$, then (3.20) holds.*

Another simple situation arises when considering the problem

$$(3.21) \quad u' + Au \ni 0, \quad u(0) = x.$$

We know that the solution u_λ of the discretization $\mathbf{D}_A(0, \lambda, 2\lambda, \ldots, N\lambda: 0, \ldots, 0)$ which satisfies $u_\lambda(0) = x$ (if it exists) is given by $u_\lambda(i\lambda) = J_\lambda^i x$ (see (2.3) and (2.4)). Thus if $A \in \mathbf{A}(\omega)$ and

$$(3.22) \quad R(I + \lambda A) \supset D(A) \quad \text{for small } \lambda > 0,$$

then (3.21) has a mild solution u for every $x \in D(A)$. Moreover, u is given by

$$(3.23) \quad J_\lambda^i x \to u(t) \quad \text{as } \lambda \to 0 \text{ and } i\lambda \to t.$$

When $A \in \mathbf{A}(\omega)$ satisfies (3.22), then its closure satisfies the stronger condition

$$(3.24) \quad R(I + \lambda A) \supset \overline{D(A)} \quad \text{for small } \lambda > 0,$$

and (3.21) is solvable for $x \in \overline{D(A)}$ and (3.23) still holds. In particular, if the *range condition* (3.24) holds, we have the *exponential formula*

$$(3.25) \quad S_A(t)x = \lim_{n \to \infty} (I + (t/n)A)^{-n} x$$

for the semigroup generated by $-A$ on $\overline{D(A)}$.

4. A quasilinear equation. In this section we sketch an application of the results of §3 to a (oversimplified) quasilinear problem related to those previously considered by Kato. This section, while self-contained, is primarily intended for readers with some knowledge of Kato's theory. It demonstrates a strong relationship between his existence results and the results we have sketched so far.

Here X and Y denote reflexive Banach spaces with Y densely and continuously imbedded in X. The norms of X and Y are denoted by $\| \ \|_X$ and $\| \ \|_Y$ respectively. The problem of interest has the form

(4.1) $$u' + B(u)u = 0, \quad u(0) = \varphi,$$

in which $B(u)$ is a linear operator in X for each suitable u. We detail properties of B shortly, but first we must introduce a little notation.

In what follows linear operators are single-valued. If $C: D(C) \subset X \to X$ is a linear operator, C_Y, the part of C in Y, is the restriction of C to $\{y \in D(C) \cap Y: Cy \in Y\}$. If Z is a Banach space and C is a densely defined linear operator in Z such that $C + \omega I$ is m-accretive we write $C \in N(\omega, Z)$. The Hille–Yosida theorem (which we will not use) implies that $C \in N(\omega, Z)$ exactly when $-C$ is the infinitesimal generator of a continuous semigroup e^{-Ct} of linear operators which satisfies $\|e^{-Ct}\|_Z \leq e^{\omega t}$.

In the assumptions below $r > 0$, $w_0 \in Y$ and

$$W = \{ y \in Y: \|y - w_0\|_Y \leq r \}$$

is the closed r-ball centered at w_0 in Y. We assume that

(B1) There is a $\theta \in \mathbf{R}$ and for each $w \in W$ an operator $B(w) \in N(\theta, X)$ such that $D(B(w)) \supset Y$ and $B(w)_Y \in N(\theta, Y)$.

(B2) There are $L, \gamma > 0$ such that, for $w, \hat{w} \in W$ and $y \in Y$,

$$\|(B(w) - B(\hat{w}))y\|_X \leq L\|w - \hat{w}\|_X \|y\|_Y \quad \text{and} \quad \|B(w)y\|_X \leq \gamma \|y\|_Y.$$

(B3) There is a $\mu > 0$ such that if $w \in W$ then

$$B(w)w_0 \in Y \quad \text{and} \quad \|B(w)w_0\|_Y \leq \mu.$$

In Kato's theory, (B1) is deducible from more subtle assumptions involving a linear isomorphism $S: Y \to X$ and conditions on $B(w)$ and the commutators $SB(w) - B(w)S$. Let us try to solve (4.1) via the discretization

(4.2) $$(x_i - x_{i-1})/\lambda + B(x_{i-1})x_i = 0, \quad i = 1,\ldots,N,$$
$$x_0 = \varphi.$$

Assume that $T > 0$ and (4.2) is solvable for each small $\lambda > 0$ for $x_i \in W$ for $i = 0, 1, \ldots, N$ where $T \leq N\lambda$. Then put

(4.3) $$u_\lambda(t) = x_i \quad \text{for } (i-1)\lambda < t \leq i\lambda, \quad i = 1,\ldots,N, \text{ and}$$
$$u_\lambda(0) = \varphi.$$

We claim that then u_λ converges strongly in X and weakly in Y uniformly on $[0, T]$ to the function $u: [0, T] \to W$ which is Lipschitz continuous into X and weakly continuous into Y. Moreover, u is weakly continuously differentiable into X and satisfies $u'(t) + B(u(t))u(t) = 0$ for $0 \leq t \leq T$. In particular, it is a strong solution of $u' + A(u) = 0$ where $A(u) = B(u)u$. We sketch the proof of these claims and then the proof that (4.3) is solvable.

For each small $\lambda > 0$, let (4.2) be satisfied, $T \leq N\lambda$, $x_i \in W$, and u_λ be given by (4.3). Then, by (B2) and $x_i \in W$,

(4.4) $$\|x_i - x_{i-1}\|_X = \lambda \|B(x_{i-1})x_i\|_X \leq \lambda\gamma \|x_i\|_Y \leq \lambda\gamma(r + \|w_0\|_Y).$$

Now put
(4.5) $$A(x) = B(x)x \quad \text{for } x \in D(A) = W.$$
Clearly u_λ is a solution of the discretization $\mathbf{D}_A(0, \lambda, \ldots, N\lambda: \varepsilon_1, \ldots, \varepsilon_N)$ of $u' + Au \ni 0$ where
(4.6) $$\varepsilon_i = (B(x_i) - B(x_{i-1}))x_i.$$
Using (4.4) and (B2) again, we see that the "errors" ε_i satisfy
$$\|\varepsilon_i\|_X \leq L\|x_i - x_{i-1}\|_X \|x_i\|_Y \leq \lambda\gamma(r + \|w_0\|_Y)^2,$$
and thus tend to zero in L^∞ and a fortiori in L^1. Finally, we check that $A + \omega I$ is accretive in X where $\omega = \theta + L(r + \|w_0\|_Y)$. Indeed, by (B1) and (B2), if x and $\hat{x} \in D(A) = W$,

(4.7)
$$\begin{aligned}
&\|x - \hat{x} + \lambda(A(x) - A(\hat{x}))\|_X \\
&= \|x - \hat{x} + \lambda B(x)(x - \hat{x}) + \lambda(B(x) - B(\hat{x}))\hat{x}\|_X \\
&\geq \|x - \hat{x} + \lambda B(x)(x - \hat{x})\|_X - \lambda\|(B(x) - B(\hat{x}))\hat{x}\|_X \\
&\geq (1 - \lambda\theta)\|x - \hat{x}\|_X - \lambda L\|x - \hat{x}\|_X\|\hat{x}\|_Y \\
&\geq (1 - \lambda(\theta + L(r + \|w_0\|_Y)))\|x - \hat{x}\|_X,
\end{aligned}$$

and the claim is proved. The convergence of u_λ in X uniformly in t to a continuous limit u now follows from the results described in §3. Since each u_λ takes values in W, which is weakly closed in Y, and convergence in X boundedly in Y implies weak convergence in Y by the assumptions on X and Y, the convergence claims are established. Clearly $u(t)$ is weakly continuous into Y and $B(u(t))u(t)$ is weakly continuous into X. Moreover, it is easy to pass to the weak (in X) limit as $\lambda \to 0$ and $j\lambda \to t$ in the relation
$$u_\lambda(j\lambda) = \varphi + \int_\lambda^{j\lambda} B(u_\lambda(s - \lambda))u_\lambda(s)\,ds,$$
which follows from summing (4.2) from $i = 1$ to j, to find
$$u(t) = \varphi + \int_0^t B(u(s))u(s)\,ds,$$
which proves the claims about u being a weakly continuously differentiable strong solution of (4.1).

It remains to discuss the solvability of (4.2). By the assumption (B1), if $\lambda > 0$ and $\lambda\theta < 1$, then given $x_{i-1} \in W$ and any z in X we can uniquely solve
(4.8) $$x + \lambda B(x_{i-1})x = z$$
for $x = (I + \lambda B(x_{i-1}))^{-1}z$ and $x \in Y$ if $z \in Y$ and
(4.9) $$\|(I + \lambda B(x_{i-1}))^{-1}\|_Z \leq (1 - \lambda\theta)^{-1} \quad \text{for } Z = X \text{ or } Z = Y.$$

Hence, so long as we keep x_{i-1} in W we can compute x_i in Y. We estimate the range of i for which this is possible. Without loss of generality assume $0 \leq \theta$. We will keep $\lambda\theta < 1/2$ so that $(1 - \lambda\theta)^{-1} \leq e^{2\lambda\theta}$.

By (4.9) and (B3), so long as $x_{i-1} \in W$,

$$\|x_i - w_0\|_Y \leq (1 - \lambda\theta)^{-1}\|x_i - w_0 + \lambda(B(x_{i-1})x_i - B(x_{i-1})w_0)\|_Y$$
$$= (1 - \lambda\theta)^{-1}\|x_{i-1} - (w_0 + \lambda B(x_{i-1})w_0)\|_Y$$
$$\leq (1 - \lambda\theta)^{-1}(\|x_{i-1} - w_0\|_Y + \lambda\|B(x_{i-1})w_0\|_Y)$$
$$\leq (1 - \lambda\theta)^{-1}(\|x_{i-1} - w_0\|_Y + \lambda\mu).$$

Using the above inequalities and $x_0 = \varphi$, one easily finds that

$$\|x_i - w_0\|_Y \leq e^{2i\lambda}(\|\varphi - w_0\|_Y + i\lambda\mu),$$

so we conclude that if

$$e^{2(T+\lambda)\theta}(\|\varphi - w_0\|_Y + (T+\lambda)\mu) \leq r,$$

which will hold for λ and T small enough provided that φ lies in the interior of W, then (4.2) is solvable for $x_i \in W$ when $T \leq N\lambda \leq T + \lambda$, completing the discussion.

5. Generation theorems and Kobayashi's existence criterion. In this section we introduce results of two kinds. On the one hand, if we are given a mapping from data (x, f) with the properties expected of the solution operator of the problem

$$(\text{IVP})_{x,f} \qquad u' + Au \ni f, \qquad u(0) = x,$$

when $A + \omega I$ is accretive, we ask if that mapping indeed arises from an A in this way—this is generation theory in the spirit of the first section. Secondly, we will discuss more refined questions concerning the solvability of $(\text{IVP})_{x,0}$ than have been posed so far.

To begin, let us recall that if $A + \omega I$ were m-accretive, $f \in L^1_{\text{loc}}([0, \infty): X)$ and $x \in \overline{D(A)}$, then $(\text{IVP})_{x,f}$ would have a unique mild solution $u \in C([0, \infty): X)$ which we will denote by $u = E_A(x, f)$. Moreover, E_A would enjoy certain properties which we now abstract. Let K be a closed and nonempty subset of X, L be a translation invariant subspace of $L^1_{\text{loc}}([0, \infty): X)$ and consider the following properties of a mapping $E: K \times L \to C([0, \infty): X)$:

(E1) For $x \in K$ and $f \in L$, $E(x, f)(0) = x$ and $E(x, f)(t) \in K$ for $0 \leq t$.

(E2) E is translation invariant in the sense that

$$E(x, f)(t + \tau) = E(E(x, f)(\tau), f(\cdot + \tau))(t) \quad \text{for } x \in K, f \in L \text{ and } 0 \leq t, \tau.$$

(E3) If $x, \hat{x} \in K$, $f, \hat{f} \in L$ and $u = E(x, f)$, $\hat{u} = E(\hat{x}, \hat{f})$, then

$$\|u(t) - \hat{u}(t)\| \leq e^{\omega t}\|u(0) - \hat{u}(0)\| + \int_0^t e^{\omega(t-\tau)}[u(\tau) - \hat{u}(\tau), f(\tau) - \hat{f}(\tau)]\, d\tau$$

for $0 \leq t$.

For example, if $\omega = 0$ and $L = \{0\}$, then (E1)–(E3) reduce to the requirement that $S(t)x = E(x, 0)(t)$ defines a nonexpansive semigroup on K. Next we list some results which, under various circumstances, represent operators E satisfying (E1)–(E3) as arising from solving an initial-value problem $(\text{IVP})_{x,f}$ with $A + \omega I$ accretive. In the first result we see that if L is large enough, then the situation is rather nice.

(i) Let (E1)–(E3) hold and L contain all the constant functions. Then there is a unique A such that $\overline{A + \omega I}$ is m-accretive and E is the restriction of E_A to $K \times L$. Moreover, $\overline{D(A)} = K$.

It is also easy to see that the mapping $A \to E_A$ is one-to-one on

$$\{A: \overline{A + \omega I} \text{ is } m\text{-accretive}\}$$

(essentially because $y \in Ax$ is equivalent to the constant function x being a solution of $u' + Au \ni y$ when $\overline{A + \omega I}$ is m-accretive). When $L = L^1_{\mathrm{loc}}([0, \infty): X)$, this provides us with a biunique correspondence between mappings E with the properties (E1)–(E3) and operators A with $\overline{A + \omega I}$ m-accretive; this is a perfect result. The situation in the semigroup case, that is $L = \{0\}$, is considerably more complicated and there remain interesting unsolved problems. We will restrict our attention to the case $\omega = 0$, but all the results below remain valid in the general case. We begin with the compact case.

(ii) If S is a nonexpansive semigroup on a closed convex set $\underline{K \text{ in } X}$ and K is locally compact, then there is an accretive operator A in X with $\overline{D(A)} = K$,

$$K = \bigcap \{R(I + \lambda A): \lambda > 0\}$$

and $S = S_A$ (equivalently, S is obtained from A via the exponential formula). In particular, if $X = \mathbf{R}^N$ is finite dimensional, then every nonexpansive semigroup on a closed convex set arises in this way. However, even if $X = \mathbf{R}^2$ with the maximum norm, there are distinct m-accretive operators A and B with domains all of X for which $S_A = S_B$.

The next results do not require compactness but restrict the geometry of X instead. In the event that $X = H$ is a Hilbert space, the notion of an accretive operator coincides with another notion, that of a monotone operator. Moreover, it is known that an operator is m-accretive if and only if it is accretive and not a proper restriction of another accretive operator—i.e., if it is maximal accretive (equivalently, maximal monotone). This is the origin of the "m" in m-accretive.

(iii) If $X = H$ is a Hilbert space, K is a closed and convex subset of X and S is a nonexpansive semigroup on K, then there is a unique m-accretive operator A in X such that $S = S_A$ and $\overline{D(A)} = K$. Moreover, this correspondence is biunique, the infinitesimal generator of S_A is $-A^0$ where A^0 is the minimal section of A. That is, for $x \in D(A)$, Ax is a closed convex set and $A^0 x$ is the projection of 0 on this set (its element of least norm).

The results above provide a perfect generation theorem for nonexpansive semigroups in Hilbert spaces which is really quite rich in structure. Moreover, it is nontrivial even in the case $X = \mathbf{R}$! A generalization of (iii) holds which places less severe geometrical restrictions on X, but more on K.

(iv) If the norm of X is uniformly Gateaux differentiable and the norm of X^* is Fréchet differentiable, then the relation $S = S_A$ establishes a biunique correspondence between nonexpansive semigroups on closed convex nonexpansive retracts of X and m-accretive operators.

We will defer further remarks in this vein to the comments section. For now we content ourselves with the final remark that it is still an unsolved problem to determine whether or not an arbitrary nonexpansive semigroup S on a convex K can be represented in the form $S = S_A$ for an accretive A. It seems likely that if this is to be proved *not* true, then this will be done by presenting a nonexpansive semigroup S on a closed convex set K such that $\|S(t)x - x\|/t \to \infty$ as $t \to 0+$ for every $x \in K$. If this is to be proved *true*, it will likely involve some totally new arguments—a statement which leads us to a few comments on the arguments used to prove (i)–(iv).

In order to prove (i), one proceeds according to the following idea: Assuming that E is indeed of the form E_A, we fix $z \in X$ and try to build the solutions of $u' + Au + (\omega + 1)u \ni z$, $u(0) = x$ from E. If $A + \omega I$ is accretive, the time t mapping $x \to u(t)$ so defined is a strict contraction, and by a fixed point argument we conclude that the problem has a constant solution $u \equiv x$. Then $x \in D(A)$ and $z - (\omega + 1)x \in Ax$. This leads to the following construction of A. First extend E to $K \times P$ where P is the space of piecewise constant functions. This is easy owing to (E2); e.g., if $f = z$ on $0 \leqslant t \leqslant a$ and $f = w$ on $a \leqslant t$, put $E(x, f)(t) = E(x, z)(t)$ for $0 \leqslant t \leqslant a$ and $E(x, f)(t + a) = E(x, E(x, z)(a))(t)$ if $0 \leqslant t$. Next use (E3) and the density of P in $L^1_{\mathrm{loc}}(0, \infty: X)$ to extend E to all of $K \times L^1_{\mathrm{loc}}(0, \infty: X)$. Next fix $z \in X$ and $x \in K$ and solve

$$u = E(x, -(\omega + 1)u + z)$$

by iterating: $u_0 \equiv x$, and $u_n = E(x, -(\omega + 1)u_{n-1})$. Observe that $x \to e^t u(t)$ is nonexpansive and so there is a unique element of K fixed by the map $x \to u(t)$ for all t; that is, $x = E(x, -(\omega + 1)x + z)$ has a solution. Defining A by $z - (\omega + 1)x \in Ax$ yields an operator A with the desired properties.

In order to prove the results (ii)–(iv) a quite different path is taken. One attempts to produce A by defining

(5.1) $$(I + \lambda A)^{-1} = \lim_{t \downarrow 0} \left(I + \lambda \frac{I - S(t)}{t} \right)^{-1},$$

and the main work is to show the existence of a suitable (perhaps subsequential) limit. See the comments section.

The result (i) is a strong indication that if $A + \omega I$ is accretive and the problem (IVP)$_{x,f}$ has a mild solution on $[0, \infty)$ for every $x \in \overline{D(A)}$ and constant function f, then A is probably m-accretive. However, it does not quite say this, and there is an apparently open problem here. A variant of the question involved here is the problem of trying to give sufficient conditions and necessary conditions for the solvability of (IVP)$_{x,0}$ for arbitrary $x \in \overline{D(A)}$.

For example, the following is an interesting sufficient condition: Let $A + \omega I$ be accretive and

(5.2) $$\liminf_{\lambda \downarrow 0} \frac{d(R(I + \lambda A), x)}{\lambda} = 0 \quad \text{for } x \in \overline{D(A)},$$

where $d(C, x)$ denotes the distance from the set C to x. Then the problem
(5.3) $$u' + Au \ni 0, \quad u(0) = x$$
has a mild solution on $[0, \infty)$ for every $x \in \overline{D(A)}$. This is obviously a generalization of the range condition.. It is also a sort of tangency condition: In the event that A is a continuous function on $\overline{D(A)}$ it can be shown to be equivalent to the assumption that

(5.4) $$\liminf_{\lambda \downarrow 0} \frac{d(x - \lambda Ax, \overline{D(A)})}{\lambda} = 0 \quad \text{for } x \in \overline{D(A)}.$$

If the limit inferior is replaced by the limit above, the statement just says that departing from x in the direction of $-Ax$ will leave $\overline{D(A)}$ at zero velocity.

In fact, necessary *and* sufficient conditions are known for the solvability of (5.3). For example, the following are equivalent if $A + \omega I$ is accretive:

(a) (5.3) has a mild solution on $[0, \infty)$ for every $x \in \overline{D(A)}$.

(b) For every $\varepsilon > 0$ and $x_0 \in \overline{D(A)}$ there is a $\delta \in (0, \varepsilon]$, an integer N, and $y_i \in Ax_i$, $h_i > 0$ for $i = 1, \ldots, N$ such that

$$\sum_{i=1}^{N} h_i = \delta, \quad \sum_{i=1}^{N} \|x_i - x_{i-1} + h_i y_i\| < \varepsilon \delta.$$

That (a) $\to$ (b) is trivial. The other implication requires interesting arguments which we will not sketch here. Do notice that (5.2) is just the case (b) in the particular situation $N = 1$.

6. Regularity of mild solutions. In the generality we have been discussing, if $A \in \mathbf{A}(\omega)$, the only strong solutions of the problems $u' + Au \ni f$ which are known to exist are the trivial ones, the constants; that is, $u = x$ and $f = y$ for all t where $y \in Ax$. However, mild solutions are a satisfactory extension of the notion of strong solutions, since mild solutions are unique and strong solutions are mild. We have not addressed the other part of this consistency question, namely if a mild solution turns out to be "smooth", is it a strong solution? Similarly, we have not given conditions under which mild solutions are smooth. This we will do now. We do emphasize, before this, that even in applications one does not want to be limited to strong solutions, since there are important partial differential equations which simply do not have strong solutions.

A basic fact is the following consistency between the property $A \in \mathbf{A}(\omega)$ and differentiability of mild solutions of $u' + Au \ni f$.

THEOREM 4. *Let $A \in \mathbf{A}(\omega)$, $f \in L^1(0, T: X)$ and u be a mild solution of $u' + Au \ni f$ on $(0, T)$. Let u have a right derivative $u'_R(\tau)$ at $\tau \in (0, T)$ and*

$$\lim_{h \downarrow 0} \frac{1}{h} \int_{\tau}^{\tau+h} \|f(t) - f(\tau)\| \, dt = 0,$$

that is, τ is a right Lebesgue point of f. Then the operator $\hat{A}$ given by

(6.1) $\hat{A}x = Ax$ *for* $x \neq u(\tau)$ *and* $\hat{A}u(\tau) = Au(\tau) \cup \{f(\tau) - u'_R(\tau)\}$

satisfies $\hat{A} \in \mathbf{A}(\omega)$.

If $A + \omega I$ is m-accretive and $\hat{A} \in \mathbf{A}(\omega)$ is an extension of A (as is the case for the $\hat{A}$ given by (6.1)), then $\hat{A} = A$. This maximality property arises because, if $\hat{A}$ were strictly bigger than A, then for $\lambda > 0$ and $\lambda\omega < 1$, $(I + \lambda\hat{A})^{-1}$ would be a function strictly extending the everywhere defined $(I + \lambda A)^{-1}$, and this is impossible. It thus follows at once that if $A + \omega I$ is m-accretive and $u \in W^{1,1}(0, T: X)$ is a mild solution of $u' + Au \ni f$, then u is a strong solution. When is a mild solution in $W^{1,1}(0, T: X)$? The principal conditions guaranteeing this are given by

PROPOSITION 6. *Let $A \in \mathbf{A}(\omega)$, $f: [0, T] \to X$ be of bounded variation and $x \in D(A)$. If u is a mild solution of $u' + Au \ni f$ on $[0, T]$, then u is Lipschitz continuous. Moreover,*

$$e^{|\omega|T}\|f(0+) - y\| + V(f, T) + |\omega|\int_0^T e^{|\omega|(T-\tau)}V(f, \tau)\,d\tau,$$

where

$$V(f, t) = \limsup_{h \downarrow 0} \int_0^{t-h} \frac{\|f(\tau + h) - f(\tau)\|}{h}\,d\tau$$

is the variation of f over $[0, t]$, is a Lipschitz constant for u. If X is also reflexive, then $u \in W^{1,1}(0, T: X)$.

That is, we have a regularity of u under the stated conditions which is independent of X, namely Lipschitz continuity. However, it is only under further conditions on X (e.g., reflexivity) that this guarantees differentiability and hence $u \in W^{1,1}(0, T: X)$. In particular, we have

COROLLARY 1. *Let $A + \omega I$ be m-accretive, $f: [0, T] \to X$ be of bounded variation, $x \in D(A)$ and X be reflexive. Then the mild solution of $u' + Au \ni f$, $u(0) = x$ is a Lipschitz continuous strong solution.*

Under further restrictions on X more refined statements about regularity can be made, but they do not offer essential improvements over the information above and we omit them here. The range condition can be used to replace m-accretivity of $A + \omega I$ in the case $f = 0$ to deduce results like Corollary 1.

A question related to the regularity considerations above is the following: Since the only known strong solutions are the constants in general, are they enough to determine (somehow) the class of all mild solutions? Another motivation for this question is the observation that the definition of a mild solution is not very "checkable". That is, given a function u, how can we tell if it is a mild solution of $u' + Au \ni f$? In general, we cannot simply compute u' and see if the relation is satisfied. Since $y \in Ax$ implies $u = x$ solves $u' + Au \ni y$, we know by Theorem 3 that any mild solution of $u' + Au \ni f$ satisfies

(6.2) $\quad \|u(t) - x\| \leq e^{\omega(t-s)}\|u(s) - x\| + \int_s^t e^{\omega(t-\tau)}[u(\tau) - x, f(\tau) - y]\,d\tau,$

for $0 \leq s \leq t \leq T$, $y \in Ax$. In fact, this family of inequalities can be taken to define a class of solutions called integral solutions. However, as opposed to mild solutions, the notion depends on the norm of X via the bracket and is appropriate only if $A \in \mathbf{A}(\omega)$. Moreover, it is not a good notion in general in the sense that it is easier to be an integral solution for a restriction of A than for A itself. However, it is a uniqueness criterion provided mild solutions are known to exist (which guarantees that A is "big enough" for the notion to be satisfactory). More precisely,

THEOREM 5. *Let $A \in \mathbf{A}(\omega)$, $f \in L^1(0, T: X)$ and v be a mild solution of $v' + Av \ni f$ on $[0, T]$. If $u \in C[0, T: X]$ satisfies (6.2) for every $y \in Ax$ and $u(0) = v(0)$, then $v = u$.*

Hence if the existence of a mild solution is known, then one can determine if a given function is this mild solution or not according as to whether or not the relations (6.2) hold.

7. Auxiliary results: Continuity, Trotter products and compactness. In this section we formulate a variety of auxiliary results in the subject which give additional useful information. The first among these addresses the problem of the dependence of the solution of

$$(\text{IVP})_{x,f} \qquad u' + Au \ni f, \qquad u(0) = x,$$

on A. In order to formulate the results in a multivalued generality we recall the notion of the limit inferior of a sequence A_n of operators. The operator $\liminf A_n$ is defined by $y \in \liminf A_n x$ if and only if there is a sequence $y_n \in A_n x_n$ such that $x_n \to x$ and $y_n \to y$. We will meet the condition $A \subset \liminf A_n$ below, and it will make things a bit clearer if we recall the following equivalent condition in the m-accretive case.

PROPOSITION 7. *Let $A_n + \omega I$ be m-accretive for $n = 1, 2, \ldots, \infty$ (with ∞ explicitly included). Then $A_\infty \subset \liminf A_n$ if and only if*

$$(7.1) \qquad \lim_{n \to \infty} (I + \lambda A_n)^{-1} x = (I + \lambda A_\infty)^{-1} x \quad \text{for } x \in X$$

for $\lambda > 0$ and $\lambda \omega < 1$. Moreover, (7.1) holds for all such λ if it holds for one such λ.

We call the condition (7.1) "resolvent convergence." Now let us formulate the continuous dependence theorem in some generality.

THEOREM 6. *Let $A_n \in \mathbf{A}(\omega)$, $f_n \in L^1(0, T: X)$ for $n = 1, 2, \ldots, \infty$. Let u_n be a mild solution of $u'_n + A_n u_n \ni f_n$ on $[0, T]$ for $n = 1, 2, \ldots, \infty$. Let $A_\infty \subseteq \liminf A_n$ and*

$$\lim_{n \to \infty} \int_0^T \|f_n(t) - f_\infty(t)\| \, dt + \|u_n(0) - u_\infty(0)\| = 0.$$

Then $u_n \to u_\infty$ uniformly on $[0, T]$.

This result, in the m-accretive case, says that if initial data converge, the forcing terms converge in $L^1(0, T: X)$, and the resolvents of the A_n converge, then the solutions converge. More generally, it makes the same claim provided only that the solutions exist. The method of proof involves observing that, by definition, $A_\infty \subseteq \liminf A_n$ implies that given any neighborhood of an approximate solution of $u'_\infty + A_\infty u_\infty \ni f_\infty$ then for n large enough we can find an approximate solution of $u' + A_n u_n \ni f_n$ in this neighborhood and then using the estimates in the proof of the convergence theorem. The utility of such a result is clear. For example, one may use it to prove the approximation result described next. If $A + \omega I$ is m-accretive we may define the Yosida approximation A_η of A for small $\eta > 0$ by $A_\eta = \eta^{-1}(I - (I + \eta A)^{-1})$. Clearly A_η is Lipschitz continuous with $2\eta^{-1}$ as a Lipschitz constant and it is easy to see that $A_\eta + \omega/(1 - \eta\omega)I$ is accretive. Since, as the reader could check, $(I + \lambda A_\eta)^{-1} \to (I + \lambda A)^{-1}$ as $\eta \to 0$, the continuous dependence theorem implies that the solution u_η of $u'_\eta + A_\eta u_\eta = f$, $u_\eta(0) = x$ converges uniformly to the solution u of the corresponding problem with A replacing A_η as $\eta \to 0$. Since A_η is Lipschitz continuous, this is a natural way to approximate u by regular functions in a fashion closely related to the original problem.

Another sort of result of wide applicability can be motivated as follows: Suppose we want to solve

(7.2) $$u' + Au + Bu = 0, \quad u(0) = x,$$

and that we know the solutions of the Cauchy problems for the separate equations

(7.3) $$u' + Au = 0, \quad v' + Bv = 0,$$

in the form of the semigroups S_A and S_B and that A and B are functions. Assuming a large (and totally unreasonable) amount of regularity one computes

$$\frac{d}{dt} S_A(t) S_B(t) x \big|_{t=0} = Ax + Bx.$$

That is, infinitesimally $F(t) = S_A(t)S_B(t)$ looks like $S_{A+B}(t)$ should look. Moreover, $F(t)x$ is well behaved as a function of x. Can we not then represent S_{A+B} in terms of $F(t)$? One has the following theorem to this effect:

THEOREM 7. *Let $A \in \mathbf{A}(\omega)$ satisfy the range condition (3.24) and $C = \overline{D(A)}$ be convex. For each $t > 0$ let $F(t): C \to C$ and F satisfy*

(i) $\|F(t)x - F(t)y\| \leq e^{Lt}\|x - y\|$ *for x, $y \in C$ and $0 \leq t \leq 1$,*
(ii) $\lim_{t \downarrow 0} (I + \lambda F(t))^{-1} x = (I + \lambda A)^{-1} x$ *for $x \in C$ and $\lambda > 0$, $\lambda \max(\omega, L) < 1$.*

Then for each $x \in C$, $S_A(t)x = \lim_{n \to \infty} F(t/n)^n x$ uniformly on compact t-sets.

It is part of the proof that the inverses used in (ii) exist. This result applies in the "$A + B$" case above provided that one can verify the resolvent condition (ii) given A and B. In this event, the conclusion is called a "Trotter product" formula. However, there are many other circumstances under which one can verify the hypotheses of the theorem. The proof consists of using the estimates of the convergence theorem together with another interesting approximation result, which we state for interest's sake in a special case.

PROPOSITION 8. *Let $C \subset X$ be closed, $F: C \to C$ be nonexpansive and $h > 0$. If the initial-value problem*

$$u' + h^{-1}(I - F)u = 0, \quad u(0) = x \in C$$

has a mild solution on $[0, T]$, then

$$\|F^n x - u(t)\| \leq 2\|x - z\| + \left((n - (t/h))^2 + n\right)^{1/2}\|z - Fz\|$$

holds for every $z \in C$, $0 \leq n$ and $0 \leq t \leq T$. In particular, choosing $x = z$ and $t = nh$ we have

$$\|F^n x - u(nh)\| \leq \sqrt{n}\|x - Fx\|.$$

The last sort of auxiliary result we discuss here concerns compactness. We fix an operator A with $A + \omega I$ m-accretive and consider the initial-value problem

$$(\text{IVP})_{x, f} \qquad u' + Au \ni f, \quad u(0) = x,$$

whose mild solution u we denote by $E(x, f)$. If $f = 0$, then $u(t) = E(x, 0)(t) = S_A(t)x$ where S_A is the semigroup generated by $-A$, and we will use this notation below. We ask when the images of various sets under E are compact in various senses. The simplest question concerns the semigroup case. A function in X is called compact if it maps bounded subsets of its domain into precompact sets in X and a semigroup S is compact if each $S(t)$ is compact for $t > 0$.

THEOREM 8. *Let $A + \omega I$ be m-accretive and S the semigroup on $\overline{D(A)}$ generated by $-A$. Then $S(t)$ is compact if and only if the following two conditions are satisfied*:

(i) *For each small $\lambda > 0$ the operator J_λ is compact*;

(ii) *for each bounded subset B of $\overline{D(A)}$ and $s > 0$, $\lim_{t \to s} S(t)x = S(s)x$ holds uniformly for $x \in B$*.

In applications to partial differential equations, compactness of $S(t)$ tends to arise from regularizing properties, that is $S(t)x$ will lie in a more regular class of functions for $t > 0$ than at $t = 0$. Another sort of compactness one is interested in is the compactness of the *trajectory*

$$\text{tr}(x) = \{E(x, f)(t): 0 \leq t\}$$

of the solution of $(\text{IVP})_{x, f}$. Compactness of trajectories is useful in making dynamical systems type arguments concerning the asymptotic behaviour of u. Concerning this property one has

THEOREM 9. *Let A be m-accretive, $0 \in R(A)$, $f \in L^1([0, \infty): X)$ and $x \in \overline{D(A)}$. In addition, let $(I + \lambda A)^{-1}$ be compact for some $\lambda > 0$. Then $\text{tr}(x)$ is precompact.*

The first conditions in the theorem guarantee that $\text{tr}(x)$ is bounded and the compactness comes from the assumption on J_λ.

Next we look at the solution operator for (IVP) and consider when it is compact as a function of f for fixed $x \in \overline{D(A)}$. There arises the question of what topologies to use in the domain and range space here, and the next result contains an answer.

THEOREM 10. *Let $A + \omega I$ be m-accretive and $S(t)$ be the semigroup generated by $-A$. Fix $x \in \overline{D(A)}$ and $p > 1$. Let $Q: L^p(0,T: X) \to C[0,T: X]$ be given by $Q(f) = E(x, f)$. If $S(t)$ is a compact semigroup, then Q is a compact operator.*

This result is unsatisfactory in that it does not allow the natural generality of $f \in L^1(0,T: X)$. It is possible to treat this case if we are willing to weaken our requirements in the range space. Moreover, we can then vary x as well.

THEOREM 11. *Under the assumptions of Theorem 10, if $S(t)$ is a compact semigroup then the solution operator E is compact as a mapping*

$$E: \overline{D(A)} \times L^1(0,T: X) \to L^p(0,T: X) \quad \text{for } 1 \leq p \leq \infty.$$

8. Comments and references. In this section we amplify on the main text a bit and provide some basic references for the interested reader. No attempt has been made to be complete, and nothing like completeness has been achieved. However, the references we do quote together with the references they contain should suffice to accurately represent the situation. Let us begin by noting that there are several books in the general area. The theory in Hilbert spaces is well developed in Brezis [17]. The general case is treated in Da Prato [36], Barbu [5] and Pavel [72, 73]. The books of Martin [64] and Browder [24] also treat aspects of the theory. The references provided by these works are not subsumed here and the reader will find many applications to partial differential equations in Barbu's book.

On §1. Early attempts to represent nonexpansive semigroups were made by Neuberger [69], Oharu [70] and Komura [59, 60]. Komura's dramatic ideas were a main stimulus for the rapid development which followed (e.g., Kato [49, 50], Crandall and Pazy [34], Dorroh [41] and Browder [23]).

The bracket [,] and the duality map J are well known in functional analysis. However, nomenclature and notation are inconsistent. For example, in Reich [87] of this volume, $J(x)$ denotes what would be written $\|x\|J(x)$ in our notation. Sato [88] provides specific computations of the duality map, but the reader can work out what J and [,] are for the common spaces. Workers in this subject learned Proposition 1(ix) in Kato [49]. If J is not single-valued, a stronger condition than Definition 1(iii) arises when the conclusion is required to hold for all $x^* \in J(x - \hat{x})$. An operator with this property is sometimes called s-accretive (or "Browder accretive", since this notion was taken to define accretive in Browder [23]). If J is single-valued on $X/\{0\}$, the notions coincide. Interest in s-accretivity arises from facts such as $A + B$ is accretive whenever A and B are and at least one of them is s-accretive.

Komura [59] is sometimes cited in this subject for a proof of the fact that if X is reflexive and $f: [0, T] \to X$ is Lipschitz continuous, then $f \in W^{1,1}(0, T: X)$, but theorems of Radon–Nikodym type for reflexive spaces were already proved by Phillips [77] and Dunford and Pettis [42]. Reflexive spaces are but examples of spaces with the Radon–Nikodym property.

On §2. The notion of a mild solution is already suggested in Crandall and Liggett [32], although it was too early at that time to institutionalize the idea. The term "mild" appears in Browder [21] in another context, but in a way consistent with our usage. The notation and presentation here are taken from [13], but variants of this natural idea appeared in various places (under various names), e.g. Kenmochi and Oharu [52], Kobayashi [53], and Pierre [78]. There are many questions one can ask about mild solutions which have not been seriously approached. For example, it is known that a mild solution in the current sense cannot necessarily be approximated by solutions of discretizations with uniform steps—but it is not known if this is true when, e.g., A is accretive. It is known that if $f = 0$ and $X = R$ and A is accretive, then uniform steps are enough (unpublished results of Crandall and Pierre). In most applications these issues are not serious, as A is either m-accretive or satisfies a variant of the range condition (3.22). R. Martin [63] proved that continuous accretive operators are m-accretive.

It is also known that mild solutions defined, as we have done, in the implicit way (2.6) (i.e., A is evaluated at v_i) differ from those defined in the explicit way in which Av_i is replaced by Av_{i-1} in (2.6). Indeed, it is easy to see that $u(t)$ is the limit of solutions of explicit approximations of $u' + Au \ni 0$ iff $v(t) = u(-t)$ is a mild solution of $v' - Av \ni 0$. The case of a single conservation law, a partial differential equation whose relevant solutions are not reversible and which can be accommodated in the theory [29], provides a significant counterexample. Proposition 3 is selected from [13].

The notion of a "strong solution" is standard, but sometimes people prefer to weaken it to require the continuity of the solution on $[0, T]$ and what we have called a strong solution on $[\varepsilon, T]$ for each $\varepsilon > 0$. This accommodates more examples and still allows one to do "calculus" without undue worry about the validity of the manipulations.

Examples of badly behaved nonlinear semigroups occur in [32], Plant [82] and Webb [92, 93], but mild solutions which are not strong solutions are familiar even in the linear theory when initial data do not lie in $D(A)$ or f is not sufficiently regular (in which case the variation of parameters formula typically provides such solutions).

On §3. The first proof that solutions of difference approximations converge (in a more restricted context, but with general X) was in [32]. Rasmussen [83] provided a useful proof. Takahashi [89] gave a more general convergence proof with variable steps. Benilan [7] proved the existence and uniqueness of mild solutions for $u' + Au \ni f$ when A is m-accretive. The existence also follows from results of Crandall and Pazy [35].

The full Theorem 1 was first proved in Crandall and Evans [31] by the fun method sketched here. One finds appropriate error estimates in [31] as well. The result in the case $f = 0$ was obtained by Y. Kobayashi [53] who formulated his results for quasi-accretive operators, a notion which generalizes the accretive case and which was introduced by Takahashi [89], but for which significant nonaccretive examples are lacking. Kobayashi's method was different (and simpler) in the case $f = 0$, but it becomes more complex in the general case. See also Takahashi [89, 90]. The reader may consult K. Kobayasi [56] and K. Kobayasi, Y. Kobayashi, and S. Oharu [57] for even more general results by this method.

Indeed, there are a variety of generalizations of the above to time dependent equations of the form $u' + A(t)u \ni f$, although it is not easy to be satisfied with any particular set of technicalities or definitions in this case (as is already true in the linear setting). We mention that Kato [50] and Crandall and Liggett [32] already allowed some time dependence, while Crandall and Pazy [35] is more general. The case of "integrable" time dependence was handled in Evans [43] using the results of [31] (in essence, Theorem 2), and there is also the elegant and different approach of Pierre [79]. The recent works [57], which was mentioned above, Iwamiya, Oharu and Takahashi [47], and K. Kobayasi and S. Oharu [58] extend the theory in various significant ways. It would be interesting to know the precise relationship between the convergence assertions in these works and Theorem 2.

On §4. The theory of Kato referred to is the simplest context described in his survey [51] in this volume, and we refer the reader to this paper for appropriate references. The relationship with the nonlinear theory sketched here is noted in Crandall and Souganidis [37], and can be greatly generalized. See also Hazan [46] in this regard. If the assumptions of this section are strengthened by requiring the existence of an operator $S: Y \to X$ with the properties described in [51], then the conclusions of this section can be strengthened to assert that the difference approximations converge in the strong topology of Y uniformly in t and the proof can be adapted to prove the continuity of the solution as a Y-valued function (and the assumption (B3) dropped). This is done in Crandall and Souganidis [38]. Another work, which is in a more preliminary stage, extends these results to the variable norm setting explained in Kato's article in this volume.

On §5. The result (i) is due to Benilan [7]. The result (ii) follows from Crandall and Liggett [32]. The biunique correspondence of (iii) was proved by Crandall and Pazy [34]. The existence of an m-accretive A such that $S = S_A$ in Hilbert spaces in this generality involves Minty's theorem, which essentially states the equivalence of "maximal monotone" and "m-accretive" in Hilbert spaces, and this result fails in general. (See Crandall and Liggett [33] and Calvert [27].)

The idea to obtain A via (5.11) is Komura's [60], and so is the first proof of the existence of this limit in Hilbert spaces. This result was the hardest step in the proof of (iii). New ideas had to be introduced to extend this convergence result outside of Hilbert spaces, and this was done by Baillon [3]. Reich sharpened this line of the theory, and (iv) can be found in [84]. We refer to Reich [85, 86] for further references and discussion. See also [87, Theorem 1.6]. By the way, the results of [33] to the effect that the limit (5.11) always exist if X is two-dimensional but may fail to exist if X is three dimensional show that the success of this approach must involve geometrical considerations.

The sufficiency of (5.2) for the solvability of (5.3) was a fascinating result of [53]. This result allowed very slick proofs of m-accretivity if operators of the form $A + B$ where A is m-accretive and B is continuous and accretive, generalizing results of Martin [63], Webb [93], and Barbu [6]. The equivalence of (5.2) and (5.4) is remarked in [30]. Numerous people, including Kaplan and Yorke [48] and Takahashi [89], contributed to the development of this line of thought. The sufficiency of (b) is an unpublished result of Y. Kobayashi. He also shows (b) is equivalent to another condition related to the sufficient conditions of Pierre [80]. One also finds examples indicating the distinction between various conditions in [80].

On §6. The results on regularity of mild solutions we will regard as being "from the community", but let us mention that the main facts were not so obvious in the beginning. It was mentioned in §2 that a strong solution is a mild solution—this is not entirely obvious. Theorem 4 is the heart of results in the other direction—it implies that differentiable mild solutions satisfy the equation pointwise if A is "big enough" in the sense that the operator of (6.1) cannot properly extend A (and so mild solutions are strong if they are regular enough). Theorem 5 is a simple version of Benilan's uniqueness theorem [7].

On §7. Theorem 6, in this general formulation, may not appear in the literature (we are using a formulation from [13]). See, however, Miyadera and Kobayashi [66]; and results in this spirit in general Banach spaces go back to Benilan [7], Brezis and Pazy [20], Kurtz [62] and Goldstein [45]. For examples of substantial applications of this result in PDE see, e.g., Benilan and Crandall [10] or Kenmochi and Oharu [52].

Theorem 7, the conclusion of which is called the nonlinear Chernoff formula, is due to Brezis and Pazy [20]. An earlier version and Proposition 8 are due to Miyadera and Oharu [68]. Theorem 7 has many applications in PDE—see, e.g., Berger, Brezis and Rogers [16], Coron [28], Kenmochi and Oharu [52], and Oharu and Kobayasi [58]. There has been a fair amount of recent activity concerning results of this general type in special circumstances; see, e.g., Benilan and Ismail [15], Reich [85, 86], Kobayashi [54, 55] and their references. M. Pierre and M. Rihani [81] have recently obtained quite interesting results (both positive and negative) on the validity of more general formulae involving nonuniform steps.

One can ask to what extent the implications in Theorems 6 and 7 are reversible and be led thereby to the question of convergence versus resolvent consistency. Since the conclusion of Theorem 7 always holds if $F(t) = S(t)$, this links up with the problem of the existence of the limit (5.1). See Reich [86] for recent results and references.

The various compactness results are proved in Brezis [**18**], Dafermos and Slemrod [**39**] and Baras [**4**]. See Konishi [**61**] for an early result of this type and Brezis and Friedman [**19**] for an application in PDE.

Asymptotic behaviour. We have omitted the topic of asymptotic behaviour. The works Bruck [**25**] and Baillon [**2**] stimulated a large amount of subsequent work on these lines of research, and the area remains quite active. The survey article Bruck [**26**] is a recent source on this topic and we refer the reader to it. Other recent sources on aspects of this question include Pazy [**74, 75, 76**], Reich [**87**], Miyadera [**65**], and—in a somewhat different vein—Alikakos and Rostamian [**1**].

Regularizing effects. A final topic we mention is that of regularizing effects. These concern questions of regularity—interpreted in various ways—of $S(t)x$ for $t > 0$ that x itself may not enjoy. There is no general theory available yet, but the phenomenon is widespread and of considerable interest when it is present. On the abstract side, the best known examples are the regularizing effects of linear analytic semigroups and semigroups generated by subdifferentials of convex functions in Hilbert spaces (see [**17**]). In the general nonreflexive case we have some examples, e.g. those of Benilan [**8, 9**], Veron [**91**], Benilan and Crandall [**12**], and Crandall and Pierre [**36**]. A new regularizing semigroup is also discussed in Reich [**87**]. One wonders if there is an informative unifying point of view which might relate these various examples.

BIBLIOGRAPHY

1. N. Alikakos and R. Rostamian, *Lower bound estimates and separable solutions for homogeneous equations of evolution in Banach space*, J. Differential Equations **43** (1982), 323–344.

2. J. B. Baillon, *Un théorème de type ergodique pour les contractions nonlinéaires dans un espace de Hilbert*, C. R. Acad. Sci. Paris **280** (1975), 1511–1514.

3. _____, *Générateurs et semi-groupes dans les espaces de Banach uniforment lisses*, J. Funct. Anal. **29** (1978), 199–213.

4. P. Baras, *Compacité de l'opérateur $f \to u$ solution d'une equation nonlinéaire $u' + Au \ni f$*, C. R. Acad. Sci. Paris **286** (1978), 1113–1116.

5. V. Barbu, *Nonlinear semigroups and differential equations in Banach spaces*, Nordhoff, Leyden, 1976.

6. V. Barbu, *Continuous perturbations of nonlinear m-accretive operators in Banach spaces*, Boll. Un. Mat. Ital. **6** (1972), 270–278.

7. Ph. Benilan, *Équations d'évolution dans un espace de Banach quelconque et applications*, Thesis, Orsay, 1972.

8. _____, *Opérateurs accretifs et semi-groupes dans les espaces L^p $(1 \leq p \leq \infty)$*, Functional Analysis, Japan-France Seminar, Tokyo and Kyoto, 1976 (H. Fujita, ed.), Japan Society for the Promotion of Science, 1978.

9. _____, *A strong regularity L^p for solution of the porous media equation*, Contributions to Nonlinear Differential Equations, Res. Notes in Math., no. 89, Bardos, Damlamian, Díaz Hernández, eds., Pitman, Boston, 1983, pp. 39–58.

10. Ph. Benilan, H. Brezis, and M. G. Crandall, *A semilinear elliptic equation in $L^1(\mathbf{R})^N$*, Ann. Scuola Norm. Sup. Pisa (5) **2** (1975), 523–555.

11. Ph. Benilan and M. G. Crandall, *The continuous dependence on φ of solutions of $u_t - \Delta\varphi(u) = 0$*, Indiana Univ. Math. J. **30** (1981), 162–177.

12. _____, *Regularizing effects of homogeneous evolution equations*, Contributions to Analysis and Geometry, Supplement to Amer. J. Math. (D. N. Clark, C. Pecelli, R. Sacksteder, eds.), Johns Hopkins University Press, Baltimore, 1981, pp. 23–39.

13. Ph. Benilan, M. G. Crandall, and A. Pazy, *Nonlinear evolution governed by accretive operators* (in preparation).

14. Ph. Benilan, M. G. Crandall, and M. Pierre, *Solutions of the porous medium equation in $\mathbf{R}^N$ under optimal conditions on initial values*, Indiana Univ. Math. J. **33** (1984), 51-87.

15. Ph. Benilan and S. Ismail, *Une generalisation d'un resultat de Kato-Masuda sur la formule de Trotter* (to appear).

16. A. Berger, H. Brezis, and J. C. W. Rogers, *A numerical method for solving $u_t - \Delta f(u) = 0$*, RAIRO Anal. Numér. **13** (1979), 297-312.

17. H. Brezis, *Opérateurs maximaux monotones et semi-groupes de contractions dans les espaces de Hilbert*, North Holland, Amsterdam, 1977.

18. _____, *New results concerning monotone operators and nonlinear semigroups*, Analysis of Nonlinear Problems, RIMS, 1974, pp. 2-27.

19. H. Brezis and A. Friedman, *Nonlinear parabolic equations involving initial conditions as measures*, J. Math. Pures Appl. **62** (1983), 73-92.

20. H. Brezis and A. Pazy, *Convergence and approximation of nonlinear operators in Banach spaces*, J. Funct. Anal. **9** (1972), 63-74.

21. F. E. Browder, *Nonlinear equations of evolution*, Ann. of Math. **80** (1964), 485-523.

22. _____, *Nonlinear mappings of non-expansive and accretive type in Banach spaces*, Bull. Amer. Math. Soc. **73** (1967), 875-882.

23. _____, *Nonlinear equations of evolution and nonlinear accretive operators in Banach spaces*, Bull. Amer. Math. Soc. **73** (1967), 867-874.

24. _____, *Nonlinear operators and nonlinear equations of evolution in Banach spaces*, Proc. Sympos. Pure Math., vol. 18, part 2, Amer. Math. Soc., Providence, R. I., 1976.

25. R. E. Bruck, *Asymptotic convergence of nonlinear contraction semi-groups in Hilbert spaces*, J. Funct. Anal. **18** (1975), 15-26.

26. _____, *Asymptotic behavior of nonexpansive mappings*, in Fixed Points and Nonexpansive Mappings (R. C. Since, ed.), Contemporary Math., vol. 18, Amer. Math. Soc., Providence, R. I., 1983, 1-47.

27. B. Calvert, *Maximal accretive is not m-accretive*, Boll. Un. Mat. Ital. **6** (1970), 1042-1044.

28. J. M. Coron, *Formules de Trotter pour une équations d'evolution quasilinéaire du 1er order*, J. Math. Pures Appl. **61** (1982), 19-112.

29. M. G. Crandall, *The semigroup approach to first order quasilinear equations in several space variables*, Israel J. Math. **12** (1972), 108-132.

30. _____, *An introduction to evolution governed by accretive operators*, Dynamical Systems—An International Symposium (L. Cesari, J. Hale, J. La Salle, eds.), Academic Press, New York, 1976, pp. 131-165.

31. M. G. Crandall and L. C. Evans, *On the relation of the operator $\partial/\partial s + \partial/\partial \tau$ to evolution governed by accretive operators*, Israel J. Math. **21** (1975), 261-278.

32. M. G. Crandall and T. M. Liggett, *Generation of semi-groups of nonlinear transformations on general Banach spaces*, Amer. J. Math. **93** (1971), 265-298.

33. _____, *A theorem and counterexample in the theory of nonlinear transformations*, Trans. Amer. Math. Soc. **160** (1971), 263-278.

34. M. G. Crandall and A. Pazy, *Semi-groups of nonlinear contractions and dissipative sets*, J. Funct. Anal. **3** (1969), 376-418.

35. _____, *Nonlinear evolution equations in Banach spaces*, Israel J. Math. **11** (1972), 57-94.

36. M. G. Crandall and M. Pierre, *Regularizing effects for $u_t = A\varphi(u)$ in L^1*, J. Funct. Anal. **45** (1982), 194-212.

37. M. G. Crandall and P. Souganidis, *Quasinonlinear evolution equations*, Mathematics Research Center TSR 2352, University of Wisconsin-Madison, 1982.

38. _____, *Convergence of difference approximations of quasilinear evolution equations*, Mathematics Research Center TSR #2711, University of Wisconsin-Madison, 1984 (and to appear in J. Non. Anal. Theor. Meth. Appl.).

39. C. Dafermos and M. Slemrod, *Asymptotic behaviour of nonlinear contraction semigroups*, J. Funct. Anal. **13** (1973), 97-106.

40. G. Da Prato, *Applications croissants et équations d'évolution dans les espaces de Banach*, Institutiones Mathematicae, Vol. II, Academic Press, London, 1976.

41. R. Dorroh, *A nonlinear Hille–Yosida–Phillips theorem*, J. Funct. Anal. **3** (1969), 345–353.

42. N. Dunford and B. Pettis, *Linear operators on summable functions*, Trans. Amer. Math. Soc. **47** (1940), 323–392.

43. L. C. Evans, *Nonlinear evolution equations in an arbitrary Banach space*, Israel J. Math. **26** (1977), 1–42.

44. _____, *Application of nonlinear semigroup theory to certain partial differential equations*, Nonlinear Evolution Equations (M. G. Crandall, ed.), Academic Press, N. Y., 1978.

45. J. A. Goldstein, *Approximation of nonlinear semigroups and evolution equations*, J. Math. Soc. Japan **24** (1972), 558–573.

46. M. Hazan, *Nonlinear quasilinear evolution equations: existence, uniqueness, and comparison of solutions: rate of convergence of the difference method*, Zap. Nauchn. Sem. Leningrad. Otdel. Mat. Inst. Steklov **127** (1983), 181–199.

47. T. Iwamiya, S. Oharu, and T. Takahashi, *On the class of nonlinear evolution operators in Banach spaces*, J. Non. Anal. Theor. Meth. Appl. (to appear).

48. J. Kaplan and J. Yorke, *Toward a unification of ordinary differential equations with nonlinear semigroup theory*, Proc. Internat. Conf. Ordinary Differential Equations, University of Southern California, Academic Press, New York, 1975.

49. T. Kato, *Nonlinear semigroups and evolution equations*, J. Math. Soc. Japan **19** (1967), 508–520.

50. T. Kato, *Accretive operators and nonlinear evolution equations in Banach spaces*, in Nonlinear Functional Analysis, (F. Browder, ed.), Proc. Sympos. Pure Math. vol. 18, part 1, Amer. Math. Soc., Providence, R. I., 1970, pp. 138–161.

51. T. Kato, *Nonlinear equations of evolution in Banach spaces*, Proc. Sympos. Pure Math., vol. 45, Amer. Math. Soc., Providence, R. I., 1985 (this volume).

52. N. Kenmochi and S. Oharu, *Difference approximation of nonlinear evolution equations and semigroups of nonlinear operators*, Publ. RIMS, Kyoto Univ. **10** (1974), 147–207.

53. Y. Kobayashi, *Difference approximation of Cauchy problems for quasi-dissipative operators and generation of nonlinear semigroups*, J. Math. Soc. Japan **27** (1975), 640–665.

54. _____, *Product formula for nonlinear semigroups in Hilbert spaces*, Proc. Jap. Acad. **58** (1982), 425–428.

55. _____, *A product formula approach for solving first order quasilinear equations*, Hiroshima Math. J. **14** (1985), 408–509.

56. K. Kobayasi, *On difference approximation of time dependent nonlinear evolution equations in Banach spaces*, Mem. Sagami Inst. Technology **17** (1983), 59–69.

57. K. Kobayasi, Y. Kobayashi, and S. Oharu, *Nonlinear evolution operators in Banach spaces*, Osaka J. Math. **21** (1984), 281–310.

58. K. Kobayasi and S. Oharu, *Nonlinear evolution operators in a Fréchet space*, Japan. J. Math. (N.S.) **10** (1984), 243–270.

59. Y. Komura, *Nonlinear semigroups in Hilbert spaces*, J. Math. Soc. Japan **19** (1967), 508–520.

60. _____, *Differentiability of nonlinear semigroups*, J. Math. Soc. Japan **21** (1969), 375–402.

61. Y. Konishi, *Sur la compacite des semigroupés nonlinéaires dans les espaces de Hilbert*, Proc. Japan Acad. **48** (1972), 278–280.

62. T. Kurtz, *Convergence of sequences of semigroups of nonlinear operators with an application to gas kinetics*, Trans. Amer. Math. Soc. **186** (1973), 259–272.

63. R. H. Martin, *A global existence theorem for autonomous differential equations in a Banach space*, Proc. Amer. Math. Soc. **26** (1970), 307–314.

64. _____, *Nonlinear operators and differential equations in Banach spaces*, Wiley, New York, 1976.

65. I. Miyadera, *On the infinitesimal generators and the asymptotic behaviour of non-linear contraction semigroups*, Proc. Japan Acad. **58** (1982), 1–4.

66. I. Miyadera and Y. Kobayashi, *Convergence and approximation of nonlinear semigroups*, Functional Analysis and Numerical Analysis (H. Fujita, ed.), Japan-France Seminar, Tokyo and Kyoto, 1976, Japan Society for the Promotion of Science, 1978, pp. 272–295.

67. I. Miyadera and K. Kobayasi, *On the asymptotic behaviour of almost orbits of non-linear contraction semigroups in Banach spaces*, Nonlinear Anal. **6** (1982), 349–365.

68. I. Miyadera and S. Oharu, *Approximation of semi-groups of nonlinear operators*, Tôhoku Math. J. **22** (1970), 24–47.

69. J. W. Neuberger, *An exponential formula for one parameter semigroups of nonlinear transformations*, J. Math. Soc. Japan **18** (1966), 154–157.

70. S. Oharu, *Note on the representation of semi-groups of nonlinear operators*, Proc. Japan Acad. **42** (1967), 1149–1154.

71. S. Oharu and T. Takahashi, *A convergence theorem of nonlinear semigroups and its application to first order quasilinear equations*, J. Math. Soc. Japan **26** (1974), 124–160.

72. N. Pavel, *Ecuati Diferentiale Asociate unor Operati Nonliniari pe Spatii Banach*, Editura Acad. Bucuresti, 1977.

73. _____, *Analysis of some nonlinear problems in Banach spaces and applications* (to appear).

74. A. Pazy, *Strong convergence of semigroups of nonlinear contractions in Hilbert space*, J. d'Anal. Math. **34** (1978), 1–35.

75. _____, *Semigroups of nonlinear contractions and their asymptotic behaviour*, in Nonlinear Analysis and Mechanics (R. J. Knops, ed.), Heriot–Watt Sympos., vol. 3, Pitman Research Notes in Math. **30** (1979), 36–134.

76. _____, *The Lyapunov method for semigroups of nonlinear contractions in Banach spaces*, J. d'Anal. Math. **40** (1981), 239–262.

77. R. S. Phillips, *On linear transformations*, Trans. Amer. Math. Soc. **48** (1940), 516–541.

78. M. Pierre, *Génération et perturbation de semi-groupes de contractions nonlinéaires*, These de Docteur de 3 é cycle, University de Paris VI, 1976.

79. _____, *Enveloppe d'une famille de semigroups nonlinéaires de équations d'evolution*, Publ. Math. Univ. de Besancon, anée 1976–1977.

80. _____, *Un théoreme general de génération de semi-groupes nonlinéaires*, Israel J. Math. **23** (1976), 189–199.

81. M. Pierre and M. Rihani, *About product formulas with variable step-size*, MRC Tech. Summary Rep. 2783, Univ. Wisconsin, Madison, Wisc., 1985.

82. A. T. Plant, *Flow-invariant domains of Hölder continuity for nonlinear semigroups*, Proc. Amer. Math. Soc. **53** (1975), 83–87.

83. S. Rasmussen, *Non-linear semigroups, evolution equations and product integral representations*, Various Publication Series No. 2, Aarhus Universitet, 1971/72.

84. S. Reich, *Product formulas, nonlinear semigroups, and accretive operators*, J. Funct. Anal. **36** (1980), 147–168.

85. _____, *Nonlinear semigroups, accretive operators, and applications*, Nonlinear Phenomena in Mathematical Sciences, Academic Press, New York, 1982, pp. 831–838.

86. _____, *Convergence, resolvent consistency and the fixed point property for nonexpansive mappings*, in Fixed Points and Nonexpansive Mappings (R. C. Sine, ed.), Contemporary Math., vol. 18, Amer. Math. Soc., Providence, R. I., 1983, pp. 167–174.

87. _____, *Nonlinear semigroups, holomorphic mappings and integral equations*, Proc. Sympos. Pure Math., vol. 45, Amer. Math. Soc., Providence, R. I., 1985 (this volume).

88. K. Sato, *On the generators of non-negative contraction semi-groups in Banach lattices*, J. Math. Soc. Japan **20** (1968), 431–436.

89. T. Takahashi, *Difference approximation of Cauchy problems for quasi-dissipative operators and generation of semigroups of nonlinear contractions*, Technical Report of National Aerospace Laboratory, TR-419T, 1975, pp. 1–15.

90. _____, *Convergence of difference approximations of nonlinear evolution equations and generation of semigroups*, J. Math. Soc. Japan **28** (1976), 96–113.

91. L. Veron, *Coercivité et propriétés régularisantes des semi-groupes non linéaires dans les espaces de Banach*, Publ. Math. Fac. Sci. Besancon **3** (1977).

92. G. F. Webb, *Representation of semi-groups of nonlinear nonexpansive transformations in Banach space*, J. Math. Mech. **19** (1969), 159–170.

93. _____, *Continuous nonlinear perturbations of linear accretive operators in Banach spaces*, J. Funct. Anal. **10** (1972), 191–203.

94. _____, *Nonlinear perturbations of linear accretive operators in Banach spaces*, Israel J. Math. **12** (1972), 237–248.

UNIVERSITY OF WISCONSIN–MADISON

A Theorem of Mather and the Local Structure of Nonlinear Fredholm Maps

JAMES DAMON[1]

Let E and F be Fréchet spaces with $E = E_1 \oplus G$, $F = F_1 \oplus G$ with $\dim_{\mathbf{R}} E_1$, $F_1 < \infty$. We consider for open $U \subset E$ a smooth nonlinear Fredholm mapping $f \colon U \to F$ which has the form

(∗) $$f(x, u) = (\tilde{f}(x, u), u),$$

with $x \in E_1$ and $u \in G$. For example, if E and F are Banach spaces, then the Lyapunov–Schmidt procedure allows one to locally represent a nonlinear Fredholm map in such a form.

If $u_0 \in U$, then we shall be concerned with describing the structure of f, as a mapping, in a neighborhood of u_0. For example, the problem

(∗∗) $$\Delta u + h(u, x) = r(x) \quad \text{in } \Omega, \quad u = 0 \quad \text{on } \partial\Omega$$

(with Ω a bounded domain with smooth boundary) can be studied, under suitable hypotheses on h, by solving the equation $f(u) = r$, where $f(u) = \Delta u + h(u, x)$ is an operator $f \colon C_0^{2+\alpha}(\overline{\Omega}) \to C^\alpha(\overline{\Omega})$ (using the usual notation). It will have a solution $u_0 = 0$ for $r(x) \equiv 0$ if $h(0, x) = 0$. It would also be useful to know how many solutions near u_0 there are to (∗) when $r(x)$ is near 0 in the appropriate function space. We shall describe several results which do this, using singularity theory. Local singularity theory has been used by Golubitsky and Schaeffer (see e.g. [9] and [10]) to study imperfect bifurcation theory. They have subsequently applied their results to numerous examples. Such "local" singularity results give local information about mappings f restricted to finite-dimensional subspaces and, hence, may not allow one to conclude certain local results about f itself. On the other hand, Berger, Church, and Timourian [3, 4] have used singularity theory

---

1980 *Mathematics Subject Classification*. Primary 58C25; Secondary 58E07, 47H15, 35J65.
[1] Partially supported by a grant from the National Science Foundation.

© 1986 American Mathematical Society
0082-0717/86 $1.00 + $.25 per page

to give explicitly the local structure for the operators corresponding to certain particular forms of (**), extending earlier results of Ambrosetti–Prodi [1] and Berger–Church [3]. We shall show how a "global" theorem of Mather [12, II] allows one to quite generally deduce local information about such operators.

We shall consider two mappings to be equivalent if they agree on a neighborhood of the point u_0. Such an equivalence class is referred to as a germ of the mapping at u_0. Furthermore, by translation, we may assume $u_0 = 0$ and $f(0) = 0$; then $f: E, 0 \to F, 0$ will be used to denote the germ of f at 0. The results are described using the notion of infinitesimally stable germs $f_1: \mathbf{R}^n, 0 \to \mathbf{R}^p, 0$. Such germs possess remarkable properties (due to Mather [12, IV]), which we shall discuss in §1. It is these properties which justify the usefulness of these germs in describing the local properties of mappings such as f. If there is a finite-dimensional subspace $G_1 \subset G$ such that the germ $f_1 = (f|E_1 \oplus G_1): E_1 \oplus G_1, 0 \to F_1 \oplus G_1, 0$ is infinitesimally stable, then we shall say that f is infinitesimally stable. The first result is

THEOREM 1. *If $f: E, 0 \to F, 0$ defined above (*) (with $u_0 = 0$) is infinitesimally stable, then it is smoothly equivalent in a neighborhood of 0 to $f_1 \times \mathrm{id}_{G_2}$ (G_2 is a closed complement to G_1 in G). This means there are germs of smooth invertible mappings $\phi: E, 0 \to E, 0$ and $\psi: F, 0 \to F, 0$ so that $f = \psi \circ (f_1 \times \mathrm{id}) \circ \phi$.*

Here and in the following, smooth will be understood in the Gâteaux differentiable sense.

This has two important corollaries, which follow from the properties of infinitesimally stable germs and the proof of the theorem.

COROLLARY 1. *Let E, F, and H be Banach spaces and $f: E, 0 \to F, 0$ infinitesimally stable. Suppose $g: E \oplus H, 0 \to F \oplus H, 0$ is smooth and satisfies $g(e, h) = (\bar{g}(e, h), h)$ and $\bar{g}(e, 0) = f(e)$ for $e \in H$ in a neighborhood of 0. Then, g is smoothly equivalent to $f \times \mathrm{id}_H$ by an equivalence which is the identity when $h = 0$.*

The second corollary concerns the case when f is not infinitesimally stable. However, if $f_0 = f|E_1: E_1, 0 \to F_1, 0$, we suppose that there is a finite parameter family $f_1: E_1 \oplus V, 0 \to F_1 \oplus V, 0$, with $f_1(x, v) = (\bar{f}_1(x, v), v)$ and $\bar{f}_1(x, 0) = f_0(x)$ for $x \in E_1$, $v \in V$, so that f_1 is infinitesimally stable.

This is known to hold for all germs f_0 not in a set of infinite codimension (see Tougeron [17]).

COROLLARY 2. *If f satisfies the properties given above, then it can be obtained by the restriction of $f_1 \times \mathrm{id}_G: E \oplus V, 0 \to F \oplus V, 0$ to appropriate (nonlinear) subspaces.*

The proofs of these results will be given in §2, and they will be illustrated via several examples in §3.

The author wishes to thank both Phillip Church and Norman Dancer for helpful conversations on aspects of these problems.

1. Infinitesimal stability of mappings and germs. Let $f: N \to P$ be a smooth mapping between finite-dimensional manifolds. By a vector field ζ along f we shall mean a smooth mapping $\zeta: N \to TP$ such that $\pi \circ \zeta = f$, for $\pi: TP \to P$ the projection map of the tangent bundle of P. Then f is said to be *infinitesimally stable* if, given any such vector field ζ, there exist vector fields ξ on N and η on P so that $\zeta = \xi(f) + \eta \circ f$ (here $\xi(f)$ denotes the directional derivative of f with respect to ξ). If we consider instead the germ of f at a point $x_0 \in N$ with $f(x_0) = y_0$, then the germ of f at x_0 is infinitesimally stable if the above equation can always be solved for a germ ζ at x_0, using germs of vector fields ξ at x_0 and η at y_0. Mather has proven the important result [12, II] that if f is infinitesimally stable, then by composing f with diffeomorphisms of N and P, one obtains an entire neighborhood of f in the space of smooth functions from N to P. It is a version of this result which will be used to prove Theorem 1.

Here we shall concentrate on recalling the properties of infinitesimally stable germs, because of their importance in applying Theorem 1 and its corollaries to specific examples. We may choose local coordinates near x_0 and y_0 so that f has the form $f: \mathbf{R}^n, 0 \to \mathbf{R}^p, 0$. Furthermore, by applying the Lyapunov–Schmidt procedure to f we may choose local coordinates $(x_1, \ldots, x_s, v_1, \ldots, v_q)$ for $\mathbf{R}^n$ and $(y_1, \ldots, y_t, v_1, \ldots, v_q)$ for $\mathbf{R}^p$ so that f has the form $f(x, v) = (\bar{f}(x, v), v)$, with $\bar{f}(x, 0)$ ($= f_0(x)$) having rank 0 at 0. Here $f_0: \mathbf{R}^s, 0 \to \mathbf{R}^t, 0$. In the terminology of singularity theory, such an f is called an unfolding of f_0, and the v_i are the unfolding parameters.

For f in this form it is especially easy to verify infinitesimal stability. Let $\mathscr{E}_s$ denote the ring of smooth germs $g: \mathbf{R}^s, 0 \to \mathbf{R}$; it has a maximal ideal $\mathfrak{m}_s$ consisting of germs vanishing at 0. In $(\mathscr{E}_s)^{(t)} = \mathscr{E}_s \oplus \mathscr{E}_s \cdots \oplus \mathscr{E}_s$ (t copies), which we can view as an $\mathscr{E}_s$-module, we consider the submodule L generated by $\partial f_0 / \partial x_1, \ldots, \partial f_0 / \partial x_s$, and the submodule $(I(f_0))^{(t)}$ (t-copies of $I(f_0)$ = ideal in $\mathscr{E}_s$ generated by the coordinate functions of f_0, $\{y_i \circ f_0, 1 \leq i \leq t\}$). By $\mathfrak{m}_s^k$ we mean the ideal of $\mathscr{E}_s$ generated by monomials of degree $= k$. Then the quotient $\mathscr{E}_s / \mathfrak{m}_s^k$ is a finite-dimensional vector space with basis given by the monomials of degree $< k$. Lastly, letting $v = (v_1, \ldots, v_q)$, $\partial \bar{f} / \partial v_i |_{v=0}$ is a germ on $\mathbf{R}^s$ and, hence, via its coordinate functions, can be viewed as an element of $(\mathscr{E}_s)^{(t)}$—in fact, $(\mathfrak{m}_s)^{(t)}$. Then

1.1 *Verification criterion.* By Mather [12, IV, Proposition 1.8] f is infinitesimally stable iff $\partial \bar{f} / \partial v_1 |_{v=0}, \ldots, \partial \bar{f} / \partial v_q |_{v=0}$ span the quotient space

$$N(f_0) = (\mathfrak{m}_s)^{(t)} / \left(L + I(f_0)^{(t)} + \left(\mathfrak{m}_s^{p+1} \right)^{(t)} \right).$$

Since this space is finite dimensional, this becomes a very computable criterion, using part of the Taylor series of f.

EXAMPLE 1.2. $f: \mathbf{R}^{m-1}, 0 \to \mathbf{R}^{m-1}, 0$ defined by

(1.3) $$f(x, v_1, \ldots, v_{m-2}) = \left(x^m + \sum_{i=1}^{m-2} v_i x^i, v_1, \ldots, v_{m-2} \right).$$

Then $f_0(x) = x^m$, so $\partial f_0/\partial x = mx^{m-1}$ and $I(f_0)$ is the ideal in $\mathscr{E}_1$ generated by x^m. Then the quotient space $N(f_0) = \mathfrak{m}_1/\mathfrak{m}_1^{m-1}$. Also $\partial \bar{f}/\partial v_i = x^i$, which remains unchanged when we set $v = 0$. Then since $\{x, x^2, \ldots, x^{m-2}\}$ span the quotient space $\mathfrak{m}_1/\mathfrak{m}_1^{m-1}$, f given by (1.3) is infinitesimally stable.

An important property of infinitesimally stable germs is their classification by an algebraic invariant called the local algebra. For $f: \mathbf{R}^n, 0 \to \mathbf{R}^p, 0$ we define the local algebra $Q(f)$ as $\mathscr{E}_n/I(f)$. For example, for f given by (1.3), $I(f)$ = ideal generated by $x^m + \sum_{i=1}^{m-2} v_i x^i$, $v_1, \ldots, v_{m-2}$. This is also the ideal generated by x^m, $v_i, \ldots, v_{m-2}$. Thus,

$$\mathscr{E}_{m-1}/I(f) \simeq \mathscr{E}_1/(x^m) \simeq \mathbf{R}[x]/(x^m).$$

Then Mather proves [**12**, IV, Theorem A.].

1.4 *Classification by local algebras.* If f, $g: \mathbf{R}^n, 0 \to \mathbf{R}^p, 0$ are infinitesimally stable germs with isomorphic local algebras, then f and g are locally equivalent at 0 (i.e., $g = \psi \circ f \circ \phi$ for germs of diffeomorphisms ϕ, ψ).

Furthermore, given an algebra Q satisfying certain conditions, it is possible to construct an infinitesimally stable germ f, with $Q(f) \simeq Q$. For example, consider $Q \simeq \mathscr{E}_s/I$, with $I \subset \mathfrak{m}_s^2$ generated by $f_{01}, \ldots, f_{0t}$, and suppose, for simplicity, that $\dim_{\mathbf{R}} Q < \infty$. Then we use the f_{0i} as coordinate functions to define $f_0: \mathbf{R}^s, 0 \to \mathbf{R}^t, 0$. Then $N(f_0)$ is a subspace of a quotient of $(Q)^{(t)} \xrightarrow{\sim} (Q(f_0))^{(t)}$. Hence, there are $\phi_1, \ldots, \phi_q \in (\mathfrak{m}_s)^{(t)}$ which project to a basis for $N(f_0)$. Then $f: \mathbf{R}^{s+q}, 0 \to \mathbf{R}^{t+q}, 0$ defined by

$$(1.5) \quad f(x_1, \ldots, x_s, v_1, \ldots, v_q) = \left(f_0(x) + \sum_{i=1}^{q} v_i \phi_i, v_1, \ldots, v_q\right)$$

is infinitesimally stable by 1.1 and has a local algebra $Q(f) \xrightarrow{\sim} Q(f_0) \xrightarrow{\sim} Q$.

This leads to the third property:

1.6 *Normal form for infinitesimally stable germs.* Given an algebra Q satisfying certain conditions (which include the case $\dim_{\mathbf{R}} Q < \infty$) see [**12**, IV, Theorem B.]), then there is an infinitesimally stable germ f of the form (1.5) with local algebra $Q(f) \xrightarrow{\sim} Q$.

These three results allow one to begin with an infinitesimally stable germ g, compute its local algebra $Q(g)$, and construct a normal form f with local algebra $Q(f) \xrightarrow{\sim} Q(g)$. Then, by the classification theorem, f and g are locally equivalent, so the specifically given germ f can be used instead to study the local structure of g. We illustrate this by considering the question of determining how many solutions there are to the equation $g(x) = y$, where $g: \mathbf{R}^n, 0 \to \mathbf{R}^p, 0$ is infinitesimally stable with $n \leq p$. Then $\dim_{\mathbf{R}} Q(g) < \infty$ in this case. We denote this number by $\delta(g)$. We wish to consider solutions x near 0 for y near 0. Of course, this can vary, depending on y, so we ask whether we can at least give the maximum number of solutions that will occur for some y. A partial answer is given by a result in [**6**]:

1.7 *Solutions to equations given by infinitesimally stable germs.* For y near 0 there are at most $\delta(g)$ solutions x near 0 to the equation $g(x) = y$ (see e.g. [**8**, Chapter VI]). If $\dim_\mathbf{R} dg(0) \leqslant 2$, or $Q(g)$ satisfies other conditions in [**6**], there are points y arbitrarily close to 0 with exactly $\delta(g)$-solutions to $g(x) = y$ which are close to 0.

REMARK. If g is given as a polynomial so it could also be interpreted as a complex germ, then this number $\delta(g)$ still is the maximum number of complex solutions we could expect. Hence, unlike the case for other real germs, these infinitesimally stable germs achieve the maximum number of solutions we would expect even for their complexifications.

2. Proofs of the results

PROOF OF THE THEOREM. Since f_1 is an infinitesimally stable germ, then, by the "openness of versality" (see e.g. [**16**, 4.8.2]), there is a neighborhood U_1 of 0 in $E_1 \oplus G_1$ such that $f_1|U_1$ is infinitesimally stable as a mapping. We may assume U_1 chosen small enough so that there is a neighborhood U_2 of 0 in G_2 so that $U_1 \times U_2 \subset U$, the original neighborhood on which f was defined.

Let B_1 be a closed ball in U_1. We define a map $R: U_2 \to C^\infty(B_1, F_1 \oplus G_1)$ by $R(e)(x) = \bar{f}(x, e)$ for $x \in B_1$, $e \in U_2$.

Claim. R is a smooth map of Fréchet spaces in the sense of [**11**]. For this we must show that $d^k R(e)(v_1, \ldots, v_k)$ is defined and continuous as a function of $(e, v_1, \ldots, v_k)$ with values in $C^\infty(B_1, F_1 \oplus G_1)$. A straightforward calculation shows that, evaluated at x, it must be $d_2^k f(x, e)(v_1, \ldots, v_k)$, where $d_2^k f$ denotes the kth partial derivative with respect to G_2. Since f is smooth, by the properties of the partial derivative, to show $d_2^k f$ is the kth derivative of R, it is sufficient to show that the map $(e, v, \ldots, v_k) \mapsto d_2^k f(x, e)(v_1, \ldots, v_k)$ is continuous. However, as a function of x, the rth derivative has norm given for fixed $(e, v_1, \ldots, v_k)$ by

$$(2.1) \qquad \sup_{x \in B_1, w_i \in B} \left| d_1^r d_2^k f(x, e)(v_1, \ldots, v_k)(w_1, \ldots, w_r) \right|,$$

where B denotes the unit ball in $E_1 \oplus G_1$. By the continuity of $d^{r+k} f$ on $U \times E \times \cdots \times E$ and the compactness of $B_1 \times B \times \cdots \times B$ (r factors), (2.1) is continuous as a function of $(e, v_1, \ldots, v_k)$. This establishes the claim.

Then we use a slightly modified form of Mather's theorem that infinitesimal stability implies stability:

THEOREM (MATHER). *Let $f: N \to P$ be a smooth mapping and $M \subset N$ a compact submanifold of codimension 0 (possibly with boundaries and corners). Suppose $f|M$ is infinitesimally stable. Then given compact manifold neighborhoods K of M and L of $f(M)$, there is a neighborhood $\mathcal{U}$ of $f|M$ in $C^\infty(M, P)$, and there are continuous mappings $H_1: \mathcal{U} \to \mathrm{Diff}(N)$, $H_2: \mathcal{U} \to \mathrm{Diff}(P)$ such that*
  (i) $H_1(f|M) = \mathrm{id}$, $H_2(f|M) = \mathrm{id}$,
  (ii) $g = H_2(g) \circ f \circ H_1(g)|M$,

(iii) *the composition of each H_i with the restriction maps give smooth maps of Fréchet spaces*

$$\mathcal{U} \xrightarrow{H_1} \text{Diff}(N) \xrightarrow{r_1} C^\infty(K, N), \quad \mathcal{U} \xrightarrow{H_2} \text{Diff}(P) \xrightarrow{r_2} C^\infty(L, P).$$

REMARK. The modification occurs in conclusion (iii). This follows by a very slight modification of Mather's proof and checking that various steps in his proof give smooth mappings of Fréchet spaces. We examine these technical points in §4.

To finish the proof using Mather's theorem, we observe that $f_1|B_1$ is infinitesimally stable (since $f_1|U_1$ is). Let $B_1 \subset K \subset U_1$, with K a compact submanifold with boundary, which is a neighborhood of B_1, and similarly for $f_1(B_1) \subset L \subset F_1 \oplus G_1$. Then there is a neighborhood $\mathcal{U}$ of $f_1|B_1$ in $C^\infty(B_1, F_1 \oplus G_1)$ and mappings H_i satisfying the conclusions of Mather's theorem. Let $r_i \circ H_i = H_i'$. If B_2 is a closed ball about 0 in $F_1 \oplus G_1$ and contained in L, then we may assume that $\mathcal{U}$ is chosen small enough, so that if $\phi \in H_1'(\mathcal{U})$, then $B_1 \subset \phi(K)$, and if $\psi \in H_2'(\mathcal{U})$, then $B_2 \subset \psi(L)$. Then we may further assume that U_2 is chosen small enough so that $R(U_2) \subset \mathcal{U}$. We define $\Phi: U_1 \times U_2 \to U_1 \times U_2$ by $\Phi(x, e) = (H_1(R(e))(x), e)$. On $K \times U_2$, Φ is defined using H_1; hence, by Mather's theorem, $\Phi|K \times U_2$ is smooth. By our assumption on $\mathcal{U}$, $\Phi^{-1}|B_1 \times U_2$ is defined and smooth. Hence, Φ is a local diffeomorphism in a neighborhood of 0. Similarly, $\Psi: (F_1 \oplus G_1) \times U_2 \to (F_1 \oplus G_1) \times U_2$ is defined by $\Psi(y, e) = H_2(R(e))(y), e)$. This is smooth on $L \times U_2$ with a smooth inverse defined on $B_2 \times U_2$. Hence, Ψ is also a local diffeomorphism. Finally, on a neighborhood of $0, f = \Psi \circ (f_1 \times \text{id}_{G_2}) \circ \Phi$.

REMARK. The proof depends on two results: openness of versality and infinitesimal stability implies stability. The first result is a sheaf-theoretic result and valid in a number of general situations. Also, there are generalizations of Mather's theorem for a number of situations, including mappings equivariant with respect to a compact Lie group [14], composition of mappings [2], and certain generalizations of these [13]. Thus, there are extensions of this theorem to a number of situations where extra structure is preserved.

PROOF OF COROLLARY 1. We first apply Lyapunov-Schmidt to g to obtain a $g_1: E \oplus H, 0 \to F \oplus H, 0$ which has the form $g_1(x, e, h) = (\bar{g}_1(x, e, h), e, h)$ for $x \in E_1$, $e \in G_1$, and $h \in H$. If the standard Lyapunov-Schmidt method is used, then it is easily seen that

(2.2) $$\bar{g}_1(x, e, 0) = \bar{g}(x, e, 0) = \bar{f}(x, e).$$

Then we can apply the theorem to both f and g_1. Thus,

$$f = \Psi_f \circ (f_1 \times \text{id}_{G_2}) \circ \Phi_f, \quad g_1 = \Psi_{g_1} \circ (f_1 \times \text{id}_{G_2 \oplus H}) \circ \Phi_{g_1}.$$

Hence,

(2.3) $$g_1 = \Psi_{g_1} \circ ((\Psi_f^{-1} \times \text{id}_H) \circ (f \times \text{id}_H) \circ (\Phi_f^{-1} \times \text{id}_H)) \circ \Phi_{g_1}.$$

Furthermore, by (2.2), $R_{g_1}(e, 0) = R_f(e)$, so that (2.3) is the identity when $h = 0$. □

PROOF OF COROLLARY 2.2. In the notation of the corollary, we define g: $E \oplus V$, $0 \to F \oplus V$, 0 by

$$g(x, u, v) = (\bar{f}(x, u) - f_0(x) + f_1(x, v), u, v).$$

Then $g|E_1 \oplus V = f_1$. By Theorem 1, g is equivalent to $f_1 \times \mathrm{id}_G$. Under the equivalence, E and F are mapped to submanifolds of $E \oplus V$ and $F \oplus V$, and the restriction of $f_1 \times \mathrm{id}_G$ to a map between these subspaces is equivalent to f. □

3. Examples

EXAMPLE 3.1. Rather than consider the general form of (**) and give a complicated analysis, we consider instead

$$(3.2) \quad \Delta u + u^p + \sum_{i=2}^{p-1} a_i(x) \cdot u^i + \lambda u = r(x) \quad \text{on } \Omega, \quad u = 0 \text{ on } \partial\Omega.$$

Here Ω is a bounded domain with $\partial\Omega$ smooth and $a_i(x) \in C^\alpha(\bar{\Omega})$. Then the operator f: $C_0^{2+\alpha}(\bar{\Omega}) \to C^\alpha(\bar{\Omega})$, given by

$$f(u) = \Delta u + u^p + \sum_{i=2}^{p-1} a_i(x) u^i + \lambda u,$$

is smooth. We write the polynomial term in f as $P(u, x)$. Then a simple computation shows

(3.3)
(i) $df(u_0)(h) = \Delta h + \dfrac{\partial P}{\partial u}(u_0) \cdot h,$

(ii) $d^k f(u_0)(h_1, \ldots, h_k) = \dfrac{\partial^k P}{\partial u^k}(u_0) \cdot h_1 \cdot h_2 \cdots h_k, \quad k > 1.$

Suppose λ_1 is the first eigenvalue of Δ, so that the eigenspace has dimension $= 1$. We wish to consider f near $u_0 \equiv 0$ and $\lambda = -\lambda_1$. From (3.3)

$$df(0)(h) = (\Delta + \lambda)(h);$$

$$\begin{aligned} d^k f(0)(h_1, \ldots, h_k) &= k! a_k(x) \cdot h_1 \cdot h_2 \cdots h_k && \text{if } 1 < k < p, \\ &= k! h_1 \cdots h_k && \text{if } k = p, \\ &= 0 && \text{if } k > p. \end{aligned}$$

Let g be an eigenfunction for λ_1. We consider two cases.

Case 1. Generically chosen and sufficiently small $a_i(x)$: For $a_i(x)$ constant functions with $a_{p-1}(x) \equiv 0$, we define

$$(3.4) \quad f_1: C_0^{2+\alpha}(\bar{\Omega}) \times \mathbf{R}^{p-2} \to C^\alpha(\bar{\Omega}) \times \mathbf{R}^{p-2},$$

$$f_1(u, (a_{p-2}, \ldots, a_2, \lambda)) \to (f(u), (a_{p-2}, a_2, \ldots, \lambda)).$$

Computing the derivatives as in (3.3) at $(0, (0, \ldots, 0, -\lambda_1))$, we obtain

$$df_1|C_0^{2+\alpha}(\bar{\Omega}) = df \quad \text{and} \quad df_1|\mathbf{R}^{p-2} = \mathrm{id}_{\mathbf{R}^{p-2}}.$$

For $2 \leqslant k < p$, $d^k f_1 = 0$ except $d^k f_1(h_1, \ldots, h_{k-1}, e_{p-k}) = (k-1)! h_1 \cdot h_2 \cdots h_{k-1}$, where $e_{p-k} = (p-k)$th unit basis vector in $\mathbf{R}^{p-2}$. Lastly, $d^p f_1|C_0^{2+\alpha}(\bar{\Omega}) = d^p f$. Note that f_1 is not in the form obtained from the Lyapunov–Schmidt procedure. However, since $d^k f_1 = 0$ on $C_0^{2+\alpha}(\overline{\Omega}) \times \{0\}$, $2 \le k < p$, and f_1 is the identity on $\mathbf{R}^{p-2}$, it follows that changing coordinates by applying Lyapunov–Schmidt will not change the first p-derivatives. Then restricting and projecting onto $L \times \mathbf{R}^{p-2}$, where $L = \langle g \rangle$, gives a mapping f_2: $L \times \mathbf{R}^{p-2} \to L \times \mathbf{R}^{p-2}$ (since $g \notin \text{Im}(\Delta - \lambda_1)$).

We now wish to suppose $g^k \notin \text{Im}(\Delta - \lambda_1)$, $2 \le k < p$. Since $\Delta - \lambda_1$ is selfadjoint, it is sufficient to know $\int_{\Omega} g^{k+1} \ne 0$. For $k + 1$ even, this is clear. Suppose this is also true for $k + 1$ odd. The above derivative calculations imply by an argument analogous to that given in §1 that f_2 (and hence f) is infinitesimally stable. Thus, f is locally equivalent to $f_2 \times \text{id}$, and by the classification of infinitesimally stable germs by their local algebras, f_2 is equivalent to the germ given in §1.

$$(3.5) \qquad (x, v_1, \ldots, v_{p-2}) \mapsto \left(x^p + \sum_{i=1}^{p-2} v_i x^i, v_1, \ldots, v_{p-2} \right).$$

If now we replace the constant functions by general $a_i(x)$, this gives a mapping

$$(3.6) \quad \mathscr{F}\colon C_0^{2+\alpha}(\overline{\Omega}) \times \left(C^\alpha(\overline{\Omega}) \right)^{p-2} \times \mathbf{R} \to C^\alpha(\overline{\Omega}) \times \left(C^\alpha(\overline{\Omega}) \right)^{p-2} \times \mathbf{R},$$
$$(u, a_i(x), \lambda) \mapsto (f(u), a_i(x), \lambda).$$

By Corollary 1 this is also locally equivalent to $f_2 \times \text{id}$ near $(0, (0, \ldots 0, -\lambda_1))$. To discuss the number of solutions to (3.2), we can, in general, use either results in [6] for the number of solutions to equations involving infinitesimal germs, or, in this case, we can use our special knowledge of the solution set for (3.5). We can conclude

PROPOSITION 3.7. *Both of the mappings f_1 (3.4) and $\mathscr{F}$(3.6) are locally equivalent at $(0, (0, \ldots, 0, -\lambda_1))$ to (the polynomial (3.5)) $\times$ id. Hence, for coefficients $a_i(x)$ sufficiently small (in $C^\alpha(\overline{\Omega})$) and for $r(x)$ sufficiently small in $C^\alpha(\overline{\Omega})$, there are at most p solutions to (3.2) near $u \equiv 0$. For an open set of coefficients in $(C^\alpha(\overline{\Omega}))^{p-2} \times \mathbf{R}$ containing $(0, \ldots, 0, -\lambda_1)$ in its closure, there is an $r(x)$ with exactly p-solutions to (3.2) near $u \equiv 0$.*

REMARK. For the case $p = 3$ the result states that the mapping f is smoothly equivalent to cusp $\times$ id, where the cusp is defined by $(x, v) \mapsto (x^3 + vx, v)$. This recovers part of the results of Berger, Church, and Timourian [4]. However, we note that their results also apply to nonsmooth f, but the ones described here do not.

Case 2. Fixed $a_i(x)$, $\lambda = -\lambda_1$. This case is considerably more difficult than Case 1. This is because the calculations for infinitesimal stability based on our discussion of §1 should be carried out after Lyapunov–Schmidt has been applied. In this case this will seriously alter the derivative computations already made. Specifically, let $\tilde{f}$ denote composition of f with projection onto $\text{Im}(\Delta - \lambda_1)$ along $L\ (= \langle g \rangle)$, $\tilde{f}(h) = f(h) - (f(h), g)g$. Then by the inverse function theorem, ϕ: $C_0^{2+\alpha}(\overline{\Omega}) \to C^\alpha(\overline{\Omega})$, given by $\phi(h) = \tilde{f}(h) + (h, g) \cdot g$, is a local diffeomorphism.

Then $f \circ \phi^{-1}: C^\alpha(\overline{\Omega}) \to C^\alpha(\overline{\Omega})$ is in the desired form for Lyapunov–Schmidt.

(3.8) A *sufficient condition for* $f \circ \phi^{-1}$ *to be infinitesimally stable is that*

(i) there is an r (which may be different from p) such that
$$d^r(f \circ \phi^{-1})(0)(g,\ldots,g) \notin \operatorname{Im}(\Delta - \Delta_1).$$

(ii) There are $\{h_1,\ldots,h_{r-2}\}$, an independent set of elements in $\operatorname{Im}(\Delta - \lambda_1)$, such that
$$d^{j+1}(f \circ \phi^{-1})(0)(h_j, g,\ldots,g) \notin \operatorname{Im}(\Delta - \lambda_1), \quad 1 \leqslant j \leqslant r - 2.$$

Using formulas for derivatives of composition (e.g., Ronga [15] or Fraenkel [7]), we can compute $d^j(\phi^{-1})$ in terms of $d^i\phi$ and $(d\phi)^{-1}$ and then $d^j(f \circ \phi^{-1})$ in terms of the $d^k f$ and $d^j(\phi^{-1})$. Then infinitesimal stability may be verified using (3.8). For example,
$$d^2(\phi^{-1})(d\phi, d\phi) = -(d\phi)^{-1}d^2\phi(\,,\,),$$

so
$$d^2(f \circ \phi^{-1})(d\phi, d\phi) = d^2 f(\,,\,) - \left(df \circ (d\phi)^{-1} d^2\phi(\,,\,)\right).$$

Since $d^j \phi = d^j f$ for $j > 1$, this equals the projection of $d^2 f(\,,\,)$ onto L along $\operatorname{Im}(\Delta - \lambda_1)$. For (3.2) in the case $p = 4$, we obtain

(1) f is equivalent to a fold $(x \mapsto x^2) \times \text{id}$ if $a_2(x) \cdot g^2 \notin \operatorname{Im}(\Delta - \lambda_1)$—i.e.,

(3.8a) $$\int_{\overline{\Omega}} a_2(x) \cdot g^3 \neq 0;$$

(2) if $a_2(x) \cdot g^2 \in \operatorname{Im}(\Delta - \lambda_1)$, then we may solve $(\Delta - \lambda_1) \cdot h = 2 \cdot a_2(x) \cdot g^2$. Suppose there is also an $h_1 \in C_0^{2+\alpha}(\overline{\Omega})$ orthogonal to g (in $L^2(\overline{\Omega})$), so that $a_2(x) \cdot h_1 \cdot g \notin \operatorname{Im}(\Delta - \lambda_1)$—i.e.,

(3.8b) $$\int_{\overline{\Omega}} a_2(x) h_1 \cdot g^2 \neq 0.$$

Secondly, a computation similar to the above, using (3.3), yields
$$d^3(f \circ \phi^{-1})(g, g, g) = 6a_3(x)g^3 + 3\left(2a_2 g \cdot \left(-d\phi^{-1}(2a_2 \cdot g^2)\right)\right) + \omega$$
$$= 6\left(a_3(x)g^3 - a_2(x)g \cdot h\right) + \omega$$

for $\omega \in \operatorname{Im}(\Delta - \lambda_1)$. Again, to ensure $d^3(f \circ \phi^{-1})(g, g, g) \notin \operatorname{Im}(\Delta - \lambda_1)$, we require

(3.8c) $$\int_{\overline{\Omega}} a_3(x)g^4 - a_2(x)g^2 \cdot h \neq 0.$$

If $a_3(x) \geqslant 0$ and not identically zero, and h is contained in the closure of the subspace spanned by the eigenfunctions $\neq g$ of Δ, then (3.8c) is positive for
$$\int -2a_2(x) \cdot g^2 \cdot h = \int -(\Delta - \lambda_1)(h)h = \lambda_1(h, h) - (\Delta h, h) \geqslant (\lambda_1 - \lambda_2)\|h\|_2^2.$$

If, instead, (3.8c) = 0, then we proceed one final step. For $h_2 \in C_0^{2+\alpha}(\overline{\Omega})$, to ensure that $d^3(f \circ \phi^{-1})(g, g, d\phi(h_2)) \notin \text{Im}(\Delta - \lambda_1)$, we obtain

(3.8d) $$\int 6a_3 g^3 \cdot h_2 - 4a_2 g^2 \cdot h_2' - 2a_2 g \cdot h \cdot h_2 \neq 0,$$

where $d\phi(h_2') = 2a_2 g \cdot h_2$. Finally, the condition $d^4(f \circ \phi^{-1})(g, g, g, g) \notin \text{Im}(\Delta - \lambda_1)$ becomes

(3.8e) $$\int_\Omega 12g^5 + 18a_3 g^3 \cdot h + 4a_2 \cdot g^2 \cdot h' + 3a_2 \cdot h^2 \cdot g \neq 0,$$

where $(\Delta - \lambda_1)h' = -6(a_3 g^3 - a_2 gh)$.

We can summarize the situation for $p = 4$ with

PROPOSITION. 3.9. *For equation* (3.2) *with* $p = 4$, $\lambda = -\lambda_1$:
(i) *If* (3.8a) $\neq 0$, *then f is locally equivalent to a* (fold$(x \mapsto x^2)$) $\times$ id *near* $u_0 \equiv 0$.
(ii) *If* (3.8a) $= 0$, *but* (3.8b), (3.8c) $\neq 0$, *then f is locally equivalent to* (cusp $\times$ id) *near* $u_0 \equiv 0$.
(iii) *If* (3.8a), (3.8c) $= 0$, *and* (3.8b), (3.8d), (3.8e) $\neq 0$, *then f is locally equivalent to* (swallowtail $\times$ id) *near* $u_0 \equiv 0$. *The swallowtail is the germ* (3.5) *with* $p = 4$ (*its name, due to Thom, derives from the shape of its discriminant*).

With more space, time, and patience it would be possible to give conditions ensuring the general (3.2) (or even (**) in the introduction) is locally equivalent to (3.5) for some p. Note that parts (i) and (ii) of the proposition would also follow from [5].

EXAMPLE 3.10. Consider the special case of (3.2) for $\Omega = (0, \pi) \subset \mathbf{R}$ and $f(u) = \Delta u + u^4 + a_3(x)u^3 + a_2(x)u^2 + u = r(x)$ (i.e., $\lambda_1 = -1$ and $g = \sin x$). Then, for example, if $a_2(x) \geq 0$ and not identically zero, then f is equivalent to fold $\times$ id. If $a_2(x) = \sin kx$, $k \geq 3$, then (3.8a) $= 0$, but for $h_1 = \sin kx$, (3.8b) $\neq 0$. If $a_3(x) = \sin^2 rx$, then both terms of (3.8c) are positive, so (3.8c) > 0. Then f is equivalent to cusp $\times$ id. Lastly, if instead we let $a_3(x) = -\sin^2 rx$ and $a_2(x) = \alpha_k \sin kx$, then there is a value α_k for which (3.8c) $= 0$. Then under certain conditions on r, k—e.g., $|r - k| \geq 2$ and $|r - 2k| \geq 5$—it follows that with $h_2 = \sin(2r + 1)x$, all terms of (3.8d) and (3.8e) after the first are zero, while the first are easily seen to be not zero. Hence, f is locally equivalent to (swallowtail $\times$ id). Thus, for example, we may conclude that there is an open set of $r(x)$ near 0 in $C^\alpha[0, \pi]$, so that

$$\frac{d^2 u}{dx^2} + u^4 - (\sin^2 rx)u^3 + \alpha_k \sin kx u^2 + u = r(x), \quad u(0) = u(2\pi) = 0,$$

has 4-solutions near $u_0 \equiv 0$ (when r, k satisfy the above conditions).

Finally, we point out that the results are not limited to the case where $\dim_\mathbf{R} \ker df = 1$. For example, if we seek periodic solutions on $[0, 2\pi]$ to

(3.11) $$f(u) = \frac{d^2 u}{dx^2} + u + a(x)u^2 + h(x, u) \cdot u^3 = r(x),$$

then we consider $\tilde{C}^k[0, 2\pi]$, consisting of C^k-functions u, which (together with their first k derivatives) are periodic. If $\partial^j h/\partial u^j$ is continuous and $a(x)$, $(\partial^j h/\partial x^j)(x, u) \in \tilde{C}^0[0, 2\pi]$ (for each fixed u), then f defines at 0 a smooth Fredholm mapping $f: \tilde{C}^2[0, 2\pi] \to \tilde{C}^0[0, 2\pi]$. Both ker $df(0)$ and coker $df(0)$ are spanned by $\{\cos x, \sin x\}$.

A calculation shows that if $\{a_i, b_i\}$ denote the Fourier coefficients of $a(x)$, then, for example, if either
  (i) $(a_1, b_1) \neq (0, 0)$, $(a_3, b_3) = (0, 0)$, and $a_2 \neq b_2$, or
  (ii) $(a_1, b_1) = (0, 0)$, $(a_3, b_3) \neq 0$, and $a_2, b_2 \neq 0$,
then f is infinitesimally stable with local algebra isomorphic to
  (i) $\mathbf{R}[x, y]/(x^2 + y^2, xy)$ or
  (ii) $\mathbf{R}[x, y]/(x^2 - y^2, xy)$.
Thus f is smoothly equivalent to

$$((x, y, v, w) \to (x^2 + y^2 + vx + wy, xy, v, w)) \times \text{id}$$

or

$$((x, y, v, w) \to (x^2 - y^2 + vx + wy, xy, v, w)) \times \text{id}.$$

In both cases $\delta = 4$, and by [6] there are $r(x)$ near 0 for which there are four distinct solutions to (3.11). In the second case there are always at least two solutions for $r(x)$ near 0.

4. Technical points concerning Mather's theorem. We indicate here what modifications must be verified to obtain the slightly modified form of Mather's theorem. The main change is easy to describe.

For a topological space X with base point x_0, Mather considers a category M^X whose objects are smooth manifolds and whose morphisms between N and P consist of continuous germs $g: (X, x_0) \to C^\infty_{\text{pr}}(N, P)$. Here $C^\infty_{\text{pr}}(N, P)$ denotes the space of proper smooth mappings with the Whitney topology. We shall consider instead the case where X is a Fréchet space and g is a germ of a smooth mappings of Fréchet manifolds $g: (X, x_0) \to C^\infty_0(M, L)$, where $C^\infty_0(M, L)$ denotes the open subset of smooth mappings $h: M \to L$ with $h(M) \subset \text{int}(L)$ if L has a boundary. We only consider M and L compact (but possibly with boundaries and corners).

We wish to check that Mather's arguments and constructions involving the category M^X are still valid for this modified category. For this we refer to specific sections of Mather's paper [**12**, II]. For example, we let §M2 denote §2 of Mather's paper.

First, we recall the main outline of Mather's proof. Let

$$X = \{g \in C^\infty_{\text{pr}}(M \times I, P): g(x, 0) = f(x), \text{ all } x \in M\}.$$

Also, if $\pi_M: M \times I \to M$ denotes projection, we let $x_0 = f \circ \pi_M$. Then $\bar{f}: (X, x_0) \to C^\infty_{\text{pr}}(M \times I, P)$ denotes the inclusion. It is easily seen that X is a Fréchet submanifold of $C^\infty_{\text{pr}}(M \times I, P)$. Next, let $\Delta: C^\infty_{\text{pr}}(M \times I, P) \to C^\infty_{\text{pr}}(M \times I, TP)$

be defined by $\Delta(g) = \partial g/\partial t$. Then the key step in the proof is the existence of representatives of germs defined on a neighborhood Y of x_0,

$$\tilde{\xi}_0\colon Y \to \Gamma^\infty(\pi_M^* TM), \qquad \tilde{\eta}\colon Y \to \Gamma^\infty(\pi_P^* TP),$$

with $\tilde{\xi}_0(x_0) = 0$, $\tilde{\eta}(x_0) = 0$, so that for all $y \in Y$, if $\tilde{f}(y) = g$, then

(4.1) $$\Delta(g) = dg(\tilde{\xi}_0(y)) + \tilde{\eta}(y) \circ g.$$

Given $\tilde{\xi}_0$ and $\tilde{\eta}$ satisfying (4.1), the proof is completed using the following steps in §M7:

(1) There is a neighborhood $\mathscr{U}$ of $f|M$ in $C^\infty_{\mathrm{pr}}(M, P)$ and a continuous map $\gamma_*\colon \mathscr{U} \to C^\infty_{\mathrm{pr}}(M \times I, P)$.

(2) Seeley's extension theorem can be used to give a continuous linear mapping $E\colon \Gamma^\infty(\pi_M TM) \to \Gamma^\infty(\pi_N^* TN)$, and $\tilde{\xi}$ is defined by $\tilde{\xi} = E \circ \tilde{\xi}_0$.

(3) Integration of time-dependent vector fields gives continuous mappings $\theta_1\colon O_N \to C^\infty_{\mathrm{pr}}(N \times I, N)$ and $\theta_2\colon O_P \to C^\infty_{\mathrm{pr}}(P \times I, P)$, for O_N and O_P neighborhoods of 0 in $\Gamma^\infty(\pi_N^* TN)$ or $\Gamma^\infty(\pi_P^* TP)$. Then H_i are given by

$$H_1 = \theta_1(-\tilde{\xi}(\gamma_*(g)))_{|N \times \{1\}}, \qquad H_2 = \theta_2(\tilde{\eta}(\gamma_*(g)))_{|P \times \{1\}}.$$

In the modified case, if (4.1) can be solved using L in place of P with M and L compact, then we claim that the three additional steps can be carried through to obtain smooth mappings of Fréchet spaces: (1) γ_* is smooth because it is given by composition with a smooth function. For (2) it is necessary to apply the Seeley extension theorem to both $\tilde{\xi}_0$ and $\tilde{\eta}$. In general, $\Gamma^\infty(\pi_N^* TN)$ and $\Gamma^\infty(\pi_P^* TP)$ are not Fréchet spaces, so smoothness has no meaning. However, if $K' \supset K$ and $L' \supset L$ are compact manifold neighborhoods of K and L, then the composition of extension with restriction to K' and L' is still continuous and linear and, hence, smooth as maps between Fréchet spaces. Thirdly, suppose $\varepsilon >$ distance from K to $N \setminus K'$ and L to $P \setminus L'$. By restricting to vector fields of norm $< \varepsilon$, we see that integral curves beginning in K will remain in K', and similarly for L and L'. Then the restriction of the flows to $K \times I$ can be obtained by restricting the vector fields to K' and integrating and restricting. Thus, it is only necessary to consider extensions of $\tilde{\xi}_0$ and $\tilde{\eta}$ to K' and L', respectively. Lastly, for vector fields on K' and L', the smoothness of the θ_i follows by the estimates which Mather refers to in the proof of Lemma 2 of §M7 together with the smooth dependence of solutions on parameters. (By differentiating equation (9) of §M7 with respect to parameters, one obtains a differential equation which the variations must satisfy. Then, the same reasoning Mather uses for continuity implies that the norms of the variations are also bounded by the norms of the variations of the vector fields.) Thus, the composition of the H_i with restriction to K or L will be smooth.

It remains to justify the solution of (4.1). For this it is sufficient to verify Proposition 1 of §M6 for the modified category. This in turn depends upon the results of §M4, especially Propositions 1 and 2. Part of the results in §M4 depend upon several general results in §M2 on certain natural mappings of functions spaces being smooth. That these are smooth is contained in, e.g., [11, 1.4.4], since we consider compact source spaces. The mappings in the propositions of §M4 are not even mappings of Fréchet spaces. However, with N and P compact (in reality representing M and L), we can replace $P \times \mathbf{R}$ by a compact manifold $N' \supset P \times \mathbf{R}$. Then the maps are defined using $C^\infty(N')$ by first extending to $C^\infty(P \times \mathbf{R})$, using the Seeley extension theorem. In fact, this is exactly how these propositions are used in §6, where an extension must be made from such an N'. Then the proof of Proposition 1 in §M4 gives the smoothness of the mappings there (replacing $C^\infty(P \times \mathbf{R})$ by $C^\infty(N')$). For example, consider the mapping $(f, u) \to h_{f,u}$ of that proposition. In terms of local coordinates given on p. 271 of §M4,

$$h_{(g-f) \circ \phi'^{-1}_\alpha}(y, v) = H(g \circ \phi'^{-1}_\alpha, v).$$

Yet, by definition [12, I, §3], H together with the derivatives $(\partial|m|/\partial v^m) H$ are linear in $g \circ \phi'^{-1}_\alpha$. Thus, if $d_2^m H(g \circ \phi'^{-1}_\alpha, v)$ denotes the mth partial derivative with respect to $\mathbf{R}^p$, then it will be the desired $d^m h_{f,u}$ at $(g \circ \phi'^{-1}_\alpha, v)$ if $d_2^m H(g \circ \phi'^{-1}_\alpha, v)(w_1, \ldots, w_m)$ is continuous in $g \circ \phi'^{-1}_\alpha$, v, and the w_i. This requires estimates of the derivatives

(5.1) $$\frac{\partial^l}{\partial y^l} \frac{\partial^{|m|}}{\partial v^m} \left(H(f \circ \phi'^{-1}_\alpha, v) - H(g \circ \phi'^{-1}_\alpha, v') \right).$$

By the linearity of H in the first factor together with the smooth dependence of H on v, we may estimate (5.1) using estimate (∗∗) on p. 271 together with the continuity of $(\partial^{|m|}/\partial v^m)(H(\partial^{|l|}/\partial y^l(g \circ \phi'^{-1}_\alpha), v))$ in v. The proof for the smoothness of the second mapping considered in the proposition follows by similar considerations. Lastly, the smoothness of the mappings of Propositions 2 of §M4 follow by the discussion Mather gives together with the smoothness of all other maps already referred to. □

Bibliography

1. A. Ambrosetti and G. Prodi, *On the inversion of some differentiable maps with singularities*, Ann. Mat. **93** (1972), 231–246.

2. N. Baas, *Structural stability of composed mappings*, 1974 (preprint).

3. M. Berger and P. Church, *Complete integrability and perturbation of a nonlinear Dirichlet problem*. I, Indiana Univ. Math. J. **28** (1979), 935–952.

4. M. Berger, P. Church and J. Timourian, *An application of singularity theory to non-linear elliptic partial differential equations*, Proc. Sympos. Pure Math., vol. 40, Amer. Math. Soc., Providence, R. I., 1983, pp. 119–126.

5. _____, *Folds and cusps in Banach spaces with applications to non-linear partial differential equations*. I, Indiana Univ. Math. J. **34** (1985), 1–19.

6. J. Damon and A. Galligo, *A topological invariant for stable map germs*, Invent. Math. **32** (1976), 103–132.

7. L. Fraenkel, *Formulae for high derivatives of composite functions*, Proc. Cambridge Philos. Soc. **83** (1978), 159–165.

8. M. Golubitsky and V. Guillemin, *Stable mappings and their singularities*, Springer-Verlag, 1973,

9. M. Golubitsky and D. Schaeffer, *A theory for imperfect bifurcation via singularity theory*, Comm. Pure Appl. Math. **32** (1979), 21–98.

10. _____, *Imperfect bifurcation in the presence of symmetry*, Comm. Math. Phys. **67** (1979), 205–232.

11. R. S. Hamilton, *The inverse function theorem of Nash and Moser*, Bull. Amer. Math. Soc. (N.S.) **7** (1982), 65–222.

12. J. Mather, *The stability of C^∞-mappings*. I, II, IV, Ann. Math.(2) **87** (1968), 254–291; ibid. **89** (1969), 254–291; Inst. Hautes Études Sci. Publ. Math. **37** (1969), 223–248.

13. J. Mather and J. Damon, Preliminary version of book on stability of mappings.

14. V. Poenaru, *Singularités C^∞ en présence de symmétrie*, Lecture Notes in Math., Vol. 510, Springer-Verlag, 1976.

15. F. Ronga, *A new look at Faá de Bruno's formula for higher derivatives of composite functions and the expression of some intrinsic derivatives*, Proc. Sympos. Pure Math., vol. 40, Amer. Math. Soc., Providence, R. I., 1983, pp. 423–432.

16. B. Teissier, *The hunting of invariants in the geometry of discriminants*, Real and Complex Singularities (Oslo, 1976), Noordhoff–Sijthoff, 1977.

17. J. Cl. Tougeron, *Ideaux de fonctions differentiables*, Springer-Verlag, 1972.

UNIVERSITY OF NORTH CAROLINA AT CHAPEL HILL

Remarks on S^1 Symmetries and a Special Degree for S^1-Invariant Gradient Mappings

E. N. DANCER

The main point of this paper is to announce a special degree theory for S^1-invariant gradient mappings. We give a sketch of how the degree is defined and how it is used in applications. Full details will appear elsewhere. First, however, we explain why one cannot construct a useful degree for all S^1-invariant maps. We also discuss the difference between S^1-invariance and $O(2)$-invariance, and we discuss where S^1-invariance occurs.

We commence by making some remarks on symmetries. Assume that E is a Banach space, G is a Lie group, and $\{T_g\}_{g \in G}$ is a continuous linear representation of G on E. (This means that T_g is a bounded linear operator on E for $g \in G$, $T_{gh} = T_g T_h$ for $g, h \in G$, $T_e = I$, and the map $g \to T_g x$ is continuous for each x in E.) In this paper G will usually be the circle group S^1 or the group of 2×2 orthogonal matrices $O(2)$. If $z \in E$, define $G_z = \{g \in G: T_g z = z\}$. This is known as the isotropy group. In addition, let $G(z)$ denote the orbit $\{T_g z: g \in G\}$ and $E_G = \{x \in E: T_g x = x \text{ for all } g \text{ in } G\}$. A map $F: E \to E$ is said to be G-invariant if $F(T_g x) = T_g F(x)$ if $g \in G$ and $x \in E$. (Similarly, $F: E \times R \to E$ is G-invariant if $F(T_g x, \lambda) = T_g F(x, \lambda)$.)

Assume *now* that E is finite dimensional, $F: E \to E$ is G-invariant and C^1 and $F(z) = 0$. By the symmetries, F vanishes on $M \equiv G(z)$. By differentiating along M we easily see that $N(F'(z)) \supseteq T_z(M)$, where N denotes the kernel and $T_z(M)$ denotes the tangent space to M at z. (Note that our assumptions ensure that M is a smooth manifold.) Thus, if M is not discrete, the implicit function theorem does not apply at z.

1980 *Mathematics Subject Classification.* Primary 58E07; Secondary 47H15, 58F22.

© 1986 American Mathematical Society
0082-0717/86 $1.00 + $.25 per page

THEOREM 1. *Assume that $G = S^1$, G_z is finite, and $F(x) \neq 0$ in some deleted closed neighbourhood U of M. Then for every $\varepsilon > 0$ there is a G-invariant smooth map $\tilde{F}$ such that $\|\tilde{F}(x) - F(x)\| \leq \varepsilon$ on U and $\tilde{F}$ has no zeros on U. Moreover, if F is smooth and $N > 0$, we can choose $\tilde{F}$ such that $\|D^i\tilde{F}(x) - D^iF(x)\| \leq \varepsilon$ if $i \leq N$ and $x \in U$ (that is, $\tilde{F}$ can be chosen close to F in the strong C^∞ topology on U).*

Note that G_z is finite if $G_z \neq G$. The theorem shows that any isolated orbit of zeros of F which is not in E_G (that is, does not consist of fixed points of the group action) is unstable to S^1-invariant perturbations. Thus, in any reasonable degree theory for S^1-invariant maps, we would expect the orbit to have zero degree. Thus, it does not seem possible to have a nontrivial degree for such maps. Later, we will discuss a degree theory when we *further restrict* to *gradient* mappings. Implicit here is the failure of the above theorem if F and $\tilde{F}$ are both gradient mappings. Essentially, the difference occurs because there are vector fields on S^1 which preserve the symmetry but have no zeros. On the other hand, every gradient vector field on S^1 has a zero.

The theorem is a special case of the main result in [5]. However, for the last statement on smoothness, one has to examine the *proof* there rather more carefully.

The problem behaves quite differently if G is $O(2)$ rather than S^1. For example, assume that $G = O(2)$, $G_z = \{I, \begin{pmatrix}1 & 0 \\ 0 & -1\end{pmatrix}\}$, F is differentiable and $N(F'(z))$ is one-dimensional. Then by the main result in [5], any $O(2)$-invariant $\tilde{F}$ near F in the C^0-norm must have zeros near M. In fact, if F is C^1 and $\tilde{F}$ is C^1-close to F, then, by [6], $\tilde{F}$ must have a *unique* orbit of zeros near M. The essential reason for the difference between the two cases is that if $G = S^1$ and $G_z \subset S^1$, then $\dim N(G_z) > \dim G_z$, whereas in the above case where $G = O(2)$, we find that $\dim N(G_z) = \dim G_z$. Here $N(G_z)$ is the normalizer $\{g \in G: g^{-1}G_z g \subset G_z\}$.

Note that, in the above discussion, it is important that we look at maps which are G-invariant with respect to the same group action on the domain and codomain. The problem may behave differently in other cases.

We now want to discuss how S^1 symmetries seem to occur. First, they occur in the problem of looking for periodic solutions of fixed period T for autonomous differential equations. Here the symmetries are simply translations. Second, we consider problems which are invariant with respect to rotations in a plane (say the x-y plane). The rotations give an S^1 symmetry. However, in most of these problems, we also have invariance with respect to reflections in the x-y plane. Thus we have an $O(2)$ symmetry. However, there is one case where we do not have the extra symmetry. Assume our equations have a cross-product term $\mathbf{z} \times \mathbf{x}$, where $\mathbf{z}$ is a fixed vector perpendicular to the x-y plane. Since it is easy to see this term is invariant under rotations but not reflections, we have an S^1 symmetry but not an $O(2)$ symmetry. Physical problems of this type occur for elastic rods in magnetic fields. Wolfe [11, 12] discusses such problems. Indeed, the author's work below was motivated by the wish for a general theorem which applied to the example in [11].

We now sketch the construction of a useful index theory for S^1-invariant *gradient* maps. We call the degree we construct a normal S^1 degree. It behaves a little like the Fuller index [4] in that it is not defined for a map F on a set Ω if there is an x in $\Omega \cap E_{S^1}$ such that $x = F(x)$.

More precisely, assume that E is a finite-dimensional space and $\{T_g\}_{g \in S^1}$ is a continuous linear representation of S^1 on E. Assume now that $F: E \to E$ is an S^1-invariant gradient mapping (with respect to a scalar product $\langle \, , \, \rangle$). Without loss of generality, we may assume that $\langle T_g x, T_g y \rangle = \langle x, y \rangle$ for $g \in S^1$. Suppose now that $z \notin E_{S^1}$, $F(z) = z$ and $S^1(z)$ is an isolated orbit of zeros of $I - F$. Choose a hyperplane T transversal to $S^1(z)$ at z. By the tubular neighbourhood theorem (cf. [3]) every point x near z can be uniquely expressed as $T_g y$, where g is near e, $y \in T$, and y is near x. (We write $x = T_{g(x)} \bar{y}(x)$.) Now define a map $\bar{F}$ of a neighbourhood of z in T into T by $\bar{F}(y) = \bar{y}(F(y))$. Note that if y is near z, $F(y)$ is near $F(z) = z$. It can be shown that z is an isolated fixed point of $\bar{F}$. The key point in the proof is that $\langle F(x), \tilde{z} \rangle = 0$ if $\tilde{z} \in T_x(S^1(x))$.) Note that it is possible for there to be points x near z where x and $F(x)$ lie on the same orbit. Now define $\operatorname{index}_n^{S^1}(I - F, S^1(z))$, the normal S^1-index of $S^1(z)$ for F, to be

$$|G_z|^{-1} \operatorname{index}_T(I - \bar{F}, z),$$

where $|G_z|$ denotes the order of G_z (and $G = S^1$). The extra factor has to be included to ensure that the index behaves well under small perturbations. The index is independent of the choice of T and of the choice of z on the orbit $S^1(z)$. Moreover, if the map is perturbed (by S^1-invariant gradient mappings), one readily shows that the sum of our indices for the orbits near $S^1(z)$ remains constant. (It is for this that we need the factor $|G_z|$. The reason is that an orbit $G(w)$ of fixed points of the perturbed map will intersect T several times if $|G_w| < |G_z|$.) If Ω is an invariant bounded open subset of E such that $x \neq F(x)$ on $\partial \Omega \cup (\Omega \cap E_{S^1})$, and assuming that $I - F$ only vanishes on a finite number of orbits $\{S^1(x_i)\}_{i=1}^k$ in Ω, define

$$\deg_n^{S^1}(I - F, \Omega) = \sum_{i=1}^k \operatorname{index}_n^{S^1}(I - F, S^1(x_i)).$$

We then remove the assumption that $I - F$ only vanishes on a finite number of orbits by approximating F by maps with this property. To check that the index is well defined and to prove a reasonable homotopy invariance property now reduces to the following proposition.

PROPOSITION. *Assume that $F: \Omega \times [0, 1] \to E$ is continuous and S^1-invariant and $F(\, , t)$ is a gradient mapping for each $t \in [0, 1]$. Then we can approximate F by an S^1-invariant continuous mapping $\tilde{F}$ such that $\tilde{F}(\, , t)$ is a gradient mapping for each t in $[0, 1]$ and $x - \tilde{F}(x, t)$ only vanishes on a finite number of orbits in Ω for each t in $[0, 1]$.*

The proof is rather technical. The idea (as in [**10**]) is (i) to successively define $\tilde{F}$ for x's having different isotropy group (starting with $G_z = G = S^1$), (ii) to show any types of bad behaviour occur one at a time, and (iii) to use a bifurcation analysis to examine the local solution structure where bad behaviour occurs. We also use transversality and real analyticity in significant ways in the proof.

This degree can be extended to S^1-invariant gradient mappings on a Hilbert space of the form $I - C$, where C is compact. The idea is to construct S^1-invariant finite-dimensional orthogonal projections P_n such that $P_n \to I$ strongly as $n \to \infty$. We then define the degree by using the finite-dimensional approximating maps $(I - P_n C)|_{R(P_n)}$. In fact, provided we *assume* that a suitable family P_n exists, we can define the degree on a Banach space with a weak gradient structure (cf. [**8**]).

We now apply the above degree to bifurcation problems. Assume that A: $H \times R \to H$ is completely continuous and S^1-invariant and that $A(\ , \lambda)$ is a gradient mapping for each λ. In addition, assume that $A(0, \lambda) = 0$ for each λ and that $A(x, \lambda) = \lambda Bx + K(x, \lambda)$, where B is linear and $\|x\|^{-1}K(x, \lambda) \to 0$ as $x \to 0$ locally uniformly in λ.

THEOREM 2. *Assume that the above conditions hold and that $x \neq A(x, \lambda)$ if $\lambda \in R$ and $x \in H_{S^1} \setminus \{0\}$. Suppose that λ_i is a characteristic value of B such that $N(I - \lambda_i B) \cap H_{S^1} = \{0\}$. Then there is a connected set C in $H \times R$ of nontrivial solutions of $x = A(x, \lambda)$ bifurcating from $(0, \lambda_i)$ such that C is unbounded or $(0, \lambda_j) \in \bar{C}$, where λ_j is another characteristic value of B.*

The key step in the proof of the above result is the following. *We assume that the conditions of the theorem hold and that there is an $\varepsilon > 0$ such that $x \neq A(x, \lambda_i)$ if $0 < \|x\| \leq \varepsilon$. Choose $\mu > 0$ small and then choose δ in $(0, \varepsilon)$ such that $x \neq A(x, \lambda)$ if $\lambda = \lambda_i \pm \mu$ and $0 < \|x\| \leq \delta$. Then*

$$\deg_n^{S^1}(I - A(\ , \lambda_i + \mu), H_\varepsilon \setminus H_\delta) - \deg_n^{S^1}(I - A(\ , \lambda_i - \mu), H_\varepsilon \setminus H_\delta) \neq 0.$$

This is proved by first showing that the above difference is independent of the higher-order term K and then doing a local bifurcational analysis for a suitably chosen K.

A weaker version of Theorem 1 can be proved without the condition that $x \neq A(x, \lambda)$ if $\lambda \in R$, $x \neq 0$, and $x \in H_{S^1}$. In addition, Theorem 1 can be improved to obtain more information in the case where the set C is bounded. In our special case, Theorem 1 gives much more information than Rabinowitz's bifurcation theorem [**9**] for potential operators.

Finally, we briefly consider some applications of our ideas. First, our results imply the existence of global branches of solutions for the problems in Wolfe [**11, 12**]. The idea here is to study a suitably truncated equation by reducing it to a finite-dimensional problem. We obtain in these cases a branch of solutions which is unbounded, meets another eigenvalue or continues to solutions where the equation is not elliptic.

Secondly, we can obtain bifurcation and continuation results for periodic solutions of fixed period of autonomous Hamiltonian systems. We consider the system

$$x' = \lambda J \nabla_x H(x, \lambda)$$

on R^{2n}, where $J = \begin{pmatrix} 0 & -I \\ I & 0 \end{pmatrix}$, and look for solutions of period T. If H is C^2, we can use a standard reduction (cf. Amann and Zehnder [2]) to reduce our problem to a finite-dimensional S^1-invariant gradient problem (at least if we restrict to solutions $(x(t), \lambda)$ with $|x(t)| \leq R$ on $[0, T]$ and $|\lambda| \leq R$). Our invariant degree can thus be used to keep some global control on bifurcations provided that we avoid points where nonconstant T-periodic solutions bifurcate from constant solutions. We also obtain a variant of Theorem 2 if $\nabla_x H(0, \lambda) = 0$ for all λ. Assume, *for simplicity*, that $H(x, \lambda) = \lambda H(x)$ and that $\nabla H(x) \neq 0$ for $x \neq 0$. Suppose that $y' = \lambda_i J D^2 H(0) y$ has a nontrivial T-periodic solution and that $D^2 H(0)$ is positive definite. (In fact, the results below hold under the assumptions of Fadell and Rabinowitz [7], and, indeed, we can obtain our result under weaker conditions than those in [7].) It follows from our results that there is a continuous branch of nontrivial T-periodic solutions bifurcating at $(0, \lambda_i)$ which is unbounded in $C[0, \pi] \times R$ or meets $(0, \lambda_j)$, where $y' = \lambda_j J D^2 H(0) y$ has a nontrivial T-periodic solution. Our result generalizes one in [1]. Of course, our result can be reinterpreted as a result on periodic solutions of $x' = J \nabla H(x(t))$, and we have the usual problems with nonminimal periods. Unlike the work in [7], we obtain continuous branches of solutions (but do not obtain multiplicity results), and we do not need an additional parameter as in [1] (and many others) for non-Hamiltonian systems. That we expect to need an additional parameter for the non-Hamiltonian case follows from our earlier comments on the general S^1-invariant problem.

References

1. J. C. Alexander and J. Yorke, *Global bifurcation of periodic orbits*, Amer. J. Math. **100** (1978), 263–292.

2. H. Amann and E. Zehnder, *Non-trivial solutions for a class of non-resonance problems and applications to non-linear differential equations*, Ann. Scuola. Norm. Sup. Pisa Cl. Sci. (4) **7** (1980), 539–603.

3. G. Bredon, *Introduction to compact transformation groups*, Academic Press, New York, 1972.

4. S. Chow and J. Mallet-Paret, *The Fuller index and global bifurcation*, J. Differential Equations **29** (1978), 66–85.

5. E. N. Dancer, *Perturbation of zeros in the presence of symmetries*, J. Austral. Math. Soc. Ser. A. **36** (1984), 106–125.

6. _____, *An implicit function theorem with symmetries and its application to nonlinear eigenvalue problems*, Bull. Austral. Math. Soc. **21** (1980), 81–91.

7. E. Fadell and P. Rabinowitz, *Generalized cohomology theories for Lie group actions with an application to bifurcation questions for Hamiltonian systems*, Invent. Math. **45** (1978), 139–174.

8. J. Marsden, *Applications of global analysis in mathematical physics*, Publish or Perish, Boston, 1974.

9. P. Rabinowitz, *A bifurcation theorem for potential operators*, J. Funct. Anal. **26** (1977), 48–67.

10. G. Wasserman, *Equivariant differential topology*, Topology **8** (1969), 127–150.

11. P. Wolfe, *Equilibrium states of an elastic conductor in a magnetic field: a paradigm of bifurcation theory*, Trans. Amer. Math. Soc. **278** (1983), 377–388.

12. _____, *Rotating states of an elastic conductor* (preprint).

UNIVERSITY OF NEW ENGLAND, AUSTRALIA

Abstract Differential Equations, Maximal Regularity, and Linearization

G. DA PRATO

1. Maximal regularity. Let E be a Banach space and $A: D_A \subset E \to E$ a linear operator. We assume:

(1.1)
(i) D_A is dense in E,
(ii) $|(\lambda - A)^{-1}|_{\mathscr{L}(E)} \leq \dfrac{K}{|\lambda| + 1}$ if $\operatorname{Re} \lambda \geq 0$.

It follows that A generates an analytic semigroup e^{tA}; moreover, there exist constants M and N, depending only on K, such that

$$(1.2) \qquad |e^{tA}|_{\mathscr{L}(E)} \leq M, \qquad |tAe^{tA}|_{\mathscr{L}(E)} \leq N.$$

We make hypothesis (1.1)(i) only for the sake of simplicity; in fact, many of our results hold without density of D_A (see [3] and [14]).

We are concerned here with the Cauchy problem

$$(1.3) \qquad u'(t) = Au(t) + f(t), \qquad u(0) = x,$$

where $x \in E$ and $f \in C([0, T]; E)$.

By $C([0, T]; E)$ (resp. $C^\alpha([0, T]; E)$, $\alpha \in \,]0, 1[$) we denote the Banach space of all continuous (resp. α-Hölder continuous) mappings $[0, T] \to E$. $C([0, T]; E)$ and $C^\alpha([0, T]; E)$ are endowed with the usual norms:

$$(1.4) \qquad |u|_{C([0, T]; E)} = \sup_{t \in [0, T]} |u(t)|_E;$$

$$(1.5) \qquad |u|_{C^\alpha([0, T]; E)} = |u|_{C([0, T]; E)} + [u]_{C^\alpha([0, T]; E)},$$

1980 *Mathematics Subject Classification.* Primary 34G10, 34G20; Secondary 35A20, 47H20.

© 1986 American Mathematical Society
0082-0717/86 $1.00 + $.25 per page

where

(1.6) $$[u]_{C^\alpha([0,T];E)} = \sup_{\substack{t,s \in [0,T] \\ t \neq s}} \frac{|u(t) - u(s)|_E}{|t-s|^\alpha}.$$

Finally, we denote by $h^\alpha([0, T]; E)$ (little α-Hölder continuous functions) the closure of $C^1([0, T]; E)$ (set of all continuously differentiable mappings $[0, T] \to E$) in $C^\alpha([0, T]; E)$. Note that we have the following inclusion:

(1.7) $$h^\alpha([0, T]; E) \supset C^\beta([0, T]; E) \quad \text{if } \beta > \alpha.$$

EXAMPLE 1.1. Let Ω be a bounded set in $\mathbf{R}^n$ with regular boundary $\partial\Omega$. We denote by $E = C_0(\overline{\Omega})$ the Banach space of all continuous functions $\overline{\Omega} \to \mathbf{R}$ vanishing in $\partial\Omega$ and endowed with the usual norm.

Let A be the linear operator defined by

(1.8) $$D_A = \{u \in C_0(\overline{\Omega}); \Delta u \in C_0(\overline{\Omega}) \text{ in the sense of distributions,} \\ \Delta \text{ Laplace operator}\},$$

$$Au = \Delta u.$$

Then A verifies hypotheses (1.1) (see [16]).

Return now to Problem 1.3; it is well known that its "mild" solution is given by

(1.9) $$u(t) = e^{tA}x + \int_0^t e^{(t-s)A} f(s) \, ds.$$

We are interested in regularity properties of u; in particular, we find conditions on x and f in order for u' and Au to have the same regularity of f (*maximal regularity*). This property turns out to be useful in nonautonomous equations [5], integral equations [15], delay equations, nonlinear equations [4, 10].

We start with a well-known result (see, for instance, [12]).

PROPOSITION 1.2. *Assume that* $f \in C^\alpha([0, T]; E)$, $\alpha \in \,]0,1[$ *and* $x \in D_A$; *then Problem* (1.3) *has a unique strict solution* $u \in C^1([0, T]; E) \cap C([0, T]; D_A)$.[1] *Moreover*,

(1.10) $$u'(t) = e^{tA}(Ax + f(t)) + \int_0^t Ae^{(t-s)A}(f(s) - f(t)) \, ds.$$

Note that the integral in (1.10) is meaningful, since, from (1.2),

$$|Ae^{(t-s)A}(f(s) - f(t))|_E \leqslant N[f]_{C^\alpha([0,T];E)} |t-s|^{\alpha-1}.$$

Now consider the particular case $f = 0$; then the following result is easily proved.

PROPOSITION 1.3. *Let* $u(t) = e^{tA}x$. *Then* (i) *and* (ii) (*resp.* (iii) *and* (iv)) *are equivalent.*

(i) $u \in C^\alpha([0, +\infty[; E)$.
(ii) $\sup_{t>0} t^{-\alpha}|e^{tA}x - x|_E < +\infty$.
(iii) $u \in h^\alpha([0, +\infty[; E)$.
(iv) $\lim_{t \to 0^+} t^{-\alpha}|e^{tA}x - x|_E = 0$.

---

[1] D_A is endowed with the norm $|x|_{D_A} = |A^{-1}x|_E$.

We set

(1.11) $$|x|_\alpha = \sup_{t>0} t^{-\alpha}|e^{tA}x - x|_E, \quad \alpha \in \,]0,1[,$$

(1.12) $$D_A(\alpha, \infty) = \{x \in E; \cdot |x|_\alpha < +\infty\},$$

(1.13) $$D_A(\alpha) = \{x \in E; \lim_{t \to 0^+} t^{-\alpha}|e^{tA}x - x|_E = 0\}.$$

$D_A(\alpha, \infty)$ coincides with the Lions interpolation space $(D_A, E)_{1-\alpha, \infty}$ [9], whereas $D_A(\alpha)$ is the interpolation space $(D_A, E)_{1-\alpha}$ introduced in [4]. $D_A(\alpha)$ is the closure of D_A in $D_A(\alpha, \infty)$. Note that $D_A(\alpha, \infty)$ and $D_A(\alpha)$ depend only on the couple (D_A, E).

We are now ready to prove

THEOREM 1.4. *Assume that* $f \in C^\alpha([0, T]; E)$ *(resp.* $h^\alpha([0, T]; E))$, $x \in D_A$, *and* $Ax + f(0) \in D_A(\alpha, \infty)$ *(resp.* $D_A(\alpha))$. *Let u be the mild solution to Problem* (1.3). *Then the following properties hold*:

(i) $u', Au \in C^\alpha([0, T]; E)$ *(resp.* $h^\alpha([0, T]; E))$, *and there exists a constant* $C = C(T, K, \alpha)$ *depending only on T, K, and* α, *such that*

(1.14) $$|u'|_{C^\alpha([0, T]; E)} + |Au|_{C^\alpha([0, T]; E)} \leq C(T, K, \alpha)\big(|Ax + f(0)|_\alpha$$
$$+ |f|_{C^\alpha([0, T]; E)}\big).$$

(ii) $Au(t) + f(t) \in D_A(\alpha, \infty)$ *(resp.* $D_A(\alpha)) \; \forall t \in [0, T]$.

PROOF. We will consider only the case $f \in C^\alpha([0, T]; E)$ and $Ax + f(0) \in D_A(\alpha, \infty)$; the other case is similar.

For part (i) it suffices to show that $u' \in C^\alpha([0, T]; E)$. Set $u' = h_1 + h_2 + h_3$, where

$$h_1(t) = e^{tA}(Ax + f(0)),$$
$$h_2(t) = e^{tA}g(t), \quad g(t) = f(t) - f(0),$$
$$h_3(t) = \int_0^t Ae^{(t-s)A}(f(s) - f(t))\,ds.$$

By Proposition 1.3 we have $h_1 \in C^\alpha([0, T]; E)$; moreover, if $t \geq r \geq 0$,

$$h_2(t) - h_2(r) = e^{tA}(g(t) - g(r)) + \int_r^t Ae^{\sigma A}g(r)\,d\sigma,$$

from which

$$|h_2(t) - h_2(r)|_E \leq M[g]_{C^\alpha([0, T]; E)}|t - r|^\alpha + NT^\alpha|g(r)|\int_r^t \sigma^{\alpha-1}\,d\sigma$$

$$\leq |t - r|^\alpha(M + NT^\alpha/\alpha)[f]_{C^\alpha([0, T]; E)}.$$

Concerning h_3 we have $h_3(t) - h_3(r) = J_1 + J_2 + J_3$, where

$$J_1 = \int_r^t A e^{(t-s)A}(f(s) - f(t))\, ds,$$

$$J_2 = \int_0^r ds \int_{r-s}^{t-s} A^2 e^{\sigma A}(f(s) - f(r))\, d\sigma,$$

$$J_3 = (e^{tA} - e^{(t-r)A})(f(r) - f(t)).$$

J_1 and J_3 are clearly less than const $x|t - r|^\alpha$. Concerning J_2 we have

$$|J_2|_E \leq 4N^2 [f]_{C^\alpha([0,T];E)} \int_0^r ds \int_{r-s}^{t-s} |s - r|^\alpha \sigma^{-2}\, d\sigma$$

$$= 4N^2 [f]_{C^\alpha([0,T];E)} (t - r) \int_0^r (t - s)^{-1}(r - s)^{\alpha - 1}\, ds$$

$$= 4N^2 [f]_{C^\alpha([0,T];E)} (t - r)^\alpha \int_0^{r/(t-r)} y^{\alpha - 1}(1 + y)^{-1}\, dy$$

$$\leq \text{const}|t - r|^\alpha,$$

so that (i) is proved.

We now prove (ii), that is, $u'(t) \in D_A(\alpha, \infty)$. Clearly, $h_1(t) \in D_A(\alpha, \infty)$. Concerning h_2 we have

$$e^{sA} h_2(t) - h_2(t) = \int_t^{t+s} A e^{\sigma A} g(t)\, d\sigma,$$

and it follows that

$$|e^{sA} h_2(t) - h_2(t)|_E \leq N \int_t^{t+s} |g(t)|_E \sigma^{-1}\, d\sigma$$

$$\leq (N/\alpha)[g]_{C^\alpha([0,T];E)} s^\alpha.$$

Thus, $h_2(t) \in D_A(\alpha, \infty)$. Finally, we have

$$e^{sA} h_3(t) - h_3(t) = \int_0^t dr \int_{t-r}^{t-r+s} A^2 e^{\sigma A}(f(r) - f(t))\, d\sigma,$$

from which

$$|e^{sA} h_3(t) - h_3(t)|_E \leq 4N^2 [f]_{C^\alpha([0,T];E)} \int_0^t dr \int_r^{r+s} \sigma^{-2} r^\alpha\, d\sigma$$

$$= 4N^2 [f]_{C^\alpha([0,T];E)} s^\alpha \int_0^{t/s} u^{\alpha - 1}(1 + u)^{-1}\, du$$

$$\leq \text{const } s^\alpha.$$

So $h_3(t) \in D_A(\alpha, \infty)$, and the proof is complete.

REMARKS 1.4. (a) Part (i) of Theorem 1.3 (when $f \in C^\alpha([0, T]; E)$ and $x = f(0) = 0$ can be proved using the methods of [3]. Here we have given a simpler proof due to Sinestrari (see [13]). Part (ii) is also due to Sinestrari [14] and will play an important role in subsequent applications.

(b) It is possible to study Problem (1.3) when f belongs to $L^p(0, T; E)$ (see [3] and [6]).

EXAMPLE 1.5. In the case of Example 1.1, the interpolation spaces $D_A(\alpha, \infty)$ and $D_A(\alpha)$ have been characterized by Lunardi [11]; namely, if $\alpha \in\,]0,1[$ and $\alpha \neq 1/2$, we have

(1.15) $$D_A(\alpha, \infty) = \{ u \in C^{2\alpha}(\overline{\Omega}); u = 0 \text{ on } \partial\Omega \},$$

(1.16) $$D_A(\alpha) = \{ u \in h^{2\alpha}(\overline{\Omega}); u = 0 \text{ on } \partial\Omega \}.$$

In this case, Problem (1.3) reduces to the heat equation, and Theorem 1.4 gives regularity results. For a discussion of this example see [14]; in this paper some classical results of [8] are also proved.

2. Linear nonautonomous equations. Let $\{A(t)\}_{t \in [0, T]}$ be a family of linear operators in E. We assume:

(2.1)
(i) $\left|(\lambda - A(t))^{-1}\right|_{\mathscr{L}(E)} \leq \dfrac{K}{|\lambda| + 1}$ if $\operatorname{Re} \lambda \geq 0$.

(ii) There exists a Banach space F such that
$D_{A(t)} = F$; moreover, $A \in h^\alpha([0, T]; \mathscr{L}(F; E))$.

We consider here the problem

(2.2) $$u'(t) = A(t)u(t) + f(t), \quad u(0) = x.$$

Setting $A = A(0)$, we have

(2.3) $$D_A(\alpha) = D_{A(t)}(\alpha) = (F, E)_{1-\alpha}.$$

THEOREM 2.1. *Assume that* $f \in h^\alpha([0, T]; E)$, $x \in F$, *and* $Ax + f(0) \in D_A(\alpha)$. *Then Problem (2.2) has a unique strict solution* u. *Moreover,*

(2.4)
(i) $u', Au \in h^\alpha([0, T]; E)$,
(ii) $Au(t) + f(t) \in D_A(\alpha)$.

PROOF. Fix $u \in h^\alpha([0, T]; F)$ and consider the problem

(2.5) $$v' = Av + g, \quad v(0) = x,$$

where

(2.6) $$g(t) = (A(t) - A)u(t) + f(t).$$

Since $g \in h^\alpha([0, T]; E)$ and $Ax + g(0) = Ax + f(0) \in D_A(\alpha)$, we see, by Theorem 1.4, that the mapping $\gamma: u \to v$ maps $h^\alpha([0, T]; F)$ into itself. It is easy to prove the existence of μ such that, for each $x \in F$ and $t \in [0, T]$, $|x|_F \leq \mu |x|_{D_{A(t)}}$ (see [5]).

Moreover, if $\bar{u}, u \in h^\alpha([0, T]; F)$, we have

$$|\gamma(u) - \gamma(\bar{u})|_{h^\alpha([0, T]; F)}$$
$$\leq C(T, K, \alpha) |A(\cdot) - A(0)|_{h^\alpha([0, T]; \mathscr{L}(F; E))} |u - \bar{u}|_{h^\alpha([0, T]; F)}.$$

Since

(2.7) $$\lim_{T \to 0} |A(\cdot) - A(0)|_{h^\alpha([0, T]; \mathscr{L}(F; E))} = 0,$$

there exists $T_0 > 0$ such that

(2.8) $$|A(\cdot) - A(0)|_{h^\alpha([0,T];\,\mathscr{L}(F;E))} \leq \frac{1}{C(T,K,\alpha)\mu}.$$

Thus, by the contraction principle, Problem (2.2) has a unique solution in $[0, T_0]$.
Moreover, by Theorem 1.3(ii), we have

$$u'(T_0) = v'(T_0) \in D_A(\alpha) = D_{A(T_0)}(\alpha).$$

Thus, we can repeat the previous argument and extend the solution to the interval $[0, 2T_0] \cap [0, T]$. Then, by standard compactness arguments (note that $C(T, K, \alpha)$ is independent of T_0), the conclusion follows.

REMARKS 2.2. (a) Theorem 2.1 is proved in [5]; the present proof (which uses conclusion (ii) of Theorem 1.4) is simpler. However, in [5] we give additional results; in particular, we show the existence of a strict solution when $x \in D_A$.

(b) Assume that $A \in C^\alpha([0,T];\,\mathscr{L}(F;E))$, $f \in C^\alpha([0,T];E)$, $x \in F$, and $Ax + f(0) \in D_A(\alpha, \infty)$. Then Problem (2.2) has a unique strict solution u such that $u', Au \in C^\alpha([0,T];E)$. This is proved in [1] by another technique.

3. Integral equations.

Let A be a linear operator in E such that hypotheses (1.1) hold. Moreover, let $\{B(t)\}_{t \in [0,T]}$ be a family of linear operators in E such that

(3.1)
 (i) $D_{B(t)} = D_A$,
 (ii) $B(\cdot)x$ is measurable in E for any $x \in F$,
 (iii) $|B(t)|_{\mathscr{L}(D_A;\,E)} \leq S < +\infty \quad \forall t \in [0,T]$.

Consider the integrodifferential equation

(3.2) $$u'(t) = Au(t) + \int_0^t B(s)u(t-s)\,ds + f(t), \quad t \in [0,T],$$
$$u(0) = x.$$

THEOREM 3.1. *Assume that hypotheses* (1.1) *and* (3.1) *hold. Assume, moreover, that* $f \in C^\alpha([0,T];E)$, $x \in D_A$, *and* $Ax + f(0) \in D_A(\alpha, \infty)$. *Then Problem* (3.2) *has a unique strict solution u such that* $u', Au \in C^\alpha([0,T];E)$.

PROOF. Fix $u \in C^\alpha([0,T];D_A)$ and consider the problem

(3.3) $$v' = Av + B*u + f, \quad v(0) = x,$$

where

(3.4) $$(B*u)(t) = \int_0^t B(s)u(t-s)\,ds.$$

Now fix $\overline{T}$ in $]0, T]$ and let $\overline{T} \geq t \geq r \geq 0$; we have

$$|(B*u)(t) - (B*u)(r)|_E \leq S|t-r|\,|u|_{C([0,\overline{T}];\,D_A)}$$
$$+ Sr|t-r|^\alpha [u]_{C^\alpha([0,\overline{T}];\,D_A)}.$$

It follows that

(3.5) $$|B*u|_{C^\alpha([0,\bar T];E)} \leq S(\bar T^{1-\alpha} + \bar T)|u|_{C^\alpha([0,\bar T];D_A)}.$$

Now by Theorem 1.4 the mapping $\gamma: u \to v$ maps $C^\alpha([0, \bar T]; D_A)$ into itself. Moreover, if $u, \bar u \in C^\alpha([0, \bar T]; D_A)$ we have

$$|\gamma(u) - \gamma(\bar u)|_{C^\alpha([0,\bar T]; D_A)} \leq S(\bar T^{1-\alpha} + \bar T)C(T, K, \alpha)(1 + K)|u - \bar u|_{C^\alpha([0,\bar T]; D_A)}.$$

Choosing $\bar T$ such that

(3.6) $$S(\bar T^{1-\alpha} + \bar T)C(T, K, \alpha) < 1/2,$$

we see that γ is a contraction, and there exists a unique solution u of Problem (3.2) in $[0, \bar T]$. Moreover,

(3.7) $$u'(\bar T) \in D_A(\alpha, \infty).$$

Now consider Problem (3.2) in the interval $[\bar T, 2\bar T] \cap [0, T]$.

(3.8) $$\begin{aligned} z'(t) &= Az(t) + \int_{\bar T}^t B(t-s)z(s)\,ds + f_1(t), \\ z(\bar T_0) &= u(\bar T), \end{aligned}$$

where

(3.9) $$f_1(t) = \int_0^{\bar T} B(t-s)u(s)\,ds + f(t).$$

Since

$$Az(\bar T) + f_1(\bar T) = u'(\bar T) \in D_A(\alpha, \infty),$$

we can again apply the previous argument, so that, in a finite number of steps, the conclusion follows.

THEOREM 3.2. *Assume that hypotheses* (1.1) *and* (3.1) *hold. Assume, moreover, that* $f \in C^\alpha([0, T]; E)$, $x \in D_A$. *Then Problem* (3.2) *has a strict solution u such that* $u \in C^1([0, T]; E) \cap C([0, T]; D_A)$.

PROOF. Let z be the solution to the problem

(3.10) $$z'(t) = Az(t) + f(0), \quad z(0) = x.$$

Set $u - z = v$; then v verifies the equation

(3.11) $$v'(t) = Av(t) + \int_0^t B(t-s)v(s)\,ds + p(t), \quad v(0) = 0,$$

where

(3.12) $$p(t) = \int_0^t B(t-s)z(s)\,ds + f(t) - f(0).$$

Now let us prove that $p \in C^\alpha([0, T]; E)$; for this it is sufficient to show that $p_1 \in C^\alpha([0, T]; E)$, where

$$p_1(t) = \int_0^t B(s)e^{(t-s)A}x\,ds.$$

In fact, if $t > \tau > 0$ we have

$$P_1(t) - P_1(\tau) = \int_\tau^t B(s)e^{(t-s)A}x\, ds + \int_0^\tau B(s)\left(\int_{\tau-s}^{t-s} Ae^{\sigma A}x\, d\sigma\right) ds$$
$$= I_1 + I_2,$$

so that $|I_1| \leq \text{const}|t - \tau|$, whereas $|I_2| \leq \text{const} \int_0^\tau (ds/(\tau - s)^\alpha)|t - \tau|^\alpha$. Since $p \in C^\alpha([0, T]; E)$, we may apply Theorem 3.1 and conclude that Problem (3.11) has a strict solution v. Thus, $u = v + z$ is a strict solution of Problem 3.1.

EXAMPLE 3.2. Let A be as in Example 1.1 and $\beta \in L^\infty_{\text{loc}}(\mathbf{R})$; then we can apply Theorem 3.2 to the heat equation with memory

$$u_t(t, x) = \Delta u(t, x) + \int_0^t \beta(t - s)\Delta u(s, x)\, ds + f(t, x),$$

$$0 \leq t \leq T, x \in \overline{\Omega},$$

$$u(t, x) = 0, \quad 0 \leq t \leq T, x \in \partial\Omega,$$
$$u(0, x) = u_0(x), \quad x \in \overline{\Omega}.$$

4. Nonlinear equations. Let F be a Banach space continuously and densely embedded in E. Consider the problem

(4.1) $\qquad u'(t) = f(u(t)), \quad t \geq 0, \quad u(0) = x.$

We assume that

(a) $f \in C^2(F; E)$;
(b) there exists $K \in C(F; \mathbf{R})$ such that

(4.2) $\qquad |(\lambda - f'(x))^{-1}|_{\mathscr{L}(E)} \leq \dfrac{K(x)}{1 + |\lambda|},$

$\text{Re }\lambda \geq 0$ and $|\cdot|_F \simeq |\cdot|_{f'(x)}$;

(c) $x \in F, f(x) \in (F, E)_{1-\alpha}$, for a suitable $\alpha \in\,]0, 1[$.

Setting $A = f'(x), \tilde{f}(u) = f(x) + A(u - x) + \psi(u)$, we have $\psi(x) = 0, \psi'(x) = 0$. Problem (4.1) is equivalent to

(4.3) $\qquad u' = Au + \psi(u) + f(x) - Ax, \quad u(0) = x.$

Remember that condition (4.2)(c) is equivalent to

(4.4) $\qquad f(x) \in D_A(\alpha).$

Consider now, for each u, the linearized problem

(4.5) $\qquad v' = Av + \psi(u) + f(x) - Ax, \quad v(0) = x,$

and denote by γ the mapping $u \to v$. Clearly, Problem (4.1) is equivalent to the equation $\gamma(u) = u$. By Theorem 1.4 it follows that γ maps $h^\alpha([0, T]; F)$ into itself.

Finally, set

(4.6) $\qquad \Sigma_r = \left\{u \in h^\alpha([0, T]; F); u(0) = x, |u - x|_{h^\alpha([0, T]; F)} \leq r\right\}.$

LEMMA 4.1. *There exists $T, r > 0$ such that γ maps Σ_r into itself.*

PROOF. Let $u \in \Sigma_r$, $v = \gamma(u)$; set

(4.7) $$v - x = \alpha_1 + \alpha_2,$$

where

(4.8) $$\alpha_1(t) = A^{-1}(e^{tA} - 1)f(x), \quad \alpha_2(t) = \int_0^t e^{(t-s)A}\psi(u(s))\, ds,$$

and

(4.9) $$C_T = |\alpha_1|_{h^\alpha([0, T]; F)}.$$

Then

(4.10) $$\lim_{T \to 0} C_T = 0.$$

Moreover,

$$|\psi(u(t)) - \psi(u(s))|_E \leq M_{1,r}|u(t) - u(s)|_F,$$

where $M_{1,r} = \sup_{|y-x|_F \leq r}|\psi'(y)|_{\mathscr{L}(F; E)}$. We clearly have

(4.11) $$\lim_{r \to 0} M_{1,r} = 0,$$

and

$$|\psi(u)|_{h^\alpha([0, T]; E)} \leq |\psi(u)|_{C([0, T]; E)} + M_{1,r}[u]_{h^\alpha([0, T]; E)}$$
$$\leq M_{1,r}(1 + T^\alpha)|u|_{h^\alpha([0, T]; F)}.$$

From Theorem 1.4 we have

$$|v - x|_{h^\alpha([0, T]; F)} \leq C_T + rM_{1,r}(1 + T^\alpha) \cdot C(T, K(x), \alpha),$$

and (4.10), (4.11) imply the conclusion.

The following lemma is straightforward.

LEMMA 4.2. *For any $u, \bar{u} \in \Sigma_r$ we have*

(4.12) $$|\psi(u) - \psi(\bar{u})|_{C([0, T]; E)} \leq M_{1,r}|u - \bar{u}|_{C([0, T]; F)},$$

(4.13) $$[\psi(u) - \psi(\bar{u})]_{h^\alpha([0, T]; E)} \leq M_{1,r}[u - \bar{u}]_{h^\alpha([0, T]; F)}$$
$$+ 3rM_{2,r}|u - \bar{u}|_{C([0, T]; F)},$$

where

$$M_{2,r} = \sup_{|y-x| \leq r} |\psi''(y)|_{\mathscr{L}(F; \mathscr{L}(F; E))}.$$

THEOREM 4.3. *Assume that hypotheses (4.2) hold. Then there exists $T > 0$ such that Problem (4.1) has a unique solution $u \in h^\alpha([0, T]; F) \cap h^{1,\alpha}([0, T]; E)$. Moreover,*

(4.14) $$u(T) \in F, \quad f(u(T)) \in (F, E)_{1,\alpha}.$$

PROOF. Pick r and T such that γ maps Σ_r into itself. We now show that γ is a contraction, provided that r is sufficiently small.

Let $u, \bar{u} \in \Sigma_r$, $v = \gamma(u)$, $\bar{v} = \gamma(\bar{u})$; then

$$v(t) - \bar{v}(t) = \int_0^t e^{(t-s)A}(\psi(u(s)) - \psi(\bar{u}(s)))\,ds,$$

from which

$$|v - \bar{v}|_{h^\alpha([0,T];F)} \leq C(T, K(x), \alpha)|\psi(u) - \psi(\bar{u})|_{h^\alpha([0,T];E)}.$$

Using (4.13) we get

(4.14) $\quad |v - \bar{v}|_{h^\alpha([0,T];F)} \leq C(T, K(x), \alpha)(M_{1,r} + 3rM_{2,r})|u - \bar{u}|_{h^\alpha([0,T];F)}.$

Since $\lim_{r \to 0}(M_{1,r} + 3rM_{2,r}) = 0$, we see that γ is in fact a contraction for r small.

Thus, there exists $u \in h^\alpha([0,T]; F) \cap h^{1,\alpha}([0,T]; E)$ such that $u = \gamma(u)$; moreover, from Theorem 1.4(ii), it follows that $u'(T) \in D_A(\alpha) = (F, E)_{1-\alpha}$, and the theorem is completely proved.

REMARKS 4.4. (a) Theorem 4.3 is very similar to Theorem 5.2 in [4]. In that paper, hypothesis (4.2)(b), (c) is replaced by

(4.15) $\quad D_{f'(x)}(\alpha + 1) = \{z \in F; f'(x) \in (F, E)_{1-\alpha}\} = \text{const} = D_A(\alpha + 1),$
$\quad f \in C^1(F; E) \cap C^1(D_A(\alpha + 1), D_A(\alpha)).$

For the proof we use the same linearization argument.

(b) By (4.14) we see that we can continue, by the same procedure, the solution u in an interval $[T, T + \delta]$. Consequently, we can define the solution of Problem (4.1) in a maximal interval I which is open in $[0, +\infty[$. In addition, arguing as in [4], we can prove that if the a priori estimate

(4.16) $\quad |u(t)|_F + |f(u(t))|_{D_A(\alpha)} \leq L \quad \forall t \in I, \text{ where } \alpha' > \alpha,$

holds, then the solution u is global.

(c) Regularity results and analyticity of the solution can also be studied as in [10].

We will prove finally that if x is "small" (in a suitable sense), then the solution of Problem (4.1) is global.

THEOREM 4.5. *Assume that hypotheses* (4.2) *hold. Let $T > 0$ be fixed. Then there exists $\nu > 0$ (depending on t) such that if*

(4.17) $\quad x \in F, \quad |x|_F + |f(x)|_{D_A(\alpha)} \leq \nu,$

problem (4.1) *has a unique global solution* $u \in h^\alpha([0,T]; F) \cap h^{1+\alpha}([0,T]; E)$.

PROOF. First we prove that we can choose ν and r such that if (4.17) holds, then γ maps Σ_r into Σ_r.

Let $u \in \Sigma_r$, $v = \gamma(u)$; define α_1 and α_2 as in (4.8). We have $|\alpha_1|_{h^\alpha([0,T];F)} \leq b\nu$, where b is a constant depending only on x. It follows that

$$|v - x|_{h^\alpha([0,T];F)} \leq b\nu + rM_{1,r} \cdot (1 + T^\alpha) \cdot C(T, K(x), \alpha),$$

so that the conclusion follows. Now, by (4.14), we see that r can be chosen such that γ is a contraction.

EXAMPLE 4.6. We use the notation of Example 1.1. Let $E = C_0^\beta(\overline{\Omega})$, $F = \{u \in C_0^\beta(\overline{\Omega}); \Delta u \in C_0^\beta(\overline{\Omega})\}$. Let $\varphi \in C^3(\mathbf{R})$ such that

(4.18) $$\varphi(0) = 0, \quad \varphi'(0) \geq \varepsilon > 0,$$

and set $f(u) = \varphi(\Delta u)$; then hypotheses (4.2) are fulfilled, with $\alpha > \beta/2$. It follows that if $\alpha \in {]}0, 1[$, $\alpha \neq 1/2$, there exists $\nu > 0$ such that if

(4.19) $$|u_0|_{h^{2+2\alpha}(\overline{\Omega})} \leq \nu, \quad \Delta u_0 = 0, u_0 = 0 \quad \text{on } \partial\Omega,$$

then the problem

(4.20) $$\begin{aligned} u_t(t, x) &= \varphi(\Delta u(t, x)) \quad \text{in } [0, T] \times \Omega, \\ u(t, x) &= 0 \quad \text{in } [0, T] \times \partial\Omega, \\ u(0, x) &= u_0(x) \quad \text{in } \Omega, \end{aligned}$$

has a unique global solution u such that

$$u \in C^{1,\alpha}\big([0, T]; h^{2\alpha}(\overline{\Omega})\big) \cap C^\alpha\big([0, T]; h^{2\alpha+2}(\overline{\Omega})\big).$$

REMARKS 4.7. (a) To get global existence with an arbitrary initial datum is much more difficult. In fact, we need an a priori estimate in $h^{2\alpha+2}(\overline{\Omega})$ of the solution u, for some $\alpha > 0$. An example (one-dimensional) in which it is possible to get such an estimate is given in [4] (see also [10]).

(b) Problem (4.20) can be studied using the theory of nonlinear semigroups; in this case it is possible to show the existence of a global weak solution [2]. However, Theorems 4.3 and 4.5 also apply to more general situations, such as $f(u) = \varphi(u, \nabla u, \Delta u)$ or $f(u) = \varphi(Au)$, with A an elliptic operator of order $2m$.

REFERENCES

1. P. Acquistapace and B. Terreni, *On the abstract Cauchy problem in the case of constant domains*, Ann. Mat. Pura Appl. (to appear).
2. P. Benilan and K. S. Ha, *Equations d'évolution du type $(du/dt) + \beta\partial\phi(u) \ni 0$ dans $L^\infty(\Omega)$*, C. R. Acad. Sci. Paris **281** (1975), 947–950.
3. G. Da Prato and P. Grisvard, *Sommes d'opérateurs non linéaires et équations differentielles opérationnelles*, J. Math. Pures Appl. **54** (1975), 305–387.
4. _____, *Equations d'évolution abstraites non linéaires de type parabolique*, Ann. Mat. Pura Appl. (4) **120** (1979), 329–396.
5. G. Da Prato and E. Sinestrari, *Hölder regularity for nonautonomous abstract parabolic equations*, Israel J. Math. **42** (1982), 1–19.
6. G. Di Blasio, *Linear parabolic evolution equations in L^p spaces*, Ann. Mat. Pura Appl. **138** (1984), 55–104.
7. T. Kato, *Perturbation theory for non-linear operators*, Springer-Verlag, 1961.
8. O. A. Ladyženskaja, V. A. Solonnikov and N. N. Ural'ceva, *Linear and quasilinear equations of parabolic type*, Transl. Math. Monographs, vol. 23, Amer. Math. Soc., Providence, R. I., 1968.
9. J. L. Lions and J. Peetre, *Sur une classe d'espaces d'interpolation*, Inst. Hautes Études Sci. Publ. Math. **19** (1945), 5–68.
10. A. Lunardi, *Analyticity of the maximal solution of an abstract nonlinear parabolic equation*, Nonlinear Anal. **6** (1982), 503–521.

11. _____, *Interpolation spaces between domains of elliptic operators and spaces of continuous functions with applications to nonlinear parabolic equations*, Math. Nachr. (to appear).

12. A. Pazy, *Semigroups of linear operators and applications to partial differential equations*, Lecture Notes, Univ. of Maryland, 1977.

13. E. Sinestrari, *On the solution of the inhomogeneous evolution equations in Banach spaces*, Rend. Accad. Naz. Lincei **70** (1981).

14. _____, *On the abstract Cauchy problem of parabolic type in spaces of continuous functions*, J. Math. Anal. Appl. (to appear).

15. _____, *Continuous interpolation spaces and spatial regularity in nonlinear Volterra integrodifferential equations*, J. Integral Equations (to appear).

16. H. B. Stewart, *Generation of analytic semigroups by strongly elliptic operators*, Trans. Amer. Math. Soc. **199** (1974), 141–162.

17. K. Yosida, *Functional analysis*, Springer-Verlag, 1968.

SCUOLA NORMALE SUPERIORE, PISA, ITALY

Positive Solutions for Some Classes of Semilinear Elliptic Problems

DJAIRO G. DE FIGUEIREDO

Introduction. Let us consider the Dirichlet problem

(0.1) $\quad -\Delta u = f(u) \quad \text{and} \quad u > 0 \quad \text{in } \Omega, \quad u = 0 \quad \text{on } \partial\Omega.$

where Ω is a bounded smooth domain in R^N, $N \geq 2$, and $f: R^+ \to R$ is a function satisfying the condition below, besides other requirements that will be timely introduced as we proceed.

($f1$) $\quad f: [0, \infty) \to R$ is locally lipschitzian.

The following quantities will play an important role in the analysis conducted in this paper:

(0.2) $\quad I = \inf_{s>0} \frac{f(s)}{s}, \quad S = \sup_{s>0} \frac{f(s)}{s}.$

Clearly $I \in R \cup \{-\infty\}$ and $S \in R \cup \{+\infty\}$. It is clearly seen that a necessary condition for existence of a solution of (0.1) is

(0.3) $\quad I \leq \lambda_1 \leq S.$

(This is proved via a multiplication of the equation in (0.1) by an eigenfunction $\phi_1 > 0$ of $(-\Delta, H_0^1)$ corresponding to the first eigenvalue λ_1.) Also by this very procedure we see that if either $I = \lambda_1$ or $S = \lambda_1$, then u is necessarily a positive multiple of ϕ_1 and $f(s) = \lambda_1 s$ for s in some interval $[0, a]$. Consequently the interesting cases are when

(0.4) $\quad I < \lambda_1 < S.$

That is, (0.4) is a necessary condition for existence of a solution of (0.1) which is not trivial. By trivial we mean $u = k\phi_1$ for some positive constant k.

Condition (0.4) is achieved in classes I and II below by stating hypotheses on the behavior of $f(s)/s$ at 0 and at $+\infty$. For classes III and IV we shall need

---

1980 *Mathematics Subject Classification.* Primary 35J20, 35J25, 47H15.

assumptions on $f(s)/s$ at intermediate points between 0 and ∞. Of course, the necessary condition (0.4) is not sufficient for the existence of solutions of (0.1) in most cases as simple examples show. The object of this paper is to exhibit some further conditions on the nonlinearity f that insure the existence of solutions for (0.1). This will be done for classes III and IV.

Class I (Sublinear):
$$\liminf_{s \to 0} \frac{f(s)}{s} > \lambda_1 \quad \text{and} \quad \limsup_{s \to +\infty} \frac{f(s)}{s} < \lambda_1.$$

Class II (Superlinear):
$$\limsup_{s \to 0} \frac{f(s)}{s} < \lambda_1 \quad \text{and} \quad \liminf_{s \to +\infty} \frac{f(s)}{s} > \lambda_1.$$

Class III (Sub-super!):
$$\liminf_{s \to 0} \frac{f(s)}{s} > \lambda_1 \quad \text{and} \quad \liminf_{s \to +\infty} \frac{f(s)}{s} > \lambda_1.$$

Class IV (Super-sub!):
$$\limsup_{s \to 0} \frac{f(s)}{s} < \lambda_1 \quad \text{and} \quad \limsup_{s \to +\infty} \frac{f(s)}{s} < \lambda_1.$$

It is well known that problems (0.1) in class I have at least one solution. This can be proved by the method of monotone iterations (see, for instance, [1, 16] or the review paper [2]).

Problems (0.1) in class II do not have a solution in general. The most famous case is Pohožaev's example [3]

(0.5) $\qquad -\Delta u = u^\sigma \quad \text{and} \quad u > 0 \quad \text{in } \Omega, \qquad u = 0 \quad \text{on } \partial\Omega.$

where $\sigma = (N+2)/(N-2)$, $N \geq 3$, and Ω is a star-shaped bounded domain in R^N. It is also known [4] that if $\sigma < (N+2)/(N-2)$, $N \geq 3$, then (0.5) possesses a solution. And more generally, also [4], if

(0.6) $\qquad \displaystyle\lim_{s \to +\infty} \frac{f(s)}{s^\sigma} = 0, \quad \text{where } \sigma = \frac{N+2}{N-2}, N \geq 3,$

and if there exist numbers $s_0 > s$ and $\theta > 2$ such that

(0.7) $\qquad sf(s) \geq \theta F(s) \quad \text{for } s \geq s_0, \ F(s) = \displaystyle\int_0^s f,$

then problem (0.1) has a solution. Conditions (0.6) and (0.7) are used to prove that the Euler–Lagrange functional associated to (0.1), namely

(0.8) $\qquad \Phi(u) = \dfrac{1}{2} \displaystyle\int |\nabla u|^2 - \int F(u).$

satisfies the Palais–Smale condition in $H_0^1(\Omega)$. However it has been proved in [5] that for convex domains Ω existence of solutions of (0.1) holds without assumption (0.7), assuming only (0.6) and the fact that the problem (0.1) is in class II. This is due to fairly general a priori estimates of the L^∞ norms of ∇u on $\partial\Omega$ for solutions of u of (0.1), obtained in [5] by the use of the Gidas–Ni–Nirenberg techniques [6]. So existence holds in class II even if Φ (apparently) does not satisfy the Palais–Smale condition. Nevertheless the aforementioned results still require condition (0.6). Observe however that the Euler–Lagrange functional (0.8) is still defined in $H_0^1(\Omega)$ if one only assumes

(0.9) $\qquad |f(s)| \leqslant cs^\sigma + c, \qquad \sigma = (N+2)/(N-2), n \geqslant 3.$

The investigation of problems of this type has been conducted by Brézis–Nirenberg [12] and solutions of (0.1) have been obtained under various additional conditions of the nonlinearity f. All previous remarks were made in case $N \geqslant 3$. Similar statements can be made in the case $N = 2$, where even exponential growth of f is allowed [5].

Problems (0.1) in class III also do not have a solution in general. As observed above the graph of $f(s)$ has to be somewhere below the straight line $\lambda_1 s$. But this requirement only does not suffice. We can see that problems like

(0.10) $\qquad -\Delta u = \lambda e^u \text{ in } \Omega, \qquad u = 0 \text{ on } \partial\Omega$

may have no solution for some λ's such that $\lambda < \lambda_1/e$. (That is, $\inf\{\lambda e^s/s : s > 0\} = \lambda e < \lambda_1$.) So the graph of $f(s)$ has to have a "fairly good" crossing of the line $\lambda_1 s$. Clearly if $f(s)$ vanishes at some point s_0, this is enough to guarantee the existence of a solution for (0.1). (The method of monotone iteration readily gives the solution. Here s_0 is a supersolution of (0.1).) However the graph of $f(s)$ does not have to go below $\lambda_1 s$ that far. In fact, problem (0.10) has a solution if $\lambda > 0$ is small enough. In §1 we shall see a sufficient condition on f that provides a strict supersolution for (0.1), and consequently a solution, obtained again by the method of monotone iterations. We see then that the existence of one solution for (0.1) in the case of class III does not involve the behavior of f at $+\infty$. However if we are willing to discuss the existence of further solutions, the behavior of f at $+\infty$ becomes very useful in two respects. First, if we are planning to use topological degree methods, conditions on f at $+\infty$ will be required in such a way as to obtain a priori bounds on the solutions of (0.1). Second, if we choose to use variational methods, again conditions on f at $+\infty$ will be necessary to define the Euler–Lagrange functional (0.8) in H_0^1. Also some further conditions on f at $+\infty$ will insure that Φ satisfies the Palais–Smale condition, which will permit us to use the usual critical point theory, namely here the Mountain Pass Theorem [4]. All this will be discussed in §1.

Problems in class IV clearly do not have a solution in general, since the necessary condition (0.4) may not be satisfied. In §2 we show that the solutions of a given problem (0.1) in this class are a priori bounded in L^∞. So a truncation on the nonlinearity f may be performed and the truncated problem has the same solutions as the original problem. With this truncated f, the Euler–Lagrange functional is well defined in H_0^1, bounded below, and satisfies the Palais–Smale condition. Clearly 0 is a local minimum of this functional, which also has a global minimum. It could very well happen that 0 is this global minimum. In fact this will be the case if one does not insure that $f(s)$ goes sufficiently enough above $\lambda_1 s$. This will be achieved, for instance, by assuming a condition on the primitive F, which will guarantee that the functional is negative somewhere. Consequently the global minimum is negative and it is achieved at a point $u \neq 0$.

1. Problems (0.1) in class III. The condition imposed on $f(s)/s$ near 0 implies immediately that $\underline{u} = \varepsilon\phi_1$, for $\varepsilon > 0$ sufficiently small, is a strict subsolution. In order to use the method of monotone iteration one shall insure the existence of a supersolution $\bar{u} \geqslant \underline{u}$. For that matter one uses the solution of a linear problem

(1.1) $\qquad -\Delta v = \mu v + C \quad \text{in } \Omega, \qquad v = 0 \quad \text{on } \partial\Omega,$

$0 \leqslant \mu < \lambda_1(\Omega)$ and C is a positive constant. Since other domains besides Ω will be appearing in this exposition we use the notation $\lambda_1(\Omega)$ to denote the first eigenvalue of the Laplacian in $H_0^1(\Omega)$. The solution v of (1.1) is positive in Ω and its maximum M depends on μ, C, and Ω. If $\mu < \lambda_1(\Omega^*)$, where Ω^* is the ball with the same volume as Ω, centered at 0, then it follows that $M \leqslant \omega(0)$, where ω is the solution of

(1.2) $\qquad -\Delta\omega = \mu\omega + C \quad \text{in } \Omega^*, \qquad \omega = 0 \quad \text{on } \partial\Omega^*.$

The proof of this statement uses the facts that $v^* \leqslant \omega$ in Ω^* and $\|v\|_{L^\infty} = \|v^*\|_{L^\infty}$, where v^* represents the decreasing symmetric rearrangement of v. For proofs of these facts see P. L. Lions [7], Talenti [8], or Bandle [9]. Observe also that $\lambda_1(\Omega^*) \leqslant \lambda_1(\Omega)$. So a bound for M can be written out explictly by solving (1.2), which indeed reduces to an ordinary differential equation. If we denote by M_ρ the maximum, $z(0)$, of the solution z of

(1.3) $\qquad -\Delta z = \rho z + 1 \quad \text{in } B_1(0), \qquad z = 0 \quad \text{on } \partial B_1(0),$

where $0 \leqslant \rho < \lambda_1(B_1(0))$, then $M \leqslant CR^2 M_{\mu R^2}$, where R is the radius of the ball Ω^*. In this way the following result was established in [10].

PROPOSITION 1.1. *If there exist numbers $s_0 > 0$ and $0 \leqslant \mu < \lambda_1(\Omega^*)$ such that*

(1.4) $\qquad f(s) \leqslant \mu s + s_0/R^2 M_{\mu R^2}, \qquad 0 \leqslant s \leqslant s_0,$

then problem (0.1) possesses a strict supersolution $\bar{u}$.

Existence of a strict supersolution for problem (0.1) can also be inferred from informations on the symmetrized problem. For instance, see [10].

PROPOSITION 1.2. *Suppose that there exists a function $f_1: R^+ \to R$ which is nondecreasing, $f_1(s) \geqslant f(s)$ and $f_1(s) > f(s)$ for s in some interval $(0, a)$. Assume that the problem*

(1.5) $\qquad -\Delta v = f_1(v) \quad \text{in } \Omega^*, \qquad v = 0 \quad \text{on } \partial\Omega^*,$

has a positive solution v. Then problem (0.1) in class III *possesses a strict supersolution.*

PROOF. It suffices to prove that the problem

(1.6) $\qquad -\Delta u = f_1(u) \quad \text{in } \Omega, \qquad u = 0 \quad \text{on } \partial\Omega$

has a solution. In fact, one may assume that the solution v given in the statement of the proposition is a minimal solution. Also the solution u of (1.6) obtained next will be a minimal solution. To prove the existence of such a solution we proceed by monotone iterations. Take a subsolution of (1.6) given by $u_0 = \varepsilon\phi_1$, and a subsolution of (1.5) given by $v_0 = \delta\hat{\phi}_1$, where $\hat{\phi}_1 > 0$ is a first eigenfunction of $-\Delta$ on $H_0^1(\Omega^*)$. Choose $\varepsilon > 0$ small enough such that $u_0^* \leq v_0$. Then proceed through the two following iteration schemes:

$$-\Delta u_n = f_1(u_{n-1}) \quad \text{in } \Omega, \qquad u_n = 0 \quad \text{on } \partial\Omega,$$
$$-\Delta v_n = f_1(v_{n-1}) \quad \text{in } \Omega^*, \qquad v_n = 0 \quad \text{on } \partial\Omega^*.$$

It is then seen that $u_n^* \leq v_n$ in Ω^*. Since v_n converges to v, it follows that u_n also converges to a solution of (1.6). □

Now we assume the following basic hypotheses on the behavior of f at $+\infty$:

$(f2)$ $$\lim_{s \to +\infty} \frac{f(s)}{s^\sigma} = 0, \quad \text{where } \sigma = \frac{N+2}{N-2}, \, N \geq 3.$$

From now on (in this section) we restrict ourselves to the case $N \geq 3$. The case $N = 2$ can be treated likewise, and in view of the Sobolev imbedding theorems in R^2 the nonlinearity f can be allowed to grow even exponentially (see [5]).

Suppose in addition to $(f2)$ that Ω is convex and the technical condition

$(f3)$ $$\limsup_{s \to +\infty} \frac{s(f) - \theta F(s)}{s^2 f(s)^{2/N}} \leq 0, \quad 0 \leq \theta < \frac{2N}{N-2},$$

holds. It has been proved in [5] that the solutions of a problem (0.1) in class III are a priori bounded in L^∞, and consequently in C^1. As established in [5] the convexity hypotheses in Ω can be relaxed at the expense of some additional assumptions on f. The following result holds.

PROPOSITION 1.3. *Assume $(f2)$ and $(f3)$, and that Ω is convex. Suppose that a problem (0.1) in class* III *possesses a strict supersolution $\bar{u}$. Then it has at least two solutions u_1, u_2 and $u_1 \leq u_2$.*

PROOF. Let us prove the result in the simpler case when $f(0) > 0$. The general case is proved in [10]. Extend the function f for $s < 0$ as $f(s) = f(0)$, and consider the problem

(1.7) $$-\Delta u = f(u) \quad \text{in } \Omega, \qquad u = 0 \quad \text{on } \partial\Omega.$$

The solutions of (1.7) can be proved to be all positive and in fact they are precisely the solutions of (0.1). The a priori estimates mentioned above hold for all solutions of the following parametrized family of problems.

(1.8) $$-\Delta v = tf(v) + (1-t)(\lambda v^+ + 1) \quad \text{in } \Omega, \qquad v = 0 \quad \text{on } \partial\Omega,$$

where $0 \leq t \leq 1$ and λ is some fixed number greater than $\lambda_1(\Omega)$ (v^+ denotes the positive part of v). Hence there exists a number $R > 0$ independent of t such that (1.8) has no solution in $\|u\|_{C^1} \geq R$. In order to apply topological degree theory we

take our problem to a functional-theoretical setting. Let X be the Banach space of C^1 functions in Ω which vanish on $\partial\Omega$. The equations
$$-\Delta u = tf(v) + (1-t)(\lambda v^+ + 1) \quad \text{in } \Omega, \quad u = 0 \quad \text{on } \partial\Omega$$
define mappings $T_t: X \to X$ by $T_t v = u$, $0 \leq t \leq 1$. From the elliptic theory these mappings are compact. The question of solvability of (0.1) is equivalent to the problem of obtaining a solution $u \in X$ of $u - T_1 u = 0$. From the a priori bounds it follows that

(1.9) $$\deg(I - T_t, U_R, 0) = \text{const},$$

where $U_R = \{u \in X : \|u\|_X < R\}$. Since
$$-\Delta v = \lambda v^+ + 1 \quad \text{in } \Omega, \quad v = 0 \quad \text{on } \partial\Omega$$
has no solution, it follows that the degree in (1.9) is zero. On the other hand since the problem possesses a strict subsolution $\underline{u}$ and a strict supersolution $\bar{u}$, with $\underline{u} \leq \bar{u}$, we can construct an open set $\mathcal{O} \subset U_R$ such that $\deg(I - T_1, \mathcal{O}, 0) = 1$. So $\deg(I - T_1, U_R \setminus \bar{\mathcal{O}}, 0) = -1$, and we obtain two solutions $v_1 \in \mathcal{O}$, $v_2 \in U_R \setminus \bar{\mathcal{O}}$. On the basis of these two solutions one can indeed show that existence of two ordered solutions. □

Suppose now that the condition

($f4$) $$\liminf_{s \to +\infty} \frac{sf(s) - \theta F(s)}{sf(s)^{2/(N+1)}} \geq 0, \quad \theta > 2,$$

holds. It has been proved in [**11**] that under hypotheses ($f2$) and ($f4$) the Euler–Lagrange functional associated with (0.1) satisfies the Palais–Smale condition. Then the following result [**10**] holds true.

PROPOSITION 1.4. *Assume ($f2$) and ($f4$). Suppose that a problem (0.1) in class III possesses a strict supersolution $\bar{u}$. Then it has at least two solutions.*

PROOF. The existence of a strict subsolution $\underline{u}$ and a strict supersolution $\bar{u}$, with $\underline{u} \leq \bar{u}$, implies that the Euler–Lagrange functional (0.8) has a local minimum u_1, which is in fact in the interval $[\underline{u}, \bar{u}]$. Clearly u_1 is a solution of (0.1). Next define the functional
$$\Psi(v) = \frac{1}{2}\int |\nabla v|^2 - \int G(x, v),$$
where $G(x, s) = F(s^+ + u_1(x)) - f(u_1(x))s^+ - F(u_1(x))$. It is seen that 0 is a local minimum of Ψ and that there exists $v_0 \in H_0^1$ such that $\Psi(v_0) < 0$. The hypotheses on the growth of f imply that Ψ satisfies the Palais–Smale condition. Apply then the strong form of the Mountain Pass Theorem (which holds for the case of nonstrict minimum) to obtain another critical point $v_1 \neq 0$ of Ψ. Then $u_2 = v_1 + u_1$ is the other solution of (0.1). □

The next result is also proved in [**10**]. The proof relies on certain estimates on the gradient of u on $\partial\Omega$ which allow one to perform appropriate truncations on f. In this way one can obtain existence of pairs of solutions for (0.1) without either one of the technical conditions ($f3$) or ($f4$).

PROPOSITION 1.5. *Assume* ($f2$) *and that* Ω *is convex. Suppose that problem* (0.1) *in class* III *possesses a strict supersolution* $\bar{u}$. *Then it has at least two solutions.*

2. Problems (0.1) in class IV. We start this section by proving that, as a consequence of the behavior of f at $+\infty$, the solutions of a given problem (0.1) in class IV are a priori bounded in L^∞. This will permit a truncation of the nonlinearity f and the subsequent treatment of the problem by variational methods.

PROPOSITION 2.1. *Let f be a function satisfying* ($f1$) *and in the class* IV. *Then all* (*eventual*) *solutions of* (0.1) *are a priori bounded in L^∞.*

PROOF. (i) First we prove an a priori bound in H_0^1. As a consequence of the behavior of f at $+\infty$, there are constants $0 < \mu < \lambda_1$ and $C_\mu > 0$ such that

$$(2.1) \qquad f(s) \leq \mu s + C_\mu \quad \text{for all } s \geq 0.$$

Multiplying the equation in (0.1) by u and integrating by parts we get

$$(2.2) \qquad \lambda_1 \int u^2 \leq \int |\nabla u|^2 = \int f(u)u \leq \mu \int u^2 + C_\mu \int u.$$

We have used (2.1) to obtain the second inequality in (2.1). It follows from (2.2) that the solutions of (0.1) are a priori bounded in L^2, and then in H_0^1. Moreover this bound depends only on μ and C_μ.

(ii) Now we write $f(s) = f^+(s) - f^-(s)$ and then the equation in (0.1) can be written as

$$(2.3) \qquad -\Delta u + \frac{f^-(u)}{u} u = f^+(u).$$

(Assuming, as we shall do, that $\liminf_{s\to 0}(f(s)/s) > -\infty$, we see that, for each given solution u of (0.1), the coefficient of u in (2.3) is continuous and bounded in Ω. Maybe this assumption would not be necessary in the arguments we present in the sequel.) For each solution u of (0.1) let us denote the coefficient of u in (2.3) by c_u. Clearly $c_u \geq 0$ in Ω. Also for each given solution u of (0.1) let us denote by ω_u the solution of the problem

$$(2.4) \qquad -\Delta \omega_u = f^+(u) \quad \text{in } \Omega, \qquad \omega_u = 0 \quad \text{on } \partial\Omega.$$

Clearly ω_u is a nonnegative $C^{2,\alpha}$ function. From (2.3) and (2.4) follows that

$$(2.5) \qquad -\Delta(u - \omega_u) + c_u(u - \omega_n) = -c_u \omega_u \leq 0$$

and the maximum principle gives $u \leq \omega_u$. Once we have this last inequality we may proceed by a bootstrap argument in (2.4) to get the required L^∞ bound on the solutions of (0.1). Indeed, by part (i) of this proof and Sobolev imbedding, the solutions of (0.1) are a priori bounded in L^p for $p = 2N/(N-2)$. From (2.4) the corresponding ω_u are a priori bounded in $W^{2,p}$ and consequently in an $L^{p'}$ with $p' > p$. The inequality $u \leq \omega_u$ gives the same for u, and we proceed... □

In view of Proposition 2.1, given a problem (0.1) in class IV we may assume that f is bounded. This is achieved by a truncation of f preserving condition (2.1).

The solutions of the truncated problem are precisely the same as the original problem. For this modified (but equivalent) problem the Euler–Lagrange functional Φ defined in (0.8) is well defined in H_0^1, bounded from below, and coercive. Consequently it possesses a global minimum u_1. By the hypothesis on $f(s)$ at $s = 0, u = 0$ is a local minimum of Φ. Thus a priori u_1 could be 0. This will indeed be the case if, for instance, the graph of $f(s)$ does not cross the line $\lambda_1 s$ for some $s > 0$. A necessary and sufficient condition for $u_1 \neq 0$ is the existence of a point $v_0 \neq 0$ where $\Phi(v_0) \leq 0$. So the following result holds.

PROPOSITION 2.2. *Let f be a function satisfying ($f1$) and in class IV. Suppose that there exists $0 \neq v_0 \in H_0^1$ such that $\Phi(v_0) \leq 0$. Then problem (0.1) has two solutions.*

It suffices, in addition to the above remarks, to observe that the condition $\Phi(v_0) \leq 0$ for some $v_0 \neq 0$ implies that there exists another minimum of Φ, $u_1 \neq 0$. Then use the Mountain Pass Theorem (see Corollary 2.16 of [13] or previous results in [14]).

Next we present a sufficient condition in terms of the primitive F of f which insures that there is a $v_0 \neq 0$ such that $\Phi(v_0) < 0$.

PROPOSITION 2.3. *Suppose that there is an $s_0 > 0$ such that*

$$(2.6) \quad F(s_0) \geq \frac{1}{2} s_0^2 \frac{1}{R^2} \min\left\{ \frac{(1+t)^2}{t^2} \frac{(1+t)^N - 1}{2 - (1+t)^N} : 0 < t < 2^{1/N} - 1 \right\},$$

where R is the radius of the largest ball contained in Ω. Then the functional Φ is negative for some $u \in H_0^1$.

We may assume that the ball centered at 0 with radius R is contained in Ω. Defining, for each $t \in [0, 1]$,

$$u_t(x) = \begin{cases} s_0 & \text{if } \|x\| < \dfrac{R}{1+t}, \\ s_0\left[1 - \dfrac{1+t}{tR}\left(\|x\| - \dfrac{R}{1+t}\right)\right] & \text{if } \dfrac{R}{1+t} < \|x\| \leq R, \\ 0 & \text{if } x \in \Omega \setminus B_R(0), \end{cases}$$

the result of Proposition 2.3 follows by a simple calculation.

REMARK. It should be remarked that a problem (0.1) in class IV may have two solutions even if $\Phi(u) > 0$ for all $u \neq 0$. This would mean that the global minimum of Φ is 0. So the argument above would not work. For these cases some other arguments have to be produced! See Remark 1.12 in [15].

REFERENCES

1. H. Amann, *Fixed point equations and nonlinear eigenvalue problems in ordered Banach spaces*, SIAM Rev. **18** (1976), 620–709.

2. D. G. de Figueiredo, *Positive solutions of semilinear elliptic equations*, Lecture Notes in Math., vol. 947, Springer-Verlag, 1982, pp. 34–87.

3. S. I. Pohožaev, *Eigenfunctions of $\Delta u + \lambda f(u) = 0$*, Soviet Math. Dokl. **6** (1965), 1408–1411.

4. A. Ambrosetti and P. R. Rabinowitz, *Dual variational methods in critical points theory and applications*, J. Funct. Anal. **14** (1973), 349–381.

5. D. G. de Figueiredo, P.-L. Lions and R. D. Nussbaum, *A priori estimates and existence results for positive solutions of semilinear elliptic equations*, J. Math. Pures Appl. **61** (1982), 41–63.

6. G. Gidas, W.-M. Ni and L. Nirenberg, *Symmetry and related properties via the maximum principle*, Comm. Math. Phys. **68** (1979), 209–243.

7. P.-L. Lions, *Quelques remarques sur la symmetrization de Schwarz*, Nonlinear Partial Differential Equations and Their Applications, College de France Seminar, Vol. 1, Pitman, London, 1981, pp. 308–319.

8. G. Talenti, *Elliptic equations and rearrangements*, Ann. Scuola Norm. Sup. Pisa (4) **3** (1976), 697–718.

9. C. Bandle, *Isoperimetric inequalities and applications*, Pitman, London, 1980.

10. D. G. de Figueiredo and P.-L. Lions, *On pairs of positive solutions for a class of semilinear elliptic problems*, MRC Tech. Summary Rep. 2660, 1984.

11. D. G. de Figueiredo, *On the superlinear Ambrosetti–Prodi problem*, Nonlinear Anal. **8** (1984), 655–665.

12. H. Brézis and L. Nirenberg, *Positive solutions of nonlinear elliptic equations involving critical Sobolev exponents*, Comm. Pure Appl. Math.

13. P. R. Rabinowitz, *Some aspects of critical point theory*, MRC Tech. Summary Rep. 2645, Jan. 1983.

14. _____, *Variational methods for nonlinear eigenvalue problems*, Eigenvalues of Nonlinear Problems (G. Prodi. ed.), C.I.M.E., Edizioni Cremonese, Rome, 1974, pp. 141–195.

15. P.-L. Lions, *On the existence of positive solutions of semilinear elliptic equations*, SIAM Rev. **24** (1982), 441–457.

16. H. Berestycki and P.-L. Lions, *Some applications of the method of super and subsolutions*, Bifurcation and Nonlinear Eigenvalue Problems, Lecture Notes in Math., vol. 782, Springer-Verlag, 1982, pp. 16–41.

UNIVERSIDADE DE BRASÍLIA, BRAZIL

Elliptic and Parabolic Quasilinear Equations Giving Rise to a Free Boundary: The Boundary of the Support of the Solutions

JESUS ILDEFONSO DIAZ

Abstract. In this survey we review some results and methods concerning the study of the support of the solutions of elliptic and parabolic quasilinear equations. In many of these equations and in contrast with the linear case, the support of the solutions does not coincide with the whole domain and thus a free boundary is generated by the boundary of that support.

1. Introduction. Let us consider the nonlinear Dirichlet problem

(1) $$-\text{div}\left(|\nabla u|^{p-2}\nabla u\right) + |u|^{q-1}u = f \quad \text{on } \Omega,$$

(2) $$u = g \quad \text{on } \partial\Omega,$$

where Ω is an open set of $\mathbf{R}^N$, $N \geq 1$, $p > 1$, $q > 0$, and f, g are given functions. Equation (1) appears in many different contexts: When $p = 2$ equation (1) coincides with the semilinear equation

(3) $$-\Delta u + |u|^{q-1}u = f,$$

largely studied in the literature. For instance, it is well known that (3) arises in the study of a single, irreversible, isothermic reaction (see [3]). The parameter q is called the order of the reaction and its range of values determines the behaviour of the solutions. When $p \neq 1$, (1) is a quasilinear equation which becomes degenerated for $p > 2$. In that case the equation is not uniformly elliptic, losing its elliptic character on the set $\{x \in \Omega: \nabla u(x) = 0\}$. This type of equation appears, for instance, in the study of non-Newtonian fluids with a rheological

---

1980 *Mathematics Subject Classification.* Primary 35J60, 35K60, 35R35.

Key words and phrases. Quasilinear equations, free boundary, support of the solutions, obstacle problem, comparison principle, energy methods, porous media equation.

© 1986 American Mathematical Society
0082-0717/86 $1.00 + $.25 per page

power law. When $p > 2$ the fluids are called dilatants and for $1 < p < 2$ pseudoplastics. The case $p = 2$ corresponds to Newtonian fluids (see exact references in [37]).

Existence, uniqueness and regularity results for the Dirichlet problem (1), (2) are already well known after the important works of Ladyzhenskaya–Ural'tseva, Stampacchia, Serrin and many others (see e.g. the survey [71]). Here we are interested in putting out some qualitative properties satisfied by the solutions of such problems, which exhibit very different behaviour according to the values of p and q. More concretely, we shall fix our attention on the behaviour of the support of the solution.

A well-known fact is that when (1) is linear (i.e. $p = 2$, $q = 1$) the solution u of (1) corresponding to data, say $f \geq 0$ and $q \geq 0$, is such that $u > 0$ on Ω. This is a trivial consequence of the strong maximum principle and can also be obtained by many other arguments, e.g. the Harnack inequality.

When (1) is nonlinear, entirely different behaviour may appear. Roughly speaking, the effective power of the diffusion term $\text{div}(|\nabla u|^{p-2}\nabla u)$ and of the absorption term $|u|^{q-1}u$ vary with p and q, generating new phenomena. Thus, letting Ω be an unbounded open set and f and g with compact support, the support of the solution contains the whole domain Ω if $q \geq p - 1$, but otherwise (i.e. when $q < p - 1$) the solution u has compact support and so $u = 0$ on an unbounded region of Ω. This was first shown in [13] for equation (3) and more generally in [36, 37] for (1). The main idea in order to prove the compactness of the support, assuming $q < p - 1$, lies in the construction of adequate super and subsolutions $\bar{u}$ and $\underline{u}$ of the problem (1), (2). Such functions can be chosen with compact support and so, by a comparison argument, $\underline{u} \leq u \leq \bar{u}$ on Ω, which implies that supp u is also a compact subset.

This kind of vanishing property has, in fact, a local character and this also happens even for bounded domains Ω, in the sense that the set

$$N(u) = \{x \in \bar{\Omega} : u(x) = 0\} \qquad (N(u) = \bar{\Omega} - \text{supp } u)$$

may have a positive measure. The first result in that direction seems to be the author's memoir [32], in which a general local method is proposed, and later developed in [33], [34] and [35] (see also [72, 10, 47] and [1]).

THEOREM 1. *Assume $q < p - 1$ and let u be the solution of* (1), (2). *Then*

(4)
$$N(u) \supset \{x \in N(f) \cup N(g|_{\partial\Omega})$$
$$\text{such that } d(x, \bar{\Omega} - |N(f) \cup N(g|_{\partial\Omega})|) \geq (M/C)^{(p-1-q)/p}\}$$

with C some positive constant (explicitly known) only depending on N, p, and q, and $M = \|u\|_{L^\infty}$.

Here $N(f)$ (resp. $N(g|_{\partial\Omega})$) represents the set $\{x \in \Omega : f(x) = 0\}$ (resp. $\{x \in \partial\Omega : g(x) = 0\}$). Let us remark that conclusion (4) is not empty if Ω is unbounded and f and g have compact support (in fact the constant M in (4) can be substituted for some bound of the ess. supremum of u in the interior of the set $N(f)$). Otherwise, e.g. if Ω is bounded, Theorem 1 has implicitly the following assumption:

(5) $\quad \operatorname{meas}\{x \in N(f) \cup N(g|_{\partial\Omega}):$
$$d(x, \overline{\Omega} - N(f) \cup N(g|_{\partial\Omega})) \geq (M/C)^{(p-1-q)/p}\} > 0.$$

The main idea in proving Theorem 1 is to use the functions
$$v_{\pm}(x) = \pm C|x - x_0|^{p/(p-1-q)}$$
as local super and subsolutions of (1) when x_0 is adequately chosen. Then, by comparison arguments, one obtains $v_{-}(x) \leq u(x) \leq v_{+}(x)$ on some neighbourhood of x_0 and so $u(x_0) = 0$. (For details, see the paper of J. Hernandez in these Proceedings.) By a (not difficult) modification of that argument it is possible to show that, if for instance, $g = 0$, then under the assumptions of Theorem 1 a stronger conclusion holds:

(6) $\quad N(f) = N(u)$, ie. $\{x \in \Omega: f(x) = 0\} = \{x \in \Omega: u(x) = 0\}$,

if $f(x)$ decays to zero on $\operatorname{supp} f$ as $d(x, N(f))^{pq \wedge (p-1-q)}$ in a tubular neighborhood of $\partial N(f)$ (see [33]).

Results such as Theorem 1 can be obtained for nonlinear equations under formulations more general than (1). That is the case of symmetric invariant equations such as
$$-\operatorname{div}\left(\frac{a(|\nabla u|)}{|\nabla u|^2} \nabla u\right) + c(u) = f,$$
where a and c are assumed to be nondecreasing and satisfying some balance condition (now given by the boundedness of an improper integral) substituting the assumption $q < p - 1$. Also nonisotropic equations
$$-\sum_{i=1}^{N} \frac{\partial}{\partial x_i} a_i\left(\frac{\partial u}{\partial x_i}\right) + c(u) = f$$
and fully nonlinear elliptic equations can be considered [37, 28, 33]. To finish this section, we remark that also for Variational Inequalities, e.g. the obstacle problem ($u \geq \psi$ on Ω), the method of local super and subsolutions can be applied in order to estimate the coincidence set
$$\{x \in \Omega: u(x) = \psi(x)\} \quad (\equiv N(u - \psi))$$
(see [24, 6, 77, 7, 27], and [33]). A systematic development of these results, including many others, is the subject of the book [33].

2. An energy method. An important limitation of the scope of the method commented on in §1 is its constructive character. Recently, a new method has been introduced in Antoncev [2] and developed by the author and L. Véron in [41, 42], in order to study the behaviour of the support of the solution of the general class of second order quasilinear elliptic equations

(7) $\quad -\operatorname{div} A(x, u, \nabla u) + B(x, u, \nabla u) + C(x, u) = f.$

It also has a local character and it is based on certain estimates of some energy terms associated to the solutions of (7). The structural assumptions for the treatment of (7) are the following:

(8) $\quad |A(x, r, \xi)| \leq C_1 |\xi|^{p-1} \quad$ for some $p > 1$ and $C_1 > 0$,

(9) $\quad A(x, r, \xi) \cdot \xi \geq C_2 |\xi|^p \quad$ for some $C_2 > 0$,

(10) $\quad |B(x, r, \xi)| \leq C_3 |r|^\alpha |\xi|^\beta \quad$ for some $C_3 \geq 0$, $\alpha \geq 0$, and $\beta \geq 0$,

(11) $\quad C(x, r)r \geq C_4 |r|^{q+1} \quad$ for some $q > 0$ and $C_4 > 0$.

Functions A, B and C are assumed to be Caratheodory functions on its arguments, $x \in \Omega \subset \mathbf{R}^N$, $r \in \mathbf{R}$, and $\xi \in \mathbf{R}^N$. As in the method of local super and subsolutions, it is enough to work on the subset $N(f)$ where $f = 0$.

DEFINITION. Given an open set $G \subset \mathbf{R}^N$, a function $u \in L^1_{\text{loc}}(G)$ is called a local weak solution of

(12) $\quad -\text{div}\, A(x, u, \nabla u) + B(x, u, \nabla u) + C(x, u) = 0$

on G if (i) $\nabla u \in L^p_{\text{loc}}(G)$, $B(x, u, \nabla u) \in L^1_{\text{loc}}(G)$, and $C(x, u) \in L^1_{\text{loc}}(G)$, and (ii) for any $\phi \in C^\infty_0(G)$ we have

$$\int_G \{ A(x, u, \nabla u) \cdot \nabla \phi + (B(x, u, \nabla u) + C(x, u))\phi \} \, dx = 0.$$

Given $x_0 \in G$ such that $B\rho_0(x_0) \subset G$ and u is a local weak solution u of (12) on G, we introduce the diffusion energy on $B\rho(x_0)$, $0 < \rho < \rho_0$, by

(13) $\quad E(\rho) = \displaystyle\int_{B_\rho(x_0)} A(x, u, \nabla u) \cdot \nabla u \, dx$

as well as the absorption energy on $B\rho(x_0)$ by

(14) $\quad b(\rho) = \displaystyle\int_{B_\rho(x_0)} |u|^{q+1} \, dx.$

The main conclusion of the energy method is

THEOREM 2. *Assume* $q < p - 1$, $\beta \leq p$, $\alpha = q - \beta(q+1)/p$, *and* C_3 *small enough. Then for every* $x_0 \in G$ *such that* $B\rho_0(x_0) \subset G$ *and for every local weak solution* u *of* (12) *on* G, *there exists a positive constant*

$$C^* = C^*(N, p, qE(\rho_0), b(\rho_0))$$

such that $u(x) = 0$ *a.e. on* $B\rho_1(x_0)$ *where*

(15) $\quad \rho_1 = \rho_0 - C^*.$

Before referring to the proof of Theorem 2, we shall make some remarks about its applications. First of all we point out that no monotonicity assumptions are made on the dependence of $A(x, r, \xi)$ and $C(x, r)$ on ξ and r respectively. Hence Theorem 2 can be applied even in the absence of comparison principles, in contrast with the method in §1. In order to obtain some global consequences, consider, for instance, equation (7) on an open set $\Omega \subset \mathbf{R}^N$ with Dirichlet condition $u = 0$ on $\partial\Omega$. Assume the hypotheses of Theorem 2 and let $f \in L^{(q+1)/q}(\Omega)$. Then is not difficult to show that if $u \in W_0^{1,p}(\Omega) \cap L^{q+1}(\Omega)$ is any weak solution of (7), one has

$$(16) \qquad \int_\Omega A(x, u, \nabla u) \cdot \nabla u \, dx + \int_\Omega |u|^{q+1} \, dx \leq C \int_\Omega |f|^{(q+1)/q} \, dx$$

for some structural constant C. Therefore we can apply Theorem 2 on the set $G = N(f)$ and then, if $x_0 \in N(f)$ and $\rho_0 = d(x_0, \overline{\Omega} - N(f))$, by (16) we have $E(\rho_0) + b(\rho_0) \leq C\|f\|_{(q+1)/q}^{(q+1)/q}$, which allows to us to conclude the existence of a constant $C^{**} = C^{**}(N, p, q, \|f\|_{(q+1)/q})$ such that

$$(17) \qquad N(u) \supset \{x \in N(f) : d(x, \overline{\Omega} - N(f)) \geq C^{**}\}.$$

As in Theorem 1, conclusion (17) is not empty if Ω is unbounded and f has compact support. Otherwise, we need the assumption

$$(18) \qquad \operatorname{meas}\{x \in N(f) : d(x, \overline{\Omega} - N(f)) \geq C^{**}\} > 0.$$

Hypothesis (18) has the same nature as (5) but with the important difference that no bound on $\|u\|_{L^\infty}$ is now needed. It is also interesting to remark that, when both methods may be applied, sharper estimates are obtained by using the method of local super and subsolutions. The main reason for that is the fact that comparison functions $v_\pm$ used in the proof of Theorem 1 are, in fact, exact solutions of the homogeneous equation associated to (1) and so the estimate (4) cannot be improved in some particular cases.

We now return to the proof of Theorem 2. The main ingredients in the proof are the following technical lemmas.

LEMMA 1. *Under the hypotheses of Theorem* 2, $A(\cdot, u, \nabla u) \cdot \nabla u$, $|u|^{q+1}$, $|A(\cdot, u, \nabla u)|u$, *and* $B(\cdot, u, \nabla u)u$ *belong to* $L^1(B\rho_0(x_0))$ *and, for almost every* $\rho \in (0, \rho_0)$, *we have*

$$(19) \qquad \int_{B_\rho} A(x, u, \nabla u) \cdot \nabla u \, dx + C_4 \int_{B\rho} |u|^{q+1} \, dx + \int_{B\rho} B(x, u, \nabla u) u \, dx$$
$$\leq \int_{S\rho} A(x, u, \nabla u) \cdot \vec{v} u \, ds,$$

where $B\rho = B\rho(x_0)$, $S\rho = \partial B\rho$, *and* $\vec{v} = \vec{v}(x)$ *is the outward normal vector at* $x \in S\rho$.

LEMMA 2. *Let D be a bounded open set of* $\mathbf{R}^N$, $N \geq 1$, *with a C^1 boundary ∂D. Assume $0 \leq q \leq p - 1 < \infty$. Then there exists a constant $C = C(p, q, D)$ such that for any $v \in W^{1,p}(D)$ we have*

$$\|v\|_{L^p(\partial D)} \leq C\big(\|\nabla v\|_{L^p(D)} + \|v\|_{L^{q+1}(D)}\big)^\theta \|v\|_{L^{q+1}(D)}^{1-\theta},$$

where

$$\theta = \frac{N(p - 1 - q) + q + 1}{N(p - 1 - q) + (q + 1)p}.$$

If in particular $D = B\rho(x_0)$, then the inequality

(20) $$\|v\|_{L^p(S\rho)} \leq C\left(\|\nabla v\|_{L^p(B\rho)} + \rho^{-\delta}\|v\|_{L^{q+1}(B\rho)}\right)^\theta \|v\|_{L^{q+1}(B\rho)}^{1-\theta}$$

holds, where

$$\theta = \frac{N(p-1-q) + (q+1)p}{p(q+1)} \quad \text{and} \quad C = C(N, p, q).$$

Lemma 1 is proved by taking, in the definition of local weak solution, the test functions $\phi_{n,m}(x) = \psi_n(|x - x_0|)T_m(u(x))$ and passing to the limit in n and m. Here $\psi_n : [0, \rho_0] \to \mathbf{R}^+$ is such that $\psi_n(r) = 1$ if $r \in [0, \rho - 1/n]$, $\psi_n(r) = 0$ if $r \in [\rho, \rho_0]$, and $\psi_n(r) = -n(\rho - r)$ if $r \in [\rho - 1/n, \rho]$. T_m is a truncation function, such as, e.g., $T_m(r) = \text{sign}(r)\min(m, |r|)$. Lemma 2 is an interpolation-trace result and is the key-stone of the proof of Theorem 2. It can be proved by using the Gagliardo–Niremberg interpolation inequalities and some trace results (see the details in [42]).

PROOF OF THEOREM 2. *First step.* If u is a local weak solution of (12) then

(21) $$E(\rho) + C_4 b(\rho) + \int_{B\rho} B(x, u, \nabla u) u\, dx \geq C_5(E(\rho) + b(\rho))$$

for some constant $C_5 > 0$. Indeed, by using Young's inequality, for any $\varepsilon > 0$ and $\tau > 1$ we have

$$C_3|u|^{\alpha+1}|\nabla u|^\beta \leq \frac{\varepsilon C_3}{\tau}|u|^{\tau(\alpha+1)} + \frac{(\tau-1)}{\tau}C_3\varepsilon^{-1/(\tau-1)}|\nabla u|^{\beta\tau/(\tau-1)}.$$

If we choose $\tau = (q+1)/(\alpha+1)$, then $\beta\tau/(\tau-1) = p$. So, by (14),

$$\left|\int_{B\tau} B(x, u, \nabla u) u\, dx\right| \leq \varepsilon C_3\left(\frac{p-\beta}{p}\right)b(\rho) + \frac{\beta C_3}{C_2 p}\varepsilon^{-(p-\beta)/\beta}E(\rho).$$

Hence, if

$$C_3 < C_4\left(\frac{p}{p-\beta}\right)^{(p-\beta)/p}\left(C_2 \frac{p}{\beta}\right)^{\beta/p},$$

then it is possible to find an $\varepsilon > 0$ such that

$$\varepsilon C_3\left(\frac{p-\beta}{p}\right) < C_4 \quad \text{and} \quad \frac{\beta C_1}{C_2 p}\varepsilon^{-(p-\beta)/\beta} < 1$$

and (21) holds.

END OF THE PROOF. By Lemma 1 and (21) we have

(22) $$C_5(E(\rho) + b(\rho)) \leq \int_{S\rho} A(x, u, \nabla u) \cdot \vec{\nu} u\, ds.$$

By (8) and the Holder inequality,

$$\int_{S\rho} A(x, u, \nabla u) \cdot \vec{\nu} u\, ds \leq C_1\left(\int_{S\rho} |\nabla u|^{p-1}|u|\, ds\right)$$

$$\leq C_1\left(\int_{S\rho} |\nabla u|^p\right)^{(p-1)/p}\left(\int_{S\rho} |u|^p\right)^{1/p}.$$

On the other hand, by using spherical coordinates (ω, r) with center x_0 we have

$$E(\rho) = \int_0^\rho \int_{S^{N-1}} A(r\omega, u, \nabla u) \cdot \nabla u \, r^{N-1} \, d\omega \, dr.$$

Hence E is differentiable almost everywhere and

$$\frac{dE}{d\rho}(\rho) = \int_{S_\rho} A(r\omega, u, \nabla u) \cdot \nabla u \, d\omega$$

which, by (9), implies

$$(23) \qquad \frac{dE}{d\rho}(\rho) \geq C_2 \int_{S_\rho} |\nabla u|^p \, ds.$$

Then by (22), (23) and Lemma 2 we have

$$E(\rho) + b(\rho) \leq K \left(\frac{dE}{d\rho}\right)^{(p-1)/p} \left(E(\rho)^{Q/p} b(\rho)^{(1-Q)/(q+1)} + \rho^{-\delta Q} b(\rho)^{1/(q+1)} \right)$$

for some constant K. Then, by Young's inequality

$$(24) \qquad E(\rho) + b(\rho) \leq K_1 \rho^{-\delta Q} \left(\frac{dE}{d\rho}\right)^{(p-1)/p} (E(\rho) + b(\rho))^\omega,$$

where

$$K_1 = 2K \max\left(1, b(\rho_0)^{\theta(1/(q+1) - 1/p)}\right) \max\left(\rho_0^{\delta\theta}, 1\right),$$
$$\omega = \theta/p + (1-\theta)/(q+1).$$

Hence E satisfies the differential inequality

$$(25) \qquad K_2 \rho^{-\rho\theta p/(p-1)} \frac{dE}{d\rho}(\rho) \geq E(\rho)^{(1-\omega)p/(p-1)},$$

where $K_2 = K_1^{p/(p-1)}$. But $0 < (1-\omega)p/(p-1) < 1$ and integrating in (25) we conclude that $E(\rho_1) = 0$ if

$$\rho_1^{1+\delta\theta p/(p-1)} = \rho_0^{1+\delta\theta p/(p-1)} - \frac{K_2 E(\rho_0)^{1-(1-\omega)p/(p-1)}}{1 - (1-\omega)p/(p-1)}.$$

Then, by (24), $b(\rho_1) = 0$ and this implies $u(x) = 0$ a.e. on $B\rho_1(x_0)$.

REMARKS. The constant C^* appearing in the statement of Theorem 2 can be explicitly estimated in terms of N, p, q, $E(\rho_0)$, and $b(\rho_0)$ (see [42]). We also remark that, by a careful revision of the technical lemmas and the proof of Theorem 2, this remains true in the case $q = 0$ (which arises in Variational Inequalities), and even for $-1 < q < 0$ when we consider the global Dirichlet problem (see also [23]). Finally, we send the reader to Bernis [16] and the communication of that author in this Congress for the consideration of higher order elliptic equations via another energy method. (The support of the solutions of a fourth order variational inequality is also studied in [69] and [15].)

3. Parabolic quasilinear equations. The former energy method also applies to parabolic quasilinear equations like

$$(26) \quad \frac{\partial}{\partial t}\beta(v) - \text{div}\,\mathscr{A}(t, x, v, \nabla v) + \mathscr{B}(t, x, v, \nabla v) + \mathscr{C}(t, x, v) = f(t, x),$$

where

$$(27) \quad \beta(r) = |r|^{1/m}\,\text{sign}\,r \quad \text{for some } m > 0,$$

$$(28) \quad |\mathscr{A}(t, x, r, \xi)| \leq C_1|\xi|^{p-1} \quad \text{for some } p > 1 \text{ and } C_1 > 0,$$

$$(29) \quad \mathscr{A}(t, x, r, \xi) \cdot \xi \geq C_2|\xi|^p \quad \text{for some } C_2 > 0,$$

$$(30) \quad |\mathscr{B}(t, x, r, \xi)| \leq C_3|r|^\alpha|\xi|^\beta \quad \text{for some } C_3 \geq 0,\, \alpha \geq 0,\, \text{and } \beta \geq 0,$$

$$(31) \quad \mathscr{C}(t, x, r)r \geq C_4|r|^{q+1} \quad \text{for some } q > 0 \text{ and } C_4 \geq 0.$$

Equation (26) contains as main particular cases the generic porous media equation [12], which, for simplicity, in one dimension reads

$$(32) \quad u_t - (u^m)_{xx} + b_0 \cdot (u^\lambda)_x + c_0 \cdot u^q = 0,$$

where we are assuming $u \geq 0$, $b_0 \in \mathbf{R}$, $c_0 \geq 0$, $m > 1$, $\lambda \geq 1$, and $q > 0$. Obviously (26) appears taking $v = u^m$ and $\mathscr{A}$, $\mathscr{B}$ and $\mathscr{C}$ adequately. Equation (26) also contains the equation

$$(33) \quad u_t - \text{div}\bigl(|\nabla u|^{p-2}\nabla u\bigr) = 0, \qquad p > 1,$$

which, again, appears in non-Newtonian fluids [37]. As for the elliptic case, the energy method can be applied locally to show the following local version property referred to as the *finite speed of propagation property*:

(P) Let v be a weak local solution of (26) with $f = 0$ on the set $(0, \infty) \times B_{\rho_0}(x_0)$. Assume that $v(0, x) = 0$ for $x \in B_{\rho_0}(x_0)$. Then for every $t > 0$ there exists $\rho(t)$, $0 \leq \rho(t) < \rho_0$, such that $v(t, x) = 0$ on $B_{\rho(t)}(x_0)$. The main answer given in [41], [9] (see also [2]) is that in order to have such a property it suffices to have

$$(34) \quad m(p - 1) > 1$$

and no other assumption on q (about the term $\mathscr{B}$ we assume $0 \leq \beta \leq p$, $\alpha = (p - \beta(m + 1))/mp$ and in fact the conclusion holds only on a finite interval of time $[0, T^*]$ if $\mathscr{B} \neq 0$ and $\beta \neq p$). Now the energies are defined by

$$E(t, \rho) = \int_0^t \int_{B_\rho(x_0)} \mathscr{A}(s, x, v, \nabla v) \cdot \nabla v\, dx\, ds$$

and

$$b(t, \rho) = \operatorname*{ess\,sup}_{0 \leq s \leq t} \int_{B_\rho(x_0)} |v(s, x)|^{(m+1)/m}\, dx.$$

(Here the local weak solutions are supposed to satisfy, in particular, $\nabla v \in L^p_{\text{loc}}((0, \infty) \times B_{\rho_0}(x_0))$ and $v \in L^\infty(0, \infty: L^{(m+1)/m}(B_{\rho_0}(x_0)))$.) As in the elliptic case, it is not difficult to find global consequences. Thus, for instance, if we consider the Dirichlet problem

(35) $\begin{cases} \dfrac{\partial \beta(v)}{\partial t} - \operatorname{div} \mathscr{A}(t, x, v, \nabla v) + \mathscr{C}(t, x, v) = 0 & \text{on } (0, \infty) \times \Omega, \\ v(t, x) = 0 & \text{on } (0, \infty) \times \partial\Omega, \\ v(0, x) = v_0(x) & \text{on } \Omega, \end{cases}$

where Ω is an open set of $\mathbf{R}^N$, $N \geq 1$, and $v_0 \in L^{(m+1)/m}(\Omega)$, assuming the structural hypotheses as well as $m(p-1) > 1$, then the quantity $\rho(t)$ can be estimated independently of $x_0 \in N(v_0)$ and we find that $\rho(t) \leq Ct^\mu$ for some $\mu = \mu(N, m, p) > 0$. As a consequence, we have

(36) $\qquad N(v(t,\cdot)) \supset \{x \in N(v_0) : d(x, \overline{\Omega} - N(v_0)) \geq Ct^\mu\}.$

In particular, if Ω is unbounded and v_0 has compact support, we find that for every $t > 0$ the support of $v(t, \cdot)$ is also compact. If in addition $N = 1$, then $\operatorname{supp} v(t, \cdot) = [\zeta_1(t), \zeta_2(t)]$ for some monotone real functions ζ_i.

Naturally, much more precise information is available for concrete formulations of (26). For instance, for the porous media equation

(37) $\qquad u_t - \Delta \phi(u) = 0$

with ϕ a continuous nondecreasing function such that $\phi(0) = \phi'(0) = 0$, it is known that the solution u of the Cauchy problem has compact support for each $t > 0$ (assuming $u(0, x)$ with compact support) if and only if

(38) $\qquad \displaystyle\int_0^1 \frac{ds}{\phi^{-1}(s)} < +\infty$

(see [67, 4, 68, 30, 74, 75]). In fact for the homogeneous Dirichlet problem, and even under (38), there always exists a finite time T^* such that $u(t, x) > 0$ on Ω for every $T \geq T^*$ [18]. The exponent μ in estimate (15) can also be optimally estimated if, for instance, $\phi(s) = |s|^m \cdot \operatorname{sign} s$ [5, 73]. For other results concerning the boundary of the support of the solution of (36), we refer to the recent survey of Berstch-Peletier [17]. For the particular equation (33) some references are [9, 36, 37, 7, 8] and [53, 56]. The former property of finite speed of propagation is essentially due to the assumption $m(p-1) > 1$, which expresses when the diffusion is "slow". Nevertheless, other different behaviours appear when the action of the absorption term $\mathscr{C}(t, x, v)$ or of the convection term $\mathscr{B}(t, x, v, \nabla v)$ is taken into account.

Thus, when the absorption is large with respect to the diffusion, the support of $v(t, \cdot)$ remains in a compact region for every $t \in [0, \infty)$. That property, usually referred to as *localization*, appears, for instance, for the equation

(39) $\qquad u_t - \Delta u^m + u^s = 0$

when $m > s$ (see [54, 61, 19, 20, 63]). This can be proven by global comparison functions when the domain is unbounded and more generally for the method of local super and subsolutions [35]. Also, the energy method allows us to find such a behaviour (even locally) for the solutions of the general equation (26) assuming $\max(q, 1/m) < p - 1$ [42].

A stronger property appears when the absorption is larger. For instance, if we assume $0 < s < 1$ in (39), then there exists a finite time $T_0 < +\infty$ such that the set $N(u(t, \cdot))$ has positive measure even for strictly positive initial data and nonhomogeneous Dirichlet boundary conditions [35, 11]. Moreover, if the Dirichlet conditions are homogeneous or we are concerned with the Cauchy problem associated to (39), then, in fact, $u(t, x) \equiv 0$ for every $t \geq T_0$ and a.e. in x [54, 76, 43].

With respect to the balance between diffusion and convection the situation is quite different. When the convection is large with respect to the diffusion, then there is a kind of localization property but only in some directions according to the equation. For instance, for the one-dimensional equation

$$(40) \qquad u_t - (u^m)_{xx} + b_0 \cdot (u^\lambda)_x = 0,$$

where $m > 1$, $\lambda > 0$, and $b_0 \in \mathbf{R} - \{0\}$, if u is the solution of the Cauchy problem associated to (40) and if $\operatorname{supp} u(0, x) = [a, b]$, then, for every $t > 0$, $\operatorname{supp} u(t, \cdot) = [\zeta_1(t), \zeta_2(t)]$ with $\zeta_1(0) = a$, $\zeta_2(0) = b$ and $\zeta_1(t) \geq a - \varepsilon$ (resp. $\zeta_2(t) \leq b - \varepsilon$) $\forall t \in [0, \infty)$, assuming $m > \lambda$ and $b_0 < 0$ (resp. $b_0 < 0$) (see [55, 49, 50]). When the convection is larger, i.e. when $0 < \lambda < 1$, then there is only one (localized) interface: $\operatorname{supp} u(t, \cdot) = [\zeta_1(t), +\infty)$ (resp. $\operatorname{supp} u(t, \cdot) = (-\infty, \zeta_2(t)])$ when $b_0 > 0$ (resp. $b_0 < 0$) (see [39]). We also remark that the presence of convection and absorption terms in a nonlinear diffusion does not produce any new behaviour [44, 58].

Also for "fast" diffusion equations the boundary of the support of the solution can be considered as a free boundary. For instance for the equation

$$(41) \qquad u_t - \Delta\phi(u) = 0$$

it is well known that if $\phi(s) = |s|^m \cdot \operatorname{sign} s$ with $0 < m < 1$ [22], or more generally

$$\int_0^1 \frac{ds}{\phi(s)} < +\infty$$

[70, 29], then for every initial datum there exists a finite T_0 such that the solution u of the homogenous Dirichlet problem associated to (41) is such that $u(t, \cdot) = 0$ $\forall t \geq T_0$. In fact it is also known that $u(t, x) \neq 0$ a.e. on x if $t \in (0, T_0)$, and so $\Omega \times \{T_0\}$ is a free boundary. This behaviour also appears for the equation (33) when $1 < p < 2$ (see [9, 52 and 2]).

As in the elliptic case, there is also a large literature about the support of the solution of some parabolic variational inequalities ([14, 25, 31, 43] and [76]).

Detailed results as well as other qualitative properties of the solutions of quasilinear parabolic equations will be available in [33].

Finally, we remark that there are also some references about the study of the boundary of the support for the solutions of first order quasilinear hyperbolic equations

$$u_t - \sum_{i=1}^{N} \phi_i(u)_{x_i} + \beta(u) = f$$

(see [65, 64, 26] and more recently [40]).

References[1]

1. H. W. Alt and D. Phillips, *A free boundary problem for semilinear elliptic equations* (to appear).

2. S. N. Antoncev, *On the localization of solutions of nonlinear degenerate elliptic and parabolic equations*, Soviet Math. Dokl. **24** (1981), 420–424.

3. R. Aris, *The mathematical theory of diffusion and reaction impermeable catalysts*, Clarendon Press, Oxford, 1975.

4. D. G. Aronson, *Regularity properties of flows through porous media*, SIAM Appl. Math. **17** (1969), 461–467.

5. _____, *Regularity properties of flows through porous media: the interface*, Arch. Rational Mech. Anal. **37** (1970), 1–10.

6. R. G. Aronson, *Some properties of the interface for gas flow in porous media*, Free Boundary Problems: Theory and Applications, Pitman, New York, 1983.

7. C. Atkinson and J. E. Bouillet, *Some qualitative properties of solutions of a generalised diffusion equation*, Math. Proc. Cambridge Philos. Soc. **86** (1979), 495–510.

8. _____, *A generalized diffusion equation: Radial symmetries and comparison theorems*, J. Math. Anal. Appl. **95** (1983), 37–68.

9. A. Bamberger, *Etude d'une équations doublement non linéaire*, J. Funct. Anal. **24** (1977), 148–155.

10. C. Bandle, R. P. Sperb and I. Stakgold, *Diffusion and reaction with monotone kinetics*, Nonlinear Anal. **8** (1984), 321–333.

11. C. Bandle and I. Stakgold, *The formation of the dead core in parabolic reaction-diffusion problems*, Trans. Amer. Math. Soc. **286** (1984), 275–293.

12. J. Bear, *Dynamics of fluids in porous media*, American Elsevier, New York, 1971.

13. Ph. Benilan, H. Brezis and M. G. Crandall, *A semilinear equation in $L^1(R^N)$*, Ann. Scuola Norm. Sup. Pisa **2** (1975), 523–555.

14. A. Bensoussan and J. L. Lions, *On the support of the solution of some variational inequalities of evolution*, J. Math. Soc. Japan **28** (1976), 1–17.

15. M. F. Bidaut-Veron, *Propriété de support compact de la solution d'une équation aux derivées partielles non lineaires d'ordre 4*, C. R. Acad. Sci. Paris **287** (1975), 1005–1008.

16. F. Bernis, *Compactness of the support for some nonlinear elliptic problems of arbitrary order in dimension N*, Comm. Partial Differential Equations **9** (1984), 271–312.

17. M. Bertsch and L. A. Peletier, *Porous media type equations: an overview* (to appear).

18. _____, *A positivity property of solutions of non-linear diffusion equations*, J. Differential Equations (to appear).

19. M. Bertsch, R. Kersner and L. A. Peletier, *Sur le comportement de la frontiere libre dans une equation en theorie de la filtration*, C. R. Acad. Sci. Paris **295** (1982), 63–66.

20. _____, *Positivity versus localization in degenerate diffusion equations* (to appear).

21. M. Bertsch, P. de Mottoni and L. A. Peletier, *Degenerate diffusion and the Stefan Problem* (to appear).

22. J. G. Berryman and C. J. Holland, *Stability of the separable solution for fast diffusion*, Arch. Rational Mech. Anal. **74** (1980), 279–288.

23. C. M. Brauner and B. Nikolaenko, *On nonlinear eigenvalue problems which extend into free boundaries problems*, Bifurcation and Nonlinear Eigenvalue Problems (C. Bardos, J. M. Lasry and M. Sachtzman, eds.), Lecture Notes in Math., vol. 782, Springer, 1980.

---

[1] A more complete account of references can be found in [33].

24. H. Brezis, *Solutions of variational inequalities with compact support*, Uspekhi Mat. Nauk **129** (1974), 103–108.

25. H. Brezis and A. Friedman, *Estimates in the support of solutions of parabolic variational inequalities*, Illinois J. Math. **20** (1976), 82–97.

26. E. D. Conway, *The formation and decay of shocks for a conservation law in several dimensions*, Arch. Rational Mech. Anal. **64** (1977), 47–59.

27. G. Diaz, *Estimation de l'ensemble de coincidence de la solution des problemes d'obstacle pour les equations de Hamilton-Jacobi-Bellman*, C. R. Acad. Sci. Paris **290** (1980), 587–591.

28. _____, *Some properties of the solutions of degenerate second order PDE in non-divergence form*, Applicable Anal. (to appear).

29. G. Diaz and J. I. Diaz, *Finite extinction time for a class of nonlinear parabolic equations*, Comm. Partial Differential Equations **4** (1979), 1213–1231.

30. J. I. Diaz, *Solutions with compact support for some degenerate parabolic problems*, Nonlinear Anal. **3** (1979), 831–847.

31. _____, *Anulación de soluciones para operadores acretivos en espacios de Banach*, Rev. Real Acad. Cienc. Exact. Fís. Natur. Madrid **74** (1980), 865–880.

32. _____, *Técnica de supersoluciones locales para problems estacionarios no lineales*, Memoria no. XVI, Real Academia de Ciencias, Madrid, 1982.

33. _____, *Nonlinear partial differential equations and free boundaries*. Vol. I: *Elliptic Equations*, Research Notes in Math., no. 106, Pitman, London, 1985; Vol. II: *Parabolic and Hyperbolic Equations*, in preparation.

34. J. I. Diaz and J. Hernandez, *On the existence of a free boundary for a class of reaction-diffusion systems*, SIAM J. Math. Anal. **5** (1984), 670–685.

35. _____, *Some results on the existence of free boundaries for parabolic reaction-diffusion systems*, Trends in Theory and Practice of Nonlinear Differential Equations (Proc. 5th Conf. on Trends in Theory and Practice of Nonlinear Differential Equations, Arlington, Texas, 1982) V. Lakshmikantham, ed. Lecture Notes Pure Appl. Math., no. 90, Marcel Dekker, New York, 1984, pp. 149–158.

36. J. I. Diaz and M. A. Herrero, *Proprietés de support compact pour certaines equations elliptiques et paraboliques non linearies*, C. R. Acad. Sci. Paris **286** (1978), 815–817.

37. _____, *Estimates on the support of the solutions of some nonlinear elliptic and parabolic problems*, Proc. Roy. Soc. Edinburgh Sect. A **89** (1981), 249–258.

38. J. I. Diaz and R. Jimenez, *Boundary behaviour of solutions of Signorini type problems* (to appear).

39. J. I. Diaz and R. Kersner, *Non existence d'une des frontieres libres dans une equation degénérée en théorie de la filtration*, C. R. Acad. Sci. Paris **296** (1983), 505–508.

40. J. I. Diaz and L. Veron, *Existence theory and qualitative properties of the solutions of some first order quasilinear variational inequalities*, Indiana Univ. Math. J. **32** (1983), 319–361.

41. _____, *Compacité du support des solutions d'equations quasilinearies elliptiques ou paraboliques*, C. R. Acad. Sci. Paris **297** (1983), 149–152.

42. _____, *Local vanishing properties of solutions of elliptic and parabolic quasilinear equations*, Trans. Amer. Math. Soc. **289** (1985).

43. L. C. Evans and B. Knerr, *Instantaneous shrinking of the support of non-negative solutions to certain nonlinear parabolic equations and variational inequalities*, Illinois J. Math. **23** (1979), 153–166.

44. C. Francsis, *On the porous medium equation with lower order singular nonlinear terms* (to appear).

45. A. Friedman, *Boundary behaviour of solutions of variational inequalities for elliptic operators*, Arch. Rational Mech. Anal. **27** (1967), 95–107.

46. _____, *Variational principles and free-boundary problems*, Wiley, New York, 1982.

47. A. Friedman and P. Phillips, *The free boundary of a semilinear elliptic equation* Trans. Amer. Math. Soc. **282** (1984), 153–182.

48. V. A. Galaktionov, *A boundary value problem for the nonlinear parabolic equation $u_t = \Delta u^{\sigma+1} + u^\beta$*, J. Differential Equations **17** (1981), 551–555.

49. B. H. Gilding, *Properties of solutions of an equation in the theory of filtration*, Arch. Rational Mech. Anal. **65** (1977), 203–225.

50. _____, *A nonlinear degenerate parabolic equation*, Ann. Scuola Norm. Sup. Pisa **4** (1977), 393–432.

51. M. E. Gurtin and A. C. Pipkin, *A general theory of heat conduction with finite wave speeds*, Arch. Rational Mech. Anal. **31** (1968), 113–126.

52. M. A. Herrero and J. L. Vazquez, *Asymptotic behaviour of solutions of a strongly nonlinear parabolic problem*, Ann. Fac. Sci. Toulouse Math. **3** (1981), 113–127.

53. _____, *On the propagation properties of a nonlinear degenerate parabolic equation*, Comm. Partial Differential Equations **7** (1982), 1381–1402.

54. A. S. Kalashnikov, *The propagation of disturbances in problems of nonlinear heat conduction with absorption*, Zh. Vychisl. Mat. i Mat. Fiz. **14** (1974), 891–905.

55. _____, *On the character of the propagation of perturbations in processes described by quasilinear degenerate parabolic equations*, Trudy Sem. Petrovsky (1975), 135–144.

56. _____, *On a nonlinear equation appearing in the theory of non-stationary filtration*, Trudy Sem. Petrovsk. **4** (1978), 137–146.

57. _____, *The concept of a finite rate of propagation of a perturbation*, Russian Math. Surveys **34** (1979), 235–236.

58. S. Kamin and Ph. Rosenau, *Thermal waves in an absorbing and convecting medium* (to appear).

59. R. Kersner, *The behaviour of temperature fronts in media with nonlinear thermal conductivity under absorption*, Vestnik Moskov. Univ. Ser. I Mat. Mekh. **33** (1978), 44–51.

60. _____, *Degenerate parabolic equations with general nonlinearities*, Nonlinear Anal. **4** (1980), 1043–1061.

61. _____, *Filtration with absorption: necessary and sufficient condition for the propagation or perturbations to have finite velocity*, J. Math. Anal. Appl. (1983).

62. B. Knerr, *The porous medium equation in one dimension*, Trans. Amer. Math. Soc. **234** (1977), 381–415.

63. _____, *The behaviour of the support of solutions of the equation of nonlinear heat conduction with absorption in one dimension*, Trans. Amer. Math. Soc. **249** (1979), 409–424.

64. S. N. Kruzkov, *First order quasilinear equations in several independent variables*, Math. USSR-Sb. **10** (1970), 217–243.

65. P. D. Lax, *Hyperbolic systems of conservation laws and the mathematical theory of shock waves*, C.B.M.S. Regional Conf. Ser. in Applied Math., no. 11, SIAM, Philadelphia, 1973.

66. T. Nagai and M. Mimura, *Asymptotic behaviour of a nonlinear degenerate diffusion equation in a population dynamics*, SIAM J. Appl. Math. **43** (1983), 449–469.

67. O. A. Oleinik, A. S. Kalashnikov and Chzhou Yui-Lin, *The Cauchy problem and boundary problems for equations of the type of nonstationary filtration*, Izv. Akad. Nauk SSSR Ser. Mat. **22** (1958), 667–704.

68. L. A. Peletier, *A necessary and sufficient condition for the existence of an interface in flows through porous media*, Arch. Rational Mech. Anal. **56** (1974), 163–190.

69. R. Redheffer, *On a nonlinear functional of Berkovitz and Pollard*, Arch. Rational Mech. Anal. **50** (1973), 1–9.

70. E. S. Sabinina, *A class of nonlinear degenerating parabolic equations*, Soviet Math. Dokl. **3** (1962), 495–498.

71. J. Serrin, *The solvability of boundary value problems*, Proc. Sympos. Pure Math., vol. 28, Amer. Math. Soc., Providence, R. I., 1976, pp. 507–524.

72. I. Stakgold, *Estimates for some free boundary problems*, Ordinary and Partial Differential Equations (W. N. Everitt and B. D. Sleeman, eds.), Lecture Notes in Math., vol. 846, Springer-Verlag, 1981.

73. J. L. Vazquez, *Asymptotic behaviour and propagation properties of the one-dimensional flow of a gas in a porous medium*, Trans. Amer. Math. Soc. **277** (1983), 507–527.

74. _____, *Behaviour of the velocity of one-dimensional flows in porous media* (to appear).

75. _____, *The interfaces of one-dimensional flows in porous media* (to appear).

76. L. Veron, *Effects regularisant de semi-groupes non lineaires dans les espaces de Banach*, Ann. Fac. Sci. Toulouse Math. **1** (1979), 171–200.

77. _____, *Equations d'evolution semi-lineaires du second ordre dans L^1*, Rev. Roumaine Math. Pures. Appl. **27** (1982), 95–123.

78. Y. B. Zeldovich and Y. P. Raizer, *Physics of shock waves and high-temperature hydrodynamic phenomena*, Academic Press, New York, 1969.

UNIVERSIDAD COMPLUTENSE DE MADRID, SPAIN

An Index Theory for Periodic Solutions of Convex Hamiltonian Systems

I. EKELAND[1]

0. Introduction. This paper is mostly concerned with finding, counting and describing periodic solutions to the $2n$-dimensional system of ordinary differential equations

(1) $$\dot{x} = JH'(x).$$

Here $H: \mathbf{R}^{2n} \to \mathbf{R}$ is a C^2 function, going to $+\infty$ at infinity, and J is the matrix

(2) $$J = \begin{pmatrix} 0 & -I_n \\ I_n & 0 \end{pmatrix}.$$

It is known that H is a first integral of system (1), that is, $(d/dt)H(x(t)) = 0$ whenever $x(t)$ solves (1). It follows that we can impose an additional condition:

(3) $$H(x(t)) = h.$$

Taking for instance $h = 1$, we get the boundary-value problem for (T, x):

(4) $$\dot{x} = JH'(x), \quad x(0) = x(T), \quad H(x) = 1.$$

Interest in this problem was rekindled by a seminal paper [21] of Rabinowitz in 1978, and has not abated since. This may be because it is a variational problem, but not an optimization problem: one has to find critical points of the action functional, which is known to have neither local minima nor local maxima. So problem (4) provides a useful testing ground for the many variational methods which are developed today in nonlinear analysis. In addition, it is an old problem, where Poincaré has made significant contributions, and it is quite exciting to compare modern methods with classical results.

1980 *Mathematics Subject Classification.* Primary 34B15, 49H05, 58E10.

[1] This paper was completed during a stay of the author at the University of Chicago, May, 1984. Most of the results were announced during the AMS Symposium on Nonlinear Analysis, Berkeley, July, 1983.

In the case when H is convex, the most powerful tool has been the dual action principle found by Clarke (see [7 and 9]). It performs the almost miraculous feat of changing a bad functional (unbounded from above and from below; all critical points badly degenerate) into a nice one (may have global minimum; all critical points have finite index and nullity). As perhaps the most striking consequence, let us mention the multiplicity result of Ekeland and Lasry [15], which gives conditions under which problem (4) has at least n solutions. As noted in [4], the dual action principle can also be used to treat nonconvex problems.

The purpose of this paper is to describe a new tool in the theory of convex Hamiltonian systems: it is the *index* of a periodic solution. This is a nonnegative integer which is associated with the linearized system; together with the Floquet multipliers, it contains important information about the periodic solution under consideration.

This index was first introduced in [13], where it was used to prove that, in general, problem (4) has infinitely many solutions. This comes as a consequence of another theorem, which states that if there are only finitely many solutions, they must be in resonance.

However, the article [13] introduced the index in a rather abstract and computational way. Since then, another approach, very close to the Jacobi theory of conjugate points in the classical calculus of variations, has come to the author's attention (see [14]). It is more direct, and has had consequences of its own: it has made possible a closer description of the periodic solutions one gets from variational methods, by bounding their lengths in terms of the index, and by showing that, under certain circumstances, elliptic solutions are present.

This paper is written from this new point of view. In §I we define and study the index of a linear, positive definite Hamiltonian system. In §II we define the index of a periodic solution to a convex Hamiltonian system. We also introduce the dual action principle, describe the reduction of the fixed-energy problem to the fixed-period problem, and draw the first conclusions as to the existence of solutions to problem (4).

In §III we use index theory to study the length and the (linear) stability of the periodic solutions we get. This section is inspired by the paper [3] of Ballmann, Thorbergsson and Ziller, who studied the corresponding problem for closed geodesics on positively curved compact Riemannian manifolds. We estimate the length of a closed trajectory in terms of its index, and we give a sufficient condition for problem (4) to have at least one elliptic solution.

§IV is really an outline of the longer paper [13]. We state the two main theorems: If problem (4) has only finitely many solutions, they must be in resonance; generically in the C^∞ topology, problem (4) has infinitely many solutions. We explain in detail what these statements mean, particularly the words "generic" and "resonant". Finally, we give a rough sketch of the proof, referring to [13] for a complete exposition.

We conclude by stating two open problems.

I. The index of a linear, positive definite, Hamiltonian system. Let $A(t)$ be a $2n \times 2n$ matrix, symmetric and positive definite, depending continuously on $t \in \mathbf{R}$. We consider the system of $2n$ linear differential equations

(1) $$\dot{y} = JA(t)y.$$

Here J denotes the antisymmetric $2n \times 2n$ matrix

(2) $$J = \begin{pmatrix} 0 & I_n \\ -I_n & 0 \end{pmatrix}.$$

Note that $J^2 = -I$. System (1) is Hamiltonian, which means it can be written in the form $\dot{y} = JH'_y(t, y)$; just define

(3) $$H(t, y) = \tfrac{1}{2}(A(t)y, y).$$

Set

(4) $$B(t) = A(t)^{-1}.$$

This is again a continuous, symmetric and positive definite matrix.

With every $t > 0$ we associate the quadratic form $\overline{Q}_T$ on the space E_T defined by

(5) $$\overline{Q}_T(y) = \int_0^T \tfrac{1}{2}[(J\dot{y}, y) + (B(t)J\dot{y}, J\dot{y})]\, dt,$$

(6) $$E_T = \{y \in H^1(0, T; \mathbf{R}^{2n}) | y(0) = y(T)\}.$$

Note that, because of the periodicity condition

(7) $$\overline{Q}_T(y + \xi) = \overline{Q}_T(y) \quad \forall \xi \in \mathbf{R}^{2n},$$

the true variable in (5) is $\dot{y}$, and we can choose for y any primitive we like. The only prerequisite on $\dot{y}$ is that

(8) $$\dot{y} \in L^2(0, T; \mathbf{R}^{2n}) \quad \text{and} \quad \int_0^T \dot{y}\, dt = 0.$$

In other words, we have

(9) $$\overline{Q}_T(y) = Q_T(\dot{y}),$$

where Q_T is a quadratic form on the space $L_0^2(0, T)$ defined by

(10) $$Q_T(v) = \int_0^T \tfrac{1}{2}[(Jv, \Pi v) + (B(t)Jv, Jv)]\, dt,$$

(11) $$L_0^2(0, T) = \left\{v \in L^2(0, T; \mathbf{R}^{2n}) \Big| \int_0^T v\, dt = 0\right\}.$$

Here Πv denotes the primitive of v with mean value zero:

(12) $$(d/dt)\Pi v = v \quad \text{and} \quad \int_0^T \Pi v\, dt = 0.$$

PROPOSITION 1. *Q_T degenerates on $L_0^2(0, T)$ if and only if the problem*

(13) $$\dot{y} = JA(t)y, \qquad y(0) = y(T)$$

has a nonzero solution.

PROOF. Call $\mathbf{R}^{2n}$ the space of constant functions in $L^2(0, T; \mathbf{R}^{2n})$. Then $L_0^2(0, T)$ is the orthogonal subspace. We have

$$(14) \qquad Q_T(v) = \tfrac{1}{2}(-J\Pi v - JB(t)Jv, v)_{L^2(0,T)},$$

so Q_T will degenerate on $L_0^2(0, T)$ if and only if there is some $\xi \in \mathbf{R}^{2n}$ and some $v \in L_0^2(0, T; \mathbf{R}^{2n})$, with $v \neq 0$ and

$$(15) \qquad -J\Pi v - JB(t)Jv = \xi.$$

This we rewrite as

$$(16) \qquad \Pi v - J\xi = -B(t)Jv,$$
$$(17) \qquad -Jv = A(t)(\Pi v - J\xi).$$

This is precisely equation (1), with $y = \Pi v - J\xi$, and y is clearly nonzero and T-periodic. □

This generalizes obviously to arbitrary times t_1 and t_2. Say $t_1 < t_2$. Then the following conditions are equivalent:

(a) the problem

$$(18) \qquad \dot{y} = JA(t)y, \qquad y(t_1) = y(t_2)$$

has a nonzero solution;

(b) the quadratic form

$$(19) \qquad \int_{t_1}^{t_2} \tfrac{1}{2}[(Jv, \Pi v) + (B(t)Jv, Jv)] \, dt$$

degenerates over the space $L_0^2(t_1, t_2)$.

DEFINITION 2. We say t_2 is *conjugate to* t_1 if the conditions (a) and/or (b) are met. The *multiplicity* of t_2 is the number of linearly independent solutions of (18).

Note that the conjugacy relation is not transitive: if t_2 is conjugate to t_1 and t_3 to t_2, then t_3 need not be conjugate to t_1.

Going back to the time interval $[0, T]$, note that the quadratic form Q_T splits into a compact part and a positive definite part. It follows that it can be written

$$(20) \qquad Q_T(v) = \sum_{i=1}^{\infty} \lambda_i v_i^2, \quad \text{with } v = \sum_{i=1}^{n} v_i e_i,$$

where the e_i are an appropriate basis in $L_0^2(0, T)$ and the λ_i are an increasing sequence converging to some positive limit.

The index of Q_T is the number of i such that $\lambda_i < 0$. It is also the dimension of any maximal negative definite subspace for Q_T in $L_0^2(0, T)$. It is always finite.

DEFINITION 3. The *index* i_T of system (1) on the time interval $[0, T]$ is the index of the quadratic form Q_T on the space $L_0^2(0, T)$.

The following properties are more or less obvious.

LEMMA 4. *There is some $T_0 > 0$ such that*

$$(21) \qquad i_T = 0 \quad \text{if } T \in (0, T_0).$$

PROOF. There is some $k > 0$ such that,

(22) $$\forall t \in [0,1], \quad (B(t)y, y) \geq k\|y\|^2 \quad \forall y.$$

On the other hand, by the Wirtinger inequality,

(23) $$\forall u \in L_0^2(0,T), \quad \|\Pi u\|_{L^2} \leq (T/2\pi)\|u\|_{L^2}.$$

Hence, for all $T \leq 1$, we have

(24) $$Q_T(v) \geq \frac{1}{2}\left(-\|v\|\|\Pi v\| + k\|v\|^2\right) \geq \frac{1}{2}\left(k - \frac{T}{2\pi}\right)\|v\|^2.$$

So Q_T is positive definite for $T \leq \text{Min}(1, 2k\pi)$. □

LEMMA 5. i_T is a nondecreasing function of T:

(25) $$S > T \Rightarrow i_S \geq i_T.$$

PROOF. Define a map p: $L_0^2(0,T) \to L_0^2(0,S)$ by

(26) $$(pv)(t) = \begin{cases} v(t) & \text{if } 0 \leq t \leq T, \\ 0 & \text{if } T < t \leq S. \end{cases}$$

For any $v \in L_0^2(0,T)$, we have clearly

(27) $$Q_T(v) = Q_S(pv).$$

If $E \subset L_0^2(0,T)$ is a negative definite subspace for Q_T, then $pE \subset L_0^2(0,S)$ will be a negative definite subspace for Q_S with the same dimension. Hence $i_S \geq i_T$. □

These results are contained in the following, which is an extension of the celebrated Morse index theorem in the theory of geodesics.

THEOREM 6. *For any $T > 0$, the index i_T of system* (1) *is equal to the number of points t in the interval $(0,T)$ which are conjugate to 0, counted with their multiplicities.*

PROOF. We have just seen that i_s is an integer-valued, nondecreasing function of s, starting from the value $i_s = 0$ near $s = 0$. We will show that its points of discontinuity are precisely the values of s which are conjugate to 0; and that the amount by which it increases at such a point is equal to the multiplicity of s. The result then follows.

We are dealing with a variable quadratic form Q_s on a variable space $L_0^2(0,s)$. We begin by rescaling everything to the time interval $(0,1)$. Define a map q: $L_0^2(0,s) \to L_0^2(0,1)$ by

(28) $$(qv)(t) = v(ts).$$

Then $sQ_s(v) = Q_s^0(qv)$, where Q_s^0 is the quadratic form on $L_0^2(0,1)$ defined by:

(29) $$Q_s^0(v) = \frac{1}{2}\int_0^1 [s(Jv, \Pi v) + (B(st)Jv, Jv)] \, dt.$$

A compactness argument now shows that Q_s cannot change its index without degenerating. In other words, if $\sup_{s<s_0} i_s < \inf_{s_0<s} i_s$, then Q_{s_0} must be degenerate (and $i_s = \sup_{s<s_0} i_s$). By Definition 2(b), this means that s_0 is conjugate to 0.

Let m_0 be the dimension of the nullspace of Q_{s_0} on $L_0^2(0, s)$. We want to show that:

(30) $$\inf_{s_0<s} i_s - \sup_{s<s_0} i_s = m_0.$$

Another compactness argument yields the inequality

(31) $$\lim_{s_0<s} i_s - \lim_{s<s_0} i_s \leq m_0.$$

To prove the converse inequality, we will take m_0 linearly independent functions $v_1, \ldots, v_{m_0}$ in the nullspace of $Q_{s_0}^0$, and show that

(32) $$(d/ds) Q_s^0(v_j)|_{s=s_0} < 0 \quad \text{for } j = 1, \ldots, m_0.$$

It is enough to do it for $j = 1$. We know by Proposition 1 that $v_1(t/s)$ is $\dot{y}_1(t)$, where y_1 satisfies $\dot{y}_1 = JA(t)y_1$ and $y_1(0) = y_1(s)$. It follows that v_1 is C^0. For the sake of simplicity, we shall assume that $A(t)$ is C^1, so that v_1 is in fact C^1. The proof in the general case is less straightforward.

Differentiating formula (29), we get

(33) $$\frac{d}{ds} Q_s^0(v_1) = \frac{1}{2} \int_0^1 \left[(Jv_1, \Pi v_1) + (tB'(st) Jv_1, Jv_1) \right] dt.$$

We know that there is some constant $\xi \in \mathbf{R}^{2n}$ such that

(34) $$(1/s) B(st) Jv_1 = -(\Pi v_1 + \xi).$$

Differentiating with respect to t, we get

(35) $$B'(st) Jv_1 = -v_1 - (1/s) B(st) J\dot{v}_1.$$

Hence

(36) $$\int_0^1 (tB'(st) Jv_1, Jv_1) dt = -\int_0^1 t(v_1, Jv_1) dt - \frac{1}{s} \int_0^1 (tB(st) J\dot{v}_1, Jv_1) dt.$$

The first term on the right vanishes because J is antisymmetric. Using (34), we continue:

(37) $$\int_0^1 (tB'(st) Jv_1, Jv_1) dt = \int_0^1 (\Pi v_1 + \xi, tJ\dot{v}_1) dt$$
$$= (\Pi v_1(1) + \xi, Jv_1(1))$$
$$\quad - \int_0^1 (\Pi v_1 + \xi + tv_1, Jv_1) dt$$
$$= (\Pi v_1(1) + \xi, Jv_1(1)) - \int_0^1 (\Pi v_1, Jv_1) dt.$$

Writing (37) into (33), we get

(38) $$(d/ds) Q_s^0(v_1) = (\Pi v_1(1) + \xi, Jv_1(1))$$
$$= -(1/s)(B(s) Jv_1(1), Jv_1(1))$$
$$< 0. \qquad \square$$

There are other properties of the index which may prove useful in special circumstances: the dependence on $A(t)$, and the iteration formula in the periodic case. We deal with them in the two following propositions.

If A_1 and A_2 are two symmetric positive definite matrices, $A_1 \leq A_2$ means $(A_1 y, y) \leq (A_2 y, y)$ for all y.

PROPOSITION 7. *Consider two linear systems* $\dot{y} = JA_1(t)y$ *and* $\dot{y} = JA_2(t)y$, *with A_1 and A_2 continuous, symmetric and positive definite, and $A_1(t) \leq A_2(t)$ for all t in $[0, T]$. Then their indices i_T^1 and i_T^2 satisfy $i_T^1 \leq i_T^2$.*

PROOF. Clearly $B_1(t) \geq B_2(t)$ for all t, so that the associated quadratic forms Q_T^1 and Q_T^2 satisfy $Q_T^1 \geq Q_T^2$ on $L_0^2(0, T)$. Hence the result. □

Now consider the particular case when $A(t)$ is T-periodic, for some $T > 0$. The natural index for the system then is i_T, and one should be able to compute all the i_{kT}, $k \in \mathbf{N}$, from i_T. This is possible, but the formula requires knowledge of the Floquet multipliers. We pause to give the relevant information.

Denote by $R(t)$ the resolvent of system (1):

(39) $$\dot{R}(t) = JA(t)R(t), \quad R(0) = I.$$

Then $R(t)$ is a symplectic matrix

(40) $$\dot{R}(t)^* JR(t) = J.$$

It follows that if λ is an eigenvalue of $R(t)$, so is λ^{-1}. If $|\lambda| \neq 1$, $\lambda \notin \mathbf{R}$, we thus get four distinct eigenvalues $\lambda, \bar{\lambda}, \lambda^{-1}, \bar{\lambda}^{-1}$, while if $|\lambda| = 1$, $\lambda \notin \mathbf{R}$, we get only two, λ and $\bar{\lambda}$.

If $A(t)$ is T-periodic,

(41) $$A(t + T) = A(t),$$

the eigenvalues of $R(T)$ are called the *Floquet multipliers* of system (1).

Nonreal Floquet multipliers on the unit circle have been classified by Krein ([18]; see also [16, 20 and 27]). If $\lambda \notin \mathbf{R}$, $|\lambda| = 1$, is a simple multiplier, and if $\xi \in \mathbf{C}^{2n}$ is some eigenvector, $(1/i)(J\xi, \xi)$ never vanishes. We shall say that λ is positive if $(1/i)(J\xi, \xi) > 0$, negative otherwise. More generally, if $\lambda \in \mathbf{R}$, $|\lambda| = 1$, is a Floquet multiplier with multiplicity m, then $(1/i)(J\xi, \xi)$ is a nondegenerate Hermitian form on the subspace $\mathrm{Ker}(R(T) - \lambda I)^m$ of $\mathbf{C}^{2n}$, and we shall say that it has type (p, q) if it has p positive and q negative eigenvalues. We always have $p + q = m$. If λ has type (p, q), then $\bar{\lambda}$ has type (q, p).

Now set

(42) $$C = \{e^{i\theta} | \theta \in \mathbf{R}\},$$

(43) $$\Lambda = \{\lambda ||\lambda| = 1, \mathrm{Ker}(R(T) - \lambda I) \neq 0\}.$$

We associate with system (1) a map $j: C \to \mathbf{N}$ defined as follows:
(a) j is continuous, and hence locally constant, at all points $\omega \notin \Lambda$;
(b) $j(\bar{\omega}) = j(\omega)$ for all $\omega \in C$;

(c) if $\lambda \in \Lambda$, $\text{Re}\,\lambda > 0$ and λ has type (p, q):
$$\lim_{\substack{\varepsilon \to 0 \\ \varepsilon > 0}} \left[j(\lambda e^{i\varepsilon}) - j(\lambda e^{-i\varepsilon}) \right] = q - p;$$

(d) if $1 \notin \Lambda$, $\lim_{\varepsilon \to 0, \varepsilon > 0} j(\lambda e^{i\varepsilon}) = i_T + n$;

(e) if $1 \in \Lambda$, and $\text{Ker}(R(T) - I)$ is d-dimensional,
$$\lim_{\substack{\varepsilon \to 0 \\ \varepsilon > 0}} j(e^{i\varepsilon}) = i_T + n + d',$$

where $0 \leq d' < d$;

(f) if $\lambda \in \Lambda$, $\text{Re}\,\lambda > 0$, then $j(\lambda) = \lim_{\varepsilon \to 0; \varepsilon > 0} j(\lambda e^{-i\varepsilon})$;

(g) $j(1) = i_T$ and $j(-1) \leq \lim_{\varepsilon \to 0} j(-e^{i\varepsilon})$.

Note that the Floquet multipliers on the unit circle, and the value of d' and $j(-1)$ completely determine j. For instance, if there are no Floquet multipliers on the unit circle except $+1$, j takes only two values, i_T at $\omega = 1$ and $i_T + n + d'$ everywhere else.

We can now state our iteration formula.

PROPOSITION 8. *Assume A is T-periodic*:

(44) $$A(t + T) = A(t) \quad \forall t.$$

Then for all integers k:

(45) $$i_{kT} = \sum_{\omega^k = 1} j(\omega).$$

PROOF. See [13, §IV]. There is, however, a gap in the proof of Proposition 7, which addresses case (e) and gives the formula

(46) $$\lim_{\substack{\varepsilon \to 0 \\ \varepsilon > 0}} j(e^{i\varepsilon}) = i_T + n + E(d/2)$$

where $E(\alpha) = \text{Max}\{k \in \mathbf{N} | k \leq \alpha\}$. This mistake was pointed out to me by P. Brousseau. □

We should mention that there is in the literature another definition of the index of a linear Hamiltonian system. It is found in the Russian school (see [27] for a detailed exposition), in the work of Bott [6], Duistermaat [12], Moser [20], and has found recent applications in the work of Conley and Zehnder ([10] for instance). We shall refer to it as the *standard index*.

The standard index j_T of a linear Hamiltonian system $\dot{y} = JA(t)y$ over a time interval $[0, T]$ is well-defined, provided only that 1 is not an eigenvalue of $R(T)$. It is not required that $A(t)$ be positive definite. If, however, it is, then we have both the index i_T and the standard index j_T. It has been shown by Brousseau [5] that
$$j_T = i_T + n.$$

He has also extended this relation to the general case, by defining i_T adequately for nonpositive definite $A(t)$. The problem is that the quadratic form Q_T need no longer have finite index, so Definition 3 cannot be used without changes.

II. The index of a periodic solution to a convex Hamiltonian system.

We now consider a nonlinear Hamiltonian system:

(1) $$\dot{x} = JH'(t, x).$$

The Hamiltonian $H: \mathbf{R} \times \mathbf{R}^{2n} \to \mathbf{R}$ is C^2, and H' denotes the derivative with respect to x. It is assumed that:

(a) H is periodic in t:
$$\exists T > 0: H(t + T, x) = H(t, x) \quad \forall (t, x);$$

(b) H is strictly convex in x:
$$H''(t, x) \geq 0 \quad \forall (t, x);$$

(c) $\mathrm{Min}\{H(t, x)|t \in \mathbf{R}, \|x\| = r\} \to +\infty$ when $r \to \infty$.

Under these assumptions, a dual version of the least action principle has been found by Clarke [7]. It goes as follows:

THEOREM 1 [9]. *If $\bar{x}$ is a T-periodic solution of* (1), *then it is also an extremal of the integral*

(2) $$\overline{\Phi}(x) = \int_0^T \left[\frac{1}{2}(J\dot{x}, x) + H^*(t, -J\dot{x}) \right] dt$$

under the boundary condition:

(3) $$x(0) = x(T).$$

Conversely, if $\bar{x}$ is an extremal of (2), (3), *then there is some constant $\xi \in \mathbf{R}^{2n}$ such that $\bar{x} + \xi$ is a T-periodic solution of* (1).

Some explanations are in order before we give the proof. $H^*(t, y)$ is the Fenchel conjugate of $H(t, x)$, for fixed t:

(4) $$H^*(t, y) = \mathrm{Max}\{(x, y) - H(t, x)| x \in \mathbf{R}^{2n}\}.$$

It also enjoys properties (a), (b), (c), it is C^2, and we have the Legendre reciprocity formula

(5) $$y = H'(t, x) \Leftrightarrow x = (H^*)'(t, y).$$

In the classical terminology of the calculus of variations, a C^1 path $\bar{x}$ is an extremal of (2), (3) if

(6) $$h^{-1}(\overline{\Phi}(\bar{x} + hy) - \overline{\Phi}(\bar{x})) \to 0 \quad \text{when } h \to 0$$

for all C^1 paths y such that $y(0) = y(T)$; in other words, the first variation of Φ of $\bar{x}$ is zero.

In the case at hand, this boils down to the usual Euler–Lagrange equations.

PROOF OF THEOREM 1. An extremal x of (2), (3) is any solution of the boundary-value problem

(7) $$J\dot{x} = J\frac{d}{dt}(H^*)'(t, -J\dot{x}),$$

(8) $$x(0) = x(T).$$

Set $\bar{x}(t) = (H^*)'(t, -J\dot{x}(t))$. By the Legendre reciprocity formula, we have
(9) $$-J\dot{x} = H'(t, \bar{x}).$$
On the other hand, by equation (7), we have
(10) $$\dot{x} = d\bar{x}/dt.$$
Putting (9) and (10) together, we get our result. □

It will be noted that $\bar{\Phi}(x + \xi) = \bar{\Phi}(x)$, for all constants $\xi \in \mathbf{R}^{2n}$ and all T-periodic x. The true variable then is $\dot{x}$: We have
(11) $$\bar{\Phi}(x) = \Phi(\dot{x}),$$
where
(12) $$\Phi(u) = \int_0^T \left[\frac{1}{2}(Ju, \Pi u) + H^*(t, -Ju)\right] dt,$$
(13) $$\int_0^T u\, dt = 0.$$

Once we have found a T-periodic solution x of system (1), it is easy to define its index. We shall say that x is *admissible* if $H''(t, x(t))$ is nondegenerate for all t.

DEFINITION 2. Let x be an admissible T-periodic solution of system (1). The *index* of (T, x) is the index of the linearized Hamiltonian system
(14) $$\dot{y} = JH''(t, x(t))y$$
over the time interval $[0, T]$.

We also define conjugate points, and write Theorem I.6 in this new context.

DEFINITION 3. Let x be an admissible T-periodic solution of system (1), and let $x_1 = x(t_1)$ and $x_2 = x(t_2)$ be two points on x, with $t_1 < t_2$. We say that x_2 is *conjugate* to x_1 along the arc $x(t_1, t_2)$ if the linear problem
(15) $$\dot{y} = JH''(t, x(t))y,$$
(16) $$y(t_1) = y(t_2)$$
has a nonzero solution. The *multiplicity* of x_1 is the number of linearly independent periodic solutions.

THEOREM 4. *The index of an admissible T-periodic solution x is equal to the number of t in $(0, T)$ such that $x(t)$ is conjugate to $x(0)$ along the arc $x(0, t)$.*

It turns out that the indices of the various T-periodic solutions of system (1) can be related to each other. Indeed, as the following lemma shows, the quadratic form Q_T associated with system (14) is (formally at least) $2\Phi''(\dot{x})$, so that the index of x is just the Morse index of Φ'' as a critical point for Φ.

LEMMA 5. *For v_1 and v_2 in $L_0^\infty(0, T)$, we have*
(17) $$\frac{d}{dt}\frac{d}{ds}\Phi(\dot{x} + tv_1 + sv_2) = \int_0^T \left[(Jv_1, \Pi v_2) + \left(A(t)^{-1}Jv_1, Jv_2\right)\right] dt$$
with $A(t) = H''(t, x(t))$.

PROOF. We get easily

(18) $$\frac{d}{dt}\frac{d}{ds}\Phi(\dot{x} + tv_1 + sv_2) = \int_0^T \Big[(Jv_1, \Pi v_2) + \big(((H^*)''(t, -J\dot{x}))Jv_1, Jv_2\big)\Big] dt.$$

We know, by the Legendre reciprocity formula, that

(19) $$(H^*)'(t, H'(x)) = x.$$

Differentiating this relation, we get

(20) $$(H^*)''(t, H'(x))H''(t, x) = I.$$

Writing $-J\dot{x} = H'(x)$, we get the result. □

So Morse theory applied to Φ should enable us to write down relations between the indices of the various periodic solutions of system (1) (the so-called Morse relations between critical points of Φ), and prove existence theorems. There are technical obstacles to overcome, mainly to set up Φ as a C^2 function on an appropriate function space, but this program will be carried through in §IV, for a special class of problems which we now describe.

We are given in $\mathbf{R}^{2n}$ a C^2-hypersurface S, which bounds a convex compact set C, with 0 in its interior. If $x \in S$, let $N(x)$ denote the unit vector on the outward normal to S at x. Then $JN(x)$ is tangent to S at x, and the differential equation

(21) $$\dot{x} = JN(x)$$

defines a flow on S.

We want to know whether this flow has any closed trajectories, if so how many, and what are their properties (length, stability).

In other words, we are interested in the solutions (T, x) of the problem

(22) $$\dot{x} = JN(x), \quad x(t) \in S, \quad x(0) = x(T).$$

To put this problem into Hamiltonian form, we choose some number $\alpha > 1$, and we define a function H_α on $\mathbf{R}^{2n}$ by

(23) $$H(\lambda x) = \lambda^\alpha H(x), \quad \forall \lambda > 0, \forall x \in \mathbf{R}^{2n},$$
(24) $$x \in S \Leftrightarrow H(x) = 1.$$

The differential equations $\dot{x} = JN(x)$ and $\dot{x} = JH'_\alpha(x)$ then have the same trajectories on S (but not the same speed along these trajectories). Problem (22) then is equivalent to the following: find (T, x) such that

(25) $$\dot{x} = JH'_\alpha(x), \quad x(0) = x(T), \quad H(x(t)) = 1.$$

This is a fixed-energy problem: the (constant) value of H_α is prescribed, not the period T_α. The corresponding fixed-period problem would be

(26) $$\dot{x} = JH'_\alpha(x), \quad x(0) = x(1).$$

The relation between problems (25) and (26) is described in the following result.

LEMMA 6. *Assume $\alpha \neq 2$. If x is a solution of (26), with $H(x(t)) = h > 0$, then (x_0, T_0) is a solution of (25), with*

(27) $$x_0(t) = h^{-\alpha} x(th^{\alpha(2-\alpha)}), \qquad T_0 = h^{-\alpha(2-\alpha)}.$$

Conversely, if (x, T) is a solution of (25), then for any integer $k \geq 1$,

(28) $$x_k(t) = \frac{1}{(kT)^{2-\alpha}} x(kTt)$$

is a solution of (26).

The proof is a straightforward calculation. Note that x_k is in fact $(1/k)$-periodic. So a single solution of (25) corresponds to an infinite sequence of solutions of (26), with decreasing periods.

For $\alpha > 1$, the Hamiltonian H_α enjoys properties (a), (b), (c). It should be noted that its Fenchel conjugate H_α^* is positively homogeneous of degree β, with $\alpha^{-1} + \beta^{-1} = 1$, so that the dual action functional

(29) $$\Phi_\alpha(u) = \int_0^1 \left[\frac{1}{2} (\Pi u, Ju) + H_\alpha^*(-Ju) \right] dt$$

is naturally defined on the space:

(30) $$L_0^\beta(0, T) = \left\{ u \in L^\beta(0, 1; \mathbf{R}^{2n}) \mid \int_0^1 u\, dt = 0 \right\}.$$

From now on, we will always take $1 < \alpha < 2$, so $\beta > 2$. The function Φ_α is the sum of two terms, a compact term of degree two, and a convex term of degree β. So Φ is weakly sequentially lower semicontinuous, and $\Phi_\alpha(u) \to +\infty$ when $\|u\|_\beta \to \infty$. It follows that Φ_α attains its minimum on $L_0^\beta(0, 1)$ at some point $\bar{u}$. The necessary conditions for optimality then tell us that $\Pi \bar{u}$ is an extremal of $\overline{\Phi}_\alpha$ over all 1-periodic paths. By Theorem 1, there will be some constant ξ such that $\Pi \bar{u} + \xi$ is a solution of problem (26).

To get a solution of problem (25), we have to show that $\bar{u} \neq 0$. Indeed, 0 is a critical point of Φ_α, but plugging in $(H^*)''(0) = 0$ in Lemma 5, we get:

(31) $$\frac{d}{ds}\frac{d}{dt}\Phi(tv_1 + sv_2) = \int_0^T (Jv_1, \Pi v_2)\, dt$$

and the right side can be made negative by an appropriate choice of v_1 and v_2. So 0 cannot be a local minimum, let alone a global one. In this way, which follows the pattern of [9], we have proved the following.

PROPOSITION 7. *Problem (25) has at least one solution.*

PROPOSITION 8. *Problem (26) has always infinitely many solutions, at least one of which has index zero.*

Proposition 7 is an important result with a long history: Seifert [22], Rabinowitz [21], Weinstein [25], Clarke [7]. A multiplicity result is due to Ekeland and Lasry [15], who showed that if C could be fitted between two concentric euclidean balls

(32) $$B(x, 0, r) \subset C \subset B(x_0; R)$$

with $Rr^{-1} < \sqrt{2}$, the problem (25) has at least n solutions. See Ambrosetti-Mancini [1], Hofer [17], Berestycki-Lasry-Mancini-Ruf [4] for various improvements and the relation with Weinstein's bifurcation theorem [24].

Proposition 8 will be useful to us in the next sections; the infinitely many solutions are a consequence of Lemma 6, the solution of index 0 being the minimizer.

III. Length and stability of solutions. We now wish to use index theory for finding periodic solutions with special properties. We are interested in the problem

(1) $\qquad \dot{x} = JN(x), \qquad x(t) \in S, \qquad x(0) = x(T),$

with the assumption and notations laid out in the preceding section.

There are two main distinctive features of a solution to this problem: its length, and its stability. The shorter, and the more stable, a solution, the more interesting it is from any practical point of view. In this section, we will show that solutions with low index are shorter and more likely to be stable.

But first we must define length and stability. Let (T, x) be a solution of (1). A natural measure for its length is the integral

(2) $\qquad \dfrac{1}{2}\oint (Jx, \dot{x})\, dt = \dfrac{1}{2}\oint (Jx, dx).$

Indeed, it is invariant by reparametrization, and it is positive:

(3) $\qquad (Jx, \dot{x}) = (x, -J\dot{x}) = (x, N(x)) > 0,$

since C is convex and contains the origin in its interior.

Stability for nonlinear Hamiltonian systems is very poorly understood. What replaces it is linear stability.

DEFINITION 1. Let (T, x) be a solution of (1). Consider in S a C^1 submanifold Σ of codimension 1, transversal to the trajectory x at $x(0)$, and let $\phi: (\mathbf{R}^{2n-2}, 0) \to (\mathbf{R}^{2n-2}, 0)$ be the associated Poincaré mapping. We say x is *linearly stable* if the tangent map $T\phi(0) \in \mathcal{L}(\mathbf{R}^{2n-2})$ is stable, that is, if there is some constant C such that

(4) $\qquad \|T\phi(0)^n\| \leq C \quad \text{as } n \to \pm\infty.$

The Poincaré mapping is defined locally: if $\xi \in \Sigma$ is close enough to $x(0)$, then $\phi(\xi)$ is the first point where the trajectory starting from ξ hits Σ again. As for condition (4), it can be formulated in a different way: $T_\phi(0)$ should be diagonalizable, with all its eigenvalues lying on the unit circle.

If we represent S by H_α as in the preceding section, the solution (T, x) of problem (1) will be represented by a solution (T_α, x_α) of

(5) $\qquad \dot{x} = JH'_\alpha(x), \qquad x(0) = x(T), \qquad H_\alpha(x) = 1.$

Linearize the differential equation around x_α:

(6) $\qquad \dot{y} = JH''_\alpha(x_\alpha(t))y,$

and consider the resolvent $R(t)$. It can be shown that, in an appropriate basis, $R(T_\alpha)$ will be written

(7) $$R(T_\alpha) = \begin{pmatrix} A & B \\ 0 & T\phi(0) \end{pmatrix}, \quad \text{with } A = \begin{pmatrix} 1 & \alpha - 2 \\ 0 & 1 \end{pmatrix} \in \mathscr{L}(\mathbf{R}^2).$$

This brings the Floquet multipliers of x into the picture. They split into two batches: $(1,1)$, corresponding to A, and the eigenvalues $(\lambda_3, \ldots, \lambda_{2n})$ of $T_\phi(0)$. We refer to these last $(2n - 2)$ as the *nontrivial* Floquet multipliers. This shows that the Floquet multipliers do not depend on the way we chose α, and that the definition of linear stability does not depend on the choice of Σ.

Note that if x is linearly stable, it has to be *elliptic*, that is, all Floquet multipliers must be on the unit circle. In addition, we must be able to bring $R(T_\alpha)$ into the form

(8) $$\begin{pmatrix} 1 & \alpha - 2 & 0 & & & 0 \\ & 1 & 0 & & & \\ & & \lambda_2 & & & \vdots \\ & & & \ddots & & \\ 0 & & & & & \lambda_{2n} \end{pmatrix}$$

So much for defining length and linear stability. We now have to say something about the geometry of S. We already know it is a compact C^2 manifold of codimension 1, which bounds a convex set. So its total curvature must be everywhere nonnegative.

We make one further assumption: the total curvature shall be positive everywhere. In other words, if $\rho_1(x) \leq \cdots \leq \rho_{2n-1}(x)$ are the principal radii of curvature of S at x, we are assuming that

(9) $$\forall x \in S, \quad 0 < \rho_1(x) \leq \cdots \leq \rho_{2n-1}(x) < \infty.$$

In this case, for any x_0 in the interior of C, there will always exist numbers r and R, with $0 < r \leq R < \infty$, such that, for all $x \in S$:

(10) $$r^{-2} \leq \frac{1}{\|x - x_0\|^2}\left[\frac{\|x - x_0\|}{\rho_{2n-1}(x)\cos^3\theta(x)} + \tan^2\theta\right],$$

(11) $$\frac{1}{\|x - x_0\|^2}\left[\frac{\|x - x_0\|}{\rho_1(x)\cos^3\theta(x)} + \tan^2\theta(x)\right] \leq R^{-2},$$

(12) $$r \leq \|x - x_0\| \leq R.$$

Here $\theta(x)$ is the angle between $H'(x)$ and x.

In the sequel, we will always set $x_0 = 0$.

DEFINITION 2. If (10), (11), and (12) are satisfied, we shall say that S is (r, R)-*pinched*. Note $r = R$ if and only if S is a sphere.

Now set $S = \{x | H(x) = 1\}$, as in the preceding section, with H_α positively homogeneous of degree α:

(13) $$H_\alpha(\lambda x) = \lambda^\alpha H_\alpha(x) \quad \forall \lambda > 0.$$

Let us compute the second fundamental form $(b(x)\xi, \xi)$ of S at x. Take any smooth curve $c(t)$ on S such that $c(0) = x$. We have $H(c(t)) = 1$ for all t. Differentiating twice, we get

(14) $$(c''(0), H'(x)) + (H''(x)\dot{c}(0), \dot{c}(0)) = 0.$$

By definition, the first term is $-\|H'(x)\|^{-1}(b(x)\dot{c}(0), \dot{c}(0))$. Hence the second fundamental form:

(15) $$\forall \xi \in T_x S, \quad (b(x)\xi, \xi) = (H''(x)\xi, \xi)\|H'(x)\|^{-1}.$$

This gives us $H''(x)$ on $T_x S$. Now the vector x is transversal to $T_x S$, since 0 belongs to the interior of c. So the knowledge of $H''(x)x$ will completely determine $H''(x)$ on $\mathbf{R}^{2n}$. It turns out that:

(16) $$(H''(x)x, \xi) = 0 \qquad \forall \xi \in T_x S,$$
(17) $$(H''(x)x, x) = \alpha(\alpha - 1) \qquad \forall x \in S.$$

Both relations arise from differentiating the equation $H(\lambda x) = \lambda^\alpha H(x)$, and remembering that $(H'(x), \xi) = 0$ and $H(x) = 1$. They completely determine $H''(x)x$.

From (10), (11), (12) we derive

LEMMA 3. *Let S be (r, R)-pinched. Then, for every $R' > R$, there is some $\alpha' < 2$ such that, for all $\alpha \in (\alpha', 2)$, we have*

(18) $$2(R')^{-2} I \leq H''_\alpha(x) \leq 2r^{-2} I \qquad \forall x \in S.$$

PROOF. Take $\zeta \in \mathbf{R}^{2n}$, and write

(19) $$\zeta = \xi + \eta \|x\|^{-1} x, \quad \text{with } \xi \in T_x S.$$

We get

(20) $$(H''_\alpha(x)\zeta, \zeta) = (H''_\alpha(x)\xi, \xi) + (H''_\alpha(x)x, x)\eta^2 \|x\|^{-2}$$
$$= (b(x)\xi, \xi)\|H'_\alpha(x)\| + \alpha(\alpha - 1)\eta^2 \|x\|^{-2}.$$

If ζ is collinear to x, we have $\xi = 0$ and we get

(21) $$(H''_\alpha(x)\zeta, \zeta) = \frac{\alpha(\alpha - 1)}{\|x\|^2}\|\zeta\|^2.$$

If ζ is orthogonal to x, an elementary computation gives

(22) $$\eta = \frac{(H'_\alpha, \zeta)}{(H'_\alpha, x)}\|x\| = \|H'_\alpha\| \|\zeta\| \|x\| \alpha^{-1} \sin\theta = \|\zeta\| H \tan\theta.$$

Note that $(H'_\alpha(x), x) = \alpha$ by homogeneity, so that $\|H'_\alpha\| = \alpha(\|x\|\cos\theta)^{-1}$. We continue:

(23) $$0 = (\xi, x) + \eta\|x\|,$$
(24) $$\|\zeta\|^2 = \|\xi\|^2 + \eta^2 + 2\eta\|x\|^{-1}(x, \xi) = \|\xi\|^2 - \eta^2.$$

On the other hand, we have

(25) $$p_{2n-1}^{-1}\|\xi\|^2 \leq (b(x)\xi, \xi) \leq p_1^{-1}\|\xi\|^2.$$

Writing all this into formula (20), we get

(26) $$(H_\alpha''(x)\zeta, \zeta) \leq \frac{\alpha}{\|x\|\cos\theta} \frac{(1 + H\tan^2\theta)}{\rho_1} \|\zeta\|^2 + \frac{\alpha(\alpha-1)}{\|x\|^2} H\tan^2\theta \|\zeta\|^2,$$

(27) $$(H_\alpha''(x)\zeta, \zeta) \geq \frac{\alpha}{\|x\|\cos\theta} \frac{(1 + H\tan^2\theta)}{\rho_{2n-1}} \|\zeta\|^2 + \frac{\alpha(\alpha-1)}{\|x\|^2} H\tan^2\theta \|\zeta\|^2.$$

The result follows by letting α converge to 2, and substituting $H = 1$. □

Now we can start using index theory. The two main facts we will rely on are the following.

PROPOSITION 4. *Let x be an admissible T-periodic solution of an autonomous Hamiltonian system $\dot{x} = JH'(x)$. Then $x(0)$ and $x(T)$ are conjugate along the arc $(0, T)$.*

PROOF. It is well known that $y = \dot{x}$ solves the problem $\dot{y} = JH''(x(t))y$, with $y(0) = y(T)$. This is the result, by Definition II.3. □

PROPOSITION 5. *Fix a positive number c. The index of the linear Hamiltonian system*

(28) $$\dot{x} = cJy$$

on the time interval $[0, T]$ is given by

(29) $$i_{[0,T]} = 2kn \quad \text{if } (2\pi/c)k < T \leq (2\pi/c)(k+1).$$

PROOF. The only times conjugate to 0 are $t = 2k\pi/c$, each one with multiplicity $2n$. The result follows from Theorem II.6. □

PROPOSITION 6. *Let x_α be a nonconstant, T_α-periodic solution of $\dot{x} = JH_\alpha'(x)$, and assume condition (18) is satisfied. Let $x(t_2)$ be conjugate to $x(t_1)$ along the arc (t_1, t_2), and assume $x(t_2)$ is the first point with this property. In other words, there is no $x(t)$ conjugate to $x(t_1)$ with $t_1 < t < t_2$. Then*

(30) $$\pi r^2 \leq t_2 - t_1 \leq \pi R'^2.$$

PROOF. Consider the three linear Hamiltonian systems

(31) $$\dot{y} = 2(R')^{-2} Jy,$$

(32) $$\dot{y} = 2JH_\alpha''(x(t))y,$$

(33) $$\dot{y} = 2r^{-2} Jy,$$

and let $i_{[t_1, t]}$, $i_{[t_1, t]}$, $\bar{i}_{[t_1, t]}$ denote their indices on the interval $[t_1, t]$.

By Proposition I.7, we have

(34) $$\underline{i}_{[t_1, t]} \leq i_{[t_1, t]} \leq \bar{i}_{[t_1, t]}.$$

First take $t = t_2$. Then $i_{[t_1, t]} = 0$ by Theorem I.6. So $i_{[t_1, t_2]} = 0$, and Proposition 5 gives

(35) $$t_2 - t_1 \leq \pi R'^2.$$

Now take $t > t_2$. Then $i_{[t_1, t]} \geq 1$ by Theorem I.6 again, so $\bar{i}_{[t_1, t]} \geq 1$ and Proposition 5 gives

(36) $$t - t_1 > \pi r^2;$$

letting $t \to t_2$ we get the desired result. □

From this we shall derive a relation between the index of a periodic solution and its length.

DEFINITION 7. Let (T, x) be a solution of problem (1), and (T_α, x_α) the corresponding solution of problem (5). Denote by i_α the index of (T_α, x_α). We define the index of (T, x) to be

(37) $$i = i_2.$$

A simple compactness argument shows that

(38) $$i_2 = \lim_{\substack{\alpha \to 2 \\ \alpha < 2}} i_\alpha,$$

so that we can apply the results we just proved.

THEOREM 8. *Let (T, x) be a solution of problem* (1). *Say (T, x) has index i and S is (r, R)-pinched. The following relations then hold for integer $k \geq 1$:*

(39) $$i < 2nk \Rightarrow \frac{1}{2}\oint (Jx, dx) \leq k\pi R^2,$$

(40) $$i > 2n(k - 1) \Rightarrow \frac{1}{2}\oint (Jx, dx) > k\pi r^2.$$

In addition, we always have

(41) $$\frac{1}{2}\oint (Jx, dx) \geq \pi r^2.$$

PROOF. This theorem follows the pattern set by Ballmann, Thorbergsson and Ziller [3].

Let us prove relation (39). There is some $\alpha' < 2$ such that $i = i_\alpha$ for $\alpha' < \alpha < 2$. So the index of system (5) on the time interval $[0, T_\alpha]$ is strictly less than $2nk$. By Proposition 5, we must have $T_\alpha \leq rR'^2 k$. Letting $\alpha' \to 2$, we have $R' \to R$ by Lemma 3, and we get

(42) $$T_2 \leq \pi R^2 k.$$

Now T_2 is easily computed; we have

(43) $$(Jx, dx/dt) = (Jx, -JH_2'(x)) = (x, H_2'(x)) = 2.$$

So $T_2 = \frac{1}{2}\oint(Jx, dx)$ and we have formula (39). Formula (40) is proved in a similar fashion.

Formula (41) follows from Proposition 4. Indeed, we know that $x_\alpha(0)$ and $x_\alpha(T_\alpha)$ are conjugate along the arc $(0, T_\alpha)$, so that for any $t > T_\alpha$ we have, in the notation of Proposition 6,

$$\tag{44} \bar{i}_{[0, t]} \geq i_{[0, t]} \geq 1.$$

So $t > \pi r^2$ by Proposition 5. We have $t > \pi r^2$ whenever $t > T_\alpha$, so $T_\alpha \geq \pi r^2$, and the result follows by letting $\alpha \to 2$. □

Theorem 8 has several consequences, some of which are valid under weaker hypotheses. We point them out for the reader's convenience:

PROPOSITION 9. *Assume S is a compact* C^2 *hypersurface, which bounds a convex set and satisfies*

$$\tag{45} \forall x \in S, \quad r \leq \|x - x_0\| \leq R$$

for some $x_0 \in \mathbf{R}^{2n}$, *R and* $r > 0$. *Then*

(a) *every periodic solution of problem* (1) *satisfies*

$$\tag{46} \frac{1}{2} \oint (Jx, dx) \geq \pi r^2;$$

(b) *there is a periodic solution of problem* (1) *such that*

$$\tag{47} \frac{1}{2} \oint (Jx, dx) \leq \pi R^2.$$

PROOF. Part (a) is a theorem of Croke–Weinstein [11]. See [4] for an alternative proof.

Part (b) goes as follows. Choose $\alpha < 2$, and $T > 0$. Consider the dual action functional associated with problem (5)

$$\tag{48} \Phi_\alpha^T(u) = \int_0^T \left[\frac{1}{2}(Ju, \Pi u) + H_\alpha^*(-Ju)\right] dt.$$

We have seen in §II that there is a global minimizer u_α^T of Φ_α on $L^\beta(0, T)$, and a T-periodic solution x_α^T of $\dot{x} = JH_\alpha'(x)$ with $\dot{x}_\alpha^T = u_\alpha^T$. The value of the minimum can be computed, using the Legendre reciprocity formula and the homogeneity of H_α:

$$\tag{49} \int_0^T \left[\frac{1}{2}(J\dot{x}_\alpha^T, x_\alpha^T) + H_\alpha^*(-J\dot{x}_\alpha^T)\right] dt = -\int_0^T \left[H_\alpha(x_\alpha^T) + \frac{1}{2}(J\dot{x}_\alpha^T, x_\alpha^T)\right] dt$$

$$= (\alpha/2 - 1)TH_\alpha(x_\alpha^T).$$

If $T = T_\alpha$ is adjusted so that $H_\alpha(x_\alpha^T) = 1$, we get a solution (T_α, x_α) of problem (5), with

$$\tag{50} \operatorname{Min} \Phi_\alpha = (\alpha/2 - 1)T_\alpha.$$

On the other hand, we know that $H_\alpha(x) \geq \|x\|^\alpha R^{-\alpha}$ and hence

$$H_\alpha^*(y) \leq \frac{1}{\beta} \|y\|^\beta \left(\frac{R^\alpha}{\alpha}\right)^{\beta-1}.$$

Setting

$$\tilde{\Phi}_\alpha = \int_0^{T_\alpha} \left[\frac{1}{2}(Ju, \Pi u) + \frac{1}{\beta}\|-Ju\|^\beta \left(\frac{R^\alpha}{\alpha}\right)^{\beta-1} \right] dt, \tag{51}$$

we have

$$\operatorname{Min} \Phi_\alpha \leq \operatorname{Min} \tilde{\Phi}_\alpha. \tag{52}$$

But the right side is given by formula (49), with $H_\alpha(x)$ replaced by $\|\tilde{x}_\alpha\|^\alpha R^{-\alpha}$, namely:

$$\operatorname{Min} \tilde{\Phi}_\alpha = (\alpha/2 - 1) T_\alpha \|\tilde{x}_\alpha\|^\alpha R^{-\alpha}, \tag{53}$$

where $\tilde{x}_\alpha$ is a solution of

$$dx/dt = J(\alpha/R^\alpha)\|x\|^{\alpha-2} x, \quad x(0) = x(T_\alpha). \tag{54}$$

Putting (50), (52), and (53) together, and bearing in mind that $1 < \alpha < 2$, we get

$$\|\tilde{x}_\alpha\| \leq R, \tag{55}$$

where $\tilde{x}_\alpha$ solves problem (54). So $\tilde{x}_\alpha$ must have the form $\tilde{x}_\alpha(t) = \tilde{x}_\alpha(0) e^{\omega J t}$. Equation (54) gives us $\omega = \alpha \|\tilde{x}_\alpha\|^{\alpha-2} R^{-\alpha}$, while the periodicity condition requires that $\omega = 2k\pi/T_\alpha$, with $k \in \mathbf{Z}$. Taking $k = 1$, we get

$$\|\tilde{x}_\alpha\| = \left(\frac{2\pi}{T_\alpha} \frac{R^\alpha}{\alpha} \right)^{1/(\alpha-2)} \tag{56}$$

and relation (55) gives (bearing in mind that $\alpha < 2$)

$$\frac{2\pi}{T_\alpha} \frac{R^\alpha}{\alpha} \geq 1. \tag{57}$$

Letting $\alpha \to 2$, we get $(\pi/T_2) R^2 \geq 1$ as desired. □

We now turn to the existence of stable periodic solutions. The main result we have in this direction is the following, which is a symplectic version of a result by Ballman, Thorbergsson and Ziller in Riemannian geometry [3].

THEOREM 10. *If S is (r, R)-pinched with $Rr^{-1} < \sqrt{2}$, then problem (1) has at least one linearly stable solution.*

PROOF. Consider problem (5) with $1 < \alpha < 2$. We have seen that it always has a solution (T_α, x_α) of index 0. Letting $\alpha \to 2$, we get a T_2-periodic solution of index zero for the problem

$$\dot{x} = JH_2'(x(t)). \tag{58}$$

Consider the linearized equation around x_2,

$$\dot{y} = JH_2''(x_2(t)) y. \tag{59}$$

By Lemma 3, we have

$$2R^{-2} I \leq H_2''(x_2(t)) \leq 2r^{-2} I. \tag{60}$$

Let $s \in [0, 1]$ be a parameter, and associate with each value of s the positive definite matrix

(61) $$A_s = 2[sr^{-2} + (1-s)R^{-2}]I.$$

The Hamiltonian system $\dot{y} = JA_s y$ is linear, with constant coefficients. The resolvent $R_s(t)$ has a multiplier of Krein type $(n, 0)$ at $e^{i\theta_s t}$, with

$$\theta_s = 2[sr^{-2} + (1-s)R^{-2}],$$

and a multiplier of Krein type $(0, n)$ at $e^{-i\theta_s t}$. If

(62) $$\pi R^2/2 < t < \pi r^2$$

we will have $e^{i\theta_s t} \neq \pm 1$, and hence $e^{i\theta_s t} \neq e^{-i\theta_s t}$ for $0 \leq s \leq 1$. So all the systems $\dot{y} = JA_s y$ will be strongly stable on the interval $[0, t]$, provided (62) is satisfied (see [27, Chapter III] for the definition of strong stability and the fundamental theorem of Krein theory).

Since $R < r\sqrt{2}$, condition (62) defines a nonempty interval. By the directional convexity of stability domains (see [27, Chapter III, §6]), for all $t \in (\pi R^2/2, \pi r^2)$, system (59) must also be strongly stable on $[0, t]$. To be more precise, all the Floquet multipliers of (59) lie on the unit circle, the positive ones on the lower half-circle, in the arc between $e^{-i\theta_0 t}$ and $e^{-i\theta_1 t}$, and the negative ones on the upper half-circle, in the arc between $e^{i\theta_0 t}$ and $e^{i\theta_1 t}$.

Now consider the linear Hamiltonian system $\dot{y} = JB_s(t)y$, where

(63) $$B_s(t) = sH_2''(x_2(t)) + (1-s)2R^{-2}I$$

and $0 \leq s \leq 1$. We have

(64) $$(d/ds)B_s(t) = H_2''(x_2(t)) - 2R^{-2}I \geq 0.$$

It follows (see [27, Chapter III, §1]) that, for any fixed t, the Floquet multipliers of $\dot{y} = JB_s y$ on the interval $[0, t]$ move, as s increases, clockwise if they are Krein negative and counterclockwise if they are Krein positive, as long as they stay on the unit circle.

Now define t_0 to be the infimum of all $t > \pi R^2/2$ such that, for some $s \in [0, 1]$, the system $\dot{y} = JB_s y$ is not strongly stable on $[0, T]$.

We have $t_0 \geq \pi r^2$. Indeed, if $\pi R^2/2 < t < \pi r^2$, we know where the Floquet multipliers of $\dot{y} = JB_0 y$ and $\dot{y} = JB_1 y$ are. The Floquet multipliers of $\dot{y} = JB_s y$ lie between them, as described above, so that positive and negative multipliers are confined within separate arcs and cannot interfere. So they cannot leave the unit circle, and $\dot{y} = JB_s y$ remains strongly stable.

We have $t_0 < +\infty$. Indeed, we know that $y = \dot{x}_2$ is a T-periodic solution of system (59), so $\dot{y} = JB_1 y$ has 1 as a Floquet multiplier on $[0, T_2]$, and cannot be strongly stable on that interval. Hence $t_0 \leq T_2$.

Now take $t \in (\pi R^2/2, t_0)$. The Krein positive Floquet multipliers of $\dot{y} = JB_s y$ lie on arcs which extend counterclockwise from e^{-2it/R^2} to the positive Floquet multipliers of $\dot{y} = JH_2''(x_2)y$. These arcs cannot overlap with their conjugates, on which the negative Floquet multipliers lie, because otherwise two multipliers of different signs would meet for some $s \in [0,1]$, and the corresponding system would not be strongly stable. So the positive Floquet multipliers of $\dot{y} = JH''(x_2)y$ have to lie on the arc between e^{-2it/R^2} and 1, counterclockwise from the first extremity to the second.

When t crosses t_0, the positive and negative arcs must meet, and by continuity they can only do so at 1. So, on the interval $[0, t_0]$, the system (59) must have 1 as Floquet multiplier. But this means that 0 and t_0 are conjugate. We know that 0 and T_2 are also conjugate, and that system (59) has index 0, so T_2 is the first time conjugate to 0. By Proposition 6, we have $T_2 \geq \pi r^2$. So T_2 is also the first conjugate point to 0 after πr^2. Hence $T_2 = t_0$.

So the system (59) is strongly stable on all the intervals $[0, t]$, with $\pi R^2/2 < t < T_2$. By continuity, it has to be stable on the interval $[0, T_2]$. Hence the result. □

Note that by a result of Ekeland and Lasry [15], if $Rr^{-1} < \sqrt{2}$, problem (1) has at least n solutions $x^0, \ldots, x^{n-1}$, such that, for $i \leq j \leq n$,

$$(65) \qquad \pi r^2 \leq \oint \frac{1}{2}(Jx^j, dx^j) \leq 2\pi r^2.$$

We can give some more information about these solutions:

PROPOSITION 11. *Assume the solutions of problem* (1) *are all nondegenerate. Then there are at least n solutions* $x^0, \ldots, x^{n-1}$ *satisfying* (65), *and they are all nonhyperbolic.*

Let us first explain what we mean by nondegenerate and nonhyperbolic. A periodic solution is *nondegenerate* if 1 is not a nontrivial Floquet multiplier. It is *nonhyperbolic* if there is a nontrivial Floquet multiplier on the unit circle.

Now for the proof. By [15], we get n solutions $x^0, \ldots, x^{n-1}$, all satisfying (65), such that x^j has index $2j$.

Let us look at the second iterate of x^j, that is, the solution of problem (1) we get by running twice around x^j. Call it x_2^j. Clearly, we have

$$(66) \qquad 2\pi r^2 \leq \oint \frac{1}{2}(Jx_2^j, dx_2^j) \leq 4\pi r^2.$$

Proposition 8 then gives us an estimate for the index i_2 of x_2^j, namely,

$$(67) \qquad i_2 \leq 6n.$$

On the other hand, if x_2^j is hyperbolic, Proposition I.8 gives us an explicit formula for i_2 in terms of $i_1 = j$. We get:

$$(68) \qquad i_2 = 4j + n + 1.$$

Writing this in inequality (67) yields

$$(69) \qquad 4j + 1 \leq 5n. \qquad □$$

IV. Multiplicity of solutions. We consider a C^2-hypersurface S in $\mathbf{R}^{2n}$, bounding a convex compact set containing 0 in its interior, and we associate with it the differential equation

(1) $$\dot{x} = JN(x), \quad x \in S,$$

where $N(x)$ is the exterior normal to S at x. We want to know how many closed trajectories this flow has.

It has been proved by Ekeland and Lasry [15] that, if there are $x_0 \in \mathbf{R}^{2n}$, and numbers $0 < r \leqslant R$, with $R < r\sqrt{2}$, such that

(2) $$x \in S \Rightarrow r \leqslant \|x - x_0\| \leqslant R,$$

then (1) has at least n distinct closed trajectories. See [1, 17, 26] for shorter proofs, and [4] for a relaxation of the convexity assumption. Note that [15] only assumes S to be C^1, and that the condition $R < r\sqrt{2}$ has not been improved. What happens if $R \geqslant r\sqrt{2}$ is very much in doubt (see [1] and [4] for partial results).

There are counterexamples to show that the number n cannot be improved upon. Indeed, if we define S by

(3) $$S = \left\{ x \in \mathbf{R}^{2n} \mid \sum_{i=1}^{n} \omega_i (x_i^2 + x_{i+n}^2) = 1 \right\},$$

and if the $\omega_i > 0$ are linearly independent over the rationals, then we are dealing with n uncoupled linear oscillators with no common period, and the only way to achieve a periodic solution for the system is to keep $n - 1$ of them quiet and let the last one vibrate. There are exactly n ways to do this, hence n periodic solutions.

However, this counterexample appears to be pathological in many ways; note for instance that the corresponding system is completely integrable. So it is reasonable to ask for more closed trajectories in more general systems.

This is achieved by the following two results. The first one characterizes the systems which have only finitely many closed trajectories.

THEOREM 1. *Assume $n \geqslant 2$ and the Gaussian curvature of S does not vanish. Then one of the following properties holds true*:

(a) *either system* (1) *has infinitely many distinct closed trajectories*,

(b) *or the family of all closed trajectories of* (1) *is finite, nonempty, and resonant; in particular, they cannot all be hyperbolic.*

The second one tells us that most systems have infinitely many closed trajectories.

THEOREM 2. *Assume $n \geqslant 2$. Then the property*

(4) $$P(S) = \{\textit{system } (1) \textit{ has infinitely many closed trajectories}\}$$

is generic in the C^p-topology, $3 \leqslant p \leqslant \infty$.

We have a lot of explaining to do: "resonant" in Theorem 1 and "generic" in Theorem 2.

Let x be a closed trajectory of system (1), with Floquet multipliers $(1, 1, \lambda_3, \ldots, \lambda_{2n-1})$ (see the preceding section). It is classical to define x as *resonant* if it has some nontrivial Floquet multipliers on the unit circle, say $e^{\pm 2i\pi\theta_k}$, $1 \leq k \leq m$, and if there is a nontrivial relation with integer coefficients

(5) $$n_1\theta_1 + \cdots + n_k\theta_k = n_0,$$

(6) $$n_i \in \mathbf{Z}, 0 \leq i \leq k, \Sigma|n_i| \neq 0.$$

We want to define something similar for a finite family $x^1, \ldots, x^N$ of closed trajectories. With each x^p, $1 \leq p \leq N$, we associate:

(a) its nontrivial Floquet multipliers on the unit circle, say $e^{\pm 2i\pi\theta_k^p}$, $1 \leq k \leq m^p$ (here p is an index, not an exponent), and

(b) the number

$$\gamma^p = \frac{1}{2\pi} \int_c j^p(\omega) \, d\omega$$

(the function $j: C \to \mathbf{N}$ was defined in §I, and p is the index associating it with x^p).

We define the *normalized* Floquet multipliers of x^p to be

(7) $$\alpha_0^p = \gamma^p/2 \quad \text{and} \quad \alpha_k^p = \theta_k^p/\gamma^p \quad \text{for } 1 \leq k \leq m^p.$$

Because of the way j has been defined (see §I), the numbers $\{1, \alpha_k^p | 0 \leq k \leq m^p\}$ cannot be linearly independent over the rationals. There is always a nontrivial relation with integer coefficients $\sum_{k=0}^{m^p} n_k \alpha_k^p = n_0$. For instance, if x^p is hyperbolic, it has no nontrivial Floquet multiplier on the unit circle, and $j^p(\omega) = i^p + n + 1$, where i^p is the index of x^p. So the only normalized Floquet multiplier is $\alpha_0^p = (n + i^p + 1)/2$, which is rational. So 1 and α_0^p have corank 1 over the rationals.

Recall that the *corank* of a family of vectors is the number of linearly independent relations they satisfy. The family

(8) $$\{1, \alpha_k^p | 0 \leq k \leq m^p, 1 \leq p \leq N\}$$

has at least corank N over the rationals.

DEFINITION 3. We say that the family $x^1, \ldots, x^N$ is *nonresonant* if the set (8) has corank N over the rationals. If the corank is greater, the family is called resonant.

If a family of one is resonant, it consists of a closed trajectory which is resonant in the classical sense. A resonant family remains resonant if we remove all hyperbolic trajectories. In other words, every resonant family contains a minimal resonant subfamily, consisting only of nonhyperbolic trajectories. The statement of Theorem 1 should now be clear.

Now for the word "generic" in Theorem 2. Call Ω the set of hypersurfaces $S \subset \mathbf{R}^{2n}$, which are C^k for some $k \geq 3$, bound a compact convex set C containing 0 in its interior, and have strictly positive Gaussian curvature. Introduce its gauge $g_S: \mathbf{R}^{2n} \to \mathbf{R}$:

(9) $$g_S(s) = \inf\{\lambda > 0 | x \in \lambda C\}.$$

Consider the unit sphere Σ of $\mathbf{R}^{2n}$:

(10) $$\Sigma = \{ x | \|x\| = 1 \}.$$

With each $S \in \Omega$ we associate the function h_S on Σ defined by $h_S(x) = g_S(x)$ for all $x \in \Sigma$. The map $S \to h_S$ is one-to-one and sends Ω onto an open subset of $C^k(\Sigma; \mathbf{R})$. We endow Ω with the induced topology, which turns it into a complete metric space. So Baire's theorem will hold in Ω.

Now we are back on familiar ground: Saying that property $P(S)$ is generic means that the set of S for which $P(S)$ holds true contains the intersection of countably many dense open subsets. The statement of Theorem 2 should now be clear.

We refer to [13] for complete proofs of Theorems 1 and 2. Here, we will be content with providing guidelines for the reading of [13].

First, it is intuitively clear that Theorem 2 will follow from Theorem 1. Indeed, the content of the latter is that the closed trajectories cannot be finitely many unless their normalized Floquet multipliers satisfy certain nontrivial relations with integer coefficients. But any such relation can be destroyed by a small perturbation of system (1). It would thus appear that resonance is a pathological situation, so that system (1) should be nonresonant in general, and hence have infinitely many solutions.

This intuitive argument is made rigorous by the use of Thom's transversality theorem (see [13, §VI]). The final result is somewhat better than Theorem 2, namely, we show that the property

(11) $Q(S) = \{$system (1) has no resonant family of closed trajectories$\}$

is generic in Ω. Since $Q(S) \Rightarrow P(S)$ by Theorem 1, Theorem 2 follows.

Theorem 1 is the main result, and its proof requires several ingredients. First, as in §2, we turn our attention to the problem

(12) $$\dot{x} = JH'_\alpha(x), \quad x(0) = x(1),$$

where H_α is positively homogeneous of degree $\alpha \in (1,2)$, and S is described by the equation $H(x) = 1$. The relations between the solutions of problem (2) and the closed trajectories of system (1) are described by Lemma II.6. With every closed trajectory x of system (1) is associated a sequence x_k, $k \geq 1$, of homothetic solutions of problem (12), such that x_k has minimal period k^{-1}.

We then look at the dual action functional Φ_α on $L_0^\beta(0, T)$ (see formulas (29) and (30) in §II). We already noticed that $\Phi_\alpha(u) \to +\infty$ when $\|u\|_\beta \to \infty$, that $\Phi'_\alpha(0) = 0$, and that Φ_α has another critical point, namely its minimizer.

To find still more critical points, we use Morse theory. Denote by σ_θ the S^1-action on $L_0^\beta(0, T)$,

(13) $$\sigma_\theta(u)(t) = u(t + \theta T).$$

Clearly, Φ_α is S^1-invariant:

(14) $$\Phi_\alpha \circ \sigma_\theta = \Phi_\alpha \quad \forall \theta \in S_1.$$

It follows that if u is a critical point, so are all the points in its S^1-orbit:

(15) $$\Phi'_\alpha(u) = 0 \Rightarrow \Phi'_\alpha(\sigma_\theta u) = 0 \quad \forall \theta \in S^1.$$

The S^1-action is not free. However, if u is a critical point of Φ_α, we know that it is at least C^1 because of the equations $u = \dot{x}$ and $\dot{x} = -JH'_\alpha(x)$. So its S^1-orbit is a smooth circle in $L_0^\beta(0, T)$ (unless $u = 0$), possibly covered several times. This is enough to apply Bott's version of Morse theory for functions with critical submanifolds, set forth in [28]. The result is that, if all critical S^1-orbits of Φ_α are nondegenerate, there must be at least one for each even index $0, 2, 4, \ldots$.

Let us rephrase this. If u is a critical point of Φ_α, with $u \neq 0$, then $\sigma_\theta u$ is also a critical point for each $\theta \in S^1$. Letting $\theta \to 0$, we see that u cannot be isolated, and so u must be a degenerate critical point. In fact, the kernel of $\Phi''_\alpha(u)$ must contain $(d/d\theta)\sigma_\theta u = \dot{u}$, so the nullity of u is at least one. Clearly, all points in the S^1-orbit $S^1(u)$ have the same nullity and the same index. We say that $S^1(u)$ is nondegenerate if the nullity of u is exactly one, and the index of $S^1(u)$ is defined to be the index of u. By Bott's version of Morse theory, provided all critical S^1-orbits are nondegenerate, there must be for each even integer i a critical orbit of index i. Not that this does not account for $u = 0$, which is a (nonisolated) critical point of infinite index.

It turns out that the critical orbit $S^1(u)$ is nondegenerate if and only if the solution x of problem (12) related to u by $u = \dot{x}$ has no nontrivial Floquet multiplier. We shall say that x is a nondegenerate periodic solution.

So, if all critical S^1-orbits of Φ_α are nondegenerate, there must be infinitely many. By the dual action principle, this implies that if all the solutions of problem (12) are nondegenerate, there must be infinitely many. But we already know that problem (12) has infinitely many solutions: They are grouped into infinite sequences x_k^q, $k \geq 1$, each one of which corresponds to a single closed trajectory x^q of system (1). The question is whether there are infinitely many x^q.

Our third ingredient is the index theory for periodic solutions of convex Hamiltonian systems. This is what will enable us to recognize different patterns among the x_k^q.

With each x^q we associate a sequence of integers i_k^q, $k \geq 1$, defined by

(16) $$i_k^q = \text{index } x_k^q.$$

For each fixed q, the index i_k^q is given by the iteration formula in Proposition I.8. In other words, i_1^q and the Floquet multipliers of x^q on the unit circle completely determine the sequence i_k^q, $k \geq 1$.

Say for instance there is only one closed trajectory x of system (1), and it is hyperbolic. Then the iteration formula reads

(17) $$i_k = i_1 + (k-1)n.$$

If $n \geq 3$, this sequence cannot contain all even integers. This contradicts the results we derived from Morse theory. So our initial assumption must be impossible.

Now say that there are only finitely many closed trajectories of system (1) all of which are hyperbolic. We then have finitely many sequences $i_k^q = i_1^q + (k - 1)n$, and it is not hard to show, using appropriate prime numbers, that there are even numbers which belong to none of these sequences. So we get a contradiction again, which shows that if all solutions are hyperbolic, they must be infinitely many (note that they are automatically nonresonant and nondegenerate).

The case of nonhyperbolic solutions is much more complicated. The iteration formula then reads

$$(18) \qquad i_k = i_1 + (k - 1) j_{m+1} + 2 \sum_{p=1}^{m} (j_p - j_{p+1}) E[k\theta_p],$$

where the j_p, $1 \leq p \leq m + 1$, are appropriate integers, the $e^{\pm 2i\pi\theta_p}$, $1 \leq p \leq m$, $0 < \theta_p < \frac{1}{2}$, are the nontrivial Floquet multipliers, and $E[\rho]$ denotes the greatest integer $\leq \rho$.

It can be shown that

$$(19) \qquad i_1 + n - \overline{m} \leq i_{k+1} - i_k \leq i_1 + n + \overline{m},$$

where $\overline{m}$ is the sum of the multiplicities of the $e^{2i\pi\theta_p}$, for $1 \leq p \leq m$. Note that (19) implies

$$(20) \qquad 1 \leq i_{k+1} - i_k \leq i_1 + n + m.$$

Intuitively, because of the term $E(k\theta_p)$ in formula (18), and provided the θ_p are irrational, the sequence i_k will behave like a random sequence of integers—constrained, however, by the bounds in (19). In particular, it is to be expected that the rightmost bound will be achieved for some value k_0 of k. This means that i_k will jump by at least $n \geq 3$ at $k = k_0$, so that there will be some even integer $2p$ with $i_{k_0} < 2p < i_{k_0+1}$. Since the sequence i_k is increasing, it will follow that $i_k \neq 2p$ for all k.

If we now deal with a finite family x^q, $1 \leq q \leq N$, of closed trajectories, the i_k^q will behave like N independent random sequences, and any event of positive probability will happen in due time. So we expect that a certain even integer $2p$ will be missed by all of the N increasing sequences.

To put this kind of intuitive argument on safe ground, we first make the assumption that the x^q are a nonresonant family. Each of them then is nondegenerate, and we can apply the results of Morse theory. Using the nonresonance assumption, we then build an irrational flow on a high-dimensional torus, and we define in that torus a small region corresponding to the event "all sequences i_k^q have their maximum jump in the same region". We use the known properties of irrational flows on tori (namely, the fact that any trajectory is dense) to prove that this event actually occurs (infinitely often). So all sequences i_k^q jump over the same region, of length at least $n \geq 3$, and miss all numbers in that region, including certainly an even integer.

Apart from the computational complexities that arise from the behaviour of the index sequences i_k^q (the reader may find §V of [13] hard going), there are technical difficulties which arise at earlier stages. The function Φ_α, for instance, is not C^2 on $L_0^\beta(0, T)$, so we cannot apply Bott's result immediately (not to mention the fact that the underlying space is infinite-dimensional, and that Morse theory is usually stated in the C^3 framework). We have to build a finite-dimensional model of the map Φ_α: $L^\beta(0, T) \to \mathbf{R}$ which will be C^2, so that Bott's results can be applied to it. This construction takes up much of §III in [13].

There is even a mistake. In [13], we take $\alpha = 3/2$ (except for a small modification near the origin, the effect of which is to make 0 an isolated critical point of Φ_α), with the idea that any value of α in $(1, 2)$ will do. It is then stated that

$$\text{(21)} \qquad \lim_{\substack{\varepsilon \to 0 \\ \varepsilon > 0}} j(e^{i\varepsilon}) = i_1 + n + \begin{cases} \dfrac{d}{2} & \text{if } d \text{ is even,} \\ \dfrac{d-1}{2} & \text{if } d \text{ is odd} \end{cases}$$

(Proposition IV.7 of [13]). Actually, as we have seen in section I, this formula is incorrect. However, it can be proved that if $\alpha < 2$ is close enough to 2, this formula becomes

$$\text{(22)} \qquad \lim_{\substack{\varepsilon \to 0 \\ \varepsilon > 0}} j(e^{i\varepsilon}) = i_1 + n + 1$$

and the same argument carries through (replacing n by $n + 1$), with the only consequence that the provision $n \leq 3$ is replaced by $n \leq 2$.

As a conclusion, we wish to state some open problems for future research:

Question 1. Is it true that system (1) always has at least two distinct closed trajectories?

Question 2. Does there exist some S such that system (1) has only hyperbolic closed trajectories?

Some motivation for these problems. S is always understood to be a smooth hypersurface, bounding a convex compact set with nonempty interior. It is known that system (1) has always one solution at least, and that if some additonal metric condition is satisfied, namely, if S can be put between two concentric balls, the radii of which are not too far apart, there will be more. Such a metric condition is unnatural, because it is not invariant by canonical transformations. Question 1 asks if it is really needed.

In case of a linear system, that is, $H(x) = \sum_{i=1}^n \omega_i(x_i^2 + x_{i+n}^2)$ and S is an ellipsoid, all closed trajectories are elliptic, and it is known from classical perturbation theory that all nearby systems will have both elliptic and nonelliptic trajectories. So any convex Hamiltonian system with hyperbolic closed trajectories only must be strongly nonlinear. It is very likely that, for such a system, the set of periodic points would be dense in the phase space, and the Hamiltonian flow would be ergodic on every fixed energy level. Thus, a positive answer to

Question 2 would give us a Hamiltonian system with very interesting properties. On the other hand, a negative answer would be interesting too, since it would give us a general property of convex Hamiltonian systems. Theorem IV.2, for instance, then would be a corollary of this general fact and the Birkhoff–Lewis theorem. Indeed, any convex Hamiltonian system would have a nonhyperbolic closed trajectory, and a small perturbation of the system would ensure that the hypotheses of the Birkhoff–Lewis theorem are fulfilled (see [23]). It would then follow (see [19]) that there is an infinite sequence of closed trajectories, the periods of which increase to infinity, converging to the one we initially found.

Bibliography

1. A. Ambrosetti and G. Mancini, *On a theorem of Ekeland and Lasry concerning the number of periodic Hamiltonian trajectories*, J. Differential Equations **43** (1981), 1–6.

2. V. Arnold, *Méthodes mathématiques de la mécanique classique*, "Mir", Moscow, 1976. (French translation)

3. W. Ballmann, G. Thorbergsson and W. Ziller, *Closed geodesics on positively curved manifolds*, Ann. of Math. **116** (1982), 213–247.

4. H. Berestycki, J. M. Lasry, B. Ruf and G. Mancini, *Sur le nombre des orbites périodiques des équations d'Hamilton sur une surface d'énergie étoileé*, C.R. Acad. Sci. Paris **296** (1983), p. 15–18.

5. V. Brousseau, *L'index des systèmes hamiltoniens lineaires*, these 3eme cycle, Université Paris 9 Dauphine, 1984.

6. R. Bott, *On the iteration of closed geodesics and Sturm intersection theory*, Comm. Pure Appl. Math. **9** (1956), 176–206.

7. F. Clarke, *Periodic solutions of Hamiltonian inclusions*, J. Differential Equations **40** (1981), 1–6.

8. _____, *On Hamiltonian flows and symplectic transformations*, SIAM J. Control Optim. **20** (1982), 355–359.

9. F. Clarke and I. Ekeland, *Hamiltonian trajectories having prescribed minimal period*, Comm. Pure Appl. Math. **33** (1980), 103–116.

10. C. Conley and E. Zehnder, *The Birkhoff–Lewis fixed point theorem and a conjecture of V. I. Arnold*, Invent. Math. **73** (1983), 33–49.

11. C. Croke and A. Weinstein, *Closed curves on convex hypersurfaces and periods of nonlinear oscillations*, Invent. Math. **64** (1981), 199–202.

12. J. Duistermaat, *On the Morse index in variational calculus*, Adv. in Math. **21** (1976), 173–195.

13. I. Ekeland, *Une théorie de Morse pour les systémes hamiltoniens convexes*, Ann. Inst. H. Poincaré Anal. Non Linéaire **1** (1984), 19–78.

14. _____, *Hypersurfaces pincées et systèmes hamiltoniens*, C. R. Acad. Sci. Paris (to appear).

15. I. Ekeland and J. M. Lasry, *On the number of closed trajectories for a Hamiltonian flow on a convex energy surface*, Ann. of Math. **112** (1980), 283–319.

16. I. Gelfand and V. Lidskii, *On the structure of the regions of stability of linear canonical systems of differential equations with periodic coefficients*, Uspekhi Mat. Nauk **10** (1955), 3–40. (Russian)

17. H. Hofer, *A new proof for a result of Ekeland and Lasry concerning the number of periodic Hamiltonian trajectories on a prescribed energy surface*, Boll. Un. Mat. Ital. B (6) **1** (1982), 931–942.

18. M. Krein, *Generalization of certain investigations of A. M. Liapunov on linear differential equations with periodic coefficients*, Dokl. Akad. Nauk SSSR **73** (1950), 445–448.

19. J. Moser, *Proof of a generalized form of a fixed point theorem due to G. D. Birkhoff*, Lecture Notes in Math., vol. 597, Springer-Verlag, Berlin and New York, 1977, pp. 464–494.

20. J. Moser, *New aspects in the theory of stability of Hamiltonian systems*, Comm. Pure Appl. Math. **11** (1958), 81–114.

21. P. Rabinowitz, *Periodic solutions of Hamiltonian systems*, Comm. Pure Appl. Math. **31** (1978), 157–184.

22. H. Seifert, *Periodische Bewegungen mekanischer Systemen*, Math. Z. **51** (1948), 197–216.

23. F. Takens, *Hamiltonian systems: generic properties of closed orbits and local perturbations*, Math. Ann. **188** (1970), 304–312.
24. A. Weinstein, *Normal modes for nonlinear Hamiltonian systems*, Invent. Math. **20** (1973), 47–57.
25. _____, *Periodic orbits for convex Hamiltonian systems*, Ann. of Math. **108** (1978), 507–518.
26. M. Willem, *Lectures on critical point theory*, Brasilia, 1983.
27. V. Yakubovich and V. Starzhinskii, *Linear differential equations with periodic coefficients*, Halstedt Press, Wiley, 1980.
28. R. Bott, *Nondegenerate critical manifolds*, Ann. of Math. **60** (1954), 248–261.

CEREMADE, UNIVERSITÉ PARIS - DAUPHINE

An Extension of the Leray–Schauder Degree for Fully Nonlinear Elliptic Problems

P. M. FITZPATRICK AND JACOBO PEJSACHOWICZ

Introduction. The classical approach to the question of the existence of solutions to the Dirichlet problem for a second order quasilinear elliptic partial differential equation, introduced by Leray and Schauder in their celebrated paper [Le–Sc], can be sketched in an abstract setting as follows. They observe that the nonlinear map $f: X \to Y$, induced in the appropriate function spaces by the differential equation, has a natural representation of the form

$$(*) \qquad f(x) = L_x(x) + C(x),$$

where the map $x \mapsto L_x$ is continuous from a larger space X_1 in which X is compactly embedded and has values in the space of bounded linear maps from X to Y, and where the map $C: X \to Y$ is compact. From the Schauder estimates and the maximum principle it follows that each L_x is a bijection, so that the equation $f(x) = 0$ is equivalent to the equation $x - L_x^{-1}(C(x)) = 0$. Since the map $x \mapsto L_x^{-1}(C(x))$ is compact, existence results are then deduced from the degree theory for compact vector fields constructed in the same paper. In the second part of their work, Leray and Schauder show that the question of existence of solutions to the Dirichlet problem for second order fully nonlinear elliptic equations may also be reduced to the study of the zeros of a compact vector field by studying more general representations of the induced operator; specifically, they consider representations of the form

$$(**) \qquad f(x) = h_x(x),$$

where now $x \mapsto h_x$ is a compact mapping having values in a class of homeomorphisms.

In abstract settings this latter approach has been further developed by a number of authors (cf. [Kr–Za, Br–Nu, Br, Za–Ti]). In [Br–Nu] and [Br] one

---

1980 *Mathematics Subject Classification.* Primary 35J65, 47H17.

finds a degree theory, with values in Z, developed for mappings having representations of the form (**), where each h_x belongs to a convex family of admissible homeomorphisms. In [**Kr–Za**] one finds an alternative construction (compare with [**Ca, Sm**]) of a mod 2 degree theory for C^1 Fredholm maps.

The nonlinear Riemann–Hilbert problem was considered by Šnirel'mann in [**Šn**], where he observed that the relevant operator still had a representation of the form (*), but now each L_x is a singular integral operator which is Fredholm but not necessarily an isomorphism. In [**Šn**] there is introduced the notion of a quasilinear Fredholm map; it is a map f having a representation (*), where as before $C: X \to Y$ is compact, but now $x \mapsto L_x$ is continuous from X_1 into the set of linear Fredholm operators of index 0 from X to Y. A degree is constructed for such maps in [**Šn**] by approximating quasilinear Fredholm maps by vector bundle homomorphisms and then defining the degree in terms of the intersection number of a section of a vector bundle with the zero section. Due to the several choices involved in this construction, this degree is only defined up to sign; so that while it is sufficient for proving the existence of at least one solution of $f(x) = 0$, it does not "count" the number of solutions and hence it is inadequate for more delicate problems involving questions of multiplicity and detecting bifurcation of solutions from a trivial branch.

In this paper we will sketch the construction of an integer-valued degree theory for quasilinear Fredholm maps and illustrate the use of this degree in the study of global bifurcation and multiplicity of solutions for a quite general class of nonlinear elliptic boundary value problems which have linearizations of the type for which the generalized Schauder estimates of Agmon, Douglis and Nirenberg hold [**Ag–Do–Ni,1**]. Although, for brevity, we restrict ourselves to the case of a single equation, the results are also applicable to systems of the same type [**Ag–Do–Ni,2**].

The paper is organized as follows. In §1 we shall show that the nonlinear mapping induced by a nonlinear boundary value problem of the type referred to above is actually a quasilinear Fredholm mapping. In the second section we show that if f is a quasilinear Fredholm map represented by (*), then one can choose a map $x \to K_x$, continuous from X_1 and having values in the compact linear maps from X to Y, such that $L_x + K_x$ is invertible for each $x \in X_1$. By this device the equation $f(x) = 0$ can be reduced to the equation

$$(***) \qquad x - \overline{C}(x) = 0,$$

where $\overline{C}(x) = (L_x + K_x)^{-1}(C(x) - K_x(x))$. The map $\overline{C}$ is compact.

In §3 we extend the usual notion of orientation in finite-dimensional vector spaces by assigning ± 1 to each linear bijection from X to Y in an essentially arbitrary but as coherent as possible a manner. Using this and the Leray–Schauder degree of the reduced map (***), we obtain a well-defined degree function with values in Z and independent of representation for a quasilinear map $f: X \to Y$ over an open bounded subset $\mathcal{O}$ of X, where $0 \notin f(\partial \mathcal{O})$. When $X = Y$ and the quasilinear Fredholm map is a compact vector field, our degree coincides with that of Leray and Schauder.

In the final section we use our degree theory to obtain bifurcation and multiplicity results which are applicable to nonlinear boundary value problems. It is in this setting that one sees the advantages of having a degree defined for the map f itself instead of for a reduction (***). The representation (*) for a fully nonlinear boundary value problem is much less explicit than for a quasilinear boundary value problem; it involves integrodifferential operators. Moreover, in the absence of a Gårding inequality, there is neither a preferred nor explicit choice of compact inverting family $x \mapsto K_x$ and this makes difficult the computation of the relevant invariants from the reduced equation (***).

In our setting global bifurcation does not follow directly from a change of degree argument (cf. [**Ra**]). This is because the degree of a quasilinear mapping can change sign under an admissible homotopy. There cannot exist a degree theory which extends the Leray-Schauder degree theory, which includes quasilinear Fredholm maps and which is invariant under admissible homotopies. However we prove a weak homotopy invariance from which it follows that any change in sign of the degree is completely determined by the curve of linearized problems at the origin. More precisely, we introduce the notion of parity for a curve of linear Fredholm operators whose end points are invertible which counts, mod 2, the intersection of this curve with the singular set of noninvertible Fredholm operators. There are a number of ways in which one can compute this parity. Using the weak homotopy property and the notion of parity, the proof of a general global bifurcation theorem follows from the degree theory in the standard manner.

We shall not discuss here in detail the relationship of our degree to other degree theories. This will be done in a subsequent paper [**Fi–Pe**], where we shall also give complete proofs of the results stated here and other related results. We point out, however, that since our degree is based on a representation of the form (*) it has a formal similarity with other theories based on the form (**) that have been developed since the paper of Leray and Schauder. Our degree theory is different in that it has different properties than those previously constructed. The global bifurcation theorem we give cannot, we believe, be directly deduced from previous degree theories. It does not seem to have been previously observed that a degree theory can be constructed via a Leray-Schauder reduction which is applicable to fully nonlinear elliptic boundary value problems with general elliptic boundary conditions (cf. [**Ni**; **La–Ur**, Chapter 10]).

Finally, we add a remark concerning the degree theory for nonlinear Fredholm maps, a theory also applicable to nonlinear elliptic problems. When $\mathcal{O} \subseteq X$ is open and $f : \mathcal{O} \to Y$ is a proper C^2 Fredholm map, if X admits smooth partitions of unity and f induces a Fredholm manifold structure on $\mathcal{O}$ whose coordinate functions compose locally with f to yield compact perturbations of the identity, and if, in addition, this Fredholm structure is oriented then a degree, $\deg(f, \mathcal{O}, 0)$, is defined by Elworthy and Tromba [**El–Tr**], following ideas to be found in [**Ca**] and [**Sm**]. The absolute value of this degree is invariant under proper Fredholm homotopies. In order to determine whether or not the Fredholm degree changes sign under a proper Fredholm homotopy, it is necessary to study the manner in which the Fredholm structures on $\mathcal{O}$, induced by the slices of the homotopy, are related. This aspect of the Fredholm degree was not considered in [El–Tr].

1. Quasilinear Fredholm maps and nonlinear elliptic boundary value problems. Let X, X_1, and Y be Banach spaces, with X compactly embedded in X_1. A mapping $f: X \to Y$ is called quasilinear Fredholm if f has a representation of the form

$$(1.1) \qquad f(x) = L_x(x) + C(x),$$

where

(i) for $x \in X_1$, $L_x \in \mathcal{L}(X, Y)$ is Fredholm of index 0 and the map $x \mapsto L_x$ is continuous from X_1 to $\mathcal{L}(X, Y)$; and

(ii) $C: X \to Y$ is compact.

It is clear that a quasilinear elliptic boundary value problem, with appropriate boundary conditions, generates a quasilinear Fredholm map. More interesting is the fact that a fully nonlinear elliptic boundary value problem also does so.

We will consider nonlinear elliptic boundary value problems of the form

$$(1.2) \quad \begin{cases} f(x, u(x), \ldots, D^{2m}u(x)) = 0, & x \in \Omega, \\ b_i(x, u(x), \ldots, D^{m_i}u(x)) = 0, & x \in \partial\Omega, 1 \leq i \leq m, 0 \leq m_i < 2m, \end{cases}$$

where m is a positive integer and $\Omega \subset \mathbf{R}^n$ is an open bounded set having a smooth boundary.

Our assumptions are the following:

(H1) f and b_i, $1 \leq i \leq m$, belong to $C^{2+\mu}(\overline{\Omega} \times \mathbf{R}^M)$, $0 < \mu < 1$.

(H2) The first variation of $(f, b_1, \ldots, b_m)$, defined by

$$p(x, y)(v) = \sum_{|\alpha| \leq 2m} \frac{\partial f}{\partial(D^\alpha)}(x, y) D^\alpha v, \qquad (x, y) \in \overline{\Omega} \times \mathbf{R}^M,$$

$$q_i(x, y)(v) = \sum_{|\alpha| \leq m_i} \frac{\partial b_i}{\partial(D^\alpha)}(x, y) D^\alpha v, \qquad (x, y) \in \overline{\Omega} \times \mathbf{R}^M, 1 \leq i \leq m,$$

is such that p is properly elliptic (condition L of [**Ag–Do–Ni, 1**, p. 625]) and $\{q_i | 1 \leq i \leq m\}$ cover p in the sense of Shapiro–Lopatinskii (the complementing condition of [**Ag–Do–Ni, 1**, p. 626]).

Observe that from the estimates in [**Ag–Do–Ni, 1**] it follows from (H1) and (H2) that for a fixed $y \in \mathbf{R}^M$ the differential operators $(p, q_1, \ldots, q_m)$ induce a semi-Fredholm operator $\mathscr{L}$ from $C^{2m+\alpha}(\overline{\Omega})$ to $C^\alpha(\overline{\Omega}) \times \prod_{i=1}^m C^{2m+\alpha-m_i}(\partial\Omega)$ for any $\alpha \in (0, \mu)$. $\mathscr{L}$, in fact, has finite dimensional kernel and closed range [**Ag–Do–Ni, 1**, Theorem 7.3]. Our final assumption is that

(H3) when $y = 0$ the operator induced by $(p, q_1, \ldots, q_m)$ is Fredholm of index 0.

PROPOSITION 1.3. *Assume that* (H1)–(H3) *hold. Let*

$$X = C^{2m+2+\alpha}(\overline{\Omega}) \quad \text{and} \quad Y = C^{2+\alpha}(\overline{\Omega}) \times \prod_{i=1}^{m} C^{2m+2+\alpha-m_i}(\partial\Omega)$$

for $0 < \alpha < \mu$. *Let* $\mathscr{F}: X \to Y$ *be defined by*

$$\mathscr{F}(u)(x) = \big(f(x,\ldots,D^{2m}u(x)), b_1(x,\ldots,D^{m_1}u(x)),\ldots,b_m(x,\ldots,D^{m_m}u(x))\big).$$

Then $\mathscr{F}: X \to Y$ *is quasilinear Fredholm.*

PROOF. Let

$$W = C^{\alpha}(\overline{\Omega}) \times C^{1+\alpha}(\partial\Omega) \times \prod_{i=1}^{m} C^{2m+\alpha-m_i}(\partial\Omega)$$

and define $S: Y \to W$ by

$$S(u) = \left(\Delta u, \, u + \frac{\partial u}{\partial \eta}\bigg|_{\partial\Omega}, \, \Delta' u|_{\partial\Omega}, \ldots, \Delta' u|_{\partial\Omega} \right),$$

where Δ' denotes the Laplace–Beltrami operator on $\partial\Omega$. Since S is bijective, it is clear that to prove that $\mathscr{F}$ is quasilinear Fredholm it will suffice to show that $S \circ \mathscr{F}$ is quasilinear Fredholm. However, if we let $X_1 = C^{2m+1+\alpha}(\overline{\Omega})$, observe that

$$(S \circ \mathscr{F})(u) = L_u(u) + C(u),$$

where $C: X \to W$ is compact, and $L_u(v)$ for $u \in X_1$, $v \in X$ has first component equal to

$$\sum_{|\alpha|=2m} \frac{\partial f}{\partial (D^{\alpha})}(x,\ldots,D^{2m}u(x)) D^{\alpha}\Delta v(x), \quad x \in \Omega,$$

has second component equal to

$$\sum_{|\alpha|=2m} \frac{\partial f}{\partial (D^{\alpha})}(x,\ldots,D^{2m}u(x)) D^{\alpha}\frac{\partial v}{\partial \eta}(x), \quad x \in \partial\Omega,$$

and, for $1 \leq i \leq m$, has $(i+2)$ component equal to

$$\sum_{|\alpha|=m_i} \frac{\partial b_i}{\partial (D^{\alpha})}(x,\ldots,D^{m_i}u(x)) D^{\alpha}\Delta' v(x), \quad x \in \partial\Omega.$$

It is clear that the map $u \mapsto L_u$ is continuous from X_1 to $\mathscr{L}(X, W)$. Furthermore, it is not difficult to see that (H2) implies that the first component of L_u is a $(2m+2)$ order properly elliptic differential operator which is covered in the sense of Shapiro–Lopatinskii by the $(m+1)$ boundary operators which comprise the last $(m+1)$ components of L_u. It follows from Theorem 7.3 of [**Ag–Do–Ni, 1**] that each L_u is semi-Fredholm. From the stability of the semi-Fredholm index (see [**Ka**]), together with the assumption (H3) that L_0 is Fredholm of index 0, we conclude that L_u is Fredholm of index 0 for each $u \in X_1$. □

REMARK 1.4. Each solution of (1.2) which lies in $C^{2m+\alpha}(\overline{\Omega})$ is, in fact, in $C^{2m+2+\alpha}(\overline{\Omega})$. Thus there is no loss of generality in seeking solutions of (1.2) in $C^{2m+2+\alpha}(\overline{\Omega})$.

REMARK 1.5. In the H^s spaces a discussion similar to the proof of Proposition 1.3 may be found in Babin [**Ba**]. The device of studying a nonlinear equation by first differentiating it and then studying a quasilinear equation is, of course, not novel.

REMARK 1.6. When one considers a quasilinear elliptic boundary value problem of the form

$$\begin{cases} \sum_{|\alpha| \leqslant 2m} a_\alpha(x,\ldots,D^{2m-1}u(x))D^\alpha u(x) = 0, & x \in \overline{\Omega}, \\ \sum_{|\alpha| \leqslant m_i} b_\alpha^i(x,\ldots,D^{m_i-1}u(x))D^\alpha u(x) = 0, & x \in \partial\Omega, 1 \leqslant i \leqslant m, \end{cases}$$

one need only assume that the Nemytskii operators generating the coefficients are C^α, there is no need to use the device of composing with the Laplacian, and the operator induced by the above problem yields a quasilinear Fredholm operator from $C^{2m+\alpha}(\overline{\Omega})$ into $C^\alpha(\overline{\Omega}) \times \prod_{i=1}^m C^{2m+\alpha-m_i}(\partial\Omega)$.

2. The Leray-Schauder reduction.

Let X and Y be Banach spaces. By $\mathscr{K}(X,Y)$, $\mathrm{GL}(X,Y)$, and $\Phi_0(X,Y)$ we will denote the subsets of $\mathscr{L}(X,Y)$ consisting of the compact operators, the isomorphisms, and the operators which are Fredholm of index 0, respectively. The group of isomorphisms of the form $\mathrm{Id} - K$, where $K \in \mathscr{K}(X,X)$, will be denoted by $\mathrm{GL}_C(X)$. The properties of the Leray-Schauder degree which we cite may be found in [**Ll**].

It is well known that $\mathrm{GL}_C(X)$ has two connected components identified by

$$\mathrm{GL}_C^\pm(X) = \{\phi \in \mathrm{GL}_C(X), \deg_{\mathrm{L.S.}}(\phi, B, 0) = \pm 1\},$$

where $\deg_{\mathrm{L.S.}}(\phi, B, 0)$ is the Leray-Schauder degree of ϕ on any ball B, centered at the origin. Observe that since GL_C^+ is the component of Id, then it is a subgroup and

(2.1) ϕ_1 and ϕ_2 belong to the same component of $\mathrm{GL}_C(X)$ if and only if $\deg_{\mathrm{L.S.}}(\phi_1^{-1}\phi_2, B, 0) = 1$.

We begin by showing that the principal part of a quasilinear mapping f (i.e. the family L_x) is uniquely determined, modulo compact operators, by f. Namely, we have

PROPOSITION 2.2. *If $L_x(x) + C(x)$ and $L'_x(x) + C'(x)$ are each quasilinear representations of the same map f, then $L_{x_0} - L'_{x_0} \in \mathscr{K}(X,Y)$ for each $x_0 \in X$.*

PROOF. First consider $x_0 = 0$. A direct computation shows that the map $x \mapsto L_x(x) - L'_x(x)$ is differentiable at 0, with Fréchet derivative $L_0 - L'_0$. The compactness of $L_0 - L'_0$ follows from the identity $L_x(x) - L'_x(x) = C'(x) - C(x)$ and the fact that the derivative of a compact map is compact. For an arbitrary x_0 it suffices to apply the same argument to the translated map $x \mapsto f(x + x_0)$. □

PROPOSITION 2.3. *Let Λ be a paracompact, contractible topological space and let $L\colon \Lambda \to \Phi_0(X, Y)$ be continuous. Then there exists a continuous map $K\colon \Lambda \to \mathcal{K}(X, Y)$ with $(L + K)(\Lambda) \subseteq \mathrm{GL}(X, Y)$.*

PROOF. We may assume, without loss of generality, that $X = Y$. Let $G = \mathrm{GL}(X)/\mathrm{GL}_C(X)$. It is well known that the projection $q\colon \mathrm{GL}(X) \to G$ is a principal $\mathrm{GL}_C(X)$ bundle.

For each $L \in \Phi_0(X)$ choose $K \in \mathcal{K}(X)$ such that $L + K \in \mathrm{GL}(X)$ and define $\bar{q}(L)$ to be $q(L + K)$. Then $\bar{q}\colon \Phi_0(X) \to G$ is well defined. Let us also define $\gamma\colon \Lambda \to G$ by $\gamma(\lambda) = \bar{q}(L(\lambda))$.

From the definition of $\bar{q}$ it follows that our assertion is equivalent to the existence of a lifting $\tilde{\gamma}\colon \Lambda \to \mathrm{GL}(X)$ of $\gamma\colon \Lambda \to G$. That such a lifting exists follows from the paracompactness and contractibility of Λ together with Theorem 4.9.9 in [**Hu**]. □

REMARK 2.4. The above is essentially a well-known result. When Λ reduces to a point this is the Riesz characterization of $\Phi_0(X, Y)$. For a direct proof when Λ is compact and contractible see [**Br**] and [**Fi–Pe**]. A proof when Λ is compact and contractible and X and Y are Hilbert spaces was first given in [**Ja**]. When Λ is a convex, compact subset of a Banach space a direct proof can be found in [**Kr–Za**].

PROPOSITION 2.5. *Each quasilinear mapping $f\colon X \to Y$ admits a representation of the form $f(x) = M_x(x) + C(x)$, where $x \mapsto M_x$ is continuous from X_1 to $\mathrm{GL}(X, Y)$ and $C\colon X \to Y$ is compact. Moreover, the map $\bar{C}\colon X \to X$, defined by $\bar{C}(x) = M_x^{-1}(C(x))$, is compact.*

PROOF. Let $f(x) = L_x(x) + C'(x)$ be any quasilinear representation of f. Applying Proposition 2.3 with $\Lambda = X_1$, we obtain a continuous map $K\colon X_1 \to \mathcal{K}(X, Y)$ with $M_x \equiv L_x + K_x \in \mathrm{GL}(X, Y)$ for each $x \in X_1$. Thus $f(x) = M_x(x) + C(x)$, where $C(x) = C'(x) - K_x(x)$. The compactness of both C and $\bar{C}$ now follows from the compactness of C' and the compact embedding of X_1 in X. □

Let $f\colon X \to Y$ be a quasilinear mapping and let $\mathcal{O} \subseteq X$ be open and bounded with $0 \notin f(\partial\mathcal{O})$. From Proposition 2.5 it follows that the equation $f(x) = 0$, $x \in \mathcal{O}$, is equivalent to

$$x - \bar{C}(x) = 0, \quad x \in \mathcal{O},$$

where $\bar{C}(x) \equiv M_x^{-1}(C(x))$ is a compact mapping. In particular, if

$$\deg_{\mathrm{L.S.}} (I - \bar{C}, \mathcal{O}, 0) \neq 0,$$

then $f(x) = 0$ has a solution in $\mathcal{O}$.

We specifically mention that from the discussion in this and the preceding section it follows that the existence of solutions for nonlinear elliptic boundary value problems considered in the previous section follows from the

(i) existence of a curve, $f(t, u)$, of elliptic boundary value problems joining the given problem with a linear isomorphism, and

(ii) existence of a priori bounds for solutions of $f(t, u) = 0$, $0 \leq t \leq 1$.

3. A degree for quasilinear Fredholm maps.

When $\overline{C}$ is as in Proposition 2.5, $\deg_{L.S.}(I - \overline{C}, \mathcal{O}, 0)$ is only in absolute value independent of the representation of f. In this section we introduce a modification needed in order to obtain a well-defined degree, $\deg(f, \mathcal{O}, 0)$, with values in $\mathbf{Z}$.

DEFINITION 3.1. An *orientation* is a function $\varepsilon\colon \mathrm{GL}(X, Y) \to \{-1, +1\}$ verifying the following properties:

(i) If $M_1, M_2 \in \mathrm{GL}(X, Y)$, with $M_1 - M_2 \in \mathcal{K}(X, Y)$, then $\varepsilon(M_1) = \varepsilon(M_2)$ if and only if $M_1^{-1} M_2$ (which is an element of $\mathrm{GL}_C(X)$) belongs to $\mathrm{GL}_C^+(X)$.

(ii) If $X = Y$, then $\varepsilon(\mathrm{Id}) = 1$.

The existence of an orientation function for $\mathrm{GL}(X, Y)$ follows from the following argument: $\mathrm{GL}_C(X)$ acts on $\mathrm{GL}(X, Y)$ by multiplication on the right without fixed points. Consequently, each orbit is homeomorphic to $\mathrm{GL}_C(X)$ and hence has two connected components. Note that M_1 and M_2 belong to the same orbit if and only if $M_1 - M_2$ is compact. So an orientation ε is obtained by choosing one connected component in each orbit and assigning 1 to it: to the other component one assigns -1. In the case $X = Y$, one assigns $+1$ to $\mathrm{GL}_C^+(X)$.

REMARK 3.2. Observe that if $M_1 - M_2$ is compact and ε is any orientation of $\mathrm{GL}(X, Y)$, then

$$(3.3) \quad \varepsilon(M_1) \cdot \varepsilon(M_2) = \varepsilon(M_1^{-1} M_2) = \deg_{L.S.}(M_1^{-1} M_2, B, 0).$$

We are now able to define the degree. Suppose $f\colon X \to Y$ is quasilinear Fredholm and that $\mathcal{O} \subseteq X$ is open, bounded and $0 \notin f(\partial\mathcal{O})$. Let ε be an orientation for $\mathrm{GL}(X, Y)$. Choose any representation of f, $f(x) = M_x(x) + C(x)$, where each $M_x \in \mathrm{GL}(X, Y)$. Then letting $\overline{C}(x) = M_x^{-1}(C(x))$ we define the degree of f on $\mathcal{O}$ by

$$(3.4) \quad \deg(f, \mathcal{O}, 0) = \varepsilon(M_0) \deg_{L.S.}(\mathrm{Id} + \overline{C}, \mathcal{O}, 0).$$

PROPOSITION 3.5. $\deg(f, \mathcal{O}, 0)$ *is a well-defined function of f and of the orientation ε.*

PROOF. Suppose that f is represented both by $M_x(x) + C(x)$ and by $\tilde{M}_x(x) + \tilde{C}(x)$, where each M_x and each $\tilde{M}_x$ is in $\mathrm{GL}(X, Y)$. Let $\overline{C}(x) = M_x^{-1}(C(x))$ and $\overline{\tilde{C}}(x) = \tilde{M}_x^{-1}(\tilde{C}(x))$ for $x \in X$. Clearly we have

$$(3.6) \quad x + \overline{C}(x) = M_x^{-1} \tilde{M}_x\bigl(x + \overline{\tilde{C}}(x)\bigr), \quad x \in X.$$

But Proposition 2.2 asserts that $M_x^{-1} \tilde{M}_x \in \mathrm{GL}_C(X)$ for $x \in X$, and hence the map $H\colon [0,1] \times \overline{\mathcal{O}} \to X$, defined by

$$H(t, x) = M_{tx}^{-1} \tilde{M}_{tx}\bigl(x - \overline{\tilde{C}}(x)\bigr), \quad 0 \leq t \leq 1, x \in \overline{\mathcal{O}},$$

is an admissible homotopy of compact vector fields.

From the homotopy invariance and the product formula for the Leray-Schauder degree, together with (3.6), we conclude that

$$\deg_{L.S.}(\mathrm{Id} - \overline{C}, \mathcal{O}, 0) = \deg_{L.S.}(M_0^{-1} \tilde{M}_0, B, 0) \cdot \deg_{L.S.}(\mathrm{Id} - \overline{\tilde{C}}, \mathcal{O}, 0).$$

From (3.3) it follows that

$$\varepsilon(M_0)\deg_{\text{L.S.}}(\text{Id} - \overline{C}, \mathcal{O}, 0) = \varepsilon(\tilde{M}_0)\deg_{\text{L.S.}}(\text{Id} - \overline{\overline{C}}, \mathcal{O}, 0).$$

This concludes the proof. □

REMARK 3.7. Observe that when $M \in \text{GL}(X, Y)$, $\deg(M, B, 0) = \varepsilon(M)$.

The additivity, excision, and normalization properties of the degree follow from the corresponding properties of the Leray-Schauder degree. When $f = \text{Id} - C$, with C compact, the degree agrees with that of Leray and Schauder. The homotopy invariance property requires more care.

DEFINITION 3.8. A quasilinear Fredholm homotopy is a map $H: [0,1] \times X \to Y$ having a representation

$$H(t,x) = L_x^t(x) + C(t,x), \qquad 0 \leq t \leq 1, x \in X,$$

where the map $(t, x) \mapsto L_x^t$ is continuous from $[0,1] \times X_1$ into $\Phi_0(X, Y)$ and $C: [0,1] \times X \to Y$ is compact. When $\mathcal{O} \subset X$, H is said to be admissible with respect to $\mathcal{O}$ if $0 \notin H([0,1] \times \partial\mathcal{O})$.

Now, $\text{GL}(X)$ is connected for a large class of Banach spaces including infinite-dimensional Hilbert spaces, and when this is so $\text{Id} \in \text{GL}(X)$ can be joined by a path in $\text{GL}(X)$ to $T \in \text{GL}_C^-(X)$. Thus an admissible quasilinear homotopy may induce a change in sign in the degree. However, there is a weak homotopy invariance property which is sufficient to obtain global bifurcation and multiplicity results for quasilinear Fredholm mappings.

Suppose that $\mathcal{O} \subseteq X$ is open and bounded, and that $H: [0,1] \times X \to Y$ is a quasilinear homotopy which is admissible with respect to $\mathcal{O}$. From Proposition 2.3 we see that H may be represented in the form $H(t,x) = M_x^t(x) + C(t,x)$, where each $M_x^t \in \text{GL}(X, Y)$. Since $\overline{H}(t,x) \equiv x - (M_x^t)^{-1}(C(t,x))$ is an admissible homotopy of compact vector fields, from the homotopy invariance of the Leray-Schauder degree together with (3.4) we conclude that

(3.9) $$\deg(H_1, \mathcal{O}, 0) = \varepsilon(M_0^1)\varepsilon(M_0^0)\deg(H_0, \mathcal{O}, 0).$$

The number $\delta(H) \equiv \varepsilon(M_0^1)\varepsilon(M_0^0) \in \{-1, +1\}$ depends only on the homotopy of H and on ε, but is independent of the quasilinear representation. In the next section its significance will be made clearer.

We will call a quasilinear homotopy strongly orientation preserving provided that it has a representation as in Definition 3.8, where $L_0^t - L_0^0 \in \mathcal{K}(X, Y)$ for $0 \leq t \leq 1$. In this case a representation with principal part in $\text{GL}(X, Y)$ will, by Proposition 2.2, also have the same property. Since the curve M_0^t is contained in a single orbit it follows that $\varepsilon(M_0^0) = \varepsilon(M_0^1)$. Thus for a strongly orientation preserving quasilinear homotopy H which is admissible with respect to $\mathcal{O}$, we have

$$\deg(H_0, \mathcal{O}, 0) = \deg(H_1, \mathcal{O}, 0).$$

Finally, we have the following linearization result.

PROPOSITION 3.10. *Let $f: X \to Y$ be quasilinear Fredholm and Fréchet differentiable at $x = 0$. If $f(0) = 0$ and $f'(0)$ is invertible, then 0 is an isolated point of $f^{-1}(0)$ and $\deg(f, B(0, r), 0) = \varepsilon(f'(0))$ when r is sufficiently small.*

PROOF. Define $H: [0,1] \times X \to Y$ by

$$H(t, x) = \begin{cases} \frac{1}{t} f(tx), & t \neq 0, x \in X, \\ f'(0)(x), & t = 0, x \in X. \end{cases}$$

It is not difficult to see that H is a strongly orientation preserving quasilinear Fredholm homotopy. It is admissible with respect to $B(0, r)$ if r is sufficiently small. $\square$

4. Bifurcation and multiplicity results. For a curve of linear Fredholm operators of index 0, having invertible end points, we now define the notion of parity (cf. [**Iz**] and [**Ma**]). The parity counts algebraically, mod 2, the intersection of the curve with the singular set, $\Phi_0(X, Y) \setminus \mathrm{GL}(X, Y)$. Equivalently, it gives an obstruction in Z_2 to deforming the above curve through a homotopy in $\Phi_0(X, Y)$ to a curve in $\mathrm{GL}(X, Y)$. In our setting the parity determines bifurcation of zeros of one-parameter families of quasilinear Fredholm maps and using the degree theory one shows that the bifurcating branch is global.

Given a curve $L: I = [a, b] \to \Phi_0(X, Y)$ with L_a and L_b isomorphisms, by Proposition 2.3 there exists a curve $M: [a, b] \to \mathrm{GL}(X, Y)$ such that $M_t - L_t$ is compact for $0 \leq t \leq 1$. The parity of L on I is defined by

(4.1) $\quad \sigma(L, I) = \deg_{\mathrm{L.S.}}(M_0^{-1} L_0) \cdot \deg_{\mathrm{L.S.}}(M_1^{-1} L_1) \in Z_2 = \{+1, -1\}.$

The same calculation as that used in the proof of Proposition 3.5 shows that the parity does not depend on the choice of M. In particular if we take M with $M_a = L_a$, which is always possible, we have $\sigma(L, I) = \deg_{\mathrm{L.S.}}(M_b^{-1} L_b)$.

When the above L also has the property that $L_t - \mathrm{Id}$ is compact for each t, then clearly $\sigma(L, I) = \deg_{\mathrm{L.S.}}(L_a) \cdot \deg_{\mathrm{L.S.}}(L_b)$. In general we have the following properties.

(4.2) *Normalization*: If L is a curve in $\mathrm{GL}(X, Y)$, then $\sigma(L, I) = 1$.

(4.3) *Additivity*: Given a partition of I into $I_1, I_2, \ldots, I_n$,

$$\sigma(L, I) = \prod_{i=1}^{n} \sigma(L, I_i),$$

provided both sides are defined.

(4.4) *Composition*: If $L: I \to \Phi_0(X, Y)$ and $\tilde{L}: I \to \Phi_0(Y, Z)$ each have invertible end-points, then

$$\sigma(\tilde{L} \circ L, I) = \sigma(L, I) \cdot \sigma(\tilde{L}, I).$$

In particular, σ is invariant under composition with a curve in $\mathrm{GL}(Y, Z)$.

(4.5) *Homotopy invariance*: If $L: I \to \Phi_0(X, Y)$ and $\overline{L}: I \to \Phi_0(X, Y)$ are such that $L_a = \overline{L}_a$, $L_b = \overline{L}_b$, and L is homotopic to $\overline{L}$ in $\Phi_0(X, Y)$ by a homotopy leaving end points fixed, then $\sigma(L, I) = \sigma(\overline{L}, I)$.

(4.6) *Product formula*: If $L_1: I \to \Phi_0(X_1, Y_1)$ and $L_2: I \to \Phi_0(X_2, Y_2)$ each have invertible end points, then

$$\sigma(L_1 \oplus L_2, I) = \sigma(L_1, I) \cdot \sigma(L_2, I).$$

(4.7) *Localization*: If $L_t \in GL(X, Y)$ except for $t = t_1, \ldots, t_n$ and the parity of L at t_i, $\sigma(L, t_i)$, is defined by $\sigma(L, t_i) = \lim_{\delta \to 0} \sigma(L, [t_i - \delta, t_i + \delta])$, then

$$\sigma(L, I) = \prod_{i=1}^{n} \sigma(L, t_i).$$

(4.8) $\sigma(L, t_0) = (-1)^m$, where m is the multiplicity of the curve L at t_0 as defined in [Kr–Za], [Iz], [Ma] and [Ki], respectively, provided each of these respective multiplicities is defined. In particular, when $L_t = A - tK$, where $A \in \Phi_0(X, Y)$ and $K \in \mathcal{K}(X, Y)$, we have $\sigma(L, t_0) = (-1)^m$, where m is the dimension of the generalized null space of $\text{Id} - t_0 A^{-1} K$.

REMARK 4.9. Suppose that $H: [0, 1] \times X \to Y$ is a quasilinear Fredholm homotopy which is admissible with respect to the open bounded subset $\mathcal{O}$ of X. Suppose that $L_t = (\partial H/\partial x)(t, 0)$ exists and depends continuously on $t \in [0, 1]$, with L_0 and L_1 being invertible. Then it follows that

$$\deg(H(0, \cdot), \mathcal{O}, 0) = \varepsilon\left(\frac{\partial H}{\partial x}(0, 0)\right) \cdot \varepsilon\left(\frac{\partial H}{\partial x}(1, 0)\right) \sigma(L, [0, 1]) \deg(H(1, \cdot), \mathcal{O}, 0).$$

This follows from the definition of parity and the fact that H can be represented with principal part equal to $(\partial H/\partial x)(t, 0)$ when $x = 0$. As an illustration of the usefulness of the above formula let us further assume that $0 \in \mathcal{O}$ and $H(t, x) = 0$, with $x \in \mathcal{O}$ and $t \in \{0, 1\}$ if and only if $x = 0$. Then by Proposition 3.10 and the excision property of degree, it follows that $\deg(H(t, \cdot), \mathcal{O}, 0) = \varepsilon((\partial H/\partial x)(t, 0))$ when $t \in \{0, 1\}$. Thus we conclude that $\sigma(L, [0, 1]) = 1$. Thus it follows that if $\sigma(L, [0, 1]) = -1$, then $\mathcal{O}$ cannot be admissible and, in fact, for each bounded open subset U of $\mathcal{O}$ with $0 \in U$, there exists some $t \in [0, 1]$ and $x \in \partial U$ with $H(t, x) = 0$.

Recall that if $f: \mathbf{R} \times X \to Y$ is such that $f(t, 0) = 0$ for $t \in \mathbf{R}$, then $(t_0, 0)$ is called a bifurcation point of $f^{-1}(0)$ from $\mathbf{R} \times \{0\}$ if each neighborhood of $(t_0, 0)$ intersects $f^{-1}(0) \setminus \{\mathbf{R} \times \{0\}\}$.

PROPOSITION 4.10. *Let* $f: \mathbf{R} \times X \to Y$ *be quasilinear Fredholm, be* C^1 *and have* $f(t, 0) = 0$ *for* $t \in \mathbf{R}$. *Let* $L_t = (\partial f/\partial x)(t, 0)$. *Assume that for some* $a < b$, L_a *and* L_b *are isomorphisms and that* $\sigma(L, [a, b]) = -1$. *Then* $[a, b]$ *contains a bifurcation point of* $f^{-1}(0)$ *from* $\mathbf{R} \times \{0\}$.

PROOF. If not, $f(t, x)$ is an admissible homotopy between $f_a \equiv f(a, \cdot)$ and $f_b \equiv f(b, \cdot)$ with respect to $B(0, r)$, when r is sufficiently small. By the weak homotopy invariance property and Proposition 3.10 it follows that

$$1 = \varepsilon(M_0^b) \varepsilon(M_0^a) \varepsilon(L_a) \varepsilon(L_b).$$

But $M_t^0 - L_t$ is compact, and so the right side of the above is precisely $\sigma(L, [a, b])$. □

We also have the following multiplicity result.

PROPOSITION 4.11. *Let $f: X \to Y$ be quasilinear Fredholm and also be C^1. Suppose that $f(0) = 0$ and $f'(0) \in \mathrm{GL}(X, Y)$. Moreover, assume that there exists $M \in \mathrm{GL}(X, Y)$ such that*
  (i) $tf'(x) + (1 - t)M \in \Phi_0(X, Y)$ for $0 \leq t \leq 1$, $x \in X$,
  (ii) *for some $R > 0$, $tf(x) + (1 - t)M(x) \neq 0$ when $t \in [0, 1]$, $\|x\| \geq R$, and*
  (iii) $\sigma(L, [0, 1]) = -1$, *where $L_t = tf'(0) + (1 - t)M$, $0 \leq t \leq 1$.*
Then there exists an $x \neq 0$ with $f(x) = 0$.

PROOF. Suppose the conclusion is false. Then by the additivity of the degree together with Proposition 3.10 it follows that

$$\deg(f, B(0, R), 0) = \varepsilon(f'(0)).$$

Let $H(t, x) = tf(x) + (1 - t)M(x)$ for $0 \leq t \leq 1$, $x \in X$. From assumption (i) it follows that H is a quasilinear homotopy and assumption (ii) implies that H is admissible on $B(0, R)$. From the weak homotopy property and the definition of parity we now obtain a contradiction. □

Using the additivity, linearization, and weak homotopy properties of the degree one can in fact prove the following global bifurcation result which extends the local result, Proposition 4.10, and the Rabinowitz global bifurcation theorem [**Ra**].

PROPOSITION 4.12. *Suppose that $f: \mathbf{R} \times X \to Y$ is quasilinear Fredholm with $\mathbf{R} \times \{0\} \subseteq f^{-1}(0)$. Suppose $a, b \in \mathbf{R}$ with $a < b$ and $L_t \equiv (\partial f/\partial x)(t, 0)$ exists for $t \in [a, b]$ and is continuous, with L_a and L_b being invertible. Let $\mathcal{O} \subseteq \mathbf{R} \times X$ be open with $[a, b] \times \{0\} \subseteq \mathcal{O}$. Then, if $\sigma(L, [a, b]) = -1$, there exists a connected subset, C, of $f^{-1}(0) \setminus (\mathbf{R} \times \{0\})$, with $\overline{C} \cap \{(a, b) \times \{0\}\} \neq \varnothing$ and C has at least one of the following properties*:
  (i) C *is unbounded.*
  (ii) $C \cap \partial \mathcal{O} \neq \varnothing$.
  (iii) $(t^*, 0) \in \overline{C}$ *for some $t^* \notin [a, b]$.*

The multiplicity and bifurcation results of this section may now be used to study the fully nonlinear elliptic boundary value problem described in the first section. As already noted, in order to obtain the existence of one solution based upon a priori bounds one does not need to develop the full degree theory: the discussion in the first two sections is sufficient. Let us briefly consider a global bifurcation result.

Consider the equation

(4.13) $\quad \begin{cases} f_\lambda(x, u(x), \ldots, D^{2m}u(x)) = 0, & x \in \Omega \subseteq \mathbf{R}^n, \lambda \in \mathbf{R}, \\ b_{\lambda, i}(x, u(x), \ldots, D^{m_i}u(x)) = 0, & x \in \partial\Omega, 1 \leq i \leq m, \end{cases}$
$$0 \leq m_i \leq 2m - 1,$$

where f and the b's are $C^{3+\mu}$ for some $0 < \mu < 1$. For each λ suppose that f_λ and $\{b_{\lambda, i} | 1 \leq i \leq m\}$ satisfy (H1)–(H3) of §1. We also suppose that $u \equiv 0$ solves (4.13) for all $\lambda \in \mathbf{R}$.

Now consider the linearization at $u = 0$ of (4.13)—namely, the linear map L_λ: $X \to Y$, where X and Y are as in Proposition 2.3, induced by the linear problem

(4.14)
$$\begin{cases} \sum_{|\alpha| \leqslant 2m} \frac{\partial f_\lambda}{\partial D^\alpha}(x, 0) D^\alpha v(x) = 0, & x \in \Omega, \\ \sum_{|\alpha| \leqslant m_i} \frac{\partial b_{\lambda,i}}{\partial D^\alpha}(x, 0) D^\alpha v(x) = 0, & x \in \partial\Omega, 1 \leqslant i \leqslant m. \end{cases}$$

From our global bifurcation theorem it follows that if there exist $a < b$ with L_a, $L_b \in \mathrm{GL}(X, Y)$ and $\sigma(L[a, b]) = -1$, then there is global bifurcation in $C^{2m+2+\alpha}(\overline{\Omega})$, $0 < \alpha < \mu$, of nontrivial solutions of (4.13) from $[a, b] \times \{0\}$. Properties (4.7) and (4.8) may be used to check that $\sigma(L, [a, b]) = -1$.

References

[Ag–Do–Ni, 1] S. Agmon, A. Douglis and L. Nirenberg, *Estimates near the boundary for solutions of elliptic partial differential equations satisfying general boundary conditions*, Comm. Pure Appl. Math. **12** (1959), 623–727.

[Ag–Do–Ni, 2] _____, *Estimates near the boundary for solutions of elliptic partial differential equations satisfying general boundary conditions*, Comm. Pure Appl. Math. **17** (1964), 35–92.

[Ba] A. V. Babin, *Finite dimensionality of the kernel and cokernel of quasilinear elliptic mappings*, Math. USSR-Sb. **22** (1974), 427–454.

[Br] F. E. Browder, *Nonlinear operators and nonlinear equations of evolution in Banach spaces*, Proc. Sympos. Pure Math., Vol. 18, Part 2, Amer. Math. Soc., Providence, R.I., 1976.

[Br–Nu] F. E. Browder and R. D. Nussbaum, *The topological degree for noncompact nonlinear mappings in Banach spaces*, Bull. Amer. Math. Soc. **74** (1968), 671–676.

[Ca] R. Caccioppoli, *Sulle correspondenze funczionali inverse diramate: teoria generale e applicazioni ad alcune equazioni funzionali nonlineari e al problema di Plateau*, Opere Scelte, Vol. II, Edizioni Cremonese, Roma, 1963, pp. 157–177.

[El–Tr] K. D. Elworthy and A. J. Tromba, *Degree theory on Banach manifolds*, Proc. Sympos. Pure Math., Vol. 18, Part 1, Amer. Math. Soc. Providence, R.I., 1970, pp. 86–94.

[Fi–Pe] P. M. Fitzpatrick and J. Pejsachowicz, *Topological degree for mappings induced by nonlinear elliptic boundary volume problems* (to appear).

[Hu] Dale Husemoller, *Fiber bundles*, Springer-Verlag, 1966.

[Iz] Jorge Ize, *Bifurcation theory for Fredholm operators*, Mem. Amer. Math. Soc. No. 174 (1975).

[Ja] K. Janich, *Veklorraumbundël und der Raum der Fredholm-Operatoren*, Math. Ann. **161** (1965), 129–142.

[Ka] T. Kato, *Perturbation theory for linear operators*, 2nd ed., Grundlehren Math. Wiss., vol. 132, Springer-Verlag, 1980.

[Ki] Hansjörg Kielhöfer, *Multiple eigenvalue bifurcation for Fredholm operators*, Trans. Amer. Math. Soc. (to appear).

[Kr–Za] M. A. Krasnoselsky and P. Zabreiko, *Geometrical methods in nonlinear analysis*, Springer-Verlag, 1984.

[La–Ur] O. A. Ladyzhenskaia and N. N. Ural'tseva, *Linear and quasilinear elliptic equations*, Academic Press, New York, 1968.

[Le–Sc] J. Leray and J. Schauder, *Topologie et équations fonctionelles*, Ann. Sci. Ecóle Norm. Sup. (3) **51** (1934), 45–78.

[Ll] N. G. Lloyd, *Degree theory*, Cambridge Univ. Tracts, no. 73, Cambridge Univ. Press, New York, 1978.

[Ma] R. J. Magnus, *A generalization of multiplicity and the problem of bifurcation*, Proc. London Math. Soc. (3) **32** (1976), 251–278.

[Ni] L. Nirenberg, *Variational and topological methods in nonlinear problems*, Bull. Amer. Math. Soc. (N.S.) **4** (1981), 267–302.

[Ra] P. H. Rabinowitz, *Some global results for nonlinear eigenvalue problems*, J. Funct. Anal. **7** (1971), 487–513.

[Sm] S. Smale, *An infinite dimensional version of Sard's theorem*, Amer. J. Math. **87** (1965), 861–866.

[Šn] A. I. Šnirel'mann, *The degree of a quasi-ruled mapping and a nonlinear Hilbert problem*, Math. USSR-Sb. **18** (1972), 373–396.

[Za–Ti] P. P. Zaibreiko and V. P. Tikhonov, *Determining equations and the duality principle*, Siberian Math. J. **24** (1983), 65–72.

University of Maryland

Università Della Calabria

Nonlinear Spectral Manifolds for the Navier–Stokes Equations

C. FOIAŞ AND J. C. SAUT

0. Introduction. It is well known that solutions of Navier–Stokes equations

$$\partial u/\partial t + (u \cdot \nabla)u = \nu\Delta u + \nabla p,$$
$$\nabla \cdot u = 0, \quad u(x,0) = u_0$$

in $\Omega \times]0, \infty[$ (Ω smooth open bounded set in $\mathbb{R}^n$, $n = 2, 3$), with natural boundary conditions, are regular if $n = 2$, and become regular after a finite time (depending on the solution) if $n = 3$. Consequently, their energy

$$\int_\Omega |u(x,t)|^2 \, dx$$

and their enstrophy

$$\int_\Omega |\nabla\Lambda u(x,t)|^2 \, dx$$

decay at least exponentially in time, as $t \to \infty$.

This is essentially due to Leray [15, 16]. The aim of this paper is to show that this decay is exactly of exponential type, and is characterized by an eigenvalue of the Stokes operator. More precisely, we show that the ratio of the enstrophy over the energy has a limit as $t \to \infty$ which is an eigenvalue of the Stokes operator. This fact has direct consequences on the global structure of the Navier–Stokes equations. Indeed, this allows us to construct in the space $\mathscr{R}$ of initial data a flag of nonlinear spectral manifolds of the Navier–Stokes operator. These are analytic manifolds of finite codimension which are invariant for the Navier–Stokes equations and which completely determine the energy decay of the solution. In this paper we present the construction and some properties of these manifolds, as well as the technical, which are relevant for our presentation. For the details we

---

1980 *Mathematics Subject Classification.* Primary 35Q10, 76D05; Secondary 34G10.

refer to our papers [6, 7, 8, 10]. Further developments of this subject (in particular, for the linearization of these manifolds, as well as of the Navier–Stokes equations) were indicated in [9].

1. Notations. Review of some classical results. Let Ω be either a bounded regular open set in $\mathbb{R}^n$, or the cube $[0, L]^n$, where $n = 2, 3$. In the first case, we set

(1.1) $$\mathscr{V} = \{v \in C_0^\infty(\Omega)^n;\ \text{div } v = 0\}$$

and, in the second case, we set

(1.2) $$\mathscr{V} = \bigg\{u = \text{trigonometric polynomials with values in } \mathbb{R}^n;$$

$$\text{div } u = 0,\ \int_\Omega u\, dx = 0\bigg\}.$$

In both cases we set as usual (see [17, 18])

$$H = \text{closure of } \mathscr{V} \text{ in } L^2(\Omega)^n, \qquad V = \text{closure of } \mathscr{V} \text{ in } H^1(\Omega)^n,$$

where $H^l(\Omega)$ ($l = 1, 2, \ldots$) denotes the Sobolev space of u's in $L^2(\Omega)$ such that $D^\alpha u \in L^2(\Omega)$, $\forall |\alpha| \leq l$. We shall use the classical notation

$$|u|^2 = \int_\Omega |u(x)|^2\, dx, \qquad \|v\|^2 = \int_\Omega |\nabla v(x)|^2\, dx$$

($u \in H$, $v \in V$) for norms in H (resp. V). The corresponding scalar products will be denoted by $(\cdot, \cdot)$, (resp. $((\cdot, \cdot))$).

In the first case, the Navier–Stokes equations are completed by the boundary conditions

(1.1)' $$u|_{\partial\Omega} = 0$$

and, in the second case, by the spatial periodic condition

(1.2)' $$u(x + Le_j) \equiv u(x) \qquad \forall x \in \mathbb{R}^n,\ i \leq j \leq n,$$

where (e_j), $i \leq j \leq n$, is the canonical basis in $\mathbb{R}^n$.

Let P be the orthogonal projection on H in $L^2(\Omega)^n$ and let

$$Au = -P\Delta u, \qquad B(v, w) = P[(v \cdot \nabla)w],$$

defined for $u, v, w \in V \cap H^2(\Omega)^n$ and H-valued. We also set

$$b(u, v, w) = (B(u, v), w).$$

A and B can be extended by continuity to linear (resp. bilinear) operators from V (resp. $V \times V$) into the dual $V' \supset H \supset V$ of V. Equations (0.1) completed by (1.1)' or (1.2)' are then equivalent to

(1.3) $$du/dt + \nu Au + B(u, u) = 0.$$

It is known since Leray's basic papers that

(i) For every $u_0 \in H$, (1.3) possesses a weak solution (i.e. satisfying (1.3) in V') and which belongs to $C([0, \infty]; H_{\text{weak}}) \cap L^2_{\text{loc}}(0, \infty; V)$, such that $u(0) = u_0$.

(ii) This solution becomes regular for t large enough, say $t > t_0(\nu, u_0)$. (Let us recall that, by definition, u is regular on some interval $I \subset [0, \infty)$ if $u|_I \in C(I; V)$.)

(iii) If a solution u is regular on $I = [t_0, t_1]$, then $u|_I$ is uniquely determined by $u(t_0)$.

(iv) $t_0(\nu, u_0) = 0$ if $n = 2$, or if $n = 3$ and $u_0 \in V$, for u_0 sufficiently small (see for instance the books [17, 18]).

By these results it is sufficient for the asymptotic behaviour of solutions $u(t)$ of (1.3) as $t \to \infty$ to consider only solutions which are regular on $[0, \infty)$. Let $\mathcal{R}$ be the set of initial data of such solutions ($\mathcal{R} = V$ for $n = 2$, and $\mathcal{R}$ is an open set of V containing 0 if $n = 3$). Then $S(t): u_0 \in \mathcal{R} \mapsto u(t) \in \mathcal{R}$ defines a semigroup $(S(t))_{t \geq 0}$ of nonlinear mappings $\mathcal{R} \to \mathcal{R}$ which has nice properties of analyticity (cf. [5, 11]). In particular, if $u_0 \neq 0$, $u_0 \in \mathcal{R}$, then $S(t)u_0 \neq 0 \; \forall t \in 0$ [11].

Let us also recall the inequalities (see [11, 18])

$$(1.4) \quad \begin{cases} |B(v,w)| \leq \begin{cases} C_1 \|v\| \|w\|^{1/2} |Aw|^{1/2}, \\ C_2 |v|^{1/2} \|v\|^{1/2} |Aw| & \text{for } v \in V, w \in D(A), \end{cases} \\ \|B(v,v)\|_{V'} \leq C_3 |v|^{1/2} \|v\|^{3/2}, \quad v \in V. \end{cases}$$

(Here and in the sequel, $C_1, C_2, C_2, \ldots$ will designate constants with respect to the explicit variables of the formulas.)

From (1.3), (1.4) comes (for $u(\cdot) = S(\cdot)u_0$, $u_0 \in \mathcal{R}$)

$$(1.5) \quad d\|u\|^2/dt + 2\left(\nu - C_2|u|^{1/2}\|u\|^{1/2}\right)|Au|^2 \leq 0.$$

By considering the energy equality

$$(1.6) \quad \frac{1}{2}\frac{d}{dt}|u|^2 + \nu\|u\|^2 = 0 \quad \text{and then} \quad \nu \int_0^\infty \|u\|^2 \, dt = \frac{1}{2}|u_0|^2,$$

which are valid for $u(\cdot) = S(\cdot)u_0$, it follows easily from (1.5) that[1]

$$(1.7) \quad \lim_{t \to \infty} |u(t)| \|u(t)\| = 0$$

and

$$(1.8) \quad \int_t^\infty |Au|^2 \, ds \leq C_3 \|u(t)\|^2 \quad \text{for } t \geq C_4 |u_0|^4.$$

Finally, let us recall that A is in fact the *Stokes operator*: The equation $\nu Au = Pf$ in H is equivalent to the Stokes equations

$$-\nu \Delta u = f - \nabla p, \quad \nabla \cdot u = 0 \quad \text{in } \Omega, \quad f \in L^2(\Omega)^n,$$

completed with conditions (1.1)′ or (1.2)′.

We shall denote by $0 < \lambda_1 \leq \lambda_2 \leq \cdots$ the increasing sequence of the eigenvalues of A and by $(w_m)_{m=1}^\infty$ the orthonormal in H (or in V) basis of the associated eigenvectors $Aw_m = \lambda_m w_m$. Finally, P_m will denote the orthogonal projection in H on the linear space of $w_1, \ldots, w_m$ ($m = 1, 2, \ldots$) and $Q_m = I - P_m$. Moreover, we shall denote by $0 < \Lambda_1 = \lambda_1 < \Lambda_2 < \cdots$ the increasing sequence of distinct eigenvalues of A and R_{Λ_j} the orthogonal projection of the eigenspace of Λ_j.

---

[1] More precisely, $|u(t)|\|u(t)\| \leq C_5 e^{-\nu\lambda_1 t}$, where λ_1 is the first eigenvalue of A.

2. Behavior of the ratio $\|S(t)u_0\|^2/|S(t)u_0|^2$ as $t \to +\infty$.

For a fixed $u_0 \in \mathscr{R}$, $u_0 \neq 0$, the ratio

$$(2.1) \qquad \lambda(t) = \|S(t)u_0\|^2 \Big/ |S(t)u_0|^2$$

is (see §1) defined for every $t \geq 0$. The behavior of this ratio as $t \to \infty$ is given by the following result:

THEOREM 1. *The limit*

$$(2.2) \qquad \lim_{t \to \infty} \lambda(t) = \Lambda(u_0)$$

exists and is one of the eigenvalues of the Stokes operator.

Let us recall that for an eigenvalue λ of A we denote by R_λ the orthogonal projection of H onto the eigenspace of λ, i.e.

$$R_\lambda w = \sum_{\lambda_j = \lambda} (w, w_j) w_j.$$

Evidently, if $\lambda_m < \lambda \leq \lambda_{m+1}$ and $\lambda_M \leq \lambda < \lambda_{M+1}$, then

$$(2.3) \qquad R_\lambda = P_M - P_m.$$

The next proposition says that $u(t)$ behaves asymptotically as $R_\Lambda u_0(t)$.

PROPOSITION 1. *Under the assumptions of Theorem 1, one has*

$$(2.4) \qquad \lim_{t \to \infty} \left\| \left(I - R_{\Lambda(u_0)}\right) \frac{S(t)u_0}{|S(t)u_0|} \right\| = \lim_{t \to \infty} \left| \left(I - R_{\Lambda(u_0)}\right) \frac{S(t)u_0}{|S(t)u_0|} \right| = 0.$$

In the periodic (1.2)-(1.2)', *we have the stronger convergence*

$$(2.5) \qquad \lim_{t \to \infty} \left| A\left(I - R_{\Lambda(u_0)}\right) \frac{S(t)u_0}{|S(t)u_0|} \right| = 0.$$

The proof of Proposition 1 is based on

LEMMA 1. *One has*

$$(2.6) \qquad |u(t)| \leq C_8 e^{-\nu \Lambda(u_0) t} \qquad \forall\, t \geq 0,$$

$$(2.7) \qquad \|u(t)\| \leq C_9 e^{-\nu \Lambda(u_0) t} \qquad \forall\, t \geq 0.$$

It was observed by C. Guillopé [12] that the convergence

$$\left(1 - R_{\Lambda(u_0)}\right) \frac{S(t)u_0}{|S(t)u_0|} \to 0$$

holds also in $H^l(\Omega)^n$ for $l = 3, 4, \ldots$. This fact is very useful in the proof of the asymptotic expansion of $u(t)$ indicated in [9]. The convergence of $\lambda(t) = \|u(t)\|^2/|u(t)|^2$ to $\Lambda(u_0)$ is further illuminated by the following results:

PROPOSITION 2.

$$(2.8) \qquad \int_0^\infty |\lambda(t) - \Lambda(u_0)| \, dt < \infty.$$

The equality of energy (1.6) and (2.8) imply immediately the following lower bound on $u(t)$:

$$|u(t)| = |u(0)|e^{-\nu\Lambda t}\exp\left[-\nu\int_0^t (\lambda(s) - \Lambda)\,ds\right]$$

$$\geq |u(0)|e^{-\nu\Lambda t}\exp\left[-\nu\int_0^\infty |\lambda(s) - \Lambda|\,ds\right].$$

This improves an estimate given by Dyer and Edmunds [3] for classical solutions of the Navier–Stokes equations.

PROPOSITION 3. $\lim_{t\to\infty} e^{\nu\Lambda(u_0)t}|S(t)u_0|$ exists, is finite, and is not zero. More precisely, $e^{\nu\Lambda(u_0)t}S(t)u_0$ converges (in H and V) to a nonzero eigenvector of A associated to $\Lambda(u_0)$.

This provides a precise answer to a question raised by P. D. Lax [14].

Proposition 3 implies obviously that if, as $t \to +\infty$, $|u(t)| = O(e^{-\mu t})\ \forall \mu > 0$, then $u \equiv 0$. Edmunds [4] observed that (for classical solutions) the same conclusion holds if $\operatorname{Sup}_{x\in\bar\Omega}|\nabla u(x,t)| = O(e^{-\mu t})\ \forall \mu > 0$. He used totally different arguments.

THEOREM 2. *With the previous notations, we have*

(2.9) $\quad \lim_{t\to\infty} e^{\nu\lambda t} R_\lambda S(t) u_0 = U_\lambda$ exists for $\lambda_1 \leq \lambda < 2\Lambda$ (here $\Lambda = \Lambda(u_0)$),

(2.10) $\qquad U_\lambda = 0$ if $\lambda_1 \leq \lambda < \Lambda$, $\quad U_\Lambda \neq 0$,

(2.11)

$$R_\lambda S(t)u_0 = \begin{cases} \dfrac{e^{-2\Lambda\nu t}}{(2\Lambda - \lambda)\nu} R_\lambda B(U_\Lambda, U_\Lambda) + o(e^{-2\Lambda\nu t}) & \text{if } \lambda_1 \leq \lambda < \Lambda, \\[6pt] e^{-\nu\lambda t}U_\lambda + \dfrac{e^{-2\Lambda\nu t}}{(2\Lambda - \lambda)\nu} R_\lambda B(U_\Lambda, U_\Lambda) + o(e^{-2\Lambda\nu t}) & \text{if } \Lambda \leq \lambda < 2\Lambda, \\[6pt] -te^{-2\Lambda\nu t} R_\lambda B(U_\Lambda, U_\Lambda) + O(e^{-2\Lambda\nu t}) & \text{if } \lambda = 2\Lambda, \\[6pt] \dfrac{-e^{-2\Lambda\nu t}}{(\lambda - 2\Lambda)\nu} R_\lambda B(U_\Lambda, U_\Lambda) + o(e^{-2\Lambda\nu t}) & \text{if } 2\Lambda < \lambda. \end{cases}$$

Formula (2.11) is a particular case of the asymptotic expansion of $u(t)$ as $t \to \infty$, given in [9], sufficient for our present purposes.

Theorem 2 has the following obvious but useful

COROLLARY. *A necessary and sufficient condition for $\Lambda(u_0) \geq \Lambda_k$ ($k \geq 2$) is that*

(2.12) $\qquad \lim_{t\to\infty} e^{\nu\Lambda_j t} R_{\Lambda_j} u(t) = 0 \quad \text{for every } j = 1, 2, \ldots, k - 1.$

3. Nonlinear spectral manifolds. Let us first, instead of the nonlinear equation (1.3), consider the linear equation

(3.1) $\qquad du/dt + \nu Au = 0.$

Then the solution of (4.1) corresponding to $u(0) \in V$ (or H) is given by $e^{-\nu t A} u_0$ where $(e^{-\nu t A})_{t \geq 0}$ is the semigroup of linear mappings with $-\nu A$ as generator.

It is trivial to check that, for $u_0 \neq 0$,

$$\lim_{t \to \infty} \frac{\|e^{-\nu t A} u_0\|^2}{|e^{-\nu t A} u_0|^2} = \Lambda^{\text{lin}}(u_0),$$

where $\Lambda^{\text{lin}}(u_0)$ is the "first" eigenvalue Λ_k of A such that $R_{\Lambda_k} u_0 \neq 0$. So, in the linear case, we have $\Lambda^{\text{lin}}(u_0) \geq \Lambda_k$ if and only if u_0 belongs to the following (linear) spectral manifold of A:

$$M_k^{\text{lin}} = \{ u_0 \in V;\ R_{\Lambda_1} u_0 = \cdots = R_{\Lambda_{k-1}} u_0 = 0 \}.$$

According to this remark, the sets M_k, $k = 1, 2, \ldots$, of the next theorem can be viewed as nonlinear variants of the M_k^{lin}'s. To our knowledge, the M_k's are the first nontrivial invariants of the Navier–Stokes flow to be exhibited.

THEOREM 3. *There exist analytic sets* M_k, $k = 1, 2, \ldots$, *in* $\mathcal{R}$ *such that*

(i) $\mathcal{R} = M_0 \supset M_1 \supset M_2 \supset \cdots \supset M_k \supset M_{k+1} \supset \cdots$,

(ii) M_k *is a smooth analytic manifold of codimension* $m_1 + m_2 + \cdots + m_k$ *around the origin*,

(iii) M_k *is invariant by the semigroup* $(S(t))_{t \geq 0}$, *i.e.*, $S(t) M_k \subset M_k$,

(iv) $u_0 \in M_k \setminus M_{k+1} \Leftrightarrow \Lambda(u_0) = \Lambda_{k+1}$, $k = 0, 1, \ldots$,

(v) *the tangent space of* M_k *at 0 is* M_k^{lin} ($k = 1, 2, \ldots$).

Before proving this result, let us notice that, as it will be shown in §4, M_k is actually an everywhere smooth analytic manifold of codimension $m_1 + \cdots + m_k$ in V.

PROOF OF THEOREM 3. One has, for $u_0 \in \mathcal{R}$,

(3.2) $\quad e^{\nu \Lambda_1 t} R_{\Lambda_1} u(t) - R_{\Lambda_1} u_0 + \int_0^t e^{\nu \Lambda_1 \tau} R_{\Lambda_1} B(u(\tau), u(\tau))\, d\tau = 0.$

Consequently, by Corollary 2, $\Lambda(u_0) \geq \Lambda_2$ if and only if

(3.3) $\quad R_{\Lambda_1} u_0 = \int_0^\infty e^{\nu \Lambda_1 t} R_{\Lambda_1} B(S(t) u_0, S(t) u_0)\, dt.$

Let us now define $\Phi_1 : \mathcal{R} \to R_{\Lambda_1} V$ by

$$\Phi_1(v) = R_{\Lambda_1} v - \int_0^\infty e^{\nu \Lambda_1 t} R_{\Lambda_1} B(S(t) v, S(t) v)\, dt.$$

This mapping is well defined (since $S(t)v$ decays at least as $e^{-\nu \Lambda_1 t}$) and analytic (this is essentially due to the analyticity of the mapping $v \to S(t)v$ [2]).

We set $M_1 = \Phi_1^{-1}(\{0\})$. Since $\Phi_1'(0) = R_{\Lambda_1}$, M_1 is, in a neighborhood of 0, an analytic manifold of codimension m_1 in V (see [1]). By (2.12), $v \in M_1 \setminus \{0\}$ is equivalent to $\Lambda(v) \geq \Lambda_2$. The invariance of M_1 under $S(t)$ is clear.

To construct M_2, one considers $\Phi_2 : M_1 \to R_{\Lambda_2} V$ defined by

$$\Phi_2(v) = R_{\Lambda_2} v - \int_0^\infty e^{\nu \Lambda_2 t} R_{\Lambda_2} B(S(t)v, S(t)v)\, dt.$$

It is a well defined (since for $v \in M_1$ one has $\Lambda(v) \succcurlyeq \Lambda_2$) and analytic (cf. above) mapping. Let $M_2 = \Phi_2^{-1}(\{0\})$.

By the Corollary, it is easily seen that

$$v \in M_2 \setminus \{0\} \Leftrightarrow \Lambda(v) \succcurlyeq \Lambda_3.$$

Moreover, $\Phi_2'(0)$: $(I - R_{\Lambda_1})V \to R_{\Lambda_2}V$ is the restriction of R_{Λ_2} to $(I - R_{\Lambda_1})V$, which proves that M_2 is, in a neighborhood of 0, a smooth analytic manifold of codimension m_2 in M_1, i.e. of codimension $m_1 + m_2$ in V. Again the invariance of M_2 under $S(t)$ is clear.

The construction of the other M_k's follows the same procedure.

Let us make now some supplementary remarks on the periodic case (1.2), (1.2)'.

In this case, each M_k contains an unbounded (in V) linear submanifold L_k, of infinite dimension. This is based on the following special motions of the Navier–Stokes equations. Let $u(y, t)$, $y \in \mathbb{R}$, $t \geqslant 0$, be a scalar function. We set, for $x \in \mathbb{R}^n$,

$$U(x, t) = \big(u(k_1 x_1 + \cdots + k_n x_n, t), \ldots, u(k_1 x_1 + \cdots + k_n, x_n, t)\big) \; (n \text{ times})$$

where $k = (k_1, k_2, \ldots, k_n) \in \mathbb{Z}^n$. Choosing k such that $k_1 + k_2 + \cdots + k_n = 0$, one has $\operatorname{div} U = 0$ and $(U \cdot \nabla) U = 0$. Moreover, U satisfies the (linear and nonlinear !) Navier–Stokes equations (with $p = \text{const.}$) if and only if

$$(3.4) \qquad \frac{\partial u}{\partial t} - \nu |k|^2 \frac{\partial^2 u}{\partial x^2} = 0.$$

By taking any L-periodic (in space) solution of (3.4) we obtain a solution of (0.1), (1.2)'.

Let

$$\mathscr{U} = \{U \in V, U(x) = \big(u(k_1 x_1 + \cdots + k_n x_n), \ldots, u(k_1 x_1 + \cdots + k_n x_n)\big)$$
$$(k_1, \ldots, k_n) \in \mathbb{Z}^n, k_1 + \cdots + k_n = 0; u \; L\text{-periodic}\}.$$

Then $L_k = \mathscr{U} \cap M_k^{\text{lin}}$ is an unbounded infinite-dimensional linear submanifold of M_k.

This remark shows, in particular, that $\mathscr{R}$ is unbounded in the periodic 3-dimensional case. We do not know yet whether this is also true in the nonperiodic case (1.1), (1.1)'.

Let us also remark that if M_k for some $k = 1, 2, \ldots$ is linear then, necessarily, $M_k = M_k^{\text{lin}}$. Therefore, since M_k is invariant under $S(t)$ ($\forall\, t \geqslant 0$),

$$\big(R_{\Lambda_1} + \cdots + R_{\Lambda_k}\big) B\big((1 - R_{\Lambda_1} - \cdots - R_{\Lambda_k})u, (1 - R_{\Lambda_1} - \cdots - R_{\Lambda_k})u\big) = 0$$

for all $u \in V$. (Notice that the R_Λ's make also sense in V!) But is it easy to verify that in the periodic case

$$R_{\Lambda_1} B\big((1 - R_{\Lambda_1} - \cdots - R_{\Lambda_k})u, (1 - R_{\Lambda_1} - \cdots - R_{\Lambda_k})u\big) \neq 0$$

for all $k = 1, 2, 3, \ldots$. It follows that in this case, none of the manifolds $M_1, M_2, M_3, \ldots$ is linear. Again, we do not know whether this conclusion is also valid in the nonperiodic case (1.1), (1.1)'.

4. Smoothness of the nonlinear spectral manifolds.

The main result of this section is the

THEOREM 4. *For every* $k = 1, 2, \ldots, M_k$ *is a smooth analytic submanifold of* $\mathcal{R}$, *having codimension* $m_1 + \cdots + m_k$ *in* $\mathcal{R}$.

Let us just give a brief sketch of the proof of Theorem 4. For details, we refer to our paper [6].

One first easily finds that the derivative of ϕ_1 at $u \in \mathcal{R}$ is given by

$$(4.1) \quad \phi_1'(u) \cdot v = R_{\Lambda_1} v - \int_0^\infty e^{\nu \Lambda_1 t} R_{\Lambda_1} [B(S(t)u, w(t)) + B(w(t), S(t)u)] \, dt,$$

where $w(t)$ is the solution of the linear problem

$$(4.2) \quad \frac{dw}{dt} + \nu A w + B(S(t)u, w) + B(w, S(t)u) = 0, \quad w(0) = v.$$

We are therefore led to study linear equations of the type

$$(4.3) \quad \frac{dw}{dt} + \nu A w + B(u, w) + B(w, v) = 0, \quad w(0) = w_0,$$

where $u, v \in C(0, \infty; V) \cap L^2_{\text{loc}}(0, \infty; D(A))$ are rapidly decreasing vector fields. More precisely, we assume that (for some $\alpha > 0$)

$$(4.4) \quad \begin{aligned} \|u(t)\| &\leq c_{10} e^{-\nu \alpha t}, \quad t \geq 0, \\ \|v(t)\| &\leq c_{11} e^{-\nu \alpha t}, \quad t \geq 0, \end{aligned}$$

$$(4.5) \quad \begin{aligned} |Au(t)| &\leq c_{12} e^{-\nu \alpha t}, \quad t \geq t_0, \\ |Av(t)| &\leq c_{13} e^{-\nu \alpha t}, \quad t \geq t_0. \end{aligned}$$

One can develop for equation (4.3) an asymptotic analysis of solutions similar to the nonlinear case. In particular, one obtains the

PROPOSITION 4. (1) *The limit* $\lim_{t \to \infty} (\|w(t)\|^2 / |w(t)|^2) = \tilde{\Lambda}(w_0)$ *exists and belongs to the spectrum of Stokes operator.*

(2) $\lim_{t \to \infty} e^{\nu \tilde{\Lambda}(w_0) t} |w(t)|$ *exists (in H and V), is finite, and is not zero.*

(3) $\tilde{\Lambda}(w_0) \geq \Lambda_k$ ($k \geq 2$) *if and only if* $\lim_{t \to \infty} e^{\nu \Lambda_j t} R_{\Lambda_j} w(t) = 0, \forall j = 1, 2, \ldots, k - 1$.

Point (2) in Proposition 4 is partly based on the following lemma, which will be useful later on:

LEMMA 2. *If Λ is any eigenvalue of A which is strictly smaller than $\Lambda(w_0)$, then there exist $t(\Lambda) > 0$, and $c_{14} > 0$, both independent of w_0 such that*

$$(4.6) \quad |R_\Lambda \tilde{w}(t)| \leq \frac{c_{14}}{\nu(\delta + \alpha)} (\Lambda)^{1/2} e^{-\nu t \alpha} \quad \text{for } t \geq t(\Lambda),$$

where $\tilde{w}(t) = w(t)/|w(t)|$ and $\delta = (\tilde{\Lambda}(w_0) - \Lambda)/2$. □

We can now construct linear submanifolds of V (depending on u and v) which are the counterpart of the M_k's. Namely, let us define

$$(4.7) \quad \tilde{M}_1 = \Big\{ w \in V; R_{\Lambda_1} w - \int_0^\infty e^{\nu \Lambda_1 t} R_{\Lambda_1} [B(u(t), \tilde{S}(t)w) + B(\tilde{S}(t)w, v(t))] \, dt = 0 \Big\}$$

(here $\tilde{S}(t)$ denotes the "semigroup" associated to (4.3)), and inductively ($k \geq 2$),

$$(4.8) \quad \tilde{M}_k = \Big\{ w \in \tilde{M}_{k-1}; R_{\Lambda_k} w - \int_0^\infty e^{\nu \Lambda_k t} R_{\Lambda_k} [B(u(t), \tilde{S}(t)w) + B(\tilde{S}(t)w, v(t))] \, dt = 0 \Big\}.$$

Then,

$$\tilde{\Lambda}(w) = \Lambda_k \Leftrightarrow w \in \tilde{M}_{k-1} \setminus \tilde{M}_k,$$

for $k = 1, 2, \ldots$ (by convention, $\tilde{M}_0 = V$).

The key point for the proof of Theorem 4 is the next proposition which establishes that $\tilde{M}_k$ has maximal codimension in V.

PROPOSITION 5. *For every $k = 1, 2, \ldots$ one has*

$$(4.9) \quad \mathrm{Codim}_V \tilde{M}_k = m_1 + \cdots + m_k.$$

To prove Proposition 5, one argues by contradiction. If (4.9) is not true, one can show, using in particular Lemma 2, that, for some $t^* > 0$, $\tilde{S}(t^*)H$ is not dense in H. By a classical result of Lattès–Lions [13] this turns out to be equivalent to nonuniqueness for the backward Cauchy problem associated to the adjoint equation of (4.3). But this contradicts the backward uniqueness theorem of Bardos–Tartar [1].

Let us go back to Theorem 4. Equation (4.2) obviously belongs to the class of (4.3)–(4.5). Now, by using Proposition 5, one easily deduces that the mapping ϕ_k: $\tilde{M}_{k-1} \to R_{\Lambda_k} V$ which defines M_k is a submersion for $k = 1, 2, \ldots$. This completes the proof.

References

1. C. Bardos and L. Tartar, *Sur l'unicité rétrograde des équations paraboliques et quelques questions voisines*, Arch. Rational Math. Anal. **50** (1973), 10–25.

2. N. Bourbaki, *Variétés différentiables et analytiques*, Fascicule de Résultats, §§1–7, Hermann, Paris, 1967.

3. R. H. Dyer and D. E. Edmunds, *Lower bounds for solutions of the Navier–Stokes equations*, Proc. London Math. Soc. (3) **18** (1968), 169–178.

4. D. E. Edmunds, *Asymptotic behavior of solutions of the Navier–Stokes equations*, Arch. Rational Mech. Anal. **22** (1966), 15–21.

5. C. Foiaş, *Solutions statistiques des équations de Navier–Stokes*, Cours au Collège de France, 1974.

6. C. Foiaş and J. C. Saut, *On the smoothness of the nonlinear spectral manifolds associated to the Navier–Stokes equations*, Indiana Univ. Math. J. **33** (1984), 911–912.

7. _____, *Limite du rapport de l'enstrophie sur l'énergie pour une solution faible des équations de Navier–Stokes*, C. R. Acad. Sci. Paris Sér. I Math. **293** (1981), 241–244.

8. _____, *Asymptotic behavior as $t \to \infty$ of solutions of Navier–Stokes equations*, Séminaire de Mathématiques Appliquées du Collège de France, Vol. IV, Pitman, London, 1983, pp. 74–86.

9. _____, *Transformation fonctionnelle linéarisant les équations de Navier–Stokes*, C. R. Acad. Sci. Paris Sér. Math. I **295** (1982), 325–327.

10. _____, *Asymptotic behaviour, for $t \to \infty$, of the solutions of the Navier–Stokes equations and nonlinear spectral manifolds*, Indiana Univ. Math. J. **33** (1984), 459–477.

11. C. Foiaş and R. Temam, *Some analytic and geometric properties of the solutions of the evolution of Navier–Stokes equations*, J. Math. Pures Appl. (9) **58** (1979), 339–368.

12. C. Guillopé, *Remarques à propos du comportement, lorsque $t \to \infty$, des solutions des équations de Navier–Stokes associées à une force nulle*, Bull. Soc. Math. France **111** (1983), 151–180.

13. R. Lattès and J. L. Lions, *Méthode de quasi-réversibilité et applications*, Dunod, Paris, 1967.

14. P. D. Lax, Private communication, 1979.

15. J. Leray, *Essai sur le mouvement d'un fluide visqueux emplissant l'espace*, Acta Math. **63** (1934), 193–248.

16. _____, *Essais sur les mouvements plans d'un liquide visqueux que limitent des parois*, J. Math. Pures Appl. **13** (1934), 331–419.

17. J. L. Lions, *Quelques méthodes de résolutions des problèmes aux limites non linéaires*, Dunod-Gauthier-Villar, Paris, 1969.

18. R. Temam, *Navier–Stokes equations*, Theory and Numerical Analysis, North-Holland, Amsterdam, 1979.

INDIANA UNIVERSITY

UNIVERSITÉ PARIS - SUD, FRANCE

Regularity Criteria for Weak Solutions of the Navier–Stokes System

YOSHIKAZU GIGA[1,2]

This note improves and applies L^p-theory developed in my previous papers [4–6] to study the regularity of weak solutions of the Navier–Stokes initial value problem

(NS) $\quad\quad \partial u/\partial t - \Delta u + (u, \text{grad})u + \text{grad } p = 0,$
$\quad\quad\quad\quad \text{div } u = 0 \quad \text{in } D \times (0, \infty),$
$\quad\quad\quad\quad u = 0 \quad \text{on } \partial D \times (0, \infty), \quad u(x, 0) = a(x) \quad \text{in } D,$

where $(u, \text{grad}) = \sum_{j=1}^{n} u^j(\partial/\partial x_j)$ and D is a smoothly bounded domain in R^n ($n \geq 2$). The function $a = (a^1(x), \ldots, a^n(x))$ represents a given initial velocity. The external force is assumed to be zero for simplicity.

The existence of a global-in-time weak solution of (NS) was proved by Leray [10] and Hopf [7]. More explicitly, for any L^2-divergence free initial velocity a there exists a weak solution u of (NS) that satisfies the energy inequality

(1) $\quad \int_D |u(x, t)|^2 \, dx + 2\int_{t_0}^{t} \int_D |\nabla u(x, s)|^2 \, dx \, ds \leq \int_D |u(x, t_0)|^2 \, dx,$

$$t \geq t_0, \text{ a.e. } t_0 \geq 0 \text{ or } t_0 = 0;$$

for the definition of weak solutions, see [12]. Such solutions are called *Leray–Hopf weak solutions* (for short, LH solutions). In particular,

(2) $\quad\quad u \in L^2(0, T; H^1(D)) \cap L^\infty(0, T; L^2(D))$

for every $T > 0$. However, if the space dimension $n \geq 3$ it is not known whether u develops singularities even if a is smooth.

1980 *Mathematics Subject Classification.* Primary 35Q10, 76D05.

[1] On leave of absence from the Department of Mathematics, Nagoya University, Chikusa-ku, Nagoya 464, Japan.

[2] Partially supported by Educational Project for Japanese Mathematical Scientists and the Sakkokai Foundation.

© 1986 American Mathematical Society
0082-0717/86 $1.00 + $.25 per page

In this note we give some sufficient conditions to guarantee the regularity of a weak solution u. The details and related recent references are given in [15]. Let $L^{p,q}$ denote the space $L^q(0, T; L^p(D))$, $1 \leq p, q < \infty$, and set $k = 2 - q + nq/p$. This number k represents the scaling dimension of $\int (\int |u|^p \, dx)^{q/p} \, dt$; see [1].

THEOREM 1. *Let u be a weak solution of* (NS) *satisfying* (2). *If u is in $L^{p,q}$ with $k \leq 0$ and $p > n$, u is in $C^\infty(\overline{D} \times (0, T))$.*

This improves the results of Serrin [12] and Kaniel–Shinbrot [8]. They discuss the case $k < 0$. If $k > 0$, we do not know whether u is smooth. However, we can estimate the size of a possible time singularity set.

THEOREM 2. *Let u be an LH solution. If u is in $L^{p,q}$ with $k > 0$ and $p > n$, then there is a closed subset E of $(0, T)$ whose $k/2$-dimensional Hausdorff measure vanishes, and such that u is in $C^\infty(\overline{D} \times ((0, T) \setminus E))$.*

Theorems 1 and 2 are useful to understand the difference between the cases $n = 2$ and $n \geq 3$. When $n = 2$, (2) implies $u \in L^{p,q}$ with $k = 0$ and $p > n$, so every weak solution satisfying (2) is smooth. However, when $n = 3$, (2) implies $u \in L^{6,2}$, so $k = 1$. Theorem 2 says that the $\frac{1}{2}$-dimensional Hausdorff measure of time singularity set vanishes. Usually such results are proved without using Theorems 1 and 2; see [3, 10, 11, 13]. However, I believe Theorems 1 and 2 clarify what is necessary to prove regularity of LH solutions.

REMARK. Recently, in [1] Caffarelli, Kohn, and Nirenberg study space-time interior singularities of suitable weak solutions which are special weak solutions satisfying (2). Actually, they have proved that, for $n = 3$, every suitable weak solution v is smooth w.r.t. x in $D \times (0, T) \setminus F$ such that F is a closed subset of $D \times (0, T)$ whose 1-dimensional Hausdorff measure vanishes. Moreover, for fixed t in $(0, T) \setminus E$, v is smooth in $x \in D$, where the $\frac{1}{2}$-dimensional Hausdorff measure of E is zero. This is different from the result for time singularity mentioned in the preceding paragraph because only the interior regularity is discussed.

We next discuss the marginal case $p = n$.

THEOREM 3. *Let u be an LH solution. If u is in $L^{n,q}$, then there is a closed subset E of $(0, T)$ whose Lebesgue measure vanishes, and such that u is in*

$$C^\infty(\overline{D} \times ((0, T) \setminus E)).$$

When $n = 4$, (2) implies that $u \in L^{4,2}$, so Theorem 3 says that the possible time singularity set of LH solutions has Lebesgue measure zero. This result for LH solutions is recently proved by Kato [9] without using Theorem 3.

As a by-product we have the following result which von Wahl [14] recently proved by using a different method:

THEOREM 4. *Let u be a weak solution satisfying* (2). *If u is in $C((0, T); L^n(D))$, then u is in $C^\infty(\overline{D} \times (0, T))$.*

Below we give our strategy of proving theorems; the details will appear in [15].

1 (*Uniqueness*). If there is a strong (regular) solution with the same initial data of an LH solution, then the weak solution agrees with the strong solution. More precisely,

LEMMA 1. *Let u be an LH solution. Suppose v is a weak solution satisfying* (2) *and* $v(x,0) = u(x,0)$. *If v is in* $L^{p,q}$ *with* $k \leq 0$ *and* $p > n$, *then u agrees with v in* $\overline{D} \times (0, T)$.

This lemma is essentially due to Sather and Serrin (see p12]) for $n \leq 4$. We can prove this lemma with no restrictions on n. We use the results of [5] to estimate the nonlinear term, although we mostly follow their proof.

2 (*Local existence of strong solutions*). We construct a local-in-time strong solution for L^r initial data ($r \geq n$) and estimate the maximal time interval of existence from below when $r > n$. To state this more precisely, we recall the standard function spaces and the Stokes operator.

Let X^p be the closure in $L^p(D)$ of divergence-free vector fields with compact support in D, where $1 < p < \infty$. Let P be the usual continuous projection from $L^p(D)$ to X^p. The Stokes operator A in X_p is defined by $A = -P\Delta$ with dense domain

$$D(A_p) = X^p \cap \{u \in W^{2,p}(D); u = 0 \text{ on } \partial D\}.$$

Apply P to both sides of (NS) to get

(3) $\quad du/dt + Au = Fu, \quad u(0) = a, \quad$ where $Fu = -P(u, \text{grad})u$.

As far as u is smooth, (NS) is equivalent to (3). A function u is called a *strong solution* if u satisfies (3) in the strong sense. We now state our results.

LEMMA 2. (i) *Suppose that a is in* X^n. *There is a local strong solution u of* (3) *on some interval* $(0, T_0)$ *such that u is in* $C([0, T_0); X^n)$ *and* $C^\infty(\overline{D} \times (0, T_0))$.

(ii) *Suppose that a is in* X^r *and* $r > n$. *Let* $[0, T_*)$ *be the maximal interval on which a strong solution of* (3) *exists. Then the estimate*

$$T_* \geq T_0 = C/\|a\|_{L^r(D)}^s, \quad s = 2r/(r-n),$$

holds with constant C independent of a. Moreover, u is in $C([0, T_0); X^r)$.

Part (i) is already proved in [6]. To solve (3) we solve an integral form of (3),

(4) $\qquad u(t) = e^{-tA}a + \int_0^t e^{-(t-s)A} Fu(s)\, ds.$

When $D = R^3$, (ii) is proved by Leray [10]. Since $-A$ generates an analytic semigroup e^{-tA} in X^p (see e.g. [4]) and since $A^{1/2}$ behaves just like ∇ (see [5]), the method suggested by Leray [10] works even for the initial-boundary value problem.

Theorems 1 and 2 now follow from Lemmas 1 and 2(ii) by the standard argument; cf. [3, 10, 11]. The assumption $p > n$ is used to apply Lemma 1. Note that the solution u constructed in Lemma 2(ii) satisfies the assumption for v in Lemma 1.

3 ($L^{p,q}$ local strong solution). If we prove that u in Lemma 2(i) satisfies the assumption for v in Lemma 1, we also get Theorems 3 and 4 by the standard argument. In fact we have

LEMMA 3. *Suppose a is in X^n. Then there is a local strong solution u of* (3) *on some interval* $(0, T_0)$ *such that u is in* $C([0, T_0); X^n)$, $C^\infty(\overline{D} \times (0, T_0))$ *and* $L^q(0, T_0; X^p)$ *with $n < p$, q and $k = 2 - q + nq/p = 0$.*

To construct an $L^{p,q}$ solution may be useful for other parabolic equations, so we give a short

SKETCH OF THE PROOF. We solve (4) to prove Lemma 3. To construct u we use the iteration scheme

$$u_{m+1}(t) = e^{-tA}a + \int_0^t e^{-(t-s)A}Fu_m(s)\,ds \quad (m \geq 0), \quad u_0(t) = e^{-tA}a.$$

The key step to show the convergence in $L^{p,q}$ is to prove that $\{u_m\}$ is bounded in $L^{p,q}$, which is shown below. We first have

$$(5) \qquad \|u_{m+1}(t)\|_p \leq \|e^{-tA}a\|_p + \int_0^t \|e^{-(t-s)A}Fu_m\|_p\,ds,$$

where $\|f\|_p$ denotes the norm of f in X^p. Since [4, 5] gives

$$\|A^\alpha e^{-tA}f\|_p \leq C\|f\|_p/t^\alpha, \qquad \alpha, t > 0,$$

and

$$\|f\|_p \leq C\|A^{n/2p}f\|_{p/2}, \quad f \in D(A^{n/2p})_{p/2},$$

we eventually get

$$\|e^{-(t-s)A}Fu_m\|_p \leq C\|A^{-1/2}Fu_m\|_{p/2}/(t-s)^{n/2p+1/2},$$

where C denotes a constant which depends only on n, p, α, and D and may be different in each case. By an estimate for the nonlinear term (see [5, 6]) the right side is dominated by $C\|u_m\|_p^2/(t-s)^{n/2p+1/2}$. Apply this to (5) to get

$$\|u_{m+1}(t)\|_p \leq \|e^{-tA}a\|_p + \int_0^t \|u_m\|_p^2(t-s)^{-n/2p-1/2}\,ds.$$

Let $h_m(T)$ denote $\int_0^T \|u_m\|_p^q\,dt$. We next apply the Hardy–Littlewood inequality to the second term on the right side and get

$$h_{m+1}(T) \leq h_0(T) + Ch_m(T)^2;$$

here we use $p > n$. This idea is suggested by Kato [9]. If $h_0(T)$ tends to zero as $T \to 0$, the above inequality shows that, for small T,

$$(6) \qquad h_m(T) \leq C$$

is valid for all m. Fortunately, for $h_0(T)$ we have

LEMMA 4. *Let $1/r = 1/p + 2/nq$ and $q > r > 1$. We have*

$$\int_0^T \|e^{-tA}a\|_p^q\,dt < K\|a\|_r^q.$$

This lemma is proved by applying the Marcinkiewicz interpolation inequality. We apply Lemma 4 with $r = n$ to get $h_0(T) \to 0$ if $T \to 0$. This completes the proof of (6) which shows that $\{u_m\}$ is bounded in $L^{p,q}$ if T is sufficiently small.

REMARK. In [2] Fabes, Lewis, and Riviere constructed a $L^{p,q}$ strong solution with $k = 0$. However, they are forced to assume a is in X^s, $s > n$, because they do not use Lemma 4.

ACKNOWLEDGMENTS. I am grateful to Professor Tosio Kato who showed his results before publication and pointed out a technical mistake in my original draft. I am also grateful to Professor Robert Kohn for stimulating conversations.

References

1. L. Caffarelli, R. Kohn, and L. Nirenberg, *Partial regularity of suitable weak solutions of the Navier–Stokes equations*, Comm. Pure Appl. Math. **35** (1982), 771–831.

2. E. B. Fabes, J. E. Lewis, and N. M. Riviere, *Boundary value problems for the Navier-Stokes equations*, Amer. J. Math. **99** (1977), 626–668.

3. C. Foias and R. Temam, *Some analytic and geometric properties of the solutions of the Navier–Stokes equations*, J. Math. Pures Appl. (9) **58** (1979), 339–368.

4. Y. Giga, *Analyticity of the semigroup generated by the Stokes operator in L_r spaces*, Math. Z. **178** (1981), 297–329.

5. _____, *Domains of fractional powers of the Stokes operator in L_r spaces*, Arch. Rational Mech. Anal. (1985) (in press)

6. Y. Giga and T. Miyakawa, *Solutions in L_r to the Navier–Stokes initial value problem*, Arch. Rational Mech. Anal. (1985) (in press)

7. E. Hopf, *Über die Anfangwertaufgabe für die hydrodynamischen Grundgleichungen*, Math. Nachr. **4** (1951), 213–231.

8. S. Kaniel and M. Shinbrot, *Smoothness of weak solutions of the Navier–Stokes equations*, Arch. Rational Mech. Anal. **24** (1967), 302–324.

9. T. Kato, *Strong L^p solutions of the Navier–Stokes equation in $\mathbf{R}^m$ with applications to weak solutions*, Math. Z. **187** (184), 471–480.

10. J. Leray, *Sur le mouvement d'un liquide visqueux emplissant l'espace*, Acta Math. **63** (1934), 193–248.

11. V. Scheffer, *Partial regularity of solutions to the Navier–Stokes equations*, Pacific J. Math. **66** (1976), 535–552.

12. J. Serrin, *The initial-value problem for the Navier–Stokes equations*, Nonlinear Problems (R. E. Langer, ed.), University of Wisconsin Press, Madison, Wis. 1963, pp. 69–98.

13. R. Temam, *Navier–Stokes equation and nonlinear functional analysis*, SIAM, Philadelphia, Pa. 1983.

14. W. von Wahl, *Regularity of weak solutions of the Navier–Stokes equations*, Preprint. (1983)

15. Y. Giga, *Solutions for semilinear parabolic equations in L^p and regularity of weak solutions of the Navier–Stokes system*, J. Differential Equations (to appear).

COURANT INSTITUTE OF MATHEMATICAL SCIENCES

A Strongly Nonlinear Elliptic Problem in Orlicz–Sobolev Spaces

JEAN-PIERRE GOSSEZ

0. Introduction. Boundary value problems for quasilinear elliptic equations of the form

(1) $$A(u) \equiv \sum_{|\alpha| \leqslant m} (-1)^{|\alpha|} D^\alpha A_\alpha(x, u, \nabla u, \ldots, \nabla^m u) = f$$

on open subsets of $\mathbb{R}^N$ have been extensively studied since the pioneering work of Browder [5], where such equations were considered in the context of the theory of monotone operators. In the basic results, the coefficient functions A_α are supposed to satisfy some polynomial growth conditions with respect to u and its derivatives; this allows a transformation of the problem into an equation involving an everywhere defined bounded operator acting on a Sobolev space $W^{m,p}$. For a survey of these basic results, see [6, 19]; also, see [8, 17] for more recent contributions.

In the last decade several works have been concerned with the weakening of these polynomial growth restrictions on the A_α's. Two directions can be mentioned. In [7] Browder initiated the study of the so-called "strongly nonlinear" equations, i.e., equations of the form

(2) $$A(u) + g(x, u) = f,$$

where A is given by (1), with coefficients A_α having polynomial growth, and where g satisfies the sign condition $g(x, u)u \geqslant 0$ but has otherwise unrestricted growth with respect to u. Contributions of particular interest in this direction include [15] for $m = 1$ and [21, 4] for $m > 1$. The treatment of the higher-order case here is more involved, due to the lack of a simple truncation operation in the Sobolev

---

1980 *Mathematics Subject Classification.* Primary 35J65, 47H17, 46E35.

Key words and phrases. Boundary value problem, strongly nonlinear problem, Orlicz–Sobolev space, polynomial growth, sign condition, operator of monotone type.

© 1986 American Mathematical Society
0082-0717/86 $1.00 + $.25 per page

space $W^{m,p}$, $m > 1$; use is made of Hedberg's delicate approximation procedure from potential theory [14]. In another direction, equations like (1) are considered in which all the coefficients A_α, including those of order $|\alpha| = m$, are allowed to have nonpolynomial (although somehow controlled) growth. The problems here are treated in the framework of Sobolev spaces built from Orlicz spaces, and part of the difficulty comes from the fact that, in general, these spaces are not reflexive. Contributions along these lines include [9, 11, 12, 18].

It is our purpose in this paper to investigate the situation where these two different approaches are considered simultaneously; i.e., we study equations like (2) where A gives rise to a "good" operator within the class of Orlicz–Sobolev spaces, and g satisfies only a sign condition. We limit ourselves to the case $m = 1$, and part of the paper is devoted to the technicalities involved with the truncation operation in Orlicz–Sobolev spaces of order one; some of our approximation results of [13] turn out to be very useful in this respect. It is not clear at the moment whether the method of [21, 4] can be adapted to the present situation in order to deal with the higher-order case.

In §1 we extend the result of [2] on the values of some distributions to the framework of Orlicz–Sobolev spaces. In §2 we consider the Dirichlet and Neumann problems for (2) with $m = 1$.

1. Orlicz–Sobolev spaces of order one. Throughout the paper Ω will be an open (possibly unbounded) subset of $\mathbb{R}^N$ which, unless otherwise stated, satisfies the segment property.

Let $W^1 L_M(\Omega)$ and $W^1 E_M(\Omega)$ be the Orlicz–Sobolev spaces of order one on Ω corresponding to an N-function M. Standard references about these spaces include [1, 16]. They will, as usual, be identified with subspaces of the product $(L_M(\Omega))^{N+1} \equiv \prod L_M$. Denoting by $\overline{M}$ the N-function conjugate to M, we define $W_0^1 L_M(\Omega)$ (resp. $W_0^1 E_M(\Omega)$) as the $\sigma(\prod L_M, \prod E_{\overline{M}})$ (resp. norm) closure of $\mathscr{D}(\Omega)$ in $W^1 L_M(\Omega)$. We also consider $W^{-1} L_{\overline{M}}(\Omega)$ (resp. $W^{-1} E_{\overline{M}}(\Omega)$), the space of distributions on Ω which can be written as sums of derivatives up to order one of functions in $L_{\overline{M}}(\Omega)$ (resp. $E_{\overline{M}}(\Omega)$). Since Ω has the segment property, $\mathscr{D}(\Omega)$ is $\sigma(\prod L_M, \prod L_{\overline{M}})$ dense in $W_0^1 L_M(\Omega)$ (cf. [11, Theorem 1.3]), and, consequently, the value of $S \in W^{-1} L_{\overline{M}}(\Omega)$ at an element $u \in W_0^1 L_M(\Omega)$ is well defined; it will be denoted by $\langle S, u \rangle$.

THEOREM 1. *Consider $S \in W^{-1} L_{\overline{M}}(\Omega) \cap L^1_{\text{loc}}(\Omega)$ and $u \in W_1^1 L_M(\Omega)$. Suppose that for some $h \in L^1(\Omega)$, $S(x)u(x) \geq h(x)$ a.e. in Ω. Then $Su \in L^1(\Omega)$ and*

$$\langle S, u \rangle = \int_\Omega S(x) u(x)\, dx.$$

When $M(t) = |t|^p$ with $1 < p < \infty$, the conclusion of Theorem 1 was obtained in [2]. It then holds without any restriction on Ω. Also see [3] for related results and [4] for the higher-order case. The introduction here of the (mild) assumption that Ω has the segment property is somehow related to the fact that, contrary to the situation in the usual Sobolev spaces, the topology $\sigma(\Pi L_M, \Pi E_{\overline{M}})$ used in the definition of the space $W_0^1 L_M(\Omega)$ does not allow going to the limit on u in the pairing $\langle S, u \rangle$.

The following lemmas will be needed in the proof of Theorem 1.

LEMMA 2. *Let $F: \mathbb{R} \to \mathbb{R}$ be uniformly Lipschitzian with $F(0) = 0$. Let $u \in W_0^1 L_M(\Omega)$. Then $F(u) \in W_0^1 L_M(\Omega)$. Moreover, if the set D of discontinuity points of F' is finite, then*

$$(3) \qquad \frac{\partial}{\partial x_i} F(u) = \begin{cases} F'(u)(\partial u/\partial x_i) & a.e. \text{ in } \{x \in \Omega; u(x) \notin D\}, \\ 0 & a.e. \text{ in } \{x \in \Omega; u(x) \in D\}. \end{cases}$$

PROOF. We first consider the case where F is also C^1. By [13, Theorem 4] there exists a sequence $u_n \in \mathcal{D}(\Omega)$ such that, for $|\alpha| \leq 1$ and some $\lambda > 0$,

$$\int_\Omega M((D^\alpha u_n - D^\alpha u)/\lambda) \, dx \to 0 \quad \text{as } n \to \infty.$$

Passing to a subsequence, we can assume that $u_n \to u$ a.e. in Ω. From the relations $|F(s)| \leq K|s|$, where K denotes the Lipschitz constant of F, and $\partial F(u_n)/\partial x_i = F'(u_n)(\partial u_n/\partial x_i)$, we deduce that the $C_0^1(\Omega)$ functions $F(u_n)$ remain bounded in $W_0^1 L_M(\Omega)$. Thus, going to a further subsequence, we obtain $F(u_n) \to w \in W_0^1 L_M(\Omega)$ for $\sigma(\Pi L_M, \Pi E_{\overline{M}})$ and, by a local application of the compact imbedding theorem, $F(u_n) \to w$ a.e. in Ω. Consequently, $w = F(u)$ and $F(u) \in W_0^1 L_M(\Omega)$. Finally, by the usual chain rule for weak derivatives,

$$(4) \qquad \partial F(u)/\partial x_i = F'(u)(\partial u/\partial x_i)$$

a.e. in Ω.

We turn now to the general case. Taking convolutions with mollifiers, we get a sequence $F_n \in C^\infty(\mathbb{R})$ such that $F_n \to F$ uniformly on each compact set, $F_n(0) = 0$, and $|F_n'| \leq K$. For each n, $F_n(u) \in W_0^1 L_M(\Omega)$, and we have (4) with F replaced by F_n. By arguments similar to the preceding ones, we obtain, for a subsequence, $F_n(u) \to w \in W_0^1 L_M(\Omega)$ for $\sigma(\Pi L_M, \Pi E_{\overline{M}})$ and a.e. in Ω. Consequently, $w = F(u)$ and $F(u) \in W_0^1 L_M(\Omega)$. Finally, (3) follows from the generalized chain rule for weak derivatives, as given, e.g., in [10, Theorem 7.8]. Q.E.D.

LEMMA 3. *Let $u, v \in W_0^1 L_M(\Omega)$ and let $w = \min\{u, v\}$. Then $w \in W_0^1 L_M(\Omega)$ and*

$$\frac{\partial w}{\partial x_i} = \begin{cases} \dfrac{\partial u}{\partial x_i} & a.e. \text{ in } \{x \in \Omega; u(x) \leq v(x)\}, \\ \dfrac{\partial v}{\partial x_i} & a.e. \text{ in } \{x \in \Omega; u(x) > v(x)\}. \end{cases}$$

PROOF. Note that $\min\{u, v\} = u - (u - v)^+$ and apply Lemma 2 with $F(s) = s^+$. Q.E.D.

Our last lemma describes the truncation operation which will be used in the proof of Theorem 1.

LEMMA 4. *Let* $u \in W_0^1 L_M(\Omega)$. *Then there exists a sequence* u_n *such that* (i) $u_n \in W_0^1 L_M(\Omega) \cap L^\infty(\Omega)$, (ii) supp u_n *is compact in* Ω, (iii) $|u_n(x)| \leq |u(x)|$ *a.e. in* Ω, (iv) $u_n(x)u(x) \geq 0$ *a.e. in* Ω, (v) $D^\alpha u_n \to D^\alpha u$ *a.e. in* Ω, *for* $|\alpha| \leq 1$, (vi) *for some* $\lambda > 0$ *and for* $|\alpha| \leq 1$,

$$\int_\Omega M\left(\frac{D^\alpha u_n - D^\alpha u}{\lambda}\right) \to 0.$$

PROOF. Writing $u = u^+ - u^-$, we can assume without loss of generality that $u \geq 0$ a.e. in Ω. By [13, Theorem 4] there exists a sequence $v_n \in \mathscr{D}(\Omega)$ such that, for $|\alpha| \leq 1$ and some $\lambda > 0$, $\int_\Omega M((D^\alpha u - D^\alpha v_n)/\lambda)\,dx \to 0$ as $n \to \infty$; moreover, by the construction in [13], v_n can be taken ≥ 0 in Ω. Passing to a subsequence if necessary, we can assume that for $|\alpha| \leq 1$, $D^\alpha v_n \to D^\alpha u$ a.e. in Ω. Define $u_n = \min\{u, v_n\}$. Then properties (i)–(iv) clearly hold. Computing the derivatives of u_n by Lemma 3, we obtain

$$\frac{\partial u_n}{\partial x_i} = \begin{cases} \dfrac{\partial u}{\partial x_i} & \text{a.e. in } \Omega'_n = \{x \in \Omega; u(x) \leq v_n(x)\}, \\ \dfrac{\partial v_n}{\partial x_i} & \text{a.e. in } \Omega''_n = \{x \in \Omega; u(x) > v_n(x)\}. \end{cases}$$

It follows that (v) holds. To verify (vi) we write

$$\int_\Omega M\left(\left(\frac{\partial u}{\partial x_i} - \frac{\partial u_n}{\partial x_i}\right)\Big/\lambda\right) = \int_{\Omega''_n} M\left(\left(\frac{\partial u}{\partial x_i} - \frac{\partial v_n}{\partial x_i}\right)\Big/\lambda\right)$$

$$\leq \int_\Omega M\left(\left(\frac{\partial u}{\partial x_i} - \frac{\partial v_n}{\partial x_i}\right)\Big/\lambda\right).$$

The conclusion follows since the right side goes to zero. Q.E.D.

REMARK 5. Lemmas 2 and 3 hold in any open set Ω if $W_0^1 L_M(\Omega)$ is everywhere replaced by $W^1 L_M(\Omega)$. It suffices to use Theorem 1 of [13] instead of Theorem 4. As far as Lemma 4 is concerned, we only get, by using Theorem 3 of [13], that if Ω satisfies both the segment and the cone properties, then, given $u \in W^1 L_M(\Omega)$, there exists a sequence u_n such that $u_n \in W^1 L_M(\Omega) \cap L^\infty(\Omega)$, supp u_n is compact in $\overline{\Omega}$, and (iii)–(vi) hold.

PROOF OF THEOREM 1. The arguments are rather similar to those in [2], so we only sketch them. If $u \in W_0^1 L_M(\Omega) \cap L^\infty(\Omega)$ and has compact support in Ω, then the conclusion follows from the fact that the regularized function u_ε converges to u for $\sigma(\Pi L_M, \Pi L_{\overline{M}})$ (cf. [11, Lemma 1.6]). In the general case we associate to u a sequence u_n as in Lemma 4. We have

(5) $$\langle S, u_n \rangle = \int_\Omega S(x) u_n(x)\,dx.$$

By property (vi) the left side converges to $\langle S, u \rangle$. On the other hand, using (iii) and (iv), we get $S(x)u_n(x) \geq -|h(x)|$ a.e. in Ω. Fatou's lemma and (5) then imply $Su \in L^1(\Omega)$. Using (iii) and (v), we conclude, by dominated convergence, that the right side of (5) converges to $\int_\Omega S(x)u(x)\,dx$. Q.E.D.

2. Strongly nonlinear problems. We will use the notions of complementary system and of pseudomonotonicity as described, e.g., in [**11**]. Since Ω has the segment property, $(W_0^1 L_M(\Omega), W_0^1 E_M(\Omega); W^{-1}L_{\overline{M}}(\Omega), W^{-1}E_{\overline{M}}(\Omega))$ is a complementary system (cf. [**11**, §1]). It will be denoted below by $(Y, Y_0; Z, Z_0)$.

We consider the following properties for a mapping $T: D(T) \subset Y \to Z$:

(I) T is sequentially pseudomonotone with respect to any dense subspace V of a dense subspace V' of Y_0;

(II) for each $f \in Z_0$ there is a (norm) neighbourhood $\mathcal{N}$ of f in Z such that $T^{-1}\mathcal{N}$ is bounded in Y;

(III) for some $\tilde{f} \in Z_0$ and some $r \in \mathbb{R}$, $\langle Tu - \tilde{f}, u \rangle \geq 0$ for all u in V' with $\|u\|_Y \geq r$.

It is known that these three properties imply that the range of T contains Z_0 (cf. [**12**, Theorem 1]). Moreover, given an equation of the form

$$(6) \qquad A(u) \equiv \sum_{|\alpha| \leq 1} (-1)^{|\alpha|} D^\alpha A_\alpha(x, u, \nabla u) = f,$$

concrete analytical conditions on the coefficients A_α are known which guarantee that the formula

$$\int_\Omega \sum_{|\alpha| \leq 1} A_\alpha(x, u, \nabla u) D^\alpha v \, dx = \langle Tu, v \rangle$$

gives rise to a (noneverywhere defined) mapping T from $W_0^1 L_M(\Omega)$ into $W^{-1}L_{\overline{M}}(\Omega)$ with these three properties (cf. [**11, 12**] when Ω is bounded, [**18**] when Ω is unbounded). Existence results for the Dirichlet problem for (6) can thus be derived in this way.

For our purposes the following additional conditions will be needed:

(IV) for any $\tilde{f} \in Z_0$ the set $\{u \in Y; \langle Tu, u \rangle \leq \langle \tilde{f}, u \rangle\}$ is bounded in Y;

(V) Tu remains bounded in Z whenever u varies in V' with u bounded in Y and $\langle Tu, u \rangle$ bounded from above in $\mathbb{R}$.

It is easily seen that the standard analytic conditions on the coefficients A_α of (6) which imply (I)–(III) also yield (IV) and (V) (as well as the conditions (III)' and (IV)' introduced in Remark 9); this follows easily from Remarks 4.7 and 5.3 in [**11**] and the proof of the local a priori bound in [**12**]. Thus, imposing (IV) and (V) is not really a restriction from the point of view of applications.

We can now state our existence result concerning the Dirichlet problem for

$$(7) \qquad Au + g(x, u) = f.$$

THEOREM 6. *Let* $T: D(T) \subset W_0^1 L_M(\Omega) \to W^{-1}L_{\overline{M}}(\Omega)$ *satisfy* (I)–(V). *Let* $g: \Omega \times \mathbb{R} \to \mathbb{R}$ *be a Carathéodory function such that for each* $r \in \mathbb{R}$ *there exists* $h_r \in L^1(\Omega)$ *with*

$$(8) \qquad |g(x, u)| \leq h_r(x)$$

for a.e. $x \in \Omega$ and all $u \in \mathbb{R}$ with $|u| \leq r$. Assume that

(9) $$g(x, u)u \geq 0$$

for a.e. $x \in \Omega$ and all $u \in \mathbb{R}$. Then, given $f \in W^{-1}E_{\overline{M}}(\Omega)$, there exists $u \in W_0^1 L_M(\Omega)$ such that $g(x, u) \in L^1(\Omega)$, $g(x, u)u \in L^1(\Omega)$ and

(10) $$\langle Tu, v \rangle + \int_\Omega g(x, u(x))v(x)\, dx = \langle f, v \rangle$$

for all $v \in W_0^{-1}L_M(\Omega) \cap L^\infty(\Omega)$ and for $v = u$.

PROOF. We truncate g by taking $\zeta \in \mathscr{D}(\mathbb{R}^N)$ with $0 \leq \zeta \leq 1$ and $\zeta(x) = 1$ for $|x| \leq 1$ and letting

$$g_n(x, u) = \begin{cases} \zeta(x/n)g(x, u) & \text{if } |g(x, u)| \leq n, \\ \zeta(x/n)n \operatorname{sgn} g(x, u) & \text{if } |g(x, u)| > n. \end{cases}$$

Define $(G_n u)(x) = g_n(x, u(x))$ for $u \in Y$. Then the mapping $T + G_n: D(T) \subset Y \to Z$ still satisfies (I) (by [11, Proposition 2.2]) and (III) (since $g_n(x, u)u \geq 0$). Condition (II) also holds for $T + G_n$. Indeed, one easily verifies that $R(G_n)$ is relatively compact in Z_0, so that, given $\tilde{f} \in Z_0$, one can find a neighbourhood $\mathcal{N}$ of $\tilde{f} - \overline{R(G_n)}$ in Z such that $T^{-1}\mathcal{N}$ is bounded in Y; taking a neighbourhood $\mathcal{N}_1$ of $\tilde{f}$ in Z such that $\mathcal{N}_1 - \overline{R(G_n)} \subset \mathcal{N}$, we get that $(T + G_n)^{-1}\mathcal{N}_1$ is bounded in Y. Thus, by Theorem 1 of [12] there exists $u_n \in D(t)$, a solution of

(11) $$Tu_n + G_n u_n = f.$$

We now take the limit as $n \to \infty$. By (IV) and (V), u_n and Tu_n remain bounded in Y and Z, respectively. Thus, passing to a subsequence, we can assume that $u_n \to u \in Y$ for $\sigma(Y, Z_0)$ and also a.e. in Ω, and that $Tu_n \to \chi \in Z$ for $\sigma(Z, Y_0)$. Since $g_n(x, u_n)u_n \geq 0$ and $\int_\Omega g_n(x, u_n)u_n\, dx \leq k$ for some constant k, we have

$$\int_E |g_n(x, u_n)|\, dx \leq \int_E h_r(x)\, dx + \frac{k}{r}$$

for each measurable subset E of Ω and all $r \geq 0$, so that, by Vitali's convergence theorem, $g_n(x, u_n) \to g(x, u)$ in $L^1(\Omega)$. (This type of measure-theoretic argument goes back to De La Vallée-Poussin; cf. [20, p. 159].) Also, by Fatou's Lemma, $g(x, u)u \in L^1(\Omega)$ and

(12) $$\limsup \langle Tu_n, u_n \rangle \leq \langle f, u \rangle - \int_\Omega g(x, u)u\, dx.$$

Going now to the limit in (11), we get

(13) $$\int_\Omega g(x, u)v\, dx = \langle f - \chi, v \rangle$$

for all $v \in Y_0 \cap L^\infty(\Omega)$. Theorem 1 can thus be applied with $S = g(x, u(x)) = f - \chi$, which gives

(14) $$\int_\Omega g(x, u)u\, dx = \langle f - \chi, u \rangle.$$

Substituting this in (12) and using (I), we deduce that $u \in D(T)$ and $Tu = \chi$. Therefore (13) implies (10) for $v \in Y_0 \cap L^\infty(\Omega)$, and (14) implies (10) for $v = u$. To get (10) for $v \in Y \cap L^\infty(\Omega)$, we use Theorem 4 of [13] to associate with such a v a sequence $v_n \in \mathcal{D}(\Omega)$ such that $v_n \to v$ for $\sigma(Y, Z)$, a.e. in Ω, and $\|v_n\|_\infty \le (N+1)\|v\|_\infty$. The conclusion then follows by dominated convergence. Q.E.D.

REMARK 7. If g is nondecreasing in u and if u_1, u_2 are two solutions corresponding to f_1 and f_2, then

$$\langle Tu_1 - Tu_2, u_1 - u_2 \rangle + \int_\Omega (g(x, u_1) - g(x, u_2))(u_1 - u_2)\, dx$$
$$= \langle f_1 - f_2, u_1 - u_2 \rangle.$$

Indeed,

$$\int_\Omega (g(x, u_1) - g(x, u_2)) v\, dx = \langle (f_1 - f_2) - (Tu_1 - Tu_2), v \rangle$$

for all $v \in W_0^1 L_M(\Omega) \cap L^\infty(\Omega)$, and Theorem 1 implies that this equality still holds for $v = u_1 - u_2$.

REMARK 8. If h_r in (8) is only assumed to be in $L^1_{\text{loc}}(\Omega)$, then similar arguments show that there exists $u \in W_0^1 L_M(\Omega)$ such that $g(x, u) \in L^1_{\text{loc}}(\Omega)$, $g(x, u)u \in L^1(\Omega)$, which satisfies (10) for all v in $W_0^1 L_M(\Omega) \cap L^\infty(\Omega)$ with compact support in Ω and for $v = u$.

REMARK 9. If $g(x, u) = p(x, u) + q(x, u)$ with p satisfying (8) and (9) and $|q(x, u)| \le h(x)$ for some $h \in E_{\overline{M}}(\Omega) \cap L^1(\Omega)$, a.e. $x \in \Omega$ and all $u \in \mathbb{R}$, then similar arguments show that the conclusion of Theorem 6 holds, provided conditions (III) and (IV) are strengthened as follows:

(III)' for some $\tilde{f} \in Z_0$ and some $r \in \mathbb{R}$, $\langle u, Tu - f \rangle \ge 0$ for all u in V' with $\|u\|_Y \ge r$ and all f in $\tilde{f} + X_h$,

(IV)' for any $\tilde{f} \in Z_0$, $\{u \in Y; \langle Tu, u \rangle \le \langle f, u \rangle$ for some $f \in \tilde{f} + X_h\}$ is bounded in Y.

Here $X_h = \{f; |f(x)| \le h(x)$ a.e. in $\Omega\} \subset Z_0$. These conditions on g are satisfied, for instance, if g verifies (8), $g(x, u)u \ge 0$ for a.e. $x \in \Omega$ and all $u \in \mathbb{R}$ with $|u| \ge$ some R, and if $h_R \in E_{\overline{M}}(\Omega)$.

REMARK 10. The Neumann problem for (7) can also be studied along these lines by using the approximation procedure of Remark 5 instead of Theorem 1.

REFERENCES

1. R. Adams, *Sobolev spaces*, Academic Press, New York, 1975.
2. H. Brezis and F. Browder, *Sur une propriété des espaces de Sobolev*, C. R. Acad. Sci. Paris Sér. A **287** (1978), 113–115.
3. _____, *A property of Sobolev spaces*, Comm. Partial Differential Equations **9** (1979), 1077–1083.
4. _____, *Some properties of higher order Sobolev spaces*, J. Math. Pures Appl. **61** (1982), 245–259.
5. F. Browder, *Nonlinear elliptic boundary value problems*, Bull. Amer. Math. Soc. **69** (1963), 862–874.
6. _____, *Existence theorems for nonlinear partial differential equations*, Proc. Sympos. Pure Math., vol. 16, Amer. Math. Soc., 1970, pp. 1–60.

7. _____, *Existence theory for boundary value problems for quasilinear elliptic systems with strongly nonlinear lower order terms*, Proc. Sympos. Pure Math., vol. 23, Amer. Math. Soc., 1973, pp. 269–286.

8. _____, *Pseudo-monotone operators and nonlinear elliptic boundary value problems on unbounded domains*, Proc. Nat. Acad. Sci. U.S.A. **74** (1977), 2659–2661.

9. T. Donaldson, *Nonlinear elliptic boundary value problems in Orlicz–Sobolev spaces*, J. Differential Equations **10** (1971), 507–528.

10. D. Gilbarg and N. S. Trudinger, *Elliptic partial differential equations of second order*, Springer-Verlag, Berlin, 1977.

11. J.-P. Gossez, *Nonlinear elliptic boundary value problems for equations with rapidly (or slowly) increasing coefficients*, Trans. Amer. Math. Soc. **190** (1974), 163–205.

12. _____, *Surjectivity results for pseudo-monotone mappings in complementary systems*, J. Math. Anal. Appl. **53** (1976), 484–494.

13. _____, *Some approximation properties in Orlicz–Sobolev spaces*, Studia Math. **74** (1982), 17–24.

14. L. Hedberg, *Two approximation properties in function spaces*, Ark. Mat. **16** (1978), 51–81.

15. P. Hess, *A strongly nonlinear elliptic boundary value problem*, J. Math. Anal. Appl. **43** (1973), 241–249.

16. A. Kufner, O. John and S. Fuch, *Function spaces*, Academia, Praha, 1977.

17. R. Landes, *On Galerkin's method in the existence theory of quasilinear elliptic equations*, J. Funct. Anal. **39** (1980), 123–148.

18. R. Landes and V. Mustonen, *Pseudo-monotone mappings in Sobolev–Orlicz spaces and nonlinear boundary value problems on unbounded domains*, J. Math. Anal. Appl. **88** (1982), 25–36.

19. J. L. Lions, *Quelques méthodes de résolution des problèmes aux limites non linéaires*, Dunod, Paris, 1969.

20. I. P. Natanson, *Theory of functions of a real variable*, F. Ungar Publ., New York, 1955.

21. J. Webb, *Boundary value problems for strongly nonlinear elliptic equations*, J. London Math. Soc. **21** (1980), 123–132.

UNIVERSITÉ LIBRE DE BRUXELLES, BELGIUM

Nontrivial Solutions of Semilinear Elliptic Equations of Fourth Order

YONG-GENG GU

Abstract. In this paper we study nontrivial solutions of boundary value problems of semilinear biharmonic equations, using the Mountain Pass Lemma.

I. Introduction. Let Ω be a smooth, bounded domain in $\mathbf{R}^N$, $N > 4$. We will examine the problem

(1.1) $$\begin{cases} \Delta^2 u = g(x, u) & \text{in } \Omega, \\ u|_{\partial\Omega} = \left.\dfrac{\partial u}{\partial \nu}\right|_{\partial\Omega} = 0, \end{cases}$$

where the operator

$$\Delta^2 = \sum_{i=1}^{N} \frac{\partial^4}{\partial x_i^4} + \sum_{\substack{i,j=1 \\ i \neq j}}^{N} \frac{\partial^4}{\partial x_i^2 \partial x_j^2}$$

is a biharmonic operator, $\partial u/\partial \nu$ denotes the outer normal derivative of u on the boundary $\partial\Omega$, and $g(x, t)$ is a given nonlinear function which satisfies the following conditions:

(i) $g(x, t)$ is Lipschitz continuous in $\overline{\Omega} \times \mathbf{R}$,

(ii)

(1.2) $$\lim_{t \to 0} \frac{g(x, t)}{t} = 0,$$

(iii) there are constants $\theta \in (0, \tfrac{1}{2})$ and $\bar{t} > 0$ such that

(1.3) $$0 < G(x, t) \leq \theta t g(x, t) \quad \text{for } |t| > \bar{t},$$

where $G(x, t) = \int_0^t g(x, \xi)\, d\xi$,

1980 *Mathematics Subject Classification.* Primary 35J35.

© 1986 American Mathematical Society
0082-0717/86 $1.00 + $.25 per page

(iv) there are constants $C_1 \geq 0$, $C_2 > 0$ such that

(1.4) $\quad\quad |g(x,t)| < C_1 + C_2\phi(|t|) \quad \text{for all } (x,t) \in \overline{\Omega} \times \mathbf{R}.$

Here $\phi(t)$ is a continuous, increasing function that satisfies $\phi(0) = 0$, $\phi(t) > 0$ for $t > 0$, $\lim_{t\to\infty} \phi(t) = \infty$,

(1.5) $\quad\quad 2 \leq t\phi(t)/\Phi(t) \leq \beta_N < \infty \quad \text{for } t > t_0 > 0,$

and

(1.6) $\quad\quad \lim_{t\to\infty} \Phi(t)/t^{\beta_N} = 0,$

where $\beta_N = 2N/(N-4)$, t_0 is a fixed constant, and

(1.7) $\quad\quad \Phi(t) = \int_0^t \phi(\xi)\,d\xi.$

In this paper, we will study the nontrivial solutions of the problem (1.1) using the Mountain Pass Lemma and the theorem of Orlicz space. Our main results are

THEOREM 1.1. *Under the conditions* (i)–(iv), *there exists a nontrivial solution of the boundary value problem* (1.1).

THEOREM 1.2. *If the function $g(x,t)$ satisfies the conditions* (i), (iii), (iv) *and*
   (v) $g(x,t)$ *is odd in t*,
then the problem (1.1) *possesses an unbounded sequence of solutions*.

REMARK. If $N \leq 4$, the growth conditions (1.4)–(1.6) can be considerably relaxed.

II. Existence of nontrivial solutions.

DEFINITION 2.1 (PS) [8]. Let X be a real Banach space with norm $\|\cdot\|$ and $f(x) \in C^1(X; \mathbf{R})$ a continuous differentiable functional from X to $\mathbf{R}$. We say that f satisfies the so-called Palais–Smale condition (PS) if each sequence $\{x_n\}$ such that $|f(x_n)| \leq M$ and $f'(x_n) \to 0$ in norm in X^* (the dual space of X) has a strongly convergent subsequence.

Here $f'(x)$ denotes the Fréchet derivative of f at $x \in X$.

LEMMA 2.2 (MOUNTAIN PASS LEMMA) [9]. *Let f be a $C^1(X; \mathbf{R})$ functional on a Banach space X satisfying the PS condition. Suppose there is a neighborhood Q of 0 and a point $x_0 \notin \overline{Q}$ such that*

(2.1) $\quad\quad f(0), f(x_0) < C_0 \leq \inf_{x\in\partial Q} f(x).$

Then the number

(2.2) $\quad\quad C = \inf_{\Gamma} \max_{x\in\Gamma} f(x) \geq C_0$

is a critical value of f, where Γ represents any continuous path from x_0 to 0.

Now, let us introduce some function spaces. First, we define the Sobolev space $H_0^2(\Omega)$, or in short H, as

(2.3) $\qquad H_0^2(\Omega) =$ closure of $C_0^\infty(\Omega)$ in the norm

$$\|u\| = \left(\int_\Omega |\Delta u|^2 \, dx \right)^{1/2}.$$

Second, we give a definition of the Orlicz space $L_\Phi(\Omega)$ (or L_Φ) [6]. It is clear that the function $\Phi(t)$ is an N-function by (1.5). Suppose $\Psi(t)$ is the complementary N-function of $\Phi(t)$. Then we define the Orlicz spaces and their norms as follows:

$$L_\Phi(\Omega) = \left\{ u: \int_\Omega \Phi(l|u|) \, dx < \infty \text{ for some } l > 0 \right\},$$

$$L_\Psi(\Omega) = \left\{ u: \int_\Omega \Psi(l|u|) \, dx < \infty \text{ for some } l > 0 \right\},$$

and

$$\|u\|_{L_\Phi} = \inf_\alpha \left\{ \alpha: \int_\Omega \Phi\left(\frac{|u|}{\alpha}\right) dx \leq 1 \right\},$$

$$\|u\|_{L_\Psi} = \inf_\alpha \left\{ \alpha: \int_\Omega \Psi\left(\frac{|u|}{\alpha}\right) dx \leq 1 \right\}.$$

By the hypotheses (1.6) we know the N-function $\Phi(t)$ satisfies the so-called Δ_2-condition [10], i.e. there are constants $\kappa, \tilde{t} > 0$ such that

(2.4) $\qquad \Phi(2t) \leq \kappa \Phi(t) \quad \text{for all } t > \tilde{t}.$

In the same way, the N-function $\Psi(t)$ satisfies the Δ_2-condition by (1.6).

We have the following

LEMMA 2.3. *Suppose the sequence* $\{u_n\}$ *is uniformly bounded in the norm* $\|\cdot\|$, *then there exists a subsequence of* $\{u_n\}$, *still denoted by* $\{u_n\}$, *which is convergent in the space* $L_\Phi(\Omega)$.

PROOF. First, we see that the N-function $\Phi(t)$ is controlled by the function t^{β_N} due to (1.6). The embedding operator of the Sobolev space H into the Orlicz space $L_\Phi(\Omega)$ is compact because the embedding from H into L_{β_N} is bounded [1]. So this lemma holds.

PROOF OF THEOREM 1.1. At first, we consider the functional

(2.5) $\qquad J(u) = \int_\Omega \left[\tfrac{1}{2} |\Delta u|^2 - G(x, u) \right] dx$

in the space H. It may be seen that the hypotheses of the function $g(x, t)$ guarantee the functional $J(u) \in C^1(H, \mathbf{R})$, and then the boundary value problem (1.1) is the Euler–Lagrange equation of the functional $J(u)$. It is well known that the critical point $u(x)$ of the functional J is a generalized solution of problem

(1.1) in the space H, i.e., $u(x)$ satisfies the integral equality

(2.6) $$\int_\Omega [\Delta u \Delta v - g(x,u)v]\, dx = 0 \quad \forall v \in H.$$

The standard regularity theory shows that the generalized solution $u(x)$ satisfying (2.6) is sufficiently smooth if $g(x,t)$ does and hence $u(x)$ is the classical solution [2, 4]. For reasons above, we need only seek a nontrivial critical point of the functional J in the space H.

Now, we are going to discuss the existence of the nontrivial critical point of J via the Mountain Pass Lemma. First, let us check the PS condition of J. Suppose the sequence $\{u_n(x)\} \subset H$ and a constant M independent of $u_n(x)$ are such that $|J(u_n)| \leq M$ and $J'(u_n) \to 0$ (in norm) as $n \to \infty$, i.e.

(2.7) $$\frac{1}{\|v\|}\int_\Omega [\Delta u_n \Delta v - g(x,u_n)v]\, dx \to 0 \quad \text{as } n \to \infty,\ \forall v \in H.$$

Taking in particular $v = u_n$, we have

(2.8) $$\langle J'(u_n), u_n\rangle / \|u_n\| \to 0 \quad \text{as } n \to \infty,$$

where

$$\langle J'(u_n), v\rangle = \int_\Omega [\Delta u_n \Delta v - g(x,u_n)v]\, dx.$$

By (1.3) and (2.8) for any $\varepsilon > 0$, $\exists\, n_0$ such that, when $n > n_0$ we have

(2.9) $$M \geq J(u_n) - \theta\langle J'(u_n), u_n\rangle - \theta\varepsilon\|u_n\|$$
$$\geq \left(\tfrac{1}{2}-\theta\right)\|u_n\|^2 + \int_\Omega [\theta g(x,u_n)u_n - G(x,u_n)]\, dx - \theta\varepsilon\|u_n\| - M_1$$
$$\geq \left(\tfrac{1}{2}-\theta\right)\|u_n\|^2 - \theta\varepsilon\|u_n\| - M_1,$$

where M_1 is a constant independent of u_n. Note that $\theta \in (0, \tfrac{1}{2})$, so $\{u_n\}$ is uniformly bounded in H, i.e., there exists a constant C independent of $u_n(x)$ such that

(2.10) $$\|u_n\| < C.$$

Hence, there exists a subsequence of $\{u_n\}$, still denoted by $\{u_n\}$, and a function $u(x) \in H$ such that

$$u_n \xrightarrow{\text{weakly}} u \quad \text{in } H,$$

i.e.,

(2.11) $$\lim_{n\to\infty} \int_\Omega \Delta u_n \Delta v\, dx = \int_\Omega \Delta u \Delta v\, dx \quad \forall v(x) \in H.$$

Next, we consider $(u_n - u, u_n)$. For one thing, by (2.11) we have

(2.12) $$\lim_{n\to\infty}(u_n - u, u_n) = \lim_{n\to\infty}\|u_n\|^2 - \|u\|^2,$$

and for another thing, by (2.7), using the Hölder inequality in the Orlicz space we have

(2.13) $$(u_n - u, u_n) = \int_\Omega g(x, u_n)(u_n - u)\, dx$$
$$\leq \|g(x, u_n)\|_{L_\Psi} \|u_n - u\|_{L_\Phi}.$$

If we can prove $\|g(x, u_n)\|_{L_\Psi} \leq C'$ (uniformly bounded) by Lemma 2.3, it follows that

(2.14) $$\lim_{n\to\infty} (u_n - u, u_n) \leq C' \lim_{n\to\infty} \|u_n - u\|_{L_\Phi} = 0.$$

Indeed, because the N-functions $\Phi(t)$ and $\Psi(t)$ satisfy the Δ_2-condition, and from the hypotheses (iv) we see

$$\int_\Omega \Psi\bigl[|g(x, u_n)|\bigr]\, dx \leq C_3 + C_2 \int_\Omega \Psi\bigl[\phi(|u_n|)\bigr]\, dx$$
$$< C_3 + C_2 \int_\Omega |u_n|\phi(|u_n|)\, dx \leq C_3 + C_4 \int_\Omega \Phi(|u_n|)\, dx \leq C_5,$$

and then $\|g(x, u_n)\|_{L_\Psi} \leq C_6$, where the constants $C_2, \ldots, C_6$, and C' are independent of u_n.

According to (2.11), (2.12), and (2.14), we have $\|u\|^2 = \lim_{n\to\infty} \|u_n\|^2$; hence $\lim_{n\to\infty} \|u_n - u\| = 0$. Thus the PS condition holds.

Now, let us check the other conditions of the Mountain Pass Lemma. By (ii), $\forall\, \varepsilon > 0,\, \exists\, t_* > 0$ such that

(2.15) $$|g(x, t)| \leq \varepsilon |t| \quad \text{when } |t| < t_*.$$

Using (1.4), (1.6), and the embedding theorem in Sobolev–Orlicz space, we obtain

$$\left|\int_\Omega G(x, u)\, dx\right| \leq \varepsilon C_7 \|u\|^2 + C_2 \int_\Omega \phi(|u|)\, dx \leq \varepsilon C_7 \|u\|^2 + C_8 \|u\|^{\beta_N},$$

and then

(2.16) $$J(u) \geq C_9 \|u\|^2 - C_{10} \|u\|^{\beta_N},$$

where the $C_7, \ldots, C_{10} > 0$ are constants independent of u_n. Setting a small constant $\eta_0 > 0$ for any $\eta \in (0, \eta_0)$, we have

(2.17) $$J(u)|_{\|u\|=\eta} \geq C_{11} \eta^2 > 0 \quad (C_{11} = \text{constant} > 0).$$

Applying (1.3) again, it follows that

(2.18) $$|G(x, u)| \geq C_{12} |u|^{1/\theta} \quad (C_{12} = \text{constant} > 0).$$

If we choose a function $e(x) \in H$ with $\|e\| = 1$, by (2.16) and $\theta \in (0, \tfrac{1}{2})$ again, it follows that

$$J(\lambda e) \leq \tfrac{1}{2}\lambda^2 - \int_\Omega G(x, \lambda e)\, dx \leq \tfrac{1}{2}\lambda^2 - \left(C_{13} \int_\Omega |e(x)|^{1/\theta}\, dx\right) \lambda^{1/\theta} \quad \forall\, \lambda > 0.$$

Hence $\lim_{\lambda\to\infty} J(\lambda e) = -\infty$, and then there is a $\lambda_0 > 0$ such that $J(\lambda_0 e) = 0$.

Letting $u_0(x) = \lambda_0 e(x)$, we have $J(u_0) = J(0) = 0$, and then all of the conditions of the Mountain Pass Lemma hold. Theorem 1.1 is proved.

More generally we could consider the nontrivial solution of the problem

(2.19)
$$\begin{cases} \Delta^2 u = D\left[\dfrac{\kappa(|Du|)}{|Du|} Du\right] + g(x,u) & \text{in } \Omega, \\ u|_{\partial\Omega} = \left.\dfrac{\partial u}{\partial \nu}\right|_{\partial\Omega} = 0. \end{cases}$$

The assumptions of $g(x, t)$ in (2.19) are the same as (1.1). The function $\kappa(t)$ is a continuous increasing function that satisfies the following conditions:

(vi) $\kappa(0) = 0$, $\kappa(t) > 0$ for $t > 0$, $\lim_{t \to \infty} \kappa(t) = \infty$,

(vii)

(2.20) $$2 \leqslant t\kappa(t)/K(t) \leqslant \beta'_N \quad \text{for all } t > t',$$

(viii)

(2.21) $$\lim_{t \to \infty} \frac{K(t)}{t^{\beta'_N}} = 0,$$

where the constants $t' > 0$, $\beta'_N \equiv 2N/(N-2)$, $Du = \text{grad } u = (D_1 u, \ldots, D_N u)$, $|Du| = \sqrt{\sum_{i=1}^N |D_i u|^2}$, and $K(t) = \int_0^t \kappa(\xi)\, d\xi$.

In order to study the existence of the nontrivial solutions of the problem (2.19) we define a functional

$$\tilde{J}(u) = \int_\Omega \left[\tfrac{1}{2}|\Delta u|^2 - K(|Du|) - G(x,u)\right] dx,$$

and then we need only find a critical point of the functional $\tilde{J}$. Taking into account (2.20) and (2.21), the function $K(t)$ is an N-function satisfying the Δ_2-condition. By (2.21) we know the embedding of H into the Sobolev–Orlicz space $\mathring{W}^1 L_K(\Omega)$ is compact [1], where

$$\mathring{W}^1 L_K(\Omega) = \left\{u: u, |Du| \in L_K(\Omega), \text{ and } u|_{\partial\Omega} = 0\right\}$$

is a Sobolev–Orlicz space. We have the later theorem:

THEOREM 2.4. *If the function $g(x, t)$ satisfies assumptions* (i)–(iv) *and the function $\kappa(t)$ satisfies* (vi)–(viii), *then there is at least a nontrivial solution of problem* (2.17).

III. Infinitely many solutions. In this section we will study the unbounded sequence of solutions of problem (1.1). First, we need a symmetric version of the Mountain Pass Lemma.

LEMMA 3.1 (MOUNTAIN PASS LEMMA—SYMMETRIC VERSION) [9]. *Let x be a real infinite-dimensional Banach space with norm $\|\cdot\|$ and let $f \in C^1(X; \mathbf{R})$ satisfy the PS condition. Suppose*

(a) *f is even*,

(b) *for all finite-dimensional subspaces $\tilde{X} \subset X$, there exists a ball $B_\rho(\tilde{X}) = \{x \in \tilde{X}: \|x\| < \rho\}$ such that*

(3.1) $$f(x) \leqslant 0 \quad \text{for all } x \in \tilde{X} - B_\rho(\tilde{X}),$$

(c) *there are constants* $\rho, h > 0$ *and an n-dimensional subspace* $x_n \subset X$ *such that*

(3.2) $\quad f|_{\partial B_\rho \cap X_n^\perp} \geq h \quad (X_n^\perp$ *denotes the orthogonal complement of* x_n *in* $X)$.

Then f possesses an unbounded sequence of critical values.

Now, we will establish an inequality in the Orlicz space $L_M(\Omega)$.

LEMMA 3.2. *For some constant* $\tau, \alpha, \beta > 0$, *if the N-function* $M(t)$ *with the derivative* $M'(t)$ *satisfies the condition*

(3.3) $\quad\quad\quad 1 < \alpha \leq tM'(t)/M(t) \leq \beta < \infty \quad \text{for } t > \tau,$

then the inequality

(3.4) $\quad \int_\Omega M(|u|)\,dx \leq M(l)\left(\frac{\|u\|_{L_\alpha}^\alpha}{l^\alpha} + \frac{\|u\|_{L_\beta}^\beta}{l^\beta} \right) + C(\tau) \quad \forall u \in L_M(\Omega)$

is valid, where the constant $C(\tau)$ *depends on* τ *but not on* u, *and* $l \in (\tau, \infty)$ *is an arbitrary constant.*

PROOF. Using the assumption (3.3) for any $t, l > \tau$, we have

(3.5) $\quad \begin{cases} \left(\dfrac{t}{l}\right)^\beta \leq \dfrac{M(t)}{M(l)} \leq \left(\dfrac{t}{l}\right)^\alpha & \text{when } t < l, \\[1em] \left(\dfrac{t}{l}\right)^\alpha \leq \dfrac{M(t)}{M(l)} \leq \left(\dfrac{t}{l}\right)^\beta & \text{when } t \geq l. \end{cases}$

So

(3.6) $\quad\quad\quad M(t) \leq M(l)\left[\left(\dfrac{t}{l}\right)^\alpha + \left(\dfrac{t}{l}\right)^\beta \right] \quad \forall t \in (\tau, \infty).$

Moreover, for any $u \in L_M(\Omega)$, we see

(3.7) $\quad \int_\Omega M(|u|)\,dx \leq \int_{\Omega \cap \{|u| \leq \tau\}} M(|u|)\,dx + \int_{\Omega \cap \{|u| > \tau\}} M(|u|)\,dx$

$\quad\quad\quad\quad\quad \leq C(\tau) + \int_{\Omega \cap \{|u| > \tau\}} M(|u|)\,dx.$

Taking into account (3.6) and (3.7), we get inequality (3.4).

PROOF OF THEOREM 1.2. Due to conditions (i), (iv) and (v) in §I, we know the functional J defined by (2.5) lies in $C^1(H; \mathbf{R})$, is even, and satisfies the PS conditions. Moreover, using the similar method in the proof of Theorem 1.1 we can deduce that for all $u \in \tilde{H}$,

$$J(u) \to -\infty \quad \text{when } \|u\| \to \infty,$$

where $\tilde{H}$ is an arbitrary finite-dimensional subspace of H. So the conditions (a) and (b) in Lemma 3.1 are satisfied. To complete the proof, by the symmetric version Mountain Pass Lemma, we need only verify the condition (c) in Lemma 3.1.

Let us consider the eigenvalue problem

$$\Delta^2 v = \lambda v \quad \text{in } \Omega, \qquad v|_{\partial\Omega} = \left.\frac{\partial v}{\partial \nu}\right|_{\partial\Omega} = 0. \tag{3.8}$$

It is well known that the eigenvalue problem (3.8) possesses a sequence of the eigenvalues $0 < \lambda_1 \leq \lambda_2 \leq \cdots \leq \lambda_j \leq \cdots$, and a sequence of the eigenfunctions $\{\phi_1(x), \phi_2(x), \ldots, \phi_j(x), \ldots\}$ with $\int_\Omega \phi_i \phi_j \, dx = \delta_{ij}$.

Let $H_j = \{\phi_1(x), \phi_2(x), \ldots, \phi_j(x)\}$ be the eigenfunction space in the space H spanned by the first j eigenfunctions, and let $H_j^\perp$ be the orthogonal complement of H_j in H. Clearly, $H_1 \subset H_2 \subset \cdots \subset H$. Using Lemma 3.2 and hypothesis (1.5) we have

$$\int_\Omega \Phi(|u|) \, dx \leq \Phi(l)\left(\frac{\|u\|_{L_2}^2}{l^2} + \frac{\|u\|_{L_{\beta_N}}^{\beta_N}}{l^{\beta_N}}\right) + C(t_0) \tag{3.9}$$

for some arbitrary constant $l \in (t_0, \infty)$. According to (1.4) and (1.5) we see that

$$J(u) = \int_\Omega \left[\frac{1}{2}|\Delta u|^2 - G(x, u)\right] dx \tag{3.10}$$

$$\geq \frac{1}{2}\|u\|^2 - C_2 \int_{\Omega \cap \{|u| > t_0\}} \Phi(|u|) \, dx - C_3$$

$$\geq \frac{1}{2}\|u\|^2 - \Phi(l)\left(\frac{\|u\|_{L_2}^2}{l^2} + \frac{\|u\|_{L_{\beta_N}}^{\Omega_N}}{l^{\beta_N}}\right) - C_3,$$

where C_3 is a constant independent of u. Suppose $u \in H_j^\perp$; thus we can write $u(x)$ as

$$u(x) = \sum_{\kappa \geq j+1} a_\kappa \phi_\kappa(x). \tag{3.11}$$

Taking into account (3.8) and (3.11) we obtain

$$\|u\|^2 = \int_\Omega \left[\sum_{\kappa \geq j+1} a_\kappa \Delta \phi_\kappa(x)\right]^2 dx = \sum_{\kappa \geq j+1} \lambda_\kappa a_\kappa^2 \|\phi_\kappa\|_{L_2}^2 \geq \lambda_{j+1} \|u\|_{L_2}^2,$$

i.e.,

$$\|u\|_{L_2}^2 \leq \frac{1}{\lambda_{j+1}} \|u\|^2 \quad \text{for all } u \in H_j^\perp. \tag{3.12}$$

By (3.10), (3.12), and the Sobolev embedding theorem, for all $u \in H_j^\perp \cap \partial B_\rho(H)$ we have

$$J(u) \geq \left[\frac{1}{2} - C_2 \Phi(l)\left(\frac{\lambda_{j+1}^{-1}}{l^2} + \frac{\rho^{8/(N-4)}}{l^{\beta_N}}\right)\right]\rho^2 - C_3, \tag{3.13}$$

where $B_\rho(H) = \{u \in H \mid \|u\| < \rho\}$ denotes a ball in the space H. First, we choose ρ so large that

$$\rho^2/8 > C_3. \tag{3.14}$$

For the ρ chosen, we choose l sufficiently large such that

(3.15) $$C_2\rho^{8/(N-4)}\Phi(l)/l^{\beta_N} < 1/8.$$

The estimate (3.15) is valid because the N-function $\Phi(t)$ satisfies the Δ_2-condition and assumption (1.6). Finally, we choose j so large that

(3.16) $$\left(C_2\Phi(l)/l^2\right)\lambda_{j+1}^{-1} < 1/8.$$

Putting (3.14)–(3.16) into (3.13), we get

(3.17) $$I(u) \geq \rho^2/8 > 0 \quad \text{for all } u \in H_j^{\perp} \cap \partial B_\rho(H),$$

i.e., condition (c) in Lemma 3.1 is satisfied, and Theorem 1.2 is proved.

Similarly, we can get the following

THEOREM 3.3. *Under assumptions* (i), (iii)–(v), *and* (vi)–(viii), *problem* (2.19) *possesses an unbounded sequence of solutions.*

The proof of Theorem 3.3 is similar to the proof of Theorem 1.2.

EXAMPLE.

(3.18) $$\begin{cases} \Delta^2 u = f(x)|u|^{2S}u\log(1+|u|) + dD(|Du|^{2m}Du) & \text{in } \Omega, \\ u|_{\partial\Omega} = \left.\dfrac{\partial u}{\partial \nu}\right|_{\partial\Omega} = 0. \end{cases}$$

Suppose the function $f(x)$ is sufficiently smooth in $x \in \overline{\Omega}$, and the constant $d \geq 0$. If the growth indexes $S \in (0, 4/(N-4))$ and $m \in (0, 2/(N-2))$, then problem (3.18) possesses an unbounded sequence of solutions.

Professor Yao-hua Deng and Yi-ying Chen joined with the author in this work.

REFERENCES

1. R. A. Adams, *Sobolev space*, Academic Press, New York, 1975.
2. S. Agmon, *Lectures on elliptic boundary value problems*, Van Nostrand-Reinhold, Princeton, N. J., 1965.
3. A. Ambrosetti and P. H. Rabinowitz, *Dual variational methods in critical point theory and applications*, J. Funct. Anal. **14** (1973), 349–381.
4. M. S. Berger, *Nonlinearity and functional analysis*, Academic Press, New York, 1978.
5. K. C. Chang, *Variational methods for non-differentiable functionals and their application to PDE*, J. Math. Anal. Appl. **80** (1981), 102–127.
6. Xiaxi Ding, Peizhu Luo, Yong geng Gu and Huizhong Fang, *Calculus of variational with strong nonlinearity*, Sci. Sinica Ser. A **23** (1980), 945–955.
7. Yong geng Gu, *The existence of nontrivial solutions on a class of nonlinear elliptic equations* (to appear).
8. L. Nirenberg, *Variational and topological methods*, Bull. Amer. Math. Soc. **4** (1981), 267–302.
9. P. H. Rabinowitz, *The Mountain Pass Theorem: theme and variations*, Rep. Third Internat. Sympos. on Differential Geometry and Differential Equations (Changchun, China), Science Press, Beijing 1982.
10. M. A. Krasnosel'skii and Ya. B. Rutitskii, *Convex functions and Orlicz spaces*, National Press in Phys. and Math. Literature, Moscow, 1958. (Russian)

ACADEMIA SINICA, CHINA

Approximation in Sobolev Spaces and Nonlinear Potential Theory

LARS INGE HEDBERG

The purpose of this expository paper is to discuss the following result.

THEOREM 1. *Let G be an arbitrary open set in $\mathbf{R}^N$, and let $f \in W_0^{m,p}(G)$ for some positive integer m, and some p, $1 < p < \infty$. Then there exist functions ω_n, $0 \leq \omega_n \leq 1$, such that $\operatorname{supp} \omega_n \Subset G$, $\omega_n f \in L^\infty \cap W_0^{m,p}(G)$, and*

$$\lim_{n \to \infty} \|f - \omega_n f\|_{m,p} = 0.$$

Here $W_0^{m,p}(G)$ denotes the subspace of the usual Sobolev space $W^{m,p}(\mathbf{R}^N)$ obtained by taking the closure of $C_0^\infty(G)$, the test functions with support in G, and $\|\cdot\|_{m,p}$ is the Sobolev norm defined by

$$\|f\|_{m,p}^p = \sum_{|\alpha| \leq m} \int_{\mathbf{R}^N} |D^\alpha f|^p \, dx.$$

Unfortunately the proof of this deceptively simple statement is quite complicated (except in the case $m = 1$), and it occupies much of the three papers [10, 12, and 14]. It would of course be very desirable to have a simpler proof. Here we can only give a brief outline, and explain its relation to "nonlinear potential theory".

The theorem was applied by J. R. L. Webb [21] to a strongly nonlinear elliptic problem, and the main ideas behind this application were brought out by H. Brézis and F. E. Browder [2]. Further interesting applications are given by Browder in his article in these Proceedings [3].

Theorem 1 is contained in the following two results.

THEOREM 2. *Under the same assumptions as in Theorem 1 there exist functions $\omega_n \in C_0^\infty(G)$ such that $\lim_{n \to \infty} \|f - \omega_n f\|_{m,p} = 0$.*

1980 *Mathematics Subject Classification.* Primary 46E35, 31B15; Secondary 31B25, 35J60, 35J67.

© 1986 American Mathematical Society
0082-0717/86 $1.00 + $.25 per page

THEOREM 3. *Let $f \in W^{m,p}(\mathbf{R}^N)$ for some positive integer m and some p, $1 < p < \infty$. Then there are functions $\omega_n \in W^{m,p}(\mathbf{R}^N)$ such that $0 \leq \omega_n \leq 1$, supp ω_n is compact, $\lim_{n \to \infty} \|f - \omega_n f\|_{m,p} = 0$, and $\omega_n f \in L^\infty \cap W^{m,p}$. In fact, ω_n can be chosen to be zero on the set $\{x; |f(x)| > n\}$.*

Theorem 1 is obtained by applying Theorem 3 to the function $\omega_n f$ given by Theorem 2. In Theorem 3 the function ω_n cannot in general be chosen in C^∞, since $f(x)$ may be large on a dense set of positive measure.

Theorem 3 was proved in [10, Lemma 5.2]. The proof is relatively simple, and it has been exposed very clearly by Webb and Brézis-Browder in the papers mentioned above, so there is no need to discuss that theorem further here.

The main difficulty is contained in Theorem 2, and in order to explain the ideas involved in the proof it is now necessary to introduce (m, p)-capacities and to discuss nonlinear potential theory at some length. In particular we shall discuss the concept of "(m, p)-thin set" and the "Kellogg property", which are crucial to the proof of Theorem 2.

To the space $W^{m,p}(\mathbf{R}^N)$ one can associate a set function, called (m, p)-capacity, in the following way:

For a compact set $K \subset \mathbf{R}^N$, $C_{m,p}(K) = \inf\{\|\phi\|_{m,p}^p; \phi \in C_0^\infty, \phi \geq 1 \text{ on } K\}$.

For an open set G, $C_{m,p}(G) = \sup\{C_{m,p}(K); K \subset G, K \text{ compact}\}$.

For an arbitrary set E, $C_{m,p}(E) = \inf\{C_{m,p}(G); G \supset E, G \text{ open}\}$.

It was introduced in 1963 by V. G. Maz'ja [16]. We shall be interested in this capacity only when $mp \leq N$, so that $W^{m,p} \not\subset C$, and $C_{m,p}(\{x_0\}) = 0$.

In the classical case, $p = 2$, $m = 1$, the extremal function satisfies the equation $-\Delta u + u = 0$ off the set K, and the theory is essentially the same as for Laplace's equation.

For $p \neq 2$, however, the corresponding equation is nonlinear. For example, if $m = 1$, the extremal satisfies the equation $-\text{div}(|\nabla u|^{p-2} \nabla u) + |u|^{p-2} u = 0$ off K, hence it is not immediately clear how one generalizes the classical potential theory to this situation. Nevertheless, Maz'ja [17] proved the following theorem:

If G is a region in $\mathbf{R}^N$, then a point $x_0 \in \partial G$ is regular for the Dirichlet problem for the equation $\text{div}(|\nabla u|^{p-2} \nabla u) = 0$, $u \in W^{1,p}$, if

$$\int_0^1 \left(C_{1,p}(B(x_0, r) \setminus G) r^{p-N} \right)^{p'-1} r^{-1} \, dr = \infty.$$

Here and everywhere in what follows $p' = p/(p-1)$ and $B(x_0, r) = \{x; |x - x_0| \leq r\}$. We refer to the original paper for the exact statement of the theorem.

In the classical case $p = 2$ this is the Wiener Criterion, which is then known to be both necessary and sufficent for regularity. It is also known that the set of irregular boundary points always has $(1, 2)$-capacity zero. This is the Kellogg Lemma, conjectured by O. D. Kellogg [15], and proved in general by G. C. Evans [5] (in whose honor the building was named, where this conference was held).

Whether the Kellogg Lemma continues to hold for $p \neq 2$, $1 < p \leq N$, was not known until recently, and the converse to Maz'ja's theorem is still an open problem. On the other hand, R. Gariepy and W. P. Ziemer [8] were able to give a different proof for Maz'ja's theorem that works also for much more general quasilinear elliptic equations of the type $\text{div} A(x, u, \nabla u) = B(x, u, \nabla u)$.

In the late 1960s it occurred to a number of people (B. Fuglede [7], N. G. Meyers [19], V. G. Maz'ja and V. P. Havin [18], Ju. G. Rešetnjak [20]) that one could get further by defining the capacity a little differently.

It is a theorem of A. P. Calderón [4] that for $1 < p < \infty$, $W^{m,p} = J_m L^p$, where J_m denotes convolution with the "Bessel kernel" $G_m(x)$, defined as the inverse Fourier transform of $\hat{G}_m(\xi) = (1 + |\xi|^2)^{m/2}$. In other words, $f \in W^{m,p}$ if and only if $f = G_m * g$, where $g \in L^p$, and moreover there is equivalence of norms, i.e. there is a constant $A > 0$ such that $A^{-1}\|g\|_{L^p} \leq \|f\|_{m,p} \leq A\|g\|_{L^p}$.

So let us now redefine (m, p)-capacity, without changing the notation, by setting

$$C_{m,p}(K) = \inf\left\{ \int |g|^p \, dx; f = G_m * g \geq 1 \text{ on } K, f \in C_0^\infty \right\}$$

for a compact set K. By uniform convexity there exists a unique extremal function $f_K = G_m * g_K$ in $W^{m,p}$. It turns out that this extremal has a simple representation. In fact, we have

PROPOSITION 1. *There exists a positive measure μ_K, such that $\text{supp}\,\mu_K \subset K$, $\mu_K(K) = C_{m,p}(K)$, and $f_K = G_m *(G_m * \mu_K)^{p'-1}$.*

PROOF. The mapping $I: g \to g|g|^{p-2}$ is a bijective mapping from L^p to its dual $L^{p'}$. In fact, its inverse is given by $I^{-1}: h \to h|h|^{p'-2}$. Moreover, the dual space of $W^{m,p}$, which is denoted $W^{-m,p'}$, is mapped isomorphically onto $L^{p'}$ by J_m, so that any h in $L^{p'}$ can be represented as $h = G_m * T$ for some distribution $T \in W^{-m,p'}$. Thus, any f in $W^{m,p}$ can be represented as

$$f = G_m *\left((G_m * T)|G_m * T|^{p'-2}\right), \qquad T \in W^{-m,p'}.$$

The claim is that for $f = f_K$ the distribution T is positive, i.e. a measure.

Let $\phi \geq 0$ be a C_0^∞ test function, and write $\phi = G_m * \psi$. Then

$$\int |g_K + t\psi|^p \, dx \geq \int |g_K|^p \, dx = C_{m,p}(K) \quad \text{for all } t \geq 0,$$

and thus $\int |g_K|^{p-2} g_K \psi \, dx \geq 0$ for all such ψ. But

$$|g_K|^{p-2} g_K = Ig_K = G_m * T_K, \qquad T_K \in W^{-m,p'},$$

so $\int (G_m * T_K)\psi \, dx \geq 0$. But, by definition,

$$\int (G_m * T_K)\psi \, dx = T_K(G_m * \psi) = T_K(\phi),$$

so $T_K(\phi) \geq 0$ for all positive $\phi \in C_0^\infty$, which proves the claim.

If we choose ϕ with support in the complement K^c of K, we can let t take both positive and negative values. It follows that $T_K(\phi) = 0$ for all such ϕ, i.e. supp $T_K \subset K$.

Finally, by a passage to the limit, $T_K(1) = T_K(f_K)$, and as above

$$\int (G_m * T_K) g_K \, dx = \int |g_K|^p \, dx = C_{m,p}(K).$$

We denote the measure T_K by μ_K.

We now easily obtain a dual characterization of (m, p)-capacity.

PROPOSITION 2. *Let $K \subset \mathbf{R}^N$ be compact. Then*

$$C_{m,p}(K)^{1/p} = \sup\{\mu(K); \, \mu \geq 0, \|G_m * \mu\|_{p'} \geq 1\}.$$

PROOF. Let supp $\mu \subset K$, and let $f = G_m * g$, $g \in L^p$, satisfy $f \geq 1$ on K. Then, by Fubini's theorem, and Hölder's inequality

$$\mu(K) \leq \int (G_m * g) \, d\mu = \int (G_m * \mu) g \, dx \leq \|G_m * \mu\|_{p'} \|g\|_p.$$

Passing to the limit we find

$$\frac{\mu(K)}{\|G_m * \mu\|_{p'}} \leq \inf_g \|g\|_p = C_{m,p}(K)^{1/p}.$$

On the other hand, choosing $\mu = \mu_K$, we have equality.

If a statement is true for all x except those belonging to a set E with $C_{m,p}(E) = 0$, we say that it is true (m, p)-quasieverywhere $((m, p)$-q.e.$)$.

One can prove that the extremal, f_K, which we denote

$$V^{\mu_K} = G_m * (G_m * \mu_K)^{p'-1},$$

has the following properties.

PROPOSITION 3. (a) $V^{\mu_K}(x) \geq 1$ (m, p)-q.e. on K.
(b) $V^{\mu_K}(x) \leq 1$ on supp $\mu_K \subset K$.
(c) $V^{\mu_K}(x) \leq M$ everywhere, $M = M(p, N)$.

The function $G_m * (G_m * \mu)^{p'-1} = V^\mu$ is called a nonlinear potential of the measure μ. For $p = p' = 2$ it is actually linear, and then it is the classical potential $G_m * (G_m * \mu) = G_{2m} * \mu$.

The above approach to nonlinear potential theory is due to Maz'ja and Havin [18], and Rešetnjak [20]. A different approach, due to Fuglede, is based on the Minimax Theorem (see Fuglede [7], Meyers [19]).

In studying the continuity properties of nonlinear potentials, one is led to define the concept of an "(m, p)-thin set." We denote the "relative (m, p)-capacity" by $c_{m,p}(K, x, r) = C_{m,p}(K \cap B(x, r)) r^{mp-N}$, and of several possible ways of defining a thin set we choose the following.

DEFINITION. A set K is (m, p)-thin at a point x_0 if $mp \leq N$, and

$$\int_0^1 c_{m,p}(K, x_0, r)^{p'-1} r^{-1} \, dr < \infty.$$

If a set is not thin, we say that it is thick. We set $e_{m,p}(K) = \{x \in K; K \text{ is } (m, p)\text{-thin at } x\}$.

One can prove the following

THEOREM 4. *Let $K \in \mathbf{R}^N$ be compact, and let $x_0 \in K$.*
(a) *If $p > 2 - m/N$, then $x_0 \in e_{m,p}(K)$ if and only if $V^\mu(x_0) < 1$ for some $\mu = \mu_{K \cap B(x_0, r)}$.*
(b) *If $1 < p \leq 2 - m/N$ and $x_0 \notin e_{m,p}(K)$, then $V^\mu(x_0) = 1$ for all $\mu = \mu_{K \cap B(x_0, r)}$.*
(c) *If $1 < p < 2 - m/N$, then K can be found such that $x_0 \in e_{m,p}(K)$, but $V^\mu(x_0) = 1$ for all $\mu = \mu_{K \cap B(x_0, r)}$.*

These results are contained in more general results due to D. R. Adams and N. G. Meyers [1], and the author [9].

As a consequence one can prove the following Kellogg property [9].

COROLLARY 1. $C_{m,p}(K \cap e_{m,p}(K)) = 0$ *for $p > 2 - m/N$.*

Theorem 4 shows that there is a difficulty in extending the classical Wiener Criterion for potentials to the nonlinear case. The difficulty comes from the estimates used in proving Theorem 4.

In fact, we have the inequalities

$$A^{-1} \int_0^\infty \left(\frac{\mu(B(x,r))}{r^{N-mp}} \right)^{p'-1} e^{-c_1 r} r^{-1} \, dr \leq V^\mu(x)$$

$$\leq A \int_0^\infty \left(\frac{\mu(B(x,r))}{r^{N-mp}} \right)^{p'-1} e^{-c_2 r} r^{-1} \, dr$$

for $p > 2 - m/N$. For $1 < p \leq 2 - m/N$, however, the right inequality breaks down (choose μ as a point measure; this makes $V^\mu \equiv \infty$), whereas the left inequality continues to hold.

In view of this, the following theorem of T. H. Wolff is remarkable.

THEOREM 5. *There is a constant $A > 0$ such that*

$$\int V^\mu \, d\mu \leq A \int \left(\int_0^1 \left(\frac{\mu(B(x,r))}{r^{N-mp}} \right)^{p'-1} r^{-1} \, dr \right) d\mu.$$

The converse inequality follows from the above pointwise estimate. Wolff's proof of the theorem is found in [14]. A simpler proof was later given by John L. Lewis (private communication), and Per Nilsson (private communication) observed that the inequality can be reduced to a known inequality of C. Fefferman and E. M. Stein [6] for the "sharp function" and a "maximal inequality" of D. R. Adams (see [14]).

What is important for us here, is that, as Wolff observed, Theorem 5 implies the Kellogg property.

THEOREM 6. $C_{m,p}(K \cap e_{m,p}(K)) = 0$ for all $m > 0$ and $1 < p < \infty$.

It follows that *the irregular boundary points for the Dirichlet problem considered by Maz'ja and Gariepy–Ziemer form a set of $(1, p)$-capacity zero.*

After this long excursion into potential theory we are now prepared to return to the proof of Theorem 2.

In order to prove this theorem one needs to construct a function $\omega \in C_0^\infty(G)$ such that $\omega(x) = 1$ outside an arbitrarily chosen neighborhood V of ∂G, and such that $\int_V |D^k \omega|^p |D^{m-k} f|^p \, dx$ is small. Here D^k denotes an arbitrary derivative of order $k = 1, 2, \ldots, m$.

Let us denote ∂G by K. Without loss of generality we can assume that K is compact. By successively applying the Kellogg property (Theorem 6) we can split K into disjoint sets

$$K = E_0 \cup E_1 \cup \cdots \cup E_m,$$

with the following properties:

(a) K is $(1, p)$-thick at all $x \in E_0$.
(b) $C_{k,p}(E_k) = 0$, and K is $(k + 1, p)$-thick at all $x \in E_k$, $k = 1, \ldots, m - 1$.
(c) $C_{m,p}(E_m) = 0$.

Let us now assume, in order to bring out the idea clearly, that $E_0, \ldots, E_m$ are compact. (This is, of course, not true in general, so the argument needs to be modified.) Then the multiplier ω is constructed inductively as follows.

First construct $\omega_0 \in C_0^\infty(E_0^c)$ so that $\|f - \omega_0 f\|_{m,p} < \varepsilon$, and set $f_0 = \omega_0 f$. Then construct $\omega_1 \in C_0^\infty(E_1^c)$ so that $\|f_0 - \omega_1 f_0\|_{m,p} < \varepsilon/2$, and set $f_1 = \omega_1 f_0$. Proceeding like this we finally obtain $f_m = \omega_m f_{m-1}$, $\omega_m \in C_0^\infty(E_m^c)$, and

$$\|f_{m-1} - \omega_m f_{m-1}\|_{m,p} < \varepsilon/2^m.$$

Then $f_m = \omega f$, $\omega = \omega_0 \omega_1 \cdots \omega_m \in C_0^\infty(K^c)$, and $\|f - \omega f\|_{m,p} < 2\varepsilon$.

In order to construct ω_0 one has to estimate f and its derivatives near E_0 by means of the following Poincaré inequality (see [12, Theorem 4.1]). We set $\nabla^m = (D^\alpha)_{|\alpha|=m}$ and

$$[f]_n(x) = \left\{ 2^{nN} \int_{B(x, 2^{-n})} |f(x)|^p \, dx \right\}^{1/p}.$$

PROPOSITION 4. *Let $f \in W_0^{m,p}(K^c)$ and let $x \in K$. Then, for $p < N$, and any $n = 1, 2, \ldots$*

$$[f]_n(x) \leq A 2^{-nm} c_{1,p}(K, x, 2^{-n})^{-1/p} [\nabla^m f]_n(x).$$

By the definition of a $(1, p)$-thick set

$$\sum_1^\infty c_{1,p}(K, x, 2^{-n})^{p'-1} = \infty \quad \text{for all } x \in E_0.$$

Combining the last inequality with Proposition 4 we find

(1) $$\sum_1^\infty \left\{ \frac{2^{-nm} [\nabla^m f]_n(x)}{[f]_n(x)} \right\}^{p'} = \infty \quad \text{for all } x \in E_0.$$

It turns out that this is exactly the control we need over the behavior of f near E_0 in order to construct ω_0 with the required properties. The actual construction of ω_0 is quite complicated because of the fact that we do not have any uniformity in the divergence of the series in (1). The construction is carried out in [10, Lemma 3.2] and [12, Lemma 5.2].

In order to construct ω_1, ω_2, etc., we have to replace the estimate in Proposition 4 by (see [12, Theorem 4.2 and Corollary 4.3])

PROPOSITION 5. *Let* $f \in W_0^{m,p}(K^c)$ *and let* $x \in K$. *Then, for any* $n = 1, 2, \ldots,$

$$[f]_n(x) \leq A 2^{-n(m-k)} [\nabla^{m-k} f]_n(x)$$
$$+ A 2^{-nm} c_{k+1,p}(K, x, 2^{-n})^{-1/p} [\nabla^m f]_n(x),$$

$k = 1, 2, \ldots, m - 1$ *and* $(k + 1)p < N$.

For $(k + 1)p > N$, $c_{k+1,p}(\cdot)$ *can be replaced by a constant, and for* $(k + 1)p = N$ *there is a small modification that we omit.*

Using this proposition, one can split E_k into a set where (1) holds, and another, where, roughly speaking, $[f]_n(x)$ behaves like $2^{-n(m-k)}[\nabla^{m-k}f]_n(x)$. To the first set the construction of ω_0 applies, so we can assume that the second alternative holds everywhere on E_k. In order to construct ω_k, one exploits the fact that $C_{k,p}(E_k) = 0$ in combination with estimates of the (k, p)-capacity of the set where $[\nabla^{m-k}f]_n(x)$ is large.

Again the construction is somewhat complicated, and it depends heavily on properties of nonlinear potentials. The details are found in [12, Theorems 5.3–5.5] and in [10]. See also the expository papers [11] and [13].

REFERENCES

1. D. R. Adams and N. G. Meyers, *Thinness and Wiener criteria for nonlinear potentials*, Indiana Univ. Math. J. **22** (1972), 169–197.

2. H. Brézis and F. E. Browder, *Some properties of higher order Sobolev spaces*, J. Math. Pures Appl. **61** (1982), 245–259.

3. F. E. Browder, These Proceedings.

4. A. P. Calderón, *Lebesgue spaces of differentiable functions and distributions*, Proc. Sympos. Pure Math., vol. 4, Amer. Math. Soc., Providence, R.I., 1961, pp. 33–49.

5. G. C. Evans, *Applications of Poincaré's sweeping-out process*, Proc. Nat. Acad. Sci. U.S.A. **19** (1933), 457–461.

6. C. Fefferman and E. M. Stein, H^p-*spaces of several variables*, Acta Math. **129** (1972), 137–193.

7. B. Fuglede, *Applications du théorème minimax à l'étude de diverses capacités*, C. R. Acad. Sci. Paris Sér. A **266** (1968), 921–923.

8. R. Gariepy and W. R. Ziemer, *A regularity condition at the boundary for solutions of quasilinear elliptic equations*, Arch. Rational Mech. Anal. **67** (1977), 25–39.

9. L. I. Hedberg, *Non-linear potentials and approximation in the mean by analytic functions*, Math. Z. **129** (1972), 299–319.

10. _____, *Two approximation problems in function spaces*, Ark. Mat. **16** (1978), 51–81.

11. _____, *Spectral synthesis and stability in Sobolev spaces*, Euclidean Harmonic Analysis (Proc., Univ. of Maryland, 1979), Lecture Notes in Math., vol. 779, Springer-Verlag, 1980, pp. 73–103.

12. _____, *Spectral synthesis in Sobolev spaces, and uniqueness of solutions of the Dirichlet problem*, Acta Math. **147** (1981), 237–264.

13. _____, *On the Dirichlet problem for higher order equations*, Vol. 2, Conference on Harmonic Analysis in Honor of Anthony Zygmund (W. Beckner et al., eds.), Wadsworth, Belmont, Calif., 1983, pp. 620–633.

14. L. I. Hedberg and T. H. Wolff, *Thin sets in nonlinear potential theory*, Ann. Inst. Fourier Grenoble **33**, no. 4 (1983), pp. 161–187.

15. O. D. Kellogg, *Foundations of potential theory*, Springer, 1929. (Dover, 1954).

16. V. G. Maz'ja, *The Dirichlet problem for elliptic equations of arbitrary order in unbounded regions*, Dokl. Akad. Nauk SSSR **150** (1963), 1221–1224.

17. _____, *On the continuity at a boundary point of solutions of quasilinear elliptic equations*, Vestnik Leningrad Univ. Math. **25**, no. 13 (1970), 42–55; Correction, Ibid. **27**, no. 1 (1972), 160.

18. V. G. Maz'ja and V. P. Havin, *Non-linear potential theory*, Uspekhi Mat. Nauk **27**, no. 6 (1972), 67–138.

19. N. G. Meyers, *A theory of capacities for potentials of functions in Lebesgue classes*, Math Scand. **26** (1970), 255–292.

20. Ju. G. Rešetnjak, *On the concept of capacity in the theory of functions with generalized derivatives*, Sibirsk. Mat. Ž. **10** (1969), 1109–1138.

21. J. R. L. Webb, *Boundary value problems for strongly nonlinear elliptic equations*, J. London Math. Soc. (2) **21** (1980), 123–132.

UNIVERSITY OF STOCKHOLM, SWEDEN

Some Free Boundary Problems for Predator-Prey Systems with Nonlinear Diffusion

JESÚS HERNÁNDEZ

Abstract. We give some results about the existence of free boundaries for a stationary system modelling a predator–prey interaction with nonlinear diffusion. This represents some qualitative new phenomena with respect to the well-known case of linear diffusion. Our main tool for the proof of our results is a local comparison argument.

0. Introduction. Reaction-diffusion systems have received a great deal of attention during the last years, in particular for all which concerns the study of the asymptotic behaviour of solutions of parabolic problems and the stability of solutions of the associated elliptic system with respect to the parabolic one.

We report very briefly here about a different kind of problem related to elliptic reaction-diffusion systems, namely *free boundary* problems. Actually, the content of this paper is part of a joint work in progress with J. I. Díaz and A. Tesei about the system

$$(0.1) \quad \begin{aligned} u_t - \Delta u^m &= u[a(1-u) - v] \quad \text{in } \Omega \times (0, T], \\ v_t - d\Delta v^n &= v[r(u - \gamma) - v] \quad \text{in } \Omega \times (0, T] \end{aligned}$$

together with Dirichlet boundary conditions

$$(0.2) \quad u = \varepsilon, \quad v = \psi \quad \text{on } \partial\Omega \times [0, T]$$

and nonnegative smooth initial conditions. Here $T > 0$, Ω is a bounded domain in $\mathbf{R}^N$ ($N \geqslant 1$) with smooth boundary $\partial\Omega$, a, d, $r > 0$, $0 < \gamma \leqslant 1$, $m, n > 1$, and ε, ψ are in $C^2(\partial\Omega)$ with ε, $\psi \geqslant 0$. When $m = n = 1$ and $\varepsilon \equiv \psi \equiv 0$, problem (0.1), (0.2) and the corresponding elliptic version have been considered by Conway–Gardner–Smoller [8] as a model for predator-prey interactions. A very similar problem was studied by de Mottoni–Schiaffino–Tesei in [16]. The main

---

1980 *Mathematics Subject Classification.* Primary 35R35, 35J65; Secondary 35K65, 92A15, 35B50.

difference between our paper and the preceding ones is that we consider nonlinear diffusion, i.e., $m, n > 1$, instead of ordinary diffusion $m = n = 1$ and, as we will show, this implies a dramatically different qualitative behaviour of solutions of (0.1), (0.2) and the associate stationary problem. More precisely, it is possible, in contrast with the case $m = n = 1$, that one population (or both) can only survive on some regions of Ω, but not on the whole domain. Nonlinear diffusion arises mainly in some physical problems as, e.g., the porous media equation and plasma theory (cf. the surveys by Peletier [19] and Bertsch [4]), and it has been introduced in biological models by Gurtin–McCamy [13] (cf. also [4, 5]).

Existence of solutions of (0.1), (0.2) follows from some results by Nakao [17], but the assumptions for the uniqueness results in [17] are not satisfied by the nonlinearities of our problem, and a different argument is needed. Concerning the associated elliptic system, always with nonhomogeneous Dirichlet boundary conditions, existence can be proved by using coupled sub-supersolutions as in [14, 10]. Cf. also, for that matter, [1, 4, 15, 18]. A much more difficult problem is to prove in the case of homogeneous Dirichlet boundary conditions, i.e., if $\varepsilon \equiv \psi \equiv 0$, the existence of solutions (u, v) such that $u \not\equiv 0$ *and* $v \not\equiv 0$. This problem can be nontrivial, even in the case $m = n = 1$ (cf. [8, 7]). Some results for the case of nonlinear diffusion can be found in [18].

But here we are only interested in free boundary problems for the stationary system corresponding to (0.1), (0.2). Similar problems were considered by Diaz–Hernández [10] for a system arising in combustion theory (cf. also [2] and [12]). The parabolic problem has been considered in [11] and [3], and the results of [11] include also the case of nonlinear diffusion. In this situation, roughly speaking, the fact that the nonlinear reaction term is not locally Lipschitz (it behaves like u^p, with $0 < p < 1$, near the origin) implies that, if the reaction rate is strong enough or, equivalently, if diffusion is small enough, the concentration of the reactant can be zero on an interior subdomain Ω_0 (called sometimes the *dead core*) of Ω. The boundary $\partial \Omega_0$ of Ω_0 is usually called the *free boundary*. The main results in [10] are proved by using a *local* comparison argument involving the use of local supersolutions, following an idea introduced in [9].

The rest of the paper is organized as follows. First, we give the main tool for our results, the comparison lemma in [10]. For the sake of completeness, we include the proof of this lemma. Then we apply these results to our predator-prey system. As we only want to show the qualitative nature of our results, we only treat a particular case. It is clear that many related problems (e.g., competition) can be studied in the same way. The time evolution of the free boundaries can be studied by using the methods of [3, 6] and [11].

1. Main theorems and proofs. In this section, we give first a lemma which is very useful in treating this kind of problem, it allows us to obtain a lot of information about the existence, size, location, etc., of the dead core of some nonlinear equations, i.e., about the subdomain where the solution is zero. As before, Ω is a bounded domain in $\mathbf{R}^N$ ($N \geq 1$) with smooth boundary $\partial \Omega$.

LEMMA 1.1. *Let $u \in H^2(\Omega)$ be a solution of the problem*

(1.1) $\quad -\Delta u(x) + \mu^2 f(x)|u(x)|^p \operatorname{sg} u(x) = F(x) \quad \text{in } \Omega,$

(1.2) $\quad u = \varepsilon \quad \text{on } \partial\Omega,$

where $f, F \in L^\infty(\Omega)$, $f \geq 0$ on Ω, $\varepsilon \in C^2(\partial\Omega)$, and $0 < p < 1$. If Ω_λ denotes the set

$$\Omega_\lambda = \{x \in \Omega: f(x) \geq \lambda\}, \quad \lambda > 0,$$

then we have the estimate

(1.3) $\quad \Omega_0 = \{x \in \Omega: u(x) = 0\}$
$\supset \Big\{x \in \Omega_\lambda - \operatorname{supp} F:$
$$d(x, \partial(\Omega_\lambda - \operatorname{supp} F) - (\partial\Omega - \operatorname{supp} \varepsilon)) \geq (M/K_{\lambda,\mu})^{(1-p)/2}\Big\}$$

with

$$M = \max\left\{\left(\frac{\|F\|_{L^\infty(\Omega)}}{\lambda\mu^2}\right)^{1/p}, \|\varepsilon\|_{L^\infty(\partial\Omega)}\right\}$$

and where the constant $K_{\lambda,\mu}$ is given by

(1.4) $\quad K_{\lambda,\mu} = \left(\dfrac{2N(1-p) + 4p}{\lambda\mu^2(1-p)^2}\right)^{1/(p-1)}.$

PROOF. If we denote by u_+ (resp. u_-) the solutions of the equation (1.1), (1.2) corresponding to the data F^+ and ε^+ (resp. F^- and ε^-), then a simple comparison argument yields $u_+ \geq 0$ (resp. $u_- \leq 0$) and $u_- \leq u \leq u_+$ a.e. on Ω. Hence it is clear that $\Omega_0 \supset \{x \in \Omega: u_-(x) = 0 \text{ and } u_+(x) = 0\}$. For simplicity, we shall only consider the case $F = F^+$, $\varepsilon = \varepsilon^+$, the other case being analogous. Let $u_\lambda \in H^2(\Omega)$ be such that

(1.5)
$$-\Delta u_\lambda + \lambda\mu^2|u_\lambda|^p \geq F \quad \text{in } \Omega_\lambda,$$
$$u_\lambda \geq \varepsilon \quad \text{on } \partial\Omega_\lambda \cap \partial\Omega,$$
$$u_\lambda \geq \|u\|_{L^\infty(\Omega)} \quad \text{on } \partial\Omega_\lambda - \partial\Omega.$$

We claim that $0 \leq u \leq u_\lambda$ a.e. on Ω_λ. Indeed, taking $\tilde{F}(x) = -\Delta u + \lambda\mu^2 u^p$, it is clear that $\tilde{F}(x) = F(x) + \lambda\mu^2 u^p - f(x)\mu^2 u^p$ and hence $\tilde{F} \leq F$ on Ω_λ. Moreover, we have $-\Delta u + \lambda\mu^2 u^p = \tilde{F}$ on Ω_λ, and thus by the comparison results of, e.g., [9], we obtain $0 \leq u \leq u_\lambda$. Therefore the conclusion of the lemma will follow from the construction of such functions u_λ and the sets $\{x \in \Omega_\lambda: u_\lambda(x) = 0\}$ will give the estimate (1.3) for Ω_0. We will seek functions $u_\lambda(x) = h(|x - x_0|)$ for some

$x_0 \in \Omega_\lambda$. First, we remark that for $h \in C^2(\mathbf{R})$ and any $\eta \in (0,1)$ we have

$$-\Delta h(|x-x_0|) + \lambda\mu^2 h(|x-x_0|)^p$$

$$= -h''(|x-x_0|) - \left(\frac{N-1}{|x-x_0|}\right)h'(|x-x_0|) + \lambda\mu^2 h(|x-x_0|)^p$$

$$= -h''(|x-x_0|) + \eta\lambda\mu^2 h(|x-x_0|)^p$$

$$+ (1-\eta)\lambda\mu^2 h(|x-x_0|)^p - \left(\frac{N-1}{|x-x_0|}\right)h'(|x-x_0|).$$

Now, for a fixed η, let h_η be a solution of the Cauchy problem

(1.6)
$$h_\eta''(r) = \eta\lambda\mu^2 |h_\eta(r)|^p \operatorname{sg}(h_\eta(r)),$$
$$h_\eta(0) = h_\eta'(0) = 0.$$

It is easy to check (recall that $0 < p < 1$) that

(1.7)
$$h_\eta(r) = L_\eta r^{2/(1-p)},$$

where

(1.8)
$$L_\eta = \left(\frac{2(1+p)}{\eta\lambda\mu^2(1-p)^2}\right)^{1/(p-1)}$$

is a solution of (1.6). We have

$$(1-\eta)\lambda\mu^2 h_\eta(r)^p - \frac{N-1}{r}h_\eta'(r) = L_\eta r^{2p/(1-p)}\left[(1-\eta)\lambda\mu^2 L_\eta^{p-1} - \frac{2(N-1)}{1-p}\right],$$

and choosing η such that

(1.9)
$$\eta \leq \frac{p+1}{1+p+(N-1)(1-p)}$$

leads to

$$-\Delta h_\eta(|x-x_0|) + \lambda\mu^2 h_\eta(|x-x_0|)^p \geq 0$$

for any $x \in \Omega_\lambda$.

Finally, consider the set $\tilde{\Omega} = \Omega_\lambda - \operatorname{supp} F$. (We remark that the comparison argument for solutions of (1.5) is still valid if Ω_λ is replaced by any other subset of Ω_λ). The considerations made above show that the function

$$u_\lambda(x) = K_{\lambda,\mu}|x-x_0|^{2/(1-p)}$$

with $K_{\lambda,\mu}$ given by (1.4) satisfies

$$-\Delta u_\lambda + \lambda\mu^2 u_\lambda^p \geq 0 = F(x) \quad \text{in } \tilde{\Omega},$$
$$u_\lambda \geq 0 = \varepsilon \quad \text{on } \partial\tilde{\Omega} \cap (\partial\Omega - \operatorname{supp}\varepsilon).$$

Hence it is sufficient to have

(1.10) $u_\lambda \geq \max\{\varepsilon, \|u\|_{L^\infty(\Omega)}\}$ on $\partial\tilde{\Omega} - (\partial\tilde{\Omega} \cap (\partial\Omega - \operatorname{supp}\varepsilon))$

to obtain

$$0 \leq u(x) \leq u_\lambda(x) \quad \text{on } \tilde{\Omega}.$$

But, by the Maximum Principle, we know that $u(x) \leq M$ on Ω and this implies that (1.10) is satisfied if we choose x_0 such that

(1.11) $$|x - x_0| \geq \left(\frac{M}{K_{\lambda,\mu}}\right)^{(1-p)/2}$$

for every $x \in \partial\tilde{\Omega} - (\partial\tilde{\Omega} \cap (\partial\Omega - \operatorname{supp}\varepsilon))$. The conclusion now follows trivially from (1.10) and (1.11) (we recall that $u_\lambda(x_0) = 0$).

It is clear that Lemma 1.1 gives sufficient conditions to have Ω_0 nonempty. Indeed, we remark that the constant $K_{\lambda,\mu}$ defined by (1.4) tends to $+\infty$ if μ tends to $+\infty$, for λ fixed. Then, if Ω_λ is nonempty, the estimate (1.3) shows that the dead core Ω_0 has a positive measure at least if

$$\delta(\Omega_\lambda - \operatorname{supp} F) > \left(\frac{M}{K_{\lambda,\mu}}\right)^{(1-p)/2},$$

where $\delta(\Omega_\lambda - \operatorname{supp} F)$ is the radius of the largest ball contained in $\Omega_\lambda - \operatorname{supp} F$ (always assuming $0 < p < 1$). Therefore, if Ω_λ is given, Ω_0 has a positive measure if μ is large enough. The same estimate (1.3) shows that, in some sense, the dead core Ω_0 "tends" to the whole domain Ω when μ tends to $+\infty$, if, e.g., $\Omega_\lambda = \Omega$ and $\operatorname{supp} F = \emptyset$.

In the following we shall apply our lemma to obtain some information about existence and location of free boundaries in the case of our predator-prey model with nonlinear diffusion. For the sake of simplicity, we only treat the particular case

(1.12) $$\begin{aligned} -\Delta u &= u[a(1-u) - v] & \text{in } \Omega, \\ -d\Delta v^2 &= v[r(u-\gamma) - v] & \text{in } \Omega, \\ u &= \varepsilon_1, \ v = \varepsilon_2 & \text{on } \partial\Omega, \end{aligned}$$

where, as above, $a, d, r > 0$, $0 < \gamma \leq 1$, $\varepsilon_1, \varepsilon_2 \in C^2(\partial\Omega)$, and $\varepsilon_1, \varepsilon_2 > 0$ on $\partial\Omega$. (We take $m = 1$ and $n = 2$ in (0.1)). In this model, u represents the density of preys and v the density of predators. Thus, whereas preys diffuse "linearly", predators undergo nonlinear diffusion, which supposes, it seems, a certain amount of "crowding".

By performing the change of unknown $w = v^2$, system (1.12) can be rewritten as

(1.13) $$\begin{aligned} -\Delta u + au^2 + uw^{1/2} &= au & \text{in } \Omega, \\ -d\Delta w + r(\gamma - u)w^{1/2} + w &= 0 & \text{in } \Omega, \\ u = \varepsilon_1, \ w = \varepsilon_2^2 &\equiv \psi & \text{on } \partial\Omega. \end{aligned}$$

It is very easy to check that $(u_0, u^0) - (w_0, w^0)$, where $u_0 \equiv w_0 \equiv 0$, and $u^0 = \max\{\|\varepsilon_1\|_{L^\infty}, 1\}$, $w^0 = \max\{\|\psi\|_{L^\infty}, r^2(u^0 - \gamma)^2\}$ is a coupled sub-supersolution of problem (1.13) in the sense of [14]. Hence, reasoning as in [10] and [14], we obtain the existence of at least a solution (u, w) of (1.13) such that $0 \leq u \leq u^0$, $0 \leq w \leq w^0$. In particular, if $\varepsilon_1 \leq 1$ and $\psi \leq r^2(u^0 - \gamma)^2$, then $0 \leq u \leq 1$, $0 \leq w \leq r^2(1 - \gamma)^2$. Moreover, it is an easy corollary of the Maximum Principle that $u > 0$ on $\bar{\Omega}$. This means that preys are able to survive on the whole domain Ω if $\varepsilon_1 > 0$.

We consider now the unknown w, which corresponds to the density of predators. The equation satisfied by w is

$$\text{(1.14)} \qquad -d\Delta w + r(\gamma - u)w^{1/2} + w = 0 \quad \text{in } \Omega,$$
$$w = \psi \quad \text{on } \partial\Omega.$$

First, assume that we have a (positive) solution (u, w) of (1.13) satisfying $u(x) \geq \gamma > 0$ for any $x \in \bar{\Omega}$ (this implies $\varepsilon_1 \geq \gamma$ on $\partial\Omega$). Then (1.14) can be reformulated as

$$-d\Delta w + w = r(u - \gamma)w^{1/2} \quad \text{in } \Omega,$$
$$w = \psi \quad \text{on } \partial\Omega$$

and, as both right sides are positive and $\psi > 0$, the Maximum Principle implies that $w > 0$ on $\bar{\Omega}$. The biological interpretation of this fact is that, if u is larger everywhere on the domain than the "critical" value γ, i.e., if there are "enough preys" everywhere, then predators will survive on the whole domain, and this happens for any $d > 0$.

Assume now that u does not satisfy $u(x) \geq \gamma > 0$ on $\bar{\Omega}$; this means that for some $x_0 \in \Omega$ we have $u(x_0) < \gamma$ and, by the continuity of u, we have $u(x) < \gamma$ on a neighborhood of x_0. Thus, if for $\lambda > 0$ we define the set

$$\Omega_\lambda = \{x \in \Omega : r(\gamma - u(x)) \geq \lambda > 0\}$$
$$= \left\{x \in \Omega : 0 < u(x) \leq \gamma - \frac{\lambda}{r}\right\},$$

then Ω_λ is nonempty, at least for λ sufficiently small.

Now, let $w_\lambda \in H^2(\Omega)$ such that

$$\text{(1.15)} \qquad \begin{aligned} -d\Delta w_\lambda + \lambda w_\lambda^{1/2} + w_\lambda &\geq 0 & &\text{in } \Omega_\lambda, \\ w_\lambda &\geq \psi & &\text{in } \partial\Omega_\lambda \cap \partial\Omega, \\ w_\lambda &\geq \|w\|_{L^\infty(\Omega)} & &\text{on } \partial\Omega_\lambda - \partial\Omega. \end{aligned}$$

Then, by a straightforward comparison argument, $0 \leq w(x) \leq w_\lambda(x)$ a.e. on Ω_λ. But it is obvious that, if $w_\lambda \geq 0$, w_λ is a solution of (1.15) if it satisfies

$$\text{(1.16)} \qquad \begin{aligned} -\Delta w_\lambda + \lambda d^{-1} w_\lambda^{1/2} &\geq 0 & &\text{in } \Omega_\lambda, \\ w_\lambda &\geq \psi & &\text{on } \partial\Omega_\lambda \cap \partial\Omega, \\ w_\lambda &\geq \|w\|_{L^\infty(\Omega)} & &\text{on } \partial\Omega_\lambda - \partial\Omega. \end{aligned}$$

Hence the proof of Lemma 1.1 for $\mu^2 = d^{-1}$, $F \equiv 0$, $p = 1/2$, and supp $\psi = \partial\Omega$ gives (we recall that the local supersolutions in the proof are positive) the estimate

$$(1.17) \quad \Omega_0 = \{x \in \Omega : w(x) = 0\} \supset \left\{ x \in \Omega_\lambda : d(x, \partial\Omega_\lambda) \geqslant \left(\frac{M}{K_{k,d}} \right)^{1/4} \right\},$$

where $M = \|\psi\|_{L^\infty(\partial\Omega)}$ and the constant

$$K_{\lambda,d} = \left(\frac{\lambda}{4d(N+2)} \right)^{1/2}$$

tends to $+\infty$ if $d \searrow 0$. As above, a sufficient condition to get a dead core with positive measure is

$$(1.18) \quad \delta(\Omega_\lambda) \geqslant \left(\frac{4d(N+2)M^2}{\lambda} \right)^{1/8},$$

where $\delta(\Omega_\lambda)$ is the radius of the largest ball contained in Ω_λ, and λ is fixed. It is clear now that (1.18) can be satisfied if we take $d > 0$ sufficiently small.

These results can be interpreted in the sense that if predators are subjected to nonlinear diffusion, which seems to imply some "crowding", and if $d > 0$ is too small, then they can only survive near the boundary of Ω_λ.

Finally, we consider the special case of a solution (u, w) such that $0 < u(x) < \gamma$ everywhere on $\overline{\Omega}$ (this implies $0 < \varepsilon_1 < \gamma$ on $\partial\Omega$). It is easy to see that for λ sufficiently small, we have $\Omega_\lambda = \Omega$ and then estimate (1.17) reads

$$(1.19) \quad \Omega_0 = \{x \in \Omega : w(x) = 0\} \supset \left\{ x \in \Omega : d(x, \partial\Omega) \geqslant \left(\frac{M}{K_{\lambda,d}} \right)^{1/4} \right\}.$$

Hence in this case, and for $d > 0$ small enough, predators can only survive in a thin region near the boundary (we recall that $\psi > 0$) and this region "tends to the boundary" when $d \searrow 0$.

We have thus proved

THEOREM 1.1. *Let (u, w) be a (positive) solution of the system (1.13), where the above assumptions are satisfied. If $u(x) \geqslant \gamma > 0$ on $\overline{\Omega}$, then $w > 0$ on $\overline{\Omega}$. If not, we have the estimate (1.17) for w; in particular, if (1.18) is satisfied, Ω_0 has positive measure. Moreover, if $0 < u(x) < \gamma$ on $\overline{\Omega}$, then we have the more precise estimate (1.19).*

We remark that the parameter γ plays also a "critical" role in the problem considered in [8].

References

1. H. W. Alt and S. Luckhaus, *Quasilinear elliptic-parabolic differential equations* (to appear).
2. C. Bandle, R. P. Sperb and I. Stakgold, *Diffusion-reaction with monotone kinetics* (to appear).
3. C. Bandle and I. Stakgold, *The formation of the dead core in parabolic reaction-diffusion problems* (to appear).
4. M. Bertsch, *Nonlinear diffusion problems: the large time behaviour*, Ph. D. Thesis, University of Leyden, 1983.

5. M. Bertsch and D. Hilhorst, *A density dependent diffusion equation in population dynamics: stabilization to equilibrium* (to appear).

6. M. Bertsch, T. Nanbu and L. A. Peletier, *Decay of solutions of a degenerate nonlinear diffusion equation*, Nonlinear Anal. **6** (1982), 539-554.

7. J. Blat and K. J. Brown, *Bifurcation of steady-state solutions in predator-prey and competition stystems* (to appear).

8. E. Conway, R. Gardner and J. Smoller, *Stability and bifurcation of steady-state solutions for predator-prey equations*, Adv. in Appl. Math. **3** (1982), 288-334.

9. J. I. Díaz, *Técnica de supersoluciones locales para problemas estacionarios no lineales: aplicación al cálcule de flujos subsónicos*, Memoria no. 14 de la Real Academia de Ciencias, Madrid, 1980.

10. J. I. Díaz and J. Hernández, *On the existence of a free boundary for a class of reaction-diffusion systems*, SIAM J. Math. Anal. (to appear).

11. _____, *Some results on the existence of free boundaries for parabolic reaction-diffusion systems* (to appear).

12. A. Friedman and R. Phillips, *The free boundary of a semilinear elliptic equation*, Trans. Amer. Math. Soc. **282** (1984), 153-182.

13. M. E. Gurtin and R. C. McCamy, *On the diffusion of biological populations*, Math. Bios. **33** (1977), 35-49.

14. J. Hernández, *Some existence and stability results for solutions of reaction-diffusion systems with nonlinear boundary conditions*, Nonlinear Differential Equations: Invariance, Stability and Bifurcation (P. de Mottoni and L. Salvadori, eds.), Academic Press, New York, 1981, pp. 161-173.

15. L. Maddalena, *Existence of global solutions for reaction-diffusion systems with density dependent diffusion* (to appear).

16. P. de Mottoni, A. Schiaffino and A. Tesei, *On stable space dependent stationary solutions of a competition system with diffusion* (to appear).

17. M. Nakao, *L^p-estimates of solutions of some nonlinear degenerate equations* (to appear).

18. R. dal Passo and P. de Mottoni, *Some existence, uniqueness and stability results for a class of semilinear degenerate elliptic systems* (to appear).

19. L. A. Peletier, *The porous media equation*, Applications of Nonlinear Analysis in Physical Sciences (H. Amann, N. Bazley and K. Kirschgassner, eds.), Pitman, New York, 1981, pp. 229-241.

UNIVERSIDAD AUTÓNOMA, MADRID, SPAIN

On Positive Solutions of Semilinear Periodic-Parabolic Problems

PETER HESS

Consider the question of existence of positive periodic solutions of given period $T > 0$ of the semilinear parabolic problem

$$(*) \quad \begin{cases} \mathscr{L}u = g(x, t, u) & \text{in } \Omega \times \mathbf{R}, \\ u = 0 & \text{on } \partial\Omega \times \mathbf{R}, \\ u(\cdot, 0) = u(\cdot, T) & \text{on } \overline{\Omega}, \end{cases}$$

where $\mathscr{L} := \partial/\partial t + \mathscr{A}(x, t, D)$ is a uniformly parabolic linear differential expression of second order having T-periodic coefficient functions, $g: \overline{\Omega} \times \mathbf{R} \times \mathbf{R} \to \mathbf{R}$ is sufficiently smooth and T-periodic in t, and Ω is a bounded domain in $\mathbf{R}^N$ ($N \geq 1$) having smooth boundary $\partial\Omega$. In the study of this problem, the method of sub- and supersolutions has been successfully applied (e.g. Kolesov [14], Amann [1] and the references cited therein). However, for the construction of positive sub- and supersolutions, as well as for an application of bifurcation theory to problem $(*)$, one would need a thorough understanding of the *linear eigenvalue problem*

$$(**) \quad \begin{cases} \mathscr{L}u = \lambda m(x, t)u & \text{in } \Omega \times \mathbf{R}, \\ u = 0 & \text{on } \partial\Omega \times \mathbf{R}, \\ u(\cdot, 0) = u(\cdot, T) & \text{on } \overline{\Omega}, \end{cases}$$

where $m \neq 0$ is a given (not necessarily positive) weight function. For the special case $m \equiv 1$, Lazer [16] and Castro–Lazer [4] have only recently done a first step and obtained partial results. In the first section of this note we report on the results of Beltramo and the author [3] on the existence of a principal positive eigenvalue $\lambda_1(m)$ of $(**)$ (characterized as the unique positive eigenvalue having a

---

1980 *Mathematics Subject Classification*. Primary 35K20, 35B10, 35P05; Secondary 35B50, 35P30.

positive eigenfunction): We give a condition which is *both necessary and sufficient* for its existence. In §II we outline the proofs of these results. In §III we touch the question of estimates for $\lambda_1(m)$. In §IV we then apply the results on the linear eigenvalue problem (**) in the construction of positive sub- and supersolutions for equation (*), while in §V we study the bifurcation problem for the semilinear parabolic eigenvalue problem: We indicate that results analogous to those for the elliptic eigenvalue problem (Hess [9]) can be obtained.

I. The linear parabolic-periodic eigenvalue problem: Statement of the results. Let $\mathscr{L} = \partial/\partial t + \mathscr{A}(x, t, D)$ be a uniformly parabolic differential expression with

$$\mathscr{A}(x, t, D) := -a_{jk}(x, t) D_j D_k + a_j(x, t) D_j + a_0(x, t)$$

(the summation convention is employed; $D_j = \partial/\partial x_j$). We assume that, for some $\mu \in \,]0, 1[$ and a fixed $T > 0$, the coefficient functions $a_{jk} = a_{kj}, a_j$, and $a_0 \geq 0$ belong to the real Banach space $E := \{w \in C^{\mu, \mu/2}(\overline{\Omega} \times \mathbf{R}): w \text{ is } T\text{-periodic in } t\}$, and that $\partial \Omega$ is of class $C^{2+\mu}$. Further, let $m \in E, m \neq 0$.

In order to put the eigenvalue problem (**) in a proper functional analytic setting, we let L denote the operator in the space E induced by $\mathscr{L}$ and the boundary and periodicity conditions. More precisely, $D(L) := F := \{w \in C^{2+\mu, 1+\mu/2}(\overline{\Omega} \times \mathbf{R}): w = 0 \text{ on } \partial\Omega \times \mathbf{R} \text{ and } w \text{ is } T\text{-periodic in } t\}$. Then L is a closed operator in E having a compact inverse (cf. §II). Finally, let M denote the multiplication operator in E by the function m. The eigenvalue problem (**) is then equivalent to the equation

(1.1) $$Lu = \lambda M u$$

in the space E. We provide the real Banach spaces E and F with the natural ordering given by the positive cones P_E and P_F of pointwise nonnegative functions and write $w \geq 0$ iff $w \in P$, $w > 0$ iff $w \in P \setminus \{0\}$ ($P = P_E$ or P_F).

We are primarily interested in the existence of a positive eigenvalue λ of (1.1) having a positive eigenfunction u. By the parabolic maximum principle [**17**, pp. 173–175], a necessary condition for this to occur is that $m(x, t) \not\leq 0$ on $\overline{\Omega} \times \mathbf{R}$. In contrast to the elliptic situation treated in [**11**], this condition is *not* sufficient however.

We introduce the continuous, T-periodic function $\tilde{m}$ defined on $\mathbf{R}$ by

$$\tilde{m}(t) := \max_{x \in \overline{\Omega}} m(x, t).$$

DEFINITION. We say m satisfies condition (M$^+$) provided

$$\int_0^T \tilde{m}(t) \, dt > 0.$$

THEOREM 1. *Problem* (1.1) *has a positive eigenvalue* $\lambda_1(m)$ *having a positive eigenfunction* u *if and only if condition* (M$^+$) *holds.*

In case (M$^+$) *is not satisfied, the halfplane* $\{\lambda \in \mathbf{C}: \operatorname{Re} \lambda \geq 0\}$ *belongs to the resolvent set of* $L^{-1}M \in \mathscr{L}(E)$.

In case (M$^+$) holds, $\lambda_1(m)$ is the unique positive eigenvalue with positive eigenfunction, and $u \in \text{Int}(P_F)$. Moreover,
  (i) if $\lambda \in \mathbf{C}$ is an eigenvalue of (1.1) with $\text{Re } \lambda \geq 0$, then $\text{Re } \lambda \geq \lambda_1(m)$;
  (ii) $1/\lambda_1(m)$ is an algebraically simple eigenvalue of $L^{-1}M$.

REMARK. Sharpening assertion (i), Beltramo [2] shows that if m is independent of $x \in \Omega$: $m = m(t)$, and $m_0 := \int_0^T m(t)\,dt > 0$, then (1.1) has on the axis $\text{Re } \lambda = \lambda_1(m)$ precisely the eigenvalues $\lambda_n := \lambda_1(m) + in\beta$ ($n \in \mathbf{Z}$), where $\beta = 2\pi/m_0$, with associated eigenfunctions $u_n = v_n u$, $u =$ principal eigenfunction and v_n: $v_n(t) = \exp\{in\beta \int_0^t m(\tau)\,d\tau\}$.

If m depends nontrivially on $x \in \Omega$, and (M$^+$) holds, $\lambda_1(m)$ is the only eigenvalue of (1.1) on the axis $\text{Re } \lambda = \lambda_1(m)$. This result has to be compared with the elliptic situation [7, 10].

Regarding the inhomogeneous linear problem

(1.2) $$(L - \lambda M)u = h, \quad h \in E \text{ given},$$

we have

THEOREM 2. *Suppose* (M$^+$) *is satisfied. Then*
  (i) *for* $0 \leq \lambda < \lambda_1(m)$, *equation* (1.2) *is always uniquely solvable, and* $h \geq 0$ *implies* $u \geq 0$;
  (ii) *for* $\lambda \geq \lambda_1(m)$, $h > 0$ *implies* $u \not\geq 0$ (*provided the solution u exists*).

If we introduce also the function $\underline{m}(t) := \min_{x \in \bar{\Omega}} m(x, t)$, there is in addition a unique negative eigenvalue $\lambda_{-1}(m)$ having a positive eigenfunction iff $\int_0^T \underline{m}(t)\,dt < 0$ (write (1.1) in the form $Lu = (-\lambda)(-M)u$). In particular, if m is independent of $x \in \Omega$: $m = m(t)$, and if $\int_0^T m(t)\,dt = 0$, (1.1) does not admit any eigenvalue in $\mathbf{C}$ (i.e., $L^{-1}M$ is quasi-nilpotent).

We note that Theorems 1 and 2 include the results of Kato and the author [11] on the elliptic eigenvalue problem with respect to an indefinite weight function. Suppose $\mathscr{A}$ and m are independent of t, and that $m(x_0) > 0$ for some $x_0 \in \Omega$. Then m satisfies (M$^+$) for any $T > 0$, hence by Theorem 1 there exists a unique simple eigenvalue $\lambda_1(m, T) > 0$ of (1.1) having a T-periodic positive eigenfunction. Fix $T = 1$. Since a $1/k$-periodic function ($k \in \mathbf{N}^*$) has also period 1, we have $\lambda_1(m, q) = \lambda_1(m, 1)$ for all rational periods $q > 0$ by uniqueness, with the same eigenfunction u_1. Thus, u_1 is independent of t by continuity, and is a positive eigenfunction of the associated elliptic eigenvalue problem with eigenvalue $\lambda_1(m) := \lambda_1(m, 1)$.

II. The linear parabolic-periodic eigenvalue problem: Indication of the proofs.

(A) We first sketch a new proof of existence of a positive principal eigenvalue of (1.1) for the case $m = 1$ (i.e. for the main result of [4, 16]), more in the spirit of linear evolution equations.

Fix $p > N$ and set $X := L^p(\Omega)$. For $t \in \mathbf{R}$ we consider the closed linear operator $A(t) := A(\cdot, t, D)$ in X induced by $\mathscr{A}(\cdot, t, D)$, with domain

$D := D(A(t)) := W_0^{1,p}(\Omega) \cap W^{2,p}(\Omega)$ independent of t. By the results of Sobolevskii and Tanabe (e.g. [18, §5.2]), the fundamental solution $\mathscr{U}(t, s)$ for the parabolic equation

$$\frac{du}{dt}(t) + A(t)u(t) = f(t) \quad (0 < t \leq T)$$

in X exists. For given $u_0 \in X$ and $f \in C^\sigma([0, T]; X)$ $(0 < \sigma \leq 1)$, the initial value problem

(2.1) $\quad \frac{du}{dt}(t) + A(t)u(t) = f(t) \quad (0 < t \leq T), \quad u(0) = u_0$

has a unique solution $u \in C([0, T]; X) \cap C^1(\,]0, T]; X)$ with $u(t) \in D$ $(t > 0)$, given by the formula of "variations of constants"

$$u(t) = \mathscr{U}(t,0)u_0 + \int_0^t \mathscr{U}(t, s)f(s)\,ds.$$

If $u_0 \in D$, then $u \in C^1([0, T]; X)$. Moreover, by regularity (Amann [1, Lemma 4.2]), if $u_0 \in D$ and $f \in C^{\mu,\mu/2}(\overline{\Omega} \times [0, T])$, then $u \in C^{1+\sigma,(1+\sigma)/2}(\overline{\Omega} \times [0, T]) \cap C^{2+\mu,1+\mu/2}(\overline{\Omega} \times \,]0, T])$ for some $\sigma \in \,]0, 1[\,$, hence the solution u of (2.1) is regular solution of the initial-boundary value problem

(2.2) $\quad \begin{cases} \mathscr{L}u = f & \text{in } \Omega \times \,]0, T], \\ u = 0 & \text{on } \partial\Omega \times \,]0, T], \\ u(\cdot, 0) = u_0 & \text{on } \overline{\Omega} \end{cases}$

(the converse is obvious).

We now indicate the properties of $K := \mathscr{U}(T, 0)$. (The spaces X, $C_0(\overline{\Omega}) := \{v \in C(\overline{\Omega}): v = 0 \text{ on } \partial\Omega\}$, and $C_0^1(\overline{\Omega})$ are again provided with the natural ordering.)

(i) K is a bounded *positive* operator in X. (For $u_0 \in D$, the parabolic maximum principle asserts that $Ku_0 \geq 0$ provided $u_0 \geq 0$, since $u(t) := \mathscr{U}(t, 0)u_0$ is regular solution of (2.2) with $f = 0$; D being dense, the assertion then follows for $u_0 \in X$ by continuous extension.)

(ii) K is positive and *compact* as an operator in $C_0(\overline{\Omega})$ or in $C_0^1(\overline{\Omega})$ (since K maps X boundedly into $C_0^1(\overline{\Omega})$).

(iii) $K: C_0(\overline{\Omega}) \to C_0^1(\overline{\Omega})$ is *strongly positive*: $u_0 > 0 \Rightarrow Ku_0 \in \text{Int}(\text{positive cone of } C_0^1(\overline{\Omega}))$. (For $u_0 \in D$, this is again a consequence of the parabolic maximum principle [17]; if $u_0 \in C_0(\overline{\Omega})$, $u_0 > 0$, choose $\tilde{u}_0 \in D$ with $0 < \tilde{u}_0 \leq u_0$ to conclude.)

Hence, $K: C_0(\overline{\Omega}) \to C_0(\overline{\Omega})$ has spectral radius $\gamma := \text{spr}(K) > 0$, and γ is the unique eigenvalue of K having a positive eigenfunction u_0 (in fact, since $K(C_0(\overline{\Omega})) \subset D \subset C_0^1(\overline{\Omega})$, it suffices to apply the Krein-Rutman theorem [15, Theorem 6.3] to the strongly positive, compact restriction $K/C_0^1(\overline{\Omega}): C_0^1(\overline{\Omega}) \to C_0^1(\overline{\Omega}))$. It is a simple consequence of the maximum principle that $0 < \gamma < 1$.

LEMMA 2.1. *L is a closed operator in E with compact positive inverse L^{-1} having positive spectral radius.*

PROOF. Clearly L maps F boundedly into E. We prove that $L: F \to E$ is bijective (which implies its closedness in E). Let $f \in E$ be given. Then $f \in C^{\mu/2}([0, T]; X)$, and $Lu = f$ iff

$$u_0 := u(0) = u(T) = Ku_0 + \int_0^T \mathcal{U}(T, s)f(s)\, ds.$$

Since $\mathrm{spr}(K) = \gamma < 1$, the equation

$$(I - K)u_0 = \int_0^T \mathcal{U}(T, s)f(s)\, ds$$

with right-hand side in $D \subset C_0(\overline{\Omega})$ has a unique solution $u_0 \in D \subset C_0(\overline{\Omega})$.

Now, $L^{-1}: E \to E$ is compact by the compactness of the imbedding $F \hookrightarrow E$, and $f > 0$ (in E) implies $L^{-1}f \in \mathrm{Int}(P_F)$ by the parabolic maximum principle. Thus $\mathrm{spr}(L^{-1}) > 0$ (consider the compact strongly positive restriction L^{-1}/F: $F \to F$). □

There is an immediate connection between the principal eigenvalue γ of K and the principal eigenvalue μ of L with respect to the weight function $m = \mathbf{1}$:

LEMMA 2.2. $\gamma = \mathrm{spr}(K)$ *is the principal eigenvalue of* K *with eigenfunction* $u_0 > 0$ *iff* $\mu = -(1/T)\log \gamma$ *is an eigenvalue of* L *with eigenfunction* $u > 0$. *Moreover,* $u(t) = e^{\mu t}\mathcal{U}(t, 0)u_0$.

The proof consists in a simple verification.

(B) We now turn to the case of a general $m \in E$. We indicate the different steps in the proof of Theorem 1; for details see [3].

Without loss of generality we may assume $|m| < 1$ on $\overline{\Omega} \times \mathbf{R}$. For $\lambda \geqslant 0$ we consider the eigenvalue problem

(2.3) $$(L - \lambda M)u = \mu u$$

in E. Setting $L_\lambda := L + \lambda(I - M)$, (2.3) is equivalent to the problem

$$L_\lambda u = (\mu + \lambda)u,$$

which is of the form as considered in part (A) (i.e., the coefficient function of the 0th order term of L_λ is nonnegative). Hence for each $\lambda \geqslant 0$ there exists the (uniquely determined) principal eigenvalue $\mu(\lambda)$ of (2.3), with associated eigenfunction $u(\lambda) > 0$.

We first study the dependence of $\mu(\lambda)$ and $u(\lambda)$ of $\lambda \geqslant 0$.

LEMMA 2.3. $\mu(\lambda)$ *is an analytic function of* λ, *and also the mapping* $\lambda \geqslant 0 \to u(\lambda) \in F$ *can be chosen to be analytic.*

This follows from the perturbation result [5, Lemma 1.3] of Crandall–Rabinowitz once we have proven that, if J denotes the injection mapping $F \hookrightarrow E$, $\mu(\lambda)$ is a J-simple eigenvalue of $L - \lambda M$, i.e. (considering L, M, J as operators in $B(F, E)$) $\dim N(L - \lambda M - \mu(\lambda)J) = \mathrm{codim}\, R(L - \lambda M - \mu(\lambda)J) = 1$ and, if $N(L - \lambda M - \mu(\lambda)J) = \mathrm{span}[u(\lambda)]$, then $Ju(\lambda) \notin R(L - \lambda M - \mu(\lambda)J)$.

LEMMA 2.4. $\mu(\lambda)$ *is a concave function of* $\lambda \geqslant 0$.

This follows from an abstract result of Kato [**13**, Theorem 6.1]. In [**3**] a direct proof is given which rests upon an observation of Berestycki and P. L. Lions (private communication).

Obviously, $\lambda > 0$ is an eigenvalue of (1.1) with positive eigenfunction iff $\mu(\lambda) = 0$. Since $\mu(0) > 0$ by Lemma 2.2, the existence of a unique positive eigenvalue $\lambda(m)$ of (1.1) with positive eigenfunction follows by Lemma 2.4 if and only if $\mu(\lambda)$ becomes negative for large $\lambda > 0$.

We first study the asymptotic behaviour of $\mu(\lambda)$ in a special case.

LEMMA 2.5. *Suppose* $m \in E$ *is independent of* $x \in \Omega$: $m = m(t)$, *and set* $m_0 := \int_0^T m(t)\,dt$. *Let* $\mu_0 > 0$ *be the principal eigenvalue of* L (*with respect to the weight function* **1**) *as guaranteed by Lemma* 2.2. *Then*

$$(2.4) \qquad \mu(\lambda) = \mu_0 - \lambda T^{-1} m_0 \qquad (\lambda \geq 0).$$

(In particular, the main assertion of Theorem 1 holds in this special case.)

PROOF. Let $u_0 \in F$, $u_0 > 0$ be an eigenfunction associated to μ_0: $Lu_0 = \mu_0 u_0$. For $\lambda \geq 0$ set

$$u_\lambda(x,t) := \exp\left(\lambda\left[\int_0^t m(\tau)\,d\tau - \frac{m_0}{T} t\right]\right) \cdot u_0(x,t)$$

$((x,t) \in \overline{\Omega} \times \mathbf{R})$. Then $u_\lambda \in F$, $u_\lambda > 0$, and it satisfies the equation

$$(L - \lambda M) u_\lambda = (\mu_0 - \lambda m_0/T) u_\lambda.$$

Hence, by uniqueness, $\mu(\lambda) = \mu_0 - \lambda m_0 T^{-1}$ and $u(\lambda) = u_\lambda$. □

We now turn to general $m \in E$.

LEMMA 2.6. *Condition* (M^+) *implies that* $\lim_{\lambda \to +\infty} \mu(\lambda) = -\infty$.

PROOF. The *proof of Lemma* 2.6. consists in two steps.

(i) One constructs a T-periodic C^2-function $\gamma: \mathbf{R} \to \Omega$ such that

$$\int_0^T m(\gamma(t), t)\,dt > 0$$

(provided (M^+) holds).

(ii) Let $\varphi: \mathbf{R}^N \times \mathbf{R} \to \mathbf{R}^N \times \mathbf{R}$ be the C^2-diffeomorphism defined by

$$\varphi: (x,t) \mapsto (y,t) := (x - \gamma(t), t).$$

Then, $\forall (y,t) \in \varphi(\overline{\Omega} \times \mathbf{R})$, and $\forall v \in C^{2,1}(\varphi(\overline{\Omega} \times \mathbf{R}))$, we have

$$(\mathscr{L}v \circ \varphi)(\varphi^{-1}(y,t)) = \left(\frac{\partial}{\partial t} + \mathscr{A}(\varphi^{-1}(y,t), D) - \dot{\gamma}(t) \cdot \frac{\partial}{\partial y}\right) v(y,t)$$

$$:= (\mathscr{L}_\varphi v)(y,t).$$

Setting further $\hat{m} := m \circ \varphi^{-1}$, the problem $(**)$ is transformed into the equation

$$\mathscr{L}_\varphi v = \lambda \hat{m} v \quad \text{in } \varphi(\Omega \times \mathbf{R}),$$

subject to the appropriate boundary and periodicity conditions. Note that $p := \int_0^T \hat{m}(0,t)\,dt = \int_0^T m(\gamma(t), t)\,dt > 0$. There exists $\varepsilon > 0$ such that $\overline{U}_\varepsilon(0) \times \mathbf{R} \subset \varphi(\Omega \times \mathbf{R})$ and $\hat{m}(y,t) \geq c(t) := \max\{\hat{m}(0,t) - p/2T, \min_{\varphi(\overline{\Omega} \times \mathbf{R})} \hat{m}\}$, $\forall (y,t)$

$\in \overline{U}_\varepsilon(0) \times \mathbf{R}$. Since $\int_0^T c(t)\,dt > 0$, Lemma 2.5, applied to the restriction of $\mathcal{L}_\varphi$ to the small cylinder $U_\varepsilon(0) \times \mathbf{R}$ and the weight function c, guarantees the existence of $\lambda_1 > 0$ and $\tilde{w} > 0$:

$$\begin{cases} \mathcal{L}_\varphi \tilde{w} = \lambda_1 c\tilde{w} \leqslant \lambda_1 \hat{m}\tilde{w} & \text{in } U_\varepsilon(0) \times \mathbf{R}, \\ \tilde{w} = 0 & \text{on } \partial U_\varepsilon(0) \times \mathbf{R}, \\ \tilde{w}(\cdot, 0) = \tilde{w}(\cdot, T) & \text{on } \overline{U}_\varepsilon(0). \end{cases}$$

Hence,

$$\mathcal{L}(\tilde{w} \circ \varphi) \leqslant \lambda_1 m(\tilde{w} \circ \varphi) \quad \text{in } \varphi^{-1}(U_\varepsilon(0) \times \mathbf{R}).$$

The extension $w \in E, w > 0$ defined by

$$w = \begin{cases} \tilde{w} \circ \varphi & \text{in } \varphi^{-1}(U_\varepsilon(0) \times \mathbf{R}), \\ 0 & \text{in } \overline{\Omega} \times \mathbf{R} \setminus \varphi^{-1}(U_\varepsilon(0) \times \mathbf{R}), \end{cases}$$

then serves as a "comparison function" as in [11, Lemma 3] and gives $\mu(\lambda_1) \leqslant 0$.

The following result complements Lemma 2.6.

LEMMA 2.7. *We have*

$$\mu(\lambda) \geqslant \mu_0 - \lambda T^{-1} \int_0^T \tilde{m}(t)\,dt \quad \forall \lambda > 0,$$

with strict inequality provided m depends nontrivially on $x \in \Omega$.

It follows in particular that $\mu(\lambda) > 0$ for all $\lambda \geqslant 0$ if condition (M$^+$) does not hold.

PROOF. First we note that if $m_1, m_2 \in E$ with $m_1 < m_2$, then for all $\lambda > 0$, $\mu_2(\lambda) < \mu_1(\lambda)$ for the corresponding eigenvalues of (2.3).

If m is independent of $x \in \Omega$: $m = \tilde{m}$, the assertion follows from Lemma 2.5. Hence, assume $m < \tilde{m}$ in $C(\overline{\Omega} \times \mathbf{R})$. We choose $m_1 \in E$ with $m < m_1 < \tilde{m}$ in $C(\overline{\Omega} \times \mathbf{R})$. If $\mu(\lambda)$ and $\mu_1(\lambda)$ denote the eigenvalues of (2.3) with respect to m and m_1, we thus have $\mu(\lambda) > \mu_1(\lambda), \forall \lambda > 0$. We show that

$$\mu_1(\lambda) \geqslant \mu_0 - \lambda T^{-1} \int_0^T \tilde{m}(t)\,dt \quad \forall \lambda > 0.$$

Suppose that this inequality is violated for some $\lambda > 0$. Then we choose

$$\varepsilon := \frac{1}{2\lambda}\left[\mu_0 - \lambda T^{-1} \int_0^T \tilde{m}(t)\,dt - \mu_1(\lambda)\right] > 0$$

and a T-periodic C^1-function $r: \mathbf{R} \to \mathbf{R}$ with $\|r - \tilde{m}\|_{C([0,T])} < \varepsilon$. Then $m_1 < r + \varepsilon := m_2$ and $\int_0^T m_2(t)\,dt \leqslant \int_0^T \tilde{m}(t)\,dt + 2T\varepsilon$. For the associated eigenvalues we have, by Lemma 2.5, $\mu_1(\lambda) > \mu_2(\lambda) \geqslant \mu_0 - \lambda T^{-1}(\int_0^T \tilde{m}(t)\,dt + 2T\varepsilon) = \mu_1(\lambda)$, a contradiction. $\square$

LEMMA 2.8. *Let $\lambda \in \mathbf{C}$ be an eigenvalue of (1.1) with $\operatorname{Re}\lambda \geqslant 0$. Then the positive principal eigenvalue $\lambda_1(m)$ exists, and $\operatorname{Re}\lambda \geqslant \lambda_1(m)$.*

One shows that $\mu(\operatorname{Re}\lambda) \leq 0$, which implies the assertions. The proof rests upon an extension of the "Kato inequality" (cf. [11]) to parabolic second order operators.

LEMMA 2.9. $\lambda_1 := \lambda_1(m)$ *is an M-simple eigenvalue of L (which implies that $1/\lambda_1$ is an algebraically simple eigenvalue of $L^{-1}M$).*

PROOF. Differentiating (2.3),
$$(L - \lambda M)u(\lambda) = \mu(\lambda)Ju(\lambda),$$
with respect to λ, at $\lambda = \lambda_1$ we get
$$(L - \lambda_1 M)\frac{du}{d\lambda}(\lambda_1) = \left(M + \frac{d\mu}{d\lambda}(\lambda_1)J\right)u(\lambda_1).$$
Suppose now λ_1 were not an M-simple eigenvalue of L, i.e. that, for some $w \in F$,
$$(L - \lambda_1 M)w = Mu(\lambda_1).$$
Set $z := (du/d\lambda)(\lambda_1) - w$. We obtain
$$(L - \lambda_1 M)z = \frac{d\mu}{d\lambda}(\lambda_1)Ju(\lambda_1).$$
Since $(d\mu/d\lambda)(\lambda_1) < 0$ by Lemma 2.4, we are led to a contradiction of the fact that $0 = \mu(\lambda_1)$ is a J-simple eigenvalue of $L - \lambda_1 M$ (Lemma 2.3). □

The sequence of Lemmata 2.3–2.9 proves Theorem 1.

III. Estimates for the principal eigenvalue. We give an *estimate from above* for $\lambda_1(m)$. We assume for simplicity that there exists $x_0 \in \Omega$ such that $\int_0^T m(x_0, t)\, dt > 0$ (which of course implies (M$^+$)), and that the coefficient functions a_{jk} of $\mathscr{A}$ belong to $C^1(\overline{\Omega} \times \mathbf{R}) \cap E$. Let $a = a(x, t)$ denote the symmetric $(N \times N)$-matrix $(a_{jk}(x, t))$, and a^{-1} its inverse. Set $Q := \Omega \times\,]0, T[\,$.

PROPOSITION 3.1. *Let $\phi \in C_0^\infty(\Omega)$ be a (time-independent) function such that $\phi \geq 0$, $\int_\Omega \phi^2 = 1$, and $\int_\Omega m\phi^2 > 0$. Further, let $\underline{w}(\phi)$ be the $\mathbf{R}^N$-valued function whose jth component is defined on $\Omega \times \mathbf{R}$ by*
$$\underline{w}_j(\phi) := \phi(a_j + D_k a_{jk}) + 2a_{jk}D_k\phi.$$
Then
$$\lambda_1(m) \leq \Lambda(\phi) := \left(\int_Q m\phi^2\right)^{-1}\left[\frac{1}{4}\int_Q \underline{w}(\phi)a^{-1}\underline{w}(\phi) + \int_Q a_0\phi^2\right].$$

PROOF. We adapt a device due to Holland [12]. For $\lambda \geq 0$ we consider again the eigenvalue problem (2.3),
$$(L - \lambda M)u(\lambda) = \mu(\lambda)u(\lambda).$$
Since $u(\lambda) \in \operatorname{Int}(P_F)$, hence $u(\lambda)(x, t) > 0$ on $\Omega \times \mathbf{R}$, we may set $u(\lambda) = e^{-\varphi(\lambda)}$. Thus the function $\varphi(\lambda)$ is defined on $\Omega \times \mathbf{R}$ and satisfies the equation

(3.1) $\quad \dfrac{\partial}{\partial t}\varphi(\lambda) + a_{jk}D_j D_k\varphi(\lambda) - a_{jk}D_j\varphi(\lambda)D_k\varphi(\lambda)$

$$- a_j D_j\varphi(\lambda) + a_0 - \lambda m = \mu(\lambda).$$

We multiply (3.1) by ϕ^2 and add the nonnegative term
$$\left(\phi \operatorname{grad} \varphi(\lambda) + \tfrac{1}{2}a^{-1}\underline{w}(\phi)\right)a\left(\phi \operatorname{grad} \varphi(\lambda) + \tfrac{1}{2}a^{-1}\underline{w}(\phi)\right)$$
on the left side of (3.1). Taking into account the particular form of $\underline{w}(\phi)$, and integrating over Q, we arrive at

$$(3.2) \quad -\int_Q \phi^2 \frac{\partial}{\partial t}\varphi(\lambda) + \int_Q D_k\left(\phi^2 a_{jk} D_j \varphi(\lambda)\right)$$
$$+ \frac{1}{4}\int_Q \underline{w}(\phi)a^{-1}\underline{w}(\phi) + \int_Q a_0 \phi^2 - \lambda \int_Q m\phi^2 \geq \mu(\lambda).$$

Now the first term on the left side of (3.2) vanishes since ϕ is independent of t and $\varphi(\lambda)$ is T-periodic, while the second term vanishes by the Gauss divergence theorem ($\phi = 0$ on $\partial\Omega$). Thus, for $\lambda = \Lambda(\phi)$, $\mu(\lambda) \leq 0$ by (3.2), which implies that $\lambda_1(m) \leq \Lambda(\phi)$. □

Estimates from below for $\lambda_1(m)$ can be obtained in much the same way as in Gossez and Lami Dozo [7, 8]; we therefore do not go into details.

Another possibility to obtain lower bounds for $\lambda_1(m)$ is by using Lemma 2.5: If (M^+) is satisfied, the function $\nu(\lambda) := \mu_0 - \lambda T^{-1}\int_0^T \tilde{m}(t)\,dt$ has a unique zero at $\bar{\lambda} := \mu_0 T \cdot (\int_0^T \tilde{m}(t)\,dt)^{-1}$. Hence, $\lambda_1(m) \geq \bar{\lambda}$.

IV. Construction of positive sub- and supersolutions. We turn to problem $(*)$, assuming that $g: (x, t, s) \in \bar{\Omega} \times \mathbf{R} \times \mathbf{R} \to g(x, t, s) \in \mathbf{R}$ is a continuous function which is T-periodic in t, such that $g(\cdot, \cdot, s)$ is of class $C^{\mu, \mu/2}(\bar{\Omega} \times \mathbf{R})$ uniformly for s in bounded intervals. Moreover, we suppose $\partial g/\partial s$ is continuous on $\bar{\Omega} \times \mathbf{R} \times \mathbf{R}$. We denote the Nemytskii operator associated to g by G.

It is well known that if $\underline{v}$ is a subsolution and $\bar{v}$ a supersolution of $(*)$ with $\underline{v} \leq \bar{v}$, then there exists a solution u of $(*)$ with $\underline{v} \leq u \leq \bar{v}$ on $\bar{\Omega} \times [0, T]$ (e.g. Kolesov [14]).

PROPOSITION 4.1. *Suppose for some $s_0 > 0$, $g(x, t, s) \geq g_0(x, t)s$ for $0 \leq s \leq s_0$, $\forall (x, t) \in \bar{\Omega} \times \mathbf{R}$, where $g_0 \in E$ satisfies condition (M^+) and $\lambda_1(g_0) \leq 1$. Then $(*)$ admits small positive T-periodic subsolutions.*

PROOF. For $\lambda \geq 0$ consider the linear eigenvalue problem $(L - \lambda G_0)u(\lambda) = \mu(\lambda)u(\lambda)$, $u(\lambda) > 0$ (G_0 = multiplication operator by g_0). Since $\mu(\lambda_1(g_0)) = 0$ and $\lambda_1(g_0) \leq 1$, we infer by Lemma 2.4 that $\mu(1) \leq 0$. Hence, with $v := u(1) > 0$, we get, for arbitrary $\varepsilon > 0$,
$$(L - G_0)(\varepsilon v) = \mu(1)\varepsilon v \leq 0$$
and thus, for $\varepsilon > 0$ sufficiently small,
$$L(\varepsilon v) \leq G_0(\varepsilon v) \leq G(\varepsilon v)$$
by assumption. Consequently, $\underline{v} := \varepsilon v$ is positive, T-periodic subsolution. □

PROPOSITION 4.2. *Suppose $g(x, t, s) \leq g_\infty(x, t)s + c(x, t)$ for all $s \geq 0$, $\forall(x, t) \in \bar{\Omega} \times \mathbf{R}$, where $g_\infty \in E$ satisfies (M^+), $\lambda_1(g_\infty) > 1$, and $c \in E$ ($c > 0$ without loss of generality). Then there exist large T-periodic supersolutions of $(*)$.*

PROOF. Solve $(L - G_\infty)w = c$. Then $w > 0$ by Theorem 2(i), and for $0 < \varepsilon \leq 1$,

$$L\left(\frac{1}{\varepsilon}w\right) \geq c + G_\infty\left(\frac{1}{\varepsilon}w\right) \geq G\left(\frac{1}{\varepsilon}w\right).$$

Hence $\bar{v} := (1/\varepsilon)w$ is positive T-periodic supersolution. □

Note that since $w \in \text{Int}(P_F)$, we can achieve $(1/\varepsilon)w \geq \underline{v}$ for arbitrary $\underline{v} \in F$, choosing $\varepsilon > 0$ small enough.

Combining Propositions 4.1 and 4.2 we get a statement that sharpens a result obtained by de Figueiredo [6, Theorem 2.2] in the elliptic case.

V. The semilinear parabolic-periodic eigenvalue problem. Theorem 1 can immediately be applied in the study of the nonlinear eigenvalue problem

(∗∗∗) $$\begin{cases} \mathscr{L}u = \lambda g(x, t, u) & \text{in } \Omega \times \mathbf{R}, \\ u = 0 & \text{on } \partial\Omega \times \mathbf{R}, \\ u(\cdot, 0) = u(\cdot, T) & \text{on } \overline{\Omega}, \end{cases}$$

where $g: \overline{\Omega} \times \mathbf{R} \times \mathbf{R} \to \mathbf{R}$ is sufficiently smooth, T-periodic in t, and $g(\cdot, \cdot, 0) = 0$. If $G: E \supset F \to E$ denotes the Nemytskii operator associated to g, (∗∗∗) can be written in the form $Lu = \lambda G(u)$ or, equivalently, as equation

(5.1) $$u = \lambda L^{-1}G(u)$$

in the space F. (5.1) admits the line $\mathbf{R} \times \{0\} \subset \mathbf{R} \times F$ of trivial solutions. We search for positive solutions ($\lambda > 0, u > 0$) bifurcating from this line. Let Σ denote the closure (in $\mathbf{R} \times F$) of the set of positive solutions, and set m:

$$m(x, t) := \frac{\partial g}{\partial s}(x, t, 0).$$

PROPOSITION 5.1. *There is bifurcation of positive solutions of* (5.1) *from the line of trivial solutions if and only if condition* (M$^+$) *holds for the function m. If* (M$^+$) *is satisfied, Σ contains a component Σ_0 unbounded in $\mathbf{R} \times F$, with $(\lambda_1(m), 0) \in \Sigma_0$. Moreover, $(\lambda_1(m), 0)$ is the only such bifurcation point.*

In the neighbourhood of $(\lambda_1(m), 0)$, the set of solutions of (5.1) can be further described. Moreover, since the principle of linearized stability holds for (∗∗∗) [16, Theorem 4], we can study the stability properties of these solutions. The investigations parallel those for the elliptic eigenvalue problem; we refer to Hess [9, §§II and III].

REFERENCES

1. H. Amann, *Periodic solutions of semilinear parabolic equations*, Nonlinear Analysis (L. Cesari, R. Kannan and H. Weinberger, eds.), Academic Press, New York, 1978, pp. 1–29.

2. A. Beltramo, *Ueber den Haupteigenwert von periodisch-parabolischen Differentialoperatoren*, Ph. D. Thesis, Univ. of Zurich, 1984.

3. A. Beltramo and P. Hess, *On the principal eigenvalue of a periodic-parabolic operator*, Comm. Partial Differential Equations **9** (1984), 919–941.

4. A. Castro and A. C. Lazer, *Results on periodic solutions of parabolic equations suggested by elliptic theory*, Boll. Un. Mat. Ital. B (6) **1**(1982), 1089–1104.

5. M. G. Crandall and P.-H. Rabinowitz, *Bifurcation, perturbation of simple eigenvalues and linearized stability*, Arch. Rational Mech. Anal. **52** (1973), 161–180.

6. D. G. de Figueiredo, *Positive solutions of semilinear elliptic problems*, Course Latin-American School of Differential Equations, Sao Paulo, June 1981.

7. J. P. Gossez and E. Lami Dozo, *On the principal eigenvalue of a second order linear elliptic problem*, Arch. Rational Mech. Anal. (to appear).

8. _____, *On an estimate for the principal eigenvalue of a linear elliptic problem*, Portugal. Math. **41** (1982), 347–350.

9. P. Hess, *On bifurcation and stability of positive solutions of nonlinear elliptic eigenvalue problems*, Dynamical Systems II, (A. R. Bednarek and L. Cesari, eds.), Academic Press, New York, 1982, pp. 103–119.

10. P. Hess, *On the principal eigenvalue of a second order linear elliptic problem with an indefinite weight function*, Math. Z. **179** (1982), 237–239.

11. P. Hess and T. Kato, *On some linear and nonlinear eigenvalue problems with an indefinite weight function*, Comm. Partial Differential Equations **5** (1980), 999–1030.

12. C. J. Holland, *A minimum principle for the principal eigenvalue for second-order linear elliptic equations with natural boundary conditions*, Comm. Pure Appl. Math. **31** (1978), 509–519.

13. T. Kato, *Superconvexity of the spectral radius, and convexity of the spectral bound and type*, Math. Z. **180** (1982), 265–273.

14. Ju. S. Kolesov, *A test for the existence of periodic solutions to parabolic equations*, Soviet Math. Dokl. **7** (1966), 1318–1320.

15. M. G. Kreĭn and M. A. Rutman, *Linear operators leaving invariant a cone in a Banach space*, Functional Analysis and Measure Theory, Amer. Math. Soc. Transl., Vol. 10, Amer. Math. Soc., Providence, R.I., 1962, pp. 199–325.

16. A. C. Lazer, *Some remarks on periodic solutions of parabolic differential equations*, Dynamical Systems II (A. R. Bednarek and L. Cesari, eds.), Academic Press, New York, 1982, pp. 227–246.

17. M. H. Protter and H. F. Weinberger, *Maximum principles in differential equations*, Prentice-Hall, Englewood Cliffs, N.J., 1967.

18. H. Tanabe, *Equations of evolution*, Pitman, London, 1979.

UNIVERSITY OF ZURICH, SWITZERLAND

The Topological Degree at a Critical Point of Mountain-Pass Type

HELMUT HOFER

Introduction. The Mountain-Pass Theorem [1, Theorem 2.1] is an elementary but very powerful tool which has been applied in various situations in the study of potential operators. We shall show that under certain circumstances the topological degree at a critical point of mountain-pass type is -1. In fact, this seems to be not surprising at first sight. Namely, if Φ is a C^2-Hilbert space functional having a gradient of the form identity-compact, then the linearisation at a nondegenerate critical point given by the Mountain-Pass Theorem must have exactly one negative eigenvalue and consequently the local degree is -1. What happens if the critical point is degenerate? As a counterexample will show, the degree need not be -1. Nevertheless, the assertion is true if Φ is in a restricted class of functionals including those arising in the study of second order elliptic equations. The present paper is based on a recent result of the author [2]. However, in [2] a normal form was used and the proof for the result was sketched. This sketchy proof contained some mistakes. Nevertheless, the normal form is correct and we shall give here a complete proof. Independently of [2], Tian [3] and Dancer [4] obtained similar results. In contrast to [3] and [4] we give here a very geometrical proof. How the results can be used to obtain rather strong multiplicity results can be seen in [5], where potential operators preserving an order structure have been studied.

The main results. Suppose F is a real Banach space and U is a nonempty open subset of F. Assume $\Phi \in C^1(U, \mathbf{R})$ and $c, d \in \mathbf{R}$. We define

$$\mathrm{Cr}(\Phi, c) = \{u \in U \mid \Phi'(u) = 0, \Phi(u) = c\}, \quad \mathrm{Cr}(\Phi) = \bigcup_{e \in \mathbf{R}} \mathrm{Cr}(\Phi, e),$$

$$\Phi^d = \Phi^{-1}((-\infty, d]), \quad \Phi_c = \Phi^{-1}([c, +\infty)),$$

$$\dot{\Phi}^d = \Phi^{-1}((-\infty, d)), \quad \Phi_c^d = \Phi^d \cap \Phi_c.$$

1980 *Mathematics Subject Classification.* Primary 58E05; Secondary 47H15, 57H10, 34G20.

If X is a topological space and A a subset of X, we denote by $\mathrm{int}(A, X)$ the interior of A in X. We say that $\Phi \in C^1(F, \mathbf{R})$ satisfies the Palais–Smale condition if the following holds.

(PS) If for some sequence $(u_n) \subset F$ we have $\Phi'(u_n) \to 0$ and $\Phi(u_n) \to d \in \mathbf{R}$, then (u_n) is precompact.

What characterises a critical point given by the Mountain-Pass Theorem?

DEFINITION 1. Let $U \subset F$ be a nonempty open subset and $\Phi \in C^1(U, \mathbf{R})$. Assume $u_0 \in \mathrm{Cr}(\Phi, d)$ for some $d \in \mathbf{R}$. We say that u_0 is of mountain-pass type (mp type) if there exists a neighborhood $V \subset U$ of u_0 such that for all open neighbourhoods $W \subset V$ of u_0 the topological space $W \cap \Phi^d$ is nonempty and not path connected.

The first result shows the existence of critical points of mountain-pass type under very general circumstances.

THEOREM 1. *Let* $\Phi \in C^1(F, \mathbf{R})$ *satisfy* (PS) *and assume* e_0, e_1 *are distinct points in F. Define*

$$A = \{ a \in C([0,1], F) | a(i) = e_i, i = 0, 1 \},$$
$$d = \inf_{a \in A} \sup \Phi(|a|), \quad |a| = a([0,1]),$$
$$c = \max\{\Phi(e_0), \Phi(e_1)\}.$$

Then if $d > c$ *the set* $\mathrm{Cr}(\Phi, d)$ *is nonempty. If, in addition, the critical points in* $\mathrm{Cr}(\Phi, d)$ *are isolated in F, then there exists a critical point* u_0 *of mp type in* $\mathrm{Cr}(\Phi, d)$.

The hypothesis that the critical points are isolated is necessary, as simple examples in $F = \mathbf{R}$ show.

In order to study the degree at a critical point of *mp* type we have to impose the following hypotheses on F and Φ.

(Φ) Let F be a real Hilbert space and $\Phi \in C^2(U, \mathbf{R})$ for some nonempty open subset U of F. Assume the gradient Φ' has the form identity-compact. Further assume that for all $u_0 \in \mathrm{Cr}(\Phi)$ the first (smallest) eigenvalue λ_1 of the linearisation $\Phi''(u_0) \in L(F)$ at u_0 is simple provided $\lambda_1 = 0$.

We have

THEOREM 2. *Let* (Φ) *hold and assume* $u_0 \in U$ *is an isolated critical point of mp type. Then the local degree at* u_0 *is* -1.

Before we prove the main results let us give an example for a functional satisfying (Φ).

Consider the differential equation

(1) $\qquad -\Delta u = f(\cdot, u) \quad \text{in } \Omega, \quad u = 0 \quad \text{on } \partial\Omega,$

where $\Omega \subset \mathbf{R}^N$, $N \geqslant 3$, is a bounded domain having a smooth boundary $\partial\Omega$. Assume $f \in C^1(\bar{\Omega} \times \mathbf{R}, \mathbf{R})$ (the regularity of the x-dependence can be considerably weakened) and $f'(x, s) = \partial_s f(x, s)$ satisfies the growth condition

$$|f'(x, s)| \leq C\left(1 + |s|^{\sigma-1}\right)$$

for all $(x, s) \in \overline{\Omega} \times \mathbf{R}$, where $C > 0$ and $\sigma \in [1, (N+2)/(N-2))$. The solutions of (1) are exactly the critical points of the C^2-functional $\Phi \in C^2(F, \mathbf{R})$, $F = H_0^1(\Omega)$, where

$$\Phi(u) = \frac{1}{2}\|u\|^2 - \int_\Omega \hat{f}(x, u(x))\, dx,$$

$\hat{f}(x, s) = \int_0^s f(x, \tau)\, d\tau$, $(u, v) = \int_\Omega \langle \nabla u, \nabla v \rangle$, and $\|u\|^2 = (u, u)$. If u_0 is a critical point of Φ we infer that u_0 is a classical solution of (1). Denote by λ_1 the first eigenvalue of $\Phi''(u_0)$ and assume $\lambda_1 = 0$. With $b(x) = f'(x, u_0(x))$, $b \in C(\overline{\Omega})$, we find $v \neq 0$ in F such that

$$-\Delta v = bv \quad \text{in } \Omega, \qquad v = 0 \quad \text{on } \partial\Omega.$$

Hence $\int bv^2 > 0$ which implies $b(x_0) > 0$ for some $x_0 \in \Omega$. By results of Manes–Micheletti [6], or Hess–Kato [7], the eigenvalue problem $-\Delta u = \hat{\lambda} bu$ in Ω, $u = 0$ on $\partial\Omega$, possesses a smallest positive eigenvalue $\hat{\lambda}_1$ and a corresponding eigenfunction u_1 not changing sign, spanning a one-dimensional eigenspace. Now our assertion follows from the fact that $\hat{\lambda}_1 = 1$.

Before we start with the proofs we give an example that shows that without (Φ) the assertion of Theorem 2 is false. Let $F = \mathbf{R}^2$ and $\Phi \in C^\infty(F, \mathbf{R})$ be defined by

$$\Phi(x, y) = x^4 + y^4 - 8x^2y^2.$$

With $e_0 = (1, -1)$ and $e_1 = (1, 1)$ Theorem 1 gives us $d = 0$ and $c = -6$. Moreover, $\text{Cr}(\Phi) = \text{Cr}(\Phi, d) = \{0\}$. Hence $0 = (0, 0)$ is of mp type. However, its local degree is -3.

Proof of the theorems. In order to prove Theorem 1 we introduce a class of homotopies of the identity by

DEFINITION 2. $D(\Phi)$ is the set consisting of all continuous maps $\sigma: [0, 1] \times F \to F$ such that $\sigma(0, \cdot) = \text{Id}$ and $t \to \Phi(\sigma(t, u))$ is nonincreasing for all $u \in F$.

The following lemma is a trivial variant of the standard deformation lemma [8].

LEMMA 1. *Let $\Phi \in C^1(F, \mathbf{R})$ and assume* (PS) *holds. Given $\bar{\varepsilon} > 0$, $d \in \mathbf{R}$, and open neighbourhoods W and V of $\text{Cr}(\Phi, d)$, such that $W \supset \text{cl}(V)$ and distance $(\partial W, V) > 0$, there exist $\varepsilon \in (0, \bar{\varepsilon}]$ and $\sigma \in D(\Phi)$ with*

(2) $\quad \sigma(\{1\} \times (\Phi^{d+\varepsilon} \setminus V)) \subset \Phi^{d-\varepsilon},$

(3) $\quad \sigma([0,1] \times \text{cl}(V)) \subset W,$

(4) $\quad \sigma(t, u) = u \quad \text{for all } (t, u) \in [0, 1] \times (\Phi_{d+\bar{\varepsilon}} \cup \Phi^{d-\bar{\varepsilon}}).$

PROOF OF THEOREM 1. Arguing indirectly we may assume by (PS) that $\text{Cr}(\Phi, d)$ contains only a finite number of critical points all being not of mp type. Further we may assume arguing as in [1], that $\text{Cr}(\Phi, d) \neq \emptyset$. By our hypothesis all the critical points in $\text{Cr}(\Phi, d)$ are isolated in F. Let $\text{Cr}(\Phi, d) = \{u_1, u_2, \ldots, u_n\}$.

We find corresponding open neighbourhoods U_i, $u_i \in U_i$, such that $U_i \cap \dot{\Phi}^d$ is either empty or path-connected and $U = \bigcup U_i \supset \mathrm{Cr}(\Phi, d)$. Define $\delta > 0$, $\bar{\varepsilon} > 0$, W, and V by

$$\bar{\varepsilon} := 2^{-1}(d - c),$$
$$\delta := 8^{-1} \min\{\mathrm{dist}((\partial U) \cup \{e_0, e_1\}, \mathrm{Cr}(\Phi, d)),$$
$$\inf\{\mathrm{dist}(u_i, \mathrm{Cr}(\Phi) \setminus \{u_i\}) | i = 1, \ldots, n\}\},$$
$$W := \{u \in F | \mathrm{dist}(u, \mathrm{Cr}(\Phi, d)) < 2\delta\},$$
$$V := \{u \in F | \mathrm{dist}(u, \mathrm{Cr}(\Phi, d)) < \delta\}.$$

Given $\bar{\varepsilon} > 0$, W, and V as above, we find by Lemma 1 $\varepsilon \in (0, \bar{\varepsilon}]$ and $\sigma \in D(\Phi)$ satisfying (2)–(4). Choose $a \in A$ with $|a| \subset \Phi^{d+\varepsilon}$. Note that $W = \bigcup W_i$ and $V = \bigcup V_i$, where W_i and V_i, $i = 1, \ldots, n$, are open (2δ)- or δ-balls, respectively, around u_i.

Define

$$M := \{t \in [0,1] | a(t) \notin V\}, \quad \Gamma := (U \cap \dot{\Phi}^d) \cup \sigma(\{1\} \times a(M)).$$

Observe that $e_0, e_1 \in \Gamma$. Denote by $\tilde{\Gamma}$ the path-component of Γ containing e_0. We shall show that $e_1 \in \tilde{\Gamma}$. Since by construction $\tilde{\Gamma} \subset \Gamma \subset \dot{\Phi}^d$ this will contradict the definition of d. Obviously we may assume $M \neq [0,1]$ because otherwise we are done. Define

$$t_0 = \sup\{t \in M | \sigma(1, a(t)) \in \tilde{\Gamma}\}.$$

We have to show that $t_0 = 1$. Suppose $t_0 < 1$. Since $0 \in \mathrm{int}(M, [0,1])$ this implies $t_0 \in (0, 1)$. Denote by $[t^-, t^+]$ the component in M containing t_0. We must have $t_0 = t^+$ because otherwise we immediately obtain a contradiction to the definition of t_0. Therefore $a(t_0) \in \partial V$. There exists a unique $i_0 \in \{1, \ldots, n\}$ such that $\|a(t_0) - u_{i_0}\| = \delta$. Define

$$\hat{t} := \sup\{t \in [0,1] | a(t) \in \mathrm{cl}(V_{i_0})\}.$$

By the preceding discussion, since $t_0 = t^+$, we find $\hat{t} \in (t_0, 1)$. Of course $a(\hat{t}) \in \partial V_{i_0}$. Hence $\hat{t} \in M$ and moreover $g_1 := \sigma(1, a(\hat{t})) \in W_{i_0} \cap \dot{\Phi}^d \subset U_{i_0} \cap \dot{\Phi}^d$. Further, $g_2 := \sigma(1, a(t_0)) \in U_{i_0} \cap \dot{\Phi}^d$. Since $U_{i_0} \cap \dot{\Phi}^d$ is path connected g_1 and g_2 belong to the same path-component. Using that $g_2 \in \tilde{\Gamma}$ we conclude $g_1 \in \tilde{\Gamma}$. Therefore we obtain the contradiction $t_0 \geq \hat{t} > t_0$. □

In order to prove Theorem 2 we need a special normal form of Φ near a degenerate critical point. The proof uses a combination of ideas due to Gromoll and Meyer [9], Takens [10], and the author [5].

THEOREM 3. *Let F be a real Hilbert space, and $\Phi \in C^2(U, \mathbf{R})$ having a gradient of the form identity-compact. Suppose 0 is an isolated critical point of Φ with $\Phi(0) = 0$. Let $F = F^- \oplus F^0 \oplus F^+$ be the canonical decomposition associated to $\Phi''(0)$ via the spectral resolution. Then there exist an origin-preserving homeomorphism D defined on a 0-neighbourhood into F and an origin-preserving C^1-map β*

defined on a 0-neighbourhood in F^0 into $F^- \oplus F^+$ such that

(5) $\quad \Phi(Du) = -\frac{1}{2}\|x\|^2 + \frac{1}{2}\|z\|^2 + \Phi(y + \beta y),$

(6) $\quad (P^- + P^+)\Phi'(y + \beta y) = 0$

for all $u = x + y + z$, $F^- \oplus F^0 \oplus F^+$, $\|u\|$ small. Moreover we have for the local degrees the formula

(7) $\quad \deg_{\mathrm{loc}}(\Phi', 0) = (-1)^{\dim(F^-)} \deg_{\mathrm{loc}}(\Psi', 0),$

where $\Psi(y) = \Phi(y + \beta y)$. One should note that Ψ is a C^2-map despite the fact that β needs only to be a C^1-map. This can be checked by a straightforward calculation.

PROOF. By the implicit function theorem we find a constant $\delta > 0$ and a C^1-map $\beta: F^0 \cap B_\delta \to F^- \oplus F^+$ such that

$$(P^- + P^+)\Phi'(y + \beta y) = 0 \quad \forall y \in F^0 \cap B_\delta.$$

Here B_δ denotes the open ball with radius $\delta > 0$ around 0. We define a map $\tilde{\Phi}$:
$(F^0 \cap B_\delta) \times [0,1] \times (F^- \oplus F^+) \to \mathbf{R}$ by

$$\tilde{\Phi}(y, t, v) = t\Phi(y + \beta y + v) + (1 - t)\left(\Psi(y) - \frac{1}{2}\|x\|^2 + \frac{1}{2}\|z\|^2\right),$$

where $v = x + z \in F^- \oplus F^+$. We estimate, denoting by $\tilde{\Phi}_t$ and $\tilde{\Phi}_v$ the partial derivatives (or gradients) with respect to t and v,

(8) $\quad |\tilde{\Phi}_t(y, t, v)| = \left|\Phi(y + \beta y + v) - \Psi(y) + \frac{1}{2}\|x\|^2 - \frac{1}{2}\|z\|^2\right|$

$$= \left|\int_0^1 (\Phi'(y + \beta y + sv), v)\, ds + \frac{1}{2}\|x\|^2 - \frac{1}{2}\|z\|^2\right|$$

$$= \left|\int_0^1 \int_0^1 (\Phi''(y + \beta y + \tau sv)sv, v)\, d\tau\, ds + \frac{1}{2}\|x\|^2 - \frac{1}{2}\|z\|^2\right|$$

$$\leq C_1 \|v\|^2 \quad \text{for all } \|v\| < \delta_1, \|y\| < \delta_1, t \in [0,1],$$

for some constants $C_1 > 0$ and $\delta_1 \in (0, \delta]$. Moreover we compute

$$\tilde{\Phi}_v(y, t, v) = t(P^- + P^+)\Phi'(y + \beta y + v) + (1 - t)(-x + z)$$

$$= t\int_0^1 (P^- + P^+)(\Phi''(y + \beta y + sv)v)\, ds + (1 - t)(-x + z).$$

Using the above computation and the fact that $\Phi''(0)|F^- \oplus F^+$ establishes an isomorphism $F^- \oplus F^+ \to F^- \oplus F^+$ we find some positive constants $C_2 > 0, C_3 > 0$, and $\delta_2 \in (0, \delta_1]$ such that

(9) $\quad C_2\|v\| \leq \|\tilde{\Phi}_v(y, t, v)\| \leq C_3\|v\| \quad \text{for all } t \in [0,1], \|y\| < \delta_2, \|v\| < \delta_2.$

Next we compute

$$\tilde{\Phi}_{vv}(y, t, v) = t(P^- + P^+)\Phi''(y + \beta y + v) + (1 - t)(-P^- + P^+).$$

Hence, for some $C_4 > 0$,

(10) $\quad \|\tilde{\Phi}_{vv}(y, t, v)\| \leq C_4 \quad \text{for all } t \in [0,1], \|y\| < \delta_2, \|v\| < \delta_2$

and finally

$$\tilde{\Phi}_{tv}(y, t, v) = (P^- + P^+)\Phi'(y + \beta y + v) + x - z$$
$$= \int_0^1 (P^- + P^+)(\Phi''(y + \beta y + sv)v) \, ds + x - z.$$

Hence, for some $C_5 > 0$,

(11) $\qquad \|\tilde{\Phi}_{tv}(y, t, v)\| \leq C_5\|u\| \quad$ for all $t \in [0, 1]$, $\|y\| < \delta_2$, $\|u\| < \delta_2$.

Now we define a time- and parameter-dependent vectorfield $G\colon (F^0 \cap B_{\delta_2}) \times [0, 1] \times ((F^- \oplus F^+) \cap B_{\delta_2}) \to F^- \oplus F^+$ by

$$G(y, t, v) = \begin{cases} 0 & \text{if } v = 0, \\ -\tilde{\Phi}_t(y, t, v)\|\tilde{\Phi}_v(y, t, v)\|^{-2}\tilde{\Phi}_v(y, t, v) & \text{if } v \neq 0. \end{cases}$$

First of all we note that G is continuous. This is immediately clear at points where $v_0 \neq 0$. If $v_0 = 0$ we can use the estimates (8) and (9) to obtain

(12) $\qquad \|G(y, t, v)\| \leq C_1\|v\|^2 C_2^{-1}\|v\|^{-1} = C_1 C_2^{-1}\|v\|.$

Next we shall show that there exists a constant $M > 0$ such that

$$\|G(y, t, v) - G(y, t, v')\| \leq M\|v - v'\| \quad \text{for all admissible } t, y, v, v'.$$

By estimate (12) we may assume that v and v' are both nonzero. Then there exists a differentiable path $\gamma\colon [0, 1] \to ((F^- \oplus F^+) \cap B_{\delta_2}) \setminus \{0\}$, $\gamma(0) = v$, $\gamma(1) = v'$, such that

$$\int_0^1 \|\dot{\gamma}(s)\| \, ds \leq 2\|v - v'\|.$$

Hence we estimate using that $G(y, t, v)$ is differentiable with respect to v on $((F^- + F^+) \cap B_{\delta_2}) \setminus \{0\}$,

$$\|G_v(y, t, v)\| \leq \|\tilde{\Phi}_{tv}(y, t, v)\| \|\tilde{\Phi}_v(y, t, v)\|^{-1}$$
$$+ 2|\tilde{\Phi}_t(y, t, v)| \|\tilde{\Phi}_v(y, t, v)\|^{-4}$$
$$\times \|\tilde{\Phi}_{vv}(y, t, v)\tilde{\Phi}_v(y, t, v)\| \|\tilde{\Phi}_v(y, t, v)\|$$
$$+ |\tilde{\Phi}_t(y, t, v)| \|\tilde{\Phi}_v(y, t, v)\|^{-2}\|\tilde{\Phi}_{vv}(y, t, v)\|$$
$$\leq C_5 C_2^{-1} + 2C_1 C_2^{-4} C_4 C_3^2 + C_1 C_2^{-2} C_4,$$

which implies our assertion. Now we study the differential equation

(13) $\qquad \dot{\eta}_{y,v} = G(y, t, \eta_{y,v}), \qquad \eta_{y,v}(0) = v.$

By the standard existence and uniqueness theorem we find $\sigma \in (0, \delta_2)$ such that for all initial values $v \in (F^- \oplus F^+) \cap B_\sigma$ and $y \in F^0 \cap B_\sigma$ the solution of the ODE (13) exists for all $t \in [0, 1]$. We compute

(14) $\qquad \dfrac{d}{dt}\tilde{\Phi}(y, t, \eta_{y,v}(t)) = \tilde{\Phi}_t(y, t, \eta_{y,v}(t)) + (\tilde{\Phi}_v(y, t, \eta_{y,v}(t)), \dot{\eta}_{y,v}(t)) = 0.$

Hence by (14)

(15) $\tilde{\Phi}(y, 1, \eta_{y,v}(1)) = \tilde{\Phi}(y, 0, v) = \Phi(y + \beta y) - \frac{1}{2}\|x\|^2 + \frac{1}{2}\|z\|^2.$

We note further

(16) $\tilde{\Phi}(y, 1, \eta_{y,v}(1)) = \Phi(y + \beta y + \eta_{y,v}(1)).$

Now we define $D: B_\sigma \to F$ by

$$Du = y + \beta y + \eta_{y,v}(1), \quad u = x + y + z.$$

Then if D is a homeomorphism we are done and have proved the first part of the theorem. For this we note that the inverse of D is given by

$$D^{-1}u = y + \sigma_y^{-1}(v - \beta y),$$

where σ_y^{-1} is the inverse of σ_y defined by $\sigma_y(v) = \eta_{y,v}(1)$. Hence D^{-1} is continuous. To complete the proof we define a homotopy $h: [0, 3] \times \text{cl}(B_\sigma) \to F$ by

$$h(y, u) = \begin{cases} (P^- + P^+)\Phi'(u) + P^0\Phi'(t\beta y + (1-t)(x+z) + y), \\ \qquad\qquad\qquad\qquad\qquad\qquad\qquad\qquad t \in [0, 1], \\ (P^- + P^+)\Phi'(x + z + (2-t)y) + P^0\Phi'(y + \beta y), \quad t \in [1, 2], \\ (3-t)(P^- + P^+)\Phi'(x+z) + (t-2)(-x+z) + P^0\Phi'(y + \beta y), \\ \qquad\qquad\qquad\qquad\qquad\qquad\qquad\qquad t \in [2, 3]. \end{cases}$$

Since $\dim(F^- \oplus F^0) < +\infty$ one easily sees that the homotopy is of the form identity-compact. Moreover there exists $\varepsilon_0 \in (0, \sigma]$ such that there is no solution of $h(t, u) = 0$ with $\|u\| = \varepsilon$ for all $\varepsilon \in (0, \varepsilon_0]$. This completes the proof of Theorem 3. □

PROOF OF THEOREM 2. We may assume $u_0 = 0$, $\Phi(0) = 0$, and, by Theorem 3, that Φ has the form

(17) $\Phi(u) = -\frac{1}{2}\|x\|^2 + \frac{1}{2}\|z\|^2 + \Psi(y), \quad u = x + y + z,$

where 0 is an isolated critical point of Φ. Further let $W = W^- \oplus W^0 \oplus W^+$, $\text{cl}(W) \subset U$, due to our orthogonal decomposition of F such that W^-, W^+ are δ-balls around 0 and W^0 is a ball around 0 such that

(18) $|\Psi(y)| \leq \delta^2/8 \quad \text{for all } y \in W^0.$

Further, we may assume $\text{Cr}(\Phi) \cap \text{cl}(W) = \{0\}$.

Step 1. $\dim(F^-) \leq 1$.

Arguing indirectly we may assume $\dim(F^-) \geq 2$. Let $\Gamma := W \cap \dot\Phi^0$. We shall show that Γ is path connected in contradiction to our assumption that 0 is of mp type. We shall write $g \sim g'$ for $g, g' \in \Gamma$ iff they are in the same path-component of Γ. Let $g = x_1 + y_1 + z_1 \in \Gamma$. By the special form of Φ we infer

$$g \sim g_1 := x_1 + y_1.$$

Now let $x_2 \in W^-$ with $\|x_2\| > \delta/2$ and $\|tx_2 + (1-t)x_1\| \geq \|x_1\|$ for all $t \in [0,1]$. Hence, by (17),

$$g_1 \sim g_2 := x_2 + y_1.$$

Now using (18) we conclude

$$g_2 \sim g_3 =: x_2.$$

Hence, we have shown up to now:

(19) Given any $g \in \Gamma$, there exists $\hat{g} \in \tilde{\Gamma} := W^- \setminus \{0\}$ with $g \sim \hat{g}$, provided $F^- \neq \{0\}$.

If $\dim(F^-) \geq 2$ the set $\tilde{\Gamma}$ is path connected. By (19) this implies that Γ is path connected which contradicts the fact that 0 is of mp type. Hence, we have proved the assertion $\dim(F^-) \leq 1$.

Next we compute the negative and the zero-Morse index.

Step 2. $(m^-, m^0) \in (\{1\} \times \mathbf{N}) \cup \{(0,1)\}$, where $\mathbf{N} = \{0,1,2,\ldots\}$. Assume $(m^-, m^0) \notin \{1\} \times \mathbf{N}$. By the preceding discussion $m^- = 0$. If $m^0 = 0$, 0 is a local minimum which cannot be of mp type. Hence $m^0 \geq 1$. By our assumption (Φ) the first eigenvalue $\lambda_1 = 0$ is simple. This yields $m^0 = 1$. Therefore $(m^-, m^0) = (0, 1)$.

Now we come to the heart of the proof, which is divided into two parts.

Step 3. $(m^-, m^0) \in \{1\} \times \mathbf{N} \Rightarrow \deg_{\text{loc}}(\Phi', 0) = -1$. By the standard properties of the degree we have

(20) $$\deg_{\text{loc}}(\Phi', 0) = -\deg_{\text{loc}}(\Psi', 0).$$

We will show that 0 is a local minimum of Ψ. This implies by results in [11] or [12] that $\deg_{\text{loc}}(\Psi', 0) = 1$. Combining this with (20) will prove the first case. Arguing indirectly, let us assume that 0 is not a local minimum of Ψ. As in the proof of (19) we find that $\Gamma = W \cap \dot{\Phi}^0$ has at most two path-components which can be represented by elements $\hat{x}, \bar{x} \in \overline{W} \setminus B_{(2\delta/3)}$. It is enough to show that $\hat{x} \sim \bar{x}$, giving a contradiction. By our assumption we find $y \in W^0 \setminus \{0\}$ such that $\Psi(y) < 0$. We define a path $a: [0,3] \to \Gamma$ joining $\hat{x}$ and $\bar{x}$ by

$$a(t) := \begin{cases} \hat{x} + ty, & t \in [0,1], \\ (2-t)\hat{x} + (t-1)\bar{x} + y, & t \in [1,2], \\ \bar{x} + (3-t)y, & t \in [2,3]. \end{cases}$$

The remaining case is $\dim(F^-) = 0$ or equivalently $(m^-, m^0) = (0, 1)$.

Step 4. $(m^-, m^0) = (0, 1) \Rightarrow \deg_{\text{loc}}(\Phi', 0) = -1$. It is enough to show that 0 is a local maximum of Ψ. Since $\dim(F^0) = 1$ this will imply that $\deg_{\text{loc}}(\Psi', 0) = -1$. Using that 0 is an isolated critical point of Ψ, there can be only four types of behaviour of Ψ, namely $\Psi \sim y^2$, $\Psi \sim -y^2$, $\Psi \sim \text{sign}(y)y^2$, and $\Psi \sim -\text{sign}(y)y^2$. If the first case holds, 0 is a local minimum of Φ, which is impossible. In the third and fourth cases it follows easily that Γ is contractible to a point. In particular Γ is path connected. Hence the second case must hold and our assertion follows. □

REFERENCES

1. A. Ambrosetti and P. Rabinowitz, *Dual variational methods in critical point theory*, J. Funct. Anal. **14** (1973), 343–381.
2. H. Hofer, *A note on the topological degree at a critical point of mountain-pass-type*, Proc. Amer. Math. Soc. **90** (1984), 309–315.
3. G. Tian, Bulletin of Science **14** (1983), 833–835.
4. E. N. Dancer, *Degenerate critical points, homotopy indices and Morse inequalities*, J. Reine Angew. Math. **350** (1984), 1–22.
5. H. Hofer, *Variational and topological methods in partially ordered Hilbert spaces*, Math. Ann. **261** (1982), 493–514.
6. A. Manes and A. M. Micheletti, *Un' extensione della teoria variazionale classica degli autovalori per operatori elliptici del secondo ordine*, Bull. Un. Mat. Ital. **7** (1973), 285–301.
7. P. Hess and T. Kato, *On some linear and nonlinear eigenvalue problems with indefinite weight function*, Comm. Partial Differential Equations **5** (1980), 999–1030.
8. P. H. Rabinowitz, *Variational methods for nonlinear eigenvalue problems*, Eigenvalues of Nonlinear Problems (G. Prodi, ed.), C.I.M.E. Edizione Cremonese, Rome, 1975, pp. 141–195.
9. D. Gromoll and W. Meyer, *On differentiable functions with isolated critical points*, Topology **8** (1969), 316–369.
10. F. Takens, *A note on sufficiency of jets*, Invent. Math. **13** (1971), 225–231.
11. P. H. Rabinowitz, *A note on the topological degree for potential operators*, J. Math. Anal. Appl. **51** (1975), 483–492.
12. H. Amann, *A note on the degree for gradient mappings*, Proc. Amer. Math. Soc. **84** (1982), 591–595.

UNIVERSITY OF BATH, ENGLAND

On a Conjecture of Lohwater
About Asymptotic Values of Meromorphic Functions

J. S. HWANG

Abstract. In 1951 Lohwater announced the following conjecture: Let $f(z)$ be meromorphic in $|z| < 1$, and let $\lim_{r \to 1}|f(re^{i\theta})| = 1$ almost everywhere on an arc A of $|z| = 1$. If P is a singular point of $f(z)$ on A, then every value of modulus 1 that is not in the range of $f(z)$ at P is an asymptotic value of $f(z)$ at some point of A arbitrarily near P. Lohwater was able to prove this only for functions of bounded characteristic. He then posed a slightly easier problem in 1954, which we have recently solved. In this paper we prove his conjecture is true.

1. Introduction. Let $D = \{z: |z| < 1\}$ be the unit disk and $C = \{z: |z| = 1\}$ the unit circle. In 1934 [20], W. Seidel demonstrated the following asymptotic behavior of omitted values of bounded analytic functions.

THEOREM S. *Let $f(z)$ be a nonconstant bounded analytic function in D, and let $|f(e^{i\theta})| = 1$ be the modulus of the radial limits for almost all points $e^{i\theta}$ on C. If $f(z) \neq \alpha$ ($|\alpha| < 1$) in D, then α is a radial limit of $f(z)$, that is, $\alpha = \lim_{r \to 1} f(re^{i\theta})$ for some θ.*

In his dissertation [10] Lohwater extended Seidel's theorem to functions of bounded characteristic defined by R. Nevanlinna (see [5, p. 38]). He then proved the following asymptotic behavior of functions in Seidel's class [10]–[12].

THEOREM L. *Let $f(z)$ be a function of bounded characteristic in D, and let the radial limits $f(e^{i\theta}) = \lim_{r \to 1} f(re^{i\theta})$ exist and have modulus one for almost all points $e^{i\theta}$ on an arc A of C. If P is a singular point of $f(z)$ on A, then every value of modulus 1 that is not in the range of $f(z)$ at P is an asymptotic value of $f(z)$ at some point of A arbitrarily near P.*

1980 *Mathematics Subject Classification.* Primary 30C80; Secondary 30D40.

Key words and phrases. Asymptotic value, boundary behavior, meromorphic function, Seidel's class U.

© 1986 American Mathematical Society
0082-0717/86 $1.00 + $.25 per page

As usual (see [17, p. 32]), we let U be the class of all meromorphic functions in D satisfying the second hypothesis in Theorem L. In [12, p. 156] Lohwater asked whether Theorem L is still true if $f(z)$ is an arbitrary function in class U. Recently, we have answered this question affirmatively by two different methods [6, 7], as follows.

THEOREM 1. *For each function $f \in U$ on an arc A of C, Theorem L is still true.*

In [7, Theorem 4] we extended Theorem 1 to functions of more general classes. As before, we let $|U|^-$ (or $|U|^+$) be the class of meromorphic functions $f(z)$ such that, for almost all points $e^{i\theta}$, the radial limit of the modulus $|f(re^{i\theta})|$ tends to 1 from below (or above), and we let $|U|$ denote the class of functions satisfying $\lim_{r \to 1} |f(re^{i\theta})| = 1$. We proved that Theorem 1 is still true for any function in $|U|^-$ (or $|U|^+$). Unfortunately, we were not able to prove Theorem 1 to be true for functions in the general class $|U|$. In fact, this was announced by Lohwater in 1951 [9], but he was only able to prove it for functions of bounded characteristic, as in Theorem L. Several partial solutions to this long-standing problem of Lohwater have been found; see, for instance, K. Noshiro [17, pp. 46–48] and M. Ohtsuka [18]. The main purpose of this paper is to solve this problem as follows:

THEOREM 2. *Let $f \in |U|$ on an arc A of C. If P is a singular point of $f(z)$ on A, then every value of modulus 1 that is not in the range of $f(z)$ at P is an asymptotic value of $f(z)$ at some point of A arbitrarily near P.*

A remark should be made concerning the difference between Theorems 1 and 2. In our first two papers [6, 7] we have shown Theorem 1 using two different approaches, namely, the constructive method and the Schwarz reflection principle method. Both rely mainly on the existence of radial limits almost everywhere on C. However, the hypothesis in Theorem 2—that is, $\lim_{r \to 1} |f(re^{i\theta})| = 1$—can no longer guarantee the existence of radial limits. In fact, we construct an analytic function $f(z)$ in D which satisfies $\lim_{r \to 1} |f(re^{i\theta})| = 1$ for almost all $e^{i\theta}$ on C, but $f(z)$ has no radial limits almost everywhere on C [7, Theorem 2].

In the next section we prove a fundamental property of functions in $|U|$ that resembles, in a sense, the existence of radial limits on a dense subset of C. This, together with a kind of conformal invariance, allows us to prove Lohwater's conjecture in §4. We then give an extension by which we obtain three partial solutions of Lohwater's problem, due to Noshiro. Finally, we study a connection with MacLane's class, Koebe's lemma, and Seidel's theorem. Because of the importance of this long-standing conjecture, we give a complete proof in the next three sections.

2. Fundamental property. Let $f(z)$ be a function in $|U|$ on an arc A of C. As we remarked earlier, it is possible that $f(z)$ has no radial limits almost everywhere on A. Therefore, to prove Theorem 2, we cannot rely on the existence of radial limits. However, there does exist a fundamental property which is, in a sense, similar to the existence of radial limits on a dense subset of A.

LEMMA 1. *Let $f \in |U|$ on an arc A of C. Then for each subarc α of A there is an arc β in D ending at a point on α such that $f(z)$ either tends to a limit along β or $|f(z)| \geq \lambda$ for each $z \in \beta$ and some $\lambda > 1$.*

PROOF. Let α be an arbitrary subarc of A, and let R_1 and R_2 be the two radii ending at the endpoints of α. By definition of $|U|$, we may, without loss of generality (choosing a subarc of α if necessary), assume that $|f(z)| \leq m$ for each $z \in R_1 \cup R_2$ and some $m > 1$. Let S_α denote the sector bounded by α and R_1 and R_2. If $f(z)$ is bounded in S_α, then, by Fatou and Lindelöf's theorem (see [5, Theorem 2.4]) (using a conformal mapping if necessary), we see that $f(z)$ has angular limits almost everywhere on α. This proves the first part of the lemma. To finish the proof, we may therefore assume that $f(z)$ is unbounded in S_α.

To prove the second part, we consider the upper level domain $G_\lambda = \{z: |f(z)| > \lambda$ and $z \in S_\alpha\}$. Since $f(z)$ is unbounded in S_α, G_λ is nonempty for any $\lambda > 0$, and, in particular, no boundary of G_λ can meet the interior of R_1 and R_2 for any $\lambda > m$. The boundary ∂G_λ of G_λ is in fact the level set $\{z: |f(z)| = \lambda\}$ in S_α. Regarding this set, we consider two cases: ∂G_λ contains (or does not contain) an arc β tending to the arc α. Since $\lambda > m > 1$ and $f \in |U|$, the first case implies that β must end at a point on α and, furthermore, that the equality $|f(z)| = \lambda$ holds for each $z \in \beta$. This proves the desired assertion.

On the other hand, if ∂G_λ contains no arcs tending to α, then each component G_λ^n of G_λ looks like a "hole" in S_α containing some poles of $f(z)$, and each closure $\overline{G}_\lambda^n$ must be disjoint from ∂S_α. Hence, the complement $S_\alpha - G_\lambda$ contains a region (connected open set) R_λ (uniquely determined by G_λ) whose boundary includes ∂S_α. Clearly, $|f(z)| < \lambda$ for each $z \in R_\lambda$. Let λ^* be a number, $m < \lambda^* < \lambda$, such that the lower level domain $H_{\lambda^*} = \{z: |f(z)| < \lambda^*$ and $z \in R_\lambda\}$ is nonempty. Then ∂H_{λ^*} is the level set $\{z: |f(z)| = \lambda^*\}$ in G_λ. If this set contains an arc β tending to the arc α, we are done. Otherwise, each component $H_{\lambda^*}^k$ of H_{λ^*} looks like a "hole" in R_λ containing some zeros of $f(z)$, and each $\overline{H}_{\lambda^*}^k$ must be disjoint from ∂R_λ. We shall prove that the complement $R_\lambda - H_{\lambda^*}$ contains a region R_{λ^*} whose boundary includes α.

Suppose this were not true. Then there is a sequence of components $H_{\lambda^*}^k$ such that each $H_{\lambda^*}^{k+1}$ separates $H_{\lambda^*}^k$ from α in R_λ. Hence, $\partial H_{\lambda^*}^k$ converges to α. Since $|f(z)| = \lambda^* > 1$ for each $z \in \partial H_{\lambda^*}^k$, $k = 1,2,\ldots$, we have $\lim_{r \to 1} \sup|f(re^{i\theta})| \geq \lambda^* > 1$ for each $e^{i\theta} \in \alpha$, contradicting the hypothesis $f \in |U|$. This proves the existence of the desired region R_{λ^*}, which of course is unique. Clearly, the inequality $\lambda^* < |f(z)| < \lambda$ holds for each $z \in R_{\lambda^*}$. Choose β from R_{λ^*} such that β tends to α. Then again by the hypothesis $f \in |U|$, β must end at a point on α, and clearly the inequality $|f(z)| \geq \lambda^* > 1$ holds for each $z \in \beta$. This proves the lemma.

3. Conformal invariance. There is a significant difference between the two classes of functions U and $|U|$. Namely, functions in U preserve conformal invariance. To see this, consider the conformal mapping $\varphi(z) = 1/(z-1)$, which maps the disk D onto the left half-plane $\{w: \operatorname{Re} w < -1/2\}$. Let U_φ denote the class of all meromorphic functions $f(z)$ in D such that the radial limits of $\operatorname{Re} \varphi \circ f(e^{i\theta}) = -1/2$ for almost all points on an arc A of C. It is easy to see that $U = U_\varphi$, and, therefore, U is invariant under $\varphi(z)$. However, there is a function $f \in |U| - U \cup |U|^- \cup |U|^+$ [7, Theorem 2] such that $\operatorname{Re} \varphi \circ f(re^{i\theta})$ varies from $-\infty$ to ∞ for almost every $e^{i\theta}$ on C. Nevertheless, if $f \in |U|$ and $f(z)$ is bounded away from 1, we still have $\lim_{r \to 1} \operatorname{Re} \varphi \circ f(re^{i\theta}) = -1/2$ for almost all points $e^{i\theta}$ on C whether or not the radial limits exist.

We now describe another conformal invariance, due to Löwner (see [21, Theorem VIII, 30]). Let $z = z(w)$ be a conformal mapping from D_w ($|w| < 1$) into D, and let E_w be the set of all points on $|w| = 1$ such that the radial limits $z(e^{i\theta})$ exist and have modulus 1 for each $e^{i\theta} \in E_w$. If E_z denotes the image of E_w, then the measures $|E_w| \leq |E_z|$. With the help of this theorem, we can now prove the following lemma, which is needed to prove Theorem 2.

LEMMA 2. *Let $f \in |U|$ on an arc A of C, $g(z) = 1/(f(z) - 1)$, $f(z) \neq 1$ in S_A, H a component in $\{z: |g(z)| > \lambda$ and $z \in S_A\}$, and $z = z(w)$ a conformal mapping from D_w onto H^*, the smallest simply connected domain containing H. Let L be the line $\varphi(C)$ under the mapping $\varphi(z) = 1/(z - 1)$. If the function $h(w) = g(z(w))$ is bounded in D_w, then, for almost all $e^{i\theta}$ on $|w| = 1$, the radial limits $h(e^{i\theta})$ either lie on the circle $|z| = \lambda$ or on L.*

PROOF. Suppose to the contrary that there is a Borel set E_1, on $|w| = 1$, of positive measure for which $h(e^{i\theta})$ neither lie on $|z| = \lambda$ nor on L for all $e^{i\theta} \in E_1$. Since $g(z)$ is analytic at each point on the level set $\{z: |g(z)| = \lambda$ and $z \in S_A\}$, the image $E_1^* = z(E_1)$ of E_1 must lie on A. Hence, by Löwner's theorem, $|E_1^*| \geq |E_1| > 0$. For each $e^{i\theta} \in E_1$, let r_θ be the radius in D_w ending at $e^{i\theta}$, and let γ_φ be the image of r_θ under the mapping $z = z(w)$. Clearly, each γ_φ ends at a point $e^{i\varphi}$ on E_1^*, and $g(z)$ has the asymptotic value $h(e^{i\theta})$ along the arc γ_φ ending at $e^{i\varphi}$. Since the original function $f \in |U|$ on A, there is a Borel set $E_2 \subset A$ such that $|E_2| = |A|$, and the radial cluster set $C_\rho(f, e^{i\varphi})$ is a subset of the unit circle for each $e^{i\varphi} \in E_2$. Let ρ_φ be the radius ending at $e^{i\varphi}$. For convenience, we say that ρ_φ lies eventually in a domain R if, for some $0 < r < 1$, we have $\rho_\varphi \cap \{z: r < |z| < 1\} \subset R$. We prove that, except on a countable set, for every point $e^{i\varphi} \in E_2 \cap E_1^*$, the associated radius ρ_φ lies eventually in H^*.

Consider an arbitrary point $e^{i\varphi} \in E_2 \cap E_1^*$ such that ρ_φ does not lie eventually in H^*. This means that ∂H^* meets ρ_φ infinitely many times. We then choose an arc from ∂H^*, which has $e^{i\varphi}$ as a limit point, and denote it by α_φ. Remember that at each $e^{i\varphi} \in E_1^*$ there is an arc γ_φ ending at $e^{i\varphi}$ and disjoint from α_φ. Let β_φ be a radial segment defined by $\beta_\varphi = \{re^{i\varphi}: 1 \leq r \leq \beta\}$. Then the union $\alpha_\varphi \cup \beta_\varphi \cup \gamma_\varphi$ forms a plane triad, and, therefore, by a theorem of R. L. Moore [15], the set of all such $e^{i\varphi}$ must be at most countable. This shows that outside a countable set each ρ_φ lies eventually in H^*, where ρ_φ ends at $e^{i\varphi} \in E_2 \cap E_1^*$.

From the assumption that $g(z)$ is bounded in H^*, we have that, for almost every $e^{i\varphi} \in E_2 \cap E_1^*$, $C_\rho(f, e^{i\varphi})$ is bounded away from 1. Since almost every ρ_φ

lies eventually in H^*, it follows that the radial cluster set of $g(z)$ satisfies $C_\rho(g, e^{i\varphi}) \subset L$ for almost every $e^{i\varphi} \in E_2 \cap E_1^*$. Since $g(z)$ has the asymptotic value $h(e^{i\theta}) \notin L$ along an arc ending at each $e^{i\varphi} \in E_1^*$, we conclude that almost every $e^{i\varphi} \in E_2 \cap E_1^*$ is an ambiguous point, and, hence, the set $E_2 \cap E_1^*$ must be at most countable, due to a theorem of Bagemihl (see [5, p. 83]). But this contradicts the fact that $|E_2| = |A|$ and $|E_1^*| > 0$, which give $|E_2 \cap E_1^*| > 0$. The lemma is proved.

4. Proof of Theorem 2. With the help of Lemmas 1 and 2, we are now able to prove Theorem 2. Let $f \in |U|$ on an arc A of C, P a singular point of $f(z)$ on A, and $\zeta = 1$ a point not in the range of $f(z)$ at P. Then $g(z) = 1/(f(z) - 1)$ is analytic in a relative vicinity of P. Applying Lemma 1, we see that, for each subarc α of A, there is an arc β in D ending at a point P_β on α such that $g(z)$ either tends to a limit along β or $|g(z)| = 1/|f(z) - 1| \leq 1/(\lambda - 1)$ for each $z \in \beta$ and some $\lambda > 1$. We may therefore choose a sequence of such arcs β_n in D ending at points P_{β_n} on A such that P_{β_n} tend to P from both sides of P as $n \to \infty$. If β_n contains a subsequence such that $f(z)$ tends to the same limit $\zeta = 1$ along each arc of this subsequence, then there is nothing more to prove. We may therefore assume that whenever $f(z)$ tends to a limit l_n along some β_n, then l_n is different from 1. This in turn implies that $g(z)$ is bounded on each β_n, $n = 1, 2, \ldots$ (but not necessarily uniformly bounded). Since $\zeta = 1$ is not in the range of $f(z)$ at P, for all sufficiently large n there is an arc γ_n connecting β_n with β_{n+1} such that $g(z)$ is analytic in the region R_n bounded by β_n, β_{n+1}, γ_n, and the subarc α_n of A which joins the endpoints of β_n and β_{n+1} and contains P. Furthermore, we may require $f(z) \neq 1$ for each $z \in \gamma_n$, and, therefore, $|g(z)| \leq \lambda_0$ for each $z \in \beta_n \cup \beta_{n+1} \cup \gamma_n$ and some $\lambda_0 > 0$. If $g(z)$ is bounded in R_n, then the assertion follows from Lohwater's theorem [12]. We may therefore assume that $g(z)$ is not bounded in R_n.

We now choose a sequence of increasing numbers $\lambda_j \to \infty$ as $j \to \infty$, and consider the upper level domain $G_\lambda = \{z: |g(z)| > \lambda$ and $z \in D\}$. Clearly, for each $\lambda > \lambda_0$, G_λ is disjoint from the boundary of R_n. Since $g(z)$ is not bounded in R_n, we may choose a component H_λ of G_λ which lies within R_n. We shall prove that $g(z)$ cannot be bounded in H_λ. To do this, we let H_λ^* be the smallest simply connected domain containing H_λ, and we let $z = z(w)$ be a conformal mapping from D_w ($|w| < 1$) onto H_λ^*. We then consider the composite function $h(w) = g(z(w))$ in D_w. To prove the unboundedness of $g(z)$ in H_λ^*, we suppose, to the contrary, that $h(w)$ is bounded in D_w. By Lemma 2 we find that, for almost all $e^{i\theta}$ on $|w| = 1$, the radial limits $h(e^{i\theta})$ either lie on the circle C_λ ($|z| = \lambda$) or on L. Let $\partial h(D_w)$ denote the boundary of the image $h(D_w)$. Then there is a point $p \in \partial h(D_w) - C_\lambda - L$. We may choose points $q \notin h(D_w)$ and $r \in h(D_w)$ as close to p as possible such that the distance from q to r is less than half the distance from q to $C_\lambda \cup L$. Then $F(w) = 1/(h(w) - q)$ is bounded analytic on D_w and assumes the value $1/(r - q)$ in D_w, but has radial limits of modulus at most $1/(2|r - q|)$ almost everywhere on $|w| = 1$. This contradicts the strong form of the maximum principle [5, Theorem 5.3]. We thus conclude that $g(z)$ is unbounded in each H_λ, $\lambda = \lambda_j > \lambda_0$, $j = 1, 2, \ldots$. Using the standard construction (see [6, p. 540]), we can construct a path Γ_n tending to α_n such that $g(z)$ tends to ∞ along Γ_n, or, equivalently, $f(z)$ converges to the omitted value $\zeta = 1$ along Γ_n. Applying Lemma 1 again, we see that Γ_n must end at a point on α_n. Letting $n \to \infty$, we conclude that $\zeta = 1$ is an asymptotic value of $f(z)$ at some point of A arbitrarily near P. This completes the proof of Theorem 2.

5. Some remarks. In view of the first two methods in [6, 7], this proof of Theorem 2 is very lengthy; therefore it is necessary to explain the difficulty we previously met. First, we may ask why we have not considered the function

$$(*) \qquad g(z) = \exp\left(\frac{f(z) + \zeta}{f(z) - \zeta}\right)$$

introduced by Lohwater [12]. Recall that the key point in proving Theorem 2 is to show that $g(z)$ is unbounded in H_λ (cf. [7, p. 716]). When we assume the contrary, it cannot be guaranteed that the original function $f(z)$ is bounded away from the omitted value ζ if $g(z)$ is defined by $(*)$. More precisely, it is possible that $w = f(z)$ tends to ζ, say $\zeta = 1$, along a parabola, say $|w - 1|^2 = |w| - 1$, while $g(z)$ is bounded on radii, because $|g(z)| = \exp(|w|^2 - 1)/|w - 1|^2 \leq e^2$. In this case the desired contradiction cannot be obtained.

Second, if $f \in U$, then $f(z)$ has radial limits of modulus 1 a.e. on A, and, therefore, the radial limits of $g(z) = 1/(f(z) - \zeta)$ lie on L a.e. on A. It follows that the radial limits of $h(w) = g(z(w))$ lie either on $|z| = \lambda$ or L a.e. on $|w| = 1$. This key property was described in Lemma 2, which need not be true for $f \in |U|$. More precisely, if we let $\zeta = 1$, then $f(z)$ can possibly tend to 1 from both sides such that $C_\rho(f, e^{i\theta})$ equals the extended plane for almost every $e^{i\theta}$ on A. Nevertheless, we did have this key property in the proof of Theorem 2 (cf. Lemma 2) because we assumed that $g(z)$ was bounded in H_λ^*. In other words, the boundedness of $g(z)$ in H_λ^* ensures that $f(z)$ is bounded away from ζ, and, hence, $C_\rho(f, e^{i\theta}) \subset L$ for almost every $e^{i\theta}$ on $A \cap \partial H_\lambda^*$.

Third, in our proof we have not used the conformality property of the mapping $z = z(w)$ from D_w onto H_λ^* because there is no information about ∂H_λ^*. Our proof could be greatly simplified if ∂H_n^* had only a null set of infinite rotation points (see J. E. McMillan [14]). Unfortunately, we do not know whether ∂H_λ^* has such a property for functions in $|U|$.

Fourth, instead of the radial cluster set, Noshiro considered the curvilinear cluster set [16; 17, p. 46]. An extension of this kind will be made in the next section.

6. Generalizations. Let $f(z)$ be a meromorphic function in D, and let γ_θ be an arc lying in D and ending at $e^{i\theta}$. Denote the curvilinear cluster set of $f(z)$ along γ_θ by $C_{\gamma_\theta}(f, e^{i\theta})$. Let A be an arc on the unit circle C. We say that a meromorphic function $f \in U_J$ on A if there is a Jordan curve J such that, for almost every $e^{i\theta}$ on A, there is a γ_θ in D ending at $e^{i\theta}$ for which $C_{\gamma_\theta}(f, e^{i\theta}) \subset J$.

From the proof of Theorem 2 we obtain without difficulty the following extension, for which we only sketch the details.

THEOREM 3. *If $f \in U_J$ on an arc A of C, and if P is a singular point of $f(z)$ on A, then every exceptional value ζ on the Jordan curve J is an asymptotic value of $f(z)$ at some point of A arbitrarily near P.*

PROOF. We need only map, by $\varphi(w)$, the Jordan domain bounded by J onto the unit disk. Then the composite function $\varphi(f(z))$ has asymptotic limits of modulus one almost everywhere on A. Since both Lemmas 1 and 2 are true for the case of asymptotic limits, the same argument as in Theorem 2 yields the desired assertion.

Theorem 3 immediately gives the following three partial solutions, due to Noshiro [17, pp. 46–48], of Lohwater's problem.

COROLLARY 1. *Theorem 3 is true under the additional condition that the set $\{e^{i\theta}: \zeta \notin C_{\gamma_\theta}(fe^{i\theta})\}$ is everywhere dense on A.*

COROLLARY 2. *Theorem 3 is true if, in addition to the hypothesis there, we require that $f(z)$ be regular in D.*

COROLLARY 3. *Theorem 3 is true under the additional condition that, for some $\alpha \notin J$, the set of all α-points P_n of $f(z)$ (i.e. $f(p_n) = \alpha$) satisfies $\Sigma(1 - |P_n|) < \infty$.*

A remark should be made concerning the existence of functions satisfying the hypothesis of Theorem 3. For this, we state and prove the following existence theorem (cf. [7, Theorem 6]).

THEOREM 4. *Let J be an arbitrary Jordan curve, and let $\zeta \neq 1$ be an arbitrary point on J. Then there is a function $f \in U_J$ which omits the value ζ and has the asymptotic value ζ on a dense subset of C.*

PROOF. We first choose a point $\alpha \in J$, $\alpha \neq \zeta$. Then, by a theorem of W. Rudin [19], there is a nonconstant analytic function $g(z)$ in D which has radial limit α almost everywhere on C. We now define the function

$$f(z) = \frac{\zeta(1 - g(z)) - \zeta + \alpha}{1 - g(z) - \zeta + \alpha}, \quad \text{where } \zeta \neq 1.$$

Clearly, $f(z)$ has the same radial limit α whenever $g(z)$ does, so that $f \in U_J$. Furthermore, $f(z) = \zeta$ if and only if $g(z) = \infty$. Since $g(z)$ is analytic in D, $f(z)$ must omit the value ζ in D.

Finally, every point on C is a singular point of $f(z)$, otherwise $f(z)$ would be identically equal to α. It follows from Theorem 3 that $f(z)$ has the asymptotic value ζ on a dense subset of C. This completes the proof.

7. MacLane's class.
In [13] G. R. MacLane introduced three equivalent classes for functions analytic in D. We now study a connection between Seidel's and MacLane's classes. To do this, we say that $f \in M$ on an arc A of C if $f(z)$ is analytic in a relative vicinity of A and $f(z)$ has asymptotic values on a sequence

of arcs ending at a dense subset of A. With this definition we are now able to state the following relation between Seidel's and MacLane's classes.

THEOREM 5. *Let $f \in |U|$ on an arc A of C, and let $f(z)$ omit a value ζ in a relative vicinity of A. Then the function $g = 1/(f - \zeta) \in M$ on A.*

PROOF. The proof is easy. In view of Lemma 1 and its proof, we see that $g(z)$ is bounded on a sequence of arcs ending at a dense subset of A. This, together with a theorem of MacLane [13, Theorem 1], yields that $g(z)$ has asymptotic values on a sequence of arcs ending at a dense subset of A, and, hence, $g \in M$ on A.

A remark should be made about analytic and meromorphic MacLane classes. The above-mentioned theorem of MacLane is true for the analytic case but false for the meromorphic case; see Barth and Clunie [3].

8. Koebe's Lemma. A well-known lemma of Koebe (see [5, p. 42]) asserts that if $f(z)$ is analytic, bounded in D, and tends to zero along a sequence of Koebe arcs, then $f(z)$ must be identically zero. This theorem was extended to normal functions by Bagemihl and Seidel [2]. On the other hand, for each sequence of Koebe arcs, a nonzero analytic function can be constructed in D which tends to zero along the sequence (see [5, Theorem 8.11]). Naturally, we may ask about a necessary and sufficient condition for functions of this kind and its connection with Seidel's class. To this end, we let G_λ be the upper level domain previously defined and decompose it into components G_λ^n, $n = 1, 2, \ldots$. We say that a meromorphic function $f \in K$ on an arc A of C if there is a sequence of positive numbers $\lambda_j \to 0$, as $j \to \infty$, such that each closure $\overline{G}_{\lambda_j}^n$ is disjoint from A for $n, j = 1, 2, \ldots$. Functions of this kind do exist, and, in fact, if $f(z)$ is an annular function, say (see Bagemihl, Erdös, and Seidel [1])

$$f(z) = \prod_{j=1}^{\infty} \left\{ 1 - \left[\frac{z}{1 - 1/n_j} \right]^{n_j} \right\}, \quad \text{where } \frac{n_{j+1}}{n_j} \to \infty,$$

then the reciprocal function $1/f \in K$, which will be seen from the following criterion.

THEOREM 6. *Let $f(z)$ be a function meromorphic in D. Then $f \in K$ on an arc A of C if and only if there is a sequence of Koebe arcs γ_n converging to A such that $f(z)$ tends to zero along γ_n. Consequently, K and $|U|$ are disjoint.*

PROOF. We first let $f \in K$ on an arc A of C, and we construct a sequence of Koebe arcs γ_n along which $f(z)$ tends to zero. By the definition we may assume that the given sequence λ_j is monotonically decreasing to zero. We begin by considering the first number λ_1, and we let S_A be the sector bounded by A and the radii R_i, $i = 1, 2$, ending at the endpoints of A. Since the closure of each component $\overline{G}_{\lambda_1}^n$ is disjoint from A, it follows that the complement $S - \bigcup_n \overline{G}_{\lambda_1}^n$ contains a component H_1 whose boundary consists of A. Hence, we may choose a simple arc γ_1 lying in H_1 and ending at points on R_1 and R_2. By definition of upper level domain, we must have $|f(z)| \leq \lambda_1$ for each z in $S - \overline{G}_{\lambda_1}^n$ and, of course, for each z in γ_1.

Turning to the second number $\lambda_2 < \lambda_1$ and using the same argument, we see that $S - \gamma_1 - \bigcup_n \overline{G}^n_{\lambda_2}$ contains a component H_2 whose boundary consists of A, and we may choose a simple arc γ_2 from H_2 ending at points on R_1 and R_2 satisfying $|f(z)| \leq \lambda_2$ for each z in γ_2. Continuing this process, we finally obtain a sequence of disjoint simple arcs γ_j ending at points on R_1 and R_2 such that $|f(z)| \leq \lambda_j$ for each z in γ_j, $j = 1, 2, \ldots$. Since $\lambda_j \to 0$ as $j \to \infty$, it follows that $f(z)$ tends to zero along γ_j. Furthermore, $f(z)$ is nonconstant; therefore γ_j must converge to A. This proves the necessity.

Conversely, if there is a Koebe sequence γ_j converging to A such that $f(z)$ tends to zero along γ_j, we show that $f \in K$ on A. Suppose the assertion is false. Then for some $\lambda > 0$ there is a component G^n_λ whose closure meets A. It follows that $|f(z)| > \lambda$ for all z in $G^n_\lambda \cap \gamma_j$, $j = 1, 2, \ldots$, a contradiction. This proves the sufficiency.

Finally, if $f \in |U|$ on A, then $|f(z)| \geq 1/2$ for each z in a radial segment R ending at a point on A. Hence, for any Koebe sequence γ_j we must have $|f(z)| \geq 1/2$ for each z in $R \cap \gamma_j$, where $j \geq j_0$. Thus $f(z)$ cannot tend to zero along γ_j. By what we have just proved, $f(z) \notin K$-class, and the proof is complete.

9. Seidel's Theorem.
In this section we finally state and prove the following extension of Seidel's theorem, which was introduced in §1.

THEOREM 7. *Let $f(z) \not\equiv 0$ be a bounded analytic function in D, and let $|f(z_0)| < m$ for some point $z_0 \in D$ and some number $m > 0$. If $|f(e^{i\theta})| \geq m$ for almost all points on C, then 0 is an angular limit of $f(z)$.*

PROOF. We first define the function $g(z) = 1/f(z)$. Then $g(z)$ is analytic in D and $|g(e^{i\theta})| \leq 1/m$ for almost all points on C. Since $|g(z_0)| > 1/m$, it follows from the strong form of the maximum principle [5, Theorem 5.3] that $g(z)$ must be unbounded in D.

We now consider the upper level domain G_λ previously defined. Choose a sequence of increasing numbers λ_n such that $1/m < \lambda_1$ and $\lambda_n \to \infty$ as $n \to \infty$. Since $|g(e^{i\theta})| \leq 1/m$ for almost all points on C, the same argument as in the proof of Theorem 2 yields that $g(z)$ must be unbounded in any component of G_{λ_n}, $n = 1, 2, \ldots$, and that $g(z)$ tends to ∞ along a path γ ending at a point on C, say $e^{i\theta}$. Hence, $f(z)$ tends to 0 along γ. Applying a theorem of Lindelöf [5, Theorem 2.2], we conclude that $f(z)$ has angular limit 0 at $e^{i\theta}$. This proves the theorem.

As a consequence, if we let $m = 1$ and $g(z) = (f(z) - \alpha)/(1 - \bar{\alpha}f(z))$, then we obtain Theorem S as a corollary.

Note that our method here is different from Seidel's [20, Theorem 2], which is purely a measure-theoretic approach relying mainly on the constant limit $|f(e^{i\theta})| = 1$ for almost all $e^{i\theta}$. This implies that the distribution function $\sigma(\theta)$ in the Poisson representation becomes singular and, therefore, $\sigma'(\theta_0) = -\infty$ for some θ_0, which gives the desired limit $f(e^{i\theta_0}) = 0$. Theorem 7 relaxes the restriction on the constant limit.

10. Open problem. To close the paper we pose a problem regarding our recent extension of Lohwater's theorem [8, Theorem 2]. From Theorem 2 we easily obtain the following extension from U to $|U|$.

THEOREM 8. *Let $f \in |U|$ on an arc A of C, and let P be a singular point of f on A. If v and $1/\bar{v}$ are a pair of values not in the range of $f(z)$ at P, then one of them is an asymptotic value of $f(z)$ at some point of A arbitrarily near P.*

For simplicity let us pose the following

Problem. Let $f(z)$ be a function of bounded characteristic in D such that the limits $|f(e^{i\theta})| = 1$ for almost all $e^{i\theta}$, and let P be a singular point of $f(z)$ on C. If $f(z)$ omits the values 0 and ∞, is it true that one of them is an angular limit of $f(z)$ at some point of C arbitrarily near P?

We remark that if $f(z)$ omits one more value, then the problem is true even for functions of unbounded characteristic, due to Theorem 8 and a theorem of Collingwood and Cartwright [4].

References

1. F. Bagemihl, P. Erdös and W. Seidel, *Sur quelques propriétés frontieres des fonctions holomorphes définies par certains products dans le cercle-unité*, Ann. École Norm. Sup. (3) **70** (1953), 135–147.

2. F. Bagemihl and W. Seidel, *Koebe arcs and Fatou points of normal functions*, Comment. Math. Helv. **36** (1961), 9–18.

3. K. F. Barth and J. G. Clunie, *Level curves of functions of bounded characteristic*, Proc. Amer. Math. Soc. **82** (1981), 553–559.

4. E. F. Collingwood and M. L. Cartwright, *Boundary theorems for a function meromorphic in the unit circle*, Acta Math. **87** (1952), 83–146.

5. E. F. Collingwood and A. J. Lohwater, *The theory of cluster sets*, Cambridge Univ. Press, London, 1966.

6. J. S. Hwang, *On a problem of Lohwater about the asymptotic behavior of Nevanlinna's class*, Proc. Amer. Math. Soc. **81** (1981), 538–540.

7. _____, *On the Schwarz reflection principle*, Trans. Amer. Math. Soc. **272** (1982), 711–719.

8. _____, *On the generalized Seidel class U*, Trans. Amer. Math. Soc. **276** (1983), 335–346.

9. A. J. Lohwater, *On the Schwarz reflection principle*, Bull. Amer. Math. Soc. **57** (1951), 470.

10. _____, *The boundary values of a class of meromorphic functions*, Duke Math. J. **19** (1952), 243–252.

11. _____, *Les valeurs asymptotiques de quelques fonctions méromorphes dans le cercle-unité*, C. R. Acad. Sci. Paris **237** (1953), 16–18.

12. _____, *On the Schwarz reflection principle*, Michigan Math. J. **2** (1953-1954), 151–156.

13. G. R. MacLane, *Asymptotic values of holomorphic functions*, Rice Univ. Stud. **49** (1963).

14. J. E. McMillan, *Boundary behavior under conformal mapping*, Proc. NRL Conf. on Classical Function Theory (1970), Math. Res. Center, Naval Res. Lab., Washington, D. C., 1970, pp. 59–76.

15. R. L. Moore, *Concerning triads in the plane and the junction points of plane continua*, Proc. Nat. Acad. Sci. U.S.A. **14** (1928), 85–88.

16. K. Noshiro, *Cluster sets of functions meromorphic in the unit circle*, Proc. Nat. Acad. Sci. U.S.A. **41** (1955), 398–401.

17. _____, *Cluster sets*, Springer-Verlag, 1960.

18. M. Ohtsuka, *On asymptotic values of functions analytic in a circle*, Trans. Amer. Math. Soc. **78** (1955), 294–304.

19. W. Rudin, *Radial cluster sets of analytic functions*, Bull. Amer. Math. Soc. **60** (1954), 545.

20. W. Seidel, *On the distribution of values of bounded analytic functions*, Trans. Amer. Math. Soc. **36** (1934), 201–226.

21. M. Tsuji, *Potential theory in modern function theory*, Maruzen, Tokyo, 1959.

ACADEMIA SINICA, TAIWAN

Parametrix of $\Box_b$

CHISATO IWASAKI

Introduction. The aim of this note is to construct a parametrix for $\Box_b$ as a pseudodifferential operator under the condition $Y(q)$ (see §1), which was shown by Folland and Kohn [2] to be a sufficient condition for hypoellipticity of $\Box_b$.

Boutet de Monvel constructed a parametrix as a pseudodifferential operator for $\Box_b$ by applying canonical transformations essentially. He assumed that the Levi form is nondegenerate and $Y(q)$. On the other hand, Rothschild and Tartakoff [5] constructed a parametrix as an integral operator under $Y(q)$. They also show interesting regularity properties of their parametrix.

We give a more explicit construction of a parametrix, using only the symbol calculus, by using the fundamental solution for $d/dt + \Box_b$.

Definitions and statements are in §1. We give the theorem for the fundamental solution of degenerate parabolic systems, which is obtained in the same way as the theorem for a single equation in [3, §2]. We show in §3 that the above theorem can be applied to $d/dt + \Box_b$ under the condition $Y(q)$.

1. Basic definition and theorem. Let M be a CR-manifold of dimension $2l + 1$. M is a real, orientable C^∞-manifold together with a subbundle S of a complex tangent bundle satisfying the following conditions:

(i) $\dim_{\mathbf{C}} S = l$;
(ii) $S \cap \bar{S} = \{0\}$;
(iii) $[\Gamma(S), \Gamma(S)] \in \Gamma(S)$, where $\Gamma(s)$ is the space of C^∞ cross sections of S.

EXAMPLE. M is a real hypersurface in an $(l + 1)$-dimensional complex manifold V. Let S be the intersection of $\mathbf{C}TM$ and the holomorphic vector bundle of V.

1980 *Mathematics Subject Classification.* Primary 35N15; Secondary 35K65.

© 1986 American Mathematical Society
0082-0717/86 $1.00 + $.25 per page

We let F be a complexification of the line bundle of TM such that $CTM = S \oplus \bar{S} \oplus F$. We denote $(\wedge^p S^*) \otimes (\wedge^q \bar{S}^*)$ by $\wedge^{p,q}$. The operator $\bar{\partial}_b: \Gamma(\wedge^{p,q}) \to \Gamma(\wedge^{p,q+1})$ is defined by

$$\langle \bar{\partial}_b \phi, (Z_1 \wedge \cdots \wedge Z_p) \otimes (W_1 \wedge \cdots \wedge W_{q+1}) \rangle$$

$$= \sum_{j=1}^{q+1} (-1)^{j-1} W_j \langle \phi, (Z_1 \wedge \cdots \wedge Z_p) \otimes (W_1 \wedge \cdots \wedge \hat{W}_j \wedge \cdots \wedge W_{q+1}) \rangle$$

$$+ \sum_{i<j} (-1)^{i+j} \langle \phi, (Z_1 \wedge \cdots \wedge Z_p) \rangle$$

$$\otimes \langle ([W_i, W_j] \wedge W_1 \wedge \cdots \hat{W}_i \cdots \hat{W}_j \wedge \cdots \wedge W_{q+1}) \rangle.$$

If we introduce a Hermitian metric on M such that S, $\bar{S}$, and F are mutually orthogonal, we get

$$\Box_b = \bar{\partial}_b \vartheta_b + \bar{\partial}_b \vartheta_b : \Gamma(\wedge^{p,q}) \to \Gamma(\wedge^{p,q}),$$

where ϑ_b is the formal adjoint of $\bar{\partial}_b$. Let M satisfy the following condition.

$(Y(q))$ $\quad \max(\mu_+, \mu_-) \geq \max(q+1, l+1-q)$, or
$\quad \min(\mu_+, \mu_-) \geq \min(q+1, l+1-q)$,

where μ_+ (μ_-) is the number of positive (negative) eigenvalues of the Levi form.

THEOREM 1. *Let M be a CR-manifold satisfying $Y(q)$. Then the parametrix Q of $\Box_b$ is constructed as a pseudodifferential operator of class $S^{-1}_{1/2,1/2}(M; \wedge^{p,q})$. Moreover, $\bar{\partial}_b Q$ and $\vartheta_b Q$ belong to $S^{-1/2}_{1/2,1/2}(M; \wedge^{p,q})$.*

2. Fundamental solution for degenerate parabolic systems. We use the Weyl symbols in this paper: For a symbol $p(x,\xi) = \sigma(P)$ we get a pseudodifferential operator

$$p(x,D)u(x) = (2\pi)^{-n} \int_{\mathbf{R}^n \times \mathbf{R}^n} e^{i(x-y)\cdot\xi} p\left(\frac{x+y}{2}, \xi\right) u(y) \, dy \, d\xi.$$

Let $p(x,\xi) = p_m(x,\xi) I_d + q_{m-1}(x,\xi)$ be a $d \times d$ matrix satisfying the following conditions:

(2.1) $p_m(x,\xi) \geq 0$, $p_m(x,\xi) \in S^m_{1,0}$.
(2.2) $q_{m-1}(x,\xi)$ is a $d \times d$ matrix, whose elements belong to $S^{m-1}_{1,0}$, satisfying

$$q_{m-1}(x,\xi) + q^*_{m-1}(x,\xi) + \widetilde{\operatorname{tr}}(A) I_d \geq c|\xi|^{m-1} I_d$$

on the characteristic set $\Sigma = \{p_m = 0\}$ for a positive constant c, where $\widetilde{\operatorname{tr}}(A)$ is the summation of positive eigenvalues of A, $A = iJH_{p_m}$. Here H_{p_m} is the Hesse matrix of p_m and $J = \begin{pmatrix} 0 & I \\ -I & 0 \end{pmatrix}$.

THEOREM 2. *Under assumptions* (2.1) *and* (2.2) *a fundamental solution* $E(t)$ *of* $d/dt + p(x, D)$ *is obtained as a pseudodifferential operator with a symbol belonging to* $S^0_{1/2,1/2}$ *with a parameter* t. $E(t)$ *belongs to* $S^{-\infty}$ *if* t *is positive. Moreover,* $E(t)$ *has the asymptotic expansion*

$$\sigma(E(t)) = \sum_{j=0}^{N} (\exp\phi) f_j + g_N \quad \text{for any } N,$$

$$f_0 = I, \quad (\exp\phi) f_j \in S^{-\epsilon j}_{1/2,1/2},$$

$$g_N \in S^{-\epsilon(N+1)}_{1/2,1/2} \text{ for some positive } \epsilon \text{ and near } \Sigma \times \{t = 0\},$$

$$\phi = -\big[p_m t + \tfrac{1}{4}\langle bt, F(At/2) Jbt\rangle + \tfrac{1}{2}\mathrm{tr}[\log\{\cosh(At/2)\}]\big] I_d - q_{m-1},$$

where

$$b = \nabla p_m = {}^t(\partial_{x_1} p_m, \ldots, \partial_{x_n} p_m, \partial_{\xi_1} p_m, \ldots, \partial_{\xi_n} p_m),$$

and

$$F(\lambda) = (i\lambda)^{-1}(1 - \lambda^{-1}\tanh\lambda).$$

3. Proof of Theorem 1. We shall show that Theorem 2 can be applied to $d/dt + \square_b$ under condition $Y(q)$ by means of the following propositions. Let $E(t)$ be a fundamental solution for $d/dt + \square_b$. Then it is easy to see that $Q = \int_0^T E(t)\,dt$ $(T > 0)$ is a parametrix for $\square_b$.

Let $L_1, \ldots, L_l$ be a local basis for $\Gamma(S)$ over an open set $U \subset M$. Choose a nonzero local base T which belongs to $\Gamma(F \cap TM)$. Then the Levi form $L = (L_{i,j})$ defined by

$$i[L_i, \overline{L}_j] \equiv L_{i,j} T \mod(L_k, \overline{L}_k)$$

is the Hermitian matrix. If $\omega^1, \ldots, \omega^l, \tau,$ is dual basis of $L_1, \ldots, L_l, T$, for $\phi = \sum_{I,J} \phi_{I,J} \omega^I \wedge \overline{\omega}^J$ we get

$$\square_b(\phi) = -\sum_{I,J} \bigg(\sum_{m \notin \{J\}} L_m \overline{L}_m + \sum_{m \in \{J\}} \overline{L}_m L_m \bigg) \phi_{I,J} \omega^I \wedge \overline{\omega}^J$$

$$- \sum_{I,J} \sum_{j \neq m} \big(L_j \overline{L}_m - \overline{L}_m L_j \big) \phi_{I,J} \omega^I \wedge \overline{\omega}^j \lrcorner (\overline{\omega}^m \wedge \overline{\omega}^J)$$

$$+ \epsilon(L\phi, \overline{L}\phi),$$

where $\epsilon(L\phi, \overline{L}\phi)$ is the linear combination of $L_j \phi_{I,J}$, $\overline{L}_k \phi_{I,J}$, and $\phi_{I,J}$. We denote the symbol of L_j by iz_j. Then we get

$$\sigma(\square_b) \equiv p_2 I_d + q_1 \mod S^0_{1,0},$$

where

$$p_2 = \sum_{j=1}^{l} |z_j|^2 \quad \text{and} \quad d = \binom{l}{q}\binom{l}{p}.$$

Set $\sigma([L_i, \bar{L}_j]) = i\langle J\nabla z_i, \nabla \bar{z}_j\rangle = c_{i,j}$. Using the following propositions, we shall show that $Y(q)$ is equivalent to (2.2).

PROPOSITION 1.
$$\text{tr}(iJH_{p_2}) = \sum_{j=1}^{l} |\mu_j| \quad on\ \Sigma,$$

where $\{\mu_j\}_{j=1}^{l}$ are eigenvalues of $C = (c_{i,j})$.

PROPOSITION 2. *The eigenvalues of q_1 on Σ are*
$$\mu_J = \left(\sum_{j\in\{J\}} \mu_j - \sum_{j\notin\{J\}} \mu_j\right)\Big/2.$$

PROPOSITION 3. $\mu_j(x, \xi) = \nu_j(x)\tau(x, \xi)$, *where* $\tau(x, \xi)$ *is defined by* $\sigma(T) = i\tau(x, \xi)$ *and* $\{\nu_j\}_{j=1}^{l}$ *are eigenvalues of the Levi matrix L.*

PROPOSITION 4. *(1)–(3) are equivalent by Propositions 1–3.*

(1) $\quad q_1 + q_1^* + \widetilde{\text{tr}}(iJH_{p_2})I_d \geq c|\xi|^{m-1}I_d \quad on\ \Sigma.$

(2) $\quad \displaystyle\sum_{j\in\{J\}} \nu_j\tau - \sum_{j\notin\{J\}} \nu_j\tau + \sum_{j=1}^{l} |\nu_j(x)||\tau| \geq c|\tau|, \quad \tau \in R, \quad \textit{for any } J.$

(3) $\quad \begin{aligned}\max(\mu_+, \mu_-) &\geq \max(q+1, l+1-q), \quad \text{or} \\ \min(\mu_+, \mu_-) &\geq \min(q+1, l+1-q).\end{aligned}$

REFERENCES

1. L. Boutet de Monvel, *Hypoelliptic operators with double characteristics and related pseudodifferential operators*, Comm. Pure Appl. Math. **27** (1974), 585–639.

2. G. Folland and J. J. Kohn, *The Neumann problem for the Cauchy–Riemann complex*, Ann. of Math. Stud., No. 75, Princeton Univ. Press, 1972.

3. C. Iwasaki and N. Iwasaki, *Parametrix for a degenerate parabolic equation and its application to the asymptotic behavior of spectral functions*, Publ. Res. Inst. Math. Sci. **17** (1981), 577–655.

4. L. P. Rothschild and E. M. Stein, *Hypoelliptic differential operators and nilpotent groups*, Acta Math. **127** (1976), 247–320.

5. L. P. Rothschild and D. Tartakoff, *Parametrices with C^∞ error for $\square_b$ and operators of Hörmander type*, Partial Differential Equations and Geometry (Proc. Conf., Park City, Utah, 1977), Lecture Notes in Pure and Appl. Math., vol. 48, Dekker, New York, 1979, pp. 255–271.

OSAKA UNIVERSITY, JAPAN

Applications of Nash–Moser Theory to Nonlinear Cauchy Problems

NOBUHISA IWASAKI

0. Introduction. A typical problem to which Nash–Moser theory is applied is to find local solutions of the nonlinear, noncharacteristic Cauchy problem for hyperbolic equations. If a theorem for linear equations has been proven, it can immediately be extended to nonlinear equations by Nash–Moser theory. In this sense, only one condition, called strictly hyperbolic, was general and well known for a single equation until recently. Here, we state a recent result for linear equations, which has greater generality than strict hyperbolicity, and extend it to nonlinear equations, and we give another application of Nash–Moser theory. A lemma, used to prove the recent result, is demonstrated by using it to solve a simple quasilinear equation.

1. The Cauchy problem. We consider a single partial differential operator p of order m on an open set Ω of $\mathbb{R}^{n+1}$. The noncharacteristic Cauchy problem is to find a solution on Ω of the equation $pu = f$ satisfying the initial data on a hypersurface of the derivatives, up to $m - 1$, of u to the conormal direction of the surface.

DEFINITION 1.1. (1) The Cauchy problem for p is *well-posed* at a point $x^\sim$ w.r.t. a noncharacteristic direction $\theta \neq 0$ if there exist a neighborhood Ω of $x^\sim$ and an infinitely differentiable function ϕ, satisfying $\phi(x^\sim) = 0$ and $d\phi(x^\sim) = \theta$, such that the following statements hold for any small t.

(E)$_t$ For every f belonging to $C_0^{+\infty}(\Omega)$ there is a distribution u that belongs to $\mathbf{E}'(\Omega)$ and satisfies the equation $pu = f$ on Ω_t, where $\Omega_t = \{x \in \Omega : \phi(x) < t\}$.

(U)$_t$ If $u \in \mathbf{E}'(\Omega)$ satisfies $pu = 0$ on Ω_t, then u vanishes identically on Ω_t.

(2) The Cauchy problem for P is *E-well-posed* if the above distribution solution u is always infinitely differentiable.

1980 *Mathematics Subject Classification.* Primary 35L30; Secondary 35L75.

Ivrii and Petkov [1] have shown that strongly hyperbolic operators should be effectively hyperbolic. The notations appearing in this statement stand for the following ones defined for the principal parts p_m.

DEFINITION 1.2. (1) Let p_m be hyperbolic w.r.t. the direction $\theta \neq 0$. p_m is *effectively hyperbolic* (w.r.t. $\theta \neq 0$) if the fundamental matrix at any singular point of p_m has nonzero real eigenvalues. (Refer to the next section.)

(2) p_m is *strongly hyperbolic* (w.r.t. $\theta \neq 0$) if the Cauchy problem for p with any arbitrary lower-order term is well-posed (w.r.t. $\theta \neq 0$).

The author [2]–[4] has proved the converse to the proposition of Ivrii and Petkov, which is stated as Theorem 1.1. This proves the proposition that strong hyperbolicity is equivalent to effective hyperbolicity for single linear equations.

THEOREM 1.1. *Let $p \in [0, 1]$ be a single partial differential operator of order m depending smoothly on a parameter t. If p is effectively hyperbolic w.r.t. dx_0 in a neighborhood of the origin O, then p is **E**-well-posed at O.*

Moreover, there exists a neighborhood Ω of O contained in a neighborhood given earlier, such that, for an infinitely differentiable function u on Ω supported on $x_0 \geq 0$, the estimates

$$\|u\|_s \leq C_s \big(\|pu\|_{s+l} + \|a\|_{s+l}\|pu\|_l\big)$$

hold for any real $s \geq 0$, for a fixed l, and uniformly on the parameter t, as long as $\|a\|_l$ is bounded, where a is the coefficient of p, and $\|\cdot\|_s$ is the Hölder norm on Ω.

REMARK. It is important that the constant l and the domain Ω be taken in common w.r.t. small variations of the coefficients.

Let us consider a nonlinear equation of order m,

$$pu = p(x_0, x, \partial^\alpha u) = 0,$$

where $(\partial^\alpha u)_{|\alpha|=l}$ are the lth derivatives of the real-valued function u, and p is a real-valued infinitely differentiable function in the variables $(x_0, x, \eta_\alpha)_{|\alpha| \leq m}$ with multi-indices $\alpha = (\alpha_1, \ldots, \alpha_n)$. We define the principal symbol p_m of p by

$$p_m(x_0, x, \eta_\alpha, \xi) = \sum_{|\alpha|=m} \left(\frac{\partial}{\partial \eta_\alpha}\right) p(x_0, x, \eta_\alpha) \xi^\alpha.$$

DEFINITION 1.3. p is *effectively hyperbolic* at $(x_0^\sim, x^\sim, u^\sim)$ w.r.t. dx_0 if $p_m(x_0, x, \partial^\alpha u, \xi)$ is an effectively hyperbolic symbol w.r.t. dx_0 in a neighborhood of $(x_0^\sim, x^\sim, u^\sim)$.

THEOREM 1.2. *Let a nonlinear operator p be effectively hyperbolic at $(x_0, x, u) = (0, 0, 0)$ w.r.t. dx_0. Then in a small neighborhood of O there exists uniquely an infinitely differentiable solution u satisfying*

$$pu = p(x_0, x, \partial^\alpha u) = 0 \quad \text{and} \quad (\partial/\partial x_0)^j u\big|_{x_0=0} = 0 \quad (j = 0, \ldots, m-1).$$

If the estimate in Theorem 1.1 is proved, then the conclusion of Theorem 1.2 will be immediate by Nash-Moser theory. For a detailed proof see [4], and for comments on related fields, see [1] and [5].

We give an example. Consider the equation

$$pu = \partial_x^2 u - \partial_y\bigl(a(x, y, u)\partial_y u\bigr) - b(x, y, u, \partial_x u, \partial_y u) = 0,$$

$$u|_{x=0} = u_0 \quad \text{and} \quad \partial_x u|_{x=0} = u_1,$$

on $\mathbb{R}^2$, where $a(x, y, u)$ and $b(x, y, u, v, w)$ are infinitely differentiable functions, and a is nonnegative. If the function

$$\partial_x^2 a(x, y, u) + 2\partial_x\partial_u a(x, y, u)v + \partial_u^2 a(x, y, u)v^2$$

is positive at $(x, y, v, w) = (0, 0, u_0(0), u_1(0))$ when $a(0, 0, u_0(0))$ vanishes, then p is effectively hyperbolic, so that the nonlinear equation has a unique solution in a neighborhood of $(x, y) = (0, 0)$.

2. Effectively hyperbolic operators. We precisely explain effective hyperbolicity. The fundamental matrix $\mathbf{F}$ is defined on singular points of characteristics of p_m as follows:

$$\sigma(u, \mathbf{F}v) = \langle u, \nabla^2 p_m v\rangle$$

for any tangent vectors u and v, where $\nabla^2 p_m$ is the Hessian of p_m, and σ is the canonical two-form $\Sigma d\xi^j \wedge dx_j$ on $T^*\mathbb{R}^{n+1}$. By definition, $\mathbf{F}$ has nonzero real eigenvalues, so that $\nabla^2 p_m \neq 0$. Therefore the multiplicity of the characteristics is at most two. Essentially, it is equivalent to second-order hyperbolic pseudodifferential operators. Therefore we consider the case $m = 2$, that is, $p_2 = -\xi_0^2 + f(x_0, x, \xi)$, where f is an infinitely differentiable, nonnegative function in (x_0, x, ξ) [$\xi \neq 0$] homogeneous of order 2 in ξ. The effective hyperbolicity of p_2 implies that $(\partial/\partial x_0)^2 f(x_0, x, \xi) > 0$ if $f(x_0, x, \xi) = 0$. It would be better if we could solve the Cauchy problem under only these conditions. However, some difficulties are always found, because the Cauchy problem does not admit all symplectic transformation. It also causes some complexities in the proofs of both the necessary and sufficient parts. We use the following theorem to pass through these difficulties. It might not be necessary, but it makes the calculations easier.

THEOREM 2.1. *The above effectively hyperbolic operator p_2 is written locally as*

$$p_2 = -(\xi_0 + \Lambda_0)(\xi_0 - \Lambda_0) + b_2,$$

where Λ_0 and b_2 are infinitely differentiable functions in (x_0, x, ξ) of homogeneous order 1 and 2 in ξ, respectively, such that $b_2 \geq 0$, $\{\xi_0 - \Lambda_0, \xi_0 + \Lambda_0\} > 0$ at $\Lambda_0 = b_2 = 0$, and $\{\xi_0 - \Lambda_0, b_2\} + cb_2 = 0$, with an infinitely differentiable function c in (x_0, x, ξ). Here $\{\,,\,\}$ denotes the Poisson bracket.

The essential point of this theorem is to find c and Λ such that

$$\{\xi_0 - \Lambda, f - \Lambda^2\} + c[f - \Lambda^2] = 0$$

in a neighborhood of $\{f = 0\}$. By definition it is easy to get an approximate function Λ_1 such that

$$\{\xi_0 - \Lambda_1, f - \Lambda_1^2\} = \text{bilinear in } (f, \nabla f)$$

and

$$\Lambda_1 = 0, \quad (\partial/\partial x_0)\Lambda_1 > 0 \quad \text{at } \{f = 0\}.$$

So we obtain a simple quasilinear equation

$$\{\xi_0 - (\Lambda_1 + \phi), f - (\Lambda_1 + \phi)^2\} + c[f - (\Lambda_1 + \phi)^2] = 0.$$

The difficult point of this equation is that the principal part is a singular vector field. It causes difficulty when we get a right inverse of the linear part in order to apply Nash–Moser theory to this equation, because it is not always the case that right inverses of singular vector fields are operators on the space of smooth functions. However, Nash–Moser theory suggests that it is sufficient to get asymptotic right inverses. So we construct them by integrating along the integral curves of the vector fields, and we get the local solution ϕ by using Nash–Moser theory without any other change. For a precise proof see [3].

EXAMPLE. Consider

$$p_2 = -\tau^2 + (t\eta + \xi)^2 + 2^{-1}x^2\eta^2.$$

The canonical type is

$$p_2 = -2^{-1}\sigma^2 + s^2\omega^2 + \zeta^2$$

by a symplectic transformation such that $\sigma = 2\tau + x\eta$, $s = t + \xi/\eta$, $\zeta = \tau + x\eta$, $z = t/2 - \xi/2\eta$, $\omega = \eta$, and $w = y + x\xi/\eta$. The standard type of Theorem 2.1 is, however,

$$p_2 = -\tau^2 + \Lambda_0^2 + b_2, \quad \Lambda_0 = 2^{-1/2}(t + \xi/\eta - 2^{-1/2}x)|\eta|,$$

and

$$b_2 = 2^{-1}(t\eta + \xi + 2^{-1/2}x\eta)^2,$$

where they should be properly modified on a conic neighborhood of $\{\eta = 0\}$.

REFERENCES

1. V. Ya. Ivrii and V. M. Petkov, *Necessary conditions for the Cauchy problem for non-strictly hyperbolic equations to be well-posed*, Uspekhi Mat. Nauk. **29** (1974), 3–70; English transl. in Russian Math. Surveys **29** (1974), 1–70.

2. N. Iwasaki, *The Cauchy problem for effectively hyperbolic equations (a special case)*, J. Math. Kyoto Univ. **23** (1983), 503–562.

3. _____, *The Cauchy problem for effectively hyperbolic equations (a standard type)*, Publ. Res. Inst. Math. Sci. Kyoto Univ. **20** (1984), 551–592.

4. _____, *The Cauchy problem for effectively hyperbolic equations (general cases)*, J. Math. Kyoto Univ. (to appear).

5. _____, *The Cauchy problem for hyperbolic equations with double characteristics*, Publ. Res. Inst. Math. Sci. Kyoto Univ. **19** (1983), 927–942.

KYOTO UNIVERSITY, JAPAN

Nonlinear Multiparametric Equations: Structure and Topological Dimension of Global Branches of Solutions

J. IZE, I. MASSABÓ, J. PEJSACHOWICZ, AND A. VIGNOLI[1]

1. Introduction. This paper is closely related to some recent results on the covering dimension of global branches of solutions arising in nonlinear problems (see [AA; FMP,I; FMP,II; MP]).

Our main purpose here is to extend those results to a more general class of maps, using only elementary tools. In particular, we avoid techniques from algebraic topology, such as Čech cohomology theory. This approach differs from the above-mentioned papers.

We start by considering equations of a very general form. Namely, we look at the equation

$$(1.1) \qquad f(\lambda, x) = 0,$$

where $f: \mathbb{R}^n \times E \to F$ is a continuous map, E, F are Banach spaces, and $\mathbb{R}^n$ is the "parameter space".

Our goal is to give information on the behaviour and structure of the global branches of equation (1.1). We will do so by introducing an appropriate and very general class of maps. After proving the main properties of this class of maps, we give a general theorem which will include, as particular cases, the results contained in [AA; FMP,I; FMP,II; FP; LS; MP] and [R].

This paper is divided into four sections. In §2 we introduce and give the main properties of our class of maps. §3 contains some results relating our class of maps to classical dimension theory. We then proceed by stating and proving the main result of this paper. §4 is devoted to some applications of the theory

---

1980 *Mathematics Subject Classification*. Primary 58B05, 58E07; Secondary 47H15, 54F45.

[1]Work performed while the fourth author was visiting the Department of Mathematics and Mechanics, IIMAS–UNAM, Mexico.

© 1986 American Mathematical Society
0082-0717/86 $1.00 + $.25 per page

developed in the preceding sections to continuation problems of Leray–Schauder type, global implicit function theorems, and global bifurcation results. In this last section, for simplicity, we restrict our attention to maps of the form

(1.2) $$f(\lambda, x) = x - k(\lambda, x),$$

where $k\colon \mathbb{R}^n \times E \to E$ is a compact map. This is by no means the only class of maps to which our general theorem can be applied. In a forthcoming paper [**IMPV**] we shall show that our class of maps is indeed very comprehensive. It contains, in particular, any class of maps for which a classical degree theory is available (e.g., k-set-contractive and condensing vector fields; nonlinear, proper C^1 Fredholm operators; A-proper maps; etc.). It also contains the class of essential compact vector fields in the sense of A. Granas [**G**] as well as maps having generalized degree theories [**I**]. In addition, we will extend and refine considerably all of the results contained in the present work. In particular, we extend our main result to not necessarily continuous maps, and we will allow the parameter space to be infinite dimensional. Further, we will give more detailed information on the structure and topological dimension of global branches of nonlinear problems.

2. Definitions and preliminary results. In what follows, E, G are Banach spaces, U is an open, not necessarily bounded, subset of E, and S is a subset of E such that $S \cap U$ is closed in U (this last assumption is superfluous, however, in several of the properties listed below).

In this section we introduce and study two classes of maps. Namely, zero-epi and zero-essential maps on $S \cap U$. These two classes turn out to be equivalent (see Proposition 2.1). We next state some of the properties of these maps and prove the homotopy invariance, which is one of the most important properties of zero-epi and zero-essential maps on $S \cap U$.

Finally, we close this section by showing that the class of zero-epi (zero-essential) maps includes the class of compact, k-set-contractive and condensing vector fields having topological degree different from zero.

It can also be shown that the class of essential compact vector fields in the sense of Granas [**G**] is contained in the class of zero-epi maps (see [**IMPV**] for further results).

DEFINITION 2.1. Let E, G, U, and S be as above, and let $g\colon U \to G$ be a continuous map. We say that g is *admissible* on $S \cap U$ if there exists an open and bounded subset V_0 such that $g^{-1}(0) \cap S \subset V_0 \subset \overline{V}_0 \subset U$.

Note that, if $g^{-1}(0) \cap S$ is bounded, closed, and contained in U, then the existence of V_0 with the above properties follows from the normality of E.

DEFINITION 2.2. Let g be admissible on $S \cap U$. Then the map g is called *zero-epi* (*0-epi*) on $S \cap U$ if the equation $g(x) = h(x)$ *has a solution* in $S \cap U$ for any compact map $h\colon E \to G$ with $\operatorname{supp} h \equiv \overline{\{x \in E\colon h(x) \neq 0\}}$ bounded and contained in U.

DEFINITION 2.3. Let g be admissible on $S \cap U$. Then the map g is said to be *zero-essential* (*0-essential*) on $S \cap U$ if, for any open and bounded set V such that $g^{-1}(0) \cap S \subset V \subset \overline{V} \subset U$, any continuous extension $\bar{g}$: $\overline{V} \to G$ of g: $\partial V \to G$, with $g - \bar{g}$ compact on $\overline{V}$, has a zero on $S \cap V$.

We would like to point out that S may coincide with the whole Banach space E. In such a case the class of maps introduced above reduces to the concept of 0-epi maps given in [**FMV**]. Furthermore, as a warning to the reader, the following simple example shows that a map g may be 0-essential (0-epi) on $S \cap U$ and not 0-essential (0-epi) on $S \cap W$, where W is an open set containing U. In fact, take $E = S = \mathbb{R}$, and let g: $\mathbb{R} \to \mathbb{R}$ be defined by $g(x) = x^2 - 1$. The map g is 0-essential (0-epi) on $S \cap U$, with $U = \mathbb{R} \setminus \{0\}$, but it is not 0-essential (0-epi) on $S \cap W$, where $W = \mathbb{R}$.

The following proposition contains the equivalence between the notions of 0-epi and 0-essential maps on $S \cap U$.

PROPOSITION 2.1. *The map* g: $U \to G$ *is 0-epi on* $S \cap U$ *if and only if* g *is 0-essential on* $S \cap U$.

PROOF. (*If*) Let h: $E \to G$ be as in Definition 2.2 and set $V = \{x \in E: h(x) \neq 0\} \cup V_0$, where V_0 is as in Definition 2.1. Clearly, V is open, bounded, and $\overline{V} = \operatorname{supp} h \cup \overline{V}_0$. Moreover, V satisfies the properties of Definition 2.3. Since h vanishes on ∂V, $\bar{g} = (g - h)|_{\overline{V}}$ is an extension of $g|_{\partial V}$ satisfying the requirements of Definition 2.3 and, as such, $\bar{g}$ has a zero on $S \cap V$.

(*Only if*) Let V and $\bar{g}$ be as in Definition 2.3. Define h: $E \to G$ as $g - \bar{g}$ on $\overline{V}$ and zero on $E \setminus V$. Then $\operatorname{supp} h \subset \overline{V}$, and h satisfies the requirements of Definition 2.2. Hence, the equation $g(x) = h(x)$ must have a solution $x \in S \cap U$. Since x cannot be in $E \setminus \overline{V}$, it has to be in $S \cap V$. Q.E.D.

We now list some elementary properties of 0-epi (and, therefore, 0-essential) maps on $S \cap U$.

PROPERTY 2.1 (Existence). *If g is zero-epi on $S \cap U$, then $g^{-1}(0) \cap S \neq \varnothing$.*

(This follows at once by taking h to be the identically zero map on E.)

PROPERTY 2.2 (Localization with respect to the open set). *If g is zero-epi on $S \cap U$, then g is zero-epi on $S \cap V$ for any open set V such that $g^{-1}(0) \cap S \subset V \subset \overline{V} \subset U$.*

PROPERTY 2.3 (Normalization). *If i: $U \hookrightarrow E$ is the inclusion, $0 \notin \partial U$ and S is a subset of E such that $S \cap U$ is closed in U, then the map i is 0-epi on $S \cap U$ if and only if $0 \in S \cap U$ and the component of 0 in U is contained in S. In particular, if U is connected, then $S \cap U = U$.*

PROOF. (*Only if*) That $0 \in S \cap U$ follows from Property 2.1. If the second assertion is false, then there exists a point x_0 in the component, V, of U containing zero such that $x_0 \notin S$. Since U is an open subset of a Banach space, it is locally path connected. Therefore, V is path connected and open. Let $\sigma(t)$, $t \in [t_0, t_1]$, be a path (in U) from 0 to x_0. Clearly, the set $\{\sigma(t)\}$ is compact, so that there exists an ε-neighbourhood of the path, which is contained in V. Let φ

be an Urysohn function taking the value 1 on the path σ and vanishing outside of the ε-neighbourhood. Set $h(x) = \sigma(\varphi(x))$. Since $\sigma(t_0) = 0$, it follows that supp h is contained in the ε-neighbourhood, so supp h is bounded. Taking into account that the map $g(x) = x$ is 0-epi on $S \cap U$, we obtain that the equation $x - \sigma(\varphi(x)) = 0$ has a solution $\bar{x} \in S \cap U$. The equality $\bar{x} = \sigma(\varphi(\bar{x}))$ implies that $\bar{x}$ belongs to σ and, thus, $\varphi(\bar{x}) = 1$. Hence, $\bar{x} = x_0 \notin S$.

(*If*) Assume that $0 \in U$ and that the component, V, of 0 in U is contained in S. The fact that the inclusion map is zero-epi on $S \cap U$ follows from a degree argument. Indeed, under the above assumptions, the Leray–Schauder degree $\deg_{LS}(I - h, V, 0) = \deg_{LS}(I, V, 0) = 1$ (notice that $h(x) = 0$ for all $x \in \partial V \subset \partial U$). Thus $x - h(x) = 0$ is solvable in V.

PROPERTY 2.4. *Let G_i, $i = 1, 2$, be Banach spaces, and let $g_i: U \to G_i$, $i = 1, 2$, be continuous maps. Define $g: U \to G_1 \times G_2$ by $g(x) = (g_1(x), g_2(x))$. Assume that g is zero-epi on $S \cap U$. Then g_2 is zero-epi on $g_1^{-1}(0) \cap S \cap U$.*

PROOF. If $h_2: E \to G_2$ is compact with bounded support contained in U, then the map $h = (0, h_2)$ is compact on E and has bounded support contained in U. Therefore, the equations $g_2(x) = h(x)$ and $g_1(x) = 0$ are solvable in $S \cap U$. Q.E.D.

The following property is an immediate consequence of the boundary dependence property of the Leray–Schauder topological degree.

PROPERTY 2.5. *Let $E = S = G$, and let $g: U \to E$ be of the form $g(x) = x - k(x)$, with $k: U \to E$ compact. If the Leray–Schauder topological degree is defined and different from zero, then g is 0-epi on U.*

We point out that Property 2.5 holds for any class of maps g for which a classical topological degree theory is available (i.e., k-set contractive vector fields, nonlinear proper Fredholm operators, A-proper maps, etc.).

THEOREM 2.1 (HOMOTOPY PROPERTY). *Let g be 0-epi on $S \cap U$, and let $h: U \times [0,1] \to G$ be compact and such that $h(x, 0) = 0$ for all $x \in U$. Assume that the set*

$$A_0 = \{x \in S \cap U: g(x) = h(x, t) \text{ for some } t \in [0,1]\}$$

is such that $A_0 \subset V_0 \subset \overline{V_0} \subset U$, where V_0 is open and bounded. Then $g(\cdot) - h(\cdot, 1)$ is 0-epi on $S \cap U$.

PROOF. Let $k: E \to G$ be a map with bounded support contained in U. Consider the set

$$A = \{x \in S \cap U: g(x) - h(x, t) = k(x) \text{ for some } t \in [0,1]\}.$$

A is bounded, because if $x \in A \cap \overline{\text{supp } k^c}$, then $k(x) = 0$ and $x \in A_0$. Thus the following inclusions hold:

$$A \subset \text{supp } k \cup A_0 \subset \text{supp } k \cup V_0 \subset U.$$

Moreover, A is closed. Indeed, take a sequence $\{x_n\}$, $x_n \in A$, $n \in \mathbb{N}$, such that $x_n \to x$. There corresponds a sequence $\{t_n\}$ in $[0,1]$ such that $g(x_n) - h(x_n, t_n) = k(x_n)$. Without loss of generality we may assume that $\{t_n\}$ converges to some element t_0 in $[0,1]$. Taking into account the continuity of all the maps involved and the fact that $S \cap U$ is closed in U, we obtain that $x \in A$. Thus, A is closed (notice that the same argument shows that A_0 is closed as well).

Let φ be an Urysohn function such that $\varphi(A) = 1$ and $\varphi((\operatorname{supp} k)^c \cap V_0^c) = 0$ (these two sets are disjoint, because if $x \in A$, then either $k(x) \neq 0$ and $x \in \operatorname{int}(\operatorname{supp} k)$, or $k(x) = 0$ and so $x \in A_0 \subset V_0$). Now, consider the map $k(x) + h(x, \varphi(x))$, which may be considered compact on E since the equality $h(x,0) = 0$ for all $x \in U$ permits us to extend h to the whole space E by setting it identically equal to zero outside $\operatorname{supp} k \cup V_0$. The support of this map is contained in $\operatorname{supp} k \cup V_0 \subset U$. Since g is 0-epi on $S \cap U$, it follows that the equation $g(x) = k(x) + h(x, \varphi(x))$ has a solution $x \in S \cap U$. Thus $x \in A$ and $\varphi(x) = 1$; i.e., $g(x) - h(x,1) = k(x)$ is solvable in $S \cap U$. Q.E.D.

The following result, which is consequence of Theorem 2.1, gives information on the structure of the set $S \cap U$ when there is a map which is 0-epi on it.

PROPOSITION 2.2. *Let g be 0-epi on $S \cap U$. Then*

(a) *either $S \cap \partial V \neq \varnothing$, or $g(S \cap \overline{V}) = G$ for any open and bounded set V such that $g^{-1}(0) \cap S \subset V \subset \overline{V} \subset U$. In particular, if g sends bounded and closed (in E) subsets of $S \cap U$ into bounded sets of G, then $S \cap \partial V \neq \varnothing$;*

(b) *either $S \cap U$ is unbounded, or $\overline{(S \cap U)} \cap \partial U \neq \varnothing$, or there exists V as above such that $g(S \cap \overline{V}) = G$.*

PROOF. (a) Assume there is such a V for which $S \cap \partial V = \varnothing$. Then, by Property 2.2, the map g is 0-epi on the set $S \cap V$ (which is closed in V). Let $p \in G$. Consider the homotopy $h: U \times [0,1] \to G$ defined by $h(x,t) = tp$. Then the set

$$A = \{x \in S \cap V : g(x) = h(x,t) \text{ for some } t \in [0,1]\}$$

is contained properly in V, since $S \cap \partial V = \varnothing$, that is the assumptions of Theorem 2.1, with U replaced by V, are satisfied and hence $g - p$ is 0-epi on $S \cap V$. Thus, by Property 2.1, $p \in g(S \cap V)$.

(b) If $\overline{(S \cap U)} \cap \partial U = \varnothing$ and $S \cap U$ is bounded, then, by normality, there exists an open, bounded set V such that $\overline{S \cap U} \subset V \subset \overline{V} \subset U$. Since $S \cap \partial V = \varnothing$, then, from part (a), we get $g(S \cap \overline{V}) = G$. Q.E.D.

3. Covering dimension and main result. In the first part of this section we study the relation between the concept of real-valued 0-epi maps and classical dimension theory. The dimension we use throughout this paper is the covering dimension, whose definition runs as follows (cf. [**P**, p. 111]): If X is a topological space, the *order* of a family, $\{u_i\}_{i \in I}$, of subsets of X, not all empty, is the largest integer n for which there exists a subset J of I with $n+1$ elements such that $\bigcap_{i \in J} u_i$ is nonempty, or is ∞ if there is no such largest integer. A family of empty subsets has order -1. A normal topological space X has *covering dimension*, $\dim X$, equal to n if n is the least integer such that every finite open covering of X has an open refinement of order not exceeding n, or equal to ∞ if there is no such integer.

The second, and most important, part of this section contains our main result (Theorem 3.1). This theorem extends the results contained in [**AA**; **FMP, I**; **FMP, II**] and [**MP**].

Theorem 3.1 shows, in particular, that if a certain map g is 0-epi on $S \cap U$, then there is a minimal closed subset of $S \cap U$ having a number of properties related not only to its global structure but also to its covering dimension at each point (local dimension).

Note that, in most of the results given thus far, we have only used the normality of the space E (the metric of E has been used, because of our definitions, when dealing with bounded sets). However, in what follows, in order to avoid complications in the definitions, the metric of E will be used in a crucial way. In particular, we recall that if A is a subset of a metric space X, then $\dim A \leq \dim X$ provided that A is closed [**P**, p. 158].

The dimension-theory results we use are given in terms of mappings with a range in finite-dimensional spheres. Using the classical notion of essentiality, we show that the definitions in §2 allow us to use those classical results. For this we recall the following characterization of dimension of a space X (a proof can be found in [**P**, p. 123]).

PROPOSITION 3.1. *Let X be a normal space. Then $\dim X \leq n$ if and only if for each closed subset A of X any map $g: A \to S^n \subset \mathbb{R}^{n+1}$ has a continuous extension over X.*

The following is a result relating 0-epi maps having finite-dimensional range with dimension theory.

In the rest of the paper we say that a map is *bounded* if it sends bounded and closed (in E) subsets of U into bounded sets.

PROPOSITION 3.2. *Let $g: U \to \mathbb{R}^n$ be 0-epi on $S \cap U$ and bounded. Let V be any open, bounded subset of U such that $g^{-1}(0) \cap S \subset V \subset \overline{V} \subset U$. Then $\dim S \cap \overline{V} \geq n$ and $\dim S \cap \partial V \geq n - 1$ for $n > 1$; if $n = 1$, $S \cap \partial V$ has at least two components on which g changes sign.*

The proof of Proposition 3.2 is based on the following lemma (and Proposition 3.1).

LEMMA 3.1. *Let g and V be as in Proposition 3.2. Then any continuous (and bounded) extension $\bar{g}: \overline{V} \to \mathbb{R}^n$ of the restriction $g: \partial V \to \mathbb{R}^n$ has a zero in $S \cap V$ if and only if the map $g/\|g\|: S \cap \partial V \to S^{n-1} \subset \mathbb{R}^n$ is not extensible over $S \cap \overline{V}$.*

PROOF. First notice that, by Proposition 2.2, $S \cap \partial V \neq \varnothing$. Moreover, the set $S \cap \overline{V}$ is closed.

(*Only if*) Consider the map $h: (S \cap \partial V) \times [0, 1] \to \mathbb{R}^n \setminus \{0\}$ defined by $h(x, t) = g(x)/(t + (1 - t)\|g(x)\|)$ (notice that h is bounded). Clearly, $h_0 = g/\|g\|$ and $h_1 = g$. Assume that $g/\|g\|$ has continuous extension $\hat{g}: S \cap \overline{V} \to S^{n-1}$. Let

$$H: (S \cap \overline{V}) \times \{0\} \cup (S \cap \partial V) \times [0, 1] \cup \partial V \times \{1\} \to \mathbb{R}^n$$

be the map

$$H(x,0) = \hat{g}(x) \quad \text{for } x \in S \cap \overline{V},$$
$$H(x,t) = h(x,t) \quad \text{for } x \in S \cap \partial V \text{ and } t \in [0,1],$$
$$H(x,1) = g(x) \quad \text{for } x \in \partial V.$$

Observe that H is continuous and bounded, and $H(x,t) \neq 0$ for $(x,t) \in (S \cap \partial V) \times [0,1]$.

By Tietze's extension theorem let $\overline{H}$ be a bounded extension of H to $\overline{V} \times [0,1]$. Now, the boundary ∂V of V and the set $B = \{x \in S \cap \overline{V} : \overline{H}(x,t) = 0 \text{ for some } t \in [0,1]\}$ are disjoint sets. Hence, there is an Urysohn function $\varphi : \overline{V} \to [0,1]$ such that $\varphi(B) = 0$ and $\varphi(\partial V) = 1$. Now consider the map $G : \overline{V} \times [0,1] \to \mathbb{R}^n$, defined by $G(x,t) = \overline{H}(x, t\varphi(x))$. The map G is continuous (bounded), and if we define $\bar{g}(x) = G(x,1)$, it is clear that $\bar{g}(x) = g(x)$ for $x \in \partial V$ and $\bar{g}(x) \neq 0$ in $S \cap \overline{V}$.

(*If*) Assume that $g: \partial V \to \mathbb{R}^n$ has a (bounded) extension $\bar{g} : \overline{V} \to \mathbb{R}^n$ such that $\bar{g}(x) \neq 0$ on $S \cap V$. Then $\bar{g}/\|\bar{g}\|$ is an extension of $g/\|g\| : S \cap \partial V \to \mathbb{R}^n \setminus \{0\}$. Q.E.D.

We would like to add in passing that the "only if" part of Lemma 3.1 is very close to the classical Borsuk homotopy extension theorem.

PROOF OF PROPOSITION 3.2. Let V be an open bounded subset of U such that $g^{-1}(0) \cap S \subset V \subset \overline{V} \subset U$. For $n > 1$ the first inequality follows from the fact that, since g is 0-epi on $S \cap U$, Lemma 3.1 applies to the restriction $g : \partial V \to \mathbb{R}^n \setminus \{0\}$. Thus the map $g/\|g\|$ from the closed subset, $S \cap \partial V$, of $S \cap \overline{V}$ into S^{n-1} is not extensible over $S \cap \overline{V}$. Hence, by Proposition 3.1, $\dim X$ cannot be less than or equal to $n - 1$.

The second inequality follows from the fact that the set

$$A = \{x \in S \cap \partial V : g(x)/\|g(x)\| \neq p \in S^{n-1}\}$$

can be written as

$$A = \bigcup_{j=1}^{+\infty} U_j,$$

with

$$U_j = \{x \in S \cap \partial V : \|f_0(x) - f_1(x)\| > j^{-1}\},$$

where $f_0 = g/\|g\|$ and $f_1 = p$. By the Countable Sum Theorem [**P**, p. 125], $\dim A \leq \dim(S \cap \partial V)$, so that, if $\dim(S \cap \partial V) < n - 1$, then $g/\|g\|$ and the constant map p are uniformly homotopic [**P**, p. 333]. This implies that $g/\|g\|$ has a nonzero extension to $S \cap \overline{V}$. This, by Lemma 3.1, contradicts the assumption that g is 0-epi on $S \cap U$.

If $n = 1$ and g does not change sign on $S \cap \partial V$, then clearly, $g/\|g\|$ has a constant ($+1$ or -1) extension to $S \cap \overline{V}$. This contradicts, via Lemma 3.1, the assumption that is 0-epi on $S \cap U$. Q.E.D.

REMARK. By Proposition 3.2 we have $\dim(S \cap U) \geq n$ ($S \cap \overline{V}$ is a closed subset of $S \cap U$ and "dim" is monotonic).

We are now in a position to prove our main result.

At this point we further assume that S intersected with any bounded, closed (in E) subset of U is compact.

THEOREM 3.1. *Let $g: U \to \mathbb{R}^n$ be 0-epi on $S \cap U$ and bounded. Then there exists a minimal closed (in U) set $\Sigma \subset S \cap U$ such that the following hold:*

(a) *g is 0-epi on $\Sigma \cap U = \Sigma$. This implies, in particular, that $g^{-1}(0) \cap \Sigma \neq \varnothing$; Σ is either unbounded or $\overline{\Sigma} \cap \partial U \neq \varnothing$; $\dim \Sigma \geq \dim \Sigma \cap \overline{V} \geq n$ and $\dim \Sigma \cap \partial V \geq n - 1$ for any open and bounded set V such that $g^{-1}(0) \cap S \subset V \subset \overline{V} \subset U$.*

(b) *If $\Sigma = \Sigma_1 \cup \Sigma_2$, with Σ_1, Σ_2 closed (in Σ) and proper subsets of Σ, then $\dim(\Sigma_1 \cap \Sigma_2) \geq n - 1$. This implies, in particular, that Σ is connected and has dimension at least n at each point.*

(c) *Σ is minimal for any map g_1 homotopic to g.*

PROOF. Note that, since g is an admissible map, $g^{-1}(0) \cap S$ is compact. Now let
$$\mathscr{C} = \{C \subset S, C \text{ closed in } S \cap U : g \text{ is 0-epi on } S \cap U\}.$$
The family $\mathscr{C}$ is nonempty since $S \cap U \in \mathscr{C}$. Define an order in $\mathscr{C}$ by inclusion of sets, and let $\mathscr{C}'$ be a chain in $\mathscr{C}$. Consider $\Sigma = \bigcap_{C \in \mathscr{C}'} C$. Since $g^{-1}(0) \cap C$ is a descending family of compact sets (notice that $g^{-1}(0) \cap C \neq \varnothing$, by Property 2.1), $g^{-1}(0) \cap \Sigma$ is nonempty and compact.

Let $h: E \to \mathbb{R}^n$ be continuous, with bounded support contained in U, and let V_0 be an open and bounded set such that $g^{-1}(0) \cap S \subset V_0 \subset \overline{V_0} \subset U$. Set $V_1 = V_0 \cup \{x \in E: h(x) \neq 0\}$, which is bounded, open, and $\overline{V_1} \subset U$. Since S is locally compact, $(g - h)^{-1}(0) \cap S \cap \overline{V_1}$ is a compact set, and $(g - h)^{-1}(0) \cap C \cap \overline{V_1}$ is a descending family of compact sets. These sets are nonempty since g is 0-epi on $C \cap U$, and $g(x) = h(x)$ has a solution in $C \cap U$. But, $h(x) = 0$ and $g(x) \neq 0$ for all $x \in S \cap V_1^c$. Thus the solution is in $C \cap \overline{V_1}$. Hence, $(g - h)^{-1}(0) \cap \Sigma \cap \overline{V_1} \neq \varnothing$; i.e., g is 0-epi on $\Sigma \cap U$.

By Zorn's lemma $\mathscr{C}$ has a minimal element, also denoted by Σ. Since g is 0-epi on $\Sigma \cap U$, by minimality, $\Sigma \subset S \cap U$. In particular, since g is bounded, we get that the remaining part of assertion (a) follows from Property 2.1 and Propositions 2.3 and 3.2.

(b) Assume that $\Sigma = \Sigma_1 \cup \Sigma_2$, where Σ_1, Σ_2 are closed proper subsets of Σ such that $\dim(\Sigma_1 \cap \Sigma_2) < n - 1$. Since Σ_1, Σ_2 are closed proper subsets of Σ, it follows that g is *not* 0-epi on $\Sigma_i \cap U$, $i = 1, 2$. Therefore, there exist open and bounded sets V_i with $g^{-1}(0) \cap \Sigma_i \subset V_i \subset \overline{V_i} \subset U$ such that $g|_{\Sigma_i \cap \partial V_i}: \Sigma_i \cap \partial V_i \to \mathbb{R}^n \setminus \{0\}$ extends to $\tilde{g}_i: \Sigma_i \cap \overline{V_i} \to \mathbb{R}^n \setminus \{0\}$, $i = 1, 2$. Let $V = V_1 \cup V_2$, $A = \Sigma \cap \overline{V}$, $A_i = (\Sigma_i \cap \overline{V}) \cup (\Sigma \cap \partial V)$, $i = 1, 2$. Set

$$g_i = \begin{cases} g & \text{on } B_i = \Sigma_i \cap (\overline{V} \setminus V_i) \cup (\Sigma \cap \partial V), \\ \tilde{g}_i & \text{on } \Sigma_i \cap \overline{V_i}, \end{cases} \quad i = 1, 2.$$

Then g_i, $i = 1, 2$, are continuous maps from A_i into $\mathbb{R}^n \setminus \{0\}$ ($g \neq 0$ on B_i and $\tilde{g}_i \neq 0$ on $\Sigma_i \cap \overline{V_i}$). Moreover, the maps $g_1/\|g_1\|$ and $g_2/\|g_2\|$ restricted to

$A_1 \cap A_2$ differ on a subset of $\Sigma_1 \cap \Sigma_2 \cap \overline{V}$ of dimension $n - 1$. Therefore, according to [**P**, p. 334], each $g_i/\|g_i\|$: $A_i \to \mathbb{R}^n \setminus \{0\}$ has a nonvanishing extension to $A = \Sigma \cap \overline{V}$. In particular, via Lemma 3.1, $g|_{\partial V}$ has an extension over $\overline{V}$ without zeroes in $\Sigma \cap V$, contradicting the fact that g is 0-epi on Σ.

It remains to show that Σ is connected and has dimension at least n at each point.

If Σ is not connected, let Σ_1, Σ_2 be a partition of Σ. Then $\Sigma_1 \cap \Sigma_2 = \varnothing$, and so Σ_1, Σ_2 cannot be proper subsets of Σ; i.e., Σ must be connected.

Finally, let $p \in \Sigma$ and recall that $\dim_p \Sigma \leq m$ if for each neighbourhood V of p in Σ there is an open set W such that $p \in W \subset V$ and $\dim \overline{W} \leq m$ [**P**, p. 167]. Now, for any open neighbourhood W of p, let $\Sigma_1 = \Sigma \setminus W$, $\Sigma_2 = \overline{W}$. Then $\dim(\Sigma_1 \cap \Sigma_2) = \dim \partial W \geq n - 1$. Then $\dim \overline{W} \geq n$ [**P**, p. 160].

(c) Let $g(\cdot) - h(\cdot, t)$ be the homotopy joining g with $g_1 = g - h_1$. Then, by Theorem 2.1, g_1 is 0-epi on $\Sigma \cap U$. From (a) there is a minimal set $\Sigma_1 \subset \Sigma$ on which g_1 is 0-epi. Since, obviously, the homotopy is reversible, g is 0-epi on Σ_1 and, hence, $\Sigma_1 = \Sigma$. Q.E.D.

We wish to point out that the idea of the proofs of (a) and (b) of Theorem 3.1 is close to that of the existence of a *Cantor space* in a compact Hausdorff space [**P**, p. 335].

As mentioned earlier, Theorem 3.1 contains the dimension result obtained in [**AA**; **FMP, I**; **FMP, II**] and [**MP**] (see §4).

4. Applications of the main result.
In this section we return to our original motivation—the study of global branches of solutions to multiparameter nonlinear equations. We show how our main result on dimension can be applied to continuation, and global implicit function and global bifurcation theorems. We restrict ourselves to the case when degree theory can be applied and the parameter space is finite dimensional. Continuation problems with Fredholm operators of positive index, and bifurcation problems having bifurcation set of codimension larger than one—i.e., situations where the topological degree cannot be applied—will be treated in [**IMPV**].

A. *Application to continuation and global implicit function theorem.* Let $\mathcal{O}$ be an open subset of $\mathbb{R}^n \times E$, where E is a Banach space and let $f: \mathcal{O} \to E$ be of the form $f(\lambda, x) = x - k(\lambda, x)$, for $(\lambda, x) \in \mathcal{O}$, with $k: \mathcal{O} \to E$ compact.

Assume that for some $\lambda_0 \in \mathbb{R}^n$, the Leray-Schauder topological degree of the map $f_{\lambda_0}(\cdot) = f(\lambda_0, \cdot)$, relative to some open and bounded subset $\mathcal{O}'$ of $\mathcal{O}_{\lambda_0} = \{x \in E: (\lambda_0, x) \in \mathcal{O}\}$, is defined (i.e., $f_{\lambda_0}^{-1}(0) \cap \mathcal{O}'$ is compact) and different from zero: $\deg(f_{\lambda_0}, \mathcal{O}', 0) \neq 0$. Under these assumptions we have

THEOREM 4.1. *Set* $S = f^{-1}(0)$ *and* $A = \{\lambda_0\} \times (\mathcal{O}_{\lambda_0} \setminus \mathcal{O}')$. *Then the set* $S \setminus A$ *has a minimal closed subset* Σ *such that the map* $g(x, \lambda) = \lambda - \lambda_0$ *is 0-epi on* $\Sigma \cap (\mathcal{O} \setminus A)$ *with the following properties.*

(a) Σ *is connected and has dimension at each point at least* n.
(b) Σ *is either unbounded, or* $\overline{\Sigma} \cap \partial \mathcal{O} \neq \varnothing$, *or* $\overline{\Sigma} \cap A \neq \varnothing$.

(c) Σ intersects the fiber $\{\lambda = \lambda_0\}$ inside $\mathcal{O}'$ and, if $f(\lambda, x) \neq 0$ for $x \in \partial \mathcal{O}'$ and $\|\lambda - \lambda_0\| < r_0$ for some $r_0 > 0$, then Σ intersects the fiber $\{\lambda = \bar{\lambda}\}$ inside $\mathcal{O}'$ for any $\bar{\lambda}$, with $\|\bar{\lambda} - \lambda_0\| < r_0$.

PROOF. Note that, from the particular form of f, the set $S = f^{-1}(0)$ has the property that its intersection with any bounded and closed (in E) subset of $\mathcal{O}$ is compact. Set $U = \mathcal{O} \setminus A$. Then if $g: U \to \mathbb{R}^n$ denotes the map $g(\lambda, x) = \lambda - \lambda_0$, the Leray–Schauder topological degree of the map $(g, f): U \to \mathbb{R}^n \times E$ is defined (indeed, $(g, f)^{-1}|_U(0, 0) = (\{\lambda_0\} \times f_{\lambda_0}^{-1}(0)) \cap U$, which is compact by assumption) and different from zero since (by the reduction property and the product formula of the degree) it coincides with $\deg(f_{\lambda_0}, \mathcal{O}', 0) \neq 0$ by assumption. Since $(g, f): U \to \mathbb{R}^n \times E$ is a compact perturbation of the identity with Leray–Schauder topological degree different from zero, we have, from Property 2.5, that (g, f) is 0-epi on U. Furthermore, by Property 2.4, we obtain that g is 0-epi on $f^{-1}(0) \cap U = S \cap (\mathcal{O} \setminus A)$. Thus, from Theorem 3.1, one gets all of the properties of the theorem except the second part of (c).

If $f(\lambda, x) \neq 0$ on $\overline{B(r_0)} \times \partial \mathcal{O}'$, where $B(r_0) = \{\lambda \in \mathbb{R}^n: \|\lambda - \lambda_0\| < r_0\}$, then from the compactness of $S \cap \overline{B(r_0)} \times \partial \mathcal{O}'$ there exists r_1 such that $f(\lambda, x) \neq 0$ on $\overline{B(r_0)} \times (\mathcal{O}' \setminus B(r_1))$, with $B(r_1) = \{x \in E: \|x\| < r_1\}$. Let $\varphi: \mathbb{R}^+ \to \mathbb{R}^+$ be a nonincreasing function, with $\varphi(r) = 1$ for $0 \leq r \leq r_1$ and $\varphi(r) = 0$ for $r \geq r_1 + \varepsilon$. Consider the homotopy $\lambda - (1 - t)\lambda_0 - t\varphi(\|x\|)\bar{\lambda}$, with $\|\bar{\lambda} - \lambda_0\| < r_0$, which is admissible by construction. Therefore $\lambda - \varphi(\|x\|)\bar{\lambda}$ is 0-epi on Σ, and, from Property 2.1, Σ must intersect the zero set of $\lambda - \varphi(\|x\|)\bar{\lambda}$ in $\mathcal{O} \setminus A$, hence in $B(r_1)$, where $\varphi(\|x\|) = 1$. Q.E.D.

Theorem 4.1 contains as a special case a global version of the implicit function theorem. Indeed, if f is continuously Fréchet differentiable with respect to x and for some zero of f the differential $\partial f(\lambda_0, x_0)/\partial x$ is an isomorphism, then the conclusion is that, for a small neighbourhood V of (λ_0, x_0), the n-dimensional manifold $f^{-1}(0) \cap V$ may be continued globally into a connected set Σ with $\Sigma \subset f^{-1}(0)$ having local dimension at least n. Moreover, either Σ is unbounded, or $\bar{\Sigma} \cap \partial \mathcal{O} \neq \emptyset$, or Σ curls back to $\mathcal{O}_{\lambda_0}$ at some point distinct from x_0.

Another particular case of Theorem 4.1 is the following extension of the well-known Leray–Schauder continuation principle [**LS**].

Let $\mathcal{O} \subset I^n \times E$ be open and bounded, where I^n is the nth cube of $\mathbb{R}^n$ centered at some point λ_0, and let $f: \mathcal{O} \to E$ be such that $f^{-1}(0)$ is a compact subset of $\mathcal{O}$ and $\deg(f_{\lambda_0}, \mathcal{O}_{\lambda_0}, 0) \neq 0$. Then there exists a connected subset Σ of $f^{-1}(0)$ whose dimension at each point is at least n, $\Sigma \cap \mathcal{O}_{\lambda_0} \neq \emptyset$, and for each $\bar{\lambda} \in I^n$, the fiber $\{\lambda = \bar{\lambda}\}$ intersects Σ.

For further extensions and refinements of the Leray–Schauder continuation principle to the case when $\mathcal{O}$ is not necessarily bounded and the parameter space need not be finite dimensional, we refer to [**IMPV**].

B. *Application to bifurcation.* Let E be a Banach space. Let $f: \mathbb{R}^n \times E \to E$ be of the form $f(\lambda, x) = x - k(\lambda, x)$, where k is compact and $k(\lambda, 0) = 0$ for all $\lambda \in \mathbb{R}^n$. Assume there are two points $\underline{\lambda}, \bar{\lambda}$ and a positive number ε such that

(1) if $f(\lambda, x) = 0$ for $\|x\| \leq 2\varepsilon$, $\|\lambda - \underline{\lambda}\| \leq \varepsilon$, or $\|\lambda - \overline{\lambda}\| \leq \varepsilon$ on the line Γ joining $\underline{\lambda}$ with $\overline{\lambda}$, then $x = 0$;

(2) the (Leray–Schauder) index $\text{ind}(f_{\underline{\lambda}}, 0)$ of $f_{\underline{\lambda}}$ at $x = 0$ is different from $\text{ind}(f_{\overline{\lambda}}, 0)$.

Then the set of nontrivial solutions of $f(\lambda; x) = 0$ has a minimal closed subset Σ which is connected and has dimension at least n at each point. Moreover, $\Sigma \cap \{(\lambda, x): \lambda$ belongs to the line going through $\underline{\lambda}$ and $\overline{\lambda}\}$ intersects Γ in the plane $x = 0$ strictly on the segment $[\underline{\lambda}, \overline{\lambda}]$ between $\underline{\lambda}$ and $\overline{\lambda}$ and contains a closed connected subset, having local dimension at least 1, which is either unbounded or intersects Γ strictly outside $[\underline{\lambda}, \overline{\lambda}]$.

PROOF. Without loss of generality one may assume that the line Γ is represented by the one-dimensional subspace $\mathbb{R} \times \{0\} \times \cdots \times \{0\}$ of $\mathbb{R}^n$, and that $\underline{\lambda} = (-r_0, 0, \ldots, 0)$, $\overline{\lambda} = (r_0, 0, \ldots, 0)$, $r_0 \in \mathbb{R}^+$. Taking $B_2 = \{x \in E: \|x\| < 2\varepsilon\}$ and $B_1 = \{\lambda \in \mathbb{R}^n: \|\lambda\| < r_0 + \varepsilon\}$, we shall compute the Leray–Schauder topological degree of the map $h: \mathbb{R}^n \times E \to \mathbb{R}^n \times E$, defined as

$$h(\lambda, x) = (\|x\| - \varepsilon, \lambda_2, \ldots, \lambda_n, f(\lambda, x))$$

for $(\lambda_1, \lambda_2, \ldots, \lambda_n, x) \in \mathbb{R}^n \times E$ on $B_1 \times B_2$. Note first that, by assumption (1), $\deg(h, B_1 \times B_2, 0)$ is defined. Moreover, by (1), one can deform h to $(\|x\| - \varepsilon, \lambda_2, \ldots, \lambda_n, f(\lambda_1, 0, \ldots, 0, x))$ so that (by the product formula of the degree)

$$\deg(h, B_1 \times B_2, 0) = \deg(\overline{h}, B_1' \times B_2, 0),$$

where $B_1' = \{\mu \in \mathbb{R}: |\mu| < r_0 + \varepsilon\}$ and $\overline{h}: \mathbb{R} \times E \to \mathbb{R} \times E$ is the map $\overline{h}(\mu, x) = (\|x\| - \varepsilon, f(\mu, 0, \ldots, 0, x))$. Consider the following homotopy on the first component of $\overline{h}$: $(1 - t)(\|x\| - \varepsilon) + t(r_0 - |\mu|)$. It is not hard to check that the zeroes of this homotopy are in $\{|\mu| \leq r_0\} \times \{\|x\| \leq \varepsilon\}$, and so $\overline{h}$ is deformable on $B_1' \times B_2$ to $(r_0 - |\mu|, f(\mu, 0, \ldots, 0, x))$. But the only zeroes of this map are $(-r_0, 0)$ and $(r_0, 0)$, so that we can compute (using the excision property and the product formula) its degree, which turns out to be $\text{ind}(f_{\underline{\lambda}}, 0) - \text{ind}(f_{\overline{\lambda}}, 0)$. Finally, we are able to conclude that

$$\deg(h, B_1 \times B_2, 0) = \text{ind}(f_{\underline{\lambda}}, 0) - \text{ind}(f_{\overline{\lambda}}, 0) \neq 0.$$

In order to apply Theorem 3.1 we proceed as follows. Let $\varphi: \mathbb{R}^+ \to \mathbb{R}^+$ be a nonincreasing function, with $\varphi(r) = \varepsilon$ for $r \leq r_0$ and $\varphi(r) = 0$ for $|r| \geq r_0 + \varepsilon$. Consider the map $f_1(\lambda, x) = (\|x\| - \varphi(\|\lambda\|), \lambda_2, \ldots, \lambda_n, f(\lambda, x))$. The zeroes of f_1 consist of the set of zeroes of f in $B_1 \times B_2$ with $\lambda = (\lambda_1, 0, \ldots, 0)$, $\|x\| = \varepsilon$, $|\lambda_1| < r_0 - \varepsilon$, together with the set $\{(\lambda, 0) \in \mathbb{R}^n \times E: \lambda = (\lambda_1, 0, \ldots, 0), |\lambda_1| > r_0 + \varepsilon\}$. Thus, if $U = \mathbb{R}^n \times (E \setminus \{0\})$, then the degree of f_1 is defined, and, moreover,

$$\deg(f_1, U, 0) = \deg(h, B_1 \times B_2, 0) \neq 0.$$

Hence, by Property 2.5, the map f_1 is 0-epi on U. By property 2.4 the map $(\|x\| - \varphi(\|\lambda\|), \lambda_2, \ldots, \lambda_n)$ is 0-epi on $S = \{(\lambda, x): f(\lambda, x) = 0, x \neq 0\}$. Hence, by Theorem 3.1, one obtains the existence of a minimal closed subset Σ of nontrivial solutions which is connected and has local dimension at least n.

Finally, by Property 2.4, the map $\|x\| - \varphi(\|\lambda\|)$ is 0-epi on the set $S_1 \equiv \Sigma \cap \{(\lambda, x): \lambda = (\lambda_1, 0, \ldots, 0), \lambda_1 \in \mathbb{R}\}$. Therefore, again by Theorem 3.1, one gets a minimal subset Σ_1 of S_1 which is connected, with local dimension at least 1, and such that $\|x\| - \varphi(\|\lambda\|)$ is 0-epi on $\Sigma_1 \cap U$. Moreover, Σ meets the level $\|x\| = \varepsilon$ inside $B_1 \times B_2$, and Σ_1 is either unbounded, or $\overline{\Sigma}_1$ intersects $(\mathbb{R}^n \times \{0\}) \cap \overline{\Sigma}$. It remains to show that $\overline{\Sigma}_1$ intersects $B_1 \times \{0\}$ and that if Σ_1 is bounded, then $\overline{\Sigma}_1$ intersects also the set of trivial solutions outside $\overline{B}_1$. In fact, if $\overline{\Sigma}_1 \cap (B_1 \times \{0\}) = \varnothing$, then $\overline{\Sigma}_1 \cap (\overline{B}_1 \times \overline{B}_2)$ is compact and, as such, is at a positive distance d from $\overline{B}_1 \times \{0\}$. The deformation $\|x\| - t\varphi(\|\lambda\|)$ for $d/2\varepsilon \leq t \leq 1$ is valid. But then, on $\overline{\Sigma}_1 \cap (\overline{B}_1 \times \overline{B}_2)$ we have $\|x\| \neq d/2$, and this contradicts Property 2.1. On the other hand, if $\overline{\Sigma}_1$ is bounded and does not intersect $\mathbb{R}^n \times \{0\}$ outside $\overline{B}_1$, then, again by compactness, the distance d from $\overline{\Sigma}_1$ to $(\mathbb{R}^n \times \{0\}) \setminus (B_1 \times \{0\})$ is positive. Let V be the intersection of a bounded open neighbourhood of $\overline{\Sigma}_1$ with the set $\{(x, \lambda): \|x\| > \tfrac{1}{2}\min(d, \varepsilon)\}$. But on $\Sigma_1 \cap \partial V$, which is contained in $B_1 \times B_2$, the map $\|x\| - \varphi(\|\lambda\|)$ is negative, contradicting the last statement in Proposition 3.2, by which $\|x\| - \varphi(\|\lambda\|)$ must change sign on $\Sigma_1 \cap \partial V$.   Q.E.D.

REMARK. We would like to add in passing that the described results remain valid if the map f is defined only on an open set and if the line Γ going through $\underline{\lambda}$ and $\overline{\lambda}$ is replaced, for example, by a curve defined by the zeroes of a mapping from $\mathbb{R}^n \to \mathbb{R}^{n-1}$ for which zero is a regular value.

References

[AA] J. C. Alexander and S. S. Antman, *Global and local behaviour of bifurcating multidimensional continua of solutions for multiparameter nonlinear eigenvalue problems*, Arch. Rational Mech. Anal. **76** (1981), 339–355.

[FMP, I] P. M. Fitzpatrick, I. Massabó and J. Pejsachowicz, *On the covering dimension of the set of solutions of some nonlinear equations*, Trans. Amer. Math. Soc. (to appear).

[FMP, II] _____, *Global several-parameter bifurcation and continuation theorems: a unified approach via complementing maps*, Math. Ann. **263** (1983), 61–73.

[FMV] M. Furi, M. Martelli and A. Vignoli, *On the solvability of nonlinear operator equations in normed spaces*, Ann. Mat. Pura Appl. **124** (1980), 321–343.

[FP] M. Furi and M. P. Pera, *On the existence of an unbounded connected component of solutions for nonlinear equations in Banach spaces*, Rend. Accad. Naz. Lincei **57** (1979), 31–38.

[G] A. Granas, *The theory of compact vector fields and some applications to the topology of functional spaces*, Rozprawy Mat. **30** (1962).

[I] J. Ize, *Bifurcation theory for Fredholm operators*, Mem. Amer. Math. Soc. **174** (1976).

[IMPV] J. Ize, I. Massabó, J. Pejsachowicz and A. Vignoli, *Structure and dimension of global branches of solutions to multiparameter nonlinear equations*, Trans. Amer. Math. Soc. (to appear).

[LS] J. Leray and J. Schauder, *Topologie et équations fonctionelles*, Ann. Sci. École Norm. Sup. **51** (1934), 45–78.

[MP] I. Massabó and J. Pejsachowicz, *On the connectivity properties of the solution set of parametrized families of compact vector fields*, J. Funct. Anal. **59** (1984), 151–166.

[P] A. R. Pears, *Dimension theory of general spaces*, Cambridge Univ. Press, 1975.

[R] P. H. Rabinowitz, *Some global results for nonlinear eigenvalue problems*, J. Funct. Anal. **7** (1971), 487–513.

IIMAS–UNAM, MEXICO (Current address of J. Ize)

UNIVERSITÀ DELLA CALABRIA, ITALY (Current address of I. Massabó and J. Pejsachowicz)

II UNIVERSITÀ DI ROMA, ITALY (Current address of A. Vignoli)